经典原版书库

Python大学教程
面向计算机科学和数据科学
（英文版）

Intro to Python for
Computer Science and Data Science
Learning to Program with AI, Big Data and the Cloud

[美] 保罗·戴特尔　哈维·戴特尔　著
（Paul Deitel）　（Harvey Deitel）

图书在版编目（CIP）数据

Python 大学教程：面向计算机科学和数据科学（英文版）/（美）保罗·戴特尔（Paul Deitel），（美）哈维·戴特尔（Harvey Deitel）著 . -- 北京：机械工业出版社，2020.12（2025.2 重印）
（经典原版书库）
书名原文：Intro to Python for Computer Science and Data Science: Learning to Program with AI, Big Data and the Cloud
ISBN 978-7-111-67150-3

Ⅰ. ①P… Ⅱ. ①保… ②哈… Ⅲ. ①软件工具-程序设计-高等学校-教材-英文 Ⅳ. ① TP311.561

中国版本图书馆 CIP 数据核字（2020）第 265157 号

北京市版权局著作权合同登记　图字：01-2019-7259 号。

Authorized Reprint from the English language edition, entitled *Intro to Python for Computer Science and Data Science: Learning to Program with AI, Big Data and the Cloud*, Paul Deitel, Harvey Deitel, published by Pearson Education, Inc., Copyright © 2020 Pearson Education, Inc.

All rights reserved. No part of this book may be reproduced or transmitted in any form or by any means, electronic or mechanical, including photocopying, recording or by any information storage retrieval system, without permission from Pearson Education, Inc.

English language edition published by China Machine Press, Copyright © 2021.

本书英文影印版由 Pearson Education Inc. 授权机械工业出版社独家出版。未经出版者书面许可，不得以任何方式复制或抄袭本书内容。

此影印版仅限于中国大陆地区（不包括香港、澳门特别行政区及台湾地区）销售发行。

本书封面贴有 Pearson Education（培生教育出版集团）激光防伪标签，无标签者不得销售。

出版发行：机械工业出版社（北京市西城区百万庄大街 22 号　邮政编码：100037）
责任编辑：曲　熠　　　　　　　　　　　　　　责任校对：李秋荣
印　　刷：北京虎彩文化传播有限公司　　　　　版　　次：2025 年 2 月第 1 版第 4 次印刷
开　　本：186mm×240mm　1/16　　　　　　　 印　　张：53.5
书　　号：ISBN 978-7-111-67150-3　　　　　　 定　　价：169.00 元

客服电话：(010) 88361066　68326294

版权所有·侵权必究
封底无防伪标均为盗版

Deitel® Series Page

How To Program Series

Android™ How to Program, 3/E
C++ How to Program, 10/E
C How to Program, 8/E
Java™ How to Program, Early Objects Version, 11/E
Java™ How to Program, Late Objects Version, 11/E
Internet & World Wide Web How to Program, 5/E
Visual Basic® 2012 How to Program, 6/E
Visual C#® How to Program, 6/E

REVEL™ Interactive Multimedia

REVEL™ for Deitel Java™

VitalSource Web Books

http://bit.ly/DeitelOnVitalSource

Android™ How to Program, 2/E and 3/E
C++ How to Program, 9/E and 10/E
Java™ How to Program, 10/E and 11/E
Simply C++: An App-Driven Tutorial Approach
Simply Visual Basic® 2010: An App-Driven Approach, 4/E
Visual Basic® 2012 How to Program, 6/E
Visual C#® How to Program, 6/E
Visual C#® 2012 How to Program, 5/E

Deitel® Developer Series

Android™ 6 for Programmers: An App-Driven Approach, 3/E
C for Programmers with an Introduction to C11
C++11 for Programmers
C# 6 for Programmers
Java™ for Programmers, 4/E
JavaScript for Programmers
Swift™ for Programmers

LiveLessons Video Training

http://deitel.com/books/LiveLessons/

Android™ 6 App Development Fundamentals, 3/E
C++ Fundamentals
Java SE 8™ Fundamentals, 2/E
Java SE 9™ Fundamentals, 3/E
C# 6 Fundamentals
C# 2012 Fundamentals
JavaScript Fundamentals
Swift™ Fundamentals

To receive updates on Deitel publications, Resource Centers, training courses, partner offers and more, please join the Deitel communities on

- Facebook®—http://facebook.com/DeitelFan
- Twitter®—@deitel
- LinkedIn®—http://linkedin.com/company/deitel-&-associates
- YouTube™—http://youtube.com/DeitelTV
- Instagram®—http://instagram.com/DeitelFan

and register for the free *Deitel® Buzz Online* e-mail newsletter at:

http://www.deitel.com/newsletter/subscribe.html

To communicate with the authors, send e-mail to:

deitel@deitel.com

For information on programming-languages corporate training seminars offered by Deitel & Associates, Inc. worldwide, write to deitel@deitel.com or visit:

http://www.deitel.com/training/

For continuing updates on Pearson/Deitel publications visit:

http://www.deitel.com
http://www.pearson.com/deitel

前　　言

"塔尔山上有黄金。"[1]

在过去的几十年中,很多发展趋势——显现。计算机的硬件速度越来越快,价格越来越便宜,尺寸越来越小。网络带宽(携带信息的能力)越来越大,价格越来越便宜。同时软件规模越来越庞大,而"开源"运动又将它们变成完全免费或者接近免费。短时间内,"物联网"将数十亿各种各样的设备连接起来,并快速产生了大规模的数据。

几年之前,如果有人让我们写一本主题为"大数据"和"云计算"的大学阶段编程入门教材,并且在封面上绘制一只彩色的大象(象征"巨大"),我们可能会如此反应:"嗯?"如果他们继续让我们在书名中囊括 AI(人工智能),我们可能会说:"真的吗?这对于编程初学者会不会过于超前了?"

如果有人让我们在书名中加上"数据科学",我们可能会问:"计算机科学不是已经包含了数据吗?为什么我们需要为此单独分一个科目?"好吧,如今谈及程序设计,最酷的说法就是"什么都是数据"——数据科学、数据分析、大数据、关系数据库(SQL)以及 NoSQL 和 NewSQL 数据库。

如今我们真的写了这样一本书!欢迎阅读!

在本书中,你将会着手学习当今最引人入胜、最前沿的计算技术——你将会看到,它将计算机科学和数据科学轻松地结合在一起,是一门适用于这些学科和相关学科的入门课程。此外,你将使用 Python 进行编程——世界上发展速度最快、最流行的编程语言之一。在前言中,我们将展示这本书的"灵魂"。

专业程序员常常很快喜欢上 Python。他们欣赏 Python 的表现力、易读性、简洁性和交互性。他们喜欢开源软件世界,这个世界正在为广泛的应用领域不断生成可复用的软件。

无论你是教师、初学者或者有经验的专业人士,这本书都将对你有所帮助。Python 对于初学者而言是优秀的第一门编程语言,并且适用于开发工业级的应用。对于新人而言,本书前面的章节奠定了坚实的编程基础。

我们希望你在这本书中学到知识,并且发现快乐与挑战。徜徉其中,享受乐趣。

将 Python 用于计算机科学和数据科学教学

许多顶尖的美国大学使用 Python 作为介绍计算机科学的语言,"CS 学科排名前

---

1. 来源不明,通常误认为是马克·吐温。

10 的有 8 个（80%）、排名前 39 的有 27 个（69%）使用 Python"[2]。Python 如今在教学和科学计算中尤其受到欢迎[3]，最近已超过 R 语言并成为最受欢迎的数据科学编程语言[4,5,6]。

模块化体系结构

我们预计计算机科学的本科生课程将会包含数据科学部分——这本书为此而设计，并且在 Python 编程方面满足数据科学入门课程的需求。

本书的模块化体系结构帮助我们满足计算机科学、数据科学和其他相关受众的多样化需求。教师能方便地进行调整，为不同专业的学生开设系列课程。

第 1~11 章介绍传统的计算机科学编程主题。第 1~10 章每章都包含可选的、简洁的"数据科学入门"一节，介绍人工智能、基础统计、趋中和离中度量、模拟、静态/动态可视化、CSV 文件的使用、数据探索和数据整理的 pandas 库、时间序列和简单的线性回归。这会帮助你学习第 12~17 章中的数据科学、AI、大数据和云计算相关的案例研究，你能在真实世界数据集中完整地使用这些案例研究。

学完第 1~5 章中的 Python 相关知识以及第 6~7 章中的一些关键部分，你已经能够在第 12~17 章中解决数据科学、AI 和大数据案例研究中的关键用例，这对于所有的编程通识课都是实用的：

- 计算机科学方面的课程可以着重于第 1~11 章并略讲第 1~10 章的"数据科学入门"部分。教师还可以介绍第 12~17 章中的部分或者全部案例研究。
- 数据科学方面的课程可以略讲第 1~11 章，着重于大部分或者全部的第 1~10 章中的"数据科学入门"部分以及第 12~17 章中的案例研究。

前言中的"章节依赖关系"部分将展示本书的独特架构并帮助教师规划个性化的教学大纲。

第 12~17 章的内容很酷、很强大、很现代。其中包含能动手实现的案例研究，例如有监督学习、无监督学习、深度学习、强化学习（在习题中）、自然语言处理、Twitter 数据挖掘、IBM Watson 可视化计算、大数据以及其他内容。在这个过程中，你将会掌握数据科学的大量术语和概念，包括术语的定义以及在不同规模的程序中使用

---

2. Guo, Philip, " Python Is Now the Most Popular Introductory Teaching Language at Top U.S. Universities," ACM, July 07, 2014, https://cacm.acm.org/blogs/blog-cacm/176450-python-is-nowthe-most-popular-introductory-teaching-language-at-top-u-s-universities/fulltext.
3. https://www.oreilly.com/ideas/5-things-to-watch-in-python-in-2017.
4. https://www.kdnuggets.com/2017/08/python-overtakes-r-leader-analytics-datascience.html.
5. https://www.r-bloggers.com/data-science-job-report-2017-r-passes-sas-but-pythonleaves-them-both-behind/.
6. https://www.oreilly.com/ideas/5-things-to-watch-in-python-in-2017.

的概念。

本书读者对象

模块化的结构使得本书适用于以下读者：
- 所有标准的 Python 计算机科学及相关专业。首先，我们的书是一本现代的、可靠的 Python CS1 入门教材。ACM/IEEE 的计算课程建议列出了 5 个类别：计算机工程、计算机科学、信息系统、信息技术和软件工程[7]。这本书对这些类别都适用。
- 数据科学专业的本科生课程。我们的书对许多数据科学课程都是有用的。对于入门级课程而言，它遵循课程建议，整合了所有课程的关键领域。在计算机科学或者数据科学课程计划中，本书都可以作为第一本专业教材，也可用作高年级课程的 Python 参考书。
- 非计算机和数据科学专业学生的辅修课程。
- 数据科学的研究生课程。这本书可以作为入门级课程的主要教材，也可以作为高年级课程的 Python 参考书。
- 两年制学院。这些学校会为准备进入四年制学院的数据科学专业的学生开设相关课程——这本书就是一个合适的选择。
- 高中。就像出于强烈的兴趣爱好而开设计算机课程一样，很多高中已经开设了 Python 编程和数据科学课程[8]。最近在 LinkedIn 上发表的一篇文章写道："高中就应该教授数据科学，课程应该直面我们的学生将要选择的职业类型，直接关注工作和技术的发展方向。"[9] 我们相信数据科学很快就会成为一门受欢迎的大学先修课程并且最终会有数据科学的 AP 考试。
- 专业行业培训课程。

本书主要特色

保持简洁，保持短小，保持新颖

- 保持简洁（KIS）。在书的各个方面以及教师和学生资源中，我们都力求简单明了。例如，在写自然语言处理的时候，我们使用了简洁直观的 TextBlob 库而不是更为复杂的 NLTK。一般情况下，当有多个库都能完成相近的任务时，我们选

---

7. https://www.acm.org/education/curricula-recommendations.
8. http://datascience.la/introduction-to-data-science-for-high-school-students/.
9. https://www.linkedin.com/pulse/data-science-should-taught-high-school-rebeccacroucher/.

择最简单的一个。
- 保持短小（KIS）。本书的 538 个案例研究中大多数都很短小——伴随交互式 IPython 的即时反馈，通常只有几行代码。我们大约只在 40 个大型脚本和完整案例研究中使用了较长的代码示例。
- 保持新颖（KIT）。我们查阅了大量最新的 Python 编程和数据科学教材以及专业书籍，浏览、阅读或者观看了大约 15 000 篇最新的文献、研究论文、白皮书、视频、博客文章、论坛文章和文档。这让我们能够"把握"Python、计算机科学、数据科学、AI、大数据和云计算社区的脉搏，从而创建出 1566 个崭新的案例研究、习题和项目（EEP）。

IPython 的实时反馈、搜索、发现和实验教学方法

- 学习这本书的理想方法是阅读它并同时运行代码示例。在整本书中，我们使用了 IPython 解释器，它采用一种友好的、实时反馈的模式，能够在 Python 及其扩展库上快速进行搜索、发现和实验。
- 大多数代码都在小型的可交互的 IPython 会话中展示。你所写的每一个代码片段，IPython 能够立即读取然后计算并给出结果。这种即时反馈使你保持注意力，并助力学习过程、支撑原型设计和加速软件开发过程。
- 我们的书总是强调现场代码的教学方法，通过样例输入和结果显示，专注于程序的完整和可运行。IPython 的"魔力"在于将代码片段转换为实时代码，每当你输入一行时，这些代码就会"活起来"。这有助于学习并鼓励动手实验。
- IPython 是学习常见错误的报错信息的好方法。我们有时故意犯错来告知你将发生什么，因此，当我们说某件事是错误的时候，试试看会发生什么。
- 本书配有 557 道自检习题（适合于"翻转课堂"，稍后介绍）和 471 个章末习题和项目，它们中的大多数遵循了同样的实时反馈理念。

Python 编程基础

- 首先，本书是一本 Python 入门教材。我们提供了丰富的 Python 编程知识和常规的编程基础内容。
- 我们讨论了 Python 的编程模型——过程式编程、函数式编程和面向对象编程。
- 我们强调解决问题和算法设计。
- 我们为准备进入产业界的学生准备了最好的习题。
- 函数式编程贯穿全书。第 4 章中的一个图表展示了 Python 中关键的函数式编程能力及其对应的章节。

538 个案例研究以及 471 个习题和项目

- 学生通过动手实践的方法,在广泛选择的真实世界案例研究、习题和项目(EEP)中开展学习,这些内容来自计算机科学、数据科学和其他很多领域。
- 538 个案例研究的内容均围绕计算机科学、数据科学、人工智能和大数据,从单个代码片段到完整案例研究均有涉及。
- 471 个习题和项目自然地拓展了章节中的示例。每章都以一系列涵盖了各种主题的习题作为结尾,这有助于教师根据受众的需求调整课程内容并且在每个学期布置不同的作业。
- EEP 向你提供了引人入胜的、富有挑战性的、有趣的 Python 基础知识,包括可以上手实验的 AI、计算机科学和数据科学项目。
- 学生将面对令人兴奋且有趣的有关 AI、大数据和云技术应用问题的挑战,比如自然语言处理、Twitter 数据挖掘、机器学习、深度学习、Hadoop、MapReduce、Spark、IBM Watson、关键的数据科学库(NumPy、pandas、SciPy、NLTK、TextBlob、spaCy、BeautifulSoup、Textatistic、Tweepy、Scikit-learn、Keras)、关键的可视化库(Matplotlib、Seaborn、Folium)以及其他。
- 我们的 EEP 鼓励你思考未来。下述案例研究虽然只出现在前言中,但本书包含了多个发人深省的类似项目:利用深度学习、物联网以及电视摄像机(由体育赛事数据训练出来的),我们可以自动进行数据分析、回顾比赛细节以及即时回放,因此球迷不再需要忍受直播体育赛事的误判和延迟。既然如此,我们可能会产生这样的想法:可以使用这些技术来取消裁判。为什么不呢?我们已经越来越多地把自己的生命托付给基于深度学习的技术,比如机器人外科医生和自动驾驶汽车。
- 项目习题鼓励你更加深入地了解所学的知识并研究本书没有涉及的技术。真正的项目通常规模更大,需要更多的网络搜索和实现代价。
- 在教师手册中,我们提供了许多习题的答案,包括第 1~11 章中核心的 Python 代码。答案仅对教师可见,详见后文中关于 Pearson 教师资源的介绍。我们没有提供项目和研究习题的答案。
- 我们鼓励你仔细观看示例和开源代码案例研究(详见 GitHub 网页),包括课程级别项目、学期级别项目、专业方向级别项目、毕业设计项目和毕业论文。

557 道自检习题及其答案

- 平均每个小节后有三道自检习题。
- 自检习题的类型包括填空、判断和讨论,能帮你测试是否理解了所学的内容。

- IPython 交互式自检习题可帮助你不断尝试并强化所学的编程技术。
- 为快速掌握所学知识，自检习题后面都跟有答案。

避免烦琐的数学语言，多用自然语言进行解释

- 数据科学的主题与数学高度相关。这本书将用作计算机科学和数据科学第一门课的教科书，学生可能没有深厚的数学知识背景，所以我们避免了烦琐的数学语言，把数学内容留在高层次的课程中。
- 在案例研究、习题和项目中，我们关注数学的概念而不是细节。我们使用 statistics、NumPy、SciPy、pandas 等 Python 库和其他的很多库来解决问题，从而隐藏了数学复杂性。所以，学生能够直接使用数学技术（如线性回归），而不需要知道背后的数学知识。在机器学习和深度学习的案例研究中，我们专注于创建在"幕后"做数学运算的对象——这是基于对象编程的关键之一。这个做法等同于安全地驾驶一辆汽车前往目的地时不需要知道制造引擎、变速箱、动力转向和防滑刹车系统背后的数学、工程和科学知识。

可视化

- 67 张可视化结果图帮助你理解概念，包含二维或三维的、静态的、动态的、动画的和交互式的图表、图形、图片、动画等。
- 我们关注由 Matplotlib、Seaborn、pandas 和 Folium（用于交互式地图）产生的高层可视化结果。
- 我们使用可视化作为教学工具。例如，我们使用动态滚压模型和柱状图使大数定律"鲜活"起来。随着滚动数量的增加，你将看到每个面在滚动总数中所占的百分比逐渐接近 16.667%（1/6），代表百分比的柱条也趋于一致。
- 你需要了解自己的数据。一种简单的办法是直接看原始数据。即使是少量的数据，你也可能很快迷失在细节当中。对于大数据而言，可视化对于数据探索和传递可复制的研究结果尤为重要，数据规模可能是百万级的、上亿的甚至更为庞大。通常而言，一图胜千言[10]——在大数据中，一个可视化结果能够比得上数据库中数亿甚至更多的个体。
- 有时候，你需要"飞到离数据 40 000 英尺高"才能在"大范围"看到它。描述性统计当然有帮助，但是也可能产生误导。你将在习题中研究 Anscombe 四重奏，这个案例研究通过可视化直观地表明：差异显著的数据集可能产生几乎相同的描述性统计。
- 我们展示了可视化结果和动画代码以便你能够自己实现。我们也通过 Jupyter

---

10. https://en.wikipedia.org/wiki/A_picture_is_worth_a_thousand_words.

Notebook 的形式给出动画的源代码文件，便于你自定义代码和动画参数，进而重新执行动画，然后查看其带来的影响。
- 许多习题都要求你创建自己的可视化结果。

数据经验

- "数据科学本科课程建议"中提出："数据经验需要在所有课程中扮演核心角色。"[11]
- 在本书的案例研究、习题和项目中，你将使用许多真实世界数据集和数据源。网上有各种免费的开源数据集供你实验。有些我们参考的网站列出很多数据集，我们鼓励你探索这些数据集。
- 我们收集了上百份教学大纲，追踪了教师的数据集偏好，并研究了最流行的监督机器学习、无监督机器学习和深度学习的数据集。你将会使用到的许多库都附带用于实验的标准数据集。
- 你将学习如何进行数据获取和分析准备，学习使用多种技术进行数据分析、模型调整并有效交流结果，特别是通过可视化。

像开发者一样思考

- 你将以开发者为视角，使用像 GitHub 和 StackOverflow 一样的流行网站并且进行大量互联网搜索。"数据科学入门"部分和第 12~17 章中的案例研究提供了丰富的数据经验。
- GitHub 为寻找开源代码提供了一个优秀的场所，你可以把代码合并到自己的项目中（并将你的代码贡献到开源社区中）。它是软件开发者版本控制工具库中的重要组成部分，这些工具帮助开发团队管理他们的开源（和私有）项目。
- 我们鼓励你学习 GitHub 等网站上发布的代码。
- 在为计算机科学和数据科学的职业生涯做准备的过程中，你将大量使用免费且开源的 Python 和数据科学库，来自政府、工业界和学术界的真实世界数据集，以及免费、免费试用或免费增值的软件和云服务。

动手实践云计算

- 很多大数据分析发生在云端。在云端动态地度量你的应用程序需要的硬件和软件规模是比较容易的。你将会使用各种各样的云服务（某些是直接的，某些是间接的），包括 Twitter、Google 翻译、IBM Watson、Microsoft Azure、OpenMapQuest、

---

11. "Curriculum Guidelines for Undergraduate Programs in Data Science," http://www.annualreviews.org/doi/full/10.1146/annurev-statistics-060116-053930 (p. 18).

geopy、Dweet.io 和 PubNub。你将在习题和项目中了解更多。
- 我们鼓励你使用各种云服务供应商提供的免费、免费试用或免费增值的服务。我们更喜欢那些不需要信用卡的，因为谁都不想冒意外地积攒巨额账单的风险。如果你决定使用需要信用卡的服务，确保你使用的免费层不会自动跳转到支付层。

数据库、大数据和大数据基础设施

- 根据 IBM（2016 年 11 月）的数据，全球 90% 的数据是在过去的两年内产生的[12]。有证据表明，数据产生的速度正在加快。
- 根据 2016 年 3 月 *Analytics Week* 的一篇文章，五年内将有超过 500 亿台设备连接到互联网，到 2020 年前，我们将每秒为地球上的每一个人产生 1.7 兆字节的新数据[13]！
- 本书包含了对关系数据库和带有 SQLite 的 SQL 的探讨，感兴趣的读者可以选择阅读。
- 数据库是存储和操作你要处理的大量数据的关键性大数据基础设施。关系数据库处理结构化数据，它们不适用于大数据应用程序中的非结构化和半结构化数据。因此，随着大数据的发展，为了有效处理这些数据，NoSQL 和 NewSQL 数据库应运而生。本书包含对 NoSQL 和 NewSQL 的概述，以及利用 MongoDB JSON 文件数据库的动手实践的案例研究。
- 第 17 章包含关于大数据硬件与软件基础设施的细致讨论。

人工智能案例研究

- 为什么这本书没有关于人工智能的章节呢？毕竟，"AI"是印在封面上的。在第 12~16 章的案例研究中，我们介绍的人工智能主题（计算机科学与数据科学的一个关键交集）包含自然语言处理、利用数据挖掘对 Twitter 进行情感分析、利用 IBM Watson 进行认知计算、监督机器学习、无监督机器学习、深度学习和强化学习（在习题中）。第 17 章介绍了大数据硬件和软件基础设施，这些基础设施支撑了计算机科学家和数据科学家研究的前沿 AI 解决方案。

计算机科学

- 第 1~10 章的 Python 基础知识会让你像计算机科学家一样思考。第 11 章提供了一个更高的视角，其中讨论的都是经典的计算机科学话题。第 11 章强调性能问题。

---

12. https://public.dhe.ibm.com/common/ssi/ecm/wr/en/wrl12345usen/watson-customer-engagement-watson-marketing-wr-other-papers-and-reports-wrl12345usen-20170719.pdf.
13. https://analyticsweek.com/content/big-data-facts/.

内置集合：列表、元组、集合、字典

- 今天，大多数应用开发人员不再构建定制的数据结构。这是 CS2 课程的主题——严格来说，我们的范围是 CS1 和相应的数据科学课程。书中用两章的篇幅详细介绍了 Python 的内置数据结构——列表、元组、字典和集合，大多数数据结构任务都可以通过这些数据结构来完成。

使用 NumPy 数组、pandas Series 和 DataFrame 进行面向数组的编程

- 在这本书中我们专注于开源库中的三个关键数据结构——NumPy 数组、pandas Series 和 pandas DataFrame。这些库被广泛用于数据科学、计算机科学、人工智能和大数据。NumPy 的性能比内置 Python 列表高出两个数量级。
- 第 7 章中详细介绍了 NumPy 数组。pandas 等许多库都是在 NumPy 的基础上构建的。第 7~9 章的"数据科学入门"部分介绍了 pandas Series 和 DataFrame，这两个库以及 NumPy 数组将在剩余的章节中频繁使用。

文件处理和序列化

- 第 9 章介绍文本文件处理，然后演示了如何使用流行的 JSON（JavaScript 对象表示法）格式把对象序列化。JSON 是一个普遍使用的数据交换格式，在后面的数据科学章节中你会经常见到它——通常为简单起见把 JSON 的细节隐藏在库中。

基于对象编程

- 在我们为本书进行调研期间研究的所有 Python 代码中，很少遇到自定义类，而这在 Python 开发者使用的强大的库中是很常见的。
- 我们强调大量使用 Python 开源社区打包到工业标准类库中的类。你将专注于：了解有哪些库，选择你的应用程序需要的库，从现有类（通常是一行或两行代码）创建对象以及让它们"跳起来、舞起来、唱起来"。这叫作基于对象编程——它使你能够简洁地构建令人印象深刻的应用程序，这是体现 Python 吸引力的重要组成部分。
- 通过这种方法，你可以使用机器学习、深度学习、强化学习（在习题中）和其他 AI 技术来解决各种各样有趣的问题，包括语音识别和计算机视觉等认知计算方面的挑战。过去，如果仅仅学过入门级编程课程，是不可能完成这些任务的。

面向对象编程

- 对计算机科学专业的学生来说，开发自定义类是一个至关重要的面向对象编程技能，相关的继承、多态和鸭子类型也同样重要。我们将在第 10 章中讨论这些

内容。
- 面向对象编程的讨论是模块化的，所以教师可以分开介绍基础或中级部分。
- 第 10 章包括关于 doctest 单元测试的讨论，以及一个有趣的关于洗牌和切牌的模拟案例研究。
- 数据科学、人工智能、大数据和云计算相关的 6 个章节只需要一些简单的特定类定义。不希望讲授第 10 章的教师可以让学生简单地模仿我们的类定义。

隐私性

- 在习题中，你将研究更加严格的隐私法，例如美国的 HIPAA（健康保险便携性和责任法案）和欧盟的 GDPR（一般数据保护条例）。隐私的关键方面是保护用户的个人身份信息（PII），而大数据的一个重要挑战是很容易在数据库中交叉引用个人信息。我们在本书的一些地方都提到了隐私问题。

安全性

- 安全比隐私更为重要。我们有针对性地处理了一些 Python 特有的安全问题。
- 人工智能和大数据带来了独特的隐私、安全和伦理方面的挑战。在习题中，学生将会研究 OWASP Python 安全项目（http://www.pythonsecurity.org/）、异常检测、区块链（比特币和以太坊等数字加密货币背后的技术）等。

伦理观

- 伦理难题：假设利用人工智能的大数据分析预测出，一个没有犯罪记录的人有很大概率犯下严重罪行，那么他应该被逮捕吗？在习题中，你将会研究这个问题以及其他伦理问题，包括深度造假（人工智能生成的仿真图片和视频）、机器学习中的偏好和 CRISPR 基因编辑。学生还可以研究与人工智能和智能助手（例如 IBM Watson、Amazon Alexa、Apple Siri、Google Assistant 和 Microsoft Cortana）有关的隐私和伦理问题。例如，就在最近，一位法官命令亚马逊将 Alexa 的记录用于一起犯罪案件的处理[14]。

复现性

- 通常在科学研究中，数据科学尤其需要复现实验和研究的结果，并有效地展现这些结果。Jupyter Notebook 就是做这项工作的首选方式。
- 我们将分享 Jupyter Notebook 的使用经验，它可以帮助你掌握"数据科学本科课程建议"中提到的数据复现技术。

---

14. https://techcrunch.com/2018/11/14/amazon-echo-recordings-judge-murder-case/.

- 我们在关于编程技术和软件的内容中讨论了 Jupyter Notebook 和 Docker 等复现技术。

透明性

- "数据科学本科课程建议"中提到了数据透明。数据透明主要指的是数据可用性。许多政府和组织现在秉持公开数据的原则，使得任何人都可以访问其数据[15]。我们给出了一个由这些实体提供的公共数据集。
- 数据透明的其他方面包括确定数据的正确性和了解其来源（想想那些"假新闻"）。我们使用的许多数据集都与我们提供的关键库绑定在一起，例如机器学习的 Scikit-learn 和深度学习的 Keras。我们还提及了各种精挑细选的数据集仓库，例如加州大学欧文分校（UCI）的机器学习库（超过 450 个数据集）[16] 和卡内基·梅隆大学的 StatLib 数据集档案（超过 100 个数据集）[17]。

性能

- 我们在一些案例研究和习题中使用性能测试分析工具比较了执行相同任务的不同方法的性能。其他与性能相关的讨论包括生成器表达式、NumPy 数组与 Python 列表、机器学习和深度学习模型的性能以及 Hadoop 和 Spark 分布式计算的性能。

大数据和并行性

- 计算机应用程序一般擅长一次做一件事。今天更复杂的应用程序需要并行地做很多事情。人类大脑被认为拥有相当于 1 千亿个并行处理器[18]。多年来我们一直在编写程序级别的并行化，这种并行化既复杂又容易出错。
- 在本书中，你不必编写自己的并行代码，而是可以通过让 Keras 在 TensorFlow 上运行、利用 Hadoop 和 Spark 大数据工具实现并行化来开展并行计算。在这个大数据/人工智能的时代，纯粹需要处理大量数据的应用程序可以利用多核处理器、图形处理单元（GPU）、张量处理单元（TPU）和大型云端计算机集群开展真正的并行计算。一些大数据任务可能有数千个处理器并行工作，以便在合理的时间内分析大量数据。将这些处理过程顺序化是不可取的，因为会花费太长时间。

---

15. https://www.mckinsey.com/~/media/McKinsey/Business%20Functions/McKinsey%20Digital/Our%20Insights/Big%20data%20The%20next%20frontier%20for%20innovation/MGI_big_data_full_report.ashx (page 56).
16. https://archive.ics.uci.edu/ml/datasets.html.
17. http://lib.stat.cmu.edu/datasets/.
18. https://www.technologyreview.com/s/532291/fmri-data-reveals-the-number-ofparallel-processes-running-in-the-brain/.

章节依赖关系

如果你是一名设计课程大纲的教师,或者一位决定阅读哪些章节的专业人士,这部分将帮助你做出最好的决定。请阅读本书的目录,它可以帮你快速熟悉本书的独特架构。按照章节顺序教学或阅读是最为简单的。然而,第 1~10 章的"数据科学入门"中的大部分内容以及第 12~17 章的案例研究,只需要掌握下面列出的第 1~5 章的内容和第 6~10 章的小部分内容即可。

第一部分:Python 基础快速入门

我们建议所有的课程都涵盖关于 Python 的第 1~5 章:

- 第 1 章介绍在第 2~11 章的 Python 编程以及第 12~17 章的大数据、人工智能和基于云计算的案例研究中需要的基本概念。这一章也包括运用 IPython 和 Jupyter Notebook 的试用案例研究。
- 第 2 章介绍 Python 编程的基础知识,利用代码示例说明关键语言特性。
- 第 3 章介绍 Python 的控制语句,重点介绍问题求解和算法开发,以及基本的列表处理。
- 第 4 章介绍使用已有函数和自定义函数作为构建块的程序架构,以及使用随机数生成的模拟技术和元组的基本原理。
- 第 5 章介绍 Python 的内置列表和元组集合的更多细节,并开始介绍函数式编程。

第二部分:Python 数据结构、字符串和文件 [19]

下面总结了关于 Python 的第 6~9 章的章节间依赖关系,并假设你已经阅读了第 1~5 章。

- 第 6 章介绍字典和集合,"数据科学入门"部分不依赖于本章内容。
- 第 7 章介绍使用 NumPy 进行面向数组的编程,"数据科学入门"部分需要关于字典(第 6 章)和数组(第 7 章)的知识。
- 第 8 章进一步深入讨论字符串,"数据科学入门"部分需要掌握原始字符串和正则表达式(8.11~8.12 节),以及 7.14 节介绍的 pandas Series 和 DataFrame。
- 第 9 章介绍文件和异常。对于 JSON 序列化,理解关于字典的基础知识(6.2 节)是很有用的。另外,"数据科学入门"部分需要关于内置函数 open 和 with 语句(9.3 节)以及 DataFrame(7.14 节)的知识。

---

19. 我们本可以把第 5 章放在第二部分,但我们最终选择把它放在第一部分,这是因为它是所有课程都需要覆盖的章节。

第三部分：Python 高阶主题

下面总结了第 10~11 章的章节间依赖关系，并假设你已经阅读了第 1~5 章。

- 第 10 章介绍面向对象程序设计。"数据科学入门"部分需要关于 DataFrame 的知识（7.14 节）。只想讲解类和对象的教师可以讲授 10.1~10.6 节；如需讲解更高阶的主题，例如继承、多态和鸭子类型，则可以讲授 10.7~10.9 节。10.10~10.15 节提供了其他高阶视角。
- 第 11 章介绍计算机科学思维。需要的预备知识包括：创建和访问数组中的元素（第 7 章），%timeit 简介（7.6 节），字符串合并方法（8.9 节），以及 Matplotlib FuncAnimation（6.4 节）等。

第四部分：人工智能、云计算和大数据案例研究

下面总结了第 12~17 章的章节间依赖关系，并假设你已经阅读了第 1~5 章。第 12~17 章的大部分内容还需要 6.2 节中关于字典的基础知识。

- 第 12 章介绍自然语言处理，需要用到关于 DataFrame 的内容（7.14 节）。
- 第 13 章介绍 Twitter 数据挖掘，需要用到的知识包括：DataFrame（7.14 节），字符串合并方法（8.9 节），JSON 基础知识（9.5 节），TextBlob（12.2 节），以及 Word Clouds（12.3 节）。一些例子需要通过继承定义类（第 10 章），但读者可以在阅读第 10 章前简单地模仿我们提供的类定义。
- 第 14 章介绍 IBM Watson 和认知计算，使用了内置函数 open 和 with。
- 第 15 章介绍机器学习，需要用到的知识包括：NumPy 数组基础知识和 unique 方法（第 7 章），DataFrame（7.14 节），以及 Matplotlib 函数 subplots（10.6 节）。
- 第 16 章介绍深度学习，需要用到的知识包括：NumPy 数组基础知识（第 7 章），字符串合并方法（8.9 节），常用的机器学习概念（第 15 章），以及案例研究"数字数据集上的 k 近邻分类法"（第 15 章）。
- 第 17 章介绍大数据，需要用到的知识包括：使用字符串切分方法（6.2.7 节），Matplotlib FuncAnimation（6.4 节），pandas Series 和 DataFrame（7.4 节），字符串合并方法（8.9 节），JSON 模块（9.5 节），NLTK 停用词（12.2.13 节），以及 Twitter 认证、Tweepy 中用于推文流处理的 StreamListener 类、geopy 和 folium 库（第 13 章）。一些例子需要通过继承定义类（第 10 章），但读者可以在阅读第 10 章前简单地模仿我们提供的类定义。

计算和数据科学课程

我们在为写本书做准备时阅读了以下的 ACM/IEEE CS 及其相关课程文档。

- 计算机科学课程 2013[20]
- CC2020：计算课程愿景 [21]
- 信息技术课程 2017[22]
- 网络安全课程 2017[23]

以及由 NSF 和高级研究院资助的研究团队发起的 2016 年"数据科学本科课程建议"[24]。

计算课程体系

- 根据"CC2020：计算课程愿景"，课程体系"需要修订和更新，以涵盖计算方面的新兴领域，例如网络安全和数据科学"[25]。
- 数据科学包括一些关键主题（除了常见内容以外），例如机器学习、深度学习、自然语言处理、语音合成与识别以及其他一些经典的人工智能技术，因此也包括计算机科学的一些主题。

数据科学课程

- 研究生层次的数据科学课程建设情况良好，本科生层次的数据科学课程正快速增长以满足强劲的行业需求。基于新课程提议，我们的鼓励动手实践的、非数学的、面向项目的、编程密集型的方法将促进数据科学内容进入本科课程。
- 现在已经有很多本科数据科学和数据分析项目，但是它们并未有机融合。为此，2016 年组建了由 25 位成员构成的数据科学课程委员会，并发布了"数据科学本科课程建议"以及相关的 10 门本科生主修课程。
- 课程委员会认为："为了提高综合课程所能提供的潜在协同效应和效率，许多传统的计算机科学、统计学和数学课程应该为数据科学专业的学生重新设计。"[26]
- 课程委员会建议利用计算机和数学思维整合这些领域的所有课程，并表明新教

---

20. ACM/IEEE (Assoc. Comput. Mach./Inst. Electr. Electron. Eng.). 2013. Computer Science Curricula 2013: Curriculum Guidelines for Undergraduate Degree Programs in Computer Science (New York: ACM), http://ai.stanford.edu/users/sahami/CS2013/final-draft/CS2013-final-report.pdf.
21. A. Clear, A. Parrish, G. van der Veer and M. Zhang "CC2020: A Vision on Computing Curricula", https://dl.acm.org/citation.cfm?id= 3017690.
22. Information Technology Curricula 2017, http://www.acm.org/binaries/content/assets/education/it2017.pdf.
23. Cybersecurity Curricula 2017, https://cybered.hosting.acm.org/wp-content/uploads/2018/02/newcover_csec2017.pdf.
24. "Curriculum Guidelines for Undergraduate Programs in Data Science", http://www.annualreviews.org/doi/full/10.1146/annurev-statistics-060116-053930.
25. http://delivery.acm.org/10.1145/3020000/3017690/p647-clear.pdf.
26. "Curriculum Guidelines for Undergraduate Programs in Data Science," http://www.annualreviews.org/doi/full/10.1146/annurev-statistics-060116-053930 (pp. 16–17).

科书是不可或缺的 [27]——这本书就是根据课程委员会的建议而设计的。
- Python 已经迅速成为世界上最受欢迎的通用编程语言之一。对于那些只想在数据科学主修课程中教授一种语言的学校来说，选择 Python 是合情合理的。

数据科学与计算机科学的重叠 [28]

在"数据科学本科课程建议"中，包括算法设计、程序设计、计算思维、数据结构、数据库、数学、统计思维、机器学习、数据科学和很多其他的内容——这和计算机科学有明显的重合部分，特别是数据科学课程还包括一些关于 AI 的关键话题。尽管这是一本 Python 编程教材，但是我们有效地将数据科学融入各种示例、习题、项目和完整实现的案例研究中，因此本书仍然涉及推荐的 10 门数据科学课程中相关领域的内容（除了较难的数学知识以外）。

"数据科学本科课程建议"中的关键点

在这一节中，我们从"数据科学本科课程建议"[29] 及其附录 [30] 中摘选了一些关键点，并努力将其和许多其他目标结合在一起：
- 学习计算机科学课程中常见的编程基础知识，包括数据结构的使用。
- 能够通过创建算法来解决问题。
- 学习过程式、函数式和面向对象的编程。
- 全面理解计算思维和统计思维，包括通过模拟来探究概念。
- 使用开发环境（我们使用 IPython 和 Jupyter Notebook）。
- 在每门课程的实际案例研究和项目中使用真实数据。
- 获取、探索和转换（辨析）数据以进行分析。
- 创建静态、动态和交互式的数据可视化。
- 提供可复现的结果。
- 使用现有的软件和基于云计算的工具。
- 使用统计模型和机器学习模型。
- 使用高性能工具（Hadoop、Spark、MapReduce 和 NoSQL）。

---

27. " Curriculum Guidelines for Undergraduate Programs in Data Science," http://www.annualreviews.org/doi/full/10.1146/annurev-statistics-060116-053930 (pp. 16–17).
28. 这一节主要面向数据科学方向的教师。因为针对计算机科学和相关学科而刚刚推出的"2020 计算课程"可能包含一些关键的数据科学话题，所以本节也为计算机科学教师提供了重要信息。
29. " Curriculum Guidelines for Undergraduate Programs in Data Science," http://www.annualreviews.org/doi/full/10.1146/annurev-statistics-060116-053930.
30. " Appendix—Detailed Courses for a Proposed Data Science Major, " http://www.annualreviews.org/doi/suppl/10.1146/annurev-statistics-060116-053930/suppl_file/st04_de_veaux_supmat.pdf.

- 关注数据的伦理、安全、隐私、可复现和透明性问题。

需要数据科学技能的工作

2011 年，麦肯锡全球研究所发表了报告《大数据：推动创新、竞争和生产力的新浪潮》。报告中写道："仅仅在美国，对于经验丰富的数据分析技术人员，缺口为 14 万～19 万名；对于能够分析大数据并根据分析结论做出决策的经理和分析师，缺口已达到 150 万名。"[31] 现在仍然是这样的情况。2018 年 8 月的《LinkedIn 劳动力报告》指出，美国缺少超过 15 万具有数据科学技能的人[32]。IBM、Burning Glass Technologies 和商业－高等教育论坛于 2017 年发布的一份报告称，到 2020 年，美国将会有成千上万的新工作需要数据科学技术[33]。

Jupyter Notebook

为了方便起见，我们以 Python 源代码（.py）文件格式提供本书的示例，可与命令行 IPython 解释器一起使用；同时提供了 Jupyter Notebook（.ipynb）文件格式，你可以将其加载到网络浏览器中并执行。你可以使用喜欢的任意一种方式执行代码示例。

Jupyter Notebook 是一个免费的开源项目，支持结合文本、图形、音频、视频和交互式的程序编码，便于在 Web 浏览器中快速且方便地输入、编辑、执行、调试和修改代码。文章《什么是 Jupyter？》这样评价它：

Jupyter 已成为科学研究和数据分析的标准。它将计算和参数结合在一起，让使用者能够构建"计算架构"。……并且简化了向同伴和同事分发工作软件的问题[34]。

根据我们的经验，它是一个同时适合新手和经验丰富的开发人员的绝佳学习环境和快速原型开发工具。因此，我们使用 Jupyter Notebook 而非 Eclipse、Visual Studio、PyCharm 或 Spyder 等传统的集成开发环境（IDE）。学术界和专业人士已经广泛使用 Jupyter 共享研究成果。传统的开源社区[35] 机制提供了对 Jupyter Notebook 的支持（具体可参见后文）。

我们认为 Jupyter Notebook 是一种优秀的 Python 教学工具，所以大多数教师都会选择使用 Jupyter。本书的 Jupyter Notebook 包括：

---

31. https://www.mckinsey.com/~/media/McKinsey/Business%20Functions/McKinsey%20Digital/Our%20Insights/Big%20data%20The%20next%20frontier%20for%20innovation/MGI_big_data_full_report.ashx (page 3).
32. https://economicgraph.linkedin.com/resources/linkedin-workforce-report-august-2018.
33. https://www.burning-glass.com/wp-content/uploads/The_Quant_Crunch.pdf (page 3).
34. https://www.oreilly.com/ideas/what-is-jupyter.
35. https://jupyter.org/community.

- 例子。
- 自检习题。
- 所有包含代码的章末习题。
- 可视化和动画，这是本书教学方法中的关键部分。我们以 Jupyter Notebook 格式提供代码，以便学生可以方便地复现我们的结果。

有关运行本书示例的信息，请参见 1.10 节。

合作和分享成果

"数据科学本科课程建议"[36] 中重点强调了团队合作和研究结果交流，这两方面内容对将要进入数据分析相关的工业、政府、学术领域的学生而言非常重要：

- 只需要复制文件或者通过 GitHub，你就可以非常方便地和团队成员分享自己创建的笔记。
- 研究结果（包括代码和心得）可以通过 nbviewer（https://nbviewer.jupyter.org）和 GitHub 之类的工具以静态网页形式共享——两者均自动将笔记变成网页展示。

Jupyter Notebook 对复现性的支持

在数据科学以及其他所有科学中，实验和研究应具有可复现性。这个问题多年来在很多文献中被讨论到，包括：

- 高德纳（Donald Knuth）在 1992 年出版的计算机科学著作 *Literate Programming*[37]。
- 文章 "Language-Agnostic Reproducible Data Analysis Using Literate Programming"[38] 中认为："Lir（文学，可再现计算）是基于高德纳提出的 'literate programming'（文学编程）的概念。"

从本质上讲，可复现性覆盖了用于产生结果的完整环境——硬件、软件、通信、算法（尤其是代码）、数据和数据的来源（起源和沿袭）。

"数据科学本科课程建议"在四个地方提到了可复现性这一目标。文章《数据科学的五十年》中提到，"让学生学会让工作可复现，可以让他们更轻松、更深入地评估自己的工作；让他们复现他人的部分分析结果，使他们能够学习诸如'原生数据分析'之类的技能，这些技能通常实践性很强，但未被系统性讲授。对他们开展围绕可复现性的训练，将使他们毕业后完成的工作成果更加可靠。"[39]

---

36. "Curriculum Guidelines for Undergraduate Programs in Data Science," http://www.annualreviews.org/doi/full/10.1146/annurev-statistics-060116-053930 (pp. 18–19).
37. Knuth, D., "Literate Programming" (PDF), The Computer Journal, British Computer Society, 1992.
38. http://journals.plos.org/plosone/article?id=10.1371/journal.pone.0164023.
39. "50 Years of Data Science," http://courses.csail.mit.edu/18.337/2015/docs/50YearsDataScience.pdf, p. 33.

Docker

第 17 章将介绍 Docker，它是一种将软件打包到容器中的工具，该工具能够跨平台地、方便地、可复现地和可移植地将软件执行需要的内容打包起来。我们在第 17 章中使用的某些软件包需要复杂的设置和配置。对于其中许多内容，你可以下载免费的现有版本的 Docker 容器。这使你能够避免复杂的安装问题，并在台式机或笔记本电脑上本地运行软件，从而使 Docker 成为一种帮助你快速且便捷地开始使用新技术的好方法。

Docker 还有助于提高可复现性。你可以创建自定义 Docker 容器，并为其配置学习中使用的每个软件和每个库的版本。这将使其他人能够重新构建你所使用的环境，然后再现你的工作，并帮助你再现自己的结果。在第 17 章中，你将使用 Docker 下载并执行一个预先配置的容器，以供你使用 Jupyter Notebook 编写和运行大数据 Spark 应用程序。

课堂测试

在本书编写过程中，我们的一位学术评审人——圣地亚哥大学经济学系助理教授 Alison Sanchez 博士在新课程"商业分析策略"中对本书进行了测试。她评论道："（这门课程）对于来自各种教育背景和专业的 Python 初学者来说真是太棒了。在我的课堂中，商业分析专业的学生刚开始这门课程时几乎没有编程经验。除了喜欢书中内容之外，他们还可以轻松地跟随示例进行练习，并在课程结束时，有能力使用从书中学到的技术来挖掘和分析 Twitter 数据。书中清楚地给出了示例代码的详细解析，这使没有计算机科学背景的学生容易理解。模块化的章节结构、广泛的当代数据科学领域的讨论话题以及配套的 Jupyter Notebook，使本书成为各种数据科学、商业分析和计算机科学等课程的教师和学生的绝佳资源。"

"翻转课堂"

现在，许多教师正在使用"翻转课堂"[40,41]。学生上课之前（通常通过视频授课）自行学习内容，上课时间将用于诸如动手写代码、以小组为单位的工作和讨论等任务。我们的书和附录用于翻转课堂是非常合适的：

- 我们提供了内容丰富的 VideoNotes，本书作者之一 Paul Deitel 将在视频中针对 Python 核心章节讲授相关概念。有关视频获取的详细信息请参见后文。
- 有些学生通过动手实践才能获得最好的学习效果，而仅仅视频是不够的。这本

---

40. https://en.wikipedia.org/wiki/Flipped_classroom.
41. https://www.edsurge.com/news/2018-05-24-a-case-for-flipping-learning-without-videos.

书最引人注目的特色之一是交互式教学方法——配有 538 个 Python 案例研究（许多仅包含一个或几个代码片段）以及 557 道带有答案的自检习题。这些使学生能够在得到即时反馈的基础上一点一点地学习——完全适合自主掌控节奏。学生可以轻松修改"热门"代码并查看更改的效果。

- 配套的 Jupyter Notebook 补充材料为学生提供了使用代码的便捷机制。
- 我们提供 471 个习题和项目，学生可以在家中和课堂上进行练习，其中许多都适用于小组项目。
- 我们在习题和项目中提供了许多有关伦理、隐私、安全等方面的探索性问题，这些适合课堂讨论和小组工作。

特色：IBM Watson 分析和认知计算

在本书编写的初期，我们对 IBM Watson 产生了浓厚的兴趣。我们进行了详尽的服务调查，发现 Watson "免费套餐"中的"无须信用卡"政策对我们的读者来说是最友好的。

IBM Watson 是一个认知计算平台，已经用于各种实际场景。认知计算系统模拟人脑的模式识别和决策能力，在得到更多的数据后进行"学习"[42,43,44]。书中包含重要的 Watson 实践方案。我们使用免费的 Watson Developer Cloud——Python SDK，它提供了应用程序编程接口（API），你可以通过编程方式与 Watson 的服务进行交互。Watson 使用起来很有趣，并且是帮助你发挥创意的绝佳平台。你将演示或使用以下 Watson API：对话、发现、语言翻译器、自然语言分类器、自然语言理解、个人见解、语音到文本、文本到语音、音调分析器和视觉识别。

Watson 的轻量级层服务和 Watson 案例研究

IBM 通过为其 API 提供免费的轻量级层来鼓励学习和实践[45]。在第 14 章中，你将尝试许多 Watson 服务的演示程序[46]。然后，你将使用 Watson 轻量级层的文字转语音、语音转文字和翻译服务去实现"旅行者助手"翻译 App。你将用英语说一个问题，然后该 App 会将你的语音翻译成英语文本，然后再将其翻译成西班牙语，最后变成西班牙语语音。接下来，你要说一个西班牙语回答（如果你不会说西班牙语，我们给你

---

42. http://whatis.techtarget.com/definition/cognitive-computing.
43. https://en.wikipedia.org/wiki/Cognitive_computing.
44. https://www.forbes.com/sites/bernardmarr/2016/03/23/what-everyone-should-knowabout-cognitive-computing.
45. 请务必查看 IBM 网站上的最新条款，因为条款和服务可能会发生更改。
46. https://console.bluemix.net/catalog/.

提供了一个可以使用的音频文件）。然后，该 App 会将语音快速翻译成西班牙语文本，再将文本翻译成英语并说出英语回复。是不是很酷！

教学方法

本书包含来自许多领域的丰富的案例研究、习题和项目。学生基于真实世界的数据集来解决有趣的现实问题。这本书专注于遵守软件工程的基本原则，同时还强调程序的清晰性。

使用不同字体来强调重点内容

我们把关键术语设置为粗体以便于识别。此外，我们把电脑屏幕上显示的相关界面组件用 bold Helvetica 字体表示，并使用 Lucida 字体表示 Python 代码。

目标和大纲

每章的开头都是对本章目标的介绍，从而让读者知道在接下来的一章中应该期待看到什么内容，同时也给予读者一个机会，可以在读完这一章后确定是否达到了预期的目标。章节大纲使得学生能够以自顶向下的方式来理解所学内容。

538 个案例研究

本书中的 538 个案例研究包含将近 4000 行代码。对于这么厚的一本书来说，这个代码量可以说是相当少的，这主要得益于 Python 是一门表达能力很强的语言。此外，我们的代码也尽可能地使用强大的类库来完成大部分工作。

160 个表格 / 插图 / 可视化表示

本书包含丰富的表格、线图以及其他的可视化表示。这些可视化表示以 2D、3D、静态、动态和交互式等多种形式呈现。

编程的智慧

我们荟萃了本书作者们加起来 90 多年的丰富编程和教学经验，将其整合到本书对于编程智慧的讨论中，包括：

- 良好的编程实践和我们推荐的 Python 习惯用法能帮助你编写出更清晰、更容易理解和更容易维护的程序。
- 列举了常见的编程错误，降低了你以后犯这些错误的可能性。
- "防错贴士"给出了如何在程序中定位和除去 bug 的建议，其中许多贴士描述了如何在一开始就防止 bug 进入程序。

- "性能贴士"重点强调可以使程序跑得更快或者占用更少内存的方法。
- "软件工程观察"重点强调软件（尤其是大型系统）体系结构和设计上的问题。

小结

在第 2~17 章的最后都有小结部分，总结了这一章所学的内容。

本书用到的软件

在本书中，所有你需要用到的软件都可以在 Windows、macOS 和 Linux 操作系统下运行，并且可以从因特网上免费下载。我们使用免费的 Anaconda Python 发行版编写此书中的案例研究，它包含了大部分你需要用到的 Python 库、可视化库和数据科学库，以及 Python、IPython 解释器、Jupyter Notebook 和 Spyder（非常优秀的 Python 数据科学集成开发环境）——虽然我们仅使用 IPython 和 Jupyter Notebook 来开发书中的程序。

Python 文档

在阅读本书时，你会发现以下文档很有帮助：
- Python 标准库：https://docs.python.org/3/library/index.html。
- Python 语言参考：https://docs.python.org/3/reference/index.html。
- Python 文档列表：https://docs.python.org/3/。

解答你的问题

在线论坛使得你可以和其他 Python 程序员互动，并解答你在 Python 上遇到的问题。常用的 Python 编程论坛以及通用的编程论坛包括：
- python-forum.io
- StackOverflow.com
- https://www.dreamincode.net/forums/forum/29-python/

除此之外，许多供应商会为他们的工具和库提供论坛。在本书中你将使用的大部分库都在 GitHub.com 上管理和维护，这些库的维护人员会通过项目主页上的 Issues 板块提供技术支持。如果无法在网上找到相关解答，请访问我们这本书的网站来获取帮助：http://www.deitel.com/[47]。

---

47. 我们的网站正在进行重大升级。如果你没有找到需要的东西，请直接通过 deitel@deitel.com 给我们发送电子邮件。

获得关于 Jupyter 的帮助

你可以可通过以下途径获得关于 Jupyter Notebook 的技术支持：
- Project Jupyter 谷歌论坛群：https://groups.google.com/forum/#!forum/jupyter。
- Jupyter 实时聊天室：https://gitter.im/jupyter/jupyter。
- GitHub：https://github.com/jupyter/help。
- StackOverflow：https://stackoverflow.com/questions/tagged/jupyter。
- Jupyter for Education Google Group（适用于采用 Jupyter 进行教学的教师）：https://groups.google.com/forum/#!forum/jupyter-education。

学生和教师的补充资源

下列补充资源适用于学生和教师。

代码示例和入门视频

为了充分理解本书，你应该在阅读相关讨论的同时执行每个代码示例。在本书网站 http://www.deitel.com/ 上，我们提供：
- 可下载的 Python 源代码（.py 文件）和 Jupyter Notebook 源代码（.ipynb 文件），涵盖书中的代码示例、基于代码的自检习题以及包含代码描述的章末习题。
- 入门视频，展示了如何使用 IPython 和 Jupyter Notebook 运行代码示例。我们会在 1.10 节介绍这些工具。
- 博客文章和本书更新。

配套网站

本书配套网站的地址是 https://www.pearson.com/deitel。配套网站除了包含上面提到的代码外，还有丰富的视频，在这些视频中，作者之一 Paul Deitel 解释了书中核心 Python 章节的大部分案例研究。

Pearson 教师资源中心的教师资源[48]

以下补充资源仅通过 Pearson Education 的 IRC（教师资源中心，地址为 http://www.pearsonhighered.com/irc）向有资格的教师提供：

---

48. 关于教辅资源，仅提供给采用本书作为教材的教师用作课堂教学、布置作业、发布考试等。如有需要的教师，请直接联系 Pearson 北京办公室查询并填表申请。联系邮箱：Copub.Hed@pearson.com。
关于配套网站资源，大部分需要访问码，访问码只有原英文版提供，中文版无法使用。——编辑注

- PPT 幻灯片。
- 教师答案手册：包含大部分习题的解析。对于项目和研究习题，我们没有提供答案解析——这其中有许多涉及本质的问题，并且适合作为学期级项目、专业方向级项目、拔尖课程项目和论文题目。在把一道习题布置为作业之前，教师应确保在 IRC 上能找到这道题对应的解答。
- 测验文档：包含多项选择题、简答题及答案，并且这些练习都很容易使用自动化评分工具来进行评分。

请不要直接写信向我们请求拥有上述教师资源（包括习题答案）的访问权限。访问权限只对使用这本书进行教学的大学教师可用。符合条件的教师可以通过 Pearson 代理获取 IRC 的访问权限。如果你不是我们的注册教师成员，请与你的 Pearson 代理联系或者访问以下网址：https://www.pearson.com/replocator。

考试试卷副本

教师可通过 Pearson 代理索取关于本书的考试试卷的副本：https://www.pearson.com/replocator。

和作者保持联系

如果需要向我们提问、需要教学进度上的协助或者向我们报告书中的错误，请给我们发送电子邮件：deitel@deitel.com。

或者在社交媒体上和我们互动：
- Facebook（http://www.deitel.com/deitelfan）
- Twitter（@deitel）
- LinkedIn（http://linkedin.com/company/deitel-&-associates）
- YouTube（http://youtube.com/DeitelTV）

致谢

我们要感谢 Barbara Deitel 为这个项目在网络上花费了很长时间收集资料。同时，我们很幸运能够和 Pearson 出版社的专业出版团队合作。我们还要感谢 Tracy Johnson（计算机科学高等教育课件联合执行经理）的指导、智慧和付出——她在本书创作过程中的每一步都向我们提出挑战，要求我们"止于至善"。Carole Snyder 负责这本书的生产制作并且和 Pearson 许可团队沟通，迅速处理了书中的图片和引用涉及的版权问题。我们选定了封面的艺术风格，然后由 Chuti Prasertsith 完成封面的设计。

我们还要感谢学术专业评审人所付出的努力。Meghan Jacoby 和 Patricia Byron-Kimball 招募了评审专家并负责管理评审过程。在非常紧张的时间安排下，评审专家仔细审查了我们的工作，在提高表达的准确性、完整性和时效性方面提供了许多建议。

评审专家列表

提案评审人	
• Irene Bruno 博士，乔治·梅森大学信息科学与技术系副教授。	• Harvey Siy 博士，内布拉斯加大学奥马哈分校信息科学与技术计算机科学副教授。
• Lance Bryant，希彭斯堡大学数学系副教授。	• Jamie Whitacre，独立数据科学顾问。
• Daniel Chen，Lander Analytics 公司数据科学家。	**书籍评审人**
Garrett Dancik，东康涅狄格州立大学计算机科学/生物信息学系副教授。	• Daniel Chen，Lander Analytics 公司数据科学家。
• Marsha Davis 博士，东康涅狄格州立大学数学科学系主任。	• Garrett Dancik，东康涅狄格州立大学计算机科学/生物信息学系副教授。
• Roland DePratti，东康涅狄格州立大学计算机科学系兼职教授。	• Pranshu Gupta，迪西尔斯大学计算机科学系助理教授。
• Shyamal Mitra，得克萨斯大学奥斯汀分校计算机科学系高级讲师。	• David Koop，马萨诸塞大学达特茅斯分校数据科学助理教授、数据科学项目联合主任。
• Mark Pauley 博士，内布拉斯加大学奥马哈分校信息科学学院生物信息学高级研究员。	• Ramon Mata-Toledo，詹姆斯·麦迪逊大学计算机科学系教授。
• Sean Raleigh，威斯敏斯特学院数学系副教授、数据科学系主任。	• Shyamal Mitra，得克萨斯大学奥斯汀分校计算机科学系高级讲师。
• Alison Sanchez，圣地亚哥大学经济学系助理教授。	• Jamie Whitacre，独立数据科学顾问。
	• Elizabeth Wickes，伊利诺伊大学信息科学学院讲师。

特别感谢

我们要特别感谢圣地亚哥大学的 Alison Sanchez 助理教授，她在圣地亚哥大学新开设的"商业分析策略"课程中使用本书的预发布版进行了课堂测试。她看完了冗长的使用建议，在还没有见过书的情况下就决定采用这本书，并且签字成为本书的评审人员。我们真诚地感谢她在整本书的写作过程中提供的指导（和勇气）。

现在让我们开始吧！当你阅读这本书的时候，我们将非常感谢你的评论、批评、纠错和改进建议。你可以将邮件发送到 deitel@deitel.com，我们会在第一时间做出回应。

再次欢迎你来到激动人心的 Python 开源编程世界。我们希望你能享受这本书，以及它所包含的和 Python、IPython、Jupyter Notebook、AI、大数据、云技术相关的前沿计算机应用开发技术。最后，我们祝你前程似锦！

关于作者

Paul J. Deitel，Deitel & Associates 公司首席执行官兼首席技术官，毕业于麻省理工学院，在计算机领域拥有 38 年的经验。Paul 是经验丰富的编程语言培训专家，自 1992 年以来就为软件开发人员教授专业课程。他已经向来自全球的企业客户提供了数百门编程课程，包括思科、IBM、西门子、Sun Microsystems（现在为 Oracle）、戴尔、富达、肯尼迪航天中心的 NASA、国家严重风暴实验室、白沙导弹靶场、Rogue Wave 软件、波音、北电网络、彪马、iRobot 等。他和他的合作者 Harvey M. Deitel 博士是世界上畅销的编程语言教科书 / 专业书籍 / 视频作者。

Harvey M. Deitel 博士，Deitel & Associates 公司董事长兼首席战略官，在计算领域拥有 58 年的经验。Deitel 博士在麻省理工学院电气工程系获得理学学士学位和硕士学位，在波士顿大学的数学系获得博士学位——他在这些专业分离出计算机科学专业前就已经学过相关知识了。在 1991 年与儿子 Paul 创立 Deitel & Associates 公司之前，他已经获得了波士顿大学的终身职位并担任计算机科学系主任，拥有丰富的大学教学经验。Deitel 品牌的出版物赢得了国际上的广泛认可，并被翻译为日语、德语、俄语、西班牙语、法语、波兰语、意大利语、简体中文、繁体中文、韩语、葡萄牙语、希腊语、乌尔都语和土耳其语等 100 多种语言出版。Deitel 博士已为学术、公司、政府和军事客户提供了数百门编程课程。

关于 Deitel & Associates 公司

Deitel & Associates 公司由 Paul Deitel 和 Harvey Deitel 创建，是一家国际认可的计算机类著作创作和企业培训组织，专门研究计算机编程语言、对象技术、移动 App 开发以及 Internet 和 Web 软件技术。该公司的培训客户包括一些世界上的大公司、政府机构、军事部门和学术机构。该公司在世界各地的客户网站上提供关于主流编程语言和平台的有讲师指导的培训课程。

通过与 Pearson/Prentice Hall 44 年的合作，Deitel & Associates 公司以印刷物和电子书的形式出版了前沿的编程教科书和专业书籍，发布了前沿的编程方面的 LiveLessons 视频课程、Safari-Live 在线研讨会和 Revel 交互式多媒体课程。如果你需要联系 Deitel & Associates 公司和作者，或者希望给有讲师指导的现场培训课程提出建议，请发送电子邮件至 deitel@deitel.com。希望了解更多关于 Deitel 现场企业培训的信息，请访问 http://www.deitel.com/training。希望购买 Deitel 书籍的个人客户，请访问 https://www.amazon.com/。公司、政府、军队和学术机构的大宗订单请直接与 Pearson 联系。希望了解更多信息，请访问 https://www.informit.com/store/sales.aspx。

阅读前的准备工作

本部分包括读者在开始阅读本书前需要了解的信息。如有信息更新，我们会将其放在 http://www.deitel.com。

获取代码示例

在我们为本书提供的网页（http://www.deitel.com）上，通过点击 Download Examples 链接可以将 examples.zip 文件下载到本地计算机，其中包含本书的所有代码示例。大多数浏览器会将文件自动保存到用户账户的 Downloads 文件夹中。也可以通过 Pearson 的配套网站（https://pearson.com/deitel）下载本书代码示例。

下载完成后，将其中的 examples 文件夹提取到用户账户的 Documents 文件夹：
- Windows 用户：C:\Users\ 用户账户名 \Documents\examples
- macOS 或 Linux 用户：~/Documents/examples

大多数操作系统有内置的提取工具，读者也可以使用 7-Zip（www.7-zip.org）或 WinZip（www.winzip.com）等压缩工具。

examples 文件夹的结构

在本书中，读者将以下面三种形式执行示例：
- IPython 交互式环境中的单独代码段。
- 完整的应用程序，即脚本。
- Jupyter Notebook：一种基于浏览器的交互式便捷编程环境，在该环境中读者可以编写并执行代码，还可以将代码与文本、图像和视频混合在一起。

1.10 节中将给出具体的操作演示。

examples 文件夹包含了多个子文件夹，每个子文件夹对应一章。子文件夹命名为 ch##，## 是两位数字的章编号 01~16，如 ch01。除了第 14、16 和 17 章，其他章的文件夹包含以下内容：
- snippets_ipynb：包含该章 Jupyter Notebook 文件的文件夹。
- snippets_py：包含该章 Python 源代码文件的文件夹。各代码段之间以一个空行分隔。读者可以将这些代码段复制并粘贴到 IPython 或 Jupyter Notebook 中运行。
- 脚本文件及其支持文件。

第 14 章包含了一个应用程序。ch16 和 ch17 文件夹中所需文件的位置分别在第 16

章和第 17 章进行了说明。

安装 Anaconda

本书使用易于安装的 Anaconda Python 发行版。它包含了执行示例所需的绝大多数内容，包括：
- IPython 解释器。
- 本书所使用的大多数 Python 和数据科学库。
- Jupyter Notebook 本地服务器，以便读者下载并执行我们所提供的 notebook 文件。
- Spyder 集成开发环境（IDE）等其他软件包，本书中仅用到了 IPython 和 Jupyter Notebook。

从 https://www.anaconda.com/download/ 可以下载 Windows、macOS 或 Linux 的 Python 3.x Anaconda 安装程序。下载完成后，运行安装程序并根据屏幕上的提示完成操作。注意安装完成后不要移动安装好的文件位置，以确保 Anaconda 能够正常运行。

更新 Anaconda

接下来，确保 Anaconda 已更新至最新版本。按下面的方式在本地系统上打开一个命令行窗口：
- 对于 macOS，从 Applications 文件夹的 Utilities 子文件夹中打开 Terminal。
- 对于 Windows，从开始菜单中打开 Anaconda Prompt。注意，如果是为了更新 Anaconda 或安装新的软件包，则需要右键单击 Anaconda Prompt，然后选择 More>Run as administrator。（如果在开始菜单中找不到 Anaconda Prompt，在屏幕下面的 Type here to search 框中进行搜索即可。）
- 对于 Linux，打开系统的 Terminal 或 shell（不同 Linux 发行版会有所不同）。

在本地系统的命令行窗口执行下面的命令可以将 Anaconda 已安装的包更新到最新版本：

1. `conda update conda`
2. `conda update --all`

包管理器

上面使用的 conda 命令会调用 conda 包管理器，这是本书中所使用的两个重要的 Python 包管理器之一。本书所使用的另一个包管理器是 pip。软件包包含了安装特定 Python 库或工具所需的文件。在本书中，优先使用 conda 安装软件包，只有在无法使用 conda 安装软件包时，才会使用 pip。有些人喜欢使用 pip，因为目前它支持更多的软件包。读者如果在使用 conda 安装软件包时遇到问题，请尝试使用 pip。

安装 Prospector 静态代码分析工具

读者可能需要使用 Prospector 分析工具来分析 Python 代码，该工具会检查代码中的常见错误并帮助读者进行改进。要安装 Prospector 及其使用的 Python 库，请在命令行窗口中运行以下命令：

```
pip install prospector
```

安装 jupyter-matplotlib

本书使用名为 Matplotlib 的可视化库实现了一些动画。要在 Jupyter Notebook 中使用它们，必须安装一个名为 ipympl 的工具。在先前打开的终端、Anaconda 命令提示符或 shell 中，依次执行以下命令[1]：

```
conda install -c conda-forge ipympl
conda install nodejs
jupyter labextension install @jupyter-widgets/jupyterlab-manager
jupyter labextension install jupyter-matplotlib
```

安装其他包

Anaconda 提供了大约 300 种流行的 Python 和数据科学包，如 NumPy、Matplotlib、pandas、Regex、BeautifulSoup、request、Bokeh、SciPy、Scikit-learn、Seaborn、spaCy、sqlite、statsmodels 等。运行本书示例代码需要安装的其他软件包数量很少，我们将在必要时提供安装说明。当读者需要安装新的软件包时，可以参考软件包的文档完成安装。

获得 Twitter 开发者账号

如果读者要运行"Twitter 数据挖掘"一章及后续章节中任何基于 Twitter 的示例，请先申请一个 Twitter 开发者账号。Twitter 现在要求先注册才能访问其 API。要申请 Twitter 开发者账号，请在 https://developer.twitter.com/en/apply-for-access 上填写信息并提交申请。Twitter 会审核每个申请。在撰写本书时，个人开发者账号会立即通过审

---

1. https://github.com/matplotlib/jupyter-matplotlib。

批；公司账号申请则需要几天到几周的时间，且有可能无法通过审批。

部分章节需要的网络连接

使用本书时，读者需要连接互联网才能安装各种其他 Python 库。在部分章节中，读者需要注册云服务账号来使用其免费套餐，其中某些服务需要通过信用卡验证用户的身份。在一些情况下，读者会使用非免费的服务。此时，读者需要利用供应商提供的货币信用额度，从而可以免费试用其服务。注意：在完成设置后，某些云服务会产生费用。因此，当读者使用此类服务完成案例研究时，请确保立即删除分配的资源。

程序输出的细微差异

在执行代码示例时，读者可能会注意到书中给出的结果与自己运行的结果之间存在一些差异：

- 由于不同操作系统进行浮点数（如 −123.45、7.5 或 0.0236937）计算的方式不同，因此可能产生输出结果的细微变化，尤其是距离小数点右边很远的那些数字。
- 当在单独的窗口中显示输出结果时，我们会裁剪窗口以删除其边界。

目 录

第1章 计算机和Python简介 ·········· 1
- 1.1 引言 ·· 2
- 1.2 硬件和软件 ·································· 3
  - 1.2.1 摩尔定律 ····························· 4
  - 1.2.2 计算机组成 ························· 4
- 1.3 数据层级 ·· 6
- 1.4 机器语言、汇编语言和高级语言 ··· 9
- 1.5 对象技术简介 ······························ 10
- 1.6 操作系统 ······································ 13
- 1.7 Python简介 ································· 16
- 1.8 （语言）库 ·································· 18
  - 1.8.1 Python标准库 ···················· 18
  - 1.8.2 数据科学库 ························· 18
- 1.9 其他常见编程语言 ····················· 20
- 1.10 试用：使用IPython和Jupyter Notebook ···················· 21
  - 1.10.1 将IPython交互模型用作计算器 ························ 21
  - 1.10.2 使用IPython解释器执行Python程序 ··············· 23
  - 1.10.3 在Jupyter Notebook中编写和执行代码 ············ 24
- 1.11 Internet和WWW ····················· 29
  - 1.11.1 Internet：网际网 ············ 29
  - 1.11.2 WWW：用户友善的Internet ···························· 30
  - 1.11.3 计算和资源云 ·················· 30
  - 1.11.4 物联网 ······························ 31
- 1.12 软件技术 ···································· 32
- 1.13 大数据 ·· 33
  - 1.13.1 大数据分析 ······················ 38
  - 1.13.2 数据科学和大数据案例研究 ······························ 39
- 1.14 数据科学入门：大数据移动应用案例研究 ························· 40

第2章 Python程序设计简介 ·········· 49
- 2.1 引言 ·· 50
- 2.2 变量和赋值语句 ·························· 50
- 2.3 算术操作 ······································ 52
- 2.4 print函数、单引号字符串和双引号字符串 ····························· 56
- 2.5 三引号字符串 ······························ 58
- 2.6 从用户处获得输入 ······················ 59
- 2.7 判断：if语句与比较操作 ·········· 61
- 2.8 对象和动态类型 ·························· 66
- 2.9 数据科学入门：基本统计功能 ··· 68
- 2.10 小结 ·· 70

第3章 控制语句和程序设计 ·········· 73
- 3.1 引言 ·· 74
- 3.2 算法 ·· 74
- 3.3 伪代码 ·· 75
- 3.4 控制语句 ······································ 75
- 3.5 if语句 ·· 78
- 3.6 if...else和if...elif...else语句 ···· 80
- 3.7 while语句 ··································· 85
- 3.8 for语句 ·· 86
  - 3.8.1 迭代、列表和迭代器 ········· 88

3.8.2	内置 range 函数 …… 88	4.12	方法：归属于对象的函数 …… 138
3.9	增量赋值 …… 89	4.13	作用域规则 …… 138
3.10	程序设计：通过序列控制重复 …… 90	4.14	import：进一步讨论 …… 140
		4.15	给函数传递实参：进一步讨论 …… 142
3.10.1	需求声明 …… 90	4.16	函数调用栈 …… 145
3.10.2	算法的伪代码形式 …… 90	4.17	函数式程序设计 …… 146
3.10.3	在 Python 中为算法编码 …… 91	4.18	数据科学入门：数据分布的度量 …… 148
3.10.4	格式化字符串简介 …… 92	4.19	小结 …… 150
3.11	程序设计：通过哨兵控制重复 …… 93		
3.12	程序设计：嵌套控制结构 …… 97	**第 5 章**	**序列：列表和元组** …… 155
3.13	内置函数 range：进一步讨论 …… 101	5.1	引言 …… 156
		5.2	列表 …… 156
3.14	使用 Decimal 类型表达货币总量 …… 102	5.3	元组 …… 161
		5.4	序列拆包 …… 163
3.15	break 和 continue 语句 …… 105	5.5	序列切片 …… 166
3.16	布尔操作 and、or 和 not …… 106	5.6	del 语句 …… 169
3.17	数据科学入门：趋势的度量——均值、中值、众数 …… 109	5.7	给函数传递列表 …… 171
		5.8	排序列表 …… 172
3.18	小结 …… 111	5.9	搜索序列 …… 174
		5.10	其他列表方法 …… 176
第 4 章	**函数** …… 119	5.11	用列表模拟栈 …… 178
4.1	引言 …… 120	5.12	列表解析 …… 179
4.2	函数的定义 …… 120	5.13	生成器表达式 …… 181
4.3	多参数函数 …… 123	5.14	过滤器、映射和约简 …… 182
4.4	随机数生成器 …… 125	5.15	其他序列处理函数 …… 185
4.5	案例研究：机会游戏 …… 128	5.16	二维列表 …… 187
4.6	Python 标准库 …… 131	5.17	数据科学入门：模拟和静态可视化 …… 191
4.7	math 模块函数 …… 132		
4.8	使用 IPython 的 tab 补全功能 …… 133	5.17.1	600、60000 和 6000000 次掷骰子的图示 …… 191
4.9	缺省形参值 …… 135		
4.10	关键字实参 …… 136	5.17.2	掷骰子实验的序列和百分比的可视化 …… 193
4.11	任意实参表 …… 136		

5.18 小结……………………………… 199	7.8 NumPy 计算方法…………………… 250
	7.9 全局函数……………………………… 252
第6章 字典和集合……………… 209	7.10 索引和切片………………………… 254
6.1 引言………………………………… 210	7.11 视图：浅拷贝……………………… 256
6.2 字典………………………………… 210	7.12 深拷贝……………………………… 258
6.2.1 创建字典…………………… 210	7.13 转换和转置………………………… 259
6.2.2 在字典中遍历……………… 212	7.14 数据科学入门：pandas Series
6.2.3 基本的字典操作…………… 212	和 DataFrame ……………………… 262
6.2.4 字典方法 keys 和 values…… 214	7.14.1 pandas series ……………… 262
6.2.5 字典比较…………………… 216	7.14.2 DataFrame ………………… 267
6.2.6 案例研究：学生成绩字典… 217	7.15 小结………………………………… 275
6.2.7 案例研究：词计数………… 218	
6.2.8 字典方法 update …………… 220	**第8章 字符串：进一步讨论**…… 283
6.2.9 字典解析…………………… 220	8.1 引言…………………………………… 284
6.3 集合………………………………… 221	8.2 格式化字符串………………………… 285
6.3.1 集合比较…………………… 223	8.2.1 类型声明……………………… 285
6.3.2 集合的数学操作…………… 225	8.2.2 域宽和对齐…………………… 286
6.3.3 集合的可变操作和方法…… 226	8.2.3 数值格式化…………………… 287
6.3.4 集合解析…………………… 228	8.2.4 字符串 format 方法………… 288
6.4 数据科学入门：动态可视化… 228	8.3 字符串拼接和重复…………………… 289
6.4.1 了解动态可视化…………… 228	8.4 字符串空白符剥离…………………… 290
6.4.2 实现动态可视化…………… 231	8.5 改变字符的大小写…………………… 291
6.5 小结………………………………… 234	8.6 字符串比较操作……………………… 292
	8.7 子串搜索……………………………… 292
第7章 使用 NumPy 进行面向数组的编程……………………… 239	8.8 子串替换……………………………… 294
	8.9 字符串切分和合并…………………… 294
7.1 引言………………………………… 240	8.10 字符和字符测试方法……………… 297
7.2 从已有数据中创建数组…………… 241	8.11 原生字符串………………………… 298
7.3 数组属性…………………………… 242	8.12 正则表达式简介…………………… 299
7.4 用特定值填充数组………………… 244	8.12.1 re 模块和 fullmatch
7.5 使用 range 创建数组……………… 244	函数…………………………… 300
7.6 列表与数组的性能比较：	8.12.2 子串替换和串切分………… 303
%timeit 简介…………………… 246	8.12.3 其他搜索函数和匹配
7.7 数组操作…………………………… 248	处理…………………………… 304

8.13 数据科学入门：pandas、正则表达式和数据治理 ············ 307
8.14 小结 ············ 312

第9章 文件和异常 ············ 319

9.1 引言 ············ 320
9.2 文件 ············ 321
9.3 文本文件处理 ············ 321
 9.3.1 写文本文件：with 语句简介 ············ 322
 9.3.2 读文本文件 ············ 323
9.4 更新文本文件 ············ 325
9.5 JSON 序列化 ············ 327
9.6 安全问题：pickle 序列化和反序列化 ············ 330
9.7 关于文件的其他说明 ············ 330
9.8 异常处理 ············ 331
 9.8.1 除 0 异常和非法输入 ············ 332
 9.8.2 try 语句 ············ 332
 9.8.3 在 except 从句中捕捉多重异常 ············ 335
 9.8.4 函数或过程能够抛出什么异常 ············ 336
 9.8.5 try 套件应该封装什么代码 ············ 336
9.9 finally 子句 ············ 336
9.10 显式引发异常 ············ 339
9.11 （可选）栈展开和回溯 ············ 339
9.12 数据科学入门：CSV 文件的处理 ············ 342
 9.12.1 Python 标准库模块 CSV ············ 342
 9.12.2 将 CVS 文件读入 pandas DataFrame ············ 344
 9.12.3 读取 Titanic Disaster 数据库 ············ 346
 9.12.4 对 Titanic Disaster 数据库进行简单的数据分析 ············ 347
 9.12.5 乘客年龄直方图 ············ 348
9.13 小结 ············ 349

第10章 面向对象程序设计 ············ 355

10.1 引言 ············ 356
10.2 定制类 Account ············ 358
 10.2.1 试用 Account 类 ············ 358
 10.2.2 Account 类的定义 ············ 360
 10.2.3 组合：对象引用作为类的成员 ············ 361
10.3 属性的受控访问 ············ 363
10.4 数据访问的特性 ············ 364
 10.4.1 试用 Time 类 ············ 364
 10.4.2 Time 类的定义 ············ 366
 10.4.3 Time 类定义的设计要领 ············ 370
10.5 私有属性模拟 ············ 371
10.6 案例研究：洗牌和切牌 ············ 373
 10.6.1 试用 Card 和 DeckofCards 类 ············ 373
 10.6.2 Card 类属性简介 ············ 375
 10.6.3 DeckofCards 类 ············ 377
 10.6.4 在 Matplotlib 中显示扑克图片 ············ 378
10.7 继承：基类和子类 ············ 382
10.8 构建继承层次和多态简介 ············ 384
 10.8.1 基类 Commission-Employee ············ 384
 10.8.2 子类 SalariedCommission-Employee ············ 387
 10.8.3 CommissionEmployee 和 SalariedCommissionEmployee 的多态处理 ············ 391

10.8.4　关于基于对象和面向对象
　　　　程序设计的说明……………391
10.9　鸭子类型和多态………………392
10.10　操作符重载……………………393
　　10.10.1　试用 Complex 类 …………394
　　10.10.2　Complex 类的定义 ………395
10.11　异常类层次和定制异常
　　　　处理…………………………397
10.12　有名元组……………………399
10.13　Python 3.7 新数据类简介……400
　　10.13.1　创建 Card 数据类 …………401
　　10.13.2　使用 Card 数据类 …………403
　　10.13.3　数据类相较有名元组的
　　　　　　优势……………………405
　　10.13.4　数据类相较传统类的
　　　　　　优势……………………406
10.14　使用文档字符串和 doctest
　　　　进行单元测试………………406
10.15　命名空间和作用域…………411
10.16　数据科学入门：时间序列和
　　　　简单线性回归………………414
10.17　小结…………………………423

第 11 章　计算机科学思维：
　　　　 递归、搜索、排序
　　　　 和大 O 表示法 ……………431

11.1　引言……………………………432
11.2　阶乘……………………………433
11.3　阶乘的递归法…………………433
11.4　斐波那契数列的递归法………436
11.5　递归和循环……………………439
11.6　搜索和排序……………………440
11.7　线性搜索………………………440
11.8　算法效率：大 O 表示法 ………442

11.9　二叉搜索………………………444
　　11.9.1　二叉搜索的实现……………445
　　11.9.2　二叉搜索的大 O 表示法 ……447
11.10　排序算法………………………448
11.11　选择排序………………………448
　　11.11.1　选择排序的实现……………449
　　11.11.2　效用函数 print_pass ………450
　　11.11.3　选择排序的大 O 表示法 ……451
11.12　插入排序………………………451
　　11.12.1　插入排序的实现……………452
　　11.12.2　插入排序的大 O 表示法 ……453
11.13　归并排序………………………454
　　11.13.1　归并排序的实现……………454
　　11.13.2　归并排序的大 O 表示法 ……459
11.14　总结：本章算法的大 O
　　　　 表示法………………………459
11.15　可视化算法……………………460
　　11.15.1　生成函数……………………462
　　11.15.2　实现选择排序算法的
　　　　　　动画演示………………463
11.16　小结……………………………468

第 12 章　自然语言处理……………477

12.1　引言……………………………478
12.2　TextBlob………………………479
　　12.2.1　创建 TextBlob ………………481
　　12.2.2　语料化：文本的断句
　　　　　　和取词…………………482
　　12.2.3　言语分部标注………………482
　　12.2.4　提取名词短语………………483
　　12.2.5　使用 TextBlob 缺省情绪分
　　　　　　析器进行文本情绪分析……484
　　12.2.6　使用 NaiveBayesAnalyzer
　　　　　　进行文本情绪分析………486

12.2.7	语言检测和翻译 ············ 487		13.6	Tweepy ··················· 525
12.2.8	屈折辨析：多元化和		13.7	使用 Tweepy 在 Twitter 中
	单一化 ············ 489			认证 ············ 525
12.2.9	拼写检查和更正 ········ 489		13.8	从 Twitter 账户中获取信息 ··· 527
12.2.10	规范化：词根和词性还原 ··· 490		13.9	Tweepy Cursors 简介：获取
12.2.11	词频 ············ 491			账户的关注者和好友 ············ 529
12.2.12	从 WordNet 中获取定义、		13.9.1	确定账户的关注者 ············ 529
	同义词、反义词 ············ 492		13.9.2	确定账户关注了谁 ············ 532
12.2.13	停用词删除 ············ 494		13.9.3	获取用户当前的推文 ············ 532
12.2.14	n-gram 模型 ············ 496		13.10	搜索当前推文 ············ 534
12.3	使用 Bar Charts 和 Word Clouds		13.11	趋势发现：Twitter
	进行词频可视化 ············ 497			趋势 API ············ 536
12.3.1	使用 Bar Charts 进行词频		13.11.1	趋势主题位置获取 ············ 536
	可视化 ············ 497		13.11.2	趋势主题列表获取 ············ 537
12.3.2	使用 Word Clouds 进行词频		13.11.3	从趋势主题中创建词云 ············ 539
	可视化 ············ 500		13.12	推文分析前的清理 /
12.4	使用 Textatistic 进行可读性			预处理 ············ 541
	评测 ············ 503		13.13	Twitter 流处理 API ············ 542
12.5	使用 spaCy 进行有名实体		13.13.1	创建 StreamListener 的
	识别 ············ 505			子类 ············ 543
12.6	使用 spaCyn 进行相似性		13.13.2	流处理初始化 ············ 545
	评估 ············ 507		13.14	推文情感分析 ············ 547
12.7	其他 NLP 工具和库 ············ 509		13.15	地址匹配与映射 ············ 551
12.8	机器学习和深度学习的自然		13.15.1	推文的获取和映射 ············ 552
	语言应用 ············ 509		13.15.2	twetutilities.py 效用函数 ··· 556
12.9	自然语言数据集 ············ 510		13.15.3	LocationListener 类 ············ 558
12.10	小结 ············ 510		13.16	存储推文的方式 ············ 559
			13.17	Twitter 和时间序列 ············ 560
第 13 章	**Twitter 数据挖掘** ············ 515		13.18	小结 ············ 560
13.1	引言 ············ 516			
13.2	Twitter API 概述 ············ 518		**第 14 章**	**IBM Watson 和认知**
13.3	创建 Twitter 账户 ············ 519			**计算** ············ 565
13.4	获取 Twitter 证书——		14.1	引言：IBM Watson 和认知
	创建 App ············ 520			计算 ············ 566
13.5	推文中有什么 ············ 521		14.2	IBM 云账户和云控制台 ············ 568

14.3 Watson 服务 568	15.3.4 超参调优 619
14.4 附加服务和工具 572	15.4 案例研究：时间序列和简单线性回归 620
14.5 Watson 开发者云 Python SDK 573	15.5 案例研究：加州订房数据集上的多线性回归 625
14.6 案例研究：旅行者助手——翻译 App 574	15.5.1 载入数据集 626
14.6.1 运行 App 前的准备工作 575	15.5.2 用 pandas 观察数据集 628
14.6.2 试用 App 576	15.5.3 特征的可视化 630
14.6.3 SimpleLanguageTranslator.py 脚本走查 577	15.5.4 划分训练集和测试集 634
14.7 Watson 资源 587	15.5.5 训练模型 634
14.8 小结 589	15.5.6 测试模型 635
	15.5.7 预期及预测价格的可视化 636
第 15 章 机器学习：分类、回归和聚类 593	15.5.8 回归模型度量 637
15.1 机器学习入门 594	15.5.9 最佳模型选择 638
15.1.1 Scikit-learn 595	15.6 案例研究：无监督机器学习——降维 639
15.1.2 机器学习的类别 596	15.7 案例研究：无监督机器学习——k 均值聚类 642
15.1.3 Scikit-learn 数据集 598	15.7.1 载入 Iris 数据集 644
15.1.4 经典的数据科学研究步骤 599	15.7.2 观察 Iris 数据集：使用 pandas 进行统计描述 646
15.2 案例研究：数字数据集上的 k 近邻分类法（第一部分） 599	15.7.3 使用 Seaborn pairplot 可视化数据集 647
15.2.1 k 近邻算法 601	15.7.4 使用评估器 KMeans 650
15.2.2 载入数据集 602	15.7.5 使用主成分分析法进行降维 652
15.2.3 可视化数据 606	15.7.6 选择最佳聚类评估器 655
15.2.4 划分训练集和测试集 608	15.8 小结 656
15.2.5 创立模型 609	
15.2.6 训练模型 610	**第 16 章 深度学习 665**
15.2.7 预测数字类别 610	16.1 引言 666
15.3 案例研究：数字数据集上的 k 近邻分类法（第二部分） 612	16.1.1 深度学习应用 668
15.3.1 模型的准确性度量 612	16.1.2 深度学习演示程序 669
15.3.2 k 折交叉验证 616	
15.3.3 多模型寻优 617	

16.1.3 Keras 资源 ································ 669
16.2 Keras 内置数据集 ································ 669
16.3 Anaconda 定制化环境 ···················· 670
16.4 神经网络 ·· 672
16.5 张量 ·· 674
16.6 视觉处理卷积神经网络和多分类器 ································ 676
 16.6.1 载入 MNIST 数据集 ·············· 677
 16.6.2 数据探索 ································ 678
 16.6.3 数据准备 ································ 680
 16.6.4 构造神经网络 ························ 682
 16.6.5 模型训练和评价 ···················· 691
 16.6.6 模型存储和载入 ···················· 696
16.7 使用 TensorBoard 进行神经网络训练的可视化 ································ 697
16.8 ConvnetJS：基于浏览器的深度学习训练和可视化 ········· 700
16.9 序列处理中的循环神经网络和 IMDb 数据集的情感分析 ································ 701
 16.9.1 载入 IMDb 电影评论数据集 ································ 702
 16.9.2 数据探索 ································ 703
 16.9.3 数据准备 ································ 705
 16.9.4 构造神经网络 ························ 706
 16.9.5 模型训练和评价 ···················· 709
16.10 深度学习模型调参 ···················· 710
16.11 ImageNet 上的卷积网络模型预学习 ································ 711
16.12 强化学习 ································ 712
 16.12.1 深度强化学习 ························ 713
 16.12.2 OpenAI Gym ························ 713
16.13 小结 ································ 714

第 17 章 大数据：Hadoop、Spark、NoSQL 和 IoT ················ 723
17.1 引言 ·· 724
17.2 关系型数据库和结构化查询语言 ······························· 728
 17.2.1 books 数据库 ························ 730
 17.2.2 SELECT 查询 ························ 734
 17.2.3 WHERE 子句 ························ 734
 17.2.4 ORDER BY 子句 ·················· 736
 17.2.5 INNER JOIN：从多个表中合并数据 ···················· 737
 17.2.6 INSERT INTO 语句 ·············· 738
 17.2.7 UPDATE 语句 ······················ 739
 17.2.8 DELETE FROM 语句 ·········· 739
17.3 NoSQL 和 NewSQL 大数据数据库概述 ························ 741
 17.3.1 NoSQL 键值对数据库 ·········· 741
 17.3.2 NoSQL 文档数据库 ·············· 742
 17.3.3 NoSQL 列数据库 ·················· 742
 17.3.4 NoSQL 图数据库 ·················· 743
 17.3.5 NewSQL 数据库 ···················· 743
17.4 案例研究：MongoDB JSON 文档数据库 ································ 744
 17.4.1 创建 MongoDB Atlas 簇 ······ 745
 17.4.2 将推文注入 MongoDB ········· 746
17.5 Hadoop ································ 755
 17.5.1 Hadoop 概述 ························ 755
 17.5.2 使用 MapReduce 汇总 Romeo and Juliet 的词长度 ········ 758
 17.5.3 在微软 Azure HDInsight 上创建 Apache 簇 ·············· 758
 17.5.4 Hadoop Streaming ················ 760
 17.5.5 Mapper 的实现 ······················ 760

17.5.6 Reducer 的实现·················761
17.5.7 准备运行 MapReduce
案例研究···················762
17.5.8 运行 MapReduce 作业········763
17.6 Spark·································766
17.6.1 Spark 概述···················766
17.6.2 Docker 和 Jupyter
Docker 栈··················767
17.6.3 使用 Spark 进行词统计·······770
17.6.4 在微软 Azure 上进行
词统计······················773
17.7 Spark Streaming：使用 pyspark-notebook Docker 栈进行 Twitter 哈希标注统计···················777

17.7.1 将推文注入套接字··········777
17.7.2 推文哈希标注累计和
Spark SQL 简介···········780
17.8 物联网和仪表盘···················786
17.8.1 发布和订阅················788
17.8.2 使用 Freeboard 仪表盘可视化
PubNub 实时采样流·······788
17.8.3 使用 Python 模拟互联网
恒温器······················790
17.8.4 使用 Freeboard.io 创建
仪表盘······················792
17.8.5 创建 Python PubNub
订阅························794
17.9 小结·······························798

Contents

1 Introduction to Computers and Python   1
1.1 Introduction   2
1.2 Hardware and Software   3
     1.2.1 Moore's Law   4
     1.2.2 Computer Organization   4
1.3 Data Hierarchy   6
1.4 Machine Languages, Assembly Languages and High-Level Languages   9
1.5 Introduction to Object Technology   10
1.6 Operating Systems   13
1.7 Python   16
1.8 It's the Libraries!   18
     1.8.1 Python Standard Library   18
     1.8.2 Data-Science Libraries   18
1.9 Other Popular Programming Languages   20
1.10 Test-Drive: Using IPython and Jupyter Notebooks   21
     1.10.1 Using IPython Interactive Mode as a Calculator   21
     1.10.2 Executing a Python Program Using the IPython Interpreter   23
     1.10.3 Writing and Executing Code in a Jupyter Notebook   24
1.11 Internet and World Wide Web   29
     1.11.1 Internet: A Network of Networks   29
     1.11.2 World Wide Web: Making the Internet User-Friendly   30
     1.11.3 The Cloud   30
     1.11.4 Internet of Things   31
1.12 Software Technologies   32
1.13 How Big Is Big Data?   33
     1.13.1 Big Data Analytics   38
     1.13.2 Data Science and Big Data Are Making a Difference: Use Cases   39
1.14 Intro to Data Science: Case Study—A Big-Data Mobile Application   40

2 Introduction to Python Programming   49
2.1 Introduction   50
2.2 Variables and Assignment Statements   50

2.3	Arithmetic	52
2.4	Function `print` and an Intro to Single- and Double-Quoted Strings	56
2.5	Triple-Quoted Strings	58
2.6	Getting Input from the User	59
2.7	Decision Making: The `if` Statement and Comparison Operators	61
2.8	Objects and Dynamic Typing	66
2.9	Intro to Data Science: Basic Descriptive Statistics	68
2.10	Wrap-Up	70

3 Control Statements and Program Development 73

3.1	Introduction	74
3.2	Algorithms	74
3.3	Pseudocode	75
3.4	Control Statements	75
3.5	`if` Statement	78
3.6	`if…else` and `if…elif…else` Statements	80
3.7	`while` Statement	85
3.8	`for` Statement	86
	3.8.1 Iterables, Lists and Iterators	88
	3.8.2 Built-In `range` Function	88
3.9	Augmented Assignments	89
3.10	Program Development: Sequence-Controlled Repetition	90
	3.10.1 Requirements Statement	90
	3.10.2 Pseudocode for the Algorithm	90
	3.10.3 Coding the Algorithm in Python	91
	3.10.4 Introduction to Formatted Strings	92
3.11	Program Development: Sentinel-Controlled Repetition	93
3.12	Program Development: Nested Control Statements	97
3.13	Built-In Function `range`: A Deeper Look	101
3.14	Using Type `Decimal` for Monetary Amounts	102
3.15	`break` and `continue` Statements	105
3.16	Boolean Operators `and`, `or` and `not`	106
3.17	Intro to Data Science: Measures of Central Tendency—Mean, Median and Mode	109
3.18	Wrap-Up	111

4 Functions 119

4.1	Introduction	120
4.2	Defining Functions	120
4.3	Functions with Multiple Parameters	123
4.4	Random-Number Generation	125
4.5	Case Study: A Game of Chance	128
4.6	Python Standard Library	131
4.7	`math` Module Functions	132
4.8	Using IPython Tab Completion for Discovery	133

4.9	Default Parameter Values	135
4.10	Keyword Arguments	136
4.11	Arbitrary Argument Lists	136
4.12	Methods: Functions That Belong to Objects	138
4.13	Scope Rules	138
4.14	`import`: A Deeper Look	140
4.15	Passing Arguments to Functions: A Deeper Look	142
4.16	Function-Call Stack	145
4.17	Functional-Style Programming	146
4.18	Intro to Data Science: Measures of Dispersion	148
4.19	Wrap-Up	150

5 Sequences: Lists and Tuples 155

5.1	Introduction	156
5.2	Lists	156
5.3	Tuples	161
5.4	Unpacking Sequences	163
5.5	Sequence Slicing	166
5.6	`del` Statement	169
5.7	Passing Lists to Functions	171
5.8	Sorting Lists	172
5.9	Searching Sequences	174
5.10	Other List Methods	176
5.11	Simulating Stacks with Lists	178
5.12	List Comprehensions	179
5.13	Generator Expressions	181
5.14	Filter, Map and Reduce	182
5.15	Other Sequence Processing Functions	185
5.16	Two-Dimensional Lists	187
5.17	Intro to Data Science: Simulation and Static Visualizations	191
	5.17.1 Sample Graphs for 600, 60,000 and 6,000,000 Die Rolls	191
	5.17.2 Visualizing Die-Roll Frequencies and Percentages	193
5.18	Wrap-Up	199

6 Dictionaries and Sets 209

6.1	Introduction	210
6.2	Dictionaries	210
	6.2.1 Creating a Dictionary	210
	6.2.2 Iterating through a Dictionary	212
	6.2.3 Basic Dictionary Operations	212
	6.2.4 Dictionary Methods `keys` and `values`	214
	6.2.5 Dictionary Comparisons	216
	6.2.6 Example: Dictionary of Student Grades	217
	6.2.7 Example: Word Counts	218

	6.2.8	Dictionary Method `update`	220
	6.2.9	Dictionary Comprehensions	220
6.3	Sets		221
	6.3.1	Comparing Sets	223
	6.3.2	Mathematical Set Operations	225
	6.3.3	Mutable Set Operators and Methods	226
	6.3.4	Set Comprehensions	228
6.4	Intro to Data Science: Dynamic Visualizations		228
	6.4.1	How Dynamic Visualization Works	228
	6.4.2	Implementing a Dynamic Visualization	231
6.5	Wrap-Up		234

7   Array-Oriented Programming with NumPy   239

7.1	Introduction	240
7.2	Creating arrays from Existing Data	241
7.3	array Attributes	242
7.4	Filling arrays with Specific Values	244
7.5	Creating arrays from Ranges	244
7.6	List vs. array Performance: Introducing %timeit	246
7.7	array Operators	248
7.8	NumPy Calculation Methods	250
7.9	Universal Functions	252
7.10	Indexing and Slicing	254
7.11	Views: Shallow Copies	256
7.12	Deep Copies	258
7.13	Reshaping and Transposing	259
7.14	Intro to Data Science: pandas `Series` and `DataFrames`	262
	7.14.1 pandas `Series`	262
	7.14.2 `DataFrames`	267
7.15	Wrap-Up	275

8   Strings: A Deeper Look   283

8.1	Introduction	284
8.2	Formatting Strings	285
	8.2.1 Presentation Types	285
	8.2.2 Field Widths and Alignment	286
	8.2.3 Numeric Formatting	287
	8.2.4 String's `format` Method	288
8.3	Concatenating and Repeating Strings	289
8.4	Stripping Whitespace from Strings	290
8.5	Changing Character Case	291
8.6	Comparison Operators for Strings	292
8.7	Searching for Substrings	292
8.8	Replacing Substrings	294

8.9	Splitting and Joining Strings	294
8.10	Characters and Character-Testing Methods	297
8.11	Raw Strings	298
8.12	Introduction to Regular Expressions	299
	8.12.1 `re` Module and Function `fullmatch`	300
	8.12.2 Replacing Substrings and Splitting Strings	303
	8.12.3 Other Search Functions; Accessing Matches	304
8.13	Intro to Data Science: Pandas, Regular Expressions and Data Munging	307
8.14	Wrap-Up	312

9 Files and Exceptions 319

9.1	Introduction	320
9.2	Files	321
9.3	Text-File Processing	321
	9.3.1 Writing to a Text File: Introducing the `with` Statement	322
	9.3.2 Reading Data from a Text File	323
9.4	Updating Text Files	325
9.5	Serialization with JSON	327
9.6	Focus on Security: `pickle` Serialization and Deserialization	330
9.7	Additional Notes Regarding Files	330
9.8	Handling Exceptions	331
	9.8.1 Division by Zero and Invalid Input	332
	9.8.2 `try` Statements	332
	9.8.3 Catching Multiple Exceptions in One `except` Clause	335
	9.8.4 What Exceptions Does a Function or Method Raise?	336
	9.8.5 What Code Should Be Placed in a `try` Suite?	336
9.9	`finally` Clause	336
9.10	Explicitly Raising an Exception	339
9.11	(Optional) Stack Unwinding and Tracebacks	339
9.12	Intro to Data Science: Working with CSV Files	342
	9.12.1 Python Standard Library Module `csv`	342
	9.12.2 Reading CSV Files into Pandas `DataFrames`	344
	9.12.3 Reading the Titanic Disaster Dataset	346
	9.12.4 Simple Data Analysis with the Titanic Disaster Dataset	347
	9.12.5 Passenger Age Histogram	348
9.13	Wrap-Up	349

10 Object-Oriented Programming 355

10.1	Introduction	356
10.2	Custom Class Account	358
	10.2.1 Test-Driving Class Account	358
	10.2.2 Account Class Definition	360
	10.2.3 Composition: Object References as Members of Classes	361
10.3	Controlling Access to Attributes	363

10.4	Properties for Data Access	364
	10.4.1 Test-Driving Class `Time`	364
	10.4.2 Class `Time` Definition	366
	10.4.3 Class `Time` Definition Design Notes	370
10.5	Simulating "Private" Attributes	371
10.6	Case Study: Card Shuffling and Dealing Simulation	373
	10.6.1 Test-Driving Classes `Card` and `DeckOfCards`	373
	10.6.2 Class `Card`—Introducing Class Attributes	375
	10.6.3 Class `DeckOfCards`	377
	10.6.4 Displaying Card Images with Matplotlib	378
10.7	Inheritance: Base Classes and Subclasses	382
10.8	Building an Inheritance Hierarchy; Introducing Polymorphism	384
	10.8.1 Base Class `CommissionEmployee`	384
	10.8.2 Subclass `SalariedCommissionEmployee`	387
	10.8.3 Processing `CommissionEmployees` and `SalariedCommissionEmployees` Polymorphically	391
	10.8.4 A Note About Object-Based and Object-Oriented Programming	391
10.9	Duck Typing and Polymorphism	392
10.10	Operator Overloading	393
	10.10.1 Test-Driving Class `Complex`	394
	10.10.2 Class `Complex` Definition	395
10.11	Exception Class Hierarchy and Custom Exceptions	397
10.12	Named Tuples	399
10.13	A Brief Intro to Python 3.7's New Data Classes	400
	10.13.1 Creating a Card Data Class	401
	10.13.2 Using the `Card` Data Class	403
	10.13.3 Data Class Advantages over Named Tuples	405
	10.13.4 Data Class Advantages over Traditional Classes	406
10.14	Unit Testing with Docstrings and `doctest`	406
10.15	Namespaces and Scopes	411
10.16	Intro to Data Science: Time Series and Simple Linear Regression	414
10.17	Wrap-Up	423

11 Computer Science Thinking: Recursion, Searching, Sorting and Big O — 431

11.1	Introduction	432
11.2	Factorials	433
11.3	Recursive Factorial Example	433
11.4	Recursive Fibonacci Series Example	436
11.5	Recursion vs. Iteration	439
11.6	Searching and Sorting	440
11.7	Linear Search	440
11.8	Efficiency of Algorithms: Big O	442
11.9	Binary Search	444
	11.9.1 Binary Search Implementation	445

	11.9.2 Big O of the Binary Search	447
11.10	Sorting Algorithms	448
11.11	Selection Sort	448
	11.11.1 Selection Sort Implementation	449
	11.11.2 Utility Function `print_pass`	450
	11.11.3 Big O of the Selection Sort	451
11.12	Insertion Sort	451
	11.12.1 Insertion Sort Implementation	452
	11.12.2 Big O of the Insertion Sort	453
11.13	Merge Sort	454
	11.13.1 Merge Sort Implementation	454
	11.13.2 Big O of the Merge Sort	459
11.14	Big O Summary for This Chapter's Searching and Sorting Algorithms	459
11.15	Visualizing Algorithms	460
	11.15.1 Generator Functions	462
	11.15.2 Implementing the Selection Sort Animation	463
11.16	Wrap-Up	468

12 Natural Language Processing (NLP) 477

12.1	Introduction	478
12.2	TextBlob	479
	12.2.1 Create a TextBlob	481
	12.2.2 Tokenizing Text into Sentences and Words	482
	12.2.3 Parts-of-Speech Tagging	482
	12.2.4 Extracting Noun Phrases	483
	12.2.5 Sentiment Analysis with TextBlob's Default Sentiment Analyzer	484
	12.2.6 Sentiment Analysis with the `NaiveBayesAnalyzer`	486
	12.2.7 Language Detection and Translation	487
	12.2.8 Inflection: Pluralization and Singularization	489
	12.2.9 Spell Checking and Correction	489
	12.2.10 Normalization: Stemming and Lemmatization	490
	12.2.11 Word Frequencies	491
	12.2.12 Getting Definitions, Synonyms and Antonyms from WordNet	492
	12.2.13 Deleting Stop Words	494
	12.2.14 n-grams	496
12.3	Visualizing Word Frequencies with Bar Charts and Word Clouds	497
	12.3.1 Visualizing Word Frequencies with Pandas	497
	12.3.2 Visualizing Word Frequencies with Word Clouds	500
12.4	Readability Assessment with Textatistic	503
12.5	Named Entity Recognition with spaCy	505
12.6	Similarity Detection with spaCy	507
12.7	Other NLP Libraries and Tools	509
12.8	Machine Learning and Deep Learning Natural Language Applications	509
12.9	Natural Language Datasets	510
12.10	Wrap-Up	510

13 Data Mining Twitter 515

- 13.1 Introduction 516
- 13.2 Overview of the Twitter APIs 518
- 13.3 Creating a Twitter Account 519
- 13.4 Getting Twitter Credentials—Creating an App 520
- 13.5 What's in a Tweet? 521
- 13.6 Tweepy 525
- 13.7 Authenticating with Twitter Via Tweepy 525
- 13.8 Getting Information About a Twitter Account 527
- 13.9 Introduction to Tweepy Cursors: Getting an Account's Followers and Friends 529
  - 13.9.1 Determining an Account's Followers 529
  - 13.9.2 Determining Whom an Account Follows 532
  - 13.9.3 Getting a User's Recent Tweets 532
- 13.10 Searching Recent Tweets 534
- 13.11 Spotting Trends: Twitter Trends API 536
  - 13.11.1 Places with Trending Topics 536
  - 13.11.2 Getting a List of Trending Topics 537
  - 13.11.3 Create a Word Cloud from Trending Topics 539
- 13.12 Cleaning/Preprocessing Tweets for Analysis 541
- 13.13 Twitter Streaming API 542
  - 13.13.1 Creating a Subclass of `StreamListener` 543
  - 13.13.2 Initiating Stream Processing 545
- 13.14 Tweet Sentiment Analysis 547
- 13.15 Geocoding and Mapping 551
  - 13.15.1 Getting and Mapping the Tweets 552
  - 13.15.2 Utility Functions in `tweetutilities.py` 556
  - 13.15.3 Class `LocationListener` 558
- 13.16 Ways to Store Tweets 559
- 13.17 Twitter and Time Series 560
- 13.18 Wrap-Up 560

14 IBM Watson and Cognitive Computing 565

- 14.1 Introduction: IBM Watson and Cognitive Computing 566
- 14.2 IBM Cloud Account and Cloud Console 568
- 14.3 Watson Services 568
- 14.4 Additional Services and Tools 572
- 14.5 Watson Developer Cloud Python SDK 573
- 14.6 Case Study: Traveler's Companion Translation App 574
  - 14.6.1 Before You Run the App 575
  - 14.6.2 Test-Driving the App 576
  - 14.6.3 `SimpleLanguageTranslator.py` Script Walkthrough 577
- 14.7 Watson Resources 587
- 14.8 Wrap-Up 589

15 Machine Learning: Classification, Regression and Clustering — 593

- 15.1 Introduction to Machine Learning — 594
    - 15.1.1 Scikit-Learn — 595
    - 15.1.2 Types of Machine Learning — 596
    - 15.1.3 Datasets Bundled with Scikit-Learn — 598
    - 15.1.4 Steps in a Typical Data Science Study — 599
- 15.2 Case Study: Classification with k-Nearest Neighbors and the Digits Dataset, Part 1 — 599
    - 15.2.1 k-Nearest Neighbors Algorithm — 601
    - 15.2.2 Loading the Dataset — 602
    - 15.2.3 Visualizing the Data — 606
    - 15.2.4 Splitting the Data for Training and Testing — 608
    - 15.2.5 Creating the Model — 609
    - 15.2.6 Training the Model — 610
    - 15.2.7 Predicting Digit Classes — 610
- 15.3 Case Study: Classification with k-Nearest Neighbors and the Digits Dataset, Part 2 — 612
    - 15.3.1 Metrics for Model Accuracy — 612
    - 15.3.2 K-Fold Cross-Validation — 616
    - 15.3.3 Running Multiple Models to Find the Best One — 617
    - 15.3.4 Hyperparameter Tuning — 619
- 15.4 Case Study: Time Series and Simple Linear Regression — 620
- 15.5 Case Study: Multiple Linear Regression with the California Housing Dataset — 625
    - 15.5.1 Loading the Dataset — 626
    - 15.5.2 Exploring the Data with Pandas — 628
    - 15.5.3 Visualizing the Features — 630
    - 15.5.4 Splitting the Data for Training and Testing — 634
    - 15.5.5 Training the Model — 634
    - 15.5.6 Testing the Model — 635
    - 15.5.7 Visualizing the Expected vs. Predicted Prices — 636
    - 15.5.8 Regression Model Metrics — 637
    - 15.5.9 Choosing the Best Model — 638
- 15.6 Case Study: Unsupervised Machine Learning, Part 1—Dimensionality Reduction — 639
- 15.7 Case Study: Unsupervised Machine Learning, Part 2—k-Means Clustering — 642
    - 15.7.1 Loading the Iris Dataset — 644
    - 15.7.2 Exploring the Iris Dataset: Descriptive Statistics with Pandas — 646
    - 15.7.3 Visualizing the Dataset with a Seaborn `pairplot` — 647
    - 15.7.4 Using a `KMeans` Estimator — 650
    - 15.7.5 Dimensionality Reduction with Principal Component Analysis — 652
    - 15.7.6 Choosing the Best Clustering Estimator — 655
- 15.8 Wrap-Up — 656

16 Deep Learning — 665

- 16.1 Introduction — 666
  - 16.1.1 Deep Learning Applications — 668
  - 16.1.2 Deep Learning Demos — 669
  - 16.1.3 Keras Resources — 669
- 16.2 Keras Built-In Datasets — 669
- 16.3 Custom Anaconda Environments — 670
- 16.4 Neural Networks — 672
- 16.5 Tensors — 674
- 16.6 Convolutional Neural Networks for Vision; Multi-Classification with the MNIST Dataset — 676
  - 16.6.1 Loading the MNIST Dataset — 677
  - 16.6.2 Data Exploration — 678
  - 16.6.3 Data Preparation — 680
  - 16.6.4 Creating the Neural Network — 682
  - 16.6.5 Training and Evaluating the Model — 691
  - 16.6.6 Saving and Loading a Model — 696
- 16.7 Visualizing Neural Network Training with TensorBoard — 697
- 16.8 ConvnetJS: Browser-Based Deep-Learning Training and Visualization — 700
- 16.9 Recurrent Neural Networks for Sequences; Sentiment Analysis with the IMDb Dataset — 701
  - 16.9.1 Loading the IMDb Movie Reviews Dataset — 702
  - 16.9.2 Data Exploration — 703
  - 16.9.3 Data Preparation — 705
  - 16.9.4 Creating the Neural Network — 706
  - 16.9.5 Training and Evaluating the Model — 709
- 16.10 Tuning Deep Learning Models — 710
- 16.11 Convnet Models Pretrained on ImageNet — 711
- 16.12 Reinforcement Learning — 712
  - 16.12.1 Deep Q-Learning — 713
  - 16.12.2 OpenAI Gym — 713
- 16.13 Wrap-Up — 714

17 Big Data: Hadoop, Spark, NoSQL and IoT — 723

- 17.1 Introduction — 724
- 17.2 Relational Databases and Structured Query Language (SQL) — 728
  - 17.2.1 A books Database — 730
  - 17.2.2 SELECT Queries — 734
  - 17.2.3 WHERE Clause — 734
  - 17.2.4 ORDER BY Clause — 736
  - 17.2.5 Merging Data from Multiple Tables: INNER JOIN — 737
  - 17.2.6 INSERT INTO Statement — 738
  - 17.2.7 UPDATE Statement — 739
  - 17.2.8 DELETE FROM Statement — 739

17.3	NoSQL and NewSQL Big-Data Databases: A Brief Tour	741
	17.3.1 NoSQL Key–Value Databases	741
	17.3.2 NoSQL Document Databases	742
	17.3.3 NoSQL Columnar Databases	742
	17.3.4 NoSQL Graph Databases	743
	17.3.5 NewSQL Databases	743
17.4	Case Study: A MongoDB JSON Document Database	744
	17.4.1 Creating the MongoDB Atlas Cluster	745
	17.4.2 Streaming Tweets into MongoDB	746
17.5	Hadoop	755
	17.5.1 Hadoop Overview	755
	17.5.2 Summarizing Word Lengths in *Romeo and Juliet* via MapReduce	758
	17.5.3 Creating an Apache Hadoop Cluster in Microsoft Azure HDInsight	758
	17.5.4 Hadoop Streaming	760
	17.5.5 Implementing the Mapper	760
	17.5.6 Implementing the Reducer	761
	17.5.7 Preparing to Run the MapReduce Example	762
	17.5.8 Running the MapReduce Job	763
17.6	Spark	766
	17.6.1 Spark Overview	766
	17.6.2 Docker and the Jupyter Docker Stacks	767
	17.6.3 Word Count with Spark	770
	17.6.4 Spark Word Count on Microsoft Azure	773
17.7	Spark Streaming: Counting Twitter Hashtags Using the `pyspark-notebook` Docker Stack	777
	17.7.1 Streaming Tweets to a Socket	777
	17.7.2 Summarizing Tweet Hashtags; Introducing Spark SQL	780
17.8	Internet of Things and Dashboards	786
	17.8.1 Publish and Subscribe	788
	17.8.2 Visualizing a PubNub Sample Live Stream with a Freeboard Dashboard	788
	17.8.3 Simulating an Internet-Connected Thermostat in Python	790
	17.8.4 Creating the Dashboard with Freeboard.io	792
	17.8.5 Creating a Python PubNub Subscriber	794
17.9	Wrap-Up	798

Introduction to Computers and Python

1

Objectives

In this chapter you'll:

- Learn about exciting recent developments in computing.
- Learn computer hardware, software and Internet basics.
- Understand the data hierarchy from bits to databases.
- Understand the different types of programming languages.
- Understand object-oriented programming basics.
- Understand the strengths of Python and other leading programming languages.
- Understand the importance of libraries.
- Be introduced to key Python and data-science libraries you'll use in this book.
- Test-drive the IPython interpreter's interactive mode for executing Python code.
- Execute a Python script that animates a bar chart.
- Create and test-drive a web-browser-based Jupyter Notebook for executing Python code.
- Learn how big "big data" is and how quickly it's getting even bigger.
- Read a big-data case study on a mobile navigation app.
- Be introduced to artificial intelligence—at the intersection of computer science and data science.

Outline

1.1 Introduction	1.10.2 Executing a Python Program Using the IPython Interpreter
1.2 Hardware and Software	1.10.3 Writing and Executing Code in a Jupyter Notebook
1.2.1 Moore's Law	
1.2.2 Computer Organization	1.11 Internet and World Wide Web
1.3 Data Hierarchy	1.11.1 Internet: A Network of Networks
1.4 Machine Languages, Assembly Languages and High-Level Languages	1.11.2 World Wide Web: Making the Internet User-Friendly
	1.11.3 The Cloud
1.5 Introduction to Object Technology	1.11.4 Internet of Things
1.6 Operating Systems	1.12 Software Technologies
1.7 Python	1.13 How Big Is Big Data?
1.8 It's the Libraries!	1.13.1 Big Data Analytics
1.8.1 Python Standard Library	1.13.2 Data Science and Big Data Are Making a Difference: Use Cases
1.8.2 Data-Science Libraries	
1.9 Other Popular Programming Languages	1.14 Case Study—A Big-Data Mobile Application
1.10 Test-Drives: Using IPython and Jupyter Notebooks	1.15 Intro to Data Science: Artificial Intelligence—at the Intersection of CS and Data Science
1.10.1 Using IPython Interactive Mode as a Calculator	Exercises

1.1 Introduction

Welcome to Python—one of the world's most widely used computer programming languages and, according to the *Popularity of Programming Languages (PYPL) Index*, the world's most popular.[1] You're probably familiar with many of the powerful tasks computers perform. In this textbook, you'll get intensive, hands-on experience writing Python instructions that command computers to perform those and other tasks. **Software** (that is, the Python instructions you write, which are also called **code**) controls **hardware** (that is, computers and related devices).

Here, we introduce terminology and concepts that lay the groundwork for the Python programming you'll learn in Chapters 2–11 and the big-data, artificial-intelligence and cloud-based case studies we present in Chapters 12–17. We'll introduce hardware and software concepts and overview the data hierarchy—from individual bits to databases, which store the massive amounts of data companies need to implement contemporary applications such as Google Search, Waze, Uber, Airbnb and a myriad of others.

We'll discuss the types of programming languages and introduce *object-oriented programming* terminology and concepts. You'll learn why Python has become so popular. We'll introduce the Python Standard Library and various data-science libraries that help you avoid "reinventing the wheel." You'll use these libraries to create *software objects* that you'll interact with to perform significant tasks with modest numbers of instructions. We'll introduce additional software technologies that you're likely to use as you develop software.

Next, you'll work through three test-drives showing how to execute Python code:

- In the first, you'll use IPython to execute Python instructions interactively and immediately see their results.

---

1. `https://pypl.github.io/PYPL.html` (as of January 2019).

- In the second, you'll execute a substantial Python application that will display an animated bar chart summarizing rolls of a six-sided die as they occur. You'll see the "Law of Large Numbers" in action. In Chapter 6, you'll build this application with the Matplotlib visualization library.
- In the last, we'll introduce Jupyter Notebooks using JupyterLab—an interactive, web-browser-based tool in which you can conveniently write and execute Python instructions. Jupyter Notebooks enable you to include text, images, audios, videos, animations and code.

In the past, most computer applications ran on "standalone" computers (that is, not networked together). Today's applications can be written with the aim of communicating among the world's computers via the Internet. We'll introduce the Internet, the World Wide Web, the Cloud and the Internet of Things (IoT), laying the groundwork for the contemporary applications you'll develop in Chapters 12–17.

You'll learn just how big "big data" is and how quickly it's getting even bigger. Next, we'll present a big-data case study on the Waze mobile navigation app, which uses many current technologies to provide dynamic driving directions that get you to your destination as quickly and as safely as possible. As we walk through those technologies, we'll mention where you'll use many of them in this book. The chapter closes with our first Intro to Data Science section in which we discuss a key intersection between computer science and data science—artificial intelligence.

1.2 Hardware and Software

Computers can perform calculations and make logical decisions phenomenally faster than human beings can. Many of today's personal computers can perform billions of calculations in one second—more than a human can perform in a lifetime. *Supercomputers* are already performing *thousands of trillions (quadrillions)* of instructions per second! IBM has developed the IBM Summit supercomputer, which can perform over 122 quadrillion calculations per second (122 *petaflops*)![2] To put that in perspective, *the IBM Summit supercomputer can perform in one second almost 16 million calculations for every person on the planet!*[3] And supercomputing upper limits are growing quickly.

Computers process data under the control of sequences of instructions called **computer programs** (or simply **programs**). These software programs guide the computer through ordered actions specified by people called computer **programmers**.

A computer consists of various physical devices referred to as hardware (such as the keyboard, screen, mouse, solid-state disks, hard disks, memory, DVD drives and processing units). Computing costs are *dropping dramatically*, due to rapid developments in hardware and software technologies. Computers that might have filled large rooms and cost millions of dollars decades ago are now inscribed on computer chips smaller than a fingernail, costing perhaps a few dollars each. Ironically, silicon is one of the most abundant materials on Earth—it's an ingredient in common sand. Silicon-chip technology has made computing so economical that computers have become a commodity.

---

2. https://en.wikipedia.org/wiki/FLOPS.
3. For perspective on how far computing performance has come, consider this: In his early computing days, Harvey Deitel used the Digital Equipment Corporation PDP-1 (https://en.wikipedia.org/wiki/PDP-1), which was capable of performing only 93,458 operations per second.

1.2.1 Moore's Law

Every year, you probably expect to pay at least a little more for most products and services. The opposite has been the case in the computer and communications fields, especially with regard to the hardware supporting these technologies. For many decades, hardware costs have fallen rapidly.

Every year or two, the capacities of computers have approximately *doubled* inexpensively. This remarkable trend often is called **Moore's Law**, named for the person who identified it in the 1960s, Gordon Moore, co-founder of Intel—one of the leading manufacturers of the processors in today's computers and embedded systems. Moore's Law *and related observations* apply especially to

- the amount of memory that computers have for programs,
- the amount of secondary storage (such as solid-state drive storage) they have to hold programs and data over longer periods of time, and
- their processor speeds—the speeds at which they *execute* their programs (that is, do their work).

Similar growth has occurred in the communications field—costs have plummeted as enormous demand for communications *bandwidth* (that is, information-carrying capacity) has attracted intense competition. We know of no other fields in which technology improves so quickly and costs fall so rapidly. Such phenomenal improvement is truly fostering the *Information Revolution*.

1.2.2 Computer Organization

Regardless of differences in *physical* appearance, computers can be envisioned as divided into various **logical units** or sections:

Input Unit
This "receiving" section obtains information (data and computer programs) from **input devices** and places it at the disposal of the other units for processing. Most user input is entered into computers through keyboards, touch screens and mouse devices. Other forms of input include receiving voice commands, scanning images and barcodes, reading from secondary storage devices (such as hard drives, Blu-ray Disc™ drives and USB flash drives—also called "thumb drives" or "memory sticks"), receiving video from a webcam and having your computer receive information from the Internet (such as when you stream videos from YouTube® or download e-books from Amazon). Newer forms of input include position data from a GPS device, motion and orientation information from an *accelerometer* (a device that responds to up/down, left/right and forward/backward acceleration) in a smartphone or wireless game controller (such as those for Microsoft® Xbox®, Nintendo Switch™ and Sony® PlayStation®) and voice input from intelligent assistants like Apple Siri®, Amazon Echo® and Google Home®.

Output Unit
This "shipping" section takes information the computer has processed and places it on various **output devices** to make it available for use outside the computer. Most information that's output from computers today is displayed on screens (including touch screens), printed on paper ("going green" discourages this), played as audio or video on smart-

phones, tablets, PCs and giant screens in sports stadiums, transmitted over the Internet or used to control other devices, such as self-driving cars, robots and "intelligent" appliances. Information is also commonly output to secondary storage devices, such as solid-state drives (SSDs), hard drives, DVD drives and USB flash drives. Popular recent forms of output are smartphone and game-controller vibration, virtual reality devices like Oculus Rift®, Sony® PlayStation® VR and Google Daydream View™ and Samsung Gear VR®, and mixed reality devices like Magic Leap® One and Microsoft HoloLens™.

Memory Unit

This rapid-access, relatively low-capacity "warehouse" section retains information that has been entered through the input unit, making it immediately available for processing when needed. The memory unit also retains processed information until it can be placed on output devices by the output unit. Information in the memory unit is *volatile*—it's typically lost when the computer's power is turned off. The memory unit is often called either **memory, primary memory** or **RAM** (Random Access Memory). Main memories on desktop and notebook computers contain as much as 128 GB of RAM, though 8 to 16 GB is most common. GB stands for gigabytes; a gigabyte is approximately one billion bytes. A **byte** is eight bits. A bit is either a 0 or a 1.

Arithmetic and Logic Unit (ALU)

This "manufacturing" section performs *calculations*, such as addition, subtraction, multiplication and division. It also contains the *decision* mechanisms that allow the computer, for example, to compare two items from the memory unit to determine whether they're equal. In today's systems, the ALU is part of the next logical unit, the CPU.

Central Processing Unit (CPU)

This "administrative" section coordinates and supervises the operation of the other sections. The CPU tells the input unit when information should be read into the memory unit, tells the ALU when information from the memory unit should be used in calculations and tells the output unit when to send information from the memory unit to specific output devices. Most computers have **multicore processors** that implement multiple processors on a single integrated-circuit chip. Such processors can perform many operations simultaneously. A *dual-core processor* has two CPUs, a *quad-core processor* has four and an *octa-core processor* has eight. Intel has some processors with up to 72 cores. Today's desktop computers have processors that can execute billions of instructions per second.

Secondary Storage Unit

This is the long-term, high-capacity "warehousing" section. Programs or data not actively being used by the other units normally are placed on secondary storage devices (e.g., your *hard drive*) until they're again needed, possibly hours, days, months or even years later. Information on secondary storage devices is *persistent*—it's preserved even when the computer's power is turned off. Secondary storage information takes much longer to access than information in primary memory, but its cost per unit is much less. Examples of secondary storage devices include solid-state drives (SSDs), hard drives, read/write Blu-ray drives and USB flash drives. Many current drives hold terabytes (TB) of data—a **terabyte** is approximately one trillion bytes). Typical hard drives on desktop and notebook computers hold up to 4 TB, and some recent desktop-computer hard drives hold up to 15 TB.[4]

6 Introduction to Computers and Python

Self Check for Section 1.2

1 *(Fill-In)* Every year or two, the capacities of computers have approximately doubled inexpensively. This remarkable trend often is called _____.
Answer: Moore's Law.

2 *(True/False)* Information in the memory unit is *persistent*—it's preserved even when the computer's power is turned off
Answer: False. Information in the memory unit is *volatile*—it's typically lost when the computer's power is turned off.

3 *(Fill-In)* Most computers have _____ processors that implement multiple processors on a single integrated-circuit chip. Such processors can perform many operations simultaneously.
Answer: multicore.

1.3 Data Hierarchy

Data items processed by computers form a **data hierarchy** that becomes larger and more complex in structure as we progress from the simplest data items (called "bits") to richer ones, such as characters and fields. The following diagram illustrates a portion of the data hierarchy:

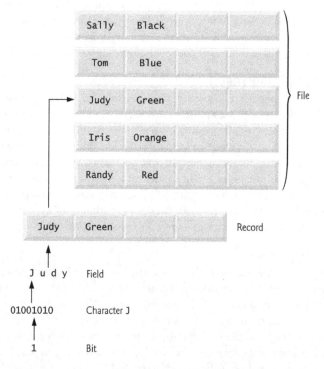

---

4. https://www.zdnet.com/article/worlds-biggest-hard-drive-meet-western-digitals-15tb-monster/.

Bits

A **bit** (short for "*b*inary dig*it*"—a digit that can assume one of *two* values) is the smallest data item in a computer. It can have the value 0 or 1. Remarkably, the impressive functions performed by computers involve only the simplest manipulations of 0s and 1s—*examining a bit's value, setting a bit's value* and *reversing a bit's value* (from 1 to 0 or from 0 to 1). Bits for the basis of the binary number system, which you can study in-depth in our online "Number Systems" appendix.

Characters

Work with data in the low-level form of bits is tedious. Instead, people prefer to work with *decimal digits* (0–9), *letters* (A–Z and a–z) and *special symbols* such as

```
$ @ % & * ( ) - + " : ; , ? /
```

Digits, letters and special symbols are known as **characters**. The computer's **character set** contains the characters used to write programs and represent data items. Computers process only 1s and 0s, so a computer's character set represents every character as a pattern of 1s and 0s. Python uses **Unicode®** characters that are composed of one, two, three or four bytes (8, 16, 24 or 32 bits, respectively)—known as **UTF-8 encoding**.[5]

Unicode contains characters for many of the world's languages. The ASCII (**American Standard Code for Information Interchange**) character set is a subset of Unicode that represents letters (a–z and A–Z), digits and some common special characters. You can view the ASCII subset of Unicode at

```
https://www.unicode.org/charts/PDF/U0000.pdf
```

The Unicode charts for all languages, symbols, emojis and more are viewable at

```
http://www.unicode.org/charts/
```

Fields

Just as characters are composed of bits, **fields** are composed of characters or bytes. A field is a group of characters or bytes that conveys meaning. For example, a field consisting of uppercase and lowercase letters can be used to represent a person's name, and a field consisting of decimal digits could represent a person's age.

Records

Several related fields can be used to compose a **record**. In a payroll system, for example, the record for an employee might consist of the following fields (possible types for these fields are shown in parentheses):

- Employee identification number (a whole number).
- Name (a string of characters).
- Address (a string of characters).
- Hourly pay rate (a number with a decimal point).
- Year-to-date earnings (a number with a decimal point).
- Amount of taxes withheld (a number with a decimal point).

---

5. `https://docs.python.org/3/howto/unicode.html`.

Thus, a record is a group of related fields. All the fields listed above belong to the same employee. A company might have many employees and a payroll record for each.

Files
A **file** is a group of related records. More generally, a file contains arbitrary data in arbitrary formats. In some operating systems, a file is viewed simply as a *sequence of bytes*—any organization of the bytes in a file, such as organizing the data into records, is a view created by the application programmer. You'll see how to do that in Chapter 9, "Files and Exceptions." It's not unusual for an organization to have many files, some containing billions, or even trillions, of characters of information.

Databases
A **database** is a collection of data organized for easy access and manipulation. The most popular model is the *relational database*, in which data is stored in simple *tables*. A table includes *records* and *fields*. For example, a table of students might include first name, last name, major, year, student ID number and grade-point-average fields. The data for each student is a record, and the individual pieces of information in each record are the fields. You can *search*, *sort* and otherwise manipulate the data, based on its relationship to multiple tables or databases. For example, a university might use data from the student database in combination with data from databases of courses, on-campus housing, meal plans, etc. We discuss databases in Chapter 17, "Big Data: Hadoop, Spark, NoSQL and IoT."

Big Data
The table below shows some common byte measurements:

Unit	Bytes	Which is approximately
1 kilobyte (KB)	1024 bytes	10^3 (1024) bytes exactly
1 megabyte (MB)	1024 kilobytes	10^6 (1,000,000) bytes
1 gigabyte (GB)	1024 megabytes	10^9 (1,000,000,000) bytes
1 terabyte (TB)	1024 gigabytes	10^{12} (1,000,000,000,000) bytes
1 petabyte (PB)	1024 terabytes	10^{15} (1,000,000,000,000,000) bytes
1 exabyte (EB)	1024 petabytes	10^{18} (1,000,000,000,000,000,000) bytes
1 zettabyte (ZB)	1024 exabytes	10^{21} (1,000,000,000,000,000,000,000) bytes

The amount of data being produced worldwide is enormous and its growth is accelerating. **Big data** applications deal with massive amounts of data. This field is growing quickly, creating lots of opportunity for software developers. Millions of IT jobs globally already are supporting big data applications. Section 1.13 discusses big data in more depth. You'll study big data and associated technologies in Chapter 17.

Self Check

1. *(Fill-In)* A(n) _____ (short for "binary digit"—a digit that can assume one of two values) is the smallest data item in a computer.
Answer: bit.

2 *(True/False)* In some operating systems, a file is viewed simply as a sequence of bytes—any organization of the bytes in a file, such as organizing the data into records, is a view created by the application programmer.
Answer: True.

3 *(Fill-In)* A database is a collection of data organized for easy access and manipulation. The most popular model is the _____ database, in which data is stored in simple tables.
Answer: *relational*

1.4 Machine Languages, Assembly Languages and High-Level Languages

Programmers write instructions in various programming languages, some directly understandable by computers and others requiring intermediate *translation* steps. Hundreds of such languages are in use today. These may be divided into three general types:

1. Machine languages.
2. Assembly languages.
3. High-level languages.

Machine Languages

Any computer can directly understand only its own **machine language**, defined by its hardware design. Machine languages generally consist of strings of numbers (ultimately reduced to 1s and 0s) that instruct computers to perform their most elementary operations one at a time. Machine languages are *machine dependent* (a particular machine language can be used on only one type of computer). Such languages are cumbersome for humans. For example, here's a section of an early machine-language payroll program that adds overtime pay to base pay and stores the result in gross pay:

```
+1300042774
+1400593419
+1200274027
```

Assembly Languages and Assemblers

Programming in machine language was simply too slow and tedious for most programmers. Instead of using the strings of numbers that computers could directly understand, programmers began using English-like abbreviations to represent elementary operations. These abbreviations formed the basis of **assembly languages**. *Translator programs* called **assemblers** were developed to convert assembly-language programs to machine language at computer speeds. The following section of an assembly-language payroll program also adds overtime pay to base pay and stores the result in gross pay:

```
load    basepay
add     overpay
store   grosspay
```

Although such code is clearer to humans, it's incomprehensible to computers until translated to machine language.

High-Level Languages and Compilers

With the advent of assembly languages, computer usage increased rapidly, but programmers still had to use numerous instructions to accomplish even the simplest tasks. To speed the programming process, **high-level languages** were developed in which single statements could be written to accomplish substantial tasks. A typical high-level-language program contains many statements, known as the program's **source code**.

Translator programs called **compilers** convert high-level-language source code into machine language. High-level languages allow you to write instructions that look almost like everyday English and contain commonly used mathematical notations. A payroll program written in a high-level language might contain a *single* statement such as

```
grossPay = basePay + overTimePay
```

From the programmer's standpoint, high-level languages are preferable to machine and assembly languages. Python is among the world's most widely used high-level programming languages.

Interpreters

Compiling a large high-level language program into machine language can take considerable computer time. *Interpreter* programs, developed to execute high-level language programs directly, avoid the delay of compilation, although they run slower than compiled programs. The most widely used Python implementation—CPython (which is written in the C programming language)—uses a clever mixture of compilation and interpretation to run programs.[6]

 Self Check

1 *(Fill-In)* Translator programs called _____ convert assembly-language programs to machine language at computer speeds.
Answer: assemblers.

2 *(Fill-In)* _____ programs, developed to execute high-level-language programs directly, avoid the delay of compilation, although they run slower than compiled programs
Answer: Interpreter.

3 *(True/False)* High-level languages allow you to write instructions that look almost like everyday English and contain commonly used mathematical notations.
Answer: True.

1.5 Introduction to Object Technology

As demands for new and more powerful software are soaring, building software quickly, correctly and economically is important. *Objects*, or more precisely, the *classes* objects come from, are essentially *reusable* software components. There are date objects, time objects, audio objects, video objects, automobile objects, people objects, etc. Almost any *noun* can be reasonably represented as a software object in terms of *attributes* (e.g., name, color and size) and *behaviors* (e.g., calculating, moving and communicating). Software-development groups can use a modular, object-oriented design-and-implementation approach to be

---

6. https://opensource.com/article/18/4/introduction-python-bytecode.

much more productive than with earlier popular techniques like "structured programming." Object-oriented programs are often easier to understand, correct and modify.

Automobile as an Object

To help you understand objects and their contents, let's begin with a simple analogy. Suppose you want to *drive a car and make it go faster by pressing its accelerator pedal*. What must happen before you can do this? Well, before you can drive a car, someone has to *design* it. A car typically begins as engineering drawings, similar to the *blueprints* that describe the design of a house. These drawings include the design for an accelerator pedal. The pedal *hides* from the driver the complex mechanisms that make the car go faster, just as the brake pedal "hides" the mechanisms that slow the car, and the steering wheel "hides" the mechanisms that turn the car. This enables people with little or no knowledge of how engines, braking and steering mechanisms work to drive a car easily.

Just as you cannot cook meals in the blueprint of a kitchen, you cannot drive a car's engineering drawings. Before you can drive a car, it must be *built* from the engineering drawings that describe it. A completed car has an *actual* accelerator pedal to make it go faster, but even that's not enough—the car won't accelerate on its own (hopefully!), so the driver must *press* the pedal to accelerate the car.

Methods and Classes

Let's use our car example to introduce some key object-oriented programming concepts. Performing a task in a program requires a **method**. The method houses the program statements that perform its tasks. The method hides these statements from its user, just as the accelerator pedal of a car hides from the driver the mechanisms of making the car go faster. In Python, a program unit called a **class** houses the set of methods that perform the class's tasks. For example, a class that represents a bank account might contain one method to *deposit* money to an account, another to *withdraw* money from an account and a third to *inquire* what the account's balance is. A class is similar in concept to a car's engineering drawings, which house the design of an accelerator pedal, steering wheel, and so on.

Instantiation

Just as someone has to *build a car* from its engineering drawings before you can drive a car, you must *build an object* of a class before a program can perform the tasks that the class's methods define. The process of doing this is called *instantiation*. An object is then referred to as an **instance** of its class.

Reuse

Just as a car's engineering drawings can be *reused* many times to build many cars, you can *reuse* a class many times to build many objects. Reuse of existing classes when building new classes and programs saves time and effort. Reuse also helps you build more reliable and effective systems because existing classes and components often have undergone extensive *testing*, *debugging* (that is, finding and removing errors) and *performance tuning*. Just as the notion of *interchangeable parts* was crucial to the Industrial Revolution, reusable classes are crucial to the software revolution that has been spurred by object technology.

In Python, you'll typically use a *building-block approach* to create your programs. To avoid reinventing the wheel, you'll use existing high-quality pieces wherever possible. This software reuse is a key benefit of object-oriented programming.

Messages and Method Calls
When you drive a car, pressing its gas pedal sends a *message* to the car to perform a task—that is, to go faster. Similarly, you *send messages to an object*. Each message is implemented as a **method call** that tells a method of the object to perform its task. For example, a program might call a bank-account object's *deposit* method to increase the account's balance.

Attributes and Instance Variables
A car, besides having capabilities to accomplish tasks, also has *attributes*, such as its color, its number of doors, the amount of gas in its tank, its current speed and its record of total miles driven (i.e., its odometer reading). Like its capabilities, the car's attributes are represented as part of its design in its engineering diagrams (which, for example, include an odometer and a fuel gauge). As you drive an actual car, these attributes are carried along with the car. Every car maintains its *own* attributes. For example, each car knows how much gas is in its own gas tank, but *not* how much is in the tanks of *other* cars.

An object, similarly, has attributes that it carries along as it's used in a program. These attributes are specified as part of the object's class. For example, a bank-account object has a *balance attribute* that represents the amount of money in the account. Each bank-account object knows the balance in the account it represents, but *not* the balances of the *other* accounts in the bank. Attributes are specified by the class's **instance variables**. A class's (and its object's) attributes and methods are intimately related, so classes wrap together their attributes and methods.

Inheritance
A new class of objects can be created conveniently by **inheritance**—the new class (called the **subclass**) starts with the characteristics of an existing class (called the **superclass**), possibly customizing them and adding unique characteristics of its own. In our car analogy, an object of class "convertible" certainly *is an* object of the more *general* class "automobile," but more *specifically*, the roof can be raised or lowered.

Object-Oriented Analysis and Design (OOAD)
Soon you'll be writing programs in Python. How will you create the **code** (i.e., the program instructions) for your programs? Perhaps, like many programmers, you'll simply turn on your computer and start typing. This approach may work for small programs (like the ones we present in the early chapters of the book), but what if you were asked to create a software system to control thousands of automated teller machines for a major bank? Or suppose you were asked to work on a team of 1,000 software developers building the next generation of the U.S. air traffic control system? For projects so large and complex, you should not simply sit down and start writing programs.

To create the best solutions, you should follow a detailed **analysis** process for determining your project's **requirements** (i.e., defining *what* the system is supposed to do), then develop a **design** that satisfies them (i.e., specifying *how* the system should do it). Ideally, you'd go through this process and carefully review the design (and have your design reviewed by other software professionals) before writing any code. If this process involves analyzing and designing your system from an object-oriented point of view, it's called an **object-oriented analysis-and-design (OOAD) process**. Languages like Python are object-oriented. Programming in such a language, called **object-oriented programming** (OOP), allows you to implement an object-oriented design as a working system.

Self Check for Section 1.5

1 *(Fill-In)* To create the best solutions, you should follow a detailed analysis process for determining your project's _____ (i.e., defining *what* the system is supposed to do) and developing a design that satisfies them (i.e., specifying *how* the system should do it).
Answer: requirements.

2 *(Fill-In)* The size, shape, color and weight of an object are _____ of the object's class.
Answer: attributes.

3 *(True/False)* Objects, or more precisely, the classes objects come from, are essentially reusable software components.
Answer: True.

1.6 Operating Systems

Operating systems are software systems that make using computers more convenient for users, application developers and system administrators. They provide services that allow each application to execute safely, efficiently and *concurrently* with other applications. The software that contains the core components of the operating system is called the **kernel**. Linux, Windows and macOS are popular desktop computer operating systems—you can use any of these with this book. The most popular mobile operating systems used in smartphones and tablets are Google's Android and Apple's iOS.

Windows—A Proprietary Operating System

In the mid-1980s, Microsoft developed the **Windows operating system**, consisting of a graphical user interface built on top of DOS (Disk Operating System)—an enormously popular personal-computer operating system that users interacted with by typing commands. Windows 10 is Microsoft's latest operating system—it includes the Cortana personal assistant for voice interactions. Windows is a *proprietary* operating system—it's controlled by Microsoft exclusively. Windows is by far the world's most widely used desktop operating system.

Linux—An Open-Source Operating System

The **Linux operating system** is among the greatest successes of the *open-source* movement. **Open-source software** departs from the *proprietary* software development style that dominated software's early years. With open-source development, individuals and companies *contribute* their efforts in developing, maintaining and evolving software in exchange for the right to use that software for their own purposes, typically at *no charge*. Open-source code is often scrutinized by a much larger audience than proprietary software, so errors often get removed faster. Open source also encourages innovation.

There are many organizations in the open-source community. Some key ones are:

- **Python Software Foundation** (responsible for Python).
- **GitHub** (provides tools for managing open-source projects—it has millions of them under development).
- The **Apache Software Foundation** (originally the creators of the Apache web server, they now oversee 350 open-source projects, including several big data infrastructure technologies we present in Chapter 17.

- The **Eclipse Foundation** (the Eclipse Integrated Development Environment helps programmers conveniently develop software)
- The **Mozilla Foundation** (creators of the Firefox web browser)
- **OpenML** (which focuses on open-source tools and data for machine learning—you'll explore machine learning in Chapter 15).
- **OpenAI** (which does research on artificial intelligence and publishes open-source tools used in AI reinforcement-learning research).
- **OpenCV** (which focuses on open-source computer-vision tools that can be used across a range of operating systems and programming languages—you'll study computer-vision applications in Chapter 16).

Rapid improvements to computing and communications, decreasing costs and open-source software have made it much easier and more economical to create software-based businesses now than just a decade ago. A great example is Facebook, which was launched from a college dorm room and built with open-source software.

The **Linux kernel** is the core of the most popular open-source, freely distributed, full-featured operating system. It's developed by a loosely organized team of volunteers and is popular in servers, personal computers and embedded systems (such as the computer systems at the heart of smartphones, smart TVs and automobile systems). Unlike that of proprietary operating systems like Microsoft's Windows and Apple's macOS, Linux source code (the program code) is available to the public for examination and modification and is free to download and install. As a result, Linux users benefit from a huge community of developers actively debugging and improving the kernel, and the ability to customize the operating system to meet specific needs.

Apple's macOS and Apple's iOS for iPhone® and iPad® Devices

Apple, founded in 1976 by Steve Jobs and Steve Wozniak, quickly became a leader in personal computing. In 1979, Jobs and several Apple employees visited Xerox PARC (Palo Alto Research Center) to learn about Xerox's desktop computer that featured a graphical user interface (GUI). That GUI served as the inspiration for the Apple Macintosh, launched in 1984.

The Objective-C programming language, created by Stepstone in the early 1980s, added capabilities for object-oriented programming (OOP) to the C programming language. Steve Jobs left Apple in 1985 and founded NeXT Inc. In 1988, NeXT licensed Objective-C from Stepstone and developed an Objective-C compiler and libraries which were used as the platform for the NeXTSTEP operating system's user interface, and Interface Builder—used to construct graphical user interfaces.

Jobs returned to Apple in 1996 when they bought NeXT. Apple's **macOS operating system** is a descendant of NeXTSTEP. Apple's proprietary operating system, **iOS**, is derived from macOS and is used in the iPhone, iPad, Apple Watch and Apple TV devices. In 2014, Apple introduced its new Swift programming language, which became open source in 2015. The iOS app-development community has largely shifted from Objective-C to Swift.

Google's Android

Android—the fastest growing mobile and smartphone operating system—is based on the Linux kernel and the Java programming language. Android is open source and free.

According to idc.com, as of 2018, Android had 86.8% of the global smartphone market share, compared to 13.2% for Apple.[7] The Android operating system is used in numerous smartphones, e-reader devices, tablets, in-store touch-screen kiosks, cars, robots, multimedia players and more.

Billions of Devices

In use today are Billions of personal computers and an even larger number of mobile devices. The following table lists many computerized devices. The explosive growth of mobile phones, tablets and other devices is creating significant opportunities for programming mobile apps. There are now various tools that enable you to use Python for Android and iOS app development, including BeeWare, Kivy, PyMob, Pythonista and others. Many are **cross-platform**, meaning that you can use them to develop apps that will run portably on Android, iOS and other platforms (like the web).

Computerized devices		
Access control systems	Airplane systems	ATMs
Automobiles	Blu-ray Disc™ players	Building controls
Cable boxes	Copiers	Credit cards
CT scanners	Desktop computers	e-Readers
Game consoles	GPS navigation systems	Home appliances
Home security systems	Internet-of-Things gateways	Light switches
Logic controllers	Lottery systems	Medical devices
Mobile phones	MRIs	Network switches
Optical sensors	Parking meters	Personal computers
Point-of-sale terminals	Printers	Robots
Routers	Servers	Smartcards
Smart meters	Smartpens	Smartphones
Tablets	Televisions	Thermostats
Transportation passes	TV set-top boxes	Vehicle diagnostic systems

Self Check for Section 1.6

1 *(Fill-In)* Windows is a(n) _____ operating system—it's controlled by Microsoft exclusively.
Answer: proprietary.

2 *(True/False)* Proprietary code is often scrutinized by a much larger audience than open-source software, so errors often get removed faster.
Answer: False. Open-source code is often scrutinized by a much larger audience than proprietary software, so errors often get removed faster.

3 *(True/False)* iOS dominates the global smartphone market over Android.
Answer: False. Android currently controls 88% of the smartphone market.

---

7. https://www.idc.com/promo/smartphone-market-share/os.

1.7 Python

Python is an object-oriented scripting language that was released publicly in 1991. It was developed by Guido van Rossum of the National Research Institute for Mathematics and Computer Science in Amsterdam.

Python has rapidly become one of the world's most popular programming languages. It's now particularly popular for educational and scientific computing,[8] and it recently surpassed the programming language R as the most popular data-science programming language.[9,10,11] Here are some reasons why Python is popular and everyone should consider learning it:[12,13,14]

- It's open source, free and widely available with a massive open-source community.
- It's easier to learn than languages like C, C++, C# and Java, enabling novices and professional developers to get up to speed quickly.
- It's easier to read than many other popular programming languages.
- It's widely used in education.[15]
- It enhances developer productivity with extensive standard libraries and *thousands* of third-party open-source libraries, so programmers can write code faster and perform complex tasks with minimal code. We'll say more about this in Section 1.8.
- There are massive numbers of free open-source Python applications.
- It's popular in web development (e.g., Django, Flask).
- It supports popular programming paradigms—procedural, functional, object-oriented and reflective.[16] We'll begin introducing functional-style programming features in Chapter 4 and use them in subsequent chapters.
- It simplifies concurrent programming—with asyncio and async/await, you're able to write single-threaded concurrent code[17], greatly simplifying the inherently complex processes of writing, debugging and maintaining that code.[18]
- There are lots of capabilities for enhancing Python performance.
- It's used to build anything from simple scripts to complex apps with massive numbers of users, such as Dropbox, YouTube, Reddit, Instagram and Quora.[19]

---

8. https://www.oreilly.com/ideas/5-things-to-watch-in-python-in-2017.
9. https://www.kdnuggets.com/2017/08/python-overtakes-r-leader-analytics-data-science.html.
10. https://www.r-bloggers.com/data-science-job-report-2017-r-passes-sas-but-python-leaves-them-both-behind/.
11. https://www.oreilly.com/ideas/5-things-to-watch-in-python-in-2017.
12. https://dbader.org/blog/why-learn-python.
13. https://simpleprogrammer.com/2017/01/18/7-reasons-why-you-should-learn-python/.
14. https://www.oreilly.com/ideas/5-things-to-watch-in-python-in-2017.
15. Tollervey, N., *Python in Education: Teach, Learn, Program* (O'Reilly Media, Inc., 2015).
16. https://en.wikipedia.org/wiki/Python_(programming_language).
17. https://docs.python.org/3/library/asyncio.html.
18. https://www.oreilly.com/ideas/5-things-to-watch-in-python-in-2017.
19. https://www.hartmannsoftware.com/Blog/Articles_from_Software_Fans/Most-Famous-Software-Programs-Written-in-Python.

- It's popular in artificial intelligence, which is enjoying explosive growth, in part because of its special relationship with data science.
- It's widely used in the financial community.[20]
- There's an extensive job market for Python programmers across many disciplines, especially in data-science-oriented positions, and Python jobs are among the highest paid of all programming jobs.[21,22]

Anaconda Python Distribution

We use the Anaconda Python distribution because it's easy to install on Windows, macOS and Linux and supports the latest versions of Python (3.7 at the time of this writing), the IPython interpreter (introduced in Section 1.10.1) and Jupyter Notebooks (introduced in Section 1.10.3). Anaconda also includes other software packages and libraries commonly used in Python programming and data science, allowing students to focus on learning Python, computer science and data science, rather than software installation issues. The IPython interpreter[23] has features that help students and professionals explore, discover and experiment with Python, the Python Standard Library and the extensive set of third-party libraries.

Zen of Python

We adhere to Tim Peters' *The Zen of Python*, which summarizes Python creator Guido van Rossum's design principles for the language. This list can be viewed in IPython with the command `import this`. The Zen of Python is defined in Python Enhancement Proposal (PEP) 20. "A PEP is a design document providing information to the Python community, or describing a new feature for Python or its processes or environment."[24]

Self Check

1 *(Fill-In)* The _____ summarizes Python creator Guido van Rossum's design principles for the Python language.
Answer: Zen of Python.

2 *(True/False)* The Python language supports popular programming paradigms—procedural, functional, object-oriented and reflective.
Answer: True.

3 *(True/False)* R is most the popular data-science programming language.
Answer: False. Python recently surpassed R as the most popular data-science programming language.

---

20. Kolanovic, M. and R. Krishnamachari, *Big Data and AI Strategies: Machine Learning and Alternative Data Approach to Investing* (J.P. Morgan, 2017).
21. https://www.infoworld.com/article/3170838/developer/get-paid-10-programming-languages-to-learn-in-2017.html.
22. https://medium.com/@ChallengeRocket/top-10-of-programming-languages-with-the-highest-salaries-in-2017-4390f468256e.
23. https://ipython.org/.
24. https://www.python.org/dev/peps/pep-0001/.

1.8 It's the Libraries!

Throughout the book, we focus on using existing libraries to help you avoid "reinventing the wheel," thus leveraging your program-development efforts. Often, rather than developing lots of original code—a costly and time-consuming process—you can simply create an object of a pre-existing library class, which takes only a single Python statement. So, libraries will help you perform significant tasks with modest amounts of code. You'll use a broad range of Python standard libraries, data-science libraries and other third-party libraries.

1.8.1 Python Standard Library

The **Python Standard Library** provides rich capabilities for text/binary data processing, mathematics, functional-style programming, file/directory access, data persistence, data compression/archiving, cryptography, operating-system services, concurrent programming, interprocess communication, networking protocols, JSON/XML/other Internet data formats, multimedia, internationalization, GUI, debugging, profiling and more. The following table lists some of the Python Standard Library modules that we use in examples or that you'll explore in the exercises.

Some of the Python Standard Library modules we use in the book	
`collections`—Additional data structures beyond lists, tuples, dictionaries and sets.	`os`—Interacting with the operating system.
`csv`—Processing comma-separated value files.	`timeit`—Performance analysis.
`datetime`, `time`—Date and time manipulations.	`queue`—First-in, first-out data structure.
`decimal`—Fixed-point and floating-point arithmetic, including monetary calculations.	`random`—Pseudorandom numbers.
`doctest`—Simple unit testing via validation tests and expected results embedded in docstrings.	`re`—Regular expressions for pattern matching.
	`sqlite3`—SQLite relational database access.
`json`—JavaScript Object Notation (JSON) processing for use with web services and NoSQL document databases.	`statistics`—Mathematical statistics functions like `mean`, `median`, `mode` and `variance`.
	`string`—String processing.
`math`—Common math constants and operations.	`sys`—Command-line argument processing; standard input, standard output and standard error streams.

1.8.2 Data-Science Libraries

Python has an enormous and rapidly growing community of open-source developers in many fields. One of the biggest reasons for Python's popularity is the extraordinary range of open-source libraries developed by the open-source community. One of our goals is to create examples, exercises, projects (EEPs) and implementation case studies that give you an engaging, challenging and entertaining introduction to Python programming, while also involving you in hands-on data science, key data-science libraries and more. You'll be amazed at the substantial tasks you can accomplish in just a few lines of code. The following table lists various popular data-science libraries. You'll use many of these as you work through our data-science examples, exercises and projects. For visualization, we'll focus primarily on Matplotlib and Seaborn, but there are many more. For a nice summary of Python visualization libraries see `http://pyviz.org/`.

Popular Python libraries used in data science

Scientific Computing and Statistics

NumPy (Numerical Python)—Python does not have a built-in array data structure. It uses lists, which are convenient but relatively slow. NumPy provides the more efficient ndarray data structure to represent lists and matrices, and it also provides routines for processing such data structures.

SciPy (Scientific Python)—Built on NumPy, SciPy adds routines for scientific processing, such as integrals, differential equations, additional matrix processing and more. scipy.org controls SciPy and NumPy.

StatsModels—Provides support for estimations of statistical models, statistical tests and statistical data exploration.

Data Manipulation and Analysis

Pandas—An extremely popular library for data manipulations. Pandas makes abundant use of NumPy's ndarray. Its two key data structures are Series (one dimensional) and DataFrames (two dimensional).

Visualization

Matplotlib—A highly customizable visualization and plotting library. Supported plots include regular, scatter, bar, contour, pie, quiver, grid, polar axis, 3D and text.

Seaborn—A higher-level visualization library built on Matplotlib. Seaborn adds a nicer look-and-feel, additional visualizations and enables you to create visualizations with less code.

Machine Learning, Deep Learning and Reinforcement Learning

scikit-learn—Top machine-learning library. Machine learning is a subset of AI. Deep learning is a subset of machine learning that focuses on neural networks.

Keras—One of the easiest to use deep-learning libraries. Keras runs on top of TensorFlow (Google), CNTK (Microsoft's cognitive toolkit for deep learning) or Theano (Université de Montréal).

TensorFlow—From Google, this is the most widely used deep learning library. TensorFlow works with GPUs (graphics processing units) or Google's custom TPUs (Tensor processing units) for performance. TensorFlow is important in AI and big data analytics—where processing demands are enormous. You'll use the version of Keras that's built into TensorFlow.

OpenAI Gym—A library and environment for developing, testing and comparing reinforcement-learning algorithms. You'll explore this in the Chapter 16 exercises.

Natural Language Processing (NLP)

NLTK (Natural Language Toolkit)—Used for natural language processing (NLP) tasks.

TextBlob—An object-oriented NLP text-processing library built on the NLTK and pattern NLP libraries. TextBlob simplifies many NLP tasks.

Gensim—Similar to NLTK. Commonly used to build an index for a collection of documents, then determine how similar another document is to each of those in the index. You'll explore this in the Chapter 12 exercises.

✓ Self Check for Section 1.8

1 *(Fill-In)* _____ help you avoid "reinventing the wheel," thus leveraging your program-development efforts.
Answer: Libraries.

2 *(Fill-In)* The _____ provides rich capabilities for many common Python programming tasks.
Answer: Python Standard Library.

1.9 Other Popular Programming Languages

The following is a brief introduction to several other popular programming languages—in the next section, we take a deeper look at Python:

- *Basic* was developed in the 1960s at Dartmouth College to familiarize novices with programming techniques. Many of its latest versions are object-oriented.

- *C* was developed in the early 1970s by Dennis Ritchie at Bell Laboratories. It initially became widely known as the UNIX operating system's development language. Today, most code for general-purpose operating systems and other performance-critical systems is written in C or C++.

- *C++*, which is based on C, was developed by Bjarne Stroustrup in the early 1980s at Bell Laboratories. C++ provides features that enhance the C language and adds capabilities for object-oriented programming.

- *Java*—Sun Microsystems in 1991 funded an internal corporate research project led by James Gosling, which resulted in the C++-based object-oriented programming language called Java. A key goal of Java is to enable developers to write programs that will run on a great variety of computer systems. This is called "write once, run anywhere." Java is used to develop enterprise applications, to enhance the functionality of web servers (the computers that provide the content to our web browsers), to provide applications for consumer devices (e.g., smartphones, tablets, television set-top boxes, appliances, automobiles and more) and for many other purposes. Java was originally the key language for developing Android smartphone and tablet apps, though several other languages are now supported.

- *C#* (based on C++ and Java) is one of Microsoft's three primary object-oriented programming languages—the other two are Visual C++ and Visual Basic. C# was developed to integrate the web into computer applications and is now widely used to develop many types of applications. As part of Microsoft's many open-source initiatives implemented over the last few years, they now offer open-source versions of C# and Visual Basic.

- *JavaScript* is the most widely used scripting language. It's primarily used to add programmability to web pages—for example, animations and interactivity with the user. All major web browsers support it. Many Python visualization libraries output JavaScript as part of visualizations that you can interact with in your web browser. Tools like NodeJS also enable JavaScript to run outside of web browsers.

- *Swift*, which was introduced in 2014, is Apple's programming language for developing iOS and macOS apps. Swift is a contemporary language that includes popular features from languages such as Objective-C, Java, C#, Ruby, Python and others. Swift is open source, so it can be used on non-Apple platforms as well.

- *R* is a popular open-source programming language for statistical applications and visualization. Python and R are the two most widely used data-science languages.

Self Check

1. *(Fill-In)* Today, most code for general-purpose operating systems and other performance-critical systems is written in _____.
Answer: C or C++.

2. *(Fill-In)* A key goal of _____ is to enable developers to write programs that will run on a great variety of computer systems and computer-controlled devices. This is sometimes called "write once, run anywhere."
Answer: Java.

1.10 Test-Drives: Using IPython and Jupyter Notebooks

In this section, you'll test-drive the IPython interpreter[25] in two modes:

- In **interactive mode**, you'll enter small bits of Python code called **snippets** and immediately see their results.
- In **script mode**, you'll execute code loaded from a file that has the .py extension (short for Python). Such files are called **scripts** or **programs**, and they're generally longer than the code snippets you'll do in interactive mode.

Then, you'll learn how to use the browser-based environment known as the Jupyter Notebook for writing and executing Python code.[26]

1.10.1 Using IPython Interactive Mode as a Calculator

Let's use IPython interactive mode to evaluate simple arithmetic expressions.

Entering IPython in Interactive Mode

First, open a command-line window on your system:

- On macOS, open a **Terminal** from the **Applications** folder's **Utilities** subfolder.
- On Windows, open the **Anaconda Command Prompt** from the start menu.
- On Linux, open your system's **Terminal** or shell (this varies by Linux distribution).

In the command-line window, type ipython, then press *Enter* (or *Return*). You'll see text like the following, this varies by platform and by IPython version:

```
Python 3.7.0 | packaged by conda-forge | (default, Jan 20 2019, 17:24:52)
Type 'copyright', 'credits' or 'license' for more information
IPython 6.5.0 -- An enhanced Interactive Python. Type '?' for help.

In [1]:
```

The text "In [1]:" is a *prompt*, indicating that IPython is waiting for your input. You can type ? for help or begin entering snippets, as you'll do momentarily.

---

25. Before reading this section, follow the instructions in the Before You Begin section to install the Anaconda Python distribution, which contains the IPython interpreter.
26. Jupyter supports many programming languages by installing their "kernels." For more information see https://github.com/jupyter/jupyter/wiki/Jupyter-kernels.

Evaluating Expressions

In interactive mode, you can evaluate expressions:

```
In [1]: 45 + 72
Out[1]: 117

In [2]:
```

After you type 45 + 72 and press *Enter*, IPython *reads* the snippet, *evaluates* it and *prints* its result in Out[1].[27] Then IPython displays the In [2] prompt to show that it's waiting for you to enter your second snippet. For each new snippet, IPython adds 1 to the number in the square brackets. Each In [1] prompt in the book indicates that we've started a new interactive session. We generally do that for each new section of a chapter.

Let's evaluate a more complex expression:

```
In [2]: 5 * (12.7 - 4) / 2
Out[2]: 21.75
```

Python uses the asterisk (*) for multiplication and the forward slash (/) for division. As in mathematics, parentheses force the evaluation order, so the parenthesized expression (12.7 - 4) evaluates first, giving 8.7. Next, 5 * 8.7 evaluates giving 43.5. Then, 43.5 / 2 evaluates, giving the result 21.75, which IPython displays in Out[2]. Whole numbers, like 5, 4 and 2, are called **integers**. Numbers with decimal points, like 12.7, 43.5 and 21.75, are called **floating-point numbers**.

Exiting Interactive Mode

To leave interactive mode, you can:

- Type the exit command at the current In [] prompt and press *Enter* to exit immediately.
- Type the key sequence <*Ctrl*> + *d* (or <*control*> + *d*). This displays the prompt "Do you really want to exit ([y]/n)?". The square brackets around y indicate that it's the default response—pressing *Enter* submits the default response and exits.
- Type <*Ctrl*> + *d* (or <*control*> + *d*) twice (macOS and Linux only).

✓ Self Check

1 *(Fill-In)* In IPython interactive mode, you'll enter small bits of Python code called _____ and immediately see their results.
Answer: snippets.

2 In IPython _____ mode, you'll execute Python code loaded from a file that has the .py extension (short for Python).
Answer: script.

3 *(IPython Session)* Evaluate the expression 5 * (3 + 4) both with and without the parentheses. Do you get the same result? Why or why not?

---

27. In the next chapter, you'll see that there are some cases in which Out[] is not displayed.

Answer: You get different results because snippet [1] first calculates 3 + 4, which is 7, then multiplies that by 5. Snippet [2] first multiplies 5 * 3, which is 15, then adds that to 4.

```
In [1]: 5 * (3 + 4)
Out[1]: 35

In [2]: 5 * 3 + 4
Out[2]: 19
```

1.10.2 Executing a Python Program Using the IPython Interpreter

In this section, you'll execute a script named RollDieDynamic.py that you'll write in Chapter 6. The **.py extension** indicates that the file contains Python source code. The script RollDieDynamic.py simulates rolling a six-sided die. It presents a colorful animated visualization that dynamically graphs the frequencies of each die face.

Changing to This Chapter's Examples Folder

You'll find the script in the book's ch01 source-code folder. In the Before You Begin section you extracted the examples folder to your user account's Documents folder. Each chapter has a folder containing that chapter's source code. The folder is named ch##, where ## is a two-digit chapter number from 01 to 17. First, open your system's command-line window. Next, use the cd ("change directory") command to change to the ch01 folder:

- On macOS/Linux, type cd ~/Documents/examples/ch01, then press *Enter*.

- On Windows, type cd C:\Users*YourAccount*\Documents\examples\ch01, then press *Enter*.

Executing the Script

To execute the script, type the following command at the command line, then press *Enter*:

```
ipython RollDieDynamic.py 6000 1
```

The script displays a window, showing the visualization. The numbers 6000 and 1 tell this script the number of times to roll dice and how many dice to roll each time. In this case, we'll update the chart 6000 times for 1 die at a time.

For a six-sided die, the values 1 through 6 should each occur with "equal likelihood"—the probability of each is 1/6th or about 16.667%. If we roll a die 6000 times, we'd expect about 1000 of each face. Like coin tossing, die rolling is *random*, so there could be some faces with fewer than 1000, some with 1000 and some with more than 1000. We took the screen captures on the next page during the script's execution. This script uses randomly generated die values, so your results will differ. Experiment with the script by changing the value 1 to 100, 1000 and 10000. Notice that as the number of die rolls gets larger, the frequencies zero in on 16.667%. This is a phenomenon of the "Law of Large Numbers."

Creating Scripts

Typically, you create your Python source code in an editor that enables you to type text. Using the editor, you type a program, make any necessary corrections and save it to your computer. **Integrated development environments (IDEs)** provide tools that support the entire software-development process, such as editors, debuggers for locating **logic errors** that cause programs to execute incorrectly and more. Some popular Python IDEs include Spyder (which comes with Anaconda), PyCharm and Visual Studio Code.

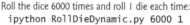

Roll the dice 6000 times and roll 1 die each time:
ipython RollDieDynamic.py 6000 1

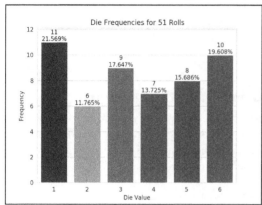
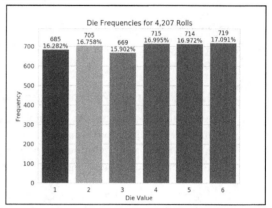

Problems That May Occur at Execution Time
Programs often do not work on the first try. For example, an executing program might try to divide by zero (an illegal operation in Python). This would cause the program to display an error message. If this occurred in a script, you'd return to the editor, make the necessary corrections and re-execute the script to determine whether the corrections fixed the problem(s).

Errors such as division by zero occur as a program runs, so they're called **runtime errors** or **execution-time errors**. **Fatal runtime errors** cause programs to terminate immediately without having successfully performed their jobs. **Non-fatal runtime errors** allow programs to run to completion, often producing incorrect results.

Self Check

1 *(Discussion)* When the example in this section finishes all 6000 rolls, does the chart show that the die faces appeared *about* 1000 times each?
Answer: Most likely, yes. This example is based on random-number generation, so the results may vary. Because of this randomness, most of the counts will be a little more than 1000 or a little less.

2 *(Discussion)* Run the example in this section again. Do the faces appear the same number of times as they did in the previous execution?
Answer: Probably not. This example uses random-number generation, so successive executions likely will produce different results. In Chapter 4, we'll show how to force Python to produce the same sequence of random numbers. This is important for *reproducibility*—a crucial data-science topic you'll investigate in the chapter exercises and throughout the book. You'll want other data scientists to be able to reproduce your results. Also, you'll want to be able to reproduce your own experimental results. This is helpful when you find and fix an error in your program and want to make sure that you've corrected it properly.

1.10.3 Writing and Executing Code in a Jupyter Notebook
The Anaconda Python Distribution that you installed in the Before You Begin section comes with the **Jupyter Notebook**—an interactive, browser-based environment in which

you can write and execute code and intermix the code with text, images and video. Jupyter Notebooks are broadly used in the data-science community in particular and the broader scientific community in general. They're the preferred means of doing Python-based data analytics studies and *reproducibly* communicating their results. The Jupyter Notebook environment actually supports many programming languages.

For your convenience, all of the book's source code also is provided in Jupyter Notebooks that you can simply load and execute. In this section, you'll use the **JupyterLab** interface, which enables you to manage your notebook files and other files that your notebooks use (like images and videos). As you'll see, JupyterLab also makes it convenient to write code, execute it, see the results, modify the code and execute it again.

You'll see that coding in a Jupyter Notebook is similar to working with IPython—in fact, Jupyter Notebooks use IPython by default. In this section, you'll create a notebook, add the code from Section 1.10.1 to it and execute that code.

Opening JupyterLab in Your Browser

To open JupyterLab, change to the `ch01` examples folder in your Terminal, shell or Anaconda Command Prompt (as in Section 1.10.2), type the following command, then press *Enter* (or *Return*):

    jupyter lab

This executes the Jupyter Notebook server on your computer and opens JupyterLab in your default web browser, showing the `ch01` folder's contents in the **File Browser** tab

at the left side of the JupyterLab interface:

26 Introduction to Computers and Python

The Jupyter Notebooks server enables you to load and run Jupyter Notebooks in your web browser. From the JupyterLab **Files** tab, you can double-click files to open them in the right side of the window where the **Launcher** tab is currently displayed. Each file you open appears as a separate tab in this part of the window. If you accidentally close your browser, you can reopen JupyterLab by entering the following address in your web browser

```
http://localhost:8888/lab
```

Creating a New Jupyter Notebook
In the **Launcher** tab under **Notebook**, click the **Python 3** button to create a new Jupyter Notebook named `Untitled.ipynb` in which you can enter and execute Python 3 code. The file extension `.ipynb` is short for IPython Notebook—the original name of the Jupyter Notebook.

Renaming the Notebook
Rename `Untitled.ipynb` as `TestDrive.ipynb`:

1. Right-click the `Untitled.ipynb` tab and select **Rename Notebook...**.
2. Change the name to `TestDrive.ipynb` and click **RENAME**.

The top of JupyterLab should now appear as follows:

Evaluating an Expression
The unit of work in a notebook is a cell in which you can enter code snippets. By default, a new notebook contains one cell—the rectangle in the `TestDrive.ipynb` notebook—but you can add more. To the cell's left, the notation [ ]: is where the Jupyter Notebook will display the cell's snippet number *after* you execute the cell. Click in the cell, then type the expression

```
45 + 72
```

To execute the current cell's code, type *Ctrl + Enter* (or *control + Enter*). JupyterLab executes the code in IPython, then displays the results below the cell:

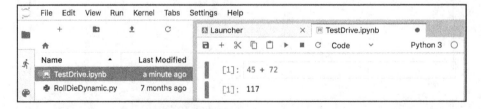

Adding and Executing Another Cell
Let's evaluate a more complex expression. First, click the **+** button in the toolbar above the notebook's first cell—this adds a new cell below the current one:

Click in the new cell, then type the expression

```
5 * (12.7 - 4) / 2
```

and execute the cell by typing *Ctrl + Enter* (or *control + Enter*):

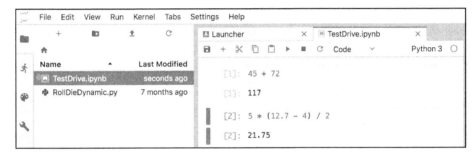

Saving the Notebook
If your notebook has unsaved changes, the **X** in the notebook's tab will change to ●. To save the notebook, select the **File** menu in JupyterLab (not at the top of your browser's window), then select **Save Notebook**.

Notebooks Provided with Each Chapter's Examples
For your convenience, each chapter's examples also are provided as ready-to-execute notebooks without their outputs. This enables you to work through them snippet-by-snippet and see the outputs appear as you execute each snippet.

So that we can show you how to load an existing notebook and execute its cells, let's reset the TestDrive.ipynb notebook to remove its output and snippet numbers. This will return it to a state like the notebooks we provide for the subsequent chapters' examples. From the **Kernel** menu select **Restart Kernel and Clear All Outputs...**, then click the **RESTART** button. The preceding command also is helpful whenever you wish to re-execute a notebook's snippets. The notebook should now appear as follows:

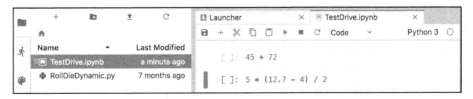

From the **File** menu, select **Save Notebook**, then click the `TestDrive.ipynb` tab's **X** button to close the notebook.

Opening and Executing an Existing Notebook
When you launch JupyterLab from a given chapter's examples folder, you'll be able to open notebooks from that folder or any of its subfolders. Once you locate a specific notebook, double-click it to open it. Open the `TestDrive.ipynb` notebook again now. Once a notebook is open, you can execute each cell individually, as you did earlier in this section, or you can execute the entire notebook at once. To do so, from the **Run** menu select **Run All Cells**. The notebook will execute the cells in order, displaying each cell's output below that cell.

Closing JupyterLab
When you're done with JupyterLab, you can close its browser tab, then in the Terminal, shell or Anaconda Command Prompt from which you launched JupyterLab, type *Ctrl + c* (or *control + c*) twice.

JupyterLab Tips
While working in JupyterLab, you might find these tips helpful:

- If you need to enter and execute many snippets, you can execute the current cell *and* add a new one below it by typing *Shift + Enter*, rather than *Ctrl + Enter* (or *control + Enter*).
- As you get into the later chapters, some of the snippets you'll enter in Jupyter Notebooks will contain many lines of code. To display line numbers within each cell, select **Show line numbers** from JupyterLab's **View** menu.

More Information on Working with JupyterLab
JupyterLab has many more features that you'll find helpful. We recommend that you read the Jupyter team's introduction to JupyterLab at:

> https://jupyterlab.readthedocs.io/en/stable/index.html

For a quick overview, click **Overview** under **GETTING STARTED**. Also, under **USER GUIDE** read the introductions to **The JupyterLab Interface**, **Working with Files**, **Text Editor** and **Notebooks** for many additional features.

 ## Self Check

1. *(True/False)* Jupyter Notebooks are the preferred means of doing Python-based data analytics studies and reproducibly communicating their results.
Answer: True.

2. *(Jupyter Notebook Session)* Ensure that JupyterLab is running, then open your `TestDrive.ipynb` notebook. Add and execute two more snippets that evaluate the expression 5 * (3 + 4) both with and without the parentheses. You should see the same results as in Section 1.10.1's Self Check Exercise 3.

Answer:

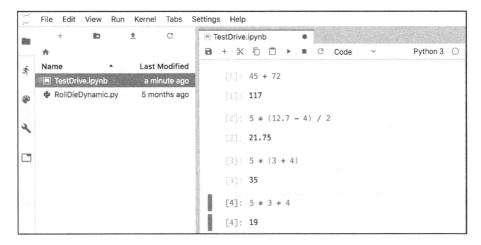

1.11 Internet and World Wide Web

In the late 1960s, ARPA—the Advanced Research Projects Agency of the United States Department of Defense—rolled out plans for networking the main computer systems of approximately a dozen ARPA-funded universities and research institutions. The computers were to be connected with communications lines operating at speeds on the order of 50,000 bits per second, a stunning rate at a time when most people (of the few who even had networking access) were connecting over telephone lines to computers at a rate of 110 bits per second. Academic research was about to take a giant leap forward. ARPA proceeded to implement what quickly became known as the ARPANET, the precursor to today's **Internet**. Today's fastest Internet speeds are on the order of billions of bits per second with trillion-bits-per-second (terabit) speeds already being tested![28]

Things worked out differently from the original plan. Although the ARPANET enabled researchers to network their computers, its main benefit proved to be the capability for quick and easy communication via what came to be known as electronic mail (e-mail). This is true even on today's Internet, with e-mail, instant messaging, file transfer and social media such as Snapchat, Instagram, Facebook and Twitter enabling billions of people worldwide to communicate quickly and easily.

The protocol (set of rules) for communicating over the ARPANET became known as the **Transmission Control Protocol (TCP)**. TCP ensured that messages, consisting of sequentially numbered pieces called *packets*, were properly delivered from sender to receiver, arrived intact and were assembled in the correct order.

1.11.1 Internet: A Network of Networks

In parallel with the early evolution of the Internet, organizations worldwide were implementing their own networks for both intra-organization (that is, within an organization) and inter-organization (that is, between organizations) communication. A huge variety of networking hardware and software appeared. One challenge was to enable these different

---

28. `https://testinternetspeed.org/blog/bt-testing-1-4-terabit-internet-connections/`.

networks to communicate with each other. ARPA accomplished this by developing the **Internet Protocol** (IP), which created a true "network of networks," the current architecture of the Internet. The combined set of protocols is now called **TCP/IP**. Each Internet-connected device has an **IP address**—a unique numerical identifier used by devices communicating via TCP/IP to locate one another on the Internet.

Businesses rapidly realized that by using the Internet, they could improve their operations and offer new and better services to their clients. Companies started spending large amounts of money to develop and enhance their Internet presence. This generated fierce competition among communications carriers and hardware and software suppliers to meet the increased infrastructure demand. As a result, **bandwidth**—the information-carrying capacity of communications lines—on the Internet has increased tremendously, while hardware costs have plummeted.

1.11.2 World Wide Web: Making the Internet User-Friendly

The **World Wide Web** (simply called "the web") is a collection of hardware and software associated with the Internet that allows computer users to locate and view documents (with various combinations of text, graphics, animations, audios and videos) on almost any subject. In 1989, Tim Berners-Lee of CERN (the European Organization for Nuclear Research) began developing **HyperText Markup Language** (HTML)—the technology for sharing information via "hyperlinked" text documents. He also wrote communication protocols such as **HyperText Transfer Protocol** (HTTP) to form the backbone of his new hypertext information system, which he referred to as the World Wide Web.

In 1994, Berners-Lee founded the **World Wide Web Consortium** (W3C, https://www.w3.org), devoted to developing web technologies. One of the W3C's primary goals is to make the web universally accessible to everyone regardless of disabilities, language or culture.

1.11.3 The Cloud

More and more computing today is done "in the cloud"—that is, distributed across the Internet worldwide. The apps you use daily are heavily dependent on various **cloud-based services** that use massive clusters of computing resources (computers, processors, memory, disk drives, etc.) and databases that communicate over the Internet with each other and the apps you use. A service that provides access to itself over the Internet is known as a **web service**. As you'll see, using cloud-based services in Python often is as simple as creating a software object and interacting with it. That object then uses web services that connect to the cloud on your behalf.

Throughout the Chapters 12–17 examples and exercises, you'll work with many cloud-based services:

- In Chapters 13 and 17, you'll use Twitter's web services (via the Python library Tweepy) to get information about specific Twitter users, search for tweets from the last seven days and to receive streams of tweets as they occur—that is, in **real time**.

- In Chapters 12 and 13, you'll use the Python library TextBlob to translate text between languages. Behind the scenes, TextBlob uses the Google Translate web service to perform those translations.

- In Chapter 14, you'll use the IBM Watson's Text to Speech, Speech to Text and Translate services. You'll implement a traveler's assistant translation app that enables you to speak a question in English, transcribes the speech to text, translates the text to Spanish and speaks the Spanish text. The app then allows you to speak a Spanish response (in case you don't speak Spanish, we provide an audio file you can use), transcribes the speech to text, translates the text to English and speaks the English response. Via IBM Watson demos, you'll also experiment with many other Watson cloud-based services in Chapter 14.

- In Chapter 17, you'll work with Microsoft Azure's HDInsight service and other Azure web services as you learn to implement big-data applications using Apache Hadoop and Spark. Azure is Microsoft's set of cloud-based services.

- In Chapter 17, you'll use the Dweet.io web service to simulate an Internet-connected thermostat that publishes temperature readings online. You'll also use a web-based service to create a "dashboard" that visualizes the temperature readings over time and warns you if the temperature gets too low or too high.

- In Chapter 17, you'll use a web-based dashboard to visualize a simulated stream of live sensor data from the PubNub web service. You'll also create a Python app that visualizes a PubNub simulated stream of live stock-price changes.

- In multiple exercises, you'll research, explore and use Wikipedia web services.

In most cases, you'll create Python objects that interact with web services on your behalf, hiding the details of how to access these services over the Internet.

Mashups

The applications-development methodology of *mashups* enables you to rapidly develop powerful software applications by combining (often free) complementary web services and other forms of information feeds—as you'll do in our IBM Watson traveler's assistant translation app. One of the first mashups combined the real-estate listings provided by http://www.craigslist.org with the mapping capabilities of Google Maps to offer maps that showed the locations of homes for sale or rent in a given area.

ProgrammableWeb (http://www.programmableweb.com/) provides a directory of over 20,750 web services and almost 8,000 mashups. They also provide how-to guides and sample code for working with web services and creating your own mashups. According to their website, some of the most widely used web services are Facebook, Google Maps, Twitter and YouTube.

1.11.4 Internet of Things

The Internet is no longer just a network of *computers*—it's an **Internet of Things** (IoT). A *thing* is any object with an IP address and the ability to send, and in some cases receive, data automatically over the Internet. Such *things* include:

- a car with a transponder for paying tolls,
- monitors for parking-space availability in a garage,
- a heart monitor implanted in a human,
- water quality monitors,

- a smart meter that reports energy usage,
- radiation detectors,
- item trackers in a warehouse,
- mobile apps that can track your movement and location,
- smart thermostats that adjust room temperatures based on weather forecasts and activity in the home, and
- intelligent home appliances.

According to `statista.com`, there are already over 23 billion IoT devices in use today, and there could be over 75 billion IoT devices in 2025.[29]

Self Check for Section 1.11

1 *(Fill-In)* The _____ was the precursor to today's Internet.
Answer: ARPANET.

2 *(Fill-In)* The _____ (simply called "the web") is a collection of hardware and software associated with the Internet that allows computer users to locate and view documents (with various combinations of text, graphics, animations, audios and videos).
Answer: World Wide Web.

3 *(Fill-In)* In the Internet of Things (IoT), a *thing* is any object with a(n) _____ and the ability to send, and in some cases receive, data automatically over the Internet.
Answer: IP address.

1.12 Software Technologies

As you learn about and work in software development, you'll frequently encounter the following buzzwords:

- **Refactoring:** Reworking programs to make them clearer and easier to maintain while preserving their correctness and functionality. Many IDEs contain built-in *refactoring tools* to do major portions of the reworking automatically.
- **Design patterns:** Proven architectures for constructing flexible and maintainable object-oriented software. The field of design patterns tries to enumerate those recurring patterns, encouraging software designers to *reuse* them to develop better-quality software using less time, money and effort.
- **Cloud computing:** You can use software and data stored in the "cloud"—i.e., accessed on remote computers (or servers) via the Internet and available on demand—rather than having it stored locally on your desktop, notebook computer or mobile device. This allows you to increase or decrease computing resources to meet your needs at any given time, which is more cost effective than purchasing hardware to provide enough storage and processing power to meet occasional peak demands. Cloud computing also saves money by shifting to the

---

29. https://www.statista.com/statistics/471264/iot-number-of-connected-devices-worldwide/.

service provider the burden of managing these apps (such as installing and upgrading the software, security, backups and disaster recovery).
- **Software Development Kits (SDKs)**—The tools and documentation that developers use to program applications. For example, in Chapter 14, you'll use the Watson Developer Cloud Python SDK to interact with IBM Watson services from a Python application.

Self Check

1 *(Fill-In)* _____ is the process of reworking programs to make them clearer and easier to maintain while preserving their correctness and functionality.
Answer: refactoring.

1.13 How Big Is Big Data?

For computer scientists and data scientists, data is now as important as writing programs. According to IBM, approximately 2.5 quintillion bytes (2.5 *exabytes*) of data are created daily,[30] and 90% of the world's data was created in the last two years.[31] According to IDC, the global data supply will reach 175 *zettabytes* (equal to 175 trillion gigabytes or 175 billion terabytes) annually by 2025.[32] Consider the following examples of various popular data measures.

Megabytes (MB)

One megabyte is about one million (actually 2^{20}) bytes. Many of the files we use on a daily basis require one or more MBs of storage. Some examples include:

- MP3 audio files—High-quality MP3s range from 1 to 2.4 MB per minute.[33]
- Photos—JPEG format photos taken on a digital camera can require about 8 to 10 MB per photo.
- Video—Smartphone cameras can record video at various resolutions. Each minute of video can require many megabytes of storage. For example, on one of our iPhones, the **Camera** settings app reports that 1080p video at 30 frames-per-second (FPS) requires 130 MB/minute and 4K video at 30 FPS requires 350 MB/minute.

Gigabytes (GB)

One gigabyte is about 1000 megabytes (actually 2^{30} bytes). A dual-layer DVD can store up to 8.5 GB[34], which translates to:

- as much as 141 hours of MP3 audio,
- approximately 1000 photos from a 16-megapixel camera,

---

30. https://www.ibm.com/blogs/watson/2016/06/welcome-to-the-world-of-a-i/.
31. https://public.dhe.ibm.com/common/ssi/ecm/wr/en/wrl12345usen/watson-customer-engagement-watson-marketing-wr-other-papers-and-reports-wrl12345usen-20170719.pdf.
32. https://www.networkworld.com/article/3325397/storage/idc-expect-175-zettabytes-of-data-worldwide-by-2025.html.
33. https://www.audiomountain.com/tech/audio-file-size.html.
34. https://en.wikipedia.org/wiki/DVD.

- approximately 7.7 minutes of 1080p video at 30 FPS, or
- approximately 2.85 minutes of 4K video at 30 FPS.

The current highest-capacity Ultra HD Blu-ray discs can store up to 100 GB of video.[35] Streaming a 4K movie can use between 7 and 10 GB per hour (highly compressed).

Terabytes (TB)
One terabyte is about 1000 gigabytes (actually 2^{40} bytes). Recent disk drives for desktop computers come in sizes up to 15 TB,[36] which is equivalent to:

- approximately 28 years of MP3 audio,
- approximately 1.68 million photos from a 16-megapixel camera,
- approximately 226 hours of 1080p video at 30 FPS and
- approximately 84 hours of 4K video at 30 FPS.

Nimbus Data now has the largest solid-state drive (SSD) at 100 TB, which can store 6.67 times the 15-TB examples of audio, photos and video listed above.[37]

Petabytes, Exabytes and Zettabytes
There are nearly four billion people online creating about 2.5 quintillion bytes of data each day[38]—that's 2500 petabytes (each petabyte is about 1000 terabytes) or 2.5 exabytes (each exabyte is about 1000 petabytes). According to a March 2016 *AnalyticsWeek* article, within five years there will be over 50 billion devices connected to the Internet (most of them through the Internet of Things, which we discuss in Sections 1.11.4 and 17.8) and by 2020 we'll be producing 1.7 megabytes of new data every second *for every person on the planet*.[39] At today's numbers (approximately 7.7 billion people[40]), that's about

- 13 petabytes of new data per second,
- 780 petabytes per minute,
- 46,800 petabytes (46.8 exabytes) per hour and
- 1,123 exabytes per day—that's 1.123 zettabytes (ZB) per day (each zettabyte is about 1000 exabytes).

That's the equivalent of over 5.5 million hours (over 600 years) of 4K video every day or approximately 116 billion photos every day!

Additional Big-Data Stats
For a real-time sense of big data, check out https://www.internetlivestats.com, with various statistics, including the numbers so far today of

- Google searches.

---

35. https://en.wikipedia.org/wiki/Ultra_HD_Blu-ray.
36. https://www.zdnet.com/article/worlds-biggest-hard-drive-meet-western-digitals-15tb-monster/.
37. https://www.cinema5d.com/nimbus-data-100tb-ssd-worlds-largest-ssd/.
38. https://public.dhe.ibm.com/common/ssi/ecm/wr/en/wrl12345usen/watson-customer-engagement-watson-marketing-wr-other-papers-and-reports-wrl12345usen-20170719.pdf.
39. https://analyticsweek.com/content/big-data-facts/.
40. https://en.wikipedia.org/wiki/World_population.

- Tweets.
- Videos viewed on YouTube.
- Photos uploaded on Instagram.

You can click each statistic to drill down for more information. For instance, they say over 250 *billion* tweets have been sent in 2018.

Some other interesting big-data facts:

- Every hour, YouTube users upload 24,000 hours of video, and almost 1 billion hours of video are watched on YouTube every day.[41]
- Every second, there are 51,773 GBs (or 51.773 TBs) of Internet traffic, 7894 tweets sent, 64,332 Google searches and 72,029 YouTube videos viewed.[42]
- On Facebook each day there are 800 million "**likes**,"[43] 60 million emojis are sent,[44] and there are over two billion searches of the more than 2.5 trillion Facebook posts since the site's inception.[45]
- In June 2017, Will Marshall, CEO of Planet, said the company has 142 satellites that image the whole planet's land mass once per day. They add one million images and seven TBs of new data each day. Together with their partners, they're using machine learning on that data to improve crop yields, see how many ships are in a given port and track deforestation. With respect to Amazon deforestation, he said: "Used to be we'd wake up after a few years and there's a big hole in the Amazon. Now we can literally count every tree on the planet every day."[46]

Domo, Inc. has a nice infographic called "Data Never Sleeps 6.0" showing how much data is generated *every minute*, including:[47]

- 473,400 tweets sent.
- 2,083,333 Snapchat photos shared.
- 97,222 hours of Netflix video viewed.
- 12,986,111 million text messages sent.
- 49,380 Instagram posts.
- 176,220 Skype calls.
- 750,000 Spotify songs streamed.
- 3,877,140 Google searches.
- 4,333,560 YouTube videos watched.

---

41. https://www.brandwatch.com/blog/youtube-stats/.
42. http://www.internetlivestats.com/one-second.
43. https://newsroom.fb.com/news/2017/06/two-billion-people-coming-together-on-facebook.
44. https://mashable.com/2017/07/17/facebook-world-emoji-day/.
45. https://techcrunch.com/2016/07/27/facebook-will-make-you-talk/.
46. https://www.bloomberg.com/news/videos/2017-06-30/learning-from-planet-s-shoe-boxed-sized-satellites-video, June 30, 2017.
47. https://www.domo.com/learn/data-never-sleeps-6.

Computing Power Over the Years
Data is getting more massive and so is the computing power for processing it. The performance of today's processors is often measured in terms of **FLOPS (floating-point operations per second)**. In the early to mid-1990s, the fastest supercomputer speeds were measured in gigaflops (10^9 FLOPS). By the late 1990s, Intel produced the first teraflop (10^{12} FLOPS) supercomputers. In the early-to-mid 2000s, speeds reached hundreds of teraflops, then in 2008, IBM released the first petaflop (10^{15} FLOPS) supercomputer. Currently, the fastest supercomputer—the IBM Summit, located at the Department of Energy's (DOE) Oak Ridge National Laboratory (ORNL)—is capable of 122.3 petaflops.[48]

Distributed computing can link thousands of personal computers via the Internet to produce even more FLOPS. In late 2016, the Folding@home network—a distributed network in which people volunteer their personal computers' resources for use in disease research and drug design[49]—was capable of over 100 petaflops.[50] Companies like IBM are now working toward supercomputers capable of exaflops (10^{18} FLOPS).[51]

The **quantum computers** now under development theoretically could operate at 18,000,000,000,000,000,000 times the speed of today's "conventional computers"![52] This number is so extraordinary that in one second, a quantum computer theoretically could do staggeringly more calculations than the total that have been done by all computers since the world's first computer appeared. This almost unimaginable computing power could wreak havoc with blockchain-based cryptocurrencies like Bitcoin. Engineers are already rethinking blockchain to prepare for such massive increases in computing power.[53]

The history of supercomputing power is that it eventually works its way down from research labs, where extraordinary amounts of money have been spent to achieve those performance numbers, into "reasonably priced" commercial computer systems and even desktop computers, laptops, tablets and smartphones.

Computing power's cost continues to decline, especially with cloud computing. People used to ask the question, "How much computing power do I need on my system to deal with my *peak* processing needs?" Today, that thinking has shifted to "Can I quickly carve out on the cloud what I need *temporarily* for my most demanding computing chores?" You pay for only what you use to accomplish a given task.

Processing the World's Data Requires Lots of Electricity
Data from the world's Internet-connected devices is exploding, and processing that data requires tremendous amounts of energy. According to a recent article, energy use for processing data in 2015 was growing at 20% per year and consuming approximately three to five percent of the world's power. The article says that total data-processing power consumption could reach 20% by 2025.[54]

---

48. https://en.wikipedia.org/wiki/FLOPS.
49. https://en.wikipedia.org/wiki/Folding@home.
50. https://en.wikipedia.org/wiki/FLOPS.
51. https://www.ibm.com/blogs/research/2017/06/supercomputing-weather-model-exascale/.
52. https://medium.com/@n.biedrzycki/only-god-can-count-that-fast-the-world-of-quantum-computing-406a0a91fcf4.
53. https://singularityhub.com/2017/11/05/is-quantum-computing-an-existential-threat-to-blockchain-technology/.
54. https://www.theguardian.com/environment/2017/dec/11/tsunami-of-data-could-consume-fifth-global-electricity-by-2025.

Another enormous electricity consumer is the blockchain-based cryptocurrency Bitcoin. Processing just one Bitcoin transaction uses approximately the same amount of energy as powering the average American home for a week. The energy use comes from the process Bitcoin "miners" use to prove that transaction data is valid.[55]

According to some estimates, a year of Bitcoin transactions consumes more energy than many countries.[56] Together, Bitcoin and Ethereum (another popular blockchain-based platform and cryptocurrency) consume more energy per year than Israel and almost as much as Greece.[57]

Morgan Stanley predicted in 2018 that "the electricity consumption required to create cryptocurrencies this year could actually outpace the firm's projected global electric vehicle demand—in 2025."[58] This situation is unsustainable, especially given the huge interest in blockchain-based applications, even beyond the cryptocurrency explosion. The blockchain community is working on fixes.[59,60]

Big-Data Opportunities

The big-data explosion is likely to continue exponentially for years to come. With 50 billion computing devices on the horizon, we can only imagine how many more there will be over the next few decades. It's crucial for businesses, governments, the military, and even individuals to get a handle on all this data.

It's interesting that some of the best writings about big data, data science, artificial intelligence and more are coming out of distinguished business organizations, such as J.P. Morgan, McKinsey and more. Big data's appeal to big business is undeniable given the rapidly accelerating accomplishments. Many companies are making significant investments and getting valuable results through technologies in this book, such as big data, machine learning, deep learning, and natural-language processing. This is forcing competitors to invest as well, rapidly increasing the need for computing professionals with data-science and computer science experience. This growth is likely to continue for many years.

✓ Self Check

1 *(Fill-In)* Today's processor performance is often measured in terms of _____.
Answer: FLOPS (floating-point operations per second).

2 *(Fill-In)* The technology that could wreak havoc with blockchain-based cryptocurrencies, like Bitcoin, and other blockchain-based technologies is _____.
Answer: quantum computers.

3 *(True/False)* With cloud computing you pay a fixed price for cloud services regardless of how much you use those services?
Answer: False. A key cloud-computing benefit is that you pay for only what you use to accomplish a given task.

---

55. https://motherboard.vice.com/en_us/article/ywbbpm/bitcoin-mining-electricity-consumption-ethereum-energy-climate-change.
56. https://digiconomist.net/bitcoin-energy-consumption.
57. https://digiconomist.net/ethereum-energy-consumption.
58. https://www.morganstanley.com/ideas/cryptocurrencies-global-utilities.
59. https://www.technologyreview.com/s/609480/bitcoin-uses-massive-amounts-of-energy-but-theres-a-plan-to-fix-it/.
60. http://mashable.com/2017/12/01/bitcoin-energy/.

1.13.1 Big Data Analytics

Data analytics is a mature and well-developed academic and professional discipline. The term "data analysis" was coined in 1962,[61] though people have been analyzing data using statistics for thousands of years going back to the ancient Egyptians.[62] Big data analytics is a more recent phenomenon—the term "big data" was coined around 2000.[63]

Consider four of the V's of big data[64,65]:

1. Volume—the amount of data the world is producing is growing exponentially.
2. Velocity—the speed at which that data is being produced, the speed at which it moves through organizations and the speed at which data changes are growing quickly.[66,67,68]
3. Variety—data used to be alphanumeric (that is, consisting of alphabetic characters, digits, punctuation and some special characters)—today it also includes images, audios, videos and data from an exploding number of Internet of Things sensors in our homes, businesses, vehicles, cities and more.
4. Veracity—the validity of the data—is it complete and accurate? Can we trust that data when making crucial decisions? Is it real?

Most data is now being created digitally in a *variety* of types, in extraordinary *volumes* and moving at astonishing *velocities*. Moore's Law and related observations have enabled us to store data economically and to process and move it faster—and all at rates growing exponentially over time. Digital data storage has become so vast in capacity, cheap and small that we can now conveniently and economically retain *all* the digital data we're creating.[69] That's big data.

The following Richard W. Hamming quote—although from 1962—sets the tone for the rest of this book:

> *"The purpose of computing is insight, not numbers."*[70]

Data science is producing new, deeper, subtler and more valuable insights at a remarkable pace. It's truly making a difference. Big data analytics is an integral part of the answer. We address big data infrastructure in Chapter 17 with hands-on case studies on NoSQL data-

---

61. https://www.forbes.com/sites/gilpress/2013/05/28/a-very-short-history-of-data-science/.
62. https://www.flydata.com/blog/a-brief-history-of-data-analysis/.
63. https://bits.blogs.nytimes.com/2013/02/01/the-origins-of-big-data-an-etymological-detective-story/.
64. https://www.ibmbigdatahub.com/infographic/four-vs-big-data.
65. There are lots of articles and papers that add many other "V-words" to this list.
66. https://www.zdnet.com/article/volume-velocity-and-variety-understanding-the-three-vs-of-big-data/.
67. https://whatis.techtarget.com/definition/3Vs.
68. https://www.forbes.com/sites/brentdykes/2017/06/28/big-data-forget-volume-and-variety-focus-on-velocity.
69. http://www.lesk.com/mlesk/ksg97/ksg.html. [The following article pointed us to this Michael Lesk article: https://www.forbes.com/sites/gilpress/2013/05/28/a-very-short-history-of-data-science/.]
70. Hamming, R. W., *Numerical Methods for Scientists and Engineers* (New York, NY., McGraw Hill, 1962). [The following article pointed us to Hamming's book and his quote that we cited: https://www.forbes.com/sites/gilpress/2013/05/28/a-very-short-history-of-data-science/.]

bases, Hadoop MapReduce programming, Spark, real-time Internet of Things (IoT) stream programming and more.

To get a sense of big data's scope in industry, government and academia, check out the high-resolution graphic.[71] You can click to zoom for easier readability:

```
http://mattturck.com/wp-content/uploads/2018/07/
    Matt_Turck_FirstMark_Big_Data_Landscape_2018_Final.png
```

1.13.2 Data Science and Big Data Are Making a Difference: Use Cases

The data-science field is growing rapidly because it's producing significant results that are making a difference. We enumerate data-science and big data use cases in the following table. We expect that the use cases and our examples, exercises and projects will inspire interesting term projects, directed-study projects, capstone-course projects and thesis research. Big-data analytics has resulted in improved profits, better customer relations, and even sports teams winning more games and championships while spending less on players.[72,73,74]

Data-science use cases		
anomaly detection	customer churn	facial recognition
assisting people with disabilities	customer experience	fitness tracking
auto-insurance risk prediction	customer retention	fraud detection
automated closed captioning	customer satisfaction	game playing
automated image captions	customer service	genomics and healthcare
automated investing	customer service agents	Geographic Information Systems (GIS)
autonomous ships	customized diets	GPS Systems
brain mapping	cybersecurity	health outcome improvement
caller identification	data mining	hospital readmission reduction
cancer diagnosis/treatment	data visualization	human genome sequencing
carbon emissions reduction	detecting new viruses	identity-theft prevention
classifying handwriting	diagnosing breast cancer	immunotherapy
computer vision	diagnosing heart disease	insurance pricing
credit scoring	diagnostic medicine	intelligent assistants
crime: predicting locations	disaster-victim identification	Internet of Things (IoT) and medical device monitoring
crime: predicting recidivism	drones	Internet of Things and weather forecasting
crime: predictive policing	dynamic driving routes	inventory control
crime: prevention	dynamic pricing	language translation
CRISPR gene editing	electronic health records	
crop-yield improvement	emotion detection	
	energy-consumption reduction	

---

71. Turck, M., and J. Hao, "Great Power, Great Responsibility: The 2018 Big Data & AI Landscape," http://mattturck.com/bigdata2018/.
72. Sawchik, T., *Big Data Baseball: Math, Miracles, and the End of a 20-Year Losing Streak* (New York, Flat Iron Books, 2015).
73. Ayres, I., *Super Crunchers* (Bantam Books, 2007), pp. 7–10.
74. Lewis, M., *Moneyball: The Art of Winning an Unfair Game* (W. W. Norton & Company, 2004).

Data-science use cases		
location-based services	predicting weather-sensitive product sales	smart thermostats
loyalty programs		smart traffic control
malware detection	predictive analytics	social analytics
mapping	preventative medicine	social graph analysis
marketing	preventing disease outbreaks	spam detection
marketing analytics	reading sign language	spatial data analysis
music generation	real-estate valuation	sports recruiting and coaching
natural-language translation	recommendation systems	stock market forecasting
new pharmaceuticals	reducing overbooking	student performance assessment
opioid abuse prevention	ride sharing	
personal assistants	risk minimization	summarizing text
personalized medicine	robo financial advisors	telemedicine
personalized shopping	security enhancements	terrorist attack prevention
phishing elimination	self-driving cars	theft prevention
pollution reduction	sentiment analysis	travel recommendations
precision medicine	sharing economy	trend spotting
predicting cancer survival	similarity detection	visual product search
predicting disease outbreaks	smart cities	voice recognition
predicting health outcomes	smart homes	voice search
predicting student enrollments	smart meters	weather forecasting

1.14 Case Study—A Big-Data Mobile Application

Google's Waze GPS navigation app, with its 90 million monthly active users,[75] is one of the most widely used big-data apps. Early GPS navigation devices and apps relied on static maps and GPS coordinates to determine the best route to your destination. They could not adjust dynamically to changing traffic situations.

Waze processes massive amounts of **crowdsourced data**—that is, the data that's continuously supplied by their users and their users' devices worldwide. They analyze this data as it arrives to determine the best route to get you to your destination in the least amount of time. To accomplish this, Waze relies on your smartphone's Internet connection. The app automatically sends location updates to their servers (assuming you allow it to). They use that data to dynamically re-route you based on current traffic conditions and to tune their maps. Users report other information, such as roadblocks, construction, obstacles, vehicles in breakdown lanes, police locations, gas prices and more. Waze then alerts other drivers in those locations.

Waze uses many technologies to provide its services. We're not privy to how Waze is implemented, but we infer below a list of technologies they probably use. You'll see many of these in Chapters 12–17. For example,

- Most apps created today use at least some open-source software. You'll take advantage of many open-source libraries and tools throughout this book.

---

75. https://www.waze.com/brands/drivers/.

- Waze communicates information over the Internet between their servers and their users' mobile devices. Today, such data typically is transmitted in JSON (JavaScript Object Notation) format, which we'll introduce in Chapter 9 and use in subsequent chapters. Often the JSON data will be hidden from you by the libraries you use.
- Waze uses speech synthesis to speak driving directions and alerts to you, and speech recognition to understand your spoken commands. We use IBM Watson's speech-synthesis and speech-recognition capabilities in Chapter 14.
- Once Waze converts a spoken natural-language command to text, it must determine the correct action to perform, which requires natural language processing (NLP). We present NLP in Chapter 12 and use it in several subsequent chapters.
- Waze displays dynamically updated visualizations such as alerts and maps. Waze also enables you to interact with the maps by moving them or zooming in and out. We create dynamic visualizations with Matplotlib and Seaborn throughout the book, and we display interactive maps with Folium in Chapters 13 and 17.
- Waze uses your phone as a streaming Internet of Things (IoT) device. Each phone is a GPS sensor that continuously streams data over the Internet to Waze. In Chapter 17, we introduce IoT and work with simulated IoT streaming sensors.
- Waze receives IoT streams from millions of phones at once. It must process, store and analyze that data immediately to update your device's maps, to display and speak relevant alerts and possibly to update your driving directions. This requires massively parallel processing capabilities implemented with clusters of computers in the cloud. In Chapter 17, we'll introduce various big-data infrastructure technologies for receiving streaming data, storing that big data in appropriate databases and processing the data with software and hardware that provide massively parallel processing capabilities.
- Waze uses artificial-intelligence capabilities to perform the data-analysis tasks that enable it to predict the best routes based on the information it receives. In Chapters 15 and 16 we use machine learning and deep learning, respectively, to analyze massive amounts of data and make predictions based on that data.
- Waze probably stores its routing information in a graph database. Such databases can efficiently calculate shortest routes. We introduce graph databases, such as Neo4J, in Chapter 17. Exercise 17.7 asks you to solve the popular "six degrees of separation" problem with Neo4j.
- Many cars are now equipped with devices that enable them to "see" cars and obstacles around them. These are used, for example, to help implement automated braking systems and are a key part of self-driving car technology. Rather than relying on users to report obstacles and stopped cars on the side of the road, navigation apps could take advantage of cameras and other sensors by using deep-learning computer-vision techniques to analyze images "on the fly" and automatically report those items. We introduce deep learning for computer vision in Chapter 16.

1.15 Intro to Data Science: Artificial Intelligence—at the Intersection of CS and Data Science

When a baby first opens its eyes, does it "see" its parent's faces? Does it understand any notion of what a face is—or even what a simple shape is? Babies must "learn" the world around them. That's what artificial intelligence (AI) is doing today. It's looking at massive amounts of data and learning from it. AI is being used to play games, implement a wide range of computer-vision applications, enable self-driving cars, enable robots to learn to perform new tasks, diagnose medical conditions, translate speech to other languages in near real time, create chatbots that can respond to arbitrary questions using massive databases of knowledge, and much more. Who'd have guessed just a few years ago that artificially intelligent self-driving cars would be allowed on our roads—or even become common? Yet, this is now a highly competitive area. The ultimate goal of all this learning is **artificial general intelligence**—an AI that can perform intelligence tasks as well as humans.

Artificial-Intelligence Milestones

Several artificial-intelligence milestones, in particular, captured people's attention and imagination, made the general public start thinking that AI is real and made businesses think about commercializing AI:

- In a 1997 match between **IBM's DeepBlue** computer system and chess Grandmaster Gary Kasparov, DeepBlue became the first computer to beat a reigning world chess champion under tournament conditions.[76] IBM loaded DeepBlue with hundreds of thousands of grandmaster chess games.[77] DeepBlue was capable of using *brute force* to evaluate up to 200 million moves per second![78] This is big data at work. IBM received the Carnegie Mellon University Fredkin Prize, which in 1980 offered $100,000 to the creators of the first computer to beat a world chess champion.[79]

- In 2011, **IBM's Watson** beat the two best human Jeopardy! players in a $1 million match. Watson simultaneously used hundreds of language-analysis techniques to locate correct answers in 200 million pages of content (including all of Wikipedia) requiring four terabytes of storage.[80,81] Watson was trained with machine learning and **reinforcement-learning techniques**.[82] Chapter 16 discusses **machine-learning** and Chapter 17's exercises introduce **reinforcement learning**.

- Go—a board game created in China thousands of years ago[83]—is widely considered to be one of the most complex games ever invented with 10^{170} possible board configurations.[84] To give you a sense of how large a number that is, it's

---

76. https://en.wikipedia.org/wiki/Deep_Blue_versus_Garry_Kasparov.
77. https://en.wikipedia.org/wiki/Deep_Blue_(chess_computer).
78. https://en.wikipedia.org/wiki/Deep_Blue_(chess_computer).
79. https://articles.latimes.com/1997/jul/30/news/mn-17696.
80. https://www.techrepublic.com/article/ibm-watson-the-inside-story-of-how-the-jeopardy-winning-supercomputer-was-born-and-what-it-wants-to-do-next/.
81. https://en.wikipedia.org/wiki/Watson_(computer).
82. https://www.aaai.org/Magazine/Watson/watson.php, *AI Magazine*, Fall 2010.
83. http://www.usgo.org/brief-history-go.
84. https://www.pbs.org/newshour/science/google-artificial-intelligence-beats-champion-at-worlds-most-complicated-board-game.

believed that there are (only) between 10^{78} and 10^{87} atoms in the known universe![85,86] In 2015, **AlphaGo**—created by Google's DeepMind group—used *deep learning with two neural networks to beat the European Go champion Fan Hui.* Go is considered to be a far more complex game than chess. Chapter 17 discusses neural networks and deep learning.

- More recently, Google generalized its AlphaGo AI to create **AlphaZero**—a game-playing AI that *teaches itself to play other games*. In December 2017, AlphaZero learned the rules of and taught itself to play chess in less than four hours using reinforcement learning. It then beat the world champion chess program, Stockfish 8, in a 100-game match—winning or drawing every game. After *training itself* in Go for just eight hours, AlphaZero was able to play Go vs. its AlphaGo predecessor, winning 60 of 100 games.[87] Chapter 17 discusses reinforcement learning.

A Personal Anecdote

When one of the authors, Harvey Deitel, was an undergraduate student at MIT in the mid-1960s, he took a graduate-level artificial-intelligence course with Marvin Minsky (to whom this book is dedicated), one of the founders of artificial intelligence (AI). Harvey:

Professor Minsky required a major term project. He told us to think about what intelligence is and to make a computer do something intelligent. Our grade in the course would be almost solely dependent on the project. No pressure!

I researched the standardized IQ tests that schools administer to help evaluate their students' intelligence capabilities. Being a mathematician at heart, I decided to tackle the popular IQ-test problem of predicting the next number in a sequence of numbers of arbitrary length and complexity. I used interactive Lisp running on an early Digital Equipment Corporation PDP-1 and was able to get my sequence predictor running on some pretty complex stuff, handling challenges well beyond what I recalled seeing on IQ tests. Lisp's ability to manipulate arbitrarily long lists recursively was exactly what I needed to meet the project's requirements. Python offers recursion (Chapter 11) and generalized list processing (Chapter 5).

I tried the sequence predictor on many of my MIT classmates. They would make up number sequences and type them into my predictor. The PDP-1 would "think" for a while—often a long while—and almost always came up with the right answer.

Then I hit a snag. One of my classmates typed in the sequence 14, 23, 34 and 42. My predictor went to work on it, and the PDP-1 chugged away for a long time, failing to predict the next number. I couldn't get it either. My classmate told me to think about it overnight, and he'd reveal the answer the next day, claiming that it was a simple sequence. My efforts were to no avail.

The following day he told me the next number was 57, but I didn't understand why. So he told me to think about it overnight again, and the following day he

---

85. https://www.universetoday.com/36302/atoms-in-the-universe/.
86. https://en.wikipedia.org/wiki/Observable_universe#Matter_content.
87. https://www.theguardian.com/technology/2017/dec/07/alphazero-google-deepmind-ai-beats-champion-program-teaching-itself-to-play-four-hours.

said the next number was 125. That didn't help a bit—I was stumped. He said that the sequence was the numbers of the two-way crosstown streets of Manhattan. I cried, "foul," but he said it met my criterion of predicting the next number in a numerical sequence. My world view was mathematics—his was broader.

Over the years, I've tried that sequence on friends, relatives and professional colleagues. A few who either lived in Manhattan or spent time there got it right. My sequence predictor needed a lot more than just mathematical knowledge to handle problems like this, requiring (a possibly vast) world knowledge.

Watson and Big Data Open New Possibilities
When Paul and I started working on this Python book, we were immediately drawn to IBM's Watson using big data and artificial-intelligence techniques like natural language processing (NLP) and machine learning to beat two of the world's best human Jeopardy! players. We realized that Watson could probably handle problems like the sequence predictor because it was loaded with the world's street maps and a whole lot more. That whet our appetite for digging in deep on big data and today's artificial-intelligence technologies.

It's notable that all of the data-science implementation case studies in Chapters 12 to 17 either are rooted in artificial intelligence technologies or discuss the big data hardware and software infrastructure that enables data scientists to implement leading-edge AI-based solutions effectively.

AI: A Field with Problems But No Solutions
For many decades, AI has been a field with problems and *no* solutions. That's because once a particular problem is solved people say, "Well, that's not intelligence, it's just a computer program that tells the computer exactly what to do." However, with machine learning (Chapter 15), deep learning (Chapter 16) and reinforcement learning (Chapter 16 exercises), we're not pre-programming solutions to *specific* problems. Instead, we're letting our computers solve problems by learning from data—and, typically, lots of it.

Many of the most interesting and challenging problems are being pursued with deep learning. Google alone has thousands of deep-learning projects underway and that number is growing quickly.[88, 89] As you work through this book, we'll introduce you to many edge-of-the-practice artificial intelligence, big data and cloud technologies and you'll work through hundreds of (often intriguing) examples, exercises and projects.

✓ Self Check

1. *(Fill-In)* The ultimate goal of AI is to produce a(n) _____.
   Answer: artificial general intelligence.

2. *(Fill-In)* IBM's Watson beat the two best human Jeopardy! players. Watson was trained using a combination of _____ learning and _____ learning techniques.
   Answer: machine, reinforcement.

---

88. http://theweek.com/speedreads/654463/google-more-than-1000-artificial-intelligence-projects-works.
89. https://www.zdnet.com/article/google-says-exponential-growth-of-ai-is-changing-nature-of-compute/.

3 *(Fill-In)* Google's _____ taught itself to play chess in less than four hours using reinforcement learning, then beat the world champion chess program, Stockfish 8, in a 100-game match—winning or drawing every game.
Answer: AlphaZero.

Exercises

1.1 *(IPython Session)* Using the techniques you learned in Section 1.10.1, execute the following expressions. Which, if any, produce a runtime error?
  a) 10 / 3
  b) 10 // 3
  c) 10 / 0
  d) 10 // 0
  e) 0 / 10
  f) 0 // 10

1.2 *(IPython Session)* Using the techniques you learned in Section 1.10.1, execute the following expressions. Which, if any, produce a runtime error?
  a) 10 / 3 + 7
  b) 10 // 3 + 7
  c) 10 / (3 + 7)
  d) 10 / 3 - 3
  e) 10 / (3 - 3)
  f) 10 // (3 - 3)

1.3 *(Creating a Jupyter Notebook)* Using the techniques you learned in Section 1.10.3, create a Jupyter Notebook containing cells for the previous exercise's expressions and execute those expressions.

1.4 *(Computer Organization)* Fill in the blanks in each of the following statements:
  a) The logical unit that receives information from outside the computer for use by the computer is the _____.
  b) _____ is a logical unit that sends information which has already been processed by the computer to various devices so that it may be used outside the computer.
  c) _____ and _____ are logical units of the computer that retain information.
  d) _____ is a logical unit of the computer that performs calculations.
  e) _____ is a logical unit of the computer that makes logical decisions.
  f) _____ is a logical unit of the computer that coordinates the activities of all the other logical units.

1.5 *(Clock as an Object)* Clocks are among the world's most common objects. Discuss how each of the following terms and concepts applies to the notion of a clock: class, object, instantiation, instance variable, reuse, method, inheritance (consider, for example, an alarm clock), superclass, subclass.

1.6 *(Gender Neutrality)* Write down the steps of a manual procedure for processing a paragraph of text and replacing gender-specific words with gender-neutral ones. Assuming that you've been given a list of gender-specific words and their gender-neutral replacements (for example, replace "wife" or "husband" with "spouse," replace "man" or "woman" with "person," replace "daughter" or "son" with "child," and so on), explain the procedure you'd use to read through a paragraph of text and manually perform these replacements. How might your procedure generate a strange term like "woperchild" and how might you modify your procedure to avoid this possibility? In Chapter 3, you'll learn that a more formal computing term for "procedure" is "algorithm," and that an algorithm specifies the *steps* to be performed and the *order* in which to perform them.

1.7 *(Self-Driving Cars)* Just a few years back the notion of driverless cars on our streets would have seemed impossible (in fact, our spell-checking software doesn't recognize the word "driverless"). Many of the technologies you'll study in this book are making self-driving cars possible. They're already common in some areas.
  a) If you hailed a taxi and a driverless taxi stopped for you, would you get into the back seat? Would you feel comfortable telling it where you want to go and trusting that it would get you there? What kinds of safety measures would you want in place? What would you do if the car headed off in the wrong direction?
  b) What if two self-driving cars approached a one-lane bridge from opposite directions? What protocol should they go through to determine which car should proceed?
  c) If a police officer pulls over a speeding self-driving car in which you're the only passenger, who—or what entity—should pay the ticket?
  d) What if you're behind a car stopped at a red light, the light turns green and the car doesn't move? You honk and nothing happens. You get out of your car and notice that there's no driver. What would you do?
  e) One serious concern with self-driving vehicles is that they could potentially be hacked. Someone could set the speed high (or low), which could be dangerous. What if they redirect you to a destination other than what you want?
  f) Imagine other scenarios that self-driving cars will encounter.

1.8 *(Research: Reproducibility)* A crucial concept in data-science studies is reproducibility, which helps others (and you) reproduce your results. Research reproducibility and list the concepts used to create reproducible results in data-science studies. Research and discuss the part that Jupyter Notebooks play in reproducibility.

1.9 *(Research: Artificial General Intelligence)* One of the most ambitious goals in the field of AI is to achieve *artificial general intelligence*—the point at which machine intelligence would equal human intelligence. Research this intriguing topic. When is this forecast to happen? What are some key ethical issues this raises? Human intelligence seems to be stable over long periods. Powerful computers with artificial general intelligence could conceivably (and quickly) evolve intelligence far beyond that of humans. Research and discuss the issues this raises.

1.10 *(Research: Intelligent Assistants)* Many companies now offer computerized intelligent assistants, such as IBM Watson, Amazon Alexa, Apple Siri, Google Assistant and Microsoft Cortana. Research these and others and list uses that can improve people's lives. Research privacy and ethics issues for intelligent assistants. Locate amusing intelligent-assistant anecdotes.

1.11 *(Research: AI in Health Care)* Research the rapidly growing field of AI big-data applications in health care. For example, suppose a diagnostic medical application had access to every x-ray that's ever been taken and the associated diagnoses—that's surely big data. As you'll see in the "Deep Learning" chapter, computer-vision applications can work with this "labeled" data to learn to diagnose medical problems. Research deep learning in diagnostic medicine and describe some of its most significant accomplishments. What are some ethical issues of having machines instead of human doctors performing medical diagnoses? Would you trust a machine-generated diagnosis? Would you ask for a second opinion?

1.12 *(Research: Big Data, AI and the Cloud—How Companies Use These Technologies)* For a major organization of your choice, research how they may be using each of the following technologies that you'll use in this book: Python, AI, big data, the cloud, mobile, natural language processing, speech recognition, speech synthesis, database, machine learning, deep learning, reinforcement learning, Hadoop, Spark, Internet of Things (IoT) and web services.

1.13 *(Research: Raspberry Pi and the Internet of Things)* It's now possible to have a computer at the heart of just about any type of device and to connect those devices to the Internet. This has led to the Internet of Things (IoT), which already interconnects tens of billions of devices. The Raspberry Pi is an economical computer which is often at the heart of IoT devices. Research the Raspberry Pi and some of the many IoT applications in which it's used.

1.14 *(Research: The Ethics of Deep Fakes)* Artificial-intelligence technologies are making it possible to create *deep fakes*—realistic fake videos of people that capture their appearance, voice, body motions and facial expressions. You can have them say and do whatever you specify. Research the ethics of deep fakes. What would happen if you turned on your TV and saw a deep-fake video of a prominent government official or newscaster reporting that a nuclear attack was about to happen? Research Orson Welles and his "War of the Worlds" radio broadcast of 1938, which created mass panic.

1.15 *(Public-Key Cryptography)* Cryptography is a crucial technology for privacy and security. Research Python's cryptography capabilities. Research online for a simple explanation of how public-key cryptography is used to implement the BitCoin cryptocurrency.

1.16 *(Blockchain: A World of Opportunity)* Cryptocurrencies like Bitcoin and Ethereum are based on a technology called blockchain that has seen explosive growth over the last few years. Research blockchain's origin, applications and how it came to be used as the basis for cryptocurrencies. Research other major applications of blockchain. Over the next many years there will be extraordinary opportunities for software developers who thoroughly understand blockchain applications development.

1.17 *(OWASP Python Security Project)* Building secure computer applications is a tremendous challenge. Many of the world's largest companies, government agencies, and military organizations have had their systems compromised. The OWASP project is concerned with "hardening" computer systems and applications to resist attacks. Research OWASP and discuss their accomplishments and current challenges.

1.18 *(IBM Watson)* We discuss IBM's Watson in Chapter 14. You'll use its cognitive computing capabilities to quickly build some intriguing applications. IBM is partnering with tens of thousands of companies—including our publisher, Pearson Education—across a wide range of industries. Research some of Watson's key accomplishments and the kinds of challenges IBM and its partners are addressing.

1.19 *(Research: Mobile App Development with Python)* Research the tools that are available for Python-based iOS and Android app development, such as BeeWare, Kivy, Py-

Mob, Pythonista and others. Which of these are cross-platform? Mobile applications development is one of the fastest growing areas of software development, and it's a great source of class projects, directed study projects, capstone exercise projects and even thesis projects. With cross-platform app-development tools, you'll be able to write your own apps and deploy them on many app stores quickly.

Introduction to Python Programming

2

Objectives

In this chapter, you'll:

- Continue using IPython interactive mode to enter code snippets and see their results immediately.
- Write simple Python statements and scripts.
- Create variables to store data for later use.
- Become familiar with built-in data types.
- Use arithmetic operators and comparison operators, and understand their precedence.
- Use single-, double- and triple-quoted strings.
- Use built-in function `print` to display text.
- Use built-in function `input` to prompt the user to enter data at the keyboard and get that data for use in the program.
- Convert text to integer values with built-in function `int`.
- Use comparison operators and the `if` statement to decide whether to execute a statement or group of statements.
- Learn about objects and Python's dynamic typing.
- Use built-in function `type` to get an object's type.

Outline

2.1 Introduction
2.2 Variables and Assignment Statements
2.3 Arithmetic
2.4 Function `print` and an Intro to Single- and Double-Quoted Strings
2.5 Triple-Quoted Strings
2.6 Getting Input from the User
2.7 Decision Making: The `if` Statement and Comparison Operators
2.8 Objects and Dynamic Typing
2.9 Intro to Data Science: Basic Descriptive Statistics
2.10 Wrap-Up
Exercises

2.1 Introduction

In this chapter, we introduce Python programming and present examples illustrating key language features. We assume you've read the IPython Test-Drive in Chapter 1, which introduced the IPython interpreter and used it to evaluate simple arithmetic expressions.

2.2 Variables and Assignment Statements

You've used IPython's interactive mode as a calculator with expressions such as

```
In [1]: 45 + 72
Out[1]: 117
```

As in algebra, Python expressions also may contain **variables**, which store values for later use in your code. Let's create a variable named x that stores the integer 7, which is the variable's **value**:

```
In [2]: x = 7
```

Snippet [2] is a **statement**. Each statement specifies a task to perform. The preceding statement creates x and uses the **assignment symbol** (=) to give x a value. The entire statement is an **assignment statement** that we read as "x is assigned the value 7." Most statements stop at the end of the line, though it's possible for statements to span more than one line. The following statement creates the variable y and assigns to it the value 3:

```
In [3]: y = 3
```

Adding Variable Values and Viewing the Result

You can now use the values of x and y in expressions:

```
In [4]: x + y
Out[4]: 10
```

The + symbol is the **addition operator**. It's a **binary operator** because it has *two* **operands** (in this case, the variables x and y) on which it performs its operation.

Calculations in Assignment Statements

You'll often save calculation results for later use. The following assignment statement adds the values of variables x and y and assigns the result to the variable `total`, which we then display:

```
In [5]: total = x + y

In [6]: total
Out[6]: 10
```

Snippet [5] is read, "total is assigned the value of x + y." The = symbol is not an operator. The right side of the = symbol always executes first, then the result is assigned to the variable on the symbol's left side.

Python Style
The *Style Guide for Python Code*[1] helps you write code that conforms to Python's coding conventions. The style guide recommends inserting one space on each side of the assignment symbol = and binary operators like + to make programs more readable.

Variable Names
A variable name, such as x, is an **identifier**. Each identifier may consist of letters, digits and underscores (_) but may not begin with a digit. Python is **case sensitive**, so number and Number are *different* identifiers because one begins with a lowercase letter and the other begins with an uppercase letter.

Types
Each value in Python has a **type** that indicates the kind of data the value represents. You can view a value's type, as in:

```
In [7]: type(x)
Out[7]: int

In [8]: type(10.5)
Out[8]: float
```

The variable x contains the integer value 7 (from snippet [2]), so Python displays int (short for integer). The value 10.5 is a **floating-point number** (that is, a number with a decimal point), so Python displays float.

Python's **type built-in function** determines a value's type. A **function** performs a task when you **call** it by writing its name, followed by parentheses, (). The parentheses contain the function's **argument**—the data that the type function needs to perform its task. You'll create *custom* functions in later chapters.

✓ Self Check

1 *(True/False)* The following are valid variable names: 3g, 87 and score_4.
Answer: False. Because they begin with a digit, 3g and 87 are invalid names.

2 *(True/False)* Python treats y and Y as the same identifier.
Answer: False. Python is case sensitive, so y and Y are different identifiers.

3 *(IPython Session)* Calculate the sum of 10.8, 12.2 and 0.2, store it in the variable total, then display total's value.
Answer:

```
In [1]: total = 10.8 + 12.2 + 0.2

In [2]: total
Out[2]: 23.1
```

---

1. https://www.python.org/dev/peps/pep-0008/.

2.3 Arithmetic

Many programs perform arithmetic calculations. The following table summarizes the **arithmetic operators**, which include some symbols not used in algebra.

Python operation	Arithmetic operator	Algebraic expression	Python expression
Addition	+	$f + 7$	f + 7
Subtraction	-	$p - c$	p - c
Multiplication	*	$b \cdot m$	b * m
Exponentiation	**	x^y	x ** y
True division	/	x/y or $\frac{x}{y}$ or $x \div y$	x / y
Floor division	//	$\lfloor x/y \rfloor$ or $\left\lfloor \frac{x}{y} \right\rfloor$ or $\lfloor x \div y \rfloor$	x // y
Remainder (modulo)	%	$r \bmod s$	r % s

Multiplication (*)

Rather than algebra's center dot (·), Python uses the **asterisk (*) multiplication operator**:

```
In [1]: 7 * 4
Out[1]: 28
```

Exponentiation (**)

The **exponentiation (**) operator** raises one value to the power of another:

```
In [2]: 2 ** 10
Out[2]: 1024
```

To calculate the square root, you can use the exponent 1/2 (that is, 0.5):

```
In [3]: 9 ** (1 / 2)
Out[3]: 3.0
```

True Division (/) vs. Floor Division (//)

True division (/) divides a numerator by a denominator and yields a floating-point number with a decimal point, as in:

```
In [4]: 7 / 4
Out[4]: 1.75
```

Floor division (//) divides a numerator by a denominator, yielding the highest *integer* that's not greater than the result. Python **truncates** (discards) the fractional part:

```
In [5]: 7 // 4
Out[5]: 1

In [6]: 3 // 5
Out[6]: 0

In [7]: 14 // 7
Out[7]: 2
```

In true division, -13 divided by 4 gives -3.25:

```
In [8]: -13 / 4
Out[8]: -3.25
```

Floor division gives the closest integer that's *not greater than* -3.25—which is -4:

```
In [9]: -13 // 4
Out[9]: -4
```

Exceptions and Tracebacks

Dividing by zero with / or // is not allowed and results in an **exception**—a sign that a problem occurred:

```
In [10]: 123 / 0
-------------------------------------------------------------------------
ZeroDivisionError                         Traceback (most recent call last)
<ipython-input-10-cd759d3fcf39> in <module>()
----> 1 123 / 0

ZeroDivisionError: division by zero
```

Python reports an exception with a **traceback**. This traceback indicates that an exception of type ZeroDivisionError occurred—most exception names end with Error. In interactive mode, the snippet number that caused the exception is specified by the 10 in the line

```
<ipython-input-10-cd759d3fcf39> in <module>()
```

The line that begins with ----> 1 shows the code that caused the exception. Sometimes snippets have more than one line of code—the 1 to the right of ----> indicates that line 1 within the snippet caused the exception. The last line shows the exception that occurred, followed by a colon (:) and an error message with more information about the exception:

```
ZeroDivisionError: division by zero
```

The "Files and Exceptions" chapter discusses exceptions in detail.

An exception also occurs if you try to use a variable that you have not yet created. The following snippet tries to add 7 to the undefined variable z, resulting in a NameError:

```
In [11]: z + 7
-------------------------------------------------------------------------
NameError                                 Traceback (most recent call last)
<ipython-input-11-f2cdbf4fe75d> in <module>()
----> 1 z + 7

NameError: name 'z' is not defined
```

Remainder Operator

Python's **remainder operator** (%) yields the remainder after the left operand is divided by the right operand:

```
In [12]: 17 % 5
Out[12]: 2
```

In this case, 17 divided by 5 yields a quotient of 3 and a remainder of 2. This operator is most commonly used with integers, but also can be used with other numeric types:

```
In [13]: 7.5 % 3.5
Out[13]: 0.5
```

In the exercises, we use the remainder operator for applications such as determining whether one number is a multiple of another—a special case of this is determining whether a number is odd or even.

Straight-Line Form

Algebraic notations such as

$$\frac{a}{b}$$

generally are not acceptable to compilers or interpreters. For this reason, algebraic expressions must be typed in **straight-line form** using Python's operators. The expression above must be written as a / b (or a // b for floor division) so that all operators and operands appear in a horizontal straight line.

Grouping Expressions with Parentheses

Parentheses group Python expressions, as they do in algebraic expressions. For example, the following code multiplies 10 times the quantity 5 + 3:

```
In [14]: 10 * (5 + 3)
Out[14]: 80
```

Without these parentheses, the result is *different*:

```
In [15]: 10 * 5 + 3
Out[15]: 53
```

The parentheses are **redundant** (unnecessary) if removing them yields the *same* result.

Operator Precedence Rules

Python applies the operators in arithmetic expressions according to the following **rules of operator precedence**. These are generally the same as those in algebra:

1. Expressions in parentheses evaluate first, so parentheses may force the order of evaluation to occur in any sequence you desire. Parentheses have the highest level of precedence. In expressions with **nested parentheses**, such as (a / (b - c)), the expression in the *innermost* parentheses (that is, b - c) evaluates first.

2. Exponentiation operations evaluate next. If an expression contains several exponentiation operations, Python applies them from right to left.

3. Multiplication, division and modulus operations evaluate next. If an expression contains several multiplication, true-division, floor-division and modulus operations, Python applies them from left to right. Multiplication, division and modulus are "on the same level of precedence."

4. Addition and subtraction operations evaluate last. If an expression contains several addition and subtraction operations, Python applies them from left to right. Addition and subtraction also have the same level of precedence.

We'll expand these rules as we introduce other operators. For the complete list of operators and their precedence (in lowest-to-highest order), see

https://docs.python.org/3/reference/expressions.html#operator-precedence

Operator Grouping

When we say that Python applies certain operators from left to right, we are referring to the operators' **grouping**. For example, in the expression

    a + b + c

the addition operators (+) group from left to right as if we parenthesized the expression as (a + b) + c. All Python operators of the same precedence group left-to-right except for the exponentiation operator (**), which groups right-to-left.

Redundant Parentheses

You can use redundant parentheses to group subexpressions to make the expression clearer. For example, the second-degree polynomial

    y = a * x ** 2 + b * x + c

can be parenthesized, for clarity, as

    y = (a * (x ** 2)) + (b * x) + c

Breaking a complex expression into a sequence of statements with shorter, simpler expressions also can promote clarity.

Operand Types

Each arithmetic operator may be used with integers and floating-point numbers. If both operands are integers, the result is an integer—except for the true-division (/) operator, which always yields a floating-point number. If both operands are floating-point numbers, the result is a floating-point number. Expressions containing an integer and a floating-point number are **mixed-type expressions**—these always produce floating-point numbers.

Self Check

1 *(Multiple Choice)* Given that $y = ax^3 + 7$, which of the following is not a correct statement for this equation?
    a) y = a * x * x * x + 7
    b) y = a * x ** 3 + 7
    c) y = a * (x * x * x) + 7
    d) y = a * x * (x * x + 7)
Answer: d is incorrect.

2 *(True/False)* In nested parentheses, the expression in the innermost pair evaluates last.
Answer: False. The expression in the innermost parentheses evaluates first.

3 *(IPython Session)* Evaluate the expression 3 * (4 - 5) with and without parentheses. Are the parentheses redundant?
Answer:

    In [1]: 3 * (4 - 5)
    Out[1]: -3

    In [2]: 3 * 4 - 5
    Out[2]: 7

The parentheses are not redundant—if you remove them the resulting value is different.

4   *(IPython Session)* Evaluate the expressions 4 ** 3 ** 2, (4 ** 3) ** 2 and 4 ** (3 ** 2). Are any of the parentheses redundant?
Answer:

```
In [3]: 4 ** 3 ** 2
Out[3]: 262144

In [4]: (4 ** 3) ** 2
Out[4]: 4096

In [5]: 4 ** (3 ** 2)
Out[5]: 262144
```

Only the parentheses in the last expression are redundant.

2.4 Function `print` and an Intro to Single- and Double-Quoted Strings

The built-in **print function** displays its argument(s) as a line of text:

```
In [1]: print('Welcome to Python!')
Welcome to Python!
```

In this case, the argument `'Welcome to Python!'` is a **string**—a sequence of characters enclosed in single quotes (`'`). Unlike when you evaluate expressions in interactive mode, the text that `print` displays here is not preceded by `Out[1]`. Also, `print` does not display a string's quotes, though we'll soon show how to display quotes in strings.

You also may enclose a string in double quotes (`"`), as in:

```
In [2]: print("Welcome to Python!")
Welcome to Python!
```

Python programmers generally prefer single quotes.

When `print` completes its task, it positions the screen cursor at the beginning of the next line. This is similar to what happens when you press the *Enter* (or *Return*) key while typing in a text editor.

Printing a Comma-Separated List of Items

The `print` function can receive a comma-separated list of arguments, as in:

```
In [3]: print('Welcome', 'to', 'Python!')
Welcome to Python!
```

The `print` function displays each argument separated from the next by a space, producing the same output as in the two preceding snippets. Here we showed a comma-separated list of strings, but the values can be of any type. We'll show in the next chapter how to prevent automatic spacing between values or use a different separator than space.

Printing Many Lines of Text with One Statement

When a backslash (\) appears in a string, it's known as the **escape character**. The backslash and the character immediately following it form an **escape sequence**. For example, \n represents the **newline character** escape sequence, which tells `print` to move the output cursor to the next line. Placing two newline characters back-to-back displays a blank line. The following snippet uses three newline characters to create many lines of output:

2.4 Function `print` and an Intro to Single- and Double-Quoted Strings

```
In [4]: print('Welcome\nto\n\nPython!')
Welcome
to

Python!
```

Other Escape Sequences

The following table shows some common escape sequences.

Escape sequence	Description
\n	Insert a newline character in a string. When the string is displayed, for each newline, move the screen cursor to the beginning of the next line.
\t	Insert a horizontal tab. When the string is displayed, for each tab, move the screen cursor to the next tab stop.
\\	Insert a backslash character in a string.
\"	Insert a double quote character in a string.
\'	Insert a single quote character in a string.

Ignoring a Line Break in a Long String

You may also split a long string (or a long statement) over several lines by using the \ **continuation character** as the last character on a line to ignore the line break:

```
In [5]: print('this is a longer string, so we \
   ...: split it over two lines')
this is a longer string, so we split it over two lines
```

The interpreter reassembles the string's parts into a single string with no line break. Though the backslash character in the preceding snippet is inside a string, it's not the escape character because another character does not follow it.

Printing the Value of an Expression

Calculations can be performed in `print` statements:

```
In [6]: print('Sum is', 7 + 3)
Sum is 10
```

Self Check

1 *(Fill-In)* The _____ function instructs the computer to display information on the screen.
Answer: print.

2 *(Fill-In)* Values of the _____ data type contain a sequence of characters.
Answer: string (type `str`).

3 *(IPython Session)* Write an expression that displays the type of `'word'`.
Answer:

```
In [1]: type('word')
Out[1]: str
```

4 *(IPython Session)* What does the following `print` statement display?

```
print('int(5.2)', 'truncates 5.2 to', int(5.2))
```

Answer:

```
In [2]: print('int(5.2)', 'truncates 5.2 to', int(5.2))
int(5.2) truncates 5.2 to 5
```

2.5 Triple-Quoted Strings

Earlier, we introduced strings delimited by a pair of single quotes (') or a pair of double quotes ("). **Triple-quoted strings** begin and end with three double quotes (""") or three single quotes ('''). The *Style Guide for Python Code* recommends three double quotes ("""). Use these to create:

- multiline strings,
- strings containing single or double quotes and
- **docstrings**, which are the recommended way to document the purposes of certain program components.

Including Quotes in Strings

In a string delimited by single quotes, you may include double-quote characters:

```
In [1]: print('Display "hi" in quotes')
Display "hi" in quotes
```

but not single quotes:

```
In [2]: print('Display 'hi' in quotes')
  File "<ipython-input-2-19bf596ccf72>", line 1
    print('Display 'hi' in quotes')
                    ^
SyntaxError: invalid syntax
```

unless you use the \' escape sequence:

```
In [3]: print('Display \'hi\' in quotes')
Display 'hi' in quotes
```

Snippet [2] displayed a **syntax error**, which is a violation of Python's language rules—in this case, a single quote inside a single-quoted string. IPython displays information about the line of code that caused the syntax error and points to the error with a ^ symbol. It also displays the message `SyntaxError: invalid syntax`.

A string delimited by double quotes may include single quote characters:

```
In [4]: print("Display the name O'Brien")
Display the name O'Brien
```

but not double quotes, unless you use the \" escape sequence:

```
In [5]: print("Display \"hi\" in quotes")
Display "hi" in quotes
```

To avoid using \' and \" inside strings, you can enclose such strings in triple quotes:

```
In [6]: print("""Display "hi" and 'bye' in quotes""")
Display "hi" and 'bye' in quotes
```

Multiline Strings

The following snippet assigns a multiline triple-quoted string to `triple_quoted_string`:

```
In [7]: triple_quoted_string = """This is a triple-quoted
   ...: string that spans two lines"""
```

IPython knows that the string is incomplete because we did not type the closing `"""` before we pressed *Enter*. So, IPython displays a **continuation prompt** `...:` at which you can input the multiline string's next line. This continues until you enter the ending `"""` and press *Enter*. The following displays `triple_quoted_string`:

```
In [8]: print(triple_quoted_string)
This is a triple-quoted
string that spans two lines
```

Python stores multiline strings with embedded newline escape sequences. When we evaluate `triple_quoted_string` rather than printing it, IPython displays the string in single quotes with a \n character where you pressed *Enter* in snippet [7]. The quotes IPython displays indicate that `triple_quoted_string` is a string—they're not part of the string's contents:

```
In [9]: triple_quoted_string
Out[9]: 'This is a triple-quoted\nstring that spans two lines'
```

✓ Self Check

1. *(Fill-In)* Multiline strings are enclosed either in _____ or in _____.
Answer: `"""` (triple double quotes) or `'''` (triple single quotes).

2. *(IPython Session)* What displays when you execute the following statement?
    ```
    print("""This is a lengthy
       multiline string containing
    a few lines \
    of text""")
    ```
Answer:
```
In [1]: print("""This is a lengthy
   ...:    multiline string containing
   ...: a few lines \
   ...: of text""")
This is a lengthy
   multiline string containing
a few lines of text
```

2.6 Getting Input from the User

The built-in **input function** requests and obtains user input:

```
In [1]: name = input("What's your name? ")
What's your name? Paul

In [2]: name
Out[2]: 'Paul'

In [3]: print(name)
Paul
```

The snippet executes as follows:

- First, input displays its string argument—called a **prompt**—to tell the user what to type and waits for the user to respond. We typed Paul (without quotes) and pressed *Enter*. We use **bold** text to distinguish the user's input from the prompt text that input displays.
- Function input then **returns** (that is, gives back) those characters as a string that the program can use. Here we assigned that string to the variable name.

Snippet [2] shows name's value. Evaluating name displays its value in single quotes as 'Paul' because it's a string. Printing name (in snippet [3]) displays the string without the quotes. If you enter quotes, they're part of the string, as in:

```
In [4]: name = input("What's your name? ")
What's your name? 'Paul'

In [5]: name
Out[5]: "'Paul'"

In [6]: print(name)
'Paul'
```

Function input Always Returns a String

Consider the following snippets that attempt to read two numbers and add them:

```
In [7]: value1 = input('Enter first number: ')
Enter first number: 7

In [8]: value2 = input('Enter second number: ')
Enter second number: 3

In [9]: value1 + value2
Out[9]: '73'
```

Rather than adding the integers 7 and 3 to produce 10, Python "adds" the *string* values '7' and '3', producing the *string* '73'. This is known as **string concatenation**. It creates a new string containing the left operand's value followed by the right operand's value.

Getting an Integer from the User

If you need an integer, convert the string to an integer using the built-in **int function**:

```
In [10]: value = input('Enter an integer: ')
Enter an integer: 7

In [11]: value = int(value)

In [12]: value
Out[12]: 7
```

We could have combined the code in snippets [10] and [11]:

```
In [13]: another_value = int(input('Enter another integer: '))
Enter another integer: 13

In [14]: another_value
Out[14]: 13
```

Variables value and another_value now contain integers. Adding them produces an integer result (rather than concatenating them):

```
In [15]: value + another_value
Out[15]: 20
```

If the string passed to int cannot be converted to an integer, a ValueError occurs:

```
In [16]: bad_value = int(input('Enter another integer: '))
Enter another integer: hello
---------------------------------------------------------------------
ValueError                                Traceback (most recent call last)
<ipython-input-16-cd36e6cf8911> in <module>()
----> 1 bad_value = int(input('Enter another integer: '))

ValueError: invalid literal for int() with base 10: 'hello'
```

Function int also can convert a floating-point value to an integer:

```
In [17]: int(10.5)
Out[17]: 10
```

To convert strings to floating-point numbers, use the built-in **float function**.

✓ Self Check

1 *(Fill-In)* The built-in _____ function converts a floating-point value to an integer value or converts a string representation of an integer to an integer value.
Answer: int.

2 *(True/False)* Built-in function get_input requests and obtains input from the user.
Answer: False. The built-in function's name is input.

3 *(IPython Session)* Use float to convert '6.2' (a string) to a floating-point value. Multiply that value by 3.3 and show the result.
Answer:

```
In [1]: float('6.2') * 3.3
Out[1]: 20.46
```

2.7 Decision Making: The if Statement and Comparison Operators

A **condition** is a Boolean expression with the value **True** or **False**. The following determines whether 7 is greater than 4 and whether 7 is less than 4:

```
In [1]: 7 > 4
Out[1]: True

In [2]: 7 < 4
Out[2]: False
```

True and False are **keywords**—words that Python reserves for its language features. Using a keyword as an identifier causes a SyntaxError. True and False are each capitalized.

You'll often create conditions using the **comparison operators** in the table at the top of the next page:

Algebraic operator	Python operator	Sample condition	Meaning
>	>	x > y	x is greater than y
<	<	x < y	x is less than y
≥	>=	x >= y	x is greater than or equal to y
≤	<=	x <= y	x is less than or equal to y
=	==	x == y	x is equal to y
≠	!=	x != y	x is not equal to y

Operators >, <, >= and <= all have the same precedence. Operators == and != both have the same precedence, which is lower than that of >, <, >= and <=. A syntax error occurs when any of the operators ==, !=, >= and <= contains spaces between its pair of symbols:

```
In [3]: 7 > = 4
  File "<ipython-input-3-5c6e2897f3b3>", line 1
    7 > = 4
        ^
SyntaxError: invalid syntax
```

Another syntax error occurs if you reverse the symbols in the operators !=, >= and <= (by writing them as =!, => and =<).

Making Decisions with the if Statement: Introducing Scripts

We now present a simple version of the **if statement**, which uses a condition to decide whether to execute a statement (or a group of statements). Here we'll read two integers from the user and compare them using six consecutive if statements, one for each comparison operator. If the condition in a given if statement is True, the corresponding print statement executes; otherwise, it's skipped.

IPython interactive mode is helpful for executing brief code snippets and seeing immediate results. When you have many statements to execute as a group, you typically write them as a **script** stored in a file with the .py (short for Python) extension—such as fig02_01.py for this example's script. Scripts are also called **programs**. For instructions on locating and executing the scripts in this book, see Chapter 1's IPython Test-Drive.

Each time you execute this script, three of the six conditions are True. To show this, we execute the script three times—once with the first integer *less than* the second, once with the *same* value for both integers and once with the first integer *greater than* the second. The three sample executions appear after the script

Figure 2.1 shows the script. Each time we present a script, we introduce it before the figure, then explain the script's code after the figure. We show line numbers for your convenience—these are not part of Python. Integrated development environments (IDEs) enable you to choose whether to display line numbers. To run this example, change to this chapter's ch02 examples folder, then enter:

```
ipython fig02_01.py
```

or, if you're in IPython already, use the command:

```
run fig02_01.py
```

2.7 Decision Making: The if Statement and Comparison Operators

```python
1   # fig02_01.py
2   """Comparing integers using if statements and comparison operators."""
3
4   print('Enter two integers, and I will tell you',
5         'the relationships they satisfy.')
6
7   # read first integer
8   number1 = int(input('Enter first integer: '))
9
10  # read second integer
11  number2 = int(input('Enter second integer: '))
12
13  if number1 == number2:
14      print(number1, 'is equal to', number2)
15
16  if number1 != number2:
17      print(number1, 'is not equal to', number2)
18
19  if number1 < number2:
20      print(number1, 'is less than', number2)
21
22  if number1 > number2:
23      print(number1, 'is greater than', number2)
24
25  if number1 <= number2:
26      print(number1, 'is less than or equal to', number2)
27
28  if number1 >= number2:
29      print(number1, 'is greater than or equal to', number2)
```

```
Enter two integers and I will tell you the relationships they satisfy.
Enter first integer: 37
Enter second integer: 42
37 is not equal to 42
37 is less than 42
37 is less than or equal to 42
```

```
Enter two integers and I will tell you the relationships they satisfy.
Enter first integer: 7
Enter second integer: 7
7 is equal to 7
7 is less than or equal to 7
7 is greater than or equal to 7
```

```
Enter two integers and I will tell you the relationships they satisfy.
Enter first integer: 54
Enter second integer: 17
54 is not equal to 17
54 is greater than 17
54 is greater than or equal to 17
```

Fig. 2.1 | Comparing integers using if statements and comparison operators.

Comments

Line 1 begins with the hash character (#), which indicates that the rest of the line is a **comment**:

```
# fig02_01.py
```

You insert comments to document your code and to improve readability. Comments also help other programmers read and understand your code. They do not cause the computer to perform any action when the code executes. For easy reference, we begin each script with a comment indicating the script's file name.

A comment also can begin to the right of the code on a given line and continue until the end of that line. Such a comment documents the code to its left.

Docstrings

The *Style Guide for Python Code* states that each script should start with a docstring that explains the script's purpose, such as the one in line 2:

```
"""Comparing integers using if statements and comparison operators."""
```

For more complex scripts, the docstring often spans many lines. In later chapters, you'll use docstrings to describe script components you define, such as new functions and new types called classes. We'll also discuss how to access docstrings with the IPython help mechanism.

Blank Lines

Line 3 is a blank line. You use blank lines and space characters to make code easier to read. Together, blank lines, space characters and tab characters are known as **white space**. Python ignores most white space—you'll see that some indentation is required.

Splitting a Lengthy Statement Across Lines
Lines 4–5

```
print('Enter two integers, and I will tell you',
      'the relationships they satisfy.')
```

display instructions to the user. These are too long to fit on one line, so we broke them into two strings. Recall that you can display several values by passing to print a comma-separated list—print separates each value from the next with a space character.

Typically, you write statements on one line. You may spread a lengthy statement over several lines with the \ continuation character. Python also allows you to split long code lines in parentheses without using continuation characters (as in lines 4–5). This is the preferred way to break long code lines according to the *Style Guide for Python Code*. Always choose breaking points that make sense, such as after a comma in the preceding call to print or before an operator in a lengthy expression.

Reading Integer Values from the User

Next, lines 8 and 11 use the built-in input and int functions to prompt for and read two integer values from the user.

2.7 Decision Making: The if Statement and Comparison Operators

if Statements
The if statement in lines 13–14

```
if number1 == number2:
    print(number1, 'is equal to', number2)
```

uses the == comparison operator to determine whether the values of variables number1 and number2 are equal. If so, the condition is True, and line 14 displays a line of text indicating that the values are equal. If any of the remaining if statements' conditions are True (lines 16, 19, 22, 25 and 28), the corresponding print displays a line of text.

Each if statement consists of the keyword if, the condition to test, and a colon (:) followed by an indented body called a **suite**. Each suite must contain one or more statements. Forgetting the colon (:) after the condition is a common syntax error.

Suite Indentation
Python requires you to indent the statements in suites. The *Style Guide for Python Code* recommends four-space indents—we use that convention throughout this book. You'll see in the next chapter that incorrect indentation can cause errors.

Confusing == and =
Using the assignment symbol (=) instead of the equality operator (==) in an if statement's condition is a common syntax error. To help avoid this, read == as "is equal to" and = as "is assigned." You'll see in the next chapter that using == in place of = in an assignment statement can lead to subtle problems.

Chaining Comparisons
You can chain comparisons to check whether a value is in a range. The following comparison determines whether x is in the range 1 through 5, inclusive:

```
In [1]: x = 3

In [2]: 1 <= x <= 5
Out[2]: True

In [3]: x = 10

In [4]: 1 <= x <= 5
Out[4]: False
```

Precedence of the Operators We've Presented So Far
The precedence of the operators introduced in this chapter is shown below:

Operators	Grouping	Type
()	left to right	parentheses
**	right to left	exponentiation
* / // %	left to right	multiplication, true division, floor division, remainder
+ -	left to right	addition, subtraction
> <= < >=	left to right	less than, less than or equal, greater than, greater than or equal
== !=	left to right	equal, not equal

The table lists the operators top-to-bottom in decreasing order of precedence. When writing expressions containing multiple operators, confirm that they evaluate in the order you expect by referring to the operator precedence chart at

> https://docs.python.org/3/reference/expressions.html#operator-precedence

✓ Self Check

1 *(Fill-In)* You use _____ to document code and improve its readability.
Answer: comments.

2 *(True/False)* The comparison operators evaluate left to right and all have the same level of precedence.
Answer: False. The operators <, <=, > and >= all have the same level of precedence and evaluate left to right. The operators == and != have the same level of precedence and evaluate left to right. Their precedence is lower than that of <, <=, > and >=.

3 *(IPython Session)* For any of the operators !=, >= or <=, show that a syntax error occurs if you reverse the symbols in a condition.
Answer:

```
In [1]: 7 =< 10
  File "<ipython-input-1-090d4004a38e>", line 1
    7 =< 10
      ^
SyntaxError: invalid syntax
```

4 *(IPython Session)* Use all six comparison operators to compare the values 5 and 9. Display the values on one line using print.
Answer:

```
In [2]: print(5 < 9, 5 <= 9, 5 > 9, 5 >= 9, 5 == 9, 5 != 9)
True True False False False True
```

2.8 Objects and Dynamic Typing

The first chapter introduced the terms *classes* and *objects* and in Section 2.2, we discussed variables, values and types. Values such as 7 (an integer), 4.1 (a floating-point number) and 'dog' are all objects. Every object has a type and a value:

```
In [1]: type(7)
Out[1]: int

In [2]: type(4.1)
Out[2]: float

In [3]: type('dog')
Out[3]: str
```

An object's value is the data stored in the object. The snippets above show objects of Python built-in types **int** (for integers), **float** (for floating-point numbers) and **str** (for strings).

Variables Refer to Objects

Assigning an object to a variable **binds** (associates) that variable's name to the object. As you've seen, you can then use the variable in your code to access the object's value:

2.8 Objects and Dynamic Typing

```
In [4]: x = 7

In [5]: x + 10
Out[5]: 17

In [6]: x
Out[6]: 7
```

After snippet [4]'s assignment, the variable x **refers** to the integer object containing 7. As shown in snippet [6], snippet [5] does not change x's value. You can change x as follows:

```
In [7]: x = x + 10

In [8]: x
Out[8]: 17
```

Dynamic Typing

Python uses **dynamic typing**—it determines the type of the object a variable refers to while executing your code. We can show this by rebinding the variable x to different objects and checking their types:

```
In [9]: type(x)
Out[9]: int

In [10]: x = 4.1

In [11]: type(x)
Out[11]: float

In [12]: x = 'dog'

In [13]: type(x)
Out[13]: str
```

Garbage Collection

Python creates objects in memory and removes them from memory as necessary. After snippet [10], the variable x now refers to a float object. The integer object from snippet [7] is no longer bound to a variable. As we'll discuss in a later chapter, Python automatically removes such objects from memory. This process—called **garbage collection**—helps ensure that memory is available for new objects you create.

Self Check

1 *(Fill-In)* Assigning an object to a variable _____ the variable's name to the object.
Answer: binds.

2 *(True/False)* A variable always references the same object.
Answer: False. You can make an existing variable refer to a different object and even one of a different type.

3 *(IPython Session)* What is the type of the expression 7.5 * 3?
Answer:

```
In [1]: type(7.5 * 3)
Out[1]: float
```

2.9 Intro to Data Science: Basic Descriptive Statistics

In data science, you'll often use statistics to describe and summarize your data. Here, we begin by introducing several such **descriptive statistics**, including:

- **minimum**—the smallest value in a collection of values.
- **maximum**—the largest value in a collection of values.
- **range**—the range of values from the minimum to the maximum.
- **count**—the number of values in a collection.
- **sum**—the total of the values in a collection.

We'll look at determining the *count* and *sum* in the next chapter. **Measures of dispersion** (also called **measures of variability**), such as *range*, help determine how spread out values are. Other measures of dispersion that we'll present in later chapters include *variance* and *standard deviation*.

Determining the Minimum of Three Values

First, let's show how to determine the minimum of three values manually. The following script prompts for and inputs three values, uses if statements to determine the minimum value, then displays it.

```
1   # fig02_02.py
2   """Find the minimum of three values."""
3
4   number1 = int(input('Enter first integer: '))
5   number2 = int(input('Enter second integer: '))
6   number3 = int(input('Enter third integer: '))
7
8   minimum = number1
9
10  if number2 < minimum:
11      minimum = number2
12
13  if number3 < minimum:
14      minimum = number3
15
16  print('Minimum value is', minimum)
```

```
Enter first integer: 12
Enter second integer: 27
Enter third integer: 36
Minimum value is 12
```

```
Enter first integer: 27
Enter second integer: 12
Enter third integer: 36
Minimum value is 12
```

Fig. 2.2 | Find the minimum of three values. (Part 1 of 2.)

```
Enter first integer: 36
Enter second integer: 27
Enter third integer: 12
Minimum value is 12
```

Fig. 2.2 | Find the minimum of three values. (Part 2 of 2.)

After inputting the three values, we process one value at a time:

- First, we assume that `number1` contains the smallest value, so line 8 assigns it to the variable `minimum`. Of course, it's possible that `number2` or `number3` contains the actual smallest value, so we still must compare each of these with `minimum`.
- The first `if` statement (lines 10–11) then tests `number2 < minimum` and if this condition is `True` assigns `number2` to `minimum`.
- The second `if` statement (lines 13–14) then tests `number3 < minimum`, and if this condition is `True` assigns `number3` to `minimum`.

Now, `minimum` contains the smallest value, so we display it. We executed the script three times to show that it always finds the smallest value regardless of whether the user enters it first, second or third.

Determining the Minimum and Maximum with Built-In Functions `min` and `max`
Python has many built-in functions for performing common tasks. Built-in functions `min` and `max` calculate the minimum and maximum, respectively, of a collection of values:

```
In [1]: min(36, 27, 12)
Out[1]: 12

In [2]: max(36, 27, 12)
Out[2]: 36
```

The functions `min` and `max` can receive any number of arguments.

Determining the Range of a Collection of Values
The *range* of values is simply the minimum through the maximum value. In this case, the range is 12 through 36. Much data science is devoted to getting to know your data. Descriptive statistics is a crucial part of that, but you also have to understand how to interpret the statistics. For example, if you have 100 numbers with a range of 12 through 36, those numbers could be distributed evenly over that range. At the opposite extreme, you could have clumping with 99 values of 12 and one 36, or one 12 and 99 values of 36.

Functional-Style Programming: Reduction
Throughout this book, we introduce various *functional-style programming* capabilities. These enable you to write code that can be more concise, clearer and easier to **debug**—that is, find and correct errors. The `min` and `max` functions are examples of a functional-style programming concept called **reduction**. They reduce a collection of values to a *single* value. Other reductions you'll see include the sum, average, variance and standard deviation of a collection of values. You'll also learn how to define custom reductions.

Upcoming Intro to Data Science Sections
In the next two chapters, we'll continue our discussion of basic descriptive statistics with *measures of central tendency*, including *mean*, *median* and *mode*, and *measures of dispersion*, including *variance* and *standard deviation*.

Self Check

1 *(Fill-In)* The range of a collection of values is a measure of _____.
Answer: dispersion.

2 *(IPython Session)* For the values 47, 95, 88, 73, 88 and 84 calculate the minimum, maximum and range.
Answer:

```
In [1]: min(47, 95, 88, 73, 88, 84)
Out[1]: 47

In [2]: max(47, 95, 88, 73, 88, 84)
Out[2]: 95

In [3]: print('Range:', min(47, 95, 88, 73, 88, 84), '-',
   ...:       max(47, 95, 88, 73, 88, 84))
   ...:
Range: 47 - 95
```

2.10 Wrap-Up

This chapter continued our discussion of arithmetic. You used variables to store values for later use. We introduced Python's arithmetic operators and showed that you must write all expressions in straight-line form. You used the built-in function `print` to display data. We created single-, double- and triple-quoted strings. You used triple-quoted strings to create multiline strings and to embed single or double quotes in strings.

You used the `input` function to prompt for and get input from the user at the keyboard. We used the functions `int` and `float` to convert strings to numeric values. We presented Python's comparison operators. Then, you used them in a script that read two integers from the user and compared their values using a series of `if` statements.

We discussed Python's dynamic typing and used the built-in function `type` to display an object's type. Finally, we introduced the basic descriptive statistics minimum and maximum and used them to calculate the range of a collection of values. In the next chapter, you'll learn Python's control statements and program development.

Exercises

Unless specified otherwise, use IPython sessions for each exercise.

2.1   *(What does this code do?)* Create the variables x = 2 and y = 3, then determine what each of the following statements displays:

    a) `print('x =', x)`
    b) `print('Value of', x, '+', x, 'is', (x + x))`
    c) `print('x =')`
    d) `print((x + y), '=', (y + x))`

2.2 *(What's wrong with this code?)* The following code should read an integer into the variable `rating`:

```
rating = input('Enter an integer rating between 1 and 10')
```

2.3 *(Fill in the missing code)* Replace *** in the following code with a statement that will print a message like `'Congratulations! Your grade of 91 earns you an A in this course'`. Your statement should print the value stored in the variable `grade`:

```
if grade >= 90:
    ***
```

2.4 *(Arithmetic)* For each of the arithmetic operators +, -, *, /, // and **, display the value of an expression with `27.5` as the left operand and `2` as the right operand.

2.5 *(Circle Area, Diameter and Circumference)* For a circle of radius 2, display the diameter, circumference and area. Use the value 3.14159 for π. Use the following formulas (r is the radius): *diameter* = $2r$, *circumference* = $2\pi r$ and *area* = πr^2. [In a later chapter, we'll introduce Python's `math` module which contains a higher-precision representation of π.]

2.6 *(Odd or Even)* Use `if` statements to determine whether an integer is odd or even. [*Hint:* Use the remainder operator. An even number is a multiple of 2. Any multiple of 2 leaves a remainder of 0 when divided by 2.]

2.7 *(Multiples)* Use `if` statements to determine whether 1024 is a multiple of 4 and whether 2 is a multiple of 10. (*Hint:* Use the remainder operator.)

2.8 *(Table of Squares and Cubes)* Write a script that calculates the squares and cubes of the numbers from 0 to 5. Print the resulting values in table format, as shown below. Use the tab escape sequence to achieve the three-column output.

```
number  square  cube
0       0       0
1       1       1
2       4       8
3       9       27
4       16      64
5       25      125
```

The next chapter shows how to "right align" numbers. You could try that as an extra challenge here. The output would be:

```
number  square  cube
     0       0     0
     1       1     1
     2       4     8
     3       9    27
     4      16    64
     5      25   125
```

2.9 *(Integer Value of a Character)* Here's a peek ahead. In this chapter, you learned about strings. Each of a string's characters has an integer representation. The set of characters a computer uses together with the characters' integer representations is called that computer's *character set*. You can indicate a character value in a program by enclosing that character in quotes, as in `'A'`. To determine a character's integer value, call the built-in function **ord**:

```
In [1]: ord('A')
Out[1]: 65
```
Display the integer equivalents of B C D b c d 0 1 2 $ * + and the space character.

2.10 *(Arithmetic, Smallest and Largest)* Write a script that inputs three integers from the user. Display the sum, average, product, smallest and largest of the numbers. Note that each of these is a reduction in functional-style programming.

2.11 *(Separating the Digits in an Integer)* Write a script that inputs a five-digit integer from the user. Separate the number into its individual digits. Print them separated by three spaces each. For example, if the user types in the number 42339, the script should print

    4   2   3   3   9

Assume that the user enters the correct number of digits. Use both the floor division and remainder operations to "pick off" each digit.

2.12 *(7% Investment Return)* Some investment advisors say that it's reasonable to expect a 7% return over the long term in the stock market. Assuming that you begin with $1000 and leave your money invested, calculate and display how much money you'll have after 10, 20 and 30 years. Use the following formula for determining these amounts:

$$a = p(1 + r)^n$$

where

    p is the original amount invested (i.e., the principal of $1000),
    r is the annual rate of return (7%),
    n is the number of years (10, 20 or 30) and
    a is the amount on deposit at the end of the nth year.

2.13 *(How Big Can Python Integers Be?)* We'll answer this question later in the book. For now, use the exponentiation operator ** with large and very large exponents to produce some huge integers and assign those to the variable number to see if Python accepts them. Did you find any integer value that Python won't accept?

2.14 *(Target Heart-Rate Calculator)* While exercising, you can use a heart-rate monitor to see that your heart rate stays within a safe range suggested by your doctors and trainers. According to the American Heart Association (AHA) (http://bit.ly/AHATargetHeart-Rates), the formula for calculating your maximum heart rate in beats per minute is 220 minus your age in years. Your target heart rate is 50–85% of your maximum heart rate. Write a script that prompts for and inputs the user's age and calculates and displays the user's maximum heart rate and the range of the user's target heart rate. [**These formulas are estimates provided by the AHA; maximum and target heart rates may vary based on the health, fitness and gender of the individual. Always consult a physician or qualified healthcare professional before beginning or modifying an exercise program.**]

2.15 *(Sort in Ascending Order)* Write a script that inputs three different floating-point numbers from the user. Display the numbers in increasing order. Recall that an if statement's suite can contain more than one statement. Prove that your script works by running it on all six possible orderings of the numbers. Does your script work with duplicate numbers? [This is challenging. In later chapters you'll do this more conveniently and with many more numbers.]

Control Statements and Program Development

3

Objectives
In this chapter, you'll:

- Decide whether to execute actions with the statements `if`, `if...else` and `if...elif...else`.
- Execute statements repeatedly with `while` and `for`.
- Shorten assignment expressions with augmented assignments.
- Use the `for` statement and the built-in `range` function to repeat actions for a sequence of values.
- Perform sentinel-controlled repetition with `while`.
- Learn problem-solving skills: understanding problem requirements, dividing problems into smaller pieces, developing algorithms to solve problems and implementing those algorithms in code.
- Develop algorithms through the process of top-down, stepwise refinement.
- Create compound conditions with the Boolean operators `and`, `or` and `not`.
- Stop looping with `break`.
- Force the next iteration of a loop with `continue`.
- Use some functional-style programming features to write scripts that are more concise, clearer, easier to debug and easier to parallelize.

Outline

3.1 Introduction
3.2 Algorithms
3.3 Pseudocode
3.4 Control Statements
3.5 `if` Statement
3.6 `if...else` and `if...elif...else` Statements
3.7 `while` Statement
3.8 `for` Statement
   3.8.1 Iterables, Lists and Iterators
   3.8.2 Built-In `range` Function
3.9 Augmented Assignments
3.10 Program Development: Sequence-Controlled Repetition
   3.10.1 Requirements Statement
   3.10.2 Pseudocode for the Algorithm
   3.10.3 Coding the Algorithm in Python
   3.10.4 Introduction to Formatted Strings
3.11 Program Development: Sentinel-Controlled Repetition
3.12 Program Development: Nested Control Statements
3.13 Built-In Function `range`: A Deeper Look
3.14 Using Type `Decimal` for Monetary Amounts
3.15 `break` and `continue` Statements
3.16 Boolean Operators `and`, `or` and `not`
3.17 Intro to Data Science: Measures of Central Tendency—Mean, Median and Mode
3.18 Wrap-Up
Exercises

3.1 Introduction

Before writing a program to solve a particular problem, you must understand the problem and have a carefully planned approach to solving it. You must also understand Python's building blocks and use proven program-construction principles.

3.2 Algorithms

You can solve any computing problem by executing a series of actions in a specific order. An **algorithm** is a *procedure* for solving a problem in terms of:

1. the **actions** to execute, and
2. the **order** in which these actions execute.

Correctly specifying the order in which the actions execute is essential. Consider the "rise-and-shine algorithm" that an executive follows for getting out of bed and going to work: (1) Get out of bed; (2) take off pajamas; (3) take a shower; (4) get dressed; (5) eat breakfast; (6) carpool to work. This routine gets the executive to work well prepared to make critical decisions. Suppose the executive performs these steps in a different order: (1) Get out of bed; (2) take off pajamas; (3) get dressed; (4) take a shower; (5) eat breakfast; (6) carpool to work. Now, our executive shows up for work soaking wet. **Program control** specifies the order in which statements (actions) execute in a program. This chapter investigates program control using Python's **control statements**.

✓ Self Check

1 *(Fill-In)* A(n) _____ is a procedure for solving a problem. It specifies the _____ to execute and the _____ in which they execute.
Answer: algorithm, actions, order.

3.3 Pseudocode

Pseudocode is an informal English-like language for "thinking out" algorithms. You write text that describes what your program should do. You then convert the pseudocode to Python by replacing pseudocode statements with their Python equivalents.

Addition-Program Pseudocode
The following pseudocode algorithm prompts the user to enter two integers, inputs them from the user at the keyboard, adds them, then stores and displays their sum:

Prompt the user to enter the first integer
Input the first integer

Prompt the user to enter the second integer
Input the second integer

Add first integer and second integer, store their sum
Display the numbers and their sum

This is the complete pseudocode algorithm. Later in the chapter, we'll show a simple process for creating a pseudocode algorithm from a *requirements statement*. The English pseudocode statements specify the actions you wish to perform and the order in which you wish to perform them.

✓ Self Check

1 *(True/False)* Pseudocode is a simple programming language.
Answer: False. Pseudocode is not a programming language. It's an artificial and informal language that helps you develop algorithms.

2 *(IPython Session)* Write Python statements that perform the tasks described by this section's pseudocode. Enter the integers 10 and 5.
Answer:

```
In [1]: number1 = int(input('Enter first integer: '))
Enter first integer: 10

In [2]: number2 = int(input('Enter second integer: '))
Enter second integer: 5

In [3]: total = number1 + number2

In [4]: print('The sum of', number1, 'and', number2, 'is', total)
The sum of 10 and 5 is 15
```

3.4 Control Statements

Usually, statements in a program execute in the order in which they're written. This is called *sequential execution*. Various Python statements enable you to specify that the next statement to execute may be *other than* the next one *in sequence*. This is called *transfer of control* and is achieved with Python *control statements*.

Forms of Control

In the 1960s, extensive use of control transfers was causing difficulty in software development. Blame was pointed at the `goto` statement. This statement allowed you to transfer control to one of many possible destinations in a program. Bohm and Jacopini's research[1] demonstrated that programs could be written without `goto` statements. The notion of *structured programming* became almost synonymous with "goto elimination." Python does not have a `goto` statement. Structured programs are clearer, easier to debug and change, and more likely to be bug-free.

Bohm and Jacopini demonstrated that all programs could be written using three forms of control—namely, **sequential execution,** the **selection statement** and the **repetition statement.** Sequential execution is simple. Python statements execute one after the other "in sequence," unless directed otherwise.

Flowcharts

A **flowchart** is a *graphical* representation of an algorithm or a part of one. You draw flowcharts using *rectangles, diamonds, rounded rectangles* and *small circles* that you connect by *arrows* called **flowlines**. Like pseudocode, flowcharts are useful for developing and representing algorithms. They clearly show how forms of control operate. Consider the following flowchart segment, which shows *sequential execution*:

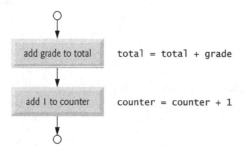

We use the **rectangle (or action) symbol** to indicate any *action,* such as a calculation or an input/output operation. The flowlines show the *order* in which the actions execute. First, the grade is added to the total, then 1 is added to the counter. We show the Python code next to each action symbol for comparison purposes. This code is not part of the flowchart.

In a flowchart for a *complete* algorithm, the first symbol is a **rounded rectangle** containing the word "Begin." The last symbol is a rounded rectangle containing the word "End." In a flowchart for only a *part* of an algorithm, we omit the rounded rectangles, instead using small circles called **connector symbols.** The most important symbol is the **decision (or diamond) symbol,** which indicates that a *decision* is to be made, such as in an `if` statement. We begin using decision symbols in the next section.

Selection Statements

Python provides three types of selection statements that execute code based on a *condition*—an expression that evaluates to either `True` or `False`:

---

1. Bohm, C., and G. Jacopini, "Flow Diagrams, Turing Machines, and Languages with Only Two Formation Rules," *Communications of the ACM,* Vol. 9, No. 5, May 1966, pp. 336–371.

- The `if` statement performs an action if a condition is `True` or skips the action if the condition is `False`.
- The **`if...else` statement** performs an action if a condition is `True` or performs a *different* action if the condition is `False`.
- The **`if...elif...else` statement** performs one of many different actions, depending on the truth or falsity of *several* conditions.

Anywhere a single action can be placed, a group of actions can be placed.

The `if` statement is called a **single-selection statement** because it selects or ignores a *single* action (or group of actions). The `if...else` statement is called a **double-selection statement** because it selects between *two different* actions (or groups of actions). The `if...elif...else` statement is called a **multiple-selection statement** because it selects one of many different actions (or groups of actions).

Repetition Statements

Python provides two repetition statements—`while` and `for`:

- The `while` statement repeats an action (or a group of actions) as long as a condition remains `True`.
- The `for` statement repeats an action (or a group of actions) for every item in a sequence of items.

Keywords

The words `if`, `elif`, `else`, `while`, `for`, `True` and `False` are keywords that Python reserves to implement its features, such as control statements. Using a keyword as an identifier such as a variable name is a syntax error. The following table lists Python's keywords.

Python keywords						
and	as	assert	async	await	break	class
continue	def	del	elif	else	except	False
finally	for	from	global	if	import	in
is	lambda	None	nonlocal	not	or	pass
raise	return	True	try	while	with	yield

Control Statements Summary

You form each Python program by combining as many control statements of each type as you need for the algorithm the program implements. With **Single-entry/single-exit** (one way in/one way out) **control statements**, the exit point of one connects to the entry point of the next. This is similar to the way a child stacks building blocks—hence, the term **control-statement stacking**. **Control-statement nesting** also connects control statements—we'll see how later in the chapter.

You can construct any Python program from only six different forms of control (sequential execution, and the `if`, `if...else`, `if...elif...else`, `while` and `for` statements). You combine these in only two ways (control-statement stacking and control-statement nesting). This is the essence of simplicity.

Self Check

1 *(Fill-In)* You can write all programs using three forms of control—_____, _____ and _____.
Answer: sequential execution, selection statements, repetition statements.

2 *(Fill-In)* A(n) _____ is a graphical representation of an algorithm.
Answer: flowchart.

3.5 if Statement

Suppose that a passing grade on an examination is 60. The pseudocode

> *If student's grade is greater than or equal to 60*
>   *Display 'Passed'*

determines whether the condition "student's grade is greater than or equal to 60" is true or false. If the condition is true, 'Passed' is displayed. Then, the next pseudocode statement in order is "performed." (Remember that pseudocode is not a real programming language.) If the condition is false, nothing is displayed, and the next pseudocode statement is "performed." The pseudocode's second line is indented. Python code requires indentation. Here it emphasizes that 'Passed' is displayed *only* if the condition is true.

Let's assign 85 to the variable grade, then show and execute the Python if statement for the pseudocode:

```
In [1]: grade = 85

In [2]: if grade >= 60:
   ...:     print('Passed')
   ...:
Passed
```

The if statement closely resembles the pseudocode. The condition grade >= 60 is True, so the indented print statement displays 'Passed'.

Suite Indentation

Indenting a suite is required; otherwise, an IndentationError syntax error occurs:

```
In [3]: if grade >= 60:
   ...: print('Passed')  # statement is not indented properly
  File "<ipython-input-3-f42783904220>", line 2
    print('Passed')  # statement is not indented properly
        ^
IndentationError: expected an indented block
```

An IndentationError also occurs if you have more than one statement in a suite and those statements do not have the *same* indentation:

```
In [4]: if grade >= 60:
   ...:         print('Passed')  # indented 4 spaces
   ...:       print('Good job!')  # incorrectly indented only two spaces
  File <ipython-input-4-8c0d75c127bf>, line 3
    print('Good job!')  # incorrectly indented only two spaces
        ^
IndentationError: unindent does not match any outer indentation level
```

Sometimes error messages may not be clear. The fact that Python calls attention to the line is usually enough for you to figure out what's wrong. Apply indentation conventions uniformly throughout your code. Programs that are not uniformly indented are hard to read.

if Statement Flowchart
The flowchart for the if statement in snippet [2] is:

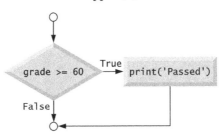

The decision (diamond) symbol contains a condition that can be either True or False. The diamond has two flowlines emerging from it:
- One indicates the direction to follow when the condition in the symbol is True. This points to the action (or group of actions) that should execute.
- The other indicates the direction to follow when the condition is False. This skips the action (or group of actions).

Every Expression Can Be Interpreted as Either True or False
You can base decisions on *any* expression. A nonzero value is True. Zero is False:

```
In [5]: if 1:
   ...:     print('Nonzero values are true, so this will print')
   ...:
Nonzero values are true, so this will print

In [6]: if 0:
   ...:     print('Zero is false, so this will not print')

In [7]:
```

Strings containing characters are True and empty strings (' ', "" or """""") are False.

An Additional Note on Confusing == and =
Using the equality operator == instead of the assignment symbol = in an assignment statement can lead to subtle problems. For example, in this session, snippet [1] defined grade with the assignment:

```
grade = 85
```

If instead we accidentally wrote:

```
grade == 85
```

then grade would be undefined and we'd get a NameError.

If grade had been defined before the preceding statement, then grade == 85 would evaluate to True or False, depending on grade's value, and not perform the intended assignment. This is a logic error.

Self Check

1 *(True/False)* If you indent a suite's statements, you will not get an IndentationError.
Answer: False. All the statements in a suite must have the *same* indentation. Otherwise, an IndentationError occurs.

2 *(IPython Session)* Redo this section's snippets [1] and [2], then change grade to 55 and repeat the if statement to show that its suite does not execute. The next section shows how to recall and re-execute earlier snippets to avoid having to re-enter the code.
Answer:

```
In [1]: grade = 85

In [2]: if grade >= 60:
   ...:     print('Passed')
   ...:
Passed

In [3]: grade = 55

In [4]: if grade >= 60:
   ...:     print('Passed')
   ...:

In [5]:
```

3.6 if...else and if...elif...else Statements

The if...else statement performs different suites, based on whether a condition is True or False. The pseudocode below displays 'Passed' if the student's grade is greater than or equal to 60; otherwise, it displays 'Failed':

> *If student's grade is greater than or equal to 60*
>   *Display 'Passed'*
> *Else*
>   *Display 'Failed'*

In either case, the next pseudocode statement in sequence after the entire *If...Else* is "performed." We indent both the *If* and *Else* suites, and by the same amount. Let's create and **initialize** (that is, give a starting value to) the variable grade, then show and execute the Python if...else statement for the preceding pseudocode:

```
In [1]: grade = 85

In [2]: if grade >= 60:
   ...:     print('Passed')
   ...: else:
   ...:     print('Failed')
   ...:
Passed
```

The condition above is True, so the if suite displays 'Passed'. Note that when you press *Enter* after typing print('Passed'), IPython indents the next line four spaces. You must delete those four spaces so that the else: suite correctly aligns under the i in if.

3.6 if…else and if…elif…else Statements

The following code assigns 57 to the variable grade, then shows the if…else statement again to demonstrate that only the else suite executes when the condition is False:

```
In [3]: grade = 57

In [4]: if grade >= 60:
   ...:     print('Passed')
   ...: else:
   ...:     print('Failed')
   ...:
Failed
```

The up and down arrow keys navigate backwards and forwards through the current interactive session's snippets. Pressing *Enter* re-executes the snippet that's displayed. Let's set grade to 99, press the up arrow key twice to recall the code from snippet [4], then press *Enter* to re-execute that code as snippet [6]. Every recalled snippet that you execute gets a new ID:

```
In [5]: grade = 99

In [6]: if grade >= 60:
   ...:     print('Passed')
   ...: else:
   ...:     print('Failed')
   ...:
Passed
```

if…else Statement Flowchart
The flowchart below shows the preceding if…else statement's flow of control:

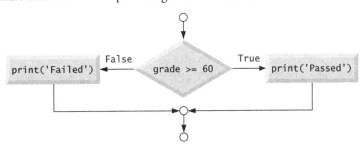

Conditional Expressions
Sometimes the suites in an if…else statement assign different values to a variable, based on a condition, as in:

```
In [7]: grade = 87

In [8]: if grade >= 60:
   ...:     result = 'Passed'
   ...: else:
   ...:     result = 'Failed'
   ...:
```

We can then print or evaluate that variable:

```
In [9]: result
Out[9]: 'Passed'
```

You can write statements like snippet [8] using a concise **conditional expression**:

```
In [10]: result = ('Passed' if grade >= 60 else 'Failed')
```

```
In [11]: result
Out[11]: 'Passed'
```

The parentheses are not required, but they make it clear that the statement assigns the conditional expression's value to result. First, Python evaluates the condition grade >= 60:

- If it's True, snippet [10] assigns to result the value of the expression to the *left* of if, namely 'Passed'. The else part does not execute.
- If it's False, snippet [10] assigns to result the value of the expression to the *right* of else, namely 'Failed'.

In interactive mode, you also can evaluate the conditional expression directly, as in:

```
In [12]: 'Passed' if grade >= 60 else 'Failed'
Out[12]: 'Passed'
```

Multiple Statements in a Suite

The following code shows two statements in the else suite of an if...else statement:

```
In [13]: grade = 49

In [14]: if grade >= 60:
   ...:     print('Passed')
   ...: else:
   ...:     print('Failed')
   ...:     print('You must take this course again')
   ...:
Failed
You must take this course again
```

In this case, grade is less than 60, so *both* statements in the else's suite execute. If you do not indent the second print, then it's not in the else's suite. So, that statement *always* executes, creating strange incorrect output:

```
In [15]: grade = 100

In [16]: if grade >= 60:
   ...:     print('Passed')
   ...: else:
   ...:     print('Failed')
   ...: print('You must take this course again')
   ...:
Passed
You must take this course again
```

if...elif...else Statement

You can test for many cases using the **if...elif...else statement**. The following pseudocode displays "A" for grades greater than or equal to 90, "B" for grades in the range 80–89, "C" for grades 70–79, "D" for grades 60–69 and "F" for all other grades:

> *If student's grade is greater than or equal to 90*
> > *Display "A"*
> *Else If student's grade is greater than or equal to 80*
> > *Display "B"*
> *Else If student's grade is greater than or equal to 70*
> > *Display "C"*
> *Else If student's grade is greater than or equal to 60*
> > *Display "D"*
> *Else*
> > *Display "F"*

Only the action for the first True condition executes. Let's show and execute the Python code for the preceding pseudocode. The pseudocode *Else If* is written with the keyword elif. Snippet [18] displays C, because grade is 77:

```
In [17]: grade = 77

In [18]: if grade >= 90:
   ...:     print('A')
   ...: elif grade >= 80:
   ...:     print('B')
   ...: elif grade >= 70:
   ...:     print('C')
   ...: elif grade >= 60:
   ...:     print('D')
   ...: else:
   ...:     print('F')
   ...:
C
```

The first condition—grade >= 90—is False, so print('A') is skipped. The second condition—grade >= 80—also is False, so print('B') is skipped. The third condition—grade >= 70—is True, so print('C') executes. Then all the remaining code in the if...elif...else statement is skipped. An if...elif...else is faster than separate if statements, because condition testing stops as soon as a condition is True.

if...elif...else Statement Flowchart

The following flowchart shows the general flow through an if...elif...else statement. It shows that, after any suite executes, control immediately exits the statement. The words to the left are not part of the flowchart. We added them to show how the flowchart corresponds to the equivalent Python code.

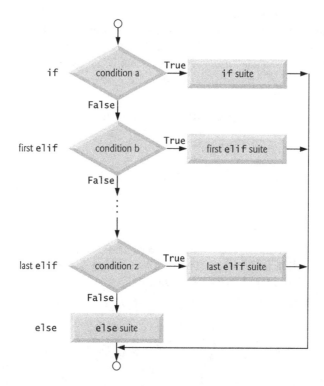

else Is Optional

The else in the if...elif...else statement is optional. Including it enables you to handle values that do not satisfy *any* of the conditions. When an if...elif statement without an else tests a value that does not make any of its conditions True, the program does not execute any of the statement's suites. The next statement in sequence after the if...elif statement executes. If you specify the else, you must place it after the last elif; otherwise, a SyntaxError occurs.

Logic Errors

The incorrectly indented code segment in snippet [16] is an example of a **nonfatal logic error**. The code executes, but it produces incorrect results. For a **fatal logic error** in a script, an exception occurs (such as a ZeroDivisionError from an attempt to divide by 0), so Python displays a traceback, then terminates the script. A fatal error in interactive mode terminates only the current snippet. Then IPython waits for your next input.

 Self Check

1 *(True/False)* A fatal logic error causes a script to produce incorrect results, then continue executing.
Answer: False. A fatal logic error causes a script to terminate.

2 *(IPython Session)* Show that a SyntaxError occurs if an if...elif statement specifies an else before the last elif.

3.7 while Statement

Answer:
```
In [1]: grade = 80

In [2]: if grade >= 90:
   ...:         print('A')
   ...: else:
   ...:         print('Not A or B')
   ...: elif grade >= 80:
  File "<ipython-input-2-033bcba40157>", line 5
    elif grade >= 80:
       ^
SyntaxError: invalid syntax
```

3.7 while Statement

The **while** statement allows you to *repeat* one or more actions while a condition remains True. Such a statement often is called a **loop**.

The following pseudocode specifies what happens when you go shopping:

While there are more items on my shopping list
Buy next item and cross it off my list

If the condition "there are more items on my shopping list" is *true*, you perform the action "Buy next item and cross it off my list." You *repeat* this action while the condition remains *true*. You stop repeating this action when the condition becomes *false*—that is, when you've crossed all items off your shopping list.

Let's use a while statement to find the first power of 3 larger than 50:

```
In [1]: product = 3

In [2]: while product <= 50:
   ...:         product = product * 3
   ...:

In [3]: product
Out[3]: 81
```

First, we create product and initialize it to 3. Then the while statement executes as follows:

1. Python tests the condition product <= 50, which is True because product is 3. The statement in the suite multiplies product by 3 and assigns the result (9) to product. One *iteration of the loop* is now complete.

2. Python again tests the condition, which is True because product is now 9. The suite's statement sets product to 27, completing the second iteration of the loop.

3. Python again tests the condition, which is True because product is now 27. The suite's statement sets product to 81, completing the third iteration of the loop.

4. Python again tests the condition, which is finally False because product is now 81. The repetition now terminates.

Snippet [3] evaluates product to see its value, 81, which is the first power of 3 larger than 50. If this while statement were part of a larger script, execution would continue with the next statement in sequence after the while.

Something in the `while` statement's suite must change `product`'s value, so the condition eventually becomes `False`. Otherwise, a logic error called an **infinite loop** occurs. Such an error prevents the `while` statement from ever terminating—the program appears to "hang." In applications executed from a Terminal, Command Prompt or shell, type *Ctrl* + *c* or *control* + *c* (depending on your keyboard) to terminate an infinite loop. IDEs typically have a toolbar button or menu option for stopping a program's execution.

`while` Statement Flowchart
The following flowchart shows the preceding `while` statement's flow of control:

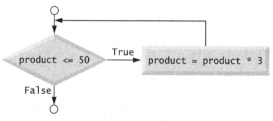

Follow the flowlines to experience the repetition. The flowline from the rectangle "closes the loop" by flowing back into the condition `product <= 50` that's tested during each iteration. When that condition becomes `False`, the `while` statement exits and control proceeds to the next statement in sequence.

✓ Self Check

1 *(True/False)* A `while` statement performs its suite while some condition remains `True`.
Answer: True.

2 *(IPython Session)* Write statements to determine the first power of 7 greater than 1000.
Answer:

```
In [1]: product = 7

In [2]: while product <= 1000:
   ...:     product = product * 7
   ...:

In [3]: product
Out[3]: 2401
```

3.8 `for` Statement

Like the `while` statement, the **`for` statement** allows you to *repeat* an action or several actions. The `for` statement performs its action(s) for each item in a **sequence** of items. For example, a string is a sequence of individual characters. Let's display `'Programming'` with its characters separated by two spaces:

```
In [1]: for character in 'Programming':
   ...:     print(character, end=' ')
   ...:
P r o g r a m m i n g
```

The for statement executes as follows:
- Upon entering the statement, it assigns the 'P' in 'Programming' to the **target** variable between keywords for and in—in this case, character.
- Next, the statement in the suite executes, displaying character's value followed by two spaces—we'll say more about this momentarily.
- After executing the suite, Python assigns to character the next item in the sequence (that is, the 'r' in 'Programming'), then executes the suite again.
- This continues while there are more items in the sequence to process. In this case, the statement terminates after displaying the letter 'g', followed by two spaces.

Using the target in the suite, as we did here to display its value, is common but not required.

for Statement Flowchart
The for statement's flowchart is similar to that of the while statement:

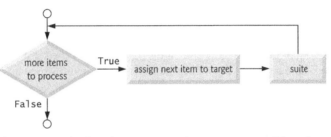

First, Python determines whether there are more items to process. If so, the for statement assigns the next item to the target, then performs the suite's action(s).

Function print's end Keyword Argument
The built-in function print displays its argument(s), then moves the cursor to the next line. You can change this behavior with the argument **end**, as in

    print(character, end=' ')

We used two spaces (' '), so each call to print displays character's value followed by two spaces. So, all the characters display horizontally on the *same* line. Python calls end a **keyword argument**, but end is not a Python keyword. The end keyword argument is *optional*. If you do not include it, print uses a newline ('\n') by default. The *Style Guide for Python Code* recommends placing no spaces around a keyword argument's =. Keyword arguments are sometimes called named arguments.

Function print's sep Keyword Argument
You can use the keyword argument **sep** (short for separator) to specify the string that appears *between* the items that print displays. When you do not specify this argument, print uses a space character by default. Let's display three numbers, each separated from the next by a comma and a space, rather than just a space:

    In [2]: print(10, 20, 30, sep=', ')
    10, 20, 30

To remove the spaces, use an **empty string** with no characters between its quotes.

3.8.1 Iterables, Lists and Iterators

The sequence to the right of the for statement's in keyword must be an **iterable**. An iterable is an object from which the for statement can take one item at a time until no more items remain. Python has other iterable sequence types besides strings. One of the most common is a **list**, which is a comma-separated collection of items enclosed in square brackets ([and]). The following code totals five integers in a list:

```
In [3]: total = 0

In [4]: for number in [2, -3, 0, 17, 9]:
   ...:     total = total + number
   ...:

In [5]: total
Out[5]: 25
```

Each sequence has an **iterator**. The for statement uses the iterator "behind the scenes" to get each consecutive item until there are no more to process. The iterator is like a bookmark—it always knows where it is in the sequence, so it can return the next item when it's called upon to do so.

For each number in the list, the suite adds the number to the total. When there are no more items to process, total contains the sum (25) of the list's items. We cover lists in detail in the "Sequences: Lists and Tuples" chapter. There, you'll see that the order of the items in a list matters and that a list's items are **mutable** (that is, modifiable).

3.8.2 Built-In range Function

Let's use a for statement and the built-in **range function** to iterate precisely 10 times, displaying the values from 0 through 9:

```
In [6]: for counter in range(10):
   ...:     print(counter, end=' ')
   ...:
0 1 2 3 4 5 6 7 8 9
```

The function call range(10) creates an iterable object that represents a sequence of consecutive integer values starting from 0 and continuing up to, but *not* including, the argument value (10). In this case, the sequence is 0, 1, 2, 3, 4, 5, 6, 7, 8, 9. The for statement exits when it finishes processing the last integer that range produces. Iterators and iterable objects are two of Python's *functional-style programming* features. We'll introduce more of these throughout the book.

Off-By-One Errors

A logic error known as an **off-by-one error** occurs when you assume that range's argument value is included in the generated sequence. For example, if you provide 9 as range's argument when trying to produce the sequence 0 through 9, range generates only 0 through 8.

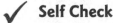 **Self Check**

1 *(Fill-In)* Function _____ generates a sequence of integers.
Answer: range.

2 *(IPython Session)* Use the range function and a for statement to calculate the total of the integers from 0 through 1,000,000.
Answer:

```
In [1]: total = 0

In [2]: for number in range(1000001):
   ...:     total = total + number
   ...:

In [3]: total
Out[3]: 500000500000
```

3.9 Augmented Assignments

Augmented assignments abbreviate assignment expressions in which the same variable name appears on the left and right of the assignment's =, as total does in:

```
for number in [1, 2, 3, 4, 5]:
    total = total + number
```

Snippet [2] reimplements this using an **addition augmented assignment (+=) statement**:

```
In [1]: total = 0

In [2]: for number in [1, 2, 3, 4, 5]:
   ...:     total += number  # add number to total
   ...:

In [3]: total
Out[3]: 15
```

The += expression in snippet [2] first adds number's value to the current total, then stores the new value in total. The table below shows sample augmented assignments:

Augmented assignment	Sample expression	Explanation	Assigns
Assume: c = 3, d = 5, e = 4, f = 2, g = 9, h = 12			
+=	c += 7	c = c + 7	10 to c
-=	d -= 4	d = d - 4	1 to d
*=	e *= 5	e = e * 5	20 to e
**=	f **= 3	f = f ** 3	8 to f
/=	g /= 2	g = g / 2	4.5 to g
//=	g //= 2	g = g // 2	4 to g
%=	h %= 9	h = h % 9	3 to h

Self Check

1. *(Fill-In)* If x is 7, the value of x after evaluating x *= 5 is _____.
Answer: 35.

2. *(IPython Session)* Create a variable x with the value 12. Use an exponentiation augmented assignment statement to square x's value. Show x's new value.
Answer:

```
In [1]: x = 12

In [2]: x **= 2

In [3]: x
Out[3]: 144
```

3.10 Program Development: Sequence-Controlled Repetition

Experience has shown that the most challenging part of solving a problem on a computer is developing an algorithm for the solution. As you'll see, once a correct algorithm has been specified, creating a working Python program from the algorithm is typically straightforward. This section and the next present problem solving and program development by creating scripts that solve two class-averaging problems.

3.10.1 Requirements Statement

A **requirements statement** describes *what* a program is supposed to do, but not *how* the program should do it. Consider the following simple requirements statement:

> *A class of ten students took a quiz. Their grades (integers in the range 0 – 100) are 98, 76, 71, 87, 83, 90, 57, 79, 82, 94. Determine the class average on the quiz.*

Once you know the problem's requirements, you can begin creating an algorithm to solve it. Then, you can implement that solution as a program.

The algorithm for solving this problem must:

1. Keep a running total of the grades.
2. Calculate the average—the total of the grades divided by the number of grades.
3. Display the result.

For this example, we'll place the 10 grades in a list. You also could input the grades from a user at the keyboard (as we'll do in the next example) or read them from a file (as you'll see how to do in the "Files and Exceptions" chapter). We also show you how to read data from SQL and NoSQL databases in later chapters.

3.10.2 Pseudocode for the Algorithm

The following pseudocode lists the actions to execute and specifies the order in which they should execute:

Set total to zero
Set grade counter to zero
Set grades to a list of the ten grades

For each grade in the grades list:
    Add the grade to the total
    Add one to the grade counter

Set the class average to the total divided by the number of grades
Display the class average

Note the mentions of *total* and *grade counter*. In Fig. 3.1's script, the variable `total` (line 5) stores the grade values' running total, and `grade_counter` (line 6) counts the number of grades we've processed. We'll use these to calculate the average. Variables for totaling and counting normally are initialized to zero before they're used, as we do in lines 5 and 6.

3.10.3 Coding the Algorithm in Python

The following script implements the pseudocode algorithm.

```
 1  # fig03_01.py
 2  """Class average program with sequence-controlled repetition."""
 3
 4  # initialization phase
 5  total = 0  # sum of grades
 6  grade_counter = 0
 7  grades = [98, 76, 71, 87, 83, 90, 57, 79, 82, 94]  # list of 10 grades
 8
 9  # processing phase
10  for grade in grades:
11      total += grade  # add current grade to the running total
12      grade_counter += 1  # indicate that one more grade was processed
13
14  # termination phase
15  average = total / grade_counter
16  print(f'Class average is {average}')
```

```
Class average is 81.7
```

Fig. 3.1 | Class average program with sequence-controlled repetition.

Execution Phases
We used blank lines and comments to break the script into three **execution phases**—initialization, processing and termination:

- The **initialization phase** creates the variables needed to process the grades and set these variables to appropriate initial values.
- The **processing phase** processes the grades, calculating the running total and counting the number of grades processed so far.
- The **termination phase** calculates and displays the class average.

Many scripts can be **decomposed** (that is, broken apart) into these three phases.

Initialization Phase

Lines 5–6 create the variables `total` and `grade_counter` and initialize each to 0. Line 7

```
grades = [98, 76, 71, 87, 83, 90, 57, 79, 82, 94]   # list of 10 grades
```

creates the variable `grades` and initializes it with a list of 10 integer grades.

Processing Phase

The `for` statement processes each `grade` in the list `grades`. Line 11 adds the current `grade` to the `total`. Then, line 12 adds 1 to the variable `grade_counter` to keep track of the number of grades processed so far. Repetition terminates when all 10 grades in the list have been processed. This is called **definite repetition** because the number of repetitions is known before the loop begins executing. In this case, it's the number of elements in the list `grades`. The *Style Guide for Python Code* recommends placing a blank line above and below each control statement (as in lines 8 and 13).

Termination Phase

When the `for` statement terminates, line 15 calculates the average and assigns it to the variable `average`. Then line 16 displays `average`. Later in this chapter, we use functional-style programming features to calculate the average of a list's items more concisely.

3.10.4 Introduction to Formatted Strings

Line 16 uses the following simple **f-string** (short for **formatted string**) to format this script's result by inserting the value of `average` into a string:

```
f'Class average is {average}'
```

The letter `f` before the string's opening quote indicates it's an f-string. You specify where to insert values by using placeholders delimited by curly braces (`{` and `}`). The placeholder

```
{average}
```

converts the variable `average`'s value to a string representation, then replaces `{average}` with that **replacement text**. Replacement-text expressions may contain values, variables or other expressions, such as calculations or function calls. In line 16, we could have used `total / grade_counter` in place of `average`, eliminating the need for line 15.

 Self Check

1 *(Fill-In)* A(n) _____ describes *what* a program is supposed to do, but not *how* the program should do it.
Answer: requirements statement.

2 *(Fill-In)* Many of the scripts you'll write can be decomposed into three phases: _____, _____ and _____.
Answer: initialization, processing, termination.

3 *(IPython Session)* Display an f-string in which you insert the values of the variables `number1` (7) and `number2` (5) and their product. The displayed string should be

```
7 times 5 is 35
```

Answer:

```
In [1]: number1 = 7

In [2]: number2 = 5

In [3]: print(f'{number1} times {number2} is {number1 * number2}')
7 times 5 is 35
```

3.11 Program Development: Sentinel-Controlled Repetition

Let's generalize the class-average problem. Consider the following requirements statement:

> *Develop a class-averaging program that processes an arbitrary number of grades each time the program executes.*

In the first class-average example, we knew in advance the 10 grades to process. The requirements statement does not state what the grades are or how many there are, so we're going to have the user enter the grades into the program. The program processes an arbitrary number of grades. How can the program determine when to stop processing grades so that it can move on to calculate and display the class average?

One way to solve this problem is to use a special value called a **sentinel value** (also called a **signal value**, a **dummy value** or a **flag value**) to indicate "end of data entry." This is a bit like the way a caboose "marks" the end of a train. The user enters grades one at a time until all the grades have been entered. The user then enters the sentinel value to indicate that there are no more grades. **Sentinel-controlled repetition** is often called **indefinite repetition** because the number of repetitions is *not* known before the loop begins executing.

A sentinel value must not be confused with any acceptable input value. Grades on a quiz are typically nonnegative integers between 0 and 100, so the value –1 is an acceptable sentinel value for this problem. Thus, a run of the class-average program might process a stream of inputs such as 95, 96, 75, 74, 89 and –1. The program would then compute and print the class average for the grades 95, 96, 75, 74 and 89. The sentinel value –1 should *not* enter into the averaging calculation.

Developing the Pseudocode Algorithm with Top-Down, Stepwise Refinement
We approach this class-average problem with a technique called **top-down, stepwise refinement**. We begin with a pseudocode representation of the *top*:

> *Determine the class average for the quiz*

The top is a single statement that conveys the program's overall function. Although it's a *complete* representation of a program, the top rarely conveys enough detail from which to write a program. The *top* specifies *what* should be done, but not *how* to implement it. So we begin the **refinement process**. We decompose the top into a sequence of smaller tasks—a process sometimes called **divide and conquer**. This results in the following **first refinement**:

> *Initialize variables*
> *Input, sum and count the quiz grades*
> *Calculate and display the class average*

Each refinement represents the *complete* algorithm—only the level of detail varies. In this refinement, the three pseudocode statements happen to correspond to the three execution phases described in the preceding section. The algorithm does not yet provide enough detail for us to write the Python program. So, we continue with the next refinement.

Second Refinement
To proceed to the **second refinement**, we commit to specific variables. The program needs to maintain

- a grade variable in which each successive user input will be stored,
- a running total of the grades,
- a count of how many grades have been processed and
- a variable that contains the calculated average.

The pseudocode statement

> *Initialize variables*

can be refined as follows:

> *Initialize total to zero*
> *Initialize grade counter to zero*

Only the variables *total* and *grade counter* need to be initialized before they're used. We do not initialize the variables for the user input and calculated average. Their values will be replaced each time we input a grade from the user and when we calculate the class average, respectively. We'll create these variables when they're needed.

The next pseudocode statement requires a loop that successively inputs each grade:

> *Input, sum and count the quiz grades*

We do not know how many grades will be entered, so we use sentinel-controlled repetition. The user enters legitimate grades successively. After the last legitimate grade has been entered, the user enters the sentinel value. The program tests for the sentinel value after each grade is input and terminates the loop when the sentinel has been entered. The second refinement of the preceding pseudocode statement is

> *Input the first grade (possibly the sentinel)*
> *While the user has not entered the sentinel*
>   *Add this grade into the running total*
>   *Add one to the grade counter*
>   *Input the next grade (possibly the sentinel)*

The pseudocode statement

> *Calculate and display the class average*

can be refined as follows:

> *If the counter is not equal to zero*
>   *Set the average to the total divided by the grade counter*
>   *Display the average*
> *Else*
>   *Display "No grades were entered"*

3.11 Program Development: Sentinel-Controlled Repetition

Notice that we're testing for the possibility of division by zero. If undetected, this would cause a fatal logic error. In the "Files and Exceptions" chapter, we discuss how to write programs that recognize such exceptions and take appropriate actions.

The following is the class-average problem's complete second refinement:

> *Initialize total to zero*
> *Initialize grade counter to zero*
>
> *Input the first grade (possibly the sentinel)*
> *While the user has not entered the sentinel*
>   *Add this grade into the running total*
>   *Add one to the grade counter*
>   *Input the next grade (possibly the sentinel)*
>
> *If the counter is not equal to zero*
>   *Set the average to the total divided by the counter*
>   *Display the average*
> *Else*
>   *Display "No grades were entered"*

Sometimes more than two refinements are necessary. You stop refining when there is enough detail for you to convert the pseudocode to Python. We include blank lines for readability. Here, they happen to separate the algorithm into the three popular execution phases.

Implementing Sentinel-Controlled Iteration

The following script implements the pseudocode algorithm and shows a sample execution in which the user enters three grades and the sentinel value.

```
 1  # fig03_02.py
 2  """Class average program with sentinel-controlled iteration."""
 3
 4  # initialization phase
 5  total = 0  # sum of grades
 6  grade_counter = 0  # number of grades entered
 7
 8  # processing phase
 9  grade = int(input('Enter grade, -1 to end: '))  # get one grade
10
11  while grade != -1:
12      total += grade
13      grade_counter += 1
14      grade = int(input('Enter grade, -1 to end: '))
15
16  # termination phase
17  if grade_counter != 0:
18      average = total / grade_counter
19      print(f'Class average is {average:.2f}')
20  else:
21      print('No grades were entered')
```

Fig. 3.2 | Class average program with sentinel-controlled iteration. (Part 1 of 2.)

```
Enter grade, -1 to end: 97
Enter grade, -1 to end: 88
Enter grade, -1 to end: 72
Enter grade, -1 to end: -1
Class average is 85.67
```

Fig. 3.2 | Class average program with sentinel-controlled iteration. (Part 2 of 2.)

Program Logic for Sentinel-Controlled Repetition
In sentinel-controlled repetition, the program reads the first value (line 9) before reaching the `while` statement. Line 9 demonstrates why we did not create the variable `grade` until we needed it in the program. If we had initialized it, that value would have been replaced immediately by this assignment.

The value input in line 9 determines whether the program's flow of control should enter the `while`'s suite (lines 12–14). If the condition in line 11 is `False`, the user entered the sentinel value (-1), so the suite does not execute because the user did not enter any grades. If the condition is `True`, the suite executes, adding the `grade` value to the `total` and incrementing the `grade_counter`. Next, line 14 inputs another grade from the user. Then, the `while`'s condition (line 11) is tested again, using the most recent `grade` entered by the user. The value of `grade` is always input immediately before the program tests the `while` condition, so we can determine whether the value just input is the sentinel before processing that value as a grade. When the sentinel value is input, the loop terminates, and the program does not add –1 to the total. In a sentinel-controlled loop that performs user input, any prompts (lines 9 and 14) should remind the user of the sentinel value.

After the loop terminates, the `if…else` statement (lines 17–21) executes. Line 17 determines whether the user entered any grades. If not, the `else` part (lines 20–21) executes and displays the message `'No grades were entered'` and the program terminates.

Formatting the Class Average with Two Decimal Places
This example formatted the class average with two digits to the right of the decimal point. In an f-string, you can optionally follow a replacement-text expression with a colon (:) and a **format specifier** that describes how to format the replacement text. The format specifier `.2f` (line 19) formats the `average` as a floating-point number (`f`) with two digits to the right of the decimal point (`.2`). In this example, the sum of the grades was 257, which, when divided by 3, yields 85.666666666…. Formatting the `average` with `.2f` *rounds* it to the hundredths position, producing the replacement text `85.67`. An average with only one digit to the right of the decimal point would be formatted with a **trailing zero** (e.g., `85.50`). The chapter "Strings: A Deeper Look" discusses many string-formatting features.

Control-Statement Stacking
In this example, notice that control statements are stacked in sequence. The `while` statement (lines 11–14) is followed immediately by an `if…else` statement (lines 17–21).

✓ Self Check

1 *(Fill-In)* Sentinel-controlled repetition is called _____ because the number of repetitions is not known before the loop begins executing.
Answer: indefinite repetition.

2 *(True/False)* Sentinel-controlled repetition uses a counter variable to control the number of times a set of instructions executes.
Answer: False. Sentinel-control repetition terminates repetition when the sentinel value is encountered.

3.12 Program Development: Nested Control Statements

Let's work through another complete problem. Once again, we plan the algorithm using pseudocode and top-down, stepwise refinement and we develop a corresponding Python script. Consider the following requirements statement:

> *A college offers a course that prepares students for the state licensing exam for real-estate brokers. Last year, several of the students who completed this course took the licensing examination. The college wants to know how well its students did on the exam. You have been asked to write a program to summarize the results. You have been given a list of these 10 students. Next to each name is written a 1 if the student passed the exam and a 2 if the student failed.*
>
> *Your program should analyze the results of the exam as follows:*
>
> *1. Input each test result (i.e., a 1 or a 2). Display the message "Enter result" each time the program requests another test result.*
>
> *2. Count the number of test results of each type.*
>
> *3. Display a summary of the test results indicating the number of students who passed and the number of students who failed.*
>
> *4. If more than eight students passed the exam, display "Bonus to instructor."*

After reading the requirements statement carefully, we make the following observations about the problem:

1. The program must process 10 test results. We'll use a for statement and the range function to control repetition.

2. Each test result is a number—either a 1 or a 2. Each time the program reads a test result, the program must determine if the number is a 1 or a 2. We test for a 1 in our algorithm. If the number is not a 1, we assume that it's a 2. (An exercise at the end of the chapter considers the consequences of this assumption.)

3. We'll use two counters—one to count the number of students who passed the exam and one to count the number of students who failed.

4. After the script processes all the results, it must decide if more than eight students passed the exam so that it can bonus the instructor.

Top-Down, Stepwise Refinement
We begin with a pseudocode representation of the top:

> *Analyze exam results and decide whether instructor should receive a bonus*

Once again, the top is a *complete* representation of the program, but several refinements are likely to be needed before the pseudocode can evolve naturally into a Python program.

First Refinement
Our first refinement is

> *Initialize variables*
> *Input the ten exam grades and count passes and failures*
> *Summarize the exam results and decide whether instructor should receive a bonus*

Here, too, even though we have a *complete* representation of the entire program, further refinement is necessary. Note again that this first refinement happens to correspond to the three-execution-phases model.

Second Refinement
We now commit to specific variables. We need counters to record the passes and failures, and a variable to store the user input. The pseudocode statement

> *Initialize variables*

can be refined as follows:

> *Initialize passes to zero*
> *Initialize failures to zero*

Only the counters for the number of passes and number of failures need to be initialized.
The pseudocode statement

> *Input the ten exam grades and count passes and failures*

requires a loop that successively inputs the result of each exam. Here it's known in advance that there are ten exam results, so the `for` statement and the `range` function are appropriate. Inside the loop (that is, *nested* within the loop), an `if...else` statement determines whether each exam result is a pass or a failure and increments the appropriate counter. The refinement of the preceding pseudocode statement is

> *For each of the ten students*
>   *Input the next exam result*
>
> *If the student passed*
>   *Add one to passes*
> *Else*
>   *Add one to failures*

The blank line before the *If...Else* improves readability.
The pseudocode statement

> *Summarize the exam results and decide whether instructor should receive a bonus*

may be refined as follows:

> *Display the number of passes*
> *Display the number of failures*
>
> *If more than eight students passed*
>   *Display "Bonus to instructor"*

Complete Pseudocode Algorithm

The pseudocode is now sufficiently refined for conversion to Python—the complete second refinement is shown below:

> *Initialize passes to zero*
> *Initialize failures to zero*
>
> *For each of the ten students*
> > *Input the next exam result*
> >
> > *If the student passed*
> > > *Add one to passes*
> >
> > *Else*
> > > *Add one to failures*
>
> *Display the number of passes*
> *Display the number of failures*
>
> *If more than eight students passed*
> > *Display "Bonus to instructor"*

Implementing the Algorithm

The following script implements the algorithm and is followed by two sample executions. Once again, notice that the Python code closely resembles the pseudocode. Lines 9–16 loop 10 times, inputting and processing one exam result each time. The if...else statement (lines 13–16) that processes each result is *nested* in the for statement—that is, it's part of the for statement's suite. If the result is 1, we add 1 to passes; otherwise, we assume the result is 2 and add 1 to failures. After inputting 10 values, the loop terminates and lines 19 and 20 display passes and failures. Lines 22–23 determine whether more than eight students passed the exam and, if so, display 'Bonus to instructor'.

```
1   # fig03_03.py
2   """Using nested control statements to analyze examination results."""
3
4   # initialize variables
5   passes = 0   # number of passes
6   failures = 0   # number of failures
7
8   # process 10 students
9   for student in range(10):
10      # get one exam result
11      result = int(input('Enter result (1=pass, 2=fail): '))
12
13      if result == 1:
14          passes = passes + 1
15      else:
16          failures = failures + 1
17
```

Fig. 3.3 | Analysis of examination results. (Part 1 of 2.)

```
18    # termination phase
19    print('Passed:', passes)
20    print('Failed:', failures)
21
22    if passes > 8:
23        print('Bonus to instructor')
```

```
Enter result (1=pass, 2=fail): 1
Enter result (1=pass, 2=fail): 2
Enter result (1=pass, 2=fail): 2
Enter result (1=pass, 2=fail): 1
Enter result (1=pass, 2=fail): 1
Enter result (1=pass, 2=fail): 1
Enter result (1=pass, 2=fail): 2
Enter result (1=pass, 2=fail): 1
Enter result (1=pass, 2=fail): 1
Enter result (1=pass, 2=fail): 2
Passed: 6
Failed: 4
```

```
Enter result (1=pass, 2=fail): 1
Enter result (1=pass, 2=fail): 1
Enter result (1=pass, 2=fail): 1
Enter result (1=pass, 2=fail): 1
Enter result (1=pass, 2=fail): 2
Enter result (1=pass, 2=fail): 1
Enter result (1=pass, 2=fail): 1
Enter result (1=pass, 2=fail): 1
Enter result (1=pass, 2=fail): 1
Enter result (1=pass, 2=fail): 1
Passed: 9
Failed: 1
Bonus to instructor
```

Fig. 3.3 | Analysis of examination results. (Part 2 of 2.)

✓ Self Check

1 *(IPython Session)* Use a for statement to input two integers. Use a nested if...else statement to display whether each value is even or odd. Enter 10 and 7 to test your code.
Answer:

```
In [1]: for count in range(2):
   ...:     value = int(input('Enter an integer: '))
   ...:     if value % 2 == 0:
   ...:         print(f'{value} is even')
   ...:     else:
   ...:         print(f'{value} is odd')
   ...:
Enter an integer: 10
10 is even
Enter an integer: 7
7 is odd
```

3.13 Built-In Function range: A Deeper Look

Function range also has two- and three-argument versions. As you've seen, range's one-argument version produces a sequence of consecutive integers from 0 up to, but not including, the argument's value. Function range's two-argument version produces a sequence of consecutive integers from its first argument's value up to, but not including, the second argument's value, as in:

```
In [1]: for number in range(5, 10):
   ...:     print(number, end=' ')
   ...:
5 6 7 8 9
```

Function range's three-argument version produces a sequence of integers from its first argument's value up to, but not including, the second argument's value, *incrementing* by the third argument's value, which is known as the **step**:

```
In [2]: for number in range(0, 10, 2):
   ...:     print(number, end=' ')
   ...:
0 2 4 6 8
```

If the third argument is negative, the sequence progresses from the first argument's value *down* to, but not including the second argument's value, *decrementing* by the third argument's value, as in:

```
In [3]: for number in range(10, 0, -2):
   ...:     print(number, end=' ')
   ...:
10 8 6 4 2
```

✓ Self Check

1 *(True/False)* Function call range(1, 10) generates the sequence 1 through 10.
Answer: False. Function call range(1, 10) generates the sequence 1 through 9.

2 *(IPython Session)* What happens if you try to print the items in range(10, 0, 2)?
Answer: Nothing displays because the step is not negative (this is not a fatal error):

```
In [1]: for number in range(10, 0, 2):
   ...:     print(number, end=' ')
   ...:

In [2]:
```

3 *(IPython Session)* Use a for statement, range and print to display on one line the sequence of values 99 88 77 66 55 44 33 22 11 0, each separated by one space.
Answer:

```
In [3]: for number in range(99, -1, -11):
   ...:     print(number, end=' ')
   ...:
99 88 77 66 55 44 33 22 11 0
```

4 *(IPython Session)* Use for and range to sum the even integers from 2 through 100, then display the sum.

Answer:

```
In [4]: total = 0

In [5]: for number in range(2, 101, 2):
   ...:     total += number
   ...:

In [6]: total
Out[6]: 2550
```

3.14 Using Type Decimal for Monetary Amounts

In this section, we introduce Decimal capabilities for precise monetary calculations. If you enter banking or other fields that require the accuracy provided by type Decimal, you should investigate Decimal's capabilities in depth.

For most scientific and other mathematical applications that use numbers with decimal points, Python's built-in floating-point numbers work well. For example, when we speak of a "normal" body temperature of 98.6, we do not need to be precise to a large number of digits. When we view the temperature on a thermometer and read it as 98.6, the actual value may be 98.5999473210643. The point here is that calling this number 98.6 is adequate for most body-temperature applications.

Floating-point values are stored in binary format (we introduced binary in the first chapter and discuss it in depth in the online "Number Systems" appendix). Some floating-point values are represented only approximately when they're converted to binary. For example, consider the variable amount with the dollars-and-cents value 112.31. If you display amount, it appears to have the exact value you assigned to it:

```
In [1]: amount = 112.31

In [2]: print(amount)
112.31
```

However, if you print amount with 20 digits of precision to the right of the decimal point, you can see that the actual floating-point value in memory is not exactly 112.31—it's only an approximation:

```
In [3]: print(f'{amount:.20f}')
112.31000000000000227374
```

Many applications require *precise* representation of numbers with decimal points. Institutions like banks that deal with millions or even billions of transactions per day have to tie out their transactions "to the penny." Floating-point numbers can represent some but not all monetary amounts with to-the-penny precision.

The **Python Standard Library**[2] provides many predefined capabilities you can use in your Python code to avoid "reinventing the wheel." For monetary calculations and other applications that require precise representation and manipulation of numbers with decimal points, the Python Standard Library provides type **Decimal**, which uses a special coding scheme to solve the problem of to-the-penny precision. That scheme requires additional memory to hold the numbers and additional processing time to perform calcu-

---

2. https://docs.python.org/3.7/library/index.html.

lations but provides the to-the-penny precision required for monetary calculations. Banks also have to deal with other issues such as using a fair rounding algorithm when they're calculating daily interest on accounts. Type Decimal offers such capabilities.[3]

Importing Type Decimal from the decimal Module
We've used several built-in types—**int** (for integers, like 10), **float** (for floating-point numbers, like 7.5) and **str** (for strings like 'Python'). The Decimal type is not built into Python. Rather, it's part of the Python Standard Library, which is divided into **modules**—groups of related capabilities. The **decimal** module defines type Decimal and its capabilities.

To use capabilities from a module, you must first **import** the entire module, as in

```
import decimal
```

and refer to the Decimal type as decimal.Decimal, or you must indicate a specific capability to import using **from...import**, as we do here:

```
In [4]: from decimal import Decimal
```

This imports only the type Decimal from the decimal module so that you can use it in your code. We'll discuss other import forms beginning in the next chapter.

Creating Decimals
You typically create a Decimal from a string:

```
In [5]: principal = Decimal('1000.00')

In [6]: principal
Out[6]: Decimal('1000.00')

In [7]: rate = Decimal('0.05')

In [8]: rate
Out[8]: Decimal('0.05')
```

We'll soon use the variables principal and rate in a compound-interest calculation.

Decimal Arithmetic
Decimals support the standard arithmetic operators +, -, *, /, //, ** and %, as well as the corresponding augmented assignments:

```
In [9]: x = Decimal('10.5')

In [10]: y = Decimal('2')

In [11]: x + y
Out[11]: Decimal('12.5')

In [12]: x // y
Out[12]: Decimal('5')

In [13]: x += y

In [14]: x
Out[14]: Decimal('12.5')
```

---

3. For more decimal module features, visit https://docs.python.org/3.7/library/decimal.html.

You may perform arithmetic between Decimals and integers, but not between Decimals and floating-point numbers.

Compound-Interest Problem Requirements Statement
Let's compute compound interest using the Decimal type for *precise* monetary calculations. Consider the following requirements statement:

> A person invests $1000 in a savings account yielding 5% interest. Assuming that the person leaves all interest on deposit in the account, calculate and display the amount of money in the account at the end of each year for 10 years. Use the following formula for determining these amounts:
>
> $a = p(1 + r)^n$
>
> where
>
> p is the original amount invested (i.e., the principal),
> r is the annual interest rate,
> n is the number of years and
> a is the amount on deposit at the end of the nth year.

Calculating Compound Interest
To solve this problem, let's use variables `principal` and `rate` that we defined in snippets [5] and [7], and a `for` statement that performs the interest calculation for each of the 10 years the money remains on deposit. For each `year`, the loop displays a formatted string containing the year number and the amount on deposit at the end of that year:

```
In [15]: for year in range(1, 11):
    ...:     amount = principal * (1 + rate) ** year
    ...:     print(f'{year:>2}{amount:>10.2f}')
    ...:
 1   1050.00
 2   1102.50
 3   1157.62
 4   1215.51
 5   1276.28
 6   1340.10
 7   1407.10
 8   1477.46
 9   1551.33
10   1628.89
```

The algebraic expression $(1 + r)^n$ from the requirements statement is written as

```
(1 + rate) ** year
```

where variable `rate` represents r and variable `year` represents n.

Formatting the Year and Amount on Deposit
The statement

```
print(f'{year:>2}{amount:>10.2f}')
```

uses an f-string with two placeholders to format the loop's output.
The placeholder

```
{year:>2}
```

uses the format specifier >2 to indicate that year's value should be **right aligned** (>) in a field of width 2—the **field width** specifies the number of character positions to use when displaying the value. For the single-digit year values 1–9, the format specifier >2 displays a space character followed by the value, thus right aligning the years in the first column. The following diagram shows the numbers 1 and 10 each formatted in a field width of 2:

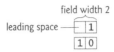

You can **left align** values with <.

The format specifier 10.2f in the placeholder

    {amount:>10.2f}

formats amount as a floating-point number (f) right aligned (>) in a field width of 10 with a decimal point and two digits to the right of the decimal point (.2). Formatting the amounts this way *aligns their decimal points vertically*, as is typical with monetary amounts. In the 10 character positions, the three rightmost characters are the number's decimal point followed by the two digits to its right. The remaining seven character positions are the leading spaces and the digits to the decimal point's left. In this example, all the dollar amounts have four digits to the left of the decimal point, so each number is formatted with three leading spaces. The following diagram shows the formatting for the value 1050.00:

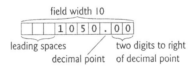

✓ Self Check

1 *(Fill-In)* A field width specifies the _____ to use when displaying a value.
Answer: number of character positions.

2 *(IPython Session)* Assume that the tax on a restaurant bill is 6.25% and that the bill amount is $37.45. Use type Decimal to calculate the bill total, then print the result with two digits to the right of the decimal point.
Answer:

    In [1]: from decimal import Decimal

    In [2]: print(f"{Decimal('37.45') * Decimal('1.0625'):.2f}")
    39.79

3.15 break and continue Statements

The break and continue statements alter a loop's flow of control. Executing a **break** statement in a while or for immediately exits that statement. In the following code, range produces the integer sequence 0–99, but the loop terminates when number is 10:

```
In [1]: for number in range(100):
   ...:     if number == 10:
   ...:         break
   ...:     print(number, end=' ')
   ...:
0 1 2 3 4 5 6 7 8 9
```

In a script, execution would continue with the next statement after the for loop. The while and for statements each have an optional else clause that executes only if the loop terminates normally—that is, not as a result of a break. We explore this in the exercises.

Executing a **continue** statement in a while or for loop skips the remainder of the loop's suite. In a while, the condition is then tested to determine whether the loop should continue executing. In a for, the loop processes the next item in the sequence (if any):

```
In [2]: for number in range(10):
   ...:     if number == 5:
   ...:         continue
   ...:     print(number, end=' ')
   ...:
0 1 2 3 4 6 7 8 9
```

3.16 Boolean Operators and, or and not

The conditional operators >, <, >=, <=, == and != can be used to form simple conditions such as grade >= 60. To form more complex conditions that combine simple conditions, use the and, or and not Boolean operators.

Boolean Operator and

To ensure that two conditions are *both* True before executing a control statement's suite, use the **Boolean and operator** to combine the conditions. The following code defines two variables, then tests a condition that's True if and only if *both* simple conditions are True—if either (or both) of the simple conditions is False, the entire and expression is False:

```
In [1]: gender = 'Female'

In [2]: age = 70

In [3]: if gender == 'Female' and age >= 65:
   ...:     print('Senior female')
   ...:
Senior female
```

The if statement has two simple conditions:

- gender == 'Female' determines whether a person is a female and
- age >= 65 determines whether that person is a senior citizen.

The simple condition to the left of the and operator evaluates first because == has higher precedence than and. If necessary, the simple condition to the right of and evaluates next, because >= has higher precedence than and. (We'll discuss shortly why the right side of an and operator evaluates *only* if the left side is True.) The entire if statement condition is True if and only if *both* of the simple conditions are True. The combined condition can be made clearer by adding redundant (unnecessary) parentheses

```
(gender == 'Female') and (age >= 65)
```

The table below summarizes the and operator by showing all four possible combinations of False and True values for *expression1* and *expression2*—such tables are called *truth tables*:

expression1	expression2	expression1 and expression2
False	False	False
False	True	False
True	False	False
True	True	True

Boolean Operator or

Use the **Boolean or operator** to test whether one *or* both of two conditions are True. The following code tests a condition that's True if either *or* both simple conditions are True—the entire condition is False only if *both* simple conditions are False:

```
In [4]: semester_average = 83

In [5]: final_exam = 95

In [6]: if semester_average >= 90 or final_exam >= 90:
   ...:     print('Student gets an A')
   ...:
Student gets an A
```

Snippet [6] also contains two simple conditions:

- `semester_average >= 90` determines whether a student's average was an A (90 or above) during the semester, and
- `final_exam >= 90` determines whether a student's final-exam grade was an A.

The truth table below summarizes the Boolean or operator. Operator and has higher precedence than or.

expression1	expression2	expression1 or expression2
False	False	False
False	True	True
True	False	True
True	True	True

Improving Performance with Short-Circuit Evaluation

Python stops evaluating an and expression as soon as it knows whether the entire condition is False. Similarly, Python stops evaluating an or expression as soon as it knows whether the entire condition is True. This is called *short-circuit evaluation*. So the condition

```
gender == 'Female' and age >= 65
```

stops evaluating immediately if gender is not equal to 'Female' because the entire expression must be False. If gender is equal to 'Female', execution continues, because the entire expression will be True if the age is greater than or equal to 65.

Similarly, the condition

```
semester_average >= 90 or final_exam >= 90
```

stops evaluating immediately if `semester_average` is greater than or equal to 90 because the entire expression must be `True`. If `semester_average` is less than 90, execution continues, because the expression could still be `True` if the `final_exam` is greater than or equal to 90.

In operator expressions that use `and`, make the condition that's more likely to be `False` the leftmost condition. In `or` operator expressions, make the condition that's more likely to be `True` the leftmost condition. These can reduce a program's execution time.

Boolean Operator not

The **Boolean not operator** "reverses" the meaning of a condition—`True` becomes `False` and `False` becomes `True`. This is a **unary operator**—it has only *one* operand. You place the not operator before a condition to choose a path of execution if the original condition (without the not operator) is `False`, such as in the following code:

```
In [7]: grade = 87

In [8]: if not grade == -1:
   ...:     print('The next grade is', grade)
   ...:
The next grade is 87
```

Often, you can avoid using `not` by expressing the condition in a more "natural" or convenient manner. For example, the preceding `if` statement can also be written as follows:

```
In [9]: if grade != -1:
   ...:     print('The next grade is', grade)
   ...:
The next grade is 87
```

The truth table below summarizes the `not` operator.

expression	not expression
False	True
True	False

The following table shows the precedence and grouping of the operators introduced so far, from top to bottom, in decreasing order of precedence.

Operators	Grouping
()	left to right
**	right to left
* / // %	left to right
+ -	left to right
< <= > >= == !=	left to right
not	left to right
and	left to right
or	left to right

Self Check

1. *(IPython Session)* Assume that i = 1, j = 2, k = 3 and m = 2. What does each of the following conditions display?
    a) (i >= 1) and (j < 4)
    b) (m <= 99) and (k < m)
    c) (j >= i) or (k == m)
    d) (k + m < j) or (3 - j >= k)
    e) not (k > m)

Answer:

```
In [1]: i = 1

In [2]: j = 2

In [3]: k = 3

In [4]: m = 2

In [5]: (i >= 1) and (j < 4)
Out[5]: True

In [6]: (m <= 99) and (k < m)
Out[6]: False

In [7]: (j >= i) or (k == m)
Out[7]: True

In [8]: (k + m < j) or (3 - j >= k)
Out[8]: False

In [9]: not (k > m)
Out[9]: False
```

3.17 Intro to Data Science: Measures of Central Tendency—Mean, Median and Mode

Here we continue our discussion of using statistics to analyze data with several additional descriptive statistics, including:

- **mean**—the average value in a set of values.
- **median**—the middle value when all the values are arranged in sorted order.
- **mode**—the most frequently occurring value.

These are **measures of central tendency**—each is a way of producing a single value that represents a "central" value in a set of values, i.e., a value which is in some sense typical of the others.

Let's calculate the mean, median and mode on a list of integers. The following session creates a list called grades, then uses the built-in **sum** and **len** functions to calculate the mean "by hand"—sum calculates the total of the grades (397) and len returns the number of grades (5):

```
In [1]: grades = [85, 93, 45, 89, 85]

In [2]: sum(grades) / len(grades)
Out[2]: 79.4
```

The previous chapter mentioned the descriptive statistics count and sum—implemented in Python as the built-in functions len and sum. Like functions min and max (introduced in the preceding chapter), sum and len are both examples of functional-style programming *reductions*—they reduce a collection of values to a single value—the sum of those values and the number of values, respectively. In Fig. 3.1's class-average example, we could have deleted lines 10–15 and replaced average in line 16 with snippet [2]'s calculation.

The Python Standard Library's **statistics module** provides functions for calculating the mean, median and mode—these, too, are reductions. To use these capabilities, first import the statistics module:

```
In [3]: import statistics
```

Then, you can access the module's functions with "statistics." followed by the name of the function to call. The following calculates the grades list's mean, median and mode, using the statistics module's **mean, median** and **mode** functions:

```
In [4]: statistics.mean(grades)
Out[4]: 79.4

In [5]: statistics.median(grades)
Out[5]: 85

In [6]: statistics.mode(grades)
Out[6]: 85
```

Each function's argument must be an *iterable*—in this case, the list grades. To confirm that the median and mode are correct, you can use the built-in **sorted function** to get a copy of grades with its values arranged in increasing order:

```
In [7]: sorted(grades)
Out[7]: [45, 85, 85, 89, 93]
```

The grades list has an odd number of values (5), so median returns the middle value (85). If the list's number of values is even, median returns the *average* of the *two* middle values. Studying the sorted values, you can see that 85 is the mode because it occurs most frequently (twice). The mode function causes a StatisticsError for lists like

```
[85, 93, 45, 89, 85, 93]
```

in which there are two or more "most frequent" values. Such a set of values is said to be **bimodal**. Here, both 85 and 93 occur twice. We'll say more about mean, median and mode in the Intro to Data Science exercises at the end of the chapter.

✓ Self Check

1 *(Fill-In)* The _____ statistic indicates the average value in a set of values.
Answer: mean.

2 *(Fill-In)* The _____ statistic indicates the most frequently occurring value in a set of values.
Answer: mode.

3 *(Fill-In)* The _____ statistic indicates the middle value in a set of values.
Answer: median.

4 *(IPython Session)* For the values 47, 95, 88, 73, 88 and 84, use the `statistics` module to calculate the mean, median and mode.
Answer:

```
In [1]: import statistics

In [2]: values = [47, 95, 88, 73, 88, 84]

In [3]: statistics.mean(values)
Out[3]: 79.16666666666667

In [4]: statistics.median(values)
Out[4]: 86.0

In [5]: statistics.mode(values)
Out[5]: 88
```

3.18 Wrap-Up

In this chapter, we discussed Python's control statements, including `if`, `if…else`, `if…elif…else`, `while`, `for`, `break` and `continue`. We used pseudocode and top-down, stepwise refinement to develop several algorithms. You saw that many simple algorithms often have three execution phases—initialization, processing and termination.

You saw that the `for` statement performs sequence-controlled iteration—it processes each item in an iterable, such as a range of integers, a string or a list. You used the built-in function `range` to generate sequences of integers from 0 up to, but not including, its argument, and to determine how many times a `for` statement iterates. You used sentinel-controlled repetition with the `while` statement to create a loop that continues executing until a sentinel value is encountered. You used built-in function `range`'s two-argument version to generate sequences of integers from the first argument's value up to, but not including, the second argument's value. You also used the three-argument version in which the third argument indicated the step between integers in a range.

We introduced the `Decimal` type for precise monetary calculations and used it to calculate compound interest. You used f-strings and various format specifiers to create formatted output. We introduced the `break` and `continue` statements for altering the flow of control in loops. We discussed the Boolean operators and, or and not for creating conditions that combine simple conditions.

Finally, we continued our discussion of descriptive statistics by introducing measures of central tendency—mean, median and mode—and calculating them with functions from the Python Standard Library's `statistics` module.

In the next chapter, you'll create custom functions and use existing functions from Python's `math` and `random` modules. We show several predefined functional-programming reductions. You'll learn more of Python's functional-programming capabilities.

Exercises

Unless specified otherwise, use IPython sessions for each exercise.

3.1 *(Validating User Input)* Modify the script of Fig. 3.3 to validate its inputs. For any input, if the value entered is other than 1 or 2, keep looping until the user enters a correct value. Use one counter to keep track of the number of passes, then calculate the number of failures after all the user's inputs have been received.

3.2 *(What's Wrong with This Code?)* What is wrong with the following code?

```
a = b = 7
print('a =', a, '\nb =', b)
```

First, answer the question, then check your work in an IPython session.

3.3 *(What Does This Code Do?)* What does the following program print?

```
for row in range(10):
    for column in range(10):
        print('<' if row % 2 == 1 else '>', end='')
    print()
```

3.4 *(Fill in the Missing Code)* In the code below

```
for ***:
    for ***:
        print('@')
    print()
```

replace the *** so that when you execute the code, it displays two rows, each containing seven @ symbols, as in:

```
@@@@@@@
@@@@@@@
```

3.5 *(if...else Statements)* Reimplement the script of Fig. 2.1 using three if...else statements rather than six if statements. [*Hint:* For example, think of == and != as "opposite" tests.]

3.6 *(Turing Test)* The great British mathematician Alan Turing proposed a simple test to determine whether machines could exhibit intelligent behavior. A user sits at a computer and does the same text chat with a human sitting at a computer and a computer operating by itself. The user doesn't know if the responses are coming back from the human or the independent computer. If the user can't distinguish which responses are coming from the human and which are coming from the computer, then it's reasonable to say that the computer is exhibiting intelligence.

Create a script that plays the part of the independent computer, giving its user a simple medical diagnosis. The script should prompt the user with 'What is your problem?' When the user answers and presses *Enter*, the script should simply ignore the user's input, then prompt the user again with 'Have you had this problem before (yes or no)?' If the user enters 'yes', print 'Well, you have it again.' If the user answers 'no', print 'Well, you have it now.'

Would this conversation convince the user that the entity at the other end exhibited intelligent behavior? Why or why not?

3.7 *(Table of Squares and Cubes)* In Exercise 2.8, you wrote a script to calculate the squares and cubes of the numbers from 0 through 5, then printed the resulting values in table format. Reimplement your script using a for loop and the f-string capabilities you learned in this chapter to produce the following table with the numbers right aligned in each column.

```
number  square  cube
     0       0     0
     1       1     1
     2       4     8
     3       9    27
     4      16    64
     5      25   125
```

3.8 *(Arithmetic, Smallest and Largest)* In Exercise 2.10, you wrote a script that input three integers, then displayed the sum, average, product, smallest and largest of those values. Reimplement your script with a loop that inputs four integers.

3.9 *(Separating the Digits in an Integer)* In Exercise 2.11, you wrote a script that separated a five-digit integer into its individual digits and displayed them. Reimplement your script to use a loop that in each iteration "picks off" one digit (left to right) using the // and % operators, then displays that digit.

3.10 *(7% Investment Return)* Reimplement Exercise 2.12 to use a loop that calculates and displays the amount of money you'll have each year at the ends of years 1 through 30.

3.11 *(Miles Per Gallon)* Drivers are concerned with the mileage obtained by their automobiles. One driver has kept track of several tankfuls of gasoline by recording miles driven and gallons used for each tankful. Develop a sentinel-controlled-repetition script that prompts the user to input the miles driven and gallons used for each tankful. The script should calculate and display the miles per gallon obtained for each tankful. After processing all input information, the script should calculate and display the combined miles per gallon obtained for all tankfuls (that is, total miles driven divided by total gallons used).

```
Enter the gallons used (-1 to end): 12.8
Enter the miles driven: 287
The miles/gallon for this tank was 22.421875
Enter the gallons used (-1 to end): 10.3
Enter the miles driven: 200
The miles/gallon for this tank was 19.417475
Enter the gallons used (-1 to end): 5
Enter the miles driven: 120
The miles/gallon for this tank was 24.000000
Enter the gallons used (-1 to end): -1
The overall average miles/gallon was 21.601423
```

3.12 *(Palindromes)* A palindrome is a number, word or text phrase that reads the same backwards or forwards. For example, each of the following five-digit integers is a palindrome: 12321, 55555, 45554 and 11611. Write a script that reads in a five-digit integer and determines whether it's a palindrome. [*Hint*: Use the // and % operators to separate the number into its digits.]

3.13 *(Factorials)* Factorial calculations are common in probability. The factorial of a nonnegative integer n is written $n!$ (pronounced "n factorial") and is defined as follows:

$$n! = n \cdot (n-1) \cdot (n-2) \cdot \ldots \cdot 1$$

for values of n greater than or equal to 1, with 0! defined to be 1. So,

$$5! = 5 \cdot 4 \cdot 3 \cdot 2 \cdot 1$$

which is 120. Factorials increase in size very rapidly. Write a script that inputs a nonnegative integer and computes and displays its factorial. Try your script on the integers 10, 20,

30 and even larger values. Did you find any integer input for which Python could not produce an integer factorial value?

3.14 *(Challenge: Approximating the Mathematical Constant π)* Write a script that computes the value of π from the following infinite series. Print a table that shows the value of π approximated by one term of this series, by two terms, by three terms, and so on. How many terms of this series do you have to use before you first get 3.14? 3.141? 3.1415? 3.14159?

$$\pi = 4 - \frac{4}{3} + \frac{4}{5} - \frac{4}{7} + \frac{4}{9} - \frac{4}{11} + \cdots$$

3.15 *(Challenge: Approximating the Mathematical Constant e)* Write a script that estimates the value of the mathematical constant *e* by using the formula below. Your script can stop after summing 10 terms.

$$e = 1 + \frac{1}{1!} + \frac{1}{2!} + \frac{1}{3!} + \cdots$$

3.16 *(Nested Control Statements)* Use a loop to find the *two* largest values of 10 numbers entered.

3.17 *(Nested Loops)* Write a script that displays the following triangle patterns separately, one below the other. Separate each pattern from the next by one blank line. Use for loops to generate the patterns. Display all asterisks (*) with a single statement of the form

```
print('*', end=' ')
```

which causes the asterisks to display side by side. [*Hint*: For the last two patterns, begin each line with zero or more space characters.]

```
(a)              (b)                  (c)                  (d)
*                * * * * * * * * * *  * * * * * * * * * *            *
* *              * * * * * * * * *      * * * * * * * * *          * *
* * *            * * * * * * * *          * * * * * * *          * * *
* * * *          * * * * * * *              * * * * * *        * * * *
* * * * *        * * * * * *                  * * * * *      * * * * *
* * * * * *      * * * * *                      * * * *    * * * * * *
* * * * * * *    * * * *                          * * *  * * * * * * *
* * * * * * * *  * * *                              * *  * * * * * * * *
* * * * * * * * **                                    *  * * * * * * * * *
* * * * * * * * * *                                      * * * * * * * * * *
```

3.18 *(Challenge: Nested Looping)* Modify your script from Exercise 3.17 to display all four patterns side-by-side (as shown above) by making clever use of nested for loops. Separate each triangle from the next by three horizontal spaces. [*Hint*: One for loop should control the row number. Its nested for loops should calculate from the row number the appropriate number of asterisks and spaces for each of the four patterns.]

3.19 *(Brute-Force Computing: Pythagorean Triples)* A right triangle can have sides that are all integers. The set of three integer values for the sides of a right triangle is called a Pythagorean triple. These three sides must satisfy the relationship that the sum of the squares of two of the sides is equal to the square of the hypotenuse. Find all Pythagorean triples for side1, side2 and hypotenuse (such as 3, 4 and 5) all no larger than 20. Use a triple-nested for-loop that tries all possibilities. This is an example of "brute-force" computing. You'll learn in more advanced computer science courses that there are many interesting problems for which there is no known algorithmic approach other than sheer brute force.

3.20 *(Binary-to-Decimal Conversion)* Input an integer containing 0s and 1s (i.e., a "binary" integer) and display its decimal equivalent. The online appendix, "Number Systems," discusses the binary number system. [*Hint*: Use the modulus and division operators to pick off the "binary" number's digits one at a time from right to left. Just as in the decimal number system, where the rightmost digit has the positional value 1 and the next digit to the left has the positional value 10, then 100, then 1000, etc., in the binary number system, the rightmost digit has the positional value 1, the next digit to the left has the positional value 2, then 4, then 8, etc. Thus, the decimal number 234 can be interpreted as 2 * 100 + 3 * 10 + 4 * 1. The decimal equivalent of binary 1101 is 1 * 8 + 1 * 4 + 0 * 2 + 1 * 1.]

3.21 *(Calculate Change Using Fewest Number of Coins)* Write a script that inputs a purchase price of a dollar or less for an item. Assume the purchaser pays with a dollar bill. Determine the amount of change the cashier should give back to the purchaser. Display the change using the *fewest* number of pennies, nickels, dimes and quarters. For example, if the purchaser is due 73 cents in change, the script would output:

```
Your change is:
2 quarters
2 dimes
3 pennies
```

3.22 *(Optional else Clause of a Loop)* The while and for statements each have an optional else clause. In a while statement, the else clause executes when the condition becomes False. In a for statement, the else clause executes when there are no more items to process. If you break out of a while or for that has an else, the else part does *not* execute. Execute the following code to see that the else clause executes only if the break statement does not:

```
for i in range(2):
    value = int(input('Enter an integer (-1 to break): '))
    print('You entered:', value)

    if value == -1:
        break
else:
    print('The loop terminated without executing the break')
```

For more information on loop else clauses, see

https://docs.python.org/3/tutorial/controlflow.html#break-and-continue-statements-and-else-clauses-on-loops

3.23 *(Validating Indentation)* The file validate_indents.py in this chapter's ch03 examples folder contains the following code with incorrect indentation:

```
grade = 93

if grade >= 90:
   print('A')
 print('Great Job!')
  print('Take a break from studying')
```

The Python Standard Library includes a code indentation validator module named **tabnanny**, which you can run as a script to check your code for proper indentation—this is one of many static code analysis tools. Execute the following command in the ch03 folder to see the results of analyzing `validate_indents.py`:

```
python -m tabnanny validate_indents.py
```

Suppose you accidentally aligned the second `print` statement under the `i` in the `if` keyword. What kind of error would that be? Would you expect `tabnanny` to flag that as an error?

3.24 *(Project: Using the **prospector** Static Code Analysis Tool)* The prospector tool runs several popular static code analysis tools to check your Python code for common errors and to help you improve your code. Check that you've installed `prospector` (see the Before You Begin section that follows the Preface). Run prospector on each of the scripts in this chapter. To do so, open the folder containing the scripts in a Terminal (macOS/Linux), Command Prompt (Windows) or shell (Linux), then run the following command from that folder:

```
prospector --strictness veryhigh --doc-warnings
```

Study the output to see the kinds of issues prospector locates in Python code. In general, run `prospector` on all new code you create.

3.25 *(Project: Using **prospector** to Analyze Open-Source Code on GitHub)* Locate a Python open-source project on GitHub, download its source code and extract it into a folder on your system. Open that folder in a Terminal (macOS/Linux), Command Prompt (Windows) or shell (Linux), then run the following command from that folder:

```
prospector --strictness veryhigh --doc-warnings
```

Study the output to see more of the kinds of issues prospector locates in Python code.

3.26 *(Research: Anscombe's Quartet)* In this book's data science case studies, we'll emphasize the importance of "getting to know your data." The *basic descriptive statistics* that you've seen in this chapter's and the previous chapter's Intro to Data Science sections certainly help you know more about your data. One caution, though, is that different datasets can have identical or nearly identical descriptive statistics and yet the data can be significantly different. For an example of this phenomenon, research *Anscombe's Quartet*. You should find four datasets and the associated visualizations. It's the visualizations that convince you the datasets are quite different. In an exercise in a later chapter, you'll create these visualizations.

3.27 *(World Population Growth)* World population has grown considerably over the centuries. Continued growth could eventually challenge the limits of breathable air, drinkable water, arable land and other limited resources. There's evidence that growth has been slowing in recent years and that world population could peak some time this century, then start to decline.

For this exercise, research world population growth issues. This is a controversial topic, so be sure to investigate various viewpoints. Get estimates for the current world population and its growth rate. Write a script that calculates world population growth each year for the next 100 years, *using the simplifying assumption that the current growth rate will stay constant*. Print the results in a table. The first column should display the year from 1 to 100. The second column should display the anticipated world population at the end of that year. The third column should display the numerical increase in the world population that would occur that year. Using your results, determine the years in which the population would be double and eventually quadruple what it is today.

3.28 *(Intro to Data Science: Mean, Median and Mode)* Calculate the mean, median and mode of the values 9, 11, 22, 34, 17, 22, 34, 22 and 40. Suppose the values included another 34. What problem might occur?

3.29 *(Intro to Data Science: Problem with the Median)* For an odd number of values, to get the median you simply arrange them in order and take the middle value. For an even number, you average the two middle values. What problem occurs if those two values are different?

3.30 *(Intro to Data Science: Outliers)* In statistics, outliers are values out of the ordinary and possibly way out of the ordinary. Sometimes, outliers are simply bad data. In the data science case studies, we'll see that outliers can distort results. Which of the three measures of central tendency we discussed—mean, median and mode—is most affected by outliers? Why? Which of these measures are not affected or least affected? Why?

3.31 *(Intro to Data Science: Categorical Data)* Mean, median and mode work well with numerical values. You can use them in calculations and arrange them in meaningful order. Categorical values are descriptive names like Boxer, Poodle, Collie, Beagle, Bulldog and Chihuahua. Normally, you don't use these in calculations nor associate an order with them. Which if any of the descriptive statistics are appropriate for categorical data?

Functions

4

Objectives

In this chapter, you'll
- Create custom functions.
- Import and use Python Standard Library modules, such as `random` and `math`, to reuse code and avoid "reinventing the wheel."
- Pass data between functions.
- Generate a range of random numbers.
- Learn simulation techniques using random-number generation.
- Pack values into a tuple and unpack values from a tuple.
- Return multiple values from a function via a tuple.
- Understand how an identifier's scope determines where in your program you can use it.
- Create functions with default parameter values.
- Call functions with keyword arguments.
- Create functions that can receive any number of arguments.
- Use methods of an object.

Outline

- 4.1 Introduction
- 4.2 Defining Functions
- 4.3 Functions with Multiple Parameters
- 4.4 Random-Number Generation
- 4.5 Case Study: A Game of Chance
- 4.6 Python Standard Library
- 4.7 `math` Module Functions
- 4.8 Using IPython Tab Completion for Discovery
- 4.9 Default Parameter Values
- 4.10 Keyword Arguments
- 4.11 Arbitrary Argument Lists
- 4.12 Methods: Functions That Belong to Objects
- 4.13 Scope Rules
- 4.14 `import`: A Deeper Look
- 4.15 Passing Arguments to Functions: A Deeper Look
- 4.16 Function-Call Stack
- 4.17 Functional-Style Programming
- 4.18 Intro to Data Science: Measures of Dispersion
- 4.19 Wrap-Up
  Exercises

4.1 Introduction

Experience has shown that the best way to develop and maintain a large program is to construct it from smaller, more manageable pieces. This technique is called **divide and conquer**. Using existing functions as building blocks for creating new programs is a key aspect of **software reusability**—it's also a major benefit of object-oriented programming. Packaging code as a function allows you to execute it from various locations in your program just by calling the function, rather than duplicating the possibly lengthy code. This also makes programs easier to modify. When you change a function's code, all calls to the function execute the updated version.

4.2 Defining Functions

You've called many built-in functions (`int`, `float`, `print`, `input`, `type`, `sum`, `len`, `min` and `max`) and a few functions from the statistics module (`mean`, `median` and `mode`). Each performed a single, well-defined task. You'll often define and call *custom* functions. The following session defines a `square` function that calculates the square of its argument. Then it calls the function twice—once to square the `int` value 7 (producing the `int` value 49) and once to square the `float` value 2.5 (producing the `float` value 6.25):

```
In [1]: def square(number):
   ...:     """Calculate the square of number."""
   ...:     return number ** 2
   ...:

In [2]: square(7)
Out[2]: 49

In [3]: square(2.5)
Out[3]: 6.25
```

The statements defining the function in the first snippet are written only once, but may be called "to do their job" from many points throughout a program and as often as you like. Calling `square` with a non-numeric argument like `'hello'` causes a `TypeError` because the exponentiation operator (`**`) works only with numeric values.

Defining a Custom Function

A **function definition** (like square in snippet [1]) begins with the **def keyword**, followed by the function name (square), a set of parentheses and a colon (:). Like variable identifiers, by convention function names should begin with a lowercase letter and in multiword names underscores should separate each word.

The required parentheses contain the function's **parameter list**—a comma-separated list of **parameters** representing the data that the function needs to perform its task. Function square has only one parameter named number—the value to be squared.

If the parentheses are empty, the function does not use parameters to perform its task. Exercise 4.7 asks you to write a parameterless date_and_time function that displays the current date and time by reading it from your computer's system clock.

The indented lines after the colon (:) are the function's **block**, which consists of an optional docstring followed by the statements that perform the function's task. We'll soon point out the difference between a function's block and a control statement's suite.

Specifying a Custom Function's Docstring

The *Style Guide for Python Code* says that the first line in a function's block should be a docstring that briefly explains the function's purpose:

```
"""Calculate the square of number."""
```

To provide more detail, you can use a multiline docstring—the style guide recommends starting with a brief explanation, followed by a blank line and the additional details.

Returning a Result to a Function's Caller

When a function finishes executing, it returns control to its caller—that is, the line of code that called the function. In square's block, the **return** statement:

```
return number ** 2
```

first squares number, then terminates the function and gives the result back to the caller. In this example, the first caller is square(7) in snippet [2], so IPython displays the result in Out[2]. Think of the return value, 49, as simply replacing the call square(7). So after the call, you'd have In [2]: 49, and that would indeed produce Out[2]: 49. The second caller square(2.5) is in snippet [3], so IPython displays the result 6.25 in Out[3].

Function calls also can be embedded in expressions. The following code calls square first, then print displays the result:

```
In [4]: print('The square of 7 is', square(7))
The square of 7 is 49
```

Here, too, think of the return value, 49, as simply replacing the call square(7), which would indeed produce the output shown above.

There are two other ways to return control from a function to its caller:

- Executing a **return** statement without an expression terminates the function and *implicitly* returns the value **None** to the caller. The Python documentation states that None represents the absence of a value. None evaluates to False in conditions.

- When there's no **return** statement in a function, it *implicitly* returns the value None after executing the last statement in the function's block.

What Happens When You Call a Function
The expression square(7) passes the argument 7 to square's parameter number. Then square calculates number ** 2 and returns the result. The parameter number exists only during the function call. It's created on each call to the function to receive the argument value, and it's destroyed when the function returns its result to the caller.

Though we did not define variables in square's block, it is possible to do so. A function's parameters and variables defined in its block are all **local variables**—they can be used only inside the function and exist only while the function is executing. Trying to access a local variable outside its function's block causes a NameError, indicating that the variable is not defined. We'll soon see how a behind-the-scenes mechanism called the *function-call stack* supports the automatic creation and destruction of a function's local variables—and helps the function return to its caller.

Accessing a Function's Docstring via IPython's Help Mechanism
IPython can help you learn about the modules and functions you intend to use in your code, as well as IPython itself. For example, to view a function's docstring to learn how to use the function, type the function's name followed by a **question mark (?)**:

```
In [5]: square?
Signature: square(number)
Docstring: Calculate the square of number.
File:      ~/Documents/examples/ch04/<ipython-input-1-7268c8ff93a9>
Type:      function
```

For our square function, the information displayed includes:

- The function's name and parameter list—known as its **signature**.
- The function's docstring.
- The name of the file containing the function's definition. For a function in an interactive session, this line shows information for the snippet that defined the function—the 1 in "<ipython-input-1-7268c8ff93a9>" means snippet [1].
- The type of the item for which you accessed IPython's help mechanism—in this case, a function.

If the function's source code is accessible from IPython—such as a function defined in the current session or imported into the session from a .py file—you can use **??** to display the function's full source-code definition:

```
In [6]: square??
Signature: square(number)
Source:
def square(number):
    """Calculate the square of number."""
    return number ** 2
File:      ~/Documents/examples/ch04/<ipython-input-1-7268c8ff93a9>
Type:      function
```

If the source code is not accessible from IPython, ?? simply shows the docstring.

If the docstring fits in the window, IPython displays the next In [] prompt. If a docstring is too long to fit, IPython indicates that there's more by displaying a colon (:) at the bottom of the window—press the *Space* key to display the next screen. You can navigate backwards and forwards through the docstring with the up and down arrow keys, respectively. IPython displays (END) at the end of the docstring. Press *q* (for "quit") at any : or the (END) prompt to return to the next In [] prompt. To get a sense of IPython's features, type ? at any In [] prompt, press *Enter*, then read the help documentation overview.

✓ Self Check

1 *(True/False)* The function body is referred to as its suite.
Answer: False. The function body is referred to as its block.

2 *(True/False)* A function's local variables exist after the function returns to its caller.
Answer: False. A function's local variables exist until the function returns to its caller.

3 *(IPython Session)* Define a function square_root that receives a number as a parameter and returns the square root of that number. Determine the square root of 6.25.
Answer:

```
In [1]: def square_root(number):
   ...:     return number ** 0.5  # or number ** (1 / 2)
   ...:

In [2]: square_root(6.25)
Out[2]: 2.5
```

4.3 Functions with Multiple Parameters

Let's define a maximum function that determines and returns the largest of three values—the following session calls the function three times with integers, floating-point numbers and strings, respectively.

```
In [1]: def maximum(value1, value2, value3):
   ...:     """Return the maximum of three values."""
   ...:     max_value = value1
   ...:     if value2 > max_value:
   ...:         max_value = value2
   ...:     if value3 > max_value:
   ...:         max_value = value3
   ...:     return max_value
   ...:

In [2]: maximum(12, 27, 36)
Out[2]: 36

In [3]: maximum(12.3, 45.6, 9.7)
Out[3]: 45.6

In [4]: maximum('yellow', 'red', 'orange')
Out[4]: 'yellow'
```

We did not place blank lines above and below the if statements, because pressing return on a blank line in interactive mode completes the function's definition.

You also may call maximum with mixed types, such as ints and floats:

```
In [5]: maximum(13.5, -3, 7)
Out[5]: 13.5
```

The call `maximum(13.5, 'hello', 7)` results in `TypeError` because strings and numbers cannot be compared to one another with the greater-than (>) operator.

Function `maximum`'s Definition
Function `maximum` specifies three parameters in a comma-separated list. Snippet [2]'s arguments 12, 27 and 36 are assigned to the parameters `value1`, `value2` and `value3`, respectively.

To determine the largest value, we process one value at a time:

- Initially, we assume that `value1` contains the largest value, so we assign it to the local variable `max_value`. Of course, it's possible that `value2` or `value3` contains the actual largest value, so we still must compare each of these with `max_value`.
- The first `if` statement then tests `value2 > max_value`, and if this condition is `True` assigns `value2` to `max_value`.
- The second `if` statement then tests `value3 > max_value`, and if this condition is `True` assigns `value3` to `max_value`.

Now, `max_value` contains the largest value, so we return it. When control returns to the caller, the parameters `value1`, `value2` and `value3` and the variable `max_value` in the function's block—which are all *local variables*—no longer exist.

Python's Built-In `max` and `min` Functions
For many common tasks, the capabilities you need already exist in Python. For example, built-in `max` and `min` functions know how to determine the largest and smallest of their two or more arguments, respectively:

```
In [6]: max('yellow', 'red', 'orange', 'blue', 'green')
Out[6]: 'yellow'

In [7]: min(15, 9, 27, 14)
Out[7]: 9
```

Each of these functions also can receive an iterable argument, such as a list or a string. Using built-in functions or functions from the Python Standard Library's modules rather than writing your own can reduce development time and increase program reliability, portability and performance. For a list of Python's built-in functions and modules, see

https://docs.python.org/3/library/index.html

✓ Self Check

1. *(Fill-In)* A function with multiple parameters specifies them in a(n) _____.
Answer: comma-separated list.

2. *(True/False)* When defining a function in IPython interactive mode, pressing *Enter* on a blank line causes IPython to display another continuation prompt so you can continue defining the function's block.
Answer: False. When defining a function in IPython interactive mode, pressing *Enter* on a blank line terminates the function definition.

3. *(IPython Session)* Call function `max` with the list `[14, 27, 5, 3]` as an argument, then call function `min` with the string `'orange'` as an argument.

Answer:

```
In [1]: max([14, 27, 5, 3])
Out[1]: 27

In [2]: min('orange')
Out[2]: 'a'
```

4.4 Random-Number Generation

We now take a brief diversion into a popular type of programming application—simulation and game playing. You can introduce the **element of chance** via the Python Standard Library's **random module**.

Rolling a Six-Sided Die
Let's produce 10 random integers in the range 1–6 to simulate rolling a six-sided die:

```
In [1]: import random

In [2]: for roll in range(10):
   ...:     print(random.randrange(1, 7), end=' ')
   ...:
4 2 5 5 4 6 4 6 1 5
```

First, we import random so we can use the module's capabilities. The **randrange** function generates an integer from the first argument value up to, but *not* including, the second argument value. Let's use the up arrow key to recall the for statement, then press *Enter* to re-execute it. Notice that *different* values are displayed:

```
In [3]: for roll in range(10):
   ...:     print(random.randrange(1, 7), end=' ')
   ...:
4 5 4 5 1 4 1 4 6 5
```

Sometimes, you may want to guarantee **reproducibility** of a random sequence—for debugging, for example. At the end of this section, we'll show how to do this with the random module's seed function.

Rolling a Six-Sided Die 6,000,000 Times
If randrange truly produces integers at random, every number in its range has an equal **probability** (or *chance* or *likelihood*) of being returned each time we call it. To show that the die faces 1–6 occur with equal likelihood, the following script simulates 6,000,000 die rolls. When you run the script, each die face should occur approximately 1,000,000 times, as in the sample output.

```
1  # fig04_01.py
2  """Roll a six-sided die 6,000,000 times."""
3  import random
4
5  # face frequency counters
6  frequency1 = 0
```

Fig. 4.1 | Roll a six-sided die 6,000,000 times. (Part 1 of 2.)

```python
7    frequency2 = 0
8    frequency3 = 0
9    frequency4 = 0
10   frequency5 = 0
11   frequency6 = 0
12
13   # 6,000,000 die rolls
14   for roll in range(6_000_000):  # note underscore separators
15       face = random.randrange(1, 7)
16
17       # increment appropriate face counter
18       if face == 1:
19           frequency1 += 1
20       elif face == 2:
21           frequency2 += 1
22       elif face == 3:
23           frequency3 += 1
24       elif face == 4:
25           frequency4 += 1
26       elif face == 5:
27           frequency5 += 1
28       elif face == 6:
29           frequency6 += 1
30
31   print(f'Face{"Frequency":>13}')
32   print(f'{1:>4}{frequency1:>13}')
33   print(f'{2:>4}{frequency2:>13}')
34   print(f'{3:>4}{frequency3:>13}')
35   print(f'{4:>4}{frequency4:>13}')
36   print(f'{5:>4}{frequency5:>13}')
37   print(f'{6:>4}{frequency6:>13}')
```

Face	Frequency
1	998686
2	1001481
3	999900
4	1000453
5	999953
6	999527

Fig. 4.1 | Roll a six-sided die 6,000,000 times. (Part 2 of 2.)

The script uses *nested* control statements (an if…elif statement nested in the for statement) to determine the number of times each die face appears. The for statement iterates 6,000,000 times. We used Python's underscore (_) digit separator to make the value 6000000 more readable. The expression range(6,000,000) would be incorrect. Commas separate arguments in function calls, so Python would treat range(6,000,000) as a call to range with the *three* arguments 6, 0 and 0.

For each die roll, the script adds 1 to the appropriate counter variable. Run the program, and observe the results. This program might take a few seconds to complete execution. As you'll see, each execution produces *different* results.

Note that we did not provide an else clause in the if…elif statement. Exercise 4.1 asks you to comment on the possible consequences of this.

Seeding the Random-Number Generator for Reproducibility

Function randrange actually generates **pseudorandom numbers**, based on an internal calculation that begins with a numeric value known as a **seed**. Repeatedly calling randrange produces a sequence of numbers that *appear* to be random, because each time you start a new interactive session or execute a script that uses the random module's functions, Python internally uses a *different* seed value.[1] When you're debugging logic errors in programs that use randomly generated data, it can be helpful to use the *same* sequence of random numbers until you've eliminated the logic errors, before testing the program with other values. To do this, you can use the random module's **seed** function to seed the **random-number generator** yourself—this forces randrange to begin calculating its pseudorandom number sequence from the seed you specify. In the following session, snippets [5] and [8] produce the same results, because snippets [4] and [7] use the same seed (32):

```
In [4]: random.seed(32)

In [5]: for roll in range(10):
   ...:     print(random.randrange(1, 7), end=' ')
   ...:
1 2 2 3 6 2 4 1 6 1
In [6]: for roll in range(10):
   ...:     print(random.randrange(1, 7), end=' ')
   ...:
1 3 5 3 1 5 6 4 3 5
In [7]: random.seed(32)

In [8]: for roll in range(10):
   ...:     print(random.randrange(1, 7), end=' ')
   ...:
1 2 2 3 6 2 4 1 6 1
```

Snippet [6] generates *different* values because it simply continues the pseudorandom number sequence that began in snippet [5].

✓ Self Check

1 *(Fill-In)* The element of chance can be introduced into computer applications using module _____.
Answer: random.

2 *(Fill-In)* The random module's _____ function enables reproducibility of random sequences.
Answer: seed.

---

1. According to the documentation, Python bases the seed value on the system clock or an operating-system-dependent randomness source. For applications requiring secure random numbers, such as cryptography, the documentation recommends using the secrets module, rather than the random module.

3 *(IPython Session)* Requirements statement: Use a for statement, randrange and a conditional expression (introduced in the preceding chapter) to simulate 20 coin flips, displaying H for heads and T for tails all on the same line, each separated by a space.
Answer:

```
In [1]: import random

In [2]: for i in range(20):
   ...:     print('H' if random.randrange(2) == 0 else 'T', end=' ')
   ...:
T H T H T T T T H T H H H T H T H H H H
```

In snippet [2]'s output, an equal number of Ts and Hs appeared—that will not always be the case with random-number generation.

4.5 Case Study: A Game of Chance

In this section, we simulate the popular dice game known as "craps." Here is the requirements statement:

> You roll two six-sided dice, each with faces containing one, two, three, four, five and six spots, respectively. When the dice come to rest, the sum of the spots on the two upward faces is calculated. If the sum is 7 or 11 on the first roll, you win. If the sum is 2, 3 or 12 on the first roll (called "craps"), you lose (i.e., the "house" wins). If the sum is 4, 5, 6, 8, 9 or 10 on the first roll, that sum becomes your "point." To win, you must continue rolling the dice until you "make your point" (i.e., roll that same point value). You lose by rolling a 7 before making your point.

The following script simulates the game and shows several sample executions, illustrating winning on the first roll, losing on the first roll, winning on a subsequent roll and losing on a subsequent roll.

```
 1  # fig04_02.py
 2  """Simulating the dice game Craps."""
 3  import random
 4
 5  def roll_dice():
 6      """Roll two dice and return their face values as a tuple."""
 7      die1 = random.randrange(1, 7)
 8      die2 = random.randrange(1, 7)
 9      return (die1, die2)  # pack die face values into a tuple
10
11  def display_dice(dice):
12      """Display one roll of the two dice."""
13      die1, die2 = dice  # unpack the tuple into variables die1 and die2
14      print(f'Player rolled {die1} + {die2} = {sum(dice)}')
15
16  die_values = roll_dice()  # first roll
17  display_dice(die_values)
18
```

Fig. 4.2 | Simulating the dice game Craps. (Part 1 of 2.)

```python
19  # determine game status and point, based on first roll
20  sum_of_dice = sum(die_values)
21
22  if sum_of_dice in (7, 11):  # win
23      game_status = 'WON'
24  elif sum_of_dice in (2, 3, 12):  # lose
25      game_status = 'LOST'
26  else:  # remember point
27      game_status = 'CONTINUE'
28      my_point = sum_of_dice
29      print('Point is', my_point)
30
31  # continue rolling until player wins or loses
32  while game_status == 'CONTINUE':
33      die_values = roll_dice()
34      display_dice(die_values)
35      sum_of_dice = sum(die_values)
36
37      if sum_of_dice == my_point:  # win by making point
38          game_status = 'WON'
39      elif sum_of_dice == 7:  # lose by rolling 7
40          game_status = 'LOST'
41
42  # display "wins" or "loses" message
43  if game_status == 'WON':
44      print('Player wins')
45  else:
46      print('Player loses')
```

```
Player rolled 2 + 5 = 7
Player wins
```

```
Player rolled 1 + 2 = 3
Player loses
```

```
Player rolled 5 + 4 = 9
Point is 9
Player rolled 4 + 4 = 8
Player rolled 2 + 3 = 5
Player rolled 5 + 4 = 9
Player wins
```

```
Player rolled 1 + 5 = 6
Point is 6
Player rolled 1 + 6 = 7
Player loses
```

Fig. 4.2 | Simulating the dice game Craps. (Part 2 of 2.)

Function `roll_dice`—Returning Multiple Values Via a Tuple

Function `roll_dice` (lines 5–9) simulates rolling two dice on each roll. The function is defined once, then called from several places in the program (lines 16 and 33). The empty parameter list indicates that `roll_dice` does not require arguments to perform its task.

The built-in and custom functions you've called so far each return one value. Sometimes it's useful to return more than one value, as in `roll_dice`, which returns both die values (line 9) as a **tuple**—an **immutable** (that is, unmodifiable) sequences of values. To create a tuple, separate its values with commas, as in line 9:

```
(die1, die2)
```

This is known as **packing a tuple**. The parentheses are optional, but we recommend using them for clarity. We discuss tuples in depth in the next chapter.

Function `display_dice`

To use a tuple's values, you can assign them to a comma-separated list of variables, which **unpacks** the tuple. To display each roll of the dice, the function `display_dice` (defined in lines 11–14 and called in lines 17 and 34) unpacks the tuple argument it receives (line 13). The number of variables to the left of = must match the number of elements in the tuple; otherwise, a `ValueError` occurs. Line 14 prints a formatted string containing both die values and their sum. We calculate the sum of the dice by passing the tuple to the built-in `sum` function—like a list, a tuple is a sequence.

Note that functions `roll_dice` and `display_dice` each begin their blocks with a docstring that states what the function does. Also, both functions contain local variables `die1` and `die2`. These variables do not "collide," because they belong to different functions' blocks. Each local variable is accessible only in the block that defined it.

First Roll

When the script begins executing, lines 16–17 roll the dice and display the results. Line 20 calculates the sum of the dice for use in lines 22–29. You can win or lose on the first roll or any subsequent roll. The variable `game_status` keeps track of the win/loss status.

The **in operator** in line 22

```
sum_of_dice in (7, 11)
```

tests whether the tuple (7, 11) contains `sum_of_dice`'s value. If this condition is `True`, you rolled a 7 or an 11. In this case, you won on the first roll, so the script sets `game_status` to `'WON'`. The operator's right operand can be any iterable. There's also a **not in** operator to determine whether a value is *not* in an iterable. The preceding concise condition is equivalent to

```
(sum_of_dice == 7) or (sum_of_dice == 11)
```

Similarly, the condition in line 24

```
sum_of_dice in (2, 3, 12)
```

tests whether the tuple (2, 3, 12) contains `sum_of_dice`'s value. If so, you lost on the first roll, so the script sets `game_status` to `'LOST'`.

For any other sum of the dice (4, 5, 6, 8, 9 or 10):

- line 27 sets `game_status` to `'CONTINUE'` so you can continue rolling

- line 28 stores the sum of the dice in my_point to keep track of what you must roll to win and
- line 29 displays my_point.

Subsequent Rolls
If game_status is equal to 'CONTINUE' (line 32), you did not win or lose, so the while statement's suite (lines 33–40) executes. Each loop iteration calls roll_dice, displays the die values and calculates their sum. If sum_of_dice is equal to my_point (line 37) or 7 (line 39), the script sets game_status to 'WON' or 'LOST', respectively, and the loop terminates. Otherwise, the while loop continues executing with the next roll.

Displaying the Final Results
When the loop terminates, the script proceeds to the if...else statement (lines 43–46), which prints 'Player wins' if game_status is 'WON', or 'Player loses' otherwise.

Self Check

1 *(Fill-In)* The _____ operator tests whether its right operand's iterable contains its left operand's value.
Answer: in.

2 *(IPython Session)* Pack a student tuple with the name 'Sue' and the list [89, 94, 85], display the tuple, then unpack it into variables name and grades, and display their values.
Answer:

```
In [1]: student = ('Sue', [89, 94, 85])

In [2]: student
Out[2]: ('Sue', [89, 94, 85])

In [3]: name, grades = student

In [4]: print(f'{name}: {grades}')
Sue: [89, 94, 85]
```

4.6 Python Standard Library

Typically, you write Python programs by combining functions and classes (that is, custom types) that you create with preexisting functions and classes defined in modules, such as those in the Python Standard Library and other libraries. A key programming goal is to avoid "reinventing the wheel."

A **module** is a file that groups related functions, data and classes. The type Decimal from the Python Standard Library's decimal module is actually a class. We introduced classes briefly in Chapter 1 and discuss them in detail in the "Object-Oriented Programming" chapter. A **package** groups related modules. In this book, you'll work with many preexisting modules and packages, and you'll create your own modules—in fact, every Python source-code (.py) file you create is a module. Creating packages is beyond this book's scope. They're typically used to organize a large library's functionality into smaller subsets that are easier to maintain and can be imported separately for convenience. For example, the matplotlib visualization library that we use in Section 5.17 has extensive functionality (its documentation is over 2300 pages), so we'll import only the subsets we need in our examples (`pyplot` and `animation`).

The Python Standard Library is provided with the core Python language. Its packages and modules contain capabilities for a wide variety of everyday programming tasks.[2] You can see a complete list of the standard library modules at

> https://docs.python.org/3/library/

You've already used capabilities from the `decimal`, `statistics` and `random` modules. In the next section, you'll use mathematics capabilities from the `math` module. You'll see many other Python Standard Library modules throughout the book's examples and exercises, including many of those in the following table:

Some popular Python Standard Library modules

`collections`—Data structures beyond lists, tuples, dictionaries and sets.
Cryptography modules—Encrypting data for secure transmission.
`csv`—Processing comma-separated value files (like those in Excel).
`datetime`—Date and time manipulations. Also modules `time` and `calendar`.
`decimal`—Fixed-point and floating-point arithmetic, including monetary calculations.
`doctest`—Embed validation tests and expected results in docstrings for simple unit testing.
`gettext` and `locale`—Internationalization and localization modules.
`json`—JavaScript Object Notation (JSON) processing used with web services and NoSQL document databases.

`math`—Common math constants and operations.
`os`—Interacting with the operating system.
`profile`, `pstats`, `timeit`—Performance analysis.
`random`—Pseudorandom numbers.
`re`—Regular expressions for pattern matching.
`sqlite3`—SQLite relational database access.
`statistics`—Mathematical statistics functions such as `mean`, `median`, `mode` and `variance`.
`string`—String processing.
`sys`—Command-line argument processing; standard input, standard output and standard error streams.
`tkinter`—Graphical user interfaces (GUIs) and canvas-based graphics.
`turtle`—Turtle graphics.
`webbrowser`—For conveniently displaying web pages in Python apps.

✓ Self Check

1 *(Fill-In)* A(n) _____ defines related functions, data and classes. A(n) _____ groups related modules.
Answer: module, package.

2 *(Fill-In)* Every Python source code (`.py`) file you create is a(n) _____.
Answer: module.

4.7 `math` Module Functions

The **`math` module** defines functions for performing various common mathematical calculations. Recall from the previous chapter that an `import` statement of the following form enables you to use a module's definitions via the module's name and a dot (`.`):

```
In [1]: import math
```

---
2. The Python Tutorial refers to this as the "batteries included" approach.

For example, the following snippet calculates the square root of 900 by calling the math module's **sqrt function**, which returns its result as a `float` value:

```
In [2]: math.sqrt(900)
Out[2]: 30.0
```

Similarly, the following snippet calculates the absolute value of -10 by calling the math module's **fabs function**, which returns its result as a `float` value:

```
In [3]: math.fabs(-10)
Out[3]: 10.0
```

Some math module functions are summarized below—you can view the complete list at

`https://docs.python.org/3/library/math.html`

Function	Description	Example
ceil(x)	Rounds x to the smallest integer not less than x	ceil(9.2) is 10.0 ceil(-9.8) is -9.0
floor(x)	Rounds x to the largest integer not greater than x	floor(9.2) is 9.0 floor(-9.8) is -10.0
sin(x)	Trigonometric sine of x (x in radians)	sin(0.0) is 0.0
cos(x)	Trigonometric cosine of x (x in radians)	cos(0.0) is 1.0
tan(x)	Trigonometric tangent of x (x in radians)	tan(0.0) is 0.0
exp(x)	Exponential function e^x	exp(1.0) is 2.718282 exp(2.0) is 7.389056
log(x)	Natural logarithm of x (base e)	log(2.718282) is 1.0 log(7.389056) is 2.0
log10(x)	Logarithm of x (base 10)	log10(10.0) is 1.0 log10(100.0) is 2.0
pow(x, y)	x raised to power y (x^y)	pow(2.0, 7.0) is 128.0 pow(9.0, .5) is 3.0
sqrt(x)	square root of x	sqrt(900.0) is 30.0 sqrt(9.0) is 3.0
fabs(x)	Absolute value of x—always returns a float. Python also has the built-in function **abs**, which returns an `int` or a `float`, based on its argument.	fabs(5.1) is 5.1 fabs(-5.1) is 5.1
fmod(x, y)	Remainder of x/y as a floating-point number	fmod(9.8, 4.0) is 1.8

4.8 Using IPython Tab Completion for Discovery

You can view a module's documentation in IPython interactive mode via **tab completion**—a **discovery** feature that speeds your coding and learning processes. After you type a portion of an identifier and press *Tab*, IPython completes the identifier for you or provides a list of identifiers that begin with what you've typed so far. This may vary based on your operating system platform and what you have imported into your IPython session:

```
In [1]: import math

In [2]: ma<Tab>
        map           %macro        %%markdown
        math          %magic        %matplotlib
        max()         %man
```

You can scroll through the identifiers with the up and down arrow keys. As you do, IPython highlights an identifier and shows it to the right of the In [] prompt.

Viewing Identifiers in a Module

To view a list of identifiers defined in a module, type the module's name and a dot (.), then press *Tab*:

```
In [3]: math.<Tab>
        acos()      atan()      copysign()   e         expm1()
        acosh()     atan2()     cos()        erf()     fabs()
        asin()      atanh()     cosh()       erfc()    factorial() >
        asinh()     ceil()      degrees()    exp()     floor()
```

If there are more identifiers to display than are currently shown, IPython displays the > symbol (on some platforms) at the right edge, in this case to the right of factorial(). You can use the up and down arrow keys to scroll through the list. In the list of identifiers:

- Those followed by parentheses are functions (or methods, as you'll see later).

- Single-word identifiers (such as Employee) that begin with an uppercase letter and multiword identifiers in which each word begins with an uppercase letter (such as CommissionEmployee) represent class names (there are none in the preceding list). This naming convention, which the *Style Guide for Python Code* recommends, is known as **CamelCase** because the uppercase letters stand out like a camel's humps.

- Lowercase identifiers without parentheses, such as pi (not shown in the preceding list) and e, are variables. The identifier pi evaluates to 3.141592653589793, and the identifier e evaluates to 2.718281828459045. In the math module, pi and e represent the mathematical constants π and e, respectively.

Python does not have *constants*, although many objects in Python are immutable (non-modifiable). So even though pi and e are real-world constants, *you must not assign new values to them*, because that would change their values. To help distinguish constants from other variables, the style guide recommends naming your custom constants with all capital letters.

Using the Currently Highlighted Function

As you navigate through the identifiers, if you wish to use a currently highlighted function, simply start typing its arguments in parentheses. IPython then hides the autocompletion list. If you need more information about the currently highlighted item, you can view its docstring by typing a question mark (?) following the name and pressing *Enter* to view the help documentation. The following shows the fabs function's docstring:

```
In [4]: math.fabs?
Docstring:
fabs(x)

Return the absolute value of the float x.
Type:      builtin_function_or_method
```

The builtin_function_or_method shown above indicates that fabs is part of a Python Standard Library module. Such modules are considered to be built into Python. In this case, fabs is a built-in function from the math module.

 Self Check

1 *(True/False)* In IPython interactive mode, to view a list of identifiers defined in a module, type the module's name and a dot (.) then press *Enter*.
Answer: False. Press *Tab*, not *Enter*.

2 *(True/False)* Python does not have constants.
Answer: True.

4.9 Default Parameter Values

When defining a function, you can specify that a parameter has a **default parameter value**. When calling the function, if you omit the argument for a parameter with a default parameter value, the default value for that parameter is automatically passed. Let's define a function rectangle_area with default parameter values:

```
In [1]: def rectangle_area(length=2, width=3):
   ...:     """Return a rectangle's area."""
   ...:     return length * width
   ...:
```

You specify a default parameter value by following a parameter's name with an = and a value—in this case, the default parameter values are 2 and 3 for length and width, respectively. Any parameters with default parameter values must appear in the parameter list to the *right* of parameters that do not have defaults.

The following call to rectangle_area has no arguments, so IPython uses both default parameter values as if you had called rectangle_area(2, 3):

```
In [2]: rectangle_area()
Out[2]: 6
```

The following call to rectangle_area has only one argument. Arguments are assigned to parameters from left to right, so 10 is used as the length. The interpreter passes the default parameter value 3 for the width as if you had called rectangle_area(10, 3):

```
In [3]: rectangle_area(10)
Out[3]: 30
```

The following call to rectangle_area has arguments for both length and width, so IPython ignores the default parameter values:

```
In [4]: rectangle_area(10, 5)
Out[4]: 50
```

136 Functions

Self Check

1 *(True/False)* When an argument with a default parameter value is omitted in a function call, the interpreter automatically passes the default parameter value in the call.
Answer: True.

2 *(True/False)* Parameters with default parameter values must be the *leftmost* arguments in a function's parameter list.
Answer: False. Parameters with default parameter values must appear to the *right* of parameters that do not have defaults.

4.10 Keyword Arguments

When calling functions, you can use **keyword arguments** to pass arguments in *any* order. To demonstrate keyword arguments, we redefine the `rectangle_area` function—this time without default parameter values:

```
In [1]: def rectangle_area(length, width):
   ...:     """Return a rectangle's area."""
   ...:     return length * width
   ...:
```

Each keyword *argument in a call* has the form *parametername=value*. The following call shows that the order of keyword arguments does not matter—they do not need to match the corresponding parameters' positions in the function definition:

```
In [2]: rectangle_area(width=5, length=10)
Out[3]: 50
```

In each function call, you must place keyword arguments *after* a function's positional arguments—that is, any arguments for which you do not specify the parameter name. Such arguments are assigned to the function's parameters left-to-right, based on the argument's positions in the argument list. Keyword arguments are also helpful for improving the readability of function calls, especially for functions with many arguments.

Self Check

1 *(True/False)* You must pass keyword arguments in the same order as their corresponding parameters in the function definition's parameter list.
Answer: False. The order of keyword arguments does not matter.

4.11 Arbitrary Argument Lists

Functions with **arbitrary argument lists**, such as built-in functions `min` and `max`, can receive *any* number of arguments. Consider the following `min` call:

```
min(88, 75, 96, 55, 83)
```

The function's documentation states that `min` has two *required* parameters (named `arg1` and `arg2`) and an optional third parameter of the form `*args`, indicating that the function can receive any number of additional arguments. The `*` before the parameter name tells Python to pack any remaining arguments into a tuple that's passed to the `args` parameter. In the call above, parameter `arg1` receives 88, parameter `arg2` receives 75 and parameter `args` receives the tuple (96, 55, 83).

Defining a Function with an Arbitrary Argument List

Let's define an average function that can receive any number of arguments:

```
In [1]: def average(*args):
   ...:     return sum(args) / len(args)
   ...:
```

The parameter name args is used by convention, but you may use any identifier. If the function has multiple parameters, the *args parameter must be the *rightmost* parameter.

Now, let's call average several times with arbitrary argument lists of different lengths:

```
In [2]: average(5, 10)
Out[2]: 7.5

In [3]: average(5, 10, 15)
Out[3]: 10.0

In [4]: average(5, 10, 15, 20)
Out[4]: 12.5
```

To calculate the average, divide the sum of the args tuple's elements (returned by built-in function sum) by the tuple's number of elements (returned by built-in function len). Note in our average definition that if the length of args is 0, a ZeroDivisionError occurs. In the next chapter, you'll see how to access a tuple's elements without unpacking them.

Passing an Iterable's Individual Elements as Function Arguments

You can unpack a tuple's, list's or other iterable's elements to pass them as individual function arguments. The * **operator**, when applied to an iterable argument in a function call, unpacks its elements. The following code creates a five-element grades list, then uses the expression *grades to unpack its elements as average's arguments:

```
In [5]: grades = [88, 75, 96, 55, 83]

In [6]: average(*grades)
Out[6]: 79.4
```

The call shown above is equivalent to average(88, 75, 96, 55, 83).

Self Check

1 *(Fill-In)* To define a function with an arbitrary argument list, specify a parameter of the form _____.
Answer: *args (again, the name args is used by convention, but is not required).

2 *(IPython Session)* Create a function named calculate_product that receives an arbitrary argument list and returns the product of all the arguments. Call the function with the arguments 10, 20 and 30, then with the sequence of integers produced by range(1, 6, 2).
Answer:

```
In [1]: def calculate_product(*args):
   ...:     product = 1
   ...:     for value in args:
   ...:         product *= value
   ...:     return product
   ...:
```

```
In [2]: calculate_product(10, 20, 30)
Out[2]: 6000

In [3]: calculate_product(*range(1, 6, 2))
Out[3]: 15
```

4.12 Methods: Functions That Belong to Objects

A **method** is simply a function that you call on an object using the form

>   *object_name*.*method_name*(*arguments*)

For example, the following session creates the string variable s and assigns it the string object 'Hello'. Then the session calls the object's **lower** and **upper** methods, which produce *new* strings containing all-lowercase and all-uppercase versions of the original string, leaving s unchanged:

```
In [1]: s = 'Hello'

In [2]: s.lower()    # call lower method on string object s
Out[2]: 'hello'

In [3]: s.upper()
Out[3]: 'HELLO'

In [4]: s
Out[4]: 'Hello'
```

The *Python Standard Library* reference at

>   https://docs.python.org/3/library/index.html

describes the methods of built-in types and the types in the Python Standard Library. In the "Object-Oriented Programming" chapter, you'll create *custom* types called classes and define custom methods that you can call on objects of those classes.

4.13 Scope Rules

Each identifier has a **scope** that determines where you can use it in your program. For that portion of the program, the identifier is said to be "in scope."

Local Scope
A local variable's identifier has **local scope**. It's "in scope" only from its definition to the end of the function's block. It "goes out of scope" when the function returns to its caller. So, a local variable can be used only inside the function that defines it.

Global Scope
Identifiers defined outside any function (or class) have **global scope**—these may include functions, variables and classes. Variables with global scope are known as **global variables**. Identifiers with global scope can be used in a .py file or interactive session anywhere after they're defined.

Accessing a Global Variable from a Function
You can access a global variable's value inside a function:
```
In [1]: x = 7

In [2]: def access_global():
   ...:     print('x printed from access_global:', x)
   ...:

In [3]: access_global()
x printed from access_global: 7
```
However, by default, you cannot *modify* a global variable in a function—when you first assign a value to a variable in a function's block, Python creates a *new* local variable:
```
In [4]: def try_to_modify_global():
   ...:     x = 3.5
   ...:     print('x printed from try_to_modify_global:', x)
   ...:

In [5]: try_to_modify_global()
x printed from try_to_modify_global: 3.5

In [6]: x
Out[6]: 7
```
In function `try_to_modify_global`'s block, the local x **shadows** the global x, making it inaccessible in the scope of the function's block. Snippet [6] shows that global variable x still exists and has its original value (7) after function `try_to_modify_global` executes.

To modify a global variable in a function's block, you must use a **global** statement to declare that the variable is defined in the global scope:
```
In [7]: def modify_global():
   ...:     global x
   ...:     x = 'hello'
   ...:     print('x printed from modify_global:', x)
   ...:

In [8]: modify_global()
x printed from modify_global: hello

In [9]: x
Out[9]: 'hello'
```

Blocks vs. Suites
You've now defined function *blocks* and control statement *suites*. When you create a variable in a block, it's *local* to that block. However, when you create a variable in a control statement's suite, the variable's scope depends on where the control statement is defined:

- If the control statement is in the global scope, then any variables defined in the control statement have global scope.
- If the control statement is in a function's block, then any variables defined in the control statement have local scope.

We'll continue our scope discussion in the "Object-Oriented Programming" chapter when we introduce custom classes.

Shadowing Functions

In the preceding chapters, when summing values, we stored the sum in a variable named `total`. The reason we did this is that `sum` is a built-in function. If you define a variable named `sum`, it *shadows* the built-in function, making it inaccessible in your code. When you execute the following assignment, Python binds the identifier `sum` to the `int` object containing 15. At this point, the identifier `sum` no longer references the built-in function. So, when you try to use `sum` as a function, a `TypeError` occurs:

```
In [10]: sum = 10 + 5

In [11]: sum
Out[11]: 15

In [12]: sum([10, 5])
-------------------------------------------------------------------------
TypeError                                 Traceback (most recent call last)
<ipython-input-12-1237d97a65fb> in <module>()
----> 1 sum([10, 5])

TypeError: 'int' object is not callable
```

Statements at Global Scope

In the scripts you've seen so far, we've written some statements outside functions at the global scope and some statements inside function blocks. Script statements at global scope execute as soon as they're encountered by the interpreter, whereas statements in a block execute only when the function is called.

Self Check

1 *(Fill-In)* An identifier's _____ describes the region of a program in which the identifier's value can be accessed.
Answer: scope.

2 *(True/False)* Once a code block terminates (e.g., when a function returns), all identifiers defined in that block "go out of scope" and can no longer be accessed.
Answer: True.

4.14 import: A Deeper Look

You've imported modules (such as `math` and `random`) with a statement like:

> `import` *module_name*

then accessed their features via each module's name and a dot (.). Also, you've imported a specific identifier from a module (such as the `decimal` module's `Decimal` type) with a statement like:

> `from` *module_name* `import` *identifier*

then used that identifier without having to precede it with the module name and a dot (.).

4.14 import: A Deeper Look

Importing Multiple Identifiers from a Module
Using the from...import statement you can import a comma-separated list of identifiers from a module then use them in your code without having to precede them with the module name and a dot (.):

```
In [1]: from math import ceil, floor

In [2]: ceil(10.3)
Out[2]: 11

In [3]: floor(10.7)
Out[3]: 10
```

Trying to use a function that's not imported causes a NameError, indicating that the name is not defined.

Caution: Avoid Wildcard Imports
You can import *all* identifiers defined in a module with a **wildcard import** of the form

```
from modulename import *
```

This makes all of the module's identifiers available for use in your code. Importing a module's identifiers with a wildcard import can lead to subtle errors—it's considered a dangerous practice that you should avoid. Consider the following snippets:

```
In [4]: e = 'hello'

In [5]: from math import *

In [6]: e
Out[6]: 2.718281828459045
```

Initially, we assign the string 'hello' to a variable named e. After executing snippet [5] though, the variable e is replaced, possibly by accident, with the math module's constant e, representing the mathematical floating-point value *e*.

Binding Names for Modules and Module Identifiers
Sometimes it's helpful to import a module and use an abbreviation for it to simplify your code. The import statement's **as** clause allows you to specify the name used to reference the module's identifiers. For example, in Section 3.17 we could have imported the statistics module and accessed its mean function as follows:

```
In [7]: import statistics as stats

In [8]: grades = [85, 93, 45, 87, 93]

In [9]: stats.mean(grades)
Out[9]: 80.6
```

As you'll see in later chapters, import...as is frequently used to import Python libraries with convenient abbreviations, like stats for the statistics module. As another example, we'll use the numpy module which typically is imported with

```
import numpy as np
```

Library documentation often mentions popular shorthand names.

Typically, when importing a module, you should use import or import...as statements, then access the module through the module name or the abbreviation following the as keyword, respectively. This ensures that you do not accidentally import an identifier that conflicts with one in your code.

 Self Check

1 *(True/False)* You must always import all the identifiers of a given module.
Answer: False. You can import only the identifiers you need by using a from...import statement.

2 *(IPython Session)* Import the decimal module with the shorthand name dec, then create a Decimal object with the value 2.5 and square its value.
Answer:

```
In [1]: import decimal as dec

In [2]: dec.Decimal('2.5') ** 2
Out[2]: Decimal('6.25')
```

4.15 Passing Arguments to Functions: A Deeper Look

Let's take a closer look at how arguments are passed to functions. In many programming languages, there are two ways to pass arguments—**pass-by-value** and **pass-by-reference** (sometimes called **call-by-value** and **call-by-reference**, respectively):

- With pass-by-value, the called function receives a *copy* of the argument's *value* and works exclusively with that copy. Changes to the function's copy do *not* affect the original variable's value in the caller.

- With pass-by-reference, the called function can access the argument's value in the caller directly and modify the value if it's mutable.

Python arguments are always passed by reference. Some people call this **pass-by-object-reference**, because "everything in Python is an object."[3] When a function call provides an argument, Python copies the argument object's *reference*—not the object itself—into the corresponding parameter. This is important for performance. Functions often manipulate large objects—frequently copying them would consume large amounts of computer memory and significantly slow program performance.

Memory Addresses, References and "Pointers"

You interact with an object via a reference, which behind the scenes is that object's address (or location) in the computer's memory—sometimes called a "pointer" in other languages. After an assignment like

    x = 7

the variable x does not actually contain the value 7. Rather, it contains a reference to an *object* containing 7 (and some other data we'll discuss in later chapters) stored *elsewhere* in

---

3. Even the functions you defined in this chapter and the classes (custom types) you'll define in later chapters are objects in Python.

memory. You might say that x "points to" (that is, references) the object containing 7, as in the diagram below:

Built-In Function id and Object Identities

Let's consider how we pass arguments to functions. First, let's create the integer variable x mentioned above—shortly we'll use x as a function argument:

```
In [1]: x = 7
```

Now x refers to (or "points to") the integer object containing 7. No two separate objects can reside at the same address in memory, so every object in memory has a *unique address*. Though we can't see an object's address, we can use the built-in **id function** to obtain a *unique* int value which identifies only that object while it remains in memory (you'll likely get a different value when you run this on your computer):

```
In [2]: id(x)
Out[2]: 4350477840
```

The integer result of calling id is known as the object's **identity**.[4] No two objects in memory can have the same *identity*. We'll use object identities to demonstrate that objects are passed by reference.

Passing an Object to a Function

Let's define a cube function that displays its parameter's identity, then returns the parameter's value cubed:

```
In [3]: def cube(number):
   ...:     print('id(number):', id(number))
   ...:     return number ** 3
   ...:
```

Next, let's call cube with the argument x, which refers to the integer object containing 7:

```
In [4]: cube(x)
id(number): 4350477840
Out[4]: 343
```

The identity displayed for cube's parameter number—4350477840—is the *same* as that displayed for x previously. Since every object has a unique identity, both the *argument* x and the *parameter* number refer to the *same object* while cube executes. So when function cube uses its parameter number in its calculation, it gets the value of number from the original object in the caller.

Testing Object Identities with the is Operator

You also can prove that the argument and the parameter refer to the same object with Python's **is operator**, which returns True if its two operands have the *same identity*:

---

4. According to the Python documentation, depending on the Python implementation you're using, an object's identity may be the object's actual memory address, but this is not required.

```
In [5]: def cube(number):
   ...:     print('number is x:', number is x)  # x is a global variable
   ...:     return number ** 3
   ...:

In [6]: cube(x)
number is x: True
Out[6]: 343
```

Immutable Objects as Arguments

When a function receives as an argument a reference to an *immutable* (unmodifiable) object—such as an int, float, string or tuple—even though you have direct access to the original object in the caller, you cannot modify the original immutable object's value. To prove this, first let's have cube display id(number) before and after assigning a new object to the parameter number via an augmented assignment:

```
In [7]: def cube(number):
   ...:     print('id(number) before modifying number:', id(number))
   ...:     number **= 3
   ...:     print('id(number) after modifying number:', id(number))
   ...:     return number
   ...:

In [8]: cube(x)
id(number) before modifying number: 4350477840
id(number) after modifying number: 4396653744
Out[8]: 343
```

When we call cube(x), the first print statement shows that id(number) initially is the same as id(x) in snippet [2]. Numeric values are immutable, so the statement

```
number **= 3
```

actually creates a *new object* containing the cubed value, then assigns that object's reference to parameter number. Recall that if there are no more references to the *original* object, it will be *garbage collected*. Function cube's second print statement shows the *new* object's identity. Object identities must be unique, so number must refer to a *different* object. To show that x was not modified, we display its value and identity again:

```
In [9]: print(f'x = {x}; id(x) = {id(x)}')
x = 7; id(x) = 4350477840
```

Mutable Objects as Arguments

In the next chapter, we'll show that when a reference to a *mutable* object like a list is passed to a function, the function *can* modify the original object in the caller.

✓ Self Check

1 *(Fill-In)* The built-in function _____ returns an object's unique identifier.
Answer: id.

2 *(True/False)* Attempts to modify mutable objects create new objects.
Answer: False. This is true for immutable objects.

3 *(IPython Session)* Create a variable width with the value 15.5, then show that modifying the variable creates a new object. Display width's identity and value before and after modifying its value.
Answer:

```
In [1]: width = 15.5

In [2]: print('id:', id(width), ' value:', width)
id: 4397553776  value: 15.5

In [3]: width = width * 3

In [4]: print('id:', id(width), ' value:', width)
id: 4397554208  value: 46.5
```

4.16 Function-Call Stack

To understand how Python performs function calls, consider a data structure (that is, a collection of related data items) known as a **stack**, which is like a pile of dishes. When you add a dish to the pile, you place it on the *top*. Similarly, when you remove a dish from the pile, you take it from the top. Stacks are known as **last-in, first-out** (LIFO) **data structures**—the last item **pushed** (that is, placed) onto the stack is the first item **popped** (that, is removed) from the stack.

Stacks and Your Web Browser's Back Button

A stack is working for you when you visit websites with your web browser. A stack of web-page addresses supports a browser's back button. For each new web page you visit, the browser *pushes* the address of the page you were viewing onto the back button's stack. This allows the browser to "remember" the web page you came from if you decide to go back to it later. Pushing onto the back button's stack may happen many times before you decide to go back to a previous web page. When you press the browser's back button, the browser *pops* the top stack element to get the prior web page's address, then displays that web page. Each time you press the back button the browser *pops* the top stack element and displays that page. This continues until the stack is empty, meaning that there are no more pages for you to go back to via the back button.

Stack Frames

Similarly, the **function-call stack** supports the function call/return mechanism. Eventually, each function must return program control to the point at which it was called. For each function call, the interpreter *pushes* an entry called a **stack frame** (or an **activation record**) onto the stack. This entry contains the *return location* that the called function needs so it can return control to its caller. When the function finishes executing, the interpreter *pops* the function's stack frame, and control transfers to the *return location* that was popped.

The *top* stack frame *always* contains the information the currently executing function needs to return control to its caller. If before a function returns it makes a call to another function, the interpreter *pushes* a stack frame for that function call onto the stack. Thus, the return address required by the newly called function to return to its caller is now on *top* of the stack.

Local Variables and Stack Frames

Most functions have one or more *parameters* and possibly *local variables* that need to:

- exist while the function is executing,
- remain active if the function makes calls to other functions, and
- "go away" when the function returns to its caller.

A called function's stack frame is the perfect place to reserve memory for the function's local variables. That stack frame is pushed when the function is called and exists while the function is executing. When that function returns, it no longer needs its local variables, so its stack frame is *popped* from the stack, and its local variables no longer exist.

Stack Overflow

Of course, the amount of memory in a computer is finite, so only a certain amount of memory can be used to store stack frames on the function-call stack. If the function-call stack runs out of memory as a result of too many function calls, a fatal error known as **stack overflow** occurs.[5] Stack overflows actually are rare unless you have a logic error that keeps calling functions that never return.

Principle of Least Privilege

The **principle of least privilege** is fundamental to good software engineering. It states that code should be granted *only* the amount of privilege and access that it needs to accomplish its designated task, but no more. An example of this is the scope of a local variable, which should not be visible when it's not needed. This is why a function's local variables are placed in stack frames on the function-call stack, so they can be used by that function while it executes and go away when it returns. Once the stack frame is popped, the memory that was occupied by it can be reused for new stack frames. Also, there is no access between stack frames, so functions cannot see each other's local variables. The principle of least privilege makes your programs more robust by preventing code from accidentally (or maliciously) modifying variable values that should not be accessible to it.

 Self Check

1 *(Fill-In)* The stack operations for adding an item to a stack and removing an item from a stack are known as _____ and _____, respectively.
Answer: push, pop.

2 *(Fill-In)* A stack's items are removed in _____ order.
Answer: last-in, first-out (LIFO).

4.17 Functional-Style Programming

Like other popular languages, such as Java and C#, Python is not a purely functional language. Rather, it offers "functional-style" features that help you write code which is less likely to contain errors, more concise and easier to read, debug and modify. Functional-style programs also can be easier to parallelize to get better performance on today's multi-

---

5. This is how the website `stackoverflow.com` got its name—a good website for getting answers to your programming questions.

core processors. The chart below lists most of Python's key functional-style programming capabilities and shows in parentheses the chapters in which we initially cover many of them.

Functional-style programming topics		
avoiding side effects (4)	generator functions (12)	lazy evaluation (5)
closures	higher-order functions (5)	list comprehensions (5)
declarative programming (4)	immutability (4)	operator module (5, 13, 17)
decorators (10)	internal iteration (4)	pure functions (4)
dictionary comprehensions (6)	iterators (3)	range function (3, 4)
filter/map/reduce (5)	itertools module (17)	reductions (3, 5)
functools module	lambda expressions (5)	set comprehensions (6)
generator expressions (5)		

We cover most of these features throughout the book—many with code examples and others from a literacy perspective. You've already used list, string and built-in function range *iterators* with the for statement, and several *reductions* (functions sum, len, min and max). We discuss declarative programming, immutability and internal iteration below.

What vs. How

As the tasks you perform get more complicated, your code can become harder to read, debug and modify, and more likely to contain errors. Specifying *how* the code works can become complex.

Functional-style programming lets you simply say *what* you want to do. It hides many details of *how* to perform each task. Typically, library code handles the *how* for you. As you'll see, this can eliminate many errors.

Consider the for statement in many other programming languages. Typically, *you* must specify all the details of counter-controlled iteration: a control variable, its initial value, how to increment it and a loop-continuation condition that uses the control variable to determine whether to continue iterating. This style of iteration is known as **external iteration** and is error-prone. For example, you might provide an incorrect initializer, increment or loop-continuation condition. External iteration **mutates** (that is, modifies) the control variable, and the for statement's suite often mutates other variables as well. Every time you modify variables you could introduce errors. Functional-style programming emphasizes **immutability**. That is, it avoids operations that modify variables' values. We'll say more in the next chapter.

Python's for statement and range function *hide* most counter-controlled iteration details. You specify *what* values range should produce and the variable that should receive each value as it's produced. Function range *knows how* to produce those values. Similarly, the for statement *knows how* to get each value from range and *how* to stop iterating when there are no more values. Specifying *what*, but not *how*, is an important aspect of **internal iteration**—a key functional-style programming concept.

The Python built-in functions sum, min and max each use internal iteration. To total the elements of the list grades, you simply declare *what* you want to do—that is, sum(grades). Function sum *knows how* to iterate through the list and add each element to the running total. Stating what you *want* done rather than programming *how* to do it is known as **declarative programming**.

Pure Functions

In pure functional programming language you focus on writing pure functions. A **pure function**'s result depends only on the argument(s) you pass to it. Also, given a particular argument (or arguments), a pure function always produces the same result. For example, built-in function sum's return value depends only on the iterable you pass to it. Given a list [1, 2, 3], sum *always* returns 6 no matter how many times you call it. Also, a pure function does not have *side effects*. For example, even if you pass a *mutable* list to a pure function, the list will contain the same values before and after the function call. When you call the pure function sum, it does not modify its argument.

```
In [1]: values = [1, 2, 3]

In [2]: sum(values)
Out[2]: 6

In [3]: sum(values)   # same call always returns same result
Out[3]: 6

In [4]: values
Out[5]: [1, 2, 3]
```

In the next chapter, we'll continue using functional-style programming concepts. Also, you'll see that *functions are objects* that you can pass to other functions as data.

4.18 Intro to Data Science: Measures of Dispersion

In our discussion of descriptive statistics, we've considered the measures of central tendency—mean, median and mode. These help us categorize typical values in a group—such as the mean height of your classmates or the most frequently purchased car brand (the mode) in a given country.

When we're talking about a group, the entire group is called the **population**. Sometimes a population is quite large, such as the people likely to vote in the next U.S. presidential election, which is a number in excess of 100,000,000 people. For practical reasons, the polling organizations trying to predict who will become the next president work with carefully selected small subsets of the population known as **samples**. Many of the polls in the 2016 election had sample sizes of about 1000 people.

In this section, we continue discussing basic descriptive statistics. We introduce **measures of dispersion** (also called **measures of variability**) that help you understand how spread out the values are. For example, in a class of students, there may be a bunch of students whose height is close to the average, with smaller numbers of students who are considerably shorter or taller.

For our purposes, we'll calculate each measure of dispersion both by hand and with functions from the module statistics, using the following population of 10 six-sided die rolls:

    1, 3, 4, 2, 6, 5, 3, 4, 5, 2

Variance

To determine the **variance**,[6] we begin with the mean of these values—3.5. You obtain this result by dividing the sum of the face values, 35, by the number of rolls, 10. Next, we subtract the mean from every die value (this produces some negative results):

    -2.5, -0.5, 0.5, -1.5, 2.5, 1.5, -0.5, 0.5, 1.5, -1.5

Then, we square each of these results (yielding only positives):

    6.25, 0.25, 0.25, 2.25, 6.25, 2.25, 0.25, 0.25, 2.25, 2.25

Finally, we calculate the mean of these squares, which is 2.25 (22.5 / 10)—this is the **population variance**. Squaring the difference between each die value and the mean of all die values emphasizes **outliers**—the values that are farthest from the mean. As we get deeper into data analytics, sometimes we'll want to pay careful attention to outliers, and sometimes we'll want to ignore them. The following code uses the `statistics` module's **pvariance** function to confirm our manual result:

```
In [1]: import statistics

In [2]: statistics.pvariance([1, 3, 4, 2, 6, 5, 3, 4, 5, 2])
Out[2]: 2.25
```

Standard Deviation

The **standard deviation** is the square root of the variance (in this case, 1.5), which tones down the effect of the outliers. The smaller the variance and standard deviation are, the closer the data values are to the mean and the less overall **dispersion** (that is, **spread**) there is between the values and the mean. The following code calculates the **population standard deviation** with the statistics module's **pstdev** function, confirming our manual result:

```
In [3]: statistics.pstdev([1, 3, 4, 2, 6, 5, 3, 4, 5, 2])
Out[3]: 1.5
```

Passing the `pvariance` function's result to the `math` module's `sqrt` function confirms our result of 1.5:

```
In [4]: import math

In [5]: math.sqrt(statistics.pvariance([1, 3, 4, 2, 6, 5, 3, 4, 5, 2]))
Out[5]: 1.5
```

Advantage of Population Standard Deviation vs. Population Variance

Suppose you've recorded the March Fahrenheit temperatures in your area. You might have 31 numbers such as 19, 32, 28 and 35. The units for these numbers are degrees. When you square your temperatures to calculate the population variance, the units of the population variance become "degrees squared." When you take the square root of the popula-

---

6. For simplicity, we're calculating the *population variance*. There is a subtle difference between the *population variance* and the *sample variance*. Instead of dividing by *n* (the number of die rolls in our example), sample variance divides by *n* – 1. The difference is pronounced for small samples and becomes insignificant as the sample size increases. The `statistics` module provides the functions pvariance and variance to calculate the population variance and sample variance, respectively. Similarly, the `statistics` module provides the functions pstdev and stdev to calculate the population standard deviation and sample standard deviation, respectively.

tion variance to calculate the population standard deviation, the units once again become degrees, which are the *same* units as your temperatures.

 Self Check

1 *(Discussion)* Why do we often work with a sample rather than the full population?
Answer: Because often the full population is unmanageably large.

2 *(True/False)* An advantage of the population variance over the population standard deviation is that its units are the same as the sample values' units.
Answer: False. This an advantage of population standard deviation over population variance.

3 *(IPython Session)* In this section, we worked with population variance and population standard deviation. There is a subtle difference between the *population variance* and the *sample variance*. In our example, instead of dividing by 10 (the number of die rolls), sample variance would divide by 9 (which is one less than the sample size). The difference is pronounced for small samples but becomes insignificant as the sample size increases. The `statistics` module provides the functions `variance` and `stdev` to calculate the sample variance and sample standard deviation, respectively. Redo the manual calculations, then use the `statistics` module's functions to confirm this difference between the two methods of calculation.
Answer:

```
In [1]: import statistics

In [2]: statistics.variance([1, 3, 4, 2, 6, 5, 3, 4, 5, 2])
Out[2]: 2.5

In [3]: statistics.stdev([1, 3, 4, 2, 6, 5, 3, 4, 5, 2])
Out[3]: 1.5811388300841898
```

4.19 Wrap-Up

In this chapter, we created custom functions. We imported capabilities from the `random` and `math` modules. We introduced random-number generation and used it to simulate rolling a six-sided die. We packed multiple values into tuples to return more than one value from a function. We also unpacked a tuple to access its values. We discussed using the Python Standard Library's modules to avoid "reinventing the wheel."

We created functions with default parameter values and called functions with keyword arguments. We also defined functions with arbitrary argument lists. We called methods of objects. We discussed how an identifier's scope determines where in your program you can use it.

You learned more about importing modules. You saw that arguments are passed-by-reference to functions, and how the function-call stack and stack frames support the function-call-and-return mechanism. We've introduced basic list and tuple capabilities over the last two chapters—in the next chapter, we'll discuss them in detail.

Finally, we continued our discussion of descriptive statistics by introducing measures of dispersion—variance and standard deviation—and calculating them with functions from the Python Standard Library's `statistics` module.

For some types of problems, it's useful to have functions call themselves. A **recursive function** calls itself, either directly or indirectly through another function. Recursion is an important topic discussed at length in upper-level computer science courses. We include a detailed treatment in the chapter "Computer Science Thinking: Recursion, Searching, Sorting and Big O."

Exercises

Unless specified otherwise, use IPython sessions for each exercise.

4.1 *(Discussion: else Clause)* In the script of Fig. 4.1, we did not include an else clause in the if…elif statement. What are the possible consequences of this choice?

4.2 *(Discussion: Function-Call Stack)* What happens if you keep pushing onto a stack, without enough popping?

4.3 *(What's Wrong with This Code?)* What is wrong with the following cube function's definition?

```
def cube(x):
    """Calculate the cube of x."""
    x ** 3

print('The cube of 2 is', cube(2))
```

4.4 *(What's Does This Code Do?)* What does the following mystery function do? Assume you pass the list [1, 2, 3, 4, 5] as an argument.

```
def mystery(x):
    y = 0

    for value in x:
        y += value ** 2

    return y
```

4.5 *(Fill in the Missing Code)* Replace the ***s in the seconds_since_midnight function so that it returns the number of seconds since midnight. The function should receive three integers representing the current time of day. Assume that the hour is a value from 0 (midnight) through 23 (11 PM) and that the minute and second are values from 0 to 59. Test your function with actual times. For example, if you call the function for 1:30:45 PM by passing 13, 30 and 45, the function should return 48645.

```
def seconds_since_midnight(***):
    hour_in_seconds = ***
    minute_in_seconds = ***
    return ***
```

4.6 *(Modified average Function)* The average function we defined in Section 4.11 can receive any number of arguments. If you call it with no arguments, however, the function causes a ZeroDivisionError. Reimplement average to receive one required argument *and* the arbitrary argument list argument *args, and update its calculation accordingly. Test your function. The function will always require at least one argument, so you'll no longer be able to get a ZeroDivisionError. When you call average with no arguments, Python should issue a TypeError indicating "average() missing 1 required positional argument."

4.7 *(Date and Time)* Python's `datetime` module contains a `datetime` type with a method `today` that returns the current date and time as a `datetime` object. Write a *parameterless* `date_and_time` function containing the following statement, then call that function to display the current date and time:

```
print(datetime.datetime.today())
```

On our system, the date and time display in the following format:

```
2018-06-08 13:04:19.214180
```

4.8 *(Rounding Numbers)* Investigate built-in function round at

```
https://docs.python.org/3/library/functions.html#round
```

then use it to round the `float` value `13.56449` to the nearest integer, tenths, hundredths and thousandths positions.

4.9 *(Temperature Conversion)* Implement a `fahrenheit` function that returns the Fahrenheit equivalent of a Celsius temperature. Use the following formula:

```
F = (9 / 5) * C + 32
```

Use this function to print a chart showing the Fahrenheit equivalents of all Celsius temperatures in the range 0–100 degrees. Use one digit of precision for the results. Print the outputs in a neat tabular format.

4.10 *(Guess the Number)* Write a script that plays "guess the number." Choose the number to be guessed by selecting a random integer in the range 1 to 1000. Do not reveal this number to the user. Display the prompt `"Guess my number between 1 and 1000 with the fewest guesses:"`. The player inputs a first guess. If the guess is incorrect, display `"Too high. Try again."` or `"Too low. Try again."` as appropriate to help the player "zero in" on the correct answer, then prompt the user for the next guess. When the user enters the correct answer, display `"Congratulations. You guessed the number!"`, and allow the user to choose whether to play again.

4.11 *(Guess-the-Number Modification)* Modify the previous exercise to count the number of guesses the player makes. If the number is 10 or fewer, display `"Either you know the secret or you got lucky!"` If the player makes more than 10 guesses, display `"You should be able to do better!"` Why should it take no more than 10 guesses? Well, with each "good guess," the player should be able to eliminate half of the numbers, then half of the remaining numbers, and so on. Doing this 10 times narrows down the possibilities to a single number. This kind of "halving" appears in many computer science applications. For example, in the "Computer Science Thinking: Recursion, Searching, Sorting and Big O" chapter, we'll present the high-speed binary search and merge sort algorithms, and you'll attempt the quicksort exercise—each of these cleverly uses halving to achieve high performance.

4.12 *(Simulation: The Tortoise and the Hare)* In this problem, you'll re-create the classic race of the tortoise and the hare. You'll use random-number generation to develop a simulation of this memorable event.

Our contenders begin the race at square 1 of 70 squares. Each square represents a position along the race course. The finish line is at square 70. The first contender to reach or pass square 70 is rewarded with a pail of fresh carrots and lettuce. The course weaves its way up the side of a slippery mountain, so occasionally the contenders lose ground.

A clock ticks once per second. With each tick of the clock, your application should adjust the position of the animals according to the rules in the table below. Use variables to keep track of the positions of the animals (i.e., position numbers are 1–70). Start each animal at position 1 (the "starting gate"). If an animal slips left before square 1, move it back to square 1.

Animal	Move type	Percentage of the time	Actual move
Tortoise	Fast plod	50%	3 squares to the right
	Slip	20%	6 squares to the left
	Slow plod	30%	1 square to the right
Hare	Sleep	20%	No move at all
	Big hop	20%	9 squares to the right
	Big slip	10%	12 squares to the left
	Small hop	30%	1 square to the right
	Small slip	20%	2 squares to the left

Create two functions that generate the percentages in the table for the tortoise and the hare, respectively, by producing a random integer i in the range $1 \leq i \leq 10$. In the function for the tortoise, perform a "fast plod" when $1 \leq i \leq 5$, a "slip" when $6 \leq i \leq 7$ or a "slow plod" when $8 \leq i \leq 10$. Use a similar technique in the function for the hare.

Begin the race by displaying

```
BANG !!!!!
AND THEY'RE OFF !!!!!
```

Then, for each tick of the clock (i.e., each iteration of a loop), display a 70-position line showing the letter "T" in the position of the tortoise and the letter "H" in the position of the hare. Occasionally, the contenders will land on the same square. In this case, the tortoise bites the hare, and your application should display "OUCH!!!" at that position. All positions other than the "T", the "H" or the "OUCH!!!" (in case of a tie) should be blank.

After each line is displayed, test for whether either animal has reached or passed square 70. If so, display the winner and terminate the simulation. If the tortoise wins, display TORTOISE WINS!!! YAY!!! If the hare wins, display Hare wins. Yuch. If both animals win on the same tick of the clock, you may want to favor the tortoise (the "underdog"), or you may want to display "It's a tie". If neither animal wins, perform the loop again to simulate the next tick of the clock. When you're ready to run your application, assemble a group of fans to watch the race. You'll be amazed at how involved your audience gets!

4.13 *(Arbitrary Argument List)* Calculate the product of a series of integers that are passed to the function `product`, which receives an arbitrary argument list. Test your function with several calls, each with a different number of arguments.

4.14 *(Computer-Assisted Instruction)* Computer-assisted instruction (CAI) refers to the use of computers in education. Write a script to help an elementary school student learn multiplication. Create a function that randomly generates and returns a tuple of two positive one-digit integers. Use that function's result in your script to prompt the user with a question, such as

    How much is 6 times 7?

For a correct answer, display the message "Very good!" and ask another multiplication question. For an incorrect answer, display the message "No. Please try again." and let the student try the same question repeatedly until the student finally gets it right.

4.15 *(Computer-Assisted Instruction: Reducing Student Fatigue)* Varying the computer's responses can help hold the student's attention. Modify the previous exercise so that various comments are displayed for each answer. Possible responses to a correct answer should include 'Very good!', 'Nice work!' and 'Keep up the good work!' Possible responses to an incorrect answer should include 'No. Please try again.', 'Wrong. Try once more.' and 'No. Keep trying.' Choose a number from 1 to 3, then use that value to select one of the three appropriate responses to each correct or incorrect answer.

4.16 *(Computer-Assisted Instruction: Difficulty Levels)* Modify the previous exercise to allow the user to enter a difficulty level. At a difficulty level of 1, the program should use only single-digit numbers in the problems and at a difficulty level of 2, numbers as large as two digits.

4.17 *(Computer-Assisted Instruction: Varying the Types of Problems)* Modify the previous exercise to allow the user to pick a type of arithmetic problem to study—1 means addition problems only, 2 means subtraction problems only, 3 means multiplication problems only, 4 means division problems only (avoid dividing by 0) and 5 means a random mixture of all these types.

4.18 *(Functional-Style Programming: Internal vs. External Iteration)* Why is internal iteration preferable to external iteration in functional-style programming?

4.19 *(Functional-Style Programming: What vs. How)* Why is programming that emphasizes "what" preferable to programming that emphasizes "how"? What is it that makes "what" programming feasible?

4.20 *(Intro to Data Science: Population Variance vs. Sample Variance)* We mentioned in the Intro to Data Science section that there's a slight difference between the way the statistics module's functions calculate the population variance and the sample variance. The same is true for the population standard deviation and the sample standard deviation. Research the reason for these differences.

Sequences: Lists and Tuples

5

Objectives

In this chapter, you'll:

- Create and initialize lists and tuples.
- Refer to elements of lists, tuples and strings.
- Sort and search lists, and search tuples.
- Pass lists and tuples to functions and methods.
- Use list methods to perform common manipulations, such as searching for items, sorting a list, inserting items and removing items.
- Use additional Python functional-style programming capabilities, including lambdas and the operations filter, map and reduce.
- Use functional-style list comprehensions to create lists quickly and easily, and use generator expressions to generate values on demand.
- Use two-dimensional lists.
- Enhance your analysis and presentation skills with the Seaborn and Matplotlib visualization libraries.

Outline

5.1 Introduction	**5.12** List Comprehensions
5.2 Lists	**5.13** Generator Expressions
5.3 Tuples	**5.14** Filter, Map and Reduce
5.4 Unpacking Sequences	**5.15** Other Sequence Processing Functions
5.5 Sequence Slicing	**5.16** Two-Dimensional Lists
5.6 `del` Statement	**5.17** Intro to Data Science: Simulation and Static Visualizations
5.7 Passing Lists to Functions	5.17.1 Sample Graphs for 600, 60,000 and 6,000,000 Die Rolls
5.8 Sorting Lists	5.17.2 Visualizing Die-Roll Frequencies and Percentages
5.9 Searching Sequences	
5.10 Other List Methods	**5.18** Wrap-Up
5.11 Simulating Stacks with Lists	Exercises

5.1 Introduction

In the last two chapters, we briefly introduced the list and tuple sequence types for representing ordered collections of items. **Collections** are prepackaged data structures consisting of related data items. Examples of collections include your favorite songs on your smartphone, your contacts list, a library's books, your cards in a card game, your favorite sports team's players, the stocks in an investment portfolio, patients in a cancer study and a shopping list. Python's built-in collections enable you to store and access data conveniently and efficiently. In this chapter, we discuss lists and tuples in more detail.

We'll demonstrate common list and tuple manipulations. You'll see that lists (which are modifiable) and tuples (which are not) have many common capabilities. Each can hold items of the same or different types. Lists can **dynamically resize** as necessary, growing and shrinking at execution time. We discuss one-dimensional and two-dimensional lists.

In the preceding chapter, we demonstrated random-number generation and simulated rolling a six-sided die. We conclude this chapter with our next Intro to Data Science section, which uses the visualization libraries Seaborn and Matplotlib to interactively develop static bar charts containing the die frequencies. In the next chapter's Intro to Data Science section, we'll present an animated visualization in which the bar chart changes *dynamically* as the number of die rolls increases—you'll see the law of large numbers "in action."

5.2 Lists

Here, we discuss lists in more detail and explain how to refer to particular list **elements**. Many of the capabilities shown in this section apply to all sequence types.

Creating a List
Lists typically store **homogeneous data**, that is, values of the *same* data type. Consider the list c, which contains five integer elements:

```
In [1]: c = [-45, 6, 0, 72, 1543]

In [2]: c
Out[2]: [-45, 6, 0, 72, 1543]
```

They also may store **heterogeneous data**, that is, data of many different types. For example, the following list contains a student's first name (a string), last name (a string), grade point average (a `float`) and graduation year (an `int`):

```
['Mary', 'Smith', 3.57, 2022]
```

Accessing Elements of a List

You reference a list element by writing the list's name followed by the element's **index** (that is, its **position number**) enclosed in square brackets (`[]`, known as the **subscription operator**). The following diagram shows the list c labeled with its element names:

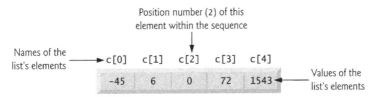

The first element in a list has the index 0. So, in the five-element list c, the first element is named c[0] and the last is c[4]:

```
In [3]: c[0]
Out[3]: -45

In [4]: c[4]
Out[4]: 1543
```

Determining a List's Length

To get a list's length, use the built-in **`len` function**:

```
In [5]: len(c)
Out[5]: 5
```

Accessing Elements from the End of the List with Negative Indices

Lists also can be accessed from the end by using *negative indices*:

So, list c's last element (c[4]), can be accessed with c[-1] and its first element with c[-5]:

```
In [6]: c[-1]
Out[6]: 1543

In [7]: c[-5]
Out[7]: -45
```

Indices Must Be Integers or Integer Expressions

An index must be an integer or integer expression (or a *slice*, as we'll soon see):

```
In [8]: a = 1

In [9]: b = 2

In [10]: c[a + b]
Out[10]: 72
```

Using a non-integer index value causes a `TypeError`.

Lists Are Mutable

Lists are mutable—their elements can be modified:

```
In [11]: c[4] = 17

In [12]: c
Out[12]: [-45, 6, 0, 72, 17]
```

You'll soon see that you also can insert and delete elements, changing the list's length.

Some Sequences Are Immutable

Python's string and tuple sequences are immutable—they cannot be modified. You can get the individual characters in a string, but attempting to assign a new value to one of the characters causes a `TypeError`:

```
In [13]: s = 'hello'

In [14]: s[0]
Out[14]: 'h'

In [15]: s[0] = 'H'
-------------------------------------------------------------------------
TypeError                                 Traceback (most recent call last)
<ipython-input-15-812ef2514689> in <module>()
----> 1 s[0] = 'H'

TypeError: 'str' object does not support item assignment
```

Attempting to Access a Nonexistent Element

Using an out-of-range list, tuple or string index causes an `IndexError`:

```
In [16]: c[100]
-------------------------------------------------------------------------
IndexError                                Traceback (most recent call last)
<ipython-input-19-9a31ea1e1a13> in <module>()
----> 1 c[100]

IndexError: list index out of range
```

Using List Elements in Expressions

List elements may be used as variables in expressions:

```
In [17]: c[0] + c[1] + c[2]
Out[17]: -39
```

Appending to a List with +=

Let's start with an *empty* list [], then use a `for` statement and `+=` to append the values 1 through 5 to the list—the list grows dynamically to accommodate each item:

```
In [18]: a_list = []

In [19]: for number in range(1, 6):
    ...:     a_list += [number]
    ...:

In [20]: a_list
Out[20]: [1, 2, 3, 4, 5]
```

When the left operand of += is a list, the right operand must be an *iterable*; otherwise, a TypeError occurs. In snippet [19]'s suite, the square brackets around number create a one-element list, which we append to a_list. If the right operand contains multiple elements, += appends them all. The following appends the characters of 'Python' to the list letters:

```
In [21]: letters = []

In [22]: letters += 'Python'

In [23]: letters
Out[23]: ['P', 'y', 't', 'h', 'o', 'n']
```

If the right operand of += is a tuple, its elements also are appended to the list. Later in the chapter, we'll use the list method append to add items to a list.

Concatenating Lists with +

You can **concatenate** two lists, two tuples or two strings using the + operator. The result is a *new* sequence of the same type containing the left operand's elements followed by the right operand's elements. The original sequences are unchanged:

```
In [24]: list1 = [10, 20, 30]

In [25]: list2 = [40, 50]

In [26]: concatenated_list = list1 + list2

In [27]: concatenated_list
Out[27]: [10, 20, 30, 40, 50]
```

A TypeError occurs if the + operator's operands are different sequence types—for example, concatenating a list and a tuple is an error.

Using for and range to Access List Indices and Values

List elements also can be accessed via their indices and the subscription operator ([]):

```
In [28]: for i in range(len(concatenated_list)):
    ...:     print(f'{i}: {concatenated_list[i]}')
    ...:
0: 10
1: 20
2: 30
3: 40
4: 50
```

The function call range(len(concatenated_list)) produces a sequence of integers representing concatenated_list's indices (in this case, 0 through 4). When looping in this manner, you must ensure that indices remain in range. Soon, we'll show a safer way to access element indices and values using built-in function enumerate.

Comparison Operators

You can compare entire lists element-by-element using comparison operators:

```
In [29]: a = [1, 2, 3]

In [30]: b = [1, 2, 3]

In [31]: c = [1, 2, 3, 4]

In [32]: a == b   # True: corresponding elements in both are equal
Out[32]: True

In [33]: a == c   # False: a and c have different elements and lengths
Out[33]: False

In [34]: a < c    # True: a has fewer elements than c
Out[34]: True

In [35]: c >= b   # True: elements 0-2 are equal but c has more elements
Out[35]: True
```

✓ Self Check

1 *(Fill-In)* Python's string and tuple sequences are _____—they cannot be modified.
Answer: immutable.

2 *(True/False)* The + operator's sequence operands may be of any sequence type.
Answer: False. The + operator's operand sequences must have the *same* type; otherwise, a TypeError occurs.

3 *(IPython Session)* Create a function cube_list that cubes each element of a list. Call the function with the list numbers containing 1 through 10. Show numbers after the call.
Answer:

```
In [1]: def cube_list(values):
   ...:     for i in range(len(values)):
   ...:         values[i] **= 3
   ...:

In [2]: numbers = [1, 2, 3, 4, 5, 6, 7, 8, 9, 10]

In [3]: cube_list(numbers)

In [4]: numbers
Out[4]: [1, 8, 27, 64, 125, 216, 343, 512, 729, 1000]
```

4 *(IPython Session)* Use an empty list named characters and a += augmented assignment statement to convert the string 'Birthday' into a list of its characters.
Answer:

```
In [5]: characters = []

In [6]: characters += 'Birthday'

In [7]: characters
Out[7]: ['B', 'i', 'r', 't', 'h', 'd', 'a', 'y']
```

5.3 Tuples

As discussed in the preceding chapter, tuples are immutable and typically store heterogeneous data, but the data can be homogeneous. A tuple's length is its number of elements and cannot change during program execution.

Creating Tuples

To create an empty tuple, use empty parentheses:

```
In [1]: student_tuple = ()

In [2]: student_tuple
Out[2]: ()

In [3]: len(student_tuple)
Out[3]: 0
```

Recall that you can pack a tuple by separating its values with commas:

```
In [4]: student_tuple = 'John', 'Green', 3.3

In [5]: student_tuple
Out[5]: ('John', 'Green', 3.3)

In [6]: len(student_tuple)
Out[6]: 3
```

When you output a tuple, Python always displays its contents in parentheses. You may surround a tuple's comma-separated list of values with optional parentheses:

```
In [7]: another_student_tuple = ('Mary', 'Red', 3.3)

In [8]: another_student_tuple
Out[8]: ('Mary', 'Red', 3.3)
```

The following code creates a one-element tuple:

```
In [9]: a_singleton_tuple = ('red',)   # note the comma

In [10]: a_singleton_tuple
Out[10]: ('red',)
```

The comma (,) that follows the string 'red' identifies a_singleton_tuple as a tuple—the parentheses are optional. If the comma were omitted, the parentheses would be redundant, and a_singleton_tuple would simply refer to the string 'red' rather than a tuple.

Accessing Tuple Elements

A tuple's elements, though related, are often of multiple types. Usually, you do not iterate over them. Rather, you access each individually. Like list indices, tuple indices start at 0. The following code creates time_tuple representing an hour, minute and second, displays the tuple, then uses its elements to calculate the number of seconds since midnight—note that we perform a *different* operation with each value in the tuple:

```
In [11]: time_tuple = (9, 16, 1)

In [12]: time_tuple
Out[12]: (9, 16, 1)
```

```
In [13]: time_tuple[0] * 3600 + time_tuple[1] * 60 + time_tuple[2]
Out[13]: 33361
```

Assigning a value to a tuple element causes a `TypeError`.

Adding Items to a String or Tuple
As with lists, the += augmented assignment statement can be used with strings and tuples, even though they're *immutable*. In the following code, after the two assignments, `tuple1` and `tuple2` refer to the *same* tuple object:

```
In [14]: tuple1 = (10, 20, 30)

In [15]: tuple2 = tuple1

In [16]: tuple2
Out[16]: (10, 20, 30)
```

Concatenating the tuple (40, 50) to `tuple1` creates a *new* tuple, then assigns a reference to it to the variable `tuple1`—`tuple2` still refers to the original tuple:

```
In [17]: tuple1 += (40, 50)

In [18]: tuple1
Out[18]: (10, 20, 30, 40, 50)

In [19]: tuple2
Out[19]: (10, 20, 30)
```

For a string or tuple, the item to the right of += must be a string or tuple, respectively—mixing types causes a `TypeError`.

Appending Tuples to Lists
You can use += to append a tuple to a list:

```
In [20]: numbers = [1, 2, 3, 4, 5]

In [21]: numbers += (6, 7)

In [22]: numbers
Out[22]: [1, 2, 3, 4, 5, 6, 7]
```

Tuples May Contain Mutable Objects
Let's create a `student_tuple` with a first name, last name and list of grades:

```
In [23]: student_tuple = ('Amanda', 'Blue', [98, 75, 87])
```

Even though the tuple is immutable, its list element is mutable:

```
In [24]: student_tuple[2][1] = 85

In [25]: student_tuple
Out[25]: ('Amanda', 'Blue', [98, 85, 87])
```

In the *double-subscripted name* `student_tuple[2][1]`, Python views `student_tuple[2]` as the element of the tuple containing the list [98, 75, 87], then uses [1] to access the list element containing 75. The assignment in snippet [24] replaces that grade with 85.

Self Check

1 *(True/False)* A += augmented assignment statement may not be used with strings and tuples, because they're immutable.
Answer: False. A += augmented assignment statement also may be used with strings and tuples, even though they're immutable. The result is a *new* string or tuple, respectively.

2 *(True/False)* Tuples can contain only immutable objects.
Answer: False. Even though a tuple is immutable, its elements can be mutable objects, such as lists.

3 *(IPython Session)* Create a single-element tuple containing 123.45, then display it.
Answer:
```
In [1]: single = (123.45,)

In [2]: single
Out[2]: (123.45,)
```

4 *(IPython Session)* Show what happens when you attempt to concatenate sequences of different types—the list [1, 2, 3] and the tuple (4, 5, 6)—using the + operator.
Answer:
```
In [3]: [1, 2, 3] + (4, 5, 6)
-------------------------------------------------------------------------
TypeError                                 Traceback (most recent call last)
<ipython-input-3-1ac3d3041bfa> in <module>()
----> 1 [1, 2, 3] + (4, 5, 6)

TypeError: can only concatenate list (not "tuple") to list
```

5.4 Unpacking Sequences

The previous chapter introduced tuple unpacking. You can unpack any sequence's elements by assigning the sequence to a comma-separated list of variables. A ValueError occurs if the number of variables to the left of the assignment symbol is not identical to the number of elements in the sequence on the right:

```
In [1]: student_tuple = ('Amanda', [98, 85, 87])

In [2]: first_name, grades = student_tuple

In [3]: first_name
Out[3]: 'Amanda'

In [4]: grades
Out[4]: [98, 85, 87]
```

The following code unpacks a string, a list and a sequence produced by range:

```
In [5]: first, second = 'hi'

In [6]: print(f'{first}  {second}')
h i
```

```
In [7]: number1, number2, number3 = [2, 3, 5]

In [8]: print(f'{number1}  {number2}  {number3}')
2  3  5

In [9]: number1, number2, number3 = range(10, 40, 10)

In [10]: print(f'{number1}  {number2}  {number3}')
10  20  30
```

Swapping Values Via Packing and Unpacking

You can swap two variables' values using sequence packing and unpacking:

```
In [11]: number1 = 99

In [12]: number2 = 22

In [13]: number1, number2 = (number2, number1)

In [14]: print(f'number1 = {number1}; number2 = {number2}')
number1 = 22; number2 = 99
```

Accessing Indices and Values Safely with Built-in Function enumerate

Earlier, we called range to produce a sequence of index values, then accessed list elements in a for loop using the index values and the subscription operator ([]). This is error-prone because you could pass the wrong arguments to range. If any value produced by range is an out-of-bounds index, using it as an index causes an IndexError.

The preferred mechanism for accessing an element's index *and* value is the built-in function **enumerate**. This function receives an iterable and creates an iterator that, for each element, returns a tuple containing the element's index and value. The following code uses the built-in function **list** to create a list containing enumerate's results:

```
In [15]: colors = ['red', 'orange', 'yellow']

In [16]: list(enumerate(colors))
Out[16]: [(0, 'red'), (1, 'orange'), (2, 'yellow')]
```

Similarly the built-in function **tuple** creates a tuple from a sequence:

```
In [17]: tuple(enumerate(colors))
Out[17]: ((0, 'red'), (1, 'orange'), (2, 'yellow'))
```

The following for loop unpacks each tuple returned by enumerate into the variables index and value and displays them:

```
In [18]: for index, value in enumerate(colors):
   ....:     print(f'{index}: {value}')
   ....:
0: red
1: orange
2: yellow
```

Creating a Primitive Bar Chart

The script in Fig. 5.1 creates a primitive **bar chart** where each bar's length is made of asterisks (*) and is proportional to the list's corresponding element value. We use the function enumerate to access the list's indices and values safely. To run this example, change to this chapter's ch05 examples folder, then enter:

```
ipython fig05_01.py
```

or, if you're in IPython already, use the command:

```
run fig05_01.py
```

```
1   # fig05_01.py
2   """Displaying a bar chart"""
3   numbers = [19, 3, 15, 7, 11]
4
5   print('\nCreating a bar chart from numbers:')
6   print(f'Index{"Value":>8}   Bar')
7
8   for index, value in enumerate(numbers):
9       print(f'{index:>5}{value:>8}   {"*" * value}')
```

```
Creating a bar chart from numbers:
Index   Value   Bar
    0      19   *******************
    1       3   ***
    2      15   ***************
    3       7   *******
    4      11   ***********
```

Fig. 5.1 | Displaying a bar chart.

The for statement uses enumerate to get each element's index and value, then displays a formatted line containing the index, the element value and the corresponding bar of asterisks. The expression

```
"*" * value
```

creates a string consisting of value asterisks. When used with a sequence, the multiplication operator (*) *repeats* the sequence—in this case, the string "*"—value times. Later in this chapter, we'll use the open-source Seaborn and Matplotlib libraries to display a publication-quality bar chart visualization.

Self Check

1 *(Fill-In)* A sequence's elements can be _____ by assigning the sequence to a comma-separated list of variables.
Answer: unpacked.

2 *(True/False)* The following expression causes an error:

```
'-' * 10
```

Answer: False: In this context, the multiplication operator (*) repeats the string ('-') 10 times.

3 *(IPython Session)* Create a tuple `high_low` representing a day of the week (a string) and its high and low temperatures (integers), display its string representation, then perform the following tasks in an interactive IPython session:
  a) Use the [] operator to access and display the `high_low` tuple's elements.
  b) Unpack the `high_low` tuple into the variables `day` and `high`. What happens and why?

Answer: For Part (b) an error occurs because you must unpack *all* the elements of a sequence.

```
In [1]: high_low = ('Monday', 87, 65)

In [2]: high_low
Out[2]: ('Monday', 87, 65)

In [3]: print(f'{high_low[0]}: High={high_low[1]}, Low={high_low[2]}')
Monday: High=87, Low=65

In [4]: day, high = high_low
-------------------------------------------------------------------------
ValueError                                Traceback (most recent call last)
<ipython-input-3-0c3ad5c97284> in <module>()
----> 1 day, high = high_low

ValueError: too many values to unpack (expected 2)
```

4 *(IPython Session)* Create the list `names` containing three name strings. Use a `for` loop and the `enumerate` function to iterate through the elements and display each element's index and value.

Answer:

```
In [4]: names = ['Amanda', 'Sam', 'David']

In [5]: for i, name in enumerate(names):
   ...:     print(f'{i}: {name}')
   ...:
0: Amanda
1: Sam
2: David
```

5.5 Sequence Slicing

You can slice sequences to create new sequences of the same type containing *subsets* of the original elements. Slice operations can modify mutable sequences—those that do *not* modify a sequence work identically for lists, tuples and strings.

Specifying a Slice with Starting and Ending Indices

Let's create a slice consisting of the elements at indices 2 through 5 of a list:

```
In [1]: numbers = [2, 3, 5, 7, 11, 13, 17, 19]

In [2]: numbers[2:6]
Out[2]: [5, 7, 11, 13]
```

The slice *copies* elements from the *starting index* to the left of the colon (2) up to, but not including, the *ending index* to the right of the colon (6). The original list is not modified.

5.5 Sequence Slicing

Specifying a Slice with Only an Ending Index
If you omit the starting index, 0 is assumed. So, the slice `numbers[:6]` is equivalent to the slice `numbers[0:6]`:

```
In [3]: numbers[:6]
Out[3]: [2, 3, 5, 7, 11, 13]

In [4]: numbers[0:6]
Out[4]: [2, 3, 5, 7, 11, 13]
```

Specifying a Slice with Only a Starting Index
If you omit the ending index, Python assumes the sequence's length (8 here), so snippet [5]'s slice contains the elements of `numbers` at indices 6 and 7:

```
In [5]: numbers[6:]
Out[5]: [17, 19]

In [6]: numbers[6:len(numbers)]
Out[6]: [17, 19]
```

Specifying a Slice with No Indices
Omitting both the start and end indices copies the entire sequence:

```
In [7]: numbers[:]
Out[7]: [2, 3, 5, 7, 11, 13, 17, 19]
```

Though slices create new objects, slices make *shallow* copies of the elements—that is, they copy the elements' references but not the objects they point to. So, in the snippet above, the new list's elements refer to the *same objects* as the original list's elements, rather than to separate copies. In the "Array-Oriented Programming with NumPy" chapter, we'll explain *deep* copying, which actually copies the referenced objects themselves, and we'll point out when deep copying is preferred.

Slicing with Steps
The following code uses a *step* of 2 to create a slice with every other element of `numbers`:

```
In [8]: numbers[::2]
Out[8]: [2, 5, 11, 17]
```

We omitted the start and end indices, so 0 and `len(numbers)` are assumed, respectively.

Slicing with Negative Indices and Steps
You can use a negative step to select slices in *reverse* order. The following code concisely creates a new list in reverse order:

```
In [9]: numbers[::-1]
Out[9]: [19, 17, 13, 11, 7, 5, 3, 2]
```

This is equivalent to:

```
In [10]: numbers[-1:-9:-1]
Out[10]: [19, 17, 13, 11, 7, 5, 3, 2]
```

Modifying Lists Via Slices
You can modify a list by assigning to a slice of it—the rest of the list is unchanged. The following code replaces `numbers`' first three elements, leaving the rest unchanged:

```
In [11]: numbers[0:3] = ['two', 'three', 'five']

In [12]: numbers
Out[12]: ['two', 'three', 'five', 7, 11, 13, 17, 19]
```

The following deletes only the first three elements of numbers by assigning an *empty* list to the three-element slice:

```
In [13]: numbers[0:3] = []

In [14]: numbers
Out[14]: [7, 11, 13, 17, 19]
```

The following assigns a list's elements to a slice of every other element of numbers:

```
In [15]: numbers = [2, 3, 5, 7, 11, 13, 17, 19]

In [16]: numbers[::2] = [100, 100, 100, 100]

In [17]: numbers
Out[17]: [100, 3, 100, 7, 100, 13, 100, 19]

In [18]: id(numbers)
Out[18]: 4434456648
```

Let's delete all the elements in numbers, leaving the *existing* list empty:

```
In [19]: numbers[:] = []

In [20]: numbers
Out[20]: []

In [21]: id(numbers)
Out[21]: 4434456648
```

Deleting numbers' contents (snippet [19]) is different from assigning numbers a *new* empty list [] (snippet [22]). To prove this, we display numbers' identity after each operation. The identities are different, so they represent separate objects in memory:

```
In [22]: numbers = []

In [23]: numbers
Out[23]: []

In [24]: id(numbers)
Out[24]: 4406030920
```

When you assign a new object to a variable (as in snippet [21]), the original object will be garbage collected if no other variables refer to it.

✓ Self Check

1 *(True/False)* Slice operations that modify a sequence work identically for lists, tuples and strings.
Answer: False. Slice operations that *do not* modify a sequence work identically for lists, tuples and strings.

2 *(Fill-In)* Assume you have a list called names. The slice expression _____ creates a new list with the elements of names in reverse order.
Answer: names[::-1]

3 *(IPython Session)* Create a list called numbers containing the values from 1 through 15, then use *slices* to perform the following operations consecutively:
   a) Select number's even integers.
   b) Replace the elements at indices 5 through 9 with 0s, then show the resulting list.
   c) Keep only the first five elements, then show the resulting list.
   d) Delete all the remaining elements by assigning to a slice. Show the resulting list.

Answer:

```
In [1]: numbers = list(range(1, 16))

In [2]: numbers
Out[2]: [1, 2, 3, 4, 5, 6, 7, 8, 9, 10, 11, 12, 13, 14, 15]

In [3]: numbers[1:len(numbers):2]
Out[3]: [2, 4, 6, 8, 10, 12, 14]

In [4]: numbers[5:10] = [0] * len(numbers[5:10])

In [5]: numbers
Out[5]: [1, 2, 3, 4, 5, 0, 0, 0, 0, 0, 11, 12, 13, 14, 15]

In [6]: numbers[5:] = []

In [7]: numbers
Out[7]: [1, 2, 3, 4, 5]

In [8]: numbers[:] = []

In [9]: numbers
Out[9]: []
```

Recall that multiplying a sequence repeats that sequence the specified number of times.

5.6 del Statement

The **del statement** also can be used to remove elements from a list and to delete variables from the interactive session. You can remove the element at any valid index or the element(s) from any valid slice.

Deleting the Element at a Specific List Index

Let's create a list, then use del to remove its last element:

```
In [1]: numbers = list(range(0, 10))

In [2]: numbers
Out[2]: [0, 1, 2, 3, 4, 5, 6, 7, 8, 9]

In [3]: del numbers[-1]

In [4]: numbers
Out[4]: [0, 1, 2, 3, 4, 5, 6, 7, 8]
```

Deleting a Slice from a List

The following deletes the list's first two elements:

```
In [5]: del numbers[0:2]

In [6]: numbers
Out[6]: [2, 3, 4, 5, 6, 7, 8]
```

The following uses a step in the slice to delete every other element from the entire list:

```
In [7]: del numbers[::2]

In [8]: numbers
Out[8]: [3, 5, 7]
```

Deleting a Slice Representing the Entire List
The following code deletes all of the list's elements:

```
In [9]: del numbers[:]

In [10]: numbers
Out[10]: []
```

Deleting a Variable from the Current Session
The del statement can delete any variable. Let's delete numbers from the interactive session, then attempt to display the variable's value, causing a NameError:

```
In [11]: del numbers

In [12]: numbers
-------------------------------------------------------------------------
NameError                                 Traceback (most recent call last)
<ipython-input-12-426f8401232b> in <module>()
----> 1 numbers

NameError: name 'numbers' is not defined
```

✓ Self Check

1 *(Fill-In)* Given a list numbers containing 1 through 10, del numbers[-2] removes the value _____ from the list.
Answer: 9.

2 *(IPython Session)* Create a list called numbers containing the values from 1 through 15, then use the del statement to perform the following operations consecutively:
   a) Delete a slice containing the first four elements, then show the resulting list.
   b) Starting with the first element, use a slice to delete every other element of the list, then show the resulting list.
Answer:

```
In [1]: numbers = list(range(1, 16))

In [2]: numbers
Out[2]: [1, 2, 3, 4, 5, 6, 7, 8, 9, 10, 11, 12, 13, 14, 15]

In [3]: del numbers[0:4]

In [4]: numbers
Out[4]: [5, 6, 7, 8, 9, 10, 11, 12, 13, 14, 15]

In [5]: del numbers[::2]

In [6]: numbers
Out[6]: [6, 8, 10, 12, 14]
```

5.7 Passing Lists to Functions

In the last chapter, we mentioned that all objects are passed by reference and demonstrated passing an immutable object as a function argument. Here, we discuss references further by examining what happens when a program passes a mutable list object to a function.

Passing an Entire List to a Function

Consider the function `modify_elements`, which receives a reference to a list and multiplies each of the list's element values by 2:

```
In [1]: def modify_elements(items):
   ...:     """Multiplies all element values in items by 2."""
   ...:     for i in range(len(items)):
   ...:         items[i] *= 2
   ...:

In [2]: numbers = [10, 3, 7, 1, 9]

In [3]: modify_elements(numbers)

In [4]: numbers
Out[4]: [20, 6, 14, 2, 18]
```

Function `modify_elements`' `items` parameter receives a reference to the *original* list, so the statement in the loop's suite modifies each element in the original list object.

Passing a Tuple to a Function

When you pass a tuple to a function, attempting to modify the tuple's immutable elements results in a `TypeError`:

```
In [5]: numbers_tuple = (10, 20, 30)

In [6]: numbers_tuple
Out[6]: (10, 20, 30)

In [7]: modify_elements(numbers_tuple)
-------------------------------------------------------------------
TypeError                                 Traceback (most recent call last)
<ipython-input-27-9339741cd595> in <module>()
----> 1 modify_elements(numbers_tuple)

<ipython-input-25-27acb8f8f44c> in modify_elements(items)
      2     """Multiplies all element values in items by 2."""
      3     for i in range(len(items)):
----> 4         items[i] *= 2
      5
      6

TypeError: 'tuple' object does not support item assignment
```

Recall that tuples may contain mutable objects, such as lists. Those objects still can be modified when a tuple is passed to a function.

A Note Regarding Tracebacks

The previous traceback shows the *two* snippets that led to the `TypeError`. The first is snippet [7]'s function call. The second is snippet [1]'s function definition. Line numbers precede each snippet's code. We've demonstrated mostly single-line snippets. When an exception occurs in such a snippet, it's always preceded by ----> 1, indicating that line 1 (the snippet's only line) caused the exception. Multiline snippets like the definition of `modify_elements` show consecutive line numbers starting at 1. The notation ----> 4 above indicates that the exception occurred in line 4 of `modify_elements`. No matter how long the traceback is, the last line of code with ----> caused the exception.

Self Check

1 *(True/False)* You cannot modify a list's contents when you pass it to a function.
Answer: False. When you pass a list (a mutable object) to a function, the function receives a reference to the original list object and can use that reference to modify the original list's contents.

2 *(True/False)* Tuples can contain lists and other mutable objects. Those mutable objects can be modified when a tuple is passed to a function.
Answer: True.

5.8 Sorting Lists

A common computing task called **sorting** enables you to arrange data either in ascending or descending order. Sorting is an intriguing problem that has attracted intense computer-science research efforts. It's studied in detail in data-structures and algorithms courses. We discuss sorting in more detail in the "Computer Science Thinking: Recursion, Searching, Sorting and Big O" chapter.

Sorting a List in Ascending Order
List method **sort** *modifies* a list to arrange its elements in ascending order:

```
In [1]: numbers = [10, 3, 7, 1, 9, 4, 2, 8, 5, 6]

In [2]: numbers.sort()

In [3]: numbers
Out[3]: [1, 2, 3, 4, 5, 6, 7, 8, 9, 10]
```

Sorting a List in Descending Order
To sort a list in descending order, call list method **sort** with the optional keyword argument **reverse** set to True (False is the default):

```
In [4]: numbers.sort(reverse=True)

In [5]: numbers
Out[5]: [10, 9, 8, 7, 6, 5, 4, 3, 2, 1]
```

Built-In Function sorted
Built-in function **sorted** *returns a new list* containing the sorted elements of its argument *sequence*—the original sequence is *unmodified*. The following code demonstrates function sorted for a list, a string and a tuple:

```
In [6]: numbers = [10, 3, 7, 1, 9, 4, 2, 8, 5, 6]

In [7]: ascending_numbers = sorted(numbers)
```

```
In [8]: ascending_numbers
Out[8]: [1, 2, 3, 4, 5, 6, 7, 8, 9, 10]

In [9]: numbers
Out[9]: [10, 3, 7, 1, 9, 4, 2, 8, 5, 6]

In [10]: letters = 'fadgchjebi'

In [11]: ascending_letters = sorted(letters)

In [12]: ascending_letters
Out[12]: ['a', 'b', 'c', 'd', 'e', 'f', 'g', 'h', 'i', 'j']

In [13]: letters
Out[13]: 'fadgchjebi'

In [14]: colors = ('red', 'orange', 'yellow', 'green', 'blue')

In [15]: ascending_colors = sorted(colors)

In [16]: ascending_colors
Out[16]: ['blue', 'green', 'orange', 'red', 'yellow']

In [17]: colors
Out[17]: ('red', 'orange', 'yellow', 'green', 'blue')
```

Use the optional keyword argument reverse with the value True to sort the elements in descending order.

Self Check

1 *(Fill-In)* To sort a list in descending order, call list method sort with the optional keyword argument _____ set to True.
Answer: reverse.

2 *(True/False)* All sequences provide a sort method.
Answer: False. Immutable sequences like tuples and strings do not provide a sort method. However, you can sort *any* sequence *without modifying it* by using built-in function sorted, which returns a *new* list containing the sorted elements of its argument sequence.

3 *(IPython Session)* Create a foods list containing 'Cookies', 'pizza', 'Grapes', 'apples', 'steak' and 'Bacon'. Use list method sort to sort the list in ascending order. Are the strings in alphabetical order?
Answer:

```
In [1]: foods = ['Cookies', 'pizza', 'Grapes',
   ...:          'apples', 'steak', 'Bacon']
   ...:

In [2]: foods.sort()

In [3]: foods
Out[3]: ['Bacon', 'Cookies', 'Grapes', 'apples', 'pizza', 'steak']
```

They're probably not in what you'd consider alphabetical order, but they are in order as defined by the underlying character set—known as **lexicographical order**. As you'll see later in the chapter, strings are compared by their character's numerical values, not their letters, and the values of uppercase letters are *lower* than the values of lowercase letters.

5.9 Searching Sequences

Often, you'll want to determine whether a sequence (such as a list, tuple or string) contains a value that matches a particular **key** value. **Searching** is the process of locating a key.

List Method index

List method **index** takes as an argument a search key—the value to locate in the list—then searches through the list from index 0 and returns the index of the *first* element that matches the search key:

```
In [1]: numbers = [3, 7, 1, 4, 2, 8, 5, 6]

In [2]: numbers.index(5)
Out[2]: 6
```

A `ValueError` occurs if the value you're searching for is not in the list.

Specifying the Starting Index of a Search

Using method `index`'s optional arguments, you can search a subset of a list's elements. You can use `*=` to *multiply a sequence*—that is, append a sequence to itself multiple times. After the following snippet, `numbers` contains two copies of the original list's contents:

```
In [3]: numbers *= 2

In [4]: numbers
Out[4]: [3, 7, 1, 4, 2, 8, 5, 6, 3, 7, 1, 4, 2, 8, 5, 6]
```

The following code searches the updated list for the value 5 starting from index 7 and continuing through the end of the list:

```
In [5]: numbers.index(5, 7)
Out[5]: 14
```

Specifying the Starting and Ending Indices of a Search

Specifying the starting and ending indices causes `index` to search from the starting index up to but not including the ending index location. The call to `index` in snippet [5]:

```
numbers.index(5, 7)
```

assumes the length of `numbers` as its optional third argument and is equivalent to:

```
numbers.index(5, 7, len(numbers))
```

The following looks for the value 7 in the range of elements with indices 0 through 3:

```
In [6]: numbers.index(7, 0, 4)
Out[6]: 1
```

Operators in and not in

Operator `in` tests whether its right operand's iterable contains the left operand's value:

```
In [7]: 1000 in numbers
Out[7]: False

In [8]: 5 in numbers
Out[8]: True
```

Similarly, operator not in tests whether its right operand's iterable does *not* contain the left operand's value:

```
In [9]: 1000 not in numbers
Out[9]: True

In [10]: 5 not in numbers
Out[10]: False
```

Using Operator in to Prevent a ValueError

You can use the operator in to ensure that calls to method index do not result in ValueErrors for search keys that are not in the corresponding sequence:

```
In [11]: key = 1000

In [12]: if key in numbers:
   ...:     print(f'found {key} at index {numbers.index(key)}')
   ...: else:
   ...:     print(f'{key} not found')
   ...:
1000 not found
```

Built-In Functions any and all

Sometimes you simply need to know whether *any* item in an iterable is True or whether *all* the items are True. The built-in function **any** returns True if any item in its iterable argument is True. The built-in function **all** returns True if all items in its iterable argument are True. Recall that nonzero values are True and 0 is False. Non-empty iterable objects also evaluate to True, whereas any empty iterable evaluates to False. Functions any and all are additional examples of internal iteration in functional-style programming.

 Self Check

1 *(Fill-In)* The _____ operator can be used to extend a list with copies of itself.
Answer: *=.

2 *(Fill-In)* Operators _____ and _____ determine whether a sequence contains or does not contain a value, respectively.
Answer: in, not in.

3 *(IPython Session)* Create a five-element list containing 67, 12, 46, 43 and 13, then use list method index to search for a 43 and 44. Ensure that no ValueError occurs when searching for 44.
Answer:

```
In [1]: numbers = [67, 12, 46, 43, 13]

In [2]: numbers.index(43)
Out[2]: 3

In [3]: if 44 in numbers:
   ...:     print(f'Found 44 at index: {numbers.index(44)}')
   ...: else:
   ...:     print('44 not found')
   ...:
44 not found
```

5.10 Other List Methods

Lists also have methods that add and remove elements. Consider the list `color_names`:

```
In [1]: color_names = ['orange', 'yellow', 'green']
```

Inserting an Element at a Specific List Index
Method **insert** adds a new item at a specified index. The following inserts `'red'` at index 0:

```
In [2]: color_names.insert(0, 'red')

In [3]: color_names
Out[3]: ['red', 'orange', 'yellow', 'green']
```

Adding an Element to the End of a List
You can add a new item to the end of a list with method **append**:

```
In [4]: color_names.append('blue')

In [5]: color_names
Out[5]: ['red', 'orange', 'yellow', 'green', 'blue']
```

Adding All the Elements of a Sequence to the End of a List
Use list method **extend** to add all the elements of another sequence to the end of a list:

```
In [6]: color_names.extend(['indigo', 'violet'])

In [7]: color_names
Out[7]: ['red', 'orange', 'yellow', 'green', 'blue', 'indigo', 'violet']
```

This is the equivalent of using +=. The following code adds all the characters of a string then all the elements of a tuple to a list:

```
In [8]: sample_list = []

In [9]: s = 'abc'

In [10]: sample_list.extend(s)

In [11]: sample_list
Out[11]: ['a', 'b', 'c']

In [12]: t = (1, 2, 3)

In [13]: sample_list.extend(t)

In [14]: sample_list
Out[14]: ['a', 'b', 'c', 1, 2, 3]
```

Rather than creating a temporary variable, like t, to store a tuple before appending it to a list, you might want to pass a tuple directly to extend. In this case, the tuple's parentheses are required, because extend expects one iterable argument:

```
In [15]: sample_list.extend((4, 5, 6))  # note the extra parentheses

In [16]: sample_list
Out[16]: ['a', 'b', 'c', 1, 2, 3, 4, 5, 6]
```

A TypeError occurs if you omit the required parentheses.

Removing the First Occurrence of an Element in a List
Method **remove** deletes the first element with a specified value—a ValueError occurs if remove's argument is not in the list:

```
In [17]: color_names.remove('green')
```

```
In [18]: color_names
Out[18]: ['red', 'orange', 'yellow', 'blue', 'indigo', 'violet']
```

Emptying a List
To delete all the elements in a list, call method **clear**:

```
In [19]: color_names.clear()
```

```
In [20]: color_names
Out[20]: []
```

This is the equivalent of the previously shown slice assignment

```
color_names[:] = []
```

Counting the Number of Occurrences of an Item
List method **count** searches for its argument and returns the number of times it is found:

```
In [21]: responses = [1, 2, 5, 4, 3, 5, 2, 1, 3, 3,
    ...:              1, 4, 3, 3, 3, 2, 3, 3, 2, 2]
    ...:
```

```
In [22]: for i in range(1, 6):
    ...:     print(f'{i} appears {responses.count(i)} times in responses')
    ...:
1 appears 3 times in responses
2 appears 5 times in responses
3 appears 8 times in responses
4 appears 2 times in responses
5 appears 2 times in responses
```

Reversing a List's Elements
List method **reverse** reverses the contents of a list in place, rather than creating a reversed copy, as we did with a slice previously:

```
In [23]: color_names = ['red', 'orange', 'yellow', 'green', 'blue']
```

```
In [24]: color_names.reverse()
```

```
In [25]: color_names
Out[25]: ['blue', 'green', 'yellow', 'orange', 'red']
```

Copying a List
List method copy returns a *new* list containing a *shallow* copy of the original list:

```
In [26]: copied_list = color_names.copy()
```

```
In [27]: copied_list
Out[27]: ['blue', 'green', 'yellow', 'orange', 'red']
```

This is equivalent to the previously demonstrated slice operation:

```
copied_list = color_names[:]
```

Self Check

1 *(Fill-In)* To add all the elements of a sequence to the end of a list, use list method _____, which is equivalent to using +=.
Answer: extend.

2 *(Fill-In)* For a list numbers, calling method _____ is equivalent to numbers[:] = [].
Answer: clear.

3 *(IPython Session)* Create a list called rainbow containing 'green', 'orange' and 'violet'. Perform the following operations consecutively using list methods and show the list's contents after each operation:
   a) Determine the index of 'violet', then use it to insert 'red' before 'violet'.
   b) Append 'yellow' to the end of the list.
   c) Reverse the list's elements.
   d) Remove the element 'orange'.
Answer:

```
In [1]: rainbow = ['green', 'orange', 'violet']

In [2]: rainbow.insert(rainbow.index('violet'), 'red')

In [3]: rainbow
Out[3]: ['green', 'orange', 'red', 'violet']

In [4]: rainbow.append('yellow')

In [5]: rainbow
Out[5]: ['green', 'orange', 'red', 'violet', 'yellow']

In [6]: rainbow.reverse()

In [7]: rainbow
Out[7]: ['yellow', 'violet', 'red', 'orange', 'green']

In [8]: rainbow.remove('orange')

In [9]: rainbow
Out[9]: ['yellow', 'violet', 'red', 'green']
```

5.11 Simulating Stacks with Lists

The preceding chapter introduced the function-call stack. Python does not have a built-in stack type, but you can think of a stack as a constrained list. You *push* using list method append, which adds a new element to the *end* of the list. You *pop* using list method **pop** with no arguments, which removes and returns the item at the *end* of the list.

Let's create an empty list called stack, push (append) two strings onto it, then pop the strings to confirm they're retrieved in last-in, first-out (LIFO) order:

```
In [1]: stack = []

In [2]: stack.append('red')

In [3]: stack
Out[3]: ['red']
```

```
In [4]: stack.append('green')

In [5]: stack
Out[5]: ['red', 'green']

In [6]: stack.pop()
Out[6]: 'green'

In [7]: stack
Out[7]: ['red']

In [8]: stack.pop()
Out[8]: 'red'

In [9]: stack
Out[9]: []

In [10]: stack.pop()
-------------------------------------------------------------------------
IndexError                                Traceback (most recent call last)
<ipython-input-10-50ea7ec13fbe> in <module>()
----> 1 stack.pop()

IndexError: pop from empty list
```

For each pop snippet, the value that pop removes and returns is displayed. Popping from an empty stack causes an IndexError, just like accessing a nonexistent list element with []. To prevent an IndexError, ensure that len(stack) is greater than 0 before calling pop. You can run out of memory if you keep pushing items faster than you pop them.

In the exercises, you'll use a list to simulate another popular collection called a **queue** in which you insert at the back and delete from the front. Items are retrieved from queues in **first-in, first-out (FIFO) order.**

✓ Self Check

1 *(Fill-In)* You can simulate a stack with a list, using methods _____ and _____ to add and remove elements, respectively, only at the end of the list.
Answer: append, pop.

2 *(Fill-In)* To prevent an IndexError when calling pop on a list, first ensure that _____.
Answer: the list's length is greater than 0.

5.12 List Comprehensions

Here, we continue discussing *functional-style* features with **list comprehensions**—a concise and convenient notation for creating new lists. List comprehensions can replace many for statements that iterate over existing sequences and create new lists, such as:

```
In [1]: list1 = []

In [2]: for item in range(1, 6):
   ...:     list1.append(item)
   ...:
```

```
In [3]: list1
Out[3]: [1, 2, 3, 4, 5]
```

Using a List Comprehension to Create a List of Integers
We can accomplish the same task in a single line of code with a list comprehension:

```
In [4]: list2 = [item for item in range(1, 6)]

In [5]: list2
Out[5]: [1, 2, 3, 4, 5]
```

Like snippet [2]'s for statement, the list comprehension's **for clause**

```
for item in range(1, 6)
```

iterates over the sequence produced by range(1, 6). For each item, the list comprehension evaluates the expression to the left of the for clause and places the expression's value (in this case, the item itself) in the new list. Snippet [4]'s particular comprehension could have been expressed more concisely using the function list:

```
list2 = list(range(1, 6))
```

Mapping: Performing Operations in a List Comprehension's Expression
A list comprehension's expression can perform tasks, such as calculations, that **map** elements to new values (possibly of different types). Mapping is a common functional-style programming operation that produces a result with the *same* number of elements as the original data being mapped. The following comprehension maps each value to its cube with the expression item ** 3:

```
In [6]: list3 = [item ** 3 for item in range(1, 6)]

In [7]: list3
Out[7]: [1, 8, 27, 64, 125]
```

Filtering: List Comprehensions with if Clauses
Another common functional-style programming operation is **filtering** elements to select only those that match a condition. This typically produces a list with *fewer* elements than the data being filtered. To do this in a list comprehension, use the **if clause**. The following includes in list4 only the even values produced by the for clause:

```
In [8]: list4 = [item for item in range(1, 11) if item % 2 == 0]

In [9]: list4
Out[9]: [2, 4, 6, 8, 10]
```

List Comprehension That Processes Another List's Elements
The for clause can process any iterable. Let's create a list of lowercase strings and use a list comprehension to create a new list containing their uppercase versions:

```
In [10]: colors = ['red', 'orange', 'yellow', 'green', 'blue']

In [11]: colors2 = [item.upper() for item in colors]

In [12]: colors2
Out[12]: ['RED', 'ORANGE', 'YELLOW', 'GREEN', 'BLUE']
```

```
In [13]: colors
Out[13]: ['red', 'orange', 'yellow', 'green', 'blue']
```

Self Check

1 *(Fill-In)* A list comprehension's _____ clause iterates over the specified sequence.
Answer: `for`.

2 *(Fill-In)* A list comprehension's _____ clause filters sequence elements to select only those that match a condition.
Answer: `if`.

3 *(IPython Session)* Use a list comprehension to create a list of tuples containing the numbers 1–5 and their cubes—that is, [(1, 1), (2, 8), (3, 27), ...]. To create tuples, place parentheses around the expression to the left of the list comprehension's `for` clause.
Answer:

```
In [1]: cubes = [(x, x ** 3) for x in range(1, 6)]

In [2]: cubes
Out[2]: [(1, 1), (2, 8), (3, 27), (4, 64), (5, 125)]
```

4 *(IPython Session)* Use a list comprehension and the `range` function with a step to create a list of the multiples of 3 that are less than 30.
Answer:

```
In [3]: multiples = [x for x in range(3, 30, 3)]

In [4]: multiples
Out[4]: [3, 6, 9, 12, 15, 18, 21, 24, 27]
```

5.13 Generator Expressions

A **generator expression** is similar to a list comprehension, but creates an iterable **generator object** that produces values *on demand*. This is known as **lazy evaluation**. List comprehensions use **greedy evaluation**—they create lists *immediately* when you execute them. For large numbers of items, creating a list can take substantial memory and time. So generator expressions can reduce your program's memory consumption and improve performance if the whole list is not needed at once.

Generator expressions have the same capabilities as list comprehensions, but you define them in parentheses instead of square brackets. The generator expression in snippet [2] squares and returns only the odd values in `numbers`:

```
In [1]: numbers = [10, 3, 7, 1, 9, 4, 2, 8, 5, 6]

In [2]: for value in (x ** 2 for x in numbers if x % 2 != 0):
   ...:     print(value, end=' ')
   ...:
9 49 1 81 25
```

To show that a generator expression does not create a list, let's assign the preceding snippet's generator expression to a variable and evaluate the variable:

```
In [3]: squares_of_odds = (x ** 2 for x in numbers if x % 2 != 0)
```

182 Sequences: Lists and Tuples

```
In [3]: squares_of_odds
Out[3]: <generator object <genexpr> at 0x1085e84c0>
```

The text "generator object <genexpr>" indicates that square_of_odds is a generator object that was created from a generator expression (genexpr).

Self Check

1 *(Fill-In)* A generator expression is _____—it produces values on demand.
Answer: lazy.

2 *(IPython Session)* Create a generator expression that cubes the even integers in a list containing 10, 3, 7, 1, 9, 4 and 2. Use function list to create a list of the results. Note that the function call's parentheses also act as the generator expression's parentheses.
Answer:

```
In [1]: list(x ** 3 for x in [10, 3, 7, 1, 9, 4, 2] if x % 2 == 0)
Out[1]: [1000, 64, 8]
```

5.14 Filter, Map and Reduce

The preceding section introduced several functional-style features—list comprehensions, filtering and mapping. Here we demonstrate the built-in filter and map functions for filtering and mapping, respectively. We continue discussing reductions in which you process a collection of elements into a *single* value, such as their count, total, product, average, minimum or maximum.

Filtering a Sequence's Values with the Built-In filter Function
Let's use built-in function **filter** to obtain the odd values in numbers:

```
In [1]: numbers = [10, 3, 7, 1, 9, 4, 2, 8, 5, 6]

In [2]: def is_odd(x):
   ...:     """Returns True only if x is odd."""
   ...:     return x % 2 != 0
   ...:

In [3]: list(filter(is_odd, numbers))
Out[3]: [3, 7, 1, 9, 5]
```

Like data, Python functions are objects that you can assign to variables, pass to other functions and return from functions. Functions that receive other functions as arguments are a functional-style capability called **higher-order functions**. For example, filter's first argument must be a function that receives one argument and returns True if the value should be included in the result. The function is_odd returns True if its argument is odd. The filter function calls is_odd once for each value in its second argument's iterable (numbers). Higher-order functions may also return a function as a result.

Function filter returns an iterator, so filter's results are not produced until you iterate through them. This is another example of lazy evaluation. In snippet [3], function list iterates through the results and creates a list containing them. We can obtain the same results as above by using a list comprehension with an `if` clause:

```
In [4]: [item for item in numbers if is_odd(item)]
Out[4]: [3, 7, 1, 9, 5]
```

Using a lambda Rather than a Function
For simple functions like `is_odd` that `return` only a *single expression's value*, you can use a **lambda expression** (or simply a **lambda**) to define the function inline where it's needed—typically as it's passed to another function:

```
In [5]: list(filter(lambda x: x % 2 != 0, numbers))
Out[5]: [3, 7, 1, 9, 5]
```

We pass `filter`'s return value (an iterator) to function `list` here to convert the results to a list and display them.

A lambda expression is an *anonymous function*—that is, a *function without a name*. In the `filter` call

```
filter(lambda x: x % 2 != 0, numbers)
```

the first argument is the lambda

```
lambda x: x % 2 != 0
```

A lambda begins with the **lambda** keyword followed by a comma-separated parameter list, a colon (:) and an expression. In this case, the parameter list has one parameter named x. A lambda *implicitly* returns its expression's value. So any simple function of the form

```
def function_name(parameter_list):
    return expression
```

may be expressed as a more concise lambda of the form

```
lambda parameter_list: expression
```

Mapping a Sequence's Values to New Values
Let's use built-in function **map** with a lambda to square each value in numbers:

```
In [6]: numbers
Out[6]: [10, 3, 7, 1, 9, 4, 2, 8, 5, 6]

In [7]: list(map(lambda x: x ** 2, numbers))
Out[7]: [100, 9, 49, 1, 81, 16, 4, 64, 25, 36]
```

Function map's first argument is a function that receives one value and returns a new value—in this case, a lambda that squares its argument. The second argument is an iterable of values to map. Function map uses lazy evaluation. So, we pass to the `list` function the iterator that map returns. This enables us to iterate through and create a list of the mapped values. Here's an equivalent list comprehension:

```
In [8]: [item ** 2 for item in numbers]
Out[8]: [100, 9, 49, 1, 81, 16, 4, 64, 25, 36]
```

Combining `filter` and `map`

You can combine the preceding `filter` and `map` operations as follows:

```
In [9]: list(map(lambda x: x ** 2,
   ...:          filter(lambda x: x % 2 != 0, numbers)))
Out[9]: [9, 49, 1, 81, 25]
```

There is a lot going on in snippet [9], so let's take a closer look at it. First, `filter` returns an iterable representing only the odd values of `numbers`. Then `map` returns an iterable representing the squares of the filtered values. Finally, `list` uses map's iterable to create the list. You might prefer the following list comprehension to the preceding snippet:

```
In [10]: [x ** 2 for x in numbers if x % 2 != 0]
Out[10]: [9, 49, 1, 81, 25]
```

For each value of x in numbers, the expression x ** 2 is performed only if the condition x % 2 != 0 is True.

Reduction: Totaling the Elements of a Sequence with `sum`

As you know reductions process a sequence's elements into a single value. You've performed reductions with the built-in functions `len`, `sum`, `min` and `max`. You also can create custom reductions using the `functools` module's `reduce` function. See https://docs.python.org/3/library/functools.html for a code example. When we investigate big data and Hadoop (introduced briefly in Chapter 1), we'll demonstrate MapReduce programming, which is based on the filter, map and reduce operations in functional-style programming.

✓ Self Check

1 *(Fill-In)* _____, _____ and _____ are common operations used in functional-style programming.
Answer: Filter, map, reduce.

2 *(Fill-In)* A(n) _____ processes a sequence's elements into a single value, such as their count, total or average.
Answer: reduction.

3 *(IPython Session)* Create a list called numbers containing 1 through 15, then perform the following tasks:
   a) Use the built-in function `filter` with a lambda to select only numbers' even elements. Create a new list containing the result.
   b) Use the built-in function `map` with a lambda to square the values of numbers' elements. Create a new list containing the result.
   c) Filter numbers' even elements, then map them to their squares. Create a new list containing the result.

Answer:

```
In [1]: numbers = list(range(1, 16))

In [2]: numbers
Out[2]: [1, 2, 3, 4, 5, 6, 7, 8, 9, 10, 11, 12, 13, 14, 15]

In [3]: list(filter(lambda x: x % 2 == 0, numbers))
```

```
Out[3]: [2, 4, 6, 8, 10, 12, 14]

In [4]: list(map(lambda x: x ** 2, numbers))
Out[4]: [1, 4, 9, 16, 25, 36, 49, 64, 81, 100, 121, 144, 169, 196, 225]

In [5]: list(map(lambda x: x**2, filter(lambda x: x % 2 == 0, numbers)))
Out[5]: [4, 16, 36, 64, 100, 144, 196]
```

4 *(IPython Session)* Map a list of the three Fahrenheit temperatures 41, 32 and 212 to a list of tuples containing the Fahrenheit temperatures and their Celsius equivalents. Convert Fahrenheit temperatures to Celsius with the following formula:

*Celsius = (Fahrenheit − 32) * (5 / 9)*

Answer:

```
In [6]: fahrenheit = [41, 32, 212]

In [7]: list(map(lambda x: (x, (x - 32) * 5 / 9), fahrenheit))
Out[7]: [(41, 5.0), (32, 0.0), (212, 100.0)]
```

The lambda's expression—(x, (x - 32) * 5 / 9)—uses parentheses to create a tuple containing the original Fahrenheit temperature (x) and the corresponding Celsius temperature, as calculated by (x - 32) * 5 / 9.

5.15 Other Sequence Processing Functions

Python provides other built-in functions for manipulating sequences.

Finding the Minimum and Maximum Values Using a Key Function

We've previously shown the built-in reduction functions min and max using arguments, such as ints or lists of ints. Sometimes you'll need to find the minimum and maximum of more complex objects, such as strings. Consider the following comparison:

```
In [1]: 'Red' < 'orange'
Out[1]: True
```

The letter 'R' "comes after" 'o' in the alphabet, so you might expect 'Red' to be less than 'orange' and the condition above to be False. However, strings are compared by their characters' underlying *numerical values*, and lowercase letters have *higher* numerical values than uppercase letters. You can confirm this with built-in function **ord**, which returns the numerical value of a character:

```
In [2]: ord('R')
Out[2]: 82

In [3]: ord('o')
Out[3]: 111
```

Consider the list colors, which contains strings with uppercase and lowercase letters:

```
In [4]: colors = ['Red', 'orange', 'Yellow', 'green', 'Blue']
```

Let's assume that we'd like to determine the minimum and maximum strings using *alphabetical* order, not *numerical* (lexicographical) order. If we arrange colors alphabetically

'Blue', 'green', 'orange', 'Red', 'Yellow'

you can see that `'Blue'` is the minimum (that is, closest to the beginning of the alphabet), and `'Yellow'` is the maximum (that is, closest to the end of the alphabet).

Since Python compares strings using numerical values, you must first convert each string to all lowercase or all uppercase letters. Then their numerical values will also represent *alphabetical* ordering. The following snippets enable `min` and `max` to determine the minimum and maximum strings alphabetically:

```
In [5]: min(colors, key=lambda s: s.lower())
Out[5]: 'Blue'

In [6]: max(colors, key=lambda s: s.lower())
Out[6]: 'Yellow'
```

The key keyword argument must be a one-parameter function that returns a value. In this case, it's a `lambda` that calls string method `lower` to get a string's lowercase version. Functions `min` and `max` call the key argument's function for each element and use the results to compare the elements.

Iterating Backward Through a Sequence

Built-in function **reversed** returns an iterator that enables you to iterate over a sequence's values backward. The following list comprehension creates a new list containing the squares of `numbers`' values in reverse order:

```
In [7]: numbers = [10, 3, 7, 1, 9, 4, 2, 8, 5, 6]

In [8]: reversed_numbers = [item ** 2 for item in reversed(numbers)]

In [9]: reversed_numbers
Out[9]: [36, 25, 64, 4, 16, 81, 1, 49, 9, 100]
```

Combining Iterables into Tuples of Corresponding Elements

Built-in function **zip** enables you to iterate over *multiple* iterables of data at the *same* time. The function receives as arguments any number of iterables and returns an iterator that produces tuples containing the elements at the same index in each. For example, snippet [12]'s call to `zip` produces the tuples (`'Bob'`, `3.5`), (`'Sue'`, `4.0`) and (`'Amanda'`, `3.75`) consisting of the elements at index 0, 1 and 2 of each list, respectively:

```
In [10]: names = ['Bob', 'Sue', 'Amanda']

In [11]: grade_point_averages = [3.5, 4.0, 3.75]

In [12]: for name, gpa in zip(names, grade_point_averages):
    ...:     print(f'Name={name}; GPA={gpa}')
    ...:
Name=Bob; GPA=3.5
Name=Sue; GPA=4.0
Name=Amanda; GPA=3.75
```

We unpack each tuple into `name` and `gpa` and display them. Function `zip`'s shortest argument determines the number of tuples produced. Here both have the same length.

Self Check

1. *(True/False)* The letter 'V' "comes after" the letter 'g' in the alphabet, so the comparison 'Violet' < 'green' yields False.
Answer: False. Strings are compared by their characters' underlying numerical values. Lowercase letters have *higher* numerical values than uppercase. So, the comparison is True.

2. *(Fill-In)* Built-in function _____ returns an iterator that enables you to iterate over a sequence's values backward.
Answer: reversed.

3. *(IPython Session)* Create the list foods containing 'Cookies', 'pizza', 'Grapes', 'apples', 'steak' and 'Bacon'. Find the smallest string with min, then reimplement the min call using the key function to ignore the strings' case. Do you get the same results? Why or why not?
Answer: The min result was different because 'apples' is the smallest string when you compare them without case sensitivity.

```
In [1]: foods = ['Cookies', 'pizza', 'Grapes',
   ...:          'apples', 'steak', 'Bacon']
   ...:

In [2]: min(foods)
Out[2]: 'Bacon'

In [3]: min(foods, key=lambda s: s.lower())
Out[3]: 'apples'
```

4. *(IPython Session)* Use zip with two integer lists to create a new list containing the sum of the elements from corresponding indices in both lists (that is, add the elements at index 0, add the elements at index 1, ...).
Answer:

```
In [4]: [(a + b) for a, b in zip([10, 20, 30], [1, 2, 3])]
Out[4]: [11, 22, 33]
```

5.16 Two-Dimensional Lists

Lists can contain other lists as elements. A typical use of such nested (or multidimensional) lists is to represent **tables** of values consisting of information arranged in **rows** and **columns**. To identify a particular table element, we specify *two* indices—by convention, the first identifies the element's row, the second the element's column.

Lists that require two indices to identify an element are called **two-dimensional lists** (or **double-indexed lists** or **double-subscripted lists**). Multidimensional lists can have more than two indices. Here, we introduce two-dimensional lists.

Creating a Two-Dimensional List

Consider a two-dimensional list with three rows and four columns (i.e., a 3-by-4 list) that might represent the grades of three students who each took four exams in a course:

```
In [1]: a = [[77, 68, 86, 73], [96, 87, 89, 81], [70, 90, 86, 81]]
```

Writing the list as follows makes its row and column tabular structure clearer:

```
a = [[77, 68, 86, 73],    # first student's grades
     [96, 87, 89, 81],    # second student's grades
     [70, 90, 86, 81]]    # third student's grades
```

Illustrating a Two-Dimensional List
The diagram below shows the list a, with its rows and columns of exam grade values:

	Column 0	Column 1	Column 2	Column 3
Row 0	77	68	86	73
Row 1	96	87	89	81
Row 2	70	90	86	81

Identifying the Elements in a Two-Dimensional List
The following diagram shows the names of list a's elements:

	Column 0	Column 1	Column 2	Column 3
Row 0	a[0][0]	a[0][1]	a[0][2]	a[0][3]
Row 1	a[1][0]	a[1][1]	a[1][2]	a[1][3]
Row 2	a[2][0]	a[2][1]	a[2][2]	a[2][3]

— Column index
— Row index
— List name

Every element is identified by a name of the form a[i][j]—a is the list's name, and i and j are the indices that uniquely identify each element's row and column, respectively. The element names in row 0 all have 0 as the first index. The element names in column 3 all have 3 as the second index.

In the two-dimensional list a:

- 77, 68, 86 and 73 initialize a[0][0], a[0][1], a[0][2] and a[0][3], respectively,
- 96, 87, 89 and 81 initialize a[1][0], a[1][1], a[1][2] and a[1][3], respectively, and
- 70, 90, 86 and 81 initialize a[2][0], a[2][1], a[2][2] and a[2][3], respectively.

A list with m rows and n columns is called an **m-by-n list** and has $m \times n$ elements.

The following nested for statement outputs the rows of the preceding two-dimensional list one row at a time:

```
In [2]: for row in a:
   ...:     for item in row:
   ...:         print(item, end=' ')
   ...:     print()
   ...:
77 68 86 73
96 87 89 81
70 90 86 81
```

How the Nested Loops Execute

Let's modify the nested loop to display the list's name and the row and column indices and value of each element:

```
In [3]: for i, row in enumerate(a):
   ...:     for j, item in enumerate(row):
   ...:         print(f'a[{i}][{j}]={item} ', end=' ')
   ...:     print()
   ...:
a[0][0]=77  a[0][1]=68  a[0][2]=86  a[0][3]=73
a[1][0]=96  a[1][1]=87  a[1][2]=89  a[1][3]=81
a[2][0]=70  a[2][1]=90  a[2][2]=86  a[2][3]=81
```

The outer for statement iterates over the two-dimensional list's rows one row at a time. During each iteration of the outer for statement, the inner for statement iterates over *each* column in the current row. So in the first iteration of the outer loop, row 0 is

[77, 68, 86, 73]

and the nested loop iterates through this list's four elements a[0][0]=77, a[0][1]=68, a[0][2]=86 and a[0][3]=73.

In the second iteration of the outer loop, row 1 is

[96, 87, 89, 81]

and the nested loop iterates through this list's four elements a[1][0]=96, a[1][1]=87, a[1][2]=89 and a[1][3]=81.

In the third iteration of the outer loop, row 2 is

[70, 90, 86, 81]

and the nested loop iterates through this list's four elements a[2][0]=70, a[2][1]=90, a[2][2]=86 and a[2][3]=81.

In the "Array-Oriented Programming with NumPy" chapter, we'll cover the NumPy library's ndarray collection and the Pandas library's DataFrame collection. These enable you to manipulate multidimensional collections more concisely and conveniently than the two-dimensional list manipulations you've seen in this section.

Self Check

1 *(Fill-In)* In a two-dimensional list, the first index by convention identifies the _____ of an element and the second index identifies the _____ of an element.
Answer: row, column.

2 *(Label the Elements)* Label the elements of the two-by-three list sales to indicate the order in which they're set to zero by the following program segment:

```
for row in range(len(sales)):
    for col in range(len(sales[row])):
        sales[row][col] = 0
```

Answer: sales[0][0], sales[0][1], sales[0][2], sales[1][0], sales[1][1], sales[1][1].

3 *(Two-Dimensional Array)* Consider a two-by-three integer list t.
 a) How many rows does t have?
 b) How many columns does t have?

c) How many elements does t have?
d) What are the names of the elements in row 1?
e) What are the names of the elements in column 2?
f) Set the element in row 0 and column 1 to 10.
g) Write a nested for statement that sets each element to the sum of its indices.

Answer:
a) 2.
b) 3.
c) 6.
d) t[1][0], t[1][1], t[1][2].
e) t[0][2], t[1][2].
f) t[0][1] = 10.
g) for row in range(len(t)):
        for column in range(len(t[row])):
            t[row][column] = row + column

4 *(IPython Session)* Given the two-by-three integer list t

   t = [[10, 7, 3], [20, 4, 17]]

a) Determine and display the average of t's elements using nested for statements to iterate through the elements.
b) Write a for statement that determines and displays the average of t's elements using the reductions sum and len to calculate the sum of each row's elements and the number of elements in each row.

Answer:

```
In [1]: t = [[10, 7, 3], [20, 4, 17]]

In [2]: total = 0

In [3]: items = 0

In [4]: for row in t:
   ...:     for item in row:
   ...:         total += item
   ...:         items += 1
   ...:

In [5]: total / items
Out[5]: 10.166666666666666

In [6]: total = 0

In [7]: items = 0

In [8]: for row in t:
   ...:     total += sum(row)
   ...:     items += len(row)
   ...:

In [9]: total / items
Out[9]: 10.166666666666666
```

5.17 Intro to Data Science: Simulation and Static Visualizations

The last few chapters' Intro to Data Science sections discussed basic descriptive statistics. Here, we focus on visualizations, which help you "get to know" your data. Visualizations give you a powerful way to understand data that goes beyond simply looking at raw data.

We use two open-source visualization libraries—Seaborn and Matplotlib—to display *static* bar charts showing the final results of a six-sided-die-rolling simulation. The **Seaborn visualization library** is built over the **Matplotlib visualization library** and simplifies many Matplotlib operations. We'll use aspects of both libraries, because some of the Seaborn operations return objects from the Matplotlib library.

In the next chapter's Intro to Data Science section, we'll make things "come alive" with *dynamic visualizations*. In this chapter's exercises, you'll use simulation techniques and explore the characteristics of some popular card and dice games.

5.17.1 Sample Graphs for 600, 60,000 and 6,000,000 Die Rolls

The screen capture below shows a vertical bar chart that for 600 die rolls summarizes the frequencies with which each of the six faces appear, and their percentages of the total. Seaborn refers to this type of graph as a **bar plot**:

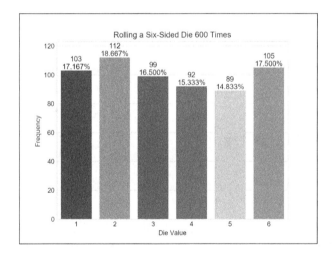

Here we expect about 100 occurrences of each die face. However, with such a small number of rolls, none of the frequencies is exactly 100 (though several are close) and most of the percentages are not close to 16.667% (about 1/6th). As we run the simulation for 60,000 die rolls, the bars will become much closer in size. At 6,000,000 die rolls, they'll appear to be exactly the same size. This is the "law of large numbers" at work. The next chapter will show the lengths of the bars changing dynamically.

We'll discuss how to control the plot's appearance and contents, including:

- the graph title inside the window (**Rolling a Six-Sided Die 600 Times**),
- the descriptive labels **Die Value** for the *x*-axis and **Frequency** for the *y*-axis,

- the text displayed above each bar, representing the *frequency* and *percentage* of the total rolls, and
- the bar colors.

We'll use various Seaborn default options. For example, Seaborn determines the text labels along the *x*-axis from the die face values 1–6 and the text labels along the *y*-axis from the actual die frequencies. Behind the scenes, Matplotlib determines the positions and sizes of the bars, based on the window size and the magnitudes of the values the bars represent. It also positions the **Frequency** axis's numeric labels based on the actual die frequencies that the bars represent. There are many more features you can customize. You should tweak these attributes to your personal preferences.

The first screen capture below shows the results for 60,000 die rolls—imagine trying to do this by hand. In this case, we expect about 10,000 of each face. The second screen capture below shows the results for 6,000,000 rolls—surely something you'd never do by hand![1] In this case, we expect about 1,000,000 of each face, and the frequency bars appear to be identical in length (they're close but not exactly the same length). Note that with more die rolls, the frequency percentages are much closer to the expected 16.667%.

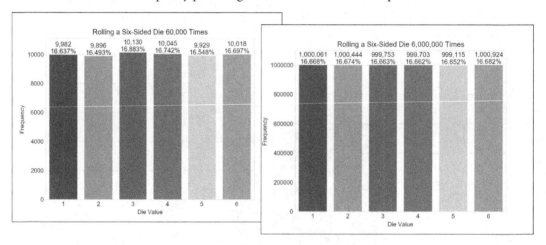

✓ Self Check

1 **(Discussion)** If you toss a coin a large odd number of times and associate the value 1 with heads and 2 with tails, what would you expect the mean to be? What would you expect the median and mode to be?
Answer: We'd expect the mean to be 1.5, which seems strange because it's not one of the possible outcomes. As the number of coin tosses increases, the percentages of heads and tails should each approach 50% of the total. However, at any given time, they are not likely to be identical. You're just as likely to have a few more heads than tails as vice versa, and as the number of rolls increases, the face with the larger number of rolls could change repeatedly. There are only two possible outcomes, so the median and mode values will be

---

1. When we taught die rolling in our first programming book in the mid-1970s, computers were so much slower that we had to limit our simulations to 6000 rolls. In writing this book's examples, we went to 6,000,000 rolls, which the program completed in a few seconds. We then went to 60,000,000 rolls, which took about a minute.

whichever value there is more of at a given time. So if there are currently more heads than tails, both the median and mode will be heads; otherwise, they'll both be tails. Similar observations apply to die rolling.

5.17.2 Visualizing Die-Roll Frequencies and Percentages

In this section, you'll interactively develop the bar plots shown in the preceding section.

Launching IPython for Interactive Matplotlib Development

IPython has built-in support for interactively developing Matplotlib graphs, which you also need to develop Seaborn graphs. Simply launch IPython with the command:

```
ipython --matplotlib
```

Importing the Libraries

First, let's import the libraries we'll use:

```
In [1]: import matplotlib.pyplot as plt

In [2]: import numpy as np

In [3]: import random

In [4]: import seaborn as sns
```

1. The **matplotlib.pyplot module** contains the Matplotlib library's graphing capabilities that we use. This module typically is imported with the name plt.

2. The NumPy (Numerical Python) library includes the function unique that we'll use to summarize the die rolls. The **numpy module** typically is imported as np.

3. The random module contains Python's random-number generation functions.

4. The **seaborn module** contains the Seaborn library's graphing capabilities we use. This module typically is imported with the name sns. Search for why this curious abbreviation was chosen.

Rolling the Die and Calculating Die Frequencies

Next, let's use a *list comprehension* to create a list of 600 random die values, then use NumPy's **unique** function to determine the unique roll values (most likely all six possible face values) and their frequencies:

```
In [5]: rolls = [random.randrange(1, 7) for i in range(600)]

In [6]: values, frequencies = np.unique(rolls, return_counts=True)
```

The NumPy library provides the high-performance **ndarray** collection, which is typically much faster than lists.[2] Though we do not use ndarray directly here, the NumPy unique function expects an ndarray argument and returns an ndarray. If you pass a list (like rolls), NumPy converts it to an ndarray for better performance. The ndarray that unique returns we'll simply assign to a variable for use by a Seaborn plotting function.

Specifying the keyword argument **return_counts=True** tells unique to count each unique value's number of occurrences. In this case, unique returns a tuple of two one-

---

2. We'll run a performance comparison in Chapter 7 where we discuss ndarray in depth.

dimensional `ndarrays` containing the sorted unique values and the corresponding frequencies, respectively. We unpack the tuple's `ndarrays` into the variables `values` and `frequencies`. If `return_counts` is `False`, only the list of unique values is returned.

Creating the Initial Bar Plot
Let's create the bar plot's title, set its style, then graph the die faces and frequencies:

```
In [7]: title = f'Rolling a Six-Sided Die {len(rolls):,} Times'

In [8]: sns.set_style('whitegrid')

In [9]: axes = sns.barplot(x=values, y=frequencies, palette='bright')
```

Snippet [7]'s f-string includes the number of die rolls in the bar plot's title. The comma (,) format specifier in

```
{len(rolls):,}
```

displays the number with *thousands separators*—so, 60000 would be displayed as 60,000.

By default, Seaborn plots graphs on a plain white background, but it provides several styles to choose from (`'darkgrid'`, `'whitegrid'`, `'dark'`, `'white'` and `'ticks'`). Snippet [8] specifies the `'whitegrid'` style, which displays light-gray horizontal lines in the vertical bar plot. These help you see more easily how each bar's height corresponds to the numeric frequency labels at the bar plot's left side.

Snippet [9] graphs the die frequencies using Seaborn's **barplot** function. When you execute this snippet, the following window appears (because you launched IPython with the `--matplotlib` option):

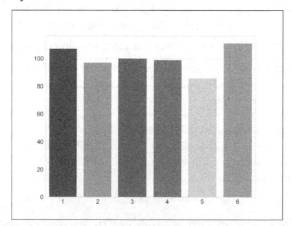

Seaborn interacts with Matplotlib to display the bars by creating a Matplotlib **Axes** object, which manages the content that appears in the window. Behind the scenes, Seaborn uses a Matplotlib **Figure** object to manage the window in which the `Axes` will appear. Function `barplot`'s first two arguments are `ndarrays` containing the *x*-axis and *y*-axis values, respectively. We used the optional `palette` keyword argument to choose Seaborn's predefined color palette `'bright'`. You can view the palette options at:

https://seaborn.pydata.org/tutorial/color_palettes.html

Function `barplot` returns the `Axes` object that it configured. We assign this to the variable `axes` so we can use it to configure other aspects of our final plot. Any changes you make to the bar plot after this point will appear *immediately* when you execute the corresponding snippet.

Setting the Window Title and Labeling the x- and y-Axes
The next two snippets add some descriptive text to the bar plot:

```
In [10]: axes.set_title(title)
Out[10]: Text(0.5,1,'Rolling a Six-Sided Die 600 Times')

In [11]: axes.set(xlabel='Die Value', ylabel='Frequency')
Out[11]: [Text(92.6667,0.5,'Frequency'), Text(0.5,58.7667,'Die Value')]
```

Snippet [10] uses the axes object's **set_title** method to display the title string centered above the plot. This method returns a Text object containing the title and its *location* in the window, which IPython simply displays as output for confirmation. You can ignore the Out[]s in the snippets above.

Snippet [11] add labels to each axis. The **set** method receives keyword arguments for the Axes object's properties to set. The method displays the xlabel text along the *x*-axis, and the ylabel text along the *y*-axis, and returns a list of Text objects containing the labels and their locations. The bar plot now appears as follows:

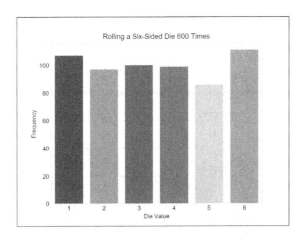

Finalizing the Bar Plot
The next two snippets complete the graph by making room for the text above each bar, then displaying it:

```
In [12]: axes.set_ylim(top=max(frequencies) * 1.10)
Out[12]: (0.0, 122.10000000000001)

In [13]: for bar, frequency in zip(axes.patches, frequencies):
   ...:     text_x = bar.get_x() + bar.get_width() / 2.0
   ...:     text_y = bar.get_height()
   ...:     text = f'{frequency:,}\n{frequency / len(rolls):.3%}'
   ...:     axes.text(text_x, text_y, text,
   ...:               fontsize=11, ha='center', va='bottom')
   ...:
```

To make room for the text above the bars, snippet [12] scales the *y*-axis by 10%. We chose this value via experimentation. The Axes object's `set_ylim` method has many optional keyword arguments. Here, we use only top to change the maximum value represented by the *y*-axis. We multiplied the largest frequency by 1.10 to ensure that the *y*-axis is 10% taller than the tallest bar.

Finally, snippet [13] displays each bar's frequency value and percentage of the total rolls. The `axes` object's `patches` collection contains two-dimensional colored shapes that represent the plot's bars. The `for` statement uses `zip` to iterate through the `patches` and their corresponding `frequency` values. Each iteration unpacks into `bar` and `frequency` one of the tuples `zip` returns. The `for` statement's suite operates as follows:

- The first statement calculates the center *x*-coordinate where the text will appear. We calculate this as the sum of the bar's left-edge *x*-coordinate (`bar.get_x()`) and half of the bar's width (`bar.get_width() / 2.0`).
- The second statement gets the *y*-coordinate where the text will appear—`bar.get_y()` represents the bar's top.
- The third statement creates a two-line string containing that bar's frequency and the corresponding percentage of the total die rolls.
- The last statement calls the `Axes` object's `text` method to display the text above the bar. This method's first two arguments specify the text's *x–y* position, and the third argument is the text to display. The keyword argument ha specifies the *horizontal alignment*—we centered text horizontally around the *x*-coordinate. The keyword argument va specifies the *vertical alignment*—we aligned the bottom of the text with at the *y*-coordinate. The final bar plot is shown below:

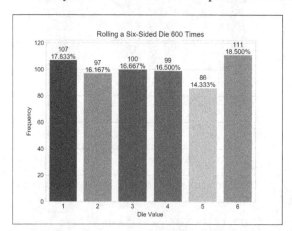

Rolling Again and Updating the Bar Plot—Introducing IPython Magics

Now that you've created a nice bar plot, you probably want to try a different number of die rolls. First, clear the existing graph by calling Matplotlib's **cla** (clear axes) function:

```
In [14]: plt.cla()
```

IPython provides special commands called **magics** for conveniently performing various tasks. Let's use the **%recall magic** to get snippet [5], which created the `rolls` list, and place the code at the next In [] prompt:

```
In [15]: %recall 5

In [16]: rolls = [random.randrange(1, 7) for i in range(600)]
```

You can now edit the snippet to change the number of rolls to 60000, then press *Enter* to create a new list:

```
In [16]: rolls = [random.randrange(1, 7) for i in range(60000)]
```

Next, recall snippets [6] through [13]. This displays all the snippets in the specified range in the next In [] prompt. Press *Enter* to re-execute these snippets:

```
In [17]: %recall 6-13

In [18]: values, frequencies = np.unique(rolls, return_counts=True)
    ...: title = f'Rolling a Six-Sided Die {len(rolls):,} Times'
    ...: sns.set_style('whitegrid')
    ...: axes = sns.barplot(x=values, y=frequencies, palette='bright')
    ...: axes.set_title(title)
    ...: axes.set(xlabel='Die Value', ylabel='Frequency')
    ...: axes.set_ylim(top=max(frequencies) * 1.10)
    ...: for bar, frequency in zip(axes.patches, frequencies):
    ...:     text_x = bar.get_x() + bar.get_width() / 2.0
    ...:     text_y = bar.get_height()
    ...:     text = f'{frequency:,}\n{frequency / len(rolls):.3%}'
    ...:     axes.text(text_x, text_y, text,
    ...:               fontsize=11, ha='center', va='bottom')
    ...:
```

The updated bar plot is shown below:

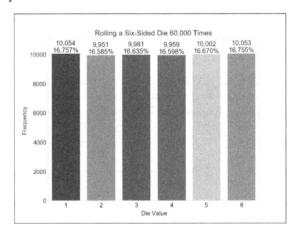

Saving Snippets to a File with the %save Magic

Once you've interactively created a plot, you may want to save the code to a file so you can turn it into a script and run it in the future. Let's use the **%save magic** to save snippets 1 through 13 to a file named RollDie.py. IPython indicates the file to which the lines were written, then displays the lines that it saved:

```
In [19]: %save RollDie.py 1-13
The following commands were written to file `RollDie.py`:
import matplotlib.pyplot as plt
import numpy as np
```

```
import random
import seaborn as sns
rolls = [random.randrange(1, 7) for i in range(600)]
values, frequencies = np.unique(rolls, return_counts=True)
title = f'Rolling a Six-Sided Die {len(rolls):,} Times'
sns.set_style("whitegrid")
axes = sns.barplot(values, frequencies, palette='bright')
axes.set_title(title)
axes.set(xlabel='Die Value', ylabel='Frequency')
axes.set_ylim(top=max(frequencies) * 1.10)
for bar, frequency in zip(axes.patches, frequencies):
    text_x = bar.get_x() + bar.get_width() / 2.0
    text_y = bar.get_height()
    text = f'{frequency:,}\n{frequency / len(rolls):.3%}'
    axes.text(text_x, text_y, text,
              fontsize=11, ha='center', va='bottom')
```

Command-Line Arguments; Displaying a Plot from a Script

Provided with this chapter's examples is an edited version of the RollDie.py file you saved above. We added comments and a two modifications so you can run the script with an argument that specifies the number of die rolls, as in:

```
ipython RollDie.py 600
```

The Python Standard Library's **sys module** enables a script to receive *command-line arguments* that are passed into the program. These include the script's name and any values that appear to the right of it when you execute the script. The sys module's **argv** list contains the arguments. In the command above, argv[0] is the *string* 'RollDie.py' and argv[1] is the *string* '600'. To control the number of die rolls with the command-line argument's value, we modified the statement that creates the rolls list as follows:

```
rolls = [random.randrange(1, 7) for i in range(int(sys.argv[1]))]
```

Note that we converted the argv[1] string to an int.

Matplotlib and Seaborn do not automatically display the plot for you when you create it in a script. So at the end of the script we added the following call to Matplotlib's **show** function, which displays the window containing the graph:

```
plt.show()
```

 Self Check

1 *(Fill-In)* The _____ format specifier indicates that a number should be displayed with thousands separators.
Answer: comma (,).

2 *(Fill-In)* A Matplotlib _____ object manages the content that appears in a Matplotlib window.
Answer: Axes.

3 *(Fill-In)* The Seaborn function _____ displays data as a bar chart.
Answer: barplot.

4 *(Fill-In)* The Matplotlib function _____ displays a plot window from a script.
Answer: show.

5 *(IPython Session)* Use the %recall magic to repeat the steps in snippets [14] through [18] to redraw the bar plot for 6,000,000 die rolls. This exercise assumes that you're continuing this section's IPython session. Notice that the heights of the six bars look the same, although each frequency is close to 1,000,000 and each percentage is close to 16.667%.
Answer:

```
In [20]: plt.cla()

In [21]: %recall 5

In [22]: rolls = [random.randrange(1, 7) for i in range(6000000)]

In [23]: %recall 6-13

In [24]: values, frequencies = np.unique(rolls, return_counts=True)
   ...: title = f'Rolling a Six-Sided Die {len(rolls):,} Times'
   ...: sns.set_style('whitegrid')
   ...: axes = sns.barplot(values, frequencies, palette='bright')
   ...: axes.set_title(title)
   ...: axes.set(xlabel='Die Value', ylabel='Frequency')
   ...: axes.set_ylim(top=max(frequencies) * 1.10)
   ...: for bar, frequency in zip(axes.patches, frequencies):
   ...:     text_x = bar.get_x() + bar.get_width() / 2.0
   ...:     text_y = bar.get_height()
   ...:     text = f'{frequency:,}\n{frequency / len(rolls):.3%}'
   ...:     axes.text(text_x, text_y, text,
   ...:               fontsize=11, ha='center', va='bottom')
   ...:
```

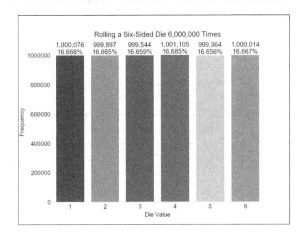

5.18 Wrap-Up

This chapter presented more details of the list and tuple sequences. You created lists, accessed their elements and determined their length. You saw that lists are mutable, so you can modify their contents, including growing and shrinking the lists as your programs execute. You saw that accessing a nonexistent element causes an IndexError. You used for statements to iterate through list elements.

We discussed tuples, which like lists are sequences, but are immutable. You unpacked a tuple's elements into separate variables. You used `enumerate` to create an iterable of tuples, each with a list index and corresponding element value.

You learned that all sequences support slicing, which creates new sequences with subsets of the original elements. You used the `del` statement to remove elements from lists and delete variables from interactive sessions. We passed lists, list elements and slices of lists to functions. You saw how to search and sort lists, and how to search tuples. We used list methods to insert, append and remove elements, and to reverse a list's elements and copy lists.

We showed how to simulate stacks with lists—in an exercise, you'll use the same list methods to simulate a queue with a list. We used the concise list-comprehension notation to create new lists. We used additional built-in methods to sum list elements, iterate backward through a list, find the minimum and maximum values, filter values and map values to new values. We showed how nested lists can represent two-dimensional tables in which data is arranged in rows and columns. You saw how nested `for` loops process two-dimensional lists.

The chapter concluded with an Intro to Data Science section that presented a die-rolling simulation and static visualizations. A detailed code example used the Seaborn and Matplotlib visualization libraries to create a *static* bar plot visualization of the simulation's final results. In the next Intro to Data Science section, we use a die-rolling simulation with a *dynamic* bar plot visualization to make the plot "come alive."

In the next chapter, "Dictionaries and Sets," we'll continue our discussion of Python's built-in collections. We'll use dictionaries to store unordered collections of key–value pairs that map immutable keys to values, just as a conventional dictionary maps words to definitions. We'll use sets to store unordered collections of unique elements.

In the "Array-Oriented Programming with NumPy" chapter, we'll discuss NumPy's `ndarray` collection in more detail. You'll see that while lists are fine for small amounts of data, they are not efficient for the large amounts of data you'll encounter in big data analytics applications. For such cases, the NumPy library's highly optimized `ndarray` collection should be used. `ndarray` (*n*-dimensional array) can be much faster than lists. We'll run Python profiling tests to see just how much faster. As you'll see, NumPy also includes many capabilities for conveniently and efficiently manipulating arrays of *many* dimensions. In big data analytics applications, the processing demands can be humongous, so everything we can do to improve performance significantly matters. In our "Big Data: Hadoop, Spark, NoSQL and IoT" chapter, you'll use one of the most popular big-data databases—MongoDB.[3]

Exercises

Use IPython sessions for each exercise where practical.

5.1 *(What's Wrong with This Code?)* What, if anything, is wrong with each of the following code segments?

    a) `day, high_temperature = ('Monday', 87, 65)`
    b) `numbers = [1, 2, 3, 4, 5]`
       `numbers[10]`

---

3. The database's name is rooted in the word "humongous."

c) ```
name = 'amanda'
name[0] = 'A'
```
d) ```
numbers = [1, 2, 3, 4, 5]
numbers[3.4]
```
e) ```
student_tuple = ('Amanda', 'Blue', [98, 75, 87])
student_tuple[0] = 'Ariana'
```
f) `('Monday', 87, 65) + 'Tuesday'`
g) `'A' += ('B', 'C')`
h) ```
x = 7
del x
print(x)
```
i) ```
numbers = [1, 2, 3, 4, 5]
numbers.index(10)
```
j) ```
numbers = [1, 2, 3, 4, 5]
numbers.extend(6, 7, 8)
```
k) ```
numbers = [1, 2, 3, 4, 5]
numbers.remove(10)
```
l) ```
values = []
values.pop()
```

5.2 *(What's Does This Code Do?)* What does the following function do, based on the sequence it receives as an argument?

```
def mystery(sequence):
    return sequence == sorted(sequence)
```

5.3 *(Fill in the Missing Code)* Replace the ***s in the following list comprehension and map function call, such that given a list of heights in inches, the code maps the list to a list of tuples containing the original height values and their corresponding values in meters. For example, if one element in the original list contains the height 69 inches, the corresponding element in the new list will contain the tuple (69, 1.7526), representing both the height in inches and the height in meters. There are 0.0254 meters per inch.

```
[*** for x in [69, 77, 54]]
list(map(lambda ***, [69, 77, 54]))
```

5.4 *(Iteration Order)* Create a 2-by-3 list, then use a nested loop to:
  a) Set each element's value to an integer indicating the order in which it was processed by the nested loop.
  b) Display the elements in tabular format. Use the column indices as headings across the top, and the row indices to the left of each row.

5.5 *(IPython Session: Slicing)* Create a string called alphabet containing 'abcdefghijklmnopqrstuvwxyz', then perform the following separate slice operations to obtain:
  a) The first half of the string using starting and ending indices.
  b) The first half of the string using only the ending index.
  c) The second half of the string using starting and ending indices.
  d) The second half of the string using only the starting index.
  e) Every second letter in the string starting with 'a'.
  f) The entire string in reverse.
  g) Every third letter of the string in reverse starting with 'z'.

5.6 *(Functions Returning Tuples)* Define a function `rotate` that receives three arguments and returns a tuple in which the first argument is at index 1, the second argument is at index 2 and the third argument is at index 0. Define variables a, b and c containing `'Doug'`, 22 and 1984. Then call the function three times. For each call, unpack its result into a, b and c, then display their values.

5.7 *(Duplicate Elimination)* Create a function that receives a list and returns a (possibly shorter) list containing only the unique values in sorted order. Test your function with a list of numbers and a list of strings.

5.8 *(Sieve of Eratosthenes)* A prime number is an integer greater than 1 that's evenly divisible only by itself and 1. The Sieve of Eratosthenes is an elegant, straightforward method of finding prime numbers. The process for finding all primes less than 1000 is:
  a) Create a 1000-element list `primes` with all elements initialized to `True`. List elements with prime indices (like 2, 3, 5, 7, 11, ...) will remain `True`. All other list elements will eventually be set to `False`.
  b) Starting with index 2, if a given element is `True` iterate through the rest of the list and set to `False` every element in `primes` whose index is a *multiple* of the index for the element we're currently processing. For list index 2, all elements beyond element 2 in the list that have indices which are multiples of 2 (i.e., 4, 6, 8, 10, ..., 998) will be set to `False`.
  c) Repeat Step (b) for the next `True` element. For list index 3 (which was initialized to `True`), all elements beyond element 3 in the list that have indices which are multiples of 3 (i.e., 6, 9, 12, 15, ..., 999) will be set to `False`; and so on. A subtle observation (think about why this is true): The square root of 999 is 31.6, you'll need to test and set to `False` only all multiples of 2, 3, 5, 7, 9, 11, 13, 17, 19, 23, 29 and 31. This will significantly improve the performance of your algorithm, especially if you decide to look for large prime numbers.

When this process completes, the list elements that are still `True` indicate that the index is a prime number. These indices can be displayed. Use a list of 1000 elements to determine and display the prime numbers less than 1000. Ignore list elements 0 and 1. [As you work through the book, you'll discover other Python capabilities that will enable you to cleverly reimplement this exercise.]

5.9 *(Palindrome Tester)* A string that's spelled identically backward and forward, like `'radar'`, is a palindrome. Write a function `is_palindrome` that takes a string and returns `True` if it's a palindrome and `False` otherwise. Use a stack (simulated with a list as we did in Section 5.11) to help determine whether a string is a palindrome. Your function should ignore case sensitivity (that is, `'a'` and `'A'` are the same), spaces and punctuation.

5.10 *(Anagrams)* An anagram of a string is another string formed by rearranging the letters in the first. Write a script that produces all possible anagrams of a given string using only techniques that you've seen to this point in the book. [The `itertools` module provides many functions, including one that produces permutations.]

5.11 *(Summarizing Letters in a String)* Write a function `summarize_letters` that receives a string and returns a list of tuples containing the unique letters and their frequencies in the string. Test your function and display each letter with its frequency. Your function should ignore case sensitivity (that is, `'a'` and `'A'` are the same) and ignore spaces and punctuation. When done, write a statement that says whether the string has all the letters of the alphabet.

5.12 *(Telephone-Number Word Generator)* You should find this exercise to be entertaining. Standard telephone keypads contain the digits zero through nine. The numbers two through nine each have three letters associated with them, as shown in the following table:

Digit	Letters	Digit	Letters	Digit	Letters
2	A B C	5	J K L	8	T U V
3	D E F	6	M N O	9	W X Y
4	G H I	7	P R S		

Many people find it difficult to memorize phone numbers, so they use the correspondence between digits and letters to develop seven-letter words (or phrases) that correspond to their phone numbers. For example, a person whose telephone number is 686-2377 might use the correspondence indicated in the preceding table to develop the seven-letter word "NUMBERS." Every seven-letter word or phrase corresponds to exactly one seven-digit telephone number. A budding data science entrepreneur might like to reserve the phone number 244-3282 ("BIGDATA").

Every seven-digit phone number without 0s or 1s corresponds to many different seven-letter words, but most of these words represent unrecognizable gibberish. A veterinarian with the phone number 738-2273 would be pleased to know that the number corresponds to the letters "PETCARE."

Write a script that, given a seven-digit number, generates every possible seven-letter word combination corresponding to that number. There are 2,187 (3^7) such combinations. Avoid phone numbers with the digits 0 and 1 (to which no letters correspond). See if your phone number corresponds to meaningful words.

5.13 *(Word or Phrase to Phone-Number Generator)* Just as people would enjoy knowing what word or phrase their phone number corresponds to, they might choose a word or phrase appropriate for their business and determine what phone numbers correspond to it. These are sometimes called vanity phone numbers, and various websites sell such phone numbers. Write a script similar to the one in the previous exercise that produces the possible phone number for the given seven-letter string.

5.14 *(Is a Sequence Sorted?)* Create a function `is_ordered` that receives a sequence and returns `True` if the elements are in sorted order. Test your function with sorted and unsorted lists, tuples and strings.

5.15 *(Tuples Representing Invoices)* When you purchase products or services from a company, you typically receive an invoice listing what you purchased and the total amount of money due. Use tuples to represent hardware store invoices that consist of four pieces of data—a part ID string, a part description string, an integer quantity of the item being purchased and, for simplicity, a `float` item price (in general, `Decimal` should be used for monetary amounts). Use the sample hardware data shown in the following table.

Part number	Part description	Quantity	Price
83	Electric sander	7	57.98
24	Power saw	18	99.99
7	Sledge hammer	11	21.50
77	Hammer	76	11.99
39	Jig saw	3	79.50

Perform the following tasks on a list of invoice tuples:

a) Use function `sorted` with a key argument to sort the tuples by part description, then display the results. To specify the element of the tuple that should be used for sorting, first import the `itemgetter` function from the `operator` module as in

```
from operator import itemgetter
```

Then, for `sorted`'s key argument specify `itemgetter(`*index*`)` where *index* specifies which element of the tuple should be used for sorting purposes.

b) Use the `sorted` function with a key argument to sort the tuples by price, then display the results.

c) Map each invoice tuple to a tuple containing the part description and quantity, sort the results by quantity, then display the results.

d) Map each invoice tuple to a tuple containing the part description and the value of the invoice (the product of the quantity and the item price), sort the results by the invoice value, then display the results.

e) Modify Part (d) to filter the results to invoice values in the range $200 to $500.

f) Calculate the total of all the invoices.

5.16 *(Sorting Letters and Removing Duplicates)* Insert 20 random letters in the range `'a'` through `'f'` into a list. Perform the following tasks and display your results:

a) Sort the list in ascending order.

b) Sort the list in descending order.

c) Get the unique values sort them in ascending order.

5.17 *(Filter/Map Performance)* With regard to the following code:

```
numbers = [10, 3, 7, 1, 9, 4, 2, 8, 5, 6]
list(map(lambda x: x ** 2,
    filter(lambda x: x % 2 != 0, numbers)))
```

a) How many times does the `filter` operation call its `lambda` argument?

b) How many times does the `map` operation call its `lambda` argument?

c) If you reverse the `filter` and `map` operations, how many times does the `map` operation call its lambda argument?

To help you answer the preceding questions, define functions that perform the same tasks as the lambdas. In each function, include a `print` statement so you can see each time the function is called. Finally, replace the lambdas in the preceding code with the names of your functions.

5.18 *(Summing the Triples of the Even Integers from 2 through 10)* Starting with a list containing 1 through 10, use `filter`, `map` and `sum` to calculate the total of the triples of the even integers from 2 through 10. Reimplement your code with list comprehensions rather than `filter` and `map`.

5.19 *(Finding the People with a Specified Last Name)* Create a list of tuples containing first and last names. Use `filter` to locate the tuples containing the last name `Jones`. Ensure that several tuples in your list have that last name.

5.20 *(Display a Two-Dimensional List in Tabular Format)* Define a function named `display_table` that receives a two-dimensional list and displays its contents in tabular format. List the column indices as headings across the top, and list the row indices at the left of each row.

5.21 *(Computer-Assisted Instruction: Reducing Student Fatigue)* Re-implement Exercise 4.15 to store the computer's responses in lists. Use random-number generation to select responses using random list indices.

5.22 *(Simulating a Queue with a List)* In this chapter, you simulated a stack using a list. You also can simulate a queue collection with a list. **Queues** represent waiting lines similar to a checkout line in a supermarket. The cashier services the person at the *front* of the line *first*. Other customers enter the line only at the end and wait for service.

In a queue, you insert items at the back (known as the **tail**) and delete items from the front (known as the **head**). For this reason, queues are first-in, first-out (FIFO) collections. The insert and remove operations are commonly known as **enqueue** and **dequeue**.

Queues have many uses in computer systems, such as sharing CPUs among a potentially large number of competing applications and the operating system itself. Applications not currently being serviced sit in a queue until a CPU becomes available. The application at the front of the queue is the next to receive service. Each application gradually advances to the front as the applications before it receive service.

Simulate a queue of integers using list methods append (to simulate *enqueue*) and **pop** with the argument 0 (to simulate *dequeue*). Enqueue the values 3, 2 and 1, then dequeue them to show that they're removed in FIFO order.

5.23 *(Functional-Style Programming: Order of `filter` and `map` Calls)* When combining `filter` and `map` operations, the order in which they're performed matters. Consider a list numbers containing 10, 3, 7, 1, 9, 4, 2, 8, 5, 6 and the following code:

```
In [1]: numbers = [10, 3, 7, 1, 9, 4, 2, 8, 5, 6]

In [2]: list(map(lambda x: x * 2,
   ...:          filter(lambda x: x % 2 == 0, numbers)))
   ...:
Out[3]: [20, 8, 4, 16, 12]
```

Reorder this code to call map first and filter second. What happens and why?

Exercises 5.24 through 5.26 are reasonably challenging. Once you've done them, you ought to be able to implement many popular card games.

5.24 *(Card Shuffling and Dealing)* In Exercises 5.24 through 5.26, you'll use lists of tuples in scripts that simulate card shuffling and dealing. Each tuple represents one card in the deck and contains a face (e.g., `'Ace'`, `'Deuce'`, `'Three'`, ..., `'Jack'`, `'Queen'`, `'King'`) and a suit (e.g., `'Hearts'`, `'Diamonds'`, `'Clubs'`, `'Spades'`). Create an `initialize_deck` function to initialize the deck of card tuples with `'Ace'` through `'King'` of each suit, as in

```
deck = [('Ace', 'Hearts'), ..., ('King', 'Hearts'),
    ('Ace', 'Diamonds'), ..., ('King', 'Diamonds'),
    ('Ace', 'Clubs'), ..., ('King', 'Clubs'),
    ('Ace', 'Spades'), ..., ('King', 'Spades')]
```

Before returning the list, use the random module's `shuffle` function to randomly order the list elements. Output the shuffled cards in the following four-column format:

```
Six of Spades       Eight of Spades     Six of Clubs        Nine of Hearts
Queen of Hearts     Seven of Clubs      Nine of Spades      King of Hearts
Three of Diamonds   Deuce of Clubs      Ace of Hearts       Ten of Spades
Four of Spades      Ace of Clubs        Seven of Diamonds   Four of Hearts
Three of Clubs      Deuce of Hearts     Five of Spades      Jack of Diamonds
King of Clubs       Ten of Hearts       Three of Hearts     Six of Diamonds
Queen of Clubs      Eight of Diamonds   Deuce of Diamonds   Ten of Diamonds
Three of Spades     King of Diamonds    Nine of Clubs       Six of Hearts
Ace of Spades       Four of Diamonds    Seven of Hearts     Eight of Clubs
Deuce of Spades     Eight of Hearts     Five of Hearts      Queen of Spades
Jack of Hearts      Seven of Spades     Four of Clubs       Nine of Diamonds
Ace of Diamonds     Queen of Diamonds   Five of Clubs       King of Spades
Five of Diamonds    Ten of Clubs        Jack of Spades      Jack of Clubs
```

5.25 *(Card Playing: Evaluating Poker Hands)* Modify Exercise 5.24 to deal a five-card poker hand as a list of five card tuples. Then create functions (i.e., is_pair, is_two_pair, is_three_of_a_kind, ...) that determine whether the hand they receive as an argument contains groups of cards, such as:

    a) one pair
    b) two pairs
    c) three of a kind (e.g., three jacks)
    d) a straight (i.e., five cards of consecutive face values)
    e) a flush (i.e., all five cards of the same suit)
    f) a full house (i.e., two cards of one face value and three cards of another)
    g) four of a kind (e.g., four aces)
    h) straight flush (i.e., a straight with all five cards of the same suit)
    i) ... and others.

See https://en.wikipedia.org/wiki/List_of_poker_hands for poker-hand types and how they rank with respect to one another. For example, three of a kind beats two pairs.

5.26 *(Card Playing: Determining the Winning Hand)* Use the methods developed in Exercise 5.25 to write a script that deals two five-card poker hands (i.e., two lists of five card tuples each), evaluates each hand and determines which wins. As each card is dealt, it should be removed from the list of tuples representing the deck.

5.27 *(Intro to Data Science: Duplicate Elimination and Counting Frequencies)* Use a list comprehension to create a list of 50 random values in the range 1 through 10. Use NumPy's `unique` function to obtain the unique values and their frequencies. Display the results.

5.28 *(Intro to Data Science: Survey Response Statistics)* Twenty students were asked to rate on a scale of 1 to 5 the quality of the food in the student cafeteria, with 1 being "awful" and 5 being "excellent." Place the 20 responses in a list

    1, 2, 5, 4, 3, 5, 2, 1, 3, 3, 1, 4, 3, 3, 3, 2, 3, 3, 2, 5

Determine and display the frequency of each rating. Use the built-in functions, statistics module functions and NumPy functions demonstrated in Section 5.17.2 to display the following response statistics: minimum, maximum, range, mean, median, mode, variance and standard deviation.

5.29 *(Intro to Data Science: Visualizing Survey Response Statistics)* Using the list in Exercise 5.28 and the techniques you learned in Section 5.17.2, display a bar chart showing the response frequencies and their percentages of the total responses.

5.30 *(Intro to Data Science: Removing the Text Above the Bars)* Modify the die-rolling simulation in Section 5.17.2 to omit displaying the frequencies and percentages above each bar. Try to minimize the number of lines of code.

5.31 *(Intro to Data Science: Coin Tossing)* Modify the die-rolling simulation in Section 5.17.2 to simulate the flipping a coin. Use randomly generated 1s and 2s to represent heads and tails, respectively. Initially, do not include the frequencies and percentages above the bars. Then modify your code to include the frequencies and percentages. Run simulations for 200, 20,000 and 200,000 coin flips. Do you get approximately 50% heads and 50% tails? Do you see the "law of large numbers" in operation here?

5.32 *(Intro to Data Science: Rolling Two Dice)* Modify the script RollDie.py that we provided with this chapter's examples to simulate rolling two dice. Calculate the sum of the two values. Each die has a value from 1 to 6, so the sum of the values will vary from 2 to 12, with 7 being the most frequent sum, and 2 and 12 the least frequent. The following diagram shows the 36 equally likely possible combinations of the two dice and their corresponding sums:

	1	2	3	4	5	6
1	2	3	4	5	6	7
2	3	4	5	6	7	8
3	4	5	6	7	8	9
4	5	6	7	8	9	10
5	6	7	8	9	10	11
6	7	8	9	10	11	12

If you roll the dice 36,000 times:
- The values 2 and 12 each occur 1/36th (2.778%) of the time, so you should expect about 1000 of each.
- The values 3 and 11 each occur 2/36ths (5.556%) of the time, so you should expect about 2000 of each, and so on.

Use a command-line argument to obtain the number of rolls. Display a bar plot summarizing the roll frequencies. The following screen captures show the final bar plots for sample executions of 360, 36,000 and 36,000,000 rolls. Use the Seaborn barplot function's optional orient keyword argument to specify a horizontal bar plot.

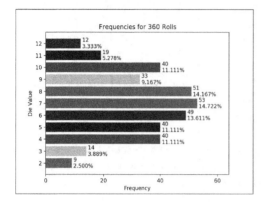
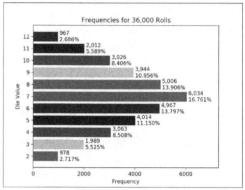
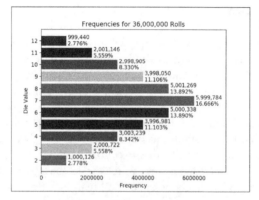

5.33 *(Intro to Data Science Challenge: Analyzing the Dice Game Craps)* In this exercise, you'll modify Chapter 4's script that simulates the dice game craps by using the techniques you learned in Section 5.17.2. The script should receive a command-line argument indicating the number of games of craps to execute and use two lists to track the total numbers of games won and lost on the first roll, second roll, third roll, etc. Summarize the results as follows:

    a) Display a horizontal bar plot indicating how many games are won and how many are lost on the first roll, second roll, third roll, etc. Since the game could continue indefinitely, you might track wins and losses through the first dozen rolls (of a pair of dice), then maintain two counters that keep track of wins and losses after 12 rolls—no matter how long the game gets. Create separate bars for wins and losses.

    b) What are the chances of winning at craps? [*Note:* You should discover that craps is one of the fairest casino games. What do you suppose this means?]

    c) What is the mean for the length of a game of craps? The median? The mode?

    d) Do the chances of winning improve with the length of the game?

Dictionaries and Sets

6

Objectives

In this chapter, you'll:

- Use dictionaries to represent unordered collections of key–value pairs.
- Use sets to represent unordered collections of unique values.
- Create, initialize and refer to elements of dictionaries and sets.
- Iterate through a dictionary's keys, values and key–value pairs.
- Add, remove and update a dictionary's key–value pairs.
- Use dictionary and set comparison operators.
- Combine sets with set operators and methods.
- Use operators in and not in to determine if a dictionary contains a key or a set contains a value.
- Use the mutable set operations to modify a set's contents.
- Use comprehensions to create dictionaries and sets quickly and conveniently.
- Learn how to build dynamic visualizations and implement more of your own in the exercises.
- Enhance your understanding of mutability and immutability.

Outline

6.1 Introduction
6.2 Dictionaries
    6.2.1 Creating a Dictionary
    6.2.2 Iterating through a Dictionary
    6.2.3 Basic Dictionary Operations
    6.2.4 Dictionary Methods `keys` and `values`
    6.2.5 Dictionary Comparisons
    6.2.6 Example: Dictionary of Student Grades
    6.2.7 Example: Word Counts
    6.2.8 Dictionary Method `update`
    6.2.9 Dictionary Comprehensions
6.3 Sets
    6.3.1 Comparing Sets
    6.3.2 Mathematical Set Operations
    6.3.3 Mutable Set Operators and Methods
    6.3.4 Set Comprehensions
6.4 Intro to Data Science: Dynamic Visualizations
    6.4.1 How Dynamic Visualization Works
    6.4.2 Implementing a Dynamic Visualization
6.5 Wrap-Up
Exercises

6.1 Introduction

We've discussed three built-in sequence collections—strings, lists and tuples. Now, we consider the built-in non-sequence collections—dictionaries and sets. A **dictionary** is an *unordered* collection which stores **key–value pairs** that map immutable keys to values, just as a conventional dictionary maps words to definitions. A **set** is an unordered collection of *unique* immutable elements.

6.2 Dictionaries

A dictionary *associates* keys with values. Each key *maps* to a specific value. The following table contains examples of dictionaries with their keys, key types, values and value types:

Keys	Key type	Values	Value type
Country names	`str`	Internet country codes	`str`
Decimal numbers	`int`	Roman numerals	`str`
States	`str`	Agricultural products	list of `str`
Hospital patients	`str`	Vital signs	tuple of `int`s and `float`s
Baseball players	`str`	Batting averages	`float`
Metric measurements	`str`	Abbreviations	`str`
Inventory codes	`str`	Quantity in stock	`int`

Unique Keys

A dictionary's keys must be *immutable* (such as strings, numbers or tuples) and *unique* (that is, no duplicates). Multiple keys can have the same value, such as two different inventory codes that have the same quantity in stock.

6.2.1 Creating a Dictionary

You can create a dictionary by enclosing in curly braces, {}, a comma-separated list of key–value pairs, each of the form *key*: *value*. You can create an empty dictionary with {}.

Let's create a dictionary with the country-name keys `'Finland'`, `'South Africa'` and `'Nepal'` and their corresponding Internet country code values `'fi'`, `'za'` and `'np'`:

```
In [1]: country_codes = {'Finland': 'fi', 'South Africa': 'za',
   ...:                  'Nepal': 'np'}
   ...:
```

```
In [2]: country_codes
Out[2]: {'Finland': 'fi', 'South Africa': 'za', 'Nepal': 'np'}
```

When you output a dictionary, its comma-separated list of key–value pairs is always enclosed in curly braces. Because dictionaries are *unordered* collections, the display order can differ from the order in which the key–value pairs were added to the dictionary. In snippet [2]'s output the key–value pairs are displayed in the order they were inserted, but do *not* write code that depends on the order of the key–value pairs.

Determining if a Dictionary Is Empty

The built-in function `len` returns the number of key–value pairs in a dictionary:

```
In [3]: len(country_codes)
Out[3]: 3
```

You can use a dictionary as a condition to determine if it's empty—a non-empty dictionary evaluates to `True`:

```
In [4]: if country_codes:
   ...:     print('country_codes is not empty')
   ...: else:
   ...:     print('country_codes is empty')
   ...:
country_codes is not empty
```

An empty dictionary evaluates to `False`. To demonstrate this, in the following code we call method **clear** to delete the dictionary's key–value pairs, then in snippet [6] we recall and re-execute snippet [4]:

```
In [5]: country_codes.clear()
```

```
In [6]: if country_codes:
   ...:     print('country_codes is not empty')
   ...: else:
   ...:     print('country_codes is empty')
   ...:
country_codes is empty
```

✓ Self Check

1 *(Fill-In)* _____ can be thought of as unordered collections in which each value is accessed through its corresponding key.
Answer: Dictionaries.

2 *(True/False)* Dictionaries may contain duplicate keys.
Answer: False. Dictionary keys must be unique. However, multiple keys may have the same value.

3 *(IPython Session)* Create a dictionary named `states` that maps three state abbreviations to their state names, then display the dictionary.
Answer:

```
In [1]: states = {'VT': 'Vermont', 'NH': 'New Hampshire',
   ...:           'MA': 'Massachusetts'}
   ...:

In [2]: states
Out[2]: {'VT': 'Vermont', 'NH': 'New Hampshire', 'MA': 'Massachusetts'}
```

6.2.2 Iterating through a Dictionary

The following dictionary maps month-name strings to `int` values representing the numbers of days in the corresponding month. Note that *multiple* keys can have the *same* value:

```
In [1]: days_per_month = {'January': 31, 'February': 28, 'March': 31}

In [2]: days_per_month
Out[2]: {'January': 31, 'February': 28, 'March': 31}
```

Again, the dictionary's string representation shows the key–value pairs in their insertion order, but this is not guaranteed because dictionaries are *unordered*. We'll show how to process keys in *sorted* order later in this chapter.

The following `for` statement iterates through `days_per_month`'s key–value pairs. Dictionary method **items** returns each key–value pair as a tuple, which we unpack into `month` and `days`:

```
In [3]: for month, days in days_per_month.items():
   ...:     print(f'{month} has {days} days')
   ...:
January has 31 days
February has 28 days
March has 31 days
```

✓ Self Check

1 *(Fill-In)* Dictionary method _____ returns each key–value pair as a tuple.
Answer: `items`.

6.2.3 Basic Dictionary Operations

For this section, let's begin by creating and displaying the dictionary `roman_numerals`. We intentionally provide the incorrect value 100 for the key `'X'`, which we'll correct shortly:

```
In [1]: roman_numerals = {'I': 1, 'II': 2, 'III': 3, 'V': 5, 'X': 100}

In [2]: roman_numerals
Out[2]: {'I': 1, 'II': 2, 'III': 3, 'V': 5, 'X': 100}
```

Accessing the Value Associated with a Key
Let's get the value associated with the key `'V'`:

```
In [3]: roman_numerals['V']
Out[3]: 5
```

Updating the Value of an Existing Key–Value Pair
You can update a key's associated value in an assignment statement, which we do here to replace the incorrect value associated with the key 'X':

```
In [4]: roman_numerals['X'] = 10

In [5]: roman_numerals
Out[5]: {'I': 1, 'II': 2, 'III': 3, 'V': 5, 'X': 10}
```

Adding a New Key–Value Pair
Assigning a value to a nonexistent key inserts the key–value pair in the dictionary:

```
In [6]: roman_numerals['L'] = 50

In [7]: roman_numerals
Out[7]: {'I': 1, 'II': 2, 'III': 3, 'V': 5, 'X': 10, 'L': 50}
```

String keys are case sensitive. Assigning to a nonexistent key inserts a new key–value pair. This may be what you intend, or it could be a logic error.

Removing a Key–Value Pair
You can delete a key–value pair from a dictionary with the del statement:

```
In [8]: del roman_numerals['III']

In [9]: roman_numerals
Out[9]: {'I': 1, 'II': 2, 'V': 5, 'X': 10, 'L': 50}
```

You also can remove a key–value pair with the dictionary method **pop**, which returns the value for the removed key:

```
In [10]: roman_numerals.pop('X')
Out[10]: 10

In [11]: roman_numerals
Out[11]: {'I': 1, 'II': 2, 'V': 5, 'L': 50}
```

Attempting to Access a Nonexistent Key
Accessing a nonexistent key results in a KeyError:

```
In [12]: roman_numerals['III']
-------------------------------------------------------------------------
KeyError                                  Traceback (most recent call last)
<ipython-input-12-ccd50c7f0c8b> in <module>()
----> 1 roman_numerals['III']

KeyError: 'III'
```

You can prevent this error by using dictionary method **get**, which normally returns its argument's corresponding value. If that key is not found, get returns None. IPython does not display anything when None is returned in snippet [13]. If you specify a second argument to get, it returns that value if the key is not found:

```
In [13]: roman_numerals.get('III')

In [14]: roman_numerals.get('III', 'III not in dictionary')
Out[14]: 'III not in dictionary'

In [15]: roman_numerals.get('V')
Out[15]: 5
```

Testing Whether a Dictionary Contains a Specified Key
Operators in and not in can determine whether a dictionary contains a specified key:

```
In [16]: 'V' in roman_numerals
Out[16]: True

In [17]: 'III' in roman_numerals
Out[17]: False

In [18]: 'III' not in roman_numerals
Out[18]: True
```

Self Check

1 *(True/False)* Assigning to a nonexistent dictionary key causes an exception.
Answer: False. Assigning to a nonexistent key inserts a new key–value pair. This may be what you intend, or it could be a logic error if you incorrectly specify the key.

2 *(Fill-In)* What does an expression of the following form do when the *key* is in the dictionary?

dictionaryName[*key*] = *value*

Answer: It updates the *value* associated with the *key*, replacing the original *value*.

3 *(IPython Session)* String dictionary keys are case sensitive. Confirm this by using the following dictionary and assigning 10 to the key 'x'—doing so adds a new key–value pair rather than correcting the value for the key 'X':

```
roman_numerals = {'I': 1, 'II': 2, 'III': 3, 'V': 5, 'X': 100}
```

Answer:

```
In [1]: roman_numerals = {'I': 1, 'II': 2, 'III': 3, 'V': 5, 'X': 100}

In [2]: roman_numerals['x'] = 10

In [3]: roman_numerals
Out[3]: {'I': 1, 'II': 2, 'III': 3, 'V': 5, 'X': 100, 'x': 10}
```

6.2.4 Dictionary Methods keys and values
Earlier, we used dictionary method items to iterate through tuples of a dictionary's key–value pairs. Similarly, methods **keys** and **values** can be used to iterate through only a dictionary's keys or values, respectively:

```
In [1]: months = {'January': 1, 'February': 2, 'March': 3}

In [2]: for month_name in months.keys():
   ...:     print(month_name, end=' ')
   ...:
January February March

In [3]: for month_number in months.values():
   ...:     print(month_number, end=' ')
   ...:
1 2 3
```

Dictionary Views

Dictionary methods `items`, `keys` and `values` each return a view of a dictionary's data. When you iterate over a **view**, it "sees" the dictionary's current contents—it does *not* have its own copy of the data.

To show that views do *not* maintain their own copies of a dictionary's data, let's first save the view returned by `keys` into the variable `months_view`, then iterate through it:

```
In [4]: months_view = months.keys()

In [5]: for key in months_view:
   ...:     print(key, end=' ')
   ...:
January February March
```

Next, let's add a new key–value pair to `months` and display the updated dictionary:

```
In [6]: months['December'] = 12

In [7]: months
Out[7]: {'January': 1, 'February': 2, 'March': 3, 'December': 12}
```

Now, let's iterate through `months_view` again. The key we added above is indeed displayed:

```
In [8]: for key in months_view:
   ...:     print(key, end=' ')
   ...:
January February March December
```

Do not modify a dictionary while iterating through a view. According to Section 4.10.1 of the Python Standard Library documentation,[1] either you'll get a `RuntimeError` or the loop might not process all of the view's values.

Converting Dictionary Keys, Values and Key–Value Pairs to Lists

You might occasionally need *lists* of a dictionary's keys, values or key–value pairs. To obtain such a list, pass the view returned by `keys`, `values` or `items` to the built-in `list` function. Modifying these lists does *not* modify the corresponding dictionary:

```
In [9]: list(months.keys())
Out[9]: ['January', 'February', 'March', 'December']

In [10]: list(months.values())
Out[10]: [1, 2, 3, 12]

In [11]: list(months.items())
Out[11]: [('January', 1), ('February', 2), ('March', 3), ('December', 12)]
```

Processing Keys in Sorted Order

To process keys in *sorted* order, you can use built-in function `sorted` as follows:

```
In [12]: for month_name in sorted(months.keys()):
   ...:     print(month_name, end=' ')
   ...:
February December January March
```

---

1. https://docs.python.org/3/library/stdtypes.html#dictionary-view-objects.

Self Check

1 *(Fill-In)* Dictionary method _____ returns an unordered list of the dictionary's keys.
Answer: keys.

2 *(True/False)* A view has its own copy of the corresponding data from the dictionary.
Answer: False. A view does *not* have its own copy of the corresponding data from the dictionary. As the dictionary changes, each view updates dynamically.

3 *(IPython Session)* For the following dictionary, create lists of its keys, values and items and show those lists.

```
roman_numerals = {'I': 1, 'II': 2, 'III': 3, 'V': 5}
```

Answer:

```
In [1]: roman_numerals = {'I': 1, 'II': 2, 'III': 3, 'V': 5}

In [2]: list(roman_numerals.keys())
Out[2]: ['I', 'II', 'III', 'V']

In [3]: list(roman_numerals.values())
Out[3]: [1, 2, 3, 5]

In [4]: list(roman_numerals.items())
Out[4]: [('I', 1), ('II', 2), ('III', 3), ('V', 5)]
```

6.2.5 Dictionary Comparisons

The comparison operators == and != can be used to determine whether two dictionaries have identical or different contents. An equals (==) comparison evaluates to True if both dictionaries have the same key–value pairs, *regardless* of the order in which those key–value pairs were added to each dictionary:

```
In [1]: country_capitals1 = {'Belgium': 'Brussels',
   ...:                      'Haiti': 'Port-au-Prince'}
   ...:

In [2]: country_capitals2 = {'Nepal': 'Kathmandu',
   ...:                      'Uruguay': 'Montevideo'}
   ...:

In [3]: country_capitals3 = {'Haiti': 'Port-au-Prince',
   ...:                      'Belgium': 'Brussels'}
   ...:

In [4]: country_capitals1 == country_capitals2
Out[4]: False

In [5]: country_capitals1 == country_capitals3
Out[5]: True

In [6]: country_capitals1 != country_capitals2
Out[6]: True
```

Self Check

1 *(True/False)* The == comparison evaluates to True only if both dictionaries have the same key–value pairs in the same order.
Answer: False. The == comparison evaluates to True if both dictionaries have the same key–value pairs, regardless of their order.

6.2.6 Example: Dictionary of Student Grades

The script in Fig. 6.1 represents an instructor's grade book as a dictionary that maps each student's name (a string) to a list of integers containing that student's grades on three exams. In each iteration of the loop that displays the data (lines 13–17), we unpack a key–value pair into the variables name and grades containing one student's name and the corresponding list of three grades. Line 14 uses built-in function sum to total a given student's grades, then line 15 calculates and displays that student's average by dividing total by the number of grades for that student (len(grades)). Lines 16–17 keep track of the total of all four students' grades and the number of grades for all the students, respectively. Line 19 prints the class average of all the students' grades on all the exams.

```
1   # fig06_01.py
2   """Using a dictionary to represent an instructor's grade book."""
3   grade_book = {
4       'Susan': [92, 85, 100],
5       'Eduardo': [83, 95, 79],
6       'Azizi': [91, 89, 82],
7       'Pantipa': [97, 91, 92]
8   }
9
10  all_grades_total = 0
11  all_grades_count = 0
12
13  for name, grades in grade_book.items():
14      total = sum(grades)
15      print(f'Average for {name} is {total/len(grades):.2f}')
16      all_grades_total += total
17      all_grades_count += len(grades)
18
19  print(f"Class's average is: {all_grades_total / all_grades_count:.2f}")
```

```
Average for Susan is 92.33
Average for Eduardo is 85.67
Average for Azizi is 87.33
Average for Pantipa is 93.33
Class's average is: 89.67
```

Fig. 6.1 | Using a dictionary to represent an instructor's grade book.

6.2.7 Example: Word Counts[2]

The script in Fig. 6.2 builds a dictionary to count the number of occurrences of each word in a string. Lines 4–5 create a string `text` that we'll break into words—a process known as **tokenizing a string**. Python automatically concatenates strings separated by whitespace in parentheses. Line 7 creates an empty dictionary. The dictionary's keys will be the unique words, and its values will be integer counts of how many times each word appears in `text`.

```
1   # fig06_02.py
2   """Tokenizing a string and counting unique words."""
3
4   text = ('this is sample text with several words '
5           'this is more sample text with some different words')
6
7   word_counts = {}
8
9   # count occurrences of each unique word
10  for word in text.split():
11      if word in word_counts:
12          word_counts[word] += 1  # update existing key-value pair
13      else:
14          word_counts[word] = 1  # insert new key-value pair
15
16  print(f'{"WORD":<12}COUNT')
17
18  for word, count in sorted(word_counts.items()):
19      print(f'{word:<12}{count}')
20
21  print('\nNumber of unique words:', len(word_counts))
```

```
WORD        COUNT
different   1
is          2
more        1
sample      2
several     1
some        1
text        2
this        2
with        2
words       2
Number of unique words: 10
```

Fig. 6.2 | Tokenizing a string and counting unique words.

Line 10 tokenizes `text` by calling string method `split`, which separates the words using the method's delimiter string argument. If you do not provide an argument, `split` uses a space. The method returns a list of tokens (that is, the words in `text`). Lines 10–14

---

2. Techniques like word frequency counting are often used to analyze published works. For example, some people believe that the works of William Shakespeare actually might have been written by Sir Francis Bacon, Christopher Marlowe or others. Comparing the word frequencies of their works with those of Shakespeare can reveal writing-style similarities. We'll look at other document-analysis techniques in the "Natural Language Processing (NLP)" chapter.

iterate through the list of words. For each word, line 11 determines whether that word (the key) is already in the dictionary. If so, line 12 increments that word's count; otherwise, line 14 inserts a new key–value pair for that word with an initial count of 1.

Lines 16–21 summarize the results in a two-column table containing each word and its corresponding count. The for statement in lines 18 and 19 iterates through the dictionary's key–value pairs. It unpacks each key and value into the variables word and count, then displays them in two columns. Line 21 displays the number of unique words.

Python Standard Library Module collections

The Python Standard Library already contains the counting functionality that we implemented using the dictionary and the loop in lines 10–14. The module **collections** contains the type **Counter**, which receives an iterable and summarizes its elements. Let's reimplement the preceding script in fewer lines of code with Counter:

```
In [1]: from collections import Counter

In [2]: text = ('this is sample text with several words '
   ...:         'this is more sample text with some different words')
   ...:

In [3]: counter = Counter(text.split())

In [4]: for word, count in sorted(counter.items()):
   ...:     print(f'{word:<12}{count}')
   ...:
different   1
is          2
more        1
sample      2
several     1
some        1
text        2
this        2
with        2
words       2

In [5]: print('Number of unique keys:', len(counter.keys()))
Number of unique keys: 10
```

Snippet [3] creates the Counter, which summarizes the list of strings returned by text.split(). In snippet [4], Counter method **items** returns each string and its associated count as a tuple. We use built-in function sorted to get a list of these tuples in ascending order. By default sorted orders the tuples by their first elements. If those are identical, then it looks at the second element, and so on. The for statement iterates over the resulting sorted list, displaying each word and count in two columns.

✓ Self Check

1 *(Fill-In)* String method _____ tokenizes a string using the delimiter provided in the method's string argument.
Answer: split.

2 *(IPython Session)* Use a comprehension to create a list of 50 random integers in the range 1–5. Summarize them with a Counter. Display the results in two-column format.

Answer:

```
In [1]: import random

In [2]: numbers = [random.randrange(1, 6) for i in range(50)]

In [3]: from collections import Counter

In [4]: counter = Counter(numbers)

In [5]: for value, count in sorted(counter.items()):
   ...:     print(f'{value:<4}{count}')
   ...:
1   9
2   6
3   13
4   10
5   12
```

6.2.8 Dictionary Method update

You may insert and update key–value pairs using dictionary method **update**. First, let's create an empty country_codes dictionary:

```
In [1]: country_codes = {}
```

The following update call receives a dictionary of key–value pairs to insert or update:

```
In [2]: country_codes.update({'South Africa': 'za'})

In [3]: country_codes
Out[3]: {'South Africa': 'za'}
```

Method update can convert keyword arguments into key–value pairs to insert. The following call automatically converts the parameter name Australia into the string key 'Australia' and associates the value 'ar' with that key:

```
In [4]: country_codes.update(Australia='ar')

In [5]: country_codes
Out[5]: {'South Africa': 'za', 'Australia': 'ar'}
```

Snippet [4] provided an incorrect country code for Australia. Let's correct this by using another keyword argument to update the value associated with 'Australia':

```
In [6]: country_codes.update(Australia='au')

In [7]: country_codes
Out[7]: {'South Africa': 'za', 'Australia': 'au'}
```

Method update also can receive an iterable object containing key–value pairs, such as a list of two-element tuples.

6.2.9 Dictionary Comprehensions

Dictionary comprehensions provide a convenient notation for quickly generating dictionaries, often by mapping one dictionary to another. For example, in a dictionary with *unique* values, you can reverse the key–value pairs:

```
In [1]: months = {'January': 1, 'February': 2, 'March': 3}

In [2]: months2 = {number: name for name, number in months.items()}

In [3]: months2
Out[3]: {1: 'January', 2: 'February', 3: 'March'}
```

Curly braces delimit a *dictionary comprehension*, and the expression to the left of the for clause specifies a key–value pair of the form *key*: *value*. The comprehension iterates through months.items(), unpacking each key–value pair tuple into the variables name and number. The expression number: name reverses the key and value, so the new dictionary maps the month numbers to the month names.

What if months contained *duplicate* values? As these become the keys in months2, attempting to insert a *duplicate* key simply updates the existing key's value. So if 'February' and 'March' both mapped to 2 originally, the preceding code would have produced

```
{1: 'January', 2: 'March'}
```

A dictionary comprehension also can map a dictionary's values to new values. The following comprehension converts a dictionary of names and lists of grades into a dictionary of names and grade-point averages. The variables k and v commonly mean *key* and *value*:

```
In [4]: grades = {'Sue': [98, 87, 94], 'Bob': [84, 95, 91]}

In [5]: grades2 = {k: sum(v) / len(v) for k, v in grades.items()}

In [6]: grades2
Out[6]: {'Sue': 93.0, 'Bob': 90.0}
```

The comprehension unpacks each tuple returned by grades.items() into k (the name) and v (the list of grades). Then, the comprehension creates a new key–value pair with the key k and the value of sum(v) / len(v), which averages the list's elements.

✓ Self Check

1 *(IPython Session)* Use a dictionary comprehension to create a dictionary of the numbers 1–5 mapped to their cubes:
Answer:
```
In [1]: {number: number ** 3 for number in range(1, 6)}
Out[1]: {1: 1, 2: 8, 3: 27, 4: 64, 5: 125}
```

6.3 Sets

A set is an unordered collection of *unique* values. Sets may contain only immutable objects, like strings, ints, floats and tuples that contain only immutable elements. Though sets are iterable, they are not sequences and do not support indexing and slicing with square brackets, []. Dictionaries also do not support slicing.

Creating a Set with Curly Braces
The following code creates a set of strings named colors:

```
In [1]: colors = {'red', 'orange', 'yellow', 'green', 'red', 'blue'}

In [2]: colors
Out[2]: {'blue', 'green', 'orange', 'red', 'yellow'}
```

Notice that the duplicate string 'red' was ignored (without causing an error). An important use of sets is **duplicate elimination**, which is automatic when creating a set. Also, the resulting set's values are *not* displayed in the same order as they were listed in snippet [1]. Though the color names are displayed in sorted order, sets are *unordered*. You should not write code that depends on the order of their elements.

Determining a Set's Length
You can determine the number of items in a set with the built-in len function:

```
In [3]: len(colors)
Out[3]: 5
```

Checking Whether a Value Is in a Set
You can check whether a set contains a particular value using the in and not in operators:

```
In [4]: 'red' in colors
Out[4]: True

In [5]: 'purple' in colors
Out[5]: False

In [6]: 'purple' not in colors
Out[6]: True
```

Iterating Through a Set
Sets are iterable, so you can process each set element with a for loop:

```
In [7]: for color in colors:
   ...:     print(color.upper(), end=' ')
   ...:
RED GREEN YELLOW BLUE ORANGE
```

Sets are *unordered*, so there's no significance to the iteration order.

Creating a Set with the Built-In set Function
You can create a set from another collection of values by using the built-in **set** function—here we create a list that contains several duplicate integer values and use that list as set's argument:

```
In [8]: numbers = list(range(10)) + list(range(5))

In [9]: numbers
Out[9]: [0, 1, 2, 3, 4, 5, 6, 7, 8, 9, 0, 1, 2, 3, 4]

In [10]: set(numbers)
Out[10]: {0, 1, 2, 3, 4, 5, 6, 7, 8, 9}
```

If you need to create an empty set, you must use the set function with empty parentheses, rather than empty braces, {}, which represent an empty dictionary:

```
In [11]: set()
Out[11]: set()
```

Python displays an empty set as set() to avoid confusion with Python's string representation of an empty dictionary ({}).

Frozenset: An Immutable Set Type

Sets are *mutable*—you can add and remove elements, but set *elements* must be *immutable*. Therefore, a set cannot have other sets as elements. A **frozenset** is an *immutable* set—it cannot be modified after you create it, so a set *can* contain frozensets as elements. The built-in function **frozenset** creates a frozenset from any iterable.

 Self Check

1 *(True/False)* Sets are collections of unique mutable and immutable objects.
Answer: False. Sets are collections of unique *immutable* objects.

2 *(Fill-In)* You can create a set from another collection of values by using the built-in _____ function.
Answer: set.

3 *(IPython Session)* Assign the following string to variable text, then split it into tokens with string method split and create a set from the results. Display the unique words in sorted order.

```
'to be or not to be that is the question'
```

Answer:

```
In [1]: text = 'to be or not to be that is the question'

In [2]: unique_words = set(text.split())

In [3]: for word in sorted(unique_words):
   ...:     print(word, end=' ')
   ...:
be is not or question that the to
```

6.3.1 Comparing Sets

Various operators and methods can be used to compare sets. The following sets contain the same values, so == returns True and != returns False.

```
In [1]: {1, 3, 5} == {3, 5, 1}
Out[1]: True

In [2]: {1, 3, 5} != {3, 5, 1}
Out[2]: False
```

The < operator tests whether the set to its left is a **proper subset** of the one to its right—that is, all the elements in the left operand are in the right operand, and the sets are not equal:

```
In [3]: {1, 3, 5} < {3, 5, 1}
Out[3]: False

In [4]: {1, 3, 5} < {7, 3, 5, 1}
Out[4]: True
```

The <= operator tests whether the set to its left is an **improper subset** of the one to its right—that is, all the elements in the left operand are in the right operand, and the sets might be equal:

```
In [5]: {1, 3, 5} <= {3, 5, 1}
Out[5]: True

In [6]: {1, 3} <= {3, 5, 1}
Out[6]: True
```

You may also check for an improper subset with the set method **issubset**:

```
In [7]: {1, 3, 5}.issubset({3, 5, 1})
Out[7]: True

In [8]: {1, 2}.issubset({3, 5, 1})
Out[8]: False
```

The > operator tests whether the set to its left is a **proper superset** of the one to its right—that is, all the elements in the right operand are in the left operand, and the left operand has more elements:

```
In [9]: {1, 3, 5} > {3, 5, 1}
Out[9]: False

In [10]: {1, 3, 5, 7} > {3, 5, 1}
Out[10]: True
```

The >= operator tests whether the set to its left is an **improper superset** of the one to its right—that is, all the elements in the right operand are in the left operand, and the sets might be equal:

```
In [11]: {1, 3, 5} >= {3, 5, 1}
Out[11]: True

In [12]: {1, 3, 5} >= {3, 1}
Out[12]: True

In [13]: {1, 3} >= {3, 1, 7}
Out[13]: False
```

You may also check for an improper superset with the set method **issuperset**:

```
In [14]: {1, 3, 5}.issuperset({3, 5, 1})
Out[14]: True

In [15]: {1, 3, 5}.issuperset({3, 2})
Out[15]: False
```

The argument to issubset or issuperset can be *any* iterable. When either of these methods receives a non-set iterable argument, it first converts the iterable to a set, then performs the operation.

✓ Self Check

1 *(True/False)* Sets may be compared with only the == and != comparison operators.
Answer: False. All the comparison operators may be used to compare sets.

2 *(Fill-In)* A subset is a(n) _____ subset of another set if all the subset's elements are in the other set and the other set has more elements.
Answer: proper.

3 *(IPython Session)* Use sets and `issuperset` to determine whether the characters of the string `'abc def ghi jkl mno'` are a superset of the characters in the string `'hi mom'`. Answer:

```
In [1]: set('abc def ghi jkl mno').issuperset('hi mom')
Out[1]: True
```

6.3.2 Mathematical Set Operations

This section presents the set type's mathematical operators |, &, - and ^ and the corresponding methods.

Union

The **union** of two sets is a set consisting of all the unique elements from both sets. You can calculate the union with the **|** **operator** or with the set type's **union** method:

```
In [1]: {1, 3, 5} | {2, 3, 4}
Out[1]: {1, 2, 3, 4, 5}

In [2]: {1, 3, 5}.union([20, 20, 3, 40, 40])
Out[2]: {1, 3, 5, 20, 40}
```

The operands of the binary set operators, like |, must both be sets. The corresponding set methods may receive any iterable object as an argument—we passed a list. When a mathematical set method receives a non-set iterable argument, it first converts the iterable to a set, then applies the mathematical operation. Again, though the new sets' string representations show the values in ascending order, you should not write code that depends on this.

Intersection

The **intersection** of two sets is a set consisting of all the unique elements that the two sets have in common. You can calculate the intersection with the **&** **operator** or with the set type's **intersection** method:

```
In [3]: {1, 3, 5} & {2, 3, 4}
Out[3]: {3}

In [4]: {1, 3, 5}.intersection([1, 2, 2, 3, 3, 4, 4])
Out[4]: {1, 3}
```

Difference

The **difference** between two sets is a set consisting of the elements in the left operand that are not in the right operand. You can calculate the difference with the **-** **operator** or with the set type's **difference** method:

```
In [5]: {1, 3, 5} - {2, 3, 4}
Out[5]: {1, 5}

In [6]: {1, 3, 5, 7}.difference([2, 2, 3, 3, 4, 4])
Out[6]: {1, 5, 7}
```

Symmetric Difference

The **symmetric difference** between two sets is a set consisting of the elements of both sets that are not in common with one another. You can calculate the symmetric difference with the **^** **operator** or with the set type's **symmetric_difference** method:

```
In [7]: {1, 3, 5} ^ {2, 3, 4}
Out[7]: {1, 2, 4, 5}

In [8]: {1, 3, 5, 7}.symmetric_difference([2, 2, 3, 3, 4, 4])
Out[8]: {1, 2, 4, 5, 7}
```

Disjoint
Two sets are **disjoint** if they do not have any common elements. You can determine this with the set type's **isdisjoint** method:

```
In [9]: {1, 3, 5}.isdisjoint({2, 4, 6})
Out[9]: True

In [10]: {1, 3, 5}.isdisjoint({4, 6, 1})
Out[10]: False
```

✓ Self Check

1 *(Fill-In)* Two sets are _____ if the sets do not have any common elements.
Answer: disjoint.

2 *(IPython Session)* Given the sets {10, 20, 30} and {5, 10, 15, 20}, use the mathematical set operators to produce the following sets:
   a) {30}
   b) {5, 15, 30}
   c) {5, 10, 15, 20, 30}
   d) {10, 20}

Answer:

```
In [1]: {10, 20, 30} - {5, 10, 15, 20}
Out[1]: {30}

In [2]: {10, 20, 30} ^ {5, 10, 15, 20}
Out[2]: {5, 15, 30}

In [3]: {10, 20, 30} | {5, 10, 15, 20}
Out[3]: {5, 10, 15, 20, 30}

In [4]: {10, 20, 30} & {5, 10, 15, 20}
Out[4]: {10, 20}
```

6.3.3 Mutable Set Operators and Methods

The operators and methods presented in the preceding section each result in a *new* set. Here we discuss operators and methods that modify an *existing* set.

Mutable Mathematical Set Operations
Like operator |, **union augmented assignment** |= performs a set union operation, but |= modifies its left operand:

```
In [1]: numbers = {1, 3, 5}

In [2]: numbers |= {2, 3, 4}

In [3]: numbers
Out[3]: {1, 2, 3, 4, 5}
```

Similarly, the set type's **update** method performs a union operation modifying the set on which it's called—the argument can be any iterable:

```
In [4]: numbers.update(range(10))

In [5]: numbers
Out[5]: {0, 1, 2, 3, 4, 5, 6, 7, 8, 9}
```

The other mutable set methods are:
- intersection augmented assignment &=
- difference augmented assignment -=
- symmetric difference augmented assignment ^=

and their corresponding methods with iterable arguments are:
- `intersection_update`
- `difference_update`
- `symmetric_difference_update`

Methods for Adding and Removing Elements

Set method **add** inserts its argument if the argument is *not* already in the set; otherwise, the set remains unchanged:

```
In [6]: numbers.add(17)

In [7]: numbers.add(3)

In [8]: numbers
Out[8]: {0, 1, 2, 3, 4, 5, 6, 7, 8, 9, 17}
```

Set method **remove** removes its argument from the set—a KeyError occurs if the value is not in the set:

```
In [9]: numbers.remove(3)

In [10]: numbers
Out[10]: {0, 1, 2, 4, 5, 6, 7, 8, 9, 17}
```

Method **discard** also removes its argument from the set but does not cause an exception if the value is not in the set.

You also can remove an *arbitrary* set element and return it with **pop**, but sets are unordered, so you do not know which element will be returned:

```
In [11]: numbers.pop()
Out[11]: 0

In [12]: numbers
Out[12]: {1, 2, 4, 5, 6, 7, 8, 9, 17}
```

A KeyError occurs if the set is empty when you call pop.

Finally, method **clear** empties the set on which it's called:

```
In [13]: numbers.clear()

In [14]: numbers
Out[14]: set()
```

Self Check

1 *(True/False)* Set method pop returns the first element added to the set.
Answer: False. Set method pop returns an *arbitrary* set element.

2 *(Fill-In)* Set method _____ performs a union operation, modifying the set on which it's called.
Answer: update.

6.3.4 Set Comprehensions

Like dictionary comprehensions, you define set comprehensions in curly braces. Let's create a new set containing only the unique even values in the list numbers:

```
In [1]: numbers = [1, 2, 2, 3, 4, 5, 6, 6, 7, 8, 9, 10, 10]

In [2]: evens = {item for item in numbers if item % 2 == 0}

In [3]: evens
Out[3]: {2, 4, 6, 8, 10}
```

6.4 Intro to Data Science: Dynamic Visualizations

The preceding chapter's Intro to Data Science section introduced visualization. We simulated rolling a six-sided die and used the Seaborn and Matplotlib visualization libraries to create a publication-quality *static* bar plot showing the frequencies and percentages of each roll value. In this section, we make things "come alive" with *dynamic visualizations*.

The Law of Large Numbers

When we introduced random-number generation, we mentioned that if the random module's randrange function indeed produces integers at random, then every number in the specified range has an equal probability (or likelihood) of being chosen each time the function is called. For a six-sided die, each value 1 through 6 should occur one-sixth of the time, so the probability of any one of these values occurring is $1/6^{th}$ or about 16.667%.

In the next section, we create and execute a *dynamic* (that is, *animated*) die-rolling simulation script. In general, you'll see that the more rolls we attempt, the closer each die value's percentage of the total rolls gets to 16.667% and the heights of the bars gradually become about the same. This is a manifestation of the *law of large numbers*.

Self Check

1 *(Fill-In)* As we toss a coin an increasing number of times, we expect the percentages of heads and tails to become closer to 50% each. This is a manifestation of _____.
Answer: the law of large numbers.

6.4.1 How Dynamic Visualization Works

The plots produced with Seaborn and Matplotlib in the previous chapter's Intro to Data Science section help you analyze the results for a fixed number of die rolls *after* the simulation completes. This section's enhances that code with the Matplotlib **animation** module's **FuncAnimation** function, which updates the bar plot *dynamically*. You'll see the bars, die frequencies and percentages "come alive," updating *continuously* as the rolls occur.

Animation Frames

`FuncAnimation` drives a **frame-by-frame animation**. Each **animation frame** specifies everything that should change during one plot update. Stringing together many of these updates over time creates the animation effect. You decide what each frame displays with a function you define and pass to `FuncAnimation`.

Each animation frame will:

- roll the dice a specified number of times (from 1 to as many as you'd like), updating die frequencies with each roll,
- clear the current plot,
- create a new set of bars representing the updated frequencies, and
- create new frequency and percentage text for each bar.

Generally, displaying more frames-per-second yields smoother animation. For example, video games with fast-moving elements try to display *at least* 30 frames-per-second and often more. Though you'll specify the number of milliseconds between animation frames, the actual number of frames-per-second can be affected by the amount of work you perform in each frame and the speed of your computer's processor. This example displays an animation frame every 33 milliseconds—yielding approximately 30 (1000 / 33) frames-per-second. Try larger and smaller values to see how they affect the animation. Experimentation is important in developing the best visualizations.

Running `RollDieDynamic.py`

In the previous chapter's Intro to Data Science section, we developed the static visualization *interactively* so you could see how the code updates the bar plot as you execute each statement. The actual bar plot with the final frequencies and percentages was drawn only once.

For this dynamic visualization, the screen results update frequently so that you can see the animation. Many things change continuously—the lengths of the bars, the frequencies and percentages above the bars, the spacing and labels on the axes and the total number of die rolls shown in the plot's title. For this reason, we present this visualization as a script, rather than interactively developing it.

The script takes two command-line arguments:

- `number_of_frames`—The number of animation frames to display. This value determines the total number of times that `FuncAnimation` updates the graph. For each animation frame, `FuncAnimation` calls a function that you define (in this example, `update`) to specify how to change the plot.
- `rolls_per_frame`—The number of times to roll the die in each animation frame. We'll use a loop to roll the die this number of times, summarize the results, then update the graph with bars and text representing the new frequencies.

To understand how we use these two values, consider the following command:

```
ipython RollDieDynamic.py 6000 1
```

In this case, `FuncAnimation` calls our `update` function 6000 times, rolling one die per frame for a total of 6000 rolls. This enables you to see the bars, frequencies and percentages update one roll at a time. On our system, this animation took about 3.33 minutes (6000 frames / 30 frames-per-second / 60 seconds-per-minute) to show you only 6000 die rolls.

Displaying animation frames to the screen is a relatively slow *input–output-bound* operation compared to the die rolls, which occur at the computer's super fast CPU speeds. If we roll only one die per animation frame, we won't be able to run a large number of rolls in a reasonable amount of time. Also, for small numbers of rolls, you're unlikely to see the die percentages converge on their expected 16.667% of the total rolls.

To see the law of large numbers in action, you can increase the execution speed by rolling the die more times per animation frame. Consider the following command:

```
ipython RollDieDynamic.py 10000 600
```

In this case, `FuncAnimation` will call our `update` function 10,000 times, performing 600 rolls-per-frame for a total of 6,000,000 rolls. On our system, this took about 5.55 minutes (10,000 frames / 30 frames-per-second / 60 seconds-per-minute), but displayed approximately 18,000 rolls-per-second (30 frames-per-second * 600 rolls-per-frame), so we could quickly see the frequencies and percentages converge on their expected values of about 1,000,000 rolls per face and 16.667% per face.

Experiment with the numbers of rolls and frames until you feel that the program is helping you visualize the results most effectively. It's fun and informative to watch it run and to tweak it until you're satisfied with the animation quality.

Sample Executions

We took the following four screen captures during each of two sample executions. In the first, the screens show the graph after just 64 die rolls, then again after 604 of the 6000 total die rolls. Run this script live to see over time how the bars update dynamically. In the second execution, the screen captures show the graph after 7200 die rolls and again after 166,200 out of the 6,000,000 rolls. With more rolls, you can see the percentages closing in on their expected values of 16.667% as predicted by the law of large numbers.

Execute 6000 animation frames rolling the die once per frame:
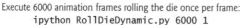

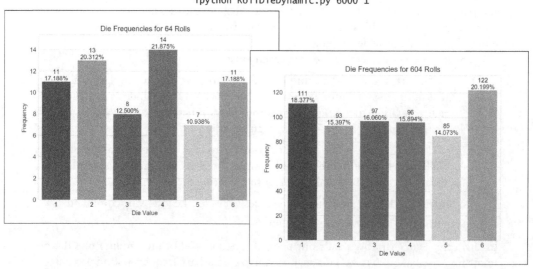

6.4 Intro to Data Science: Dynamic Visualizations

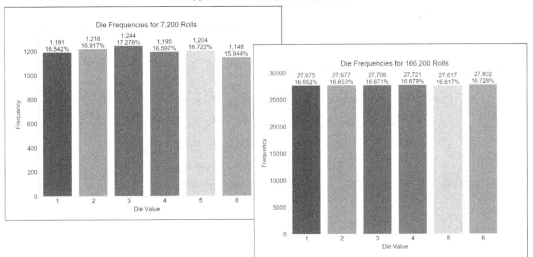

Execute 10,000 animation frames rolling the die 600 times per frame:
`ipython RollDieDynamic.py 10000 600`

✓ Self Check

1 *(Fill-In)* A(n) _____ specifies everything that should change during one plot update. Stringing together many of these over time creates the animation effect.
Answer: animation frame.

2 *(True/False)* Generally, displaying fewer frames-per-second yields smoother animation.
Answer: False. Generally, displaying *more* frames-per-second yields smoother animation.

3 *(True/False)* The actual number of frames-per-second is affected only by the millisecond interval between animation frames.
Answer: False. The actual number of frames-per-second also can be affected by the amount of work performed in each frame and the speed of your computer's processor.

6.4.2 Implementing a Dynamic Visualization

The script we present in this section uses the same Seaborn and Matplotlib features shown in the previous chapter's Intro to Data Science section. We reorganized the code for use with Matplotlib's *animation* capabilities.

Importing the Matplotlib animation Module
We focus primarily on the new features used in this example. Line 3 imports the Matplotlib animation module.

```
1   # RollDieDynamic.py
2   """Dynamically graphing frequencies of die rolls."""
3   from matplotlib import animation
4   import matplotlib.pyplot as plt
5   import random
6   import seaborn as sns
7   import sys
8
```

Dictionaries and Sets

Function update

Lines 9–27 define the update function that FuncAnimation calls once per animation frame. This function must provide at least one argument. Lines 9–10 show the beginning of the function definition. The parameters are:

- frame_number—The next value from FuncAnimation's frames argument, which we'll discuss momentarily. Though FuncAnimation requires the update function to have this parameter, we do not use it in this update function.
- rolls—The number of die rolls per animation frame.
- faces—The die face values used as labels along the graph's *x*-axis.
- frequencies—The list in which we summarize the die frequencies.

We discuss the rest of the function's body in the next several subsections.

```
 9  def update(frame_number, rolls, faces, frequencies):
10      """Configures bar plot contents for each animation frame."""
```

Function update: Rolling the Die and Updating the frequencies List

Lines 12–13 roll the die rolls times and increment the appropriate frequencies element for each roll. Note that we subtract 1 from the die value (1 through 6) before incrementing the corresponding frequencies element—as you'll see, frequencies is a six-element list (defined in line 36), so its indices are 0 through 5.

```
11      # roll die and update frequencies
12      for i in range(rolls):
13          frequencies[random.randrange(1, 7) - 1] += 1
14
```

Function update: Configuring the Bar Plot and Text

Line 16 in function update calls the matplotlib.pyplot module's **cla** (clear axes) function to remove the existing bar plot elements before drawing new ones for the current animation frame. We discussed the code in lines 17–27 in the previous chapter's Intro to Data Science section. Lines 17–20 create the bars, set the bar plot's title, set the *x*- and *y*-axis labels and scale the plot to make room for the frequency and percentage text above each bar. Lines 23–27 display the frequency and percentage text.

```
15      # reconfigure plot for updated die frequencies
16      plt.cla()  # clear old contents contents of current Figure
17      axes = sns.barplot(faces, frequencies, palette='bright')   # new bars
18      axes.set_title(f'Die Frequencies for {sum(frequencies):,} Rolls')
19      axes.set(xlabel='Die Value', ylabel='Frequency')
20      axes.set_ylim(top=max(frequencies) * 1.10)   # scale y-axis by 10%
21
22      # display frequency & percentage above each patch (bar)
23      for bar, frequency in zip(axes.patches, frequencies):
24          text_x = bar.get_x() + bar.get_width() / 2.0
25          text_y = bar.get_height()
26          text = f'{frequency:,}\n{frequency / sum(frequencies):.3%}'
27          axes.text(text_x, text_y, text, ha='center', va='bottom')
28
```

6.4 Intro to Data Science: Dynamic Visualizations

Variables Used to Configure the Graph and Maintain State

Lines 30 and 31 use the sys module's argv list to get the script's command-line arguments. Line 33 specifies the Seaborn 'whitegrid' style. Line 34 calls the matplotlib.pyplot module's **figure** function to get the Figure object in which FuncAnimation displays the animation. The function's argument is the window's title. As you'll soon see, this is one of FuncAnimation's required arguments. Line 35 creates a list containing the die face values 1–6 to display on the plot's *x*-axis. Line 36 creates the six-element frequencies list with each element initialized to 0—we update this list's counts with each die roll.

```
29  # read command-line arguments for number of frames and rolls per frame
30  number_of_frames = int(sys.argv[1])
31  rolls_per_frame = int(sys.argv[2])
32
33  sns.set_style('whitegrid')  # white background with gray grid lines
34  figure = plt.figure('Rolling a Six-Sided Die')  # Figure for animation
35  values = list(range(1, 7))  # die faces for display on x-axis
36  frequencies = [0] * 6  # six-element list of die frequencies
37
```

Calling the animation Module's FuncAnimation Function

Lines 39–41 call the Matplotlib animation module's FuncAnimation function to update the bar chart dynamically. The function returns an object representing the animation. Though this is not used explicitly, you *must* store the reference to the animation; otherwise, Python immediately terminates the animation and returns its memory to the system.

```
38  # configure and start animation that calls function update
39  die_animation = animation.FuncAnimation(
40      figure, update, repeat=False, frames=number_of_frames, interval=33,
41      fargs=(rolls_per_frame, values, frequencies))
42
43  plt.show()  # display window
```

FuncAnimation has two required arguments:

- **figure**—the Figure object in which to display the animation, and
- **update**—the function to call once per animation frame.

In this case, we also pass the following optional keyword arguments:

- **repeat**—False terminates the animation after the specified number of frames. If True (the default), when the animation completes it restarts from the beginning.
- **frames**—The total number of animation frames, which controls how many times FunctAnimation calls update. Passing an integer is equivalent to passing a range—for example, 600 means range(600). FuncAnimation passes one value from this range as the first argument in each call to update.
- **interval**—The number of milliseconds (33, in this case) between animation frames (the default is 200). After each call to update, FuncAnimation waits 33 milliseconds before making the next call.
- **fargs** (short for "function arguments")—A tuple of other arguments to pass to the function you specified in FuncAnimation's second argument. The arguments you specify in the fargs tuple correspond to update's parameters rolls, faces and frequencies (line 9).

234 Dictionaries and Sets

For a list of FuncAnimation's other optional arguments, see

> https://matplotlib.org/api/_as_gen/
> matplotlib.animation.FuncAnimation.html

Finally, line 43 displays the window.

Self Check

1 *(Fill-In)* The Matplotlib _____ module's _____ function dynamically updates a visualization.
Answer: animation, FuncAnimation.

2 *(Fill-In)* FuncAnimation's _____ keyword argument enables you to pass custom arguments to the function that's called once per animation frame.
Answer: fargs.

6.5 Wrap-Up

In this chapter, we discussed Python's dictionary and set collections. We said what a dictionary is and presented several examples. We showed the syntax of key–value pairs and showed how to use them to create dictionaries with comma-separated lists of key–value pairs in curly braces, {}. You also created dictionaries with dictionary comprehensions.

You used square brackets, [], to retrieve the value corresponding to a key, and to insert and update key–value pairs. You also used the dictionary method update to change a key's associated value. You iterated through a dictionary's keys, values and items.

You created sets of unique immutable values. You compared sets with the comparison operators, combined sets with set operators and methods, changed sets' values with the mutable set operations and created sets with set comprehensions. You saw that sets are mutable. Frozensets are immutable, so they can be used as set and frozenset elements.

In the Intro to Data Science section, we continued our visualization introduction by presenting the die-rolling simulation with a *dynamic* bar plot to make the law of large numbers "come alive." In addition, to the Seaborn and Matplotlib features shown in the previous chapter's Intro to Data Science section, we used Matplotlib's FuncAnimation function to control a frame-by-frame animation. FuncAnimation called a function we defined that specified what to display in each animation frame.

In the next chapter, we discuss array-oriented programming with the popular NumPy library. As you'll see, NumPy's ndarray collection can be up to two orders of magnitude faster than performing many of the same operations with Python's built-in lists. This power will come in handy for today's big data applications.

Exercises

Unless specified otherwise, use IPython sessions for each exercise.

6.1 *(Discussion: Dictionary Methods)* Briefly explain the operation of each of the following dictionary methods:
   a) add
   b) keys
   c) values
   d) items

6.2 *(What's Wrong with This Code?)* The following code should display the unique words in the string text and the number of occurrences of each word.

```
from collections import Counter
text = ('to be or not to be that is the question')
counter = Counter(text.split())
for word, count in sorted(counter):
    print(f'{word:<12}{count}')
```

6.3 *(What Does This Code Do?)* The dictionary temperatures contains three Fahrenheit temperature samples for each of four days. What does the for statement do?

```
temperatures = {
    'Monday': [66, 70, 74],
    'Tuesday': [50, 56, 64],
    'Wednesday': [75, 80, 83],
    'Thursday': [67, 74, 81]
}

for k, v in temperatures.items():
    print(f'{k}: {sum(v)/len(v):.2f}')
```

6.4 *(Fill in the Missing Code)* In each of the following expressions, replace the ***s with a set operator that produces the result shown in the comment. The last operation should check whether the left operand is an improper subset of the right operand. For each of the first four expressions, specify the name of the set operation that produces the specified result.

a) {1, 2, 4, 8, 16} *** {1, 4, 16, 64, 256}   # {1,2,4,8,16,64,256}
b) {1, 2, 4, 8, 16} *** {1, 4, 16, 64, 256}   # {1,4,16}
c) {1, 2, 4, 8, 16} *** {1, 4, 16, 64, 256}   # {2,8}
d) {1, 2, 4, 8, 16} *** {1, 4, 16, 64, 256}   # {2,8,64,256}
e) {1, 2, 4, 8, 16} *** {1, 4, 16, 64, 256}   # False

6.5 *(Counting Duplicate Words)* Write a script that uses a dictionary to determine and print the number of *duplicate* words in a sentence. Treat uppercase and lowercase letters the same and assume there is no punctuation in the sentence. Use the techniques you learned in Section 6.2.7. Words with counts larger than 1 have duplicates.

6.6 *(Duplicate Word Removal)* Write a function that receives a list of words, then determines and displays in alphabetical order only the unique words. Treat uppercase and lowercase letters the same. The function should use a set to get the unique words in the list. Test your function with several sentences.

6.7 *(Character Counts)* Recall that strings are sequences of characters. Use techniques similar to Fig. 6.2 to write a script that inputs a sentence from the user, then uses a dictionary to summarize the number of occurrences of each letter. Ignore case, ignore blanks and assume the user does not enter any punctuation. Display a two-column table of the letters and their counts with the letters in sorted order. Challenge: Use a set operation to determine which letters of the alphabet were not in the original string.

6.8 *(Challenge: Writing the Word Equivalent of a Check Amount)* In check-writing systems, it's crucial to prevent alteration of check amounts. One common security method requires that the amount be written in numbers and spelled out in words as well. Even if someone can alter the numerical amount of the check, it's tough to change the amount in words. Create a dictionary that maps numbers to their corresponding word equivalents. Write a script that inputs a numeric check amount that's less than 1000 and uses the dictionary to write the word equivalent of the amount. For example, the amount 112.43 should be written as

```
ONE HUNDRED TWELVE AND 43/100
```

6.9 *(Dictionary Manipulations)* Using the following dictionary, which maps country names to Internet top-level domains (TLDs):

```
tlds = {'Canada': 'ca', 'United States': 'us', 'Mexico': 'mx'}
```

perform the following tasks and display the results:
    a) Check whether the dictionary contains the key `'Canada'`.
    b) Check whether the dictionary contains the key `'France'`.
    c) Iterate through the key–value pairs and display them in two-column format.
    d) Add the key–value pair `'Sweden'` and `'sw'` (which is incorrect).
    e) Update the value for the key `'Sweden'` to `'se'`.
    f) Use a dictionary comprehension to reverse the keys and values.
    g) With the result of part (f), use a dictionary comprehension to convert the country names to all uppercase letters.

6.10 *(Set Manipulations)* Using the following sets:

```
{'red', 'green', 'blue'}
{'cyan', 'green', 'blue', 'magenta', 'red'}
```

display the results of:
    a) comparing the sets using the each of the comparison operators.
    b) combining the sets using each of the mathematical set operators.

6.11 *(Analyzing the Game of Craps)* Modify the script of Fig. 4.2 to play 1,000,000 games of craps. Use a `wins` dictionary to keep track of the number of games won for a particular number of rolls. Similarly, use a `losses` dictionary to keep track of the number of games lost for a particular number of rolls. As the simulation proceeds, keep updating the dictionaries.

A typical key–value pair in the `wins` dictionary might be

```
4: 50217
```

indicating that 50217 games were won on the 4th roll. Display a summary of the results including:
    a) the percentage of the total games played that were won.
    b) the percentage of the total games played that were lost.
    c) the percentages of the total games played that were won or lost on a given roll (column 2 of the sample output).
    d) the *cumulative* percentage of the total games played that were won or lost up to and including a given number of rolls (column 3 of the sample output).

Your output should be similar to the one below.

```
Percentage of wins: 50.2%
Percentage of losses: 49.8%
Percentage of wins/losses based on total number of rolls

         % Resolved      Cumulative %
Rolls    on this roll    of games resolved
  1         30.10%           30.10%
  2         20.80%           50.90%
  3         14.10%           65.00%
  4          9.90%           74.90%
  5          7.40%           82.30%
  6          4.60%           86.90%
  7          3.70%           90.60%
  8          2.40%           93.00%
  9          1.90%           94.90%
 10          1.10%           96.00%
 11          0.90%           96.90%
 12          0.80%           97.70%
 13          0.80%           98.50%
 14          0.30%           98.80%
 15          0.30%           99.10%
 16          0.30%           99.40%
 17          0.50%           99.90%
 25          0.10%          100.00%
```

6.12 *(Translation Dictionary)* Use an online translation tool such as Bing Microsoft Translator or Google Translate to translate English words to another language. Create a `translations` dictionary that maps the English words to their translations. Display a two-column table of translations.

6.13 *(Synonyms Dictionary)* Use an online thesaurus to look up synonyms for five words, then create a `synonyms` dictionary that maps those words to lists containing three synonyms for each word. Display the dictionary's contents as a key with an indented list of synonyms below it.

6.14 *(Intro to Data Science: Dynamic Visualization of Coin Tossing)* Modify your coin-tossing simulation from Exercise 5.31 to update the bar plot dynamically as you flip the coin. Use the techniques you learned in Section 6.4.2.

6.15 *(Intro to Data Science: Dynamic Visualization of Rolling Two Dice)* Modify your simulation of rolling two dice from Exercise 5.32 to update the bar plot dynamically as you roll the dice. Use the techniques you learned in Section 6.4.2.

6.16 *(Intro to Data Science: Dynamic Visualization of the Dice Game of Craps)* Reimplement your solution to Exercise 5.33, using the techniques you learned in Section 6.4.2 to create a dynamic bar chart showing the wins and losses on the first roll, second roll, third roll, etc.

6.17 *(Project: Cooking with Healthier Ingredients)* In the "Strings: A Deeper Look" chapter's exercises, you'll write a script that enables its user to enter ingredients from a cooking recipe, then recommends healthier replacements.[3] In preparation for that exercise, create a dictionary that maps ingredients to lists of potential replacements. Some ingredient replacements are shown below:

---

3. Always consult a healthcare professional before making significant changes to your diet.

Ingredient	Substitution
1 cup sour cream	1 cup yogurt
1 cup milk	1/2 cup evaporated milk and 1/2 cup water
1 teaspoon lemon juice	1/2 teaspoon vinegar
1 cup sugar	1/2 cup honey, 1 cup molasses or 1/4 cup agave nectar
1 cup butter	1 cup margarine or yogurt
1 cup flour	1 cup rye or rice flour
1 cup mayonnaise	1 cup cottage cheese or 1/8 cup mayonnaise and 7/8 cup yogurt
1 egg	2 tablespoons cornstarch, arrowroot flour or potato starch or 2 egg whites or 1/2 of a large banana (mashed)
1 cup milk	1 cup soy milk
1 cup oil	1 cup applesauce

Your dictionary should take into consideration that replacements are not always one-for-one. For example, if a cake recipe calls for three eggs, it might reasonably use six egg whites instead. Research conversion data for measurements and ingredient substitutes online. Your dictionary should map the ingredients to lists of potential substitutes.

Array-Oriented Programming with NumPy

7

Objectives

In this chapter you'll:

- Learn what arrays are and how they differ from lists.
- Use the numpy module's high-performance ndarrays.
- Compare list and ndarray performance with the IPython %timeit magic.
- Use ndarrays to store and retrieve data efficiently.
- Create and initialize ndarrays.
- Refer to individual ndarray elements.
- Iterate through ndarrays.
- Create and manipulate multidimensional ndarrays.
- Perform common ndarray manipulations.
- Create and manipulate pandas one-dimensional Series and two-dimensional DataFrames.
- Customize Series and DataFrame indices.
- Calculate basic descriptive statistics for data in a Series and a DataFrame.
- Customize floating-point number precision in pandas output formatting.

Outline

7.1 Introduction	**7.10** Indexing and Slicing
7.2 Creating arrays from Existing Data	**7.11** Views: Shallow Copies
7.3 array Attributes	**7.12** Deep Copies
7.4 Filling arrays with Specific Values	**7.13** Reshaping and Transposing
7.5 Creating arrays from Ranges	**7.14** Intro to Data Science: pandas Series and DataFrames
7.6 List vs. array Performance: Introducing %timeit	7.14.1 pandas Series
7.7 array Operators	7.14.2 DataFrames
7.8 NumPy Calculation Methods	**7.15** Wrap-Up
7.9 Universal Functions	Exercises

7.1 Introduction

The **NumPy (Numerical Python)** library first appeared in 2006 and is the preferred Python array implementation. It offers a high-performance, richly functional *n*-dimensional array type called **ndarray**, which from this point forward we'll refer to by its synonym, array. NumPy is one of the many open-source libraries that the Anaconda Python distribution installs. Operations on arrays are up to two orders of magnitude faster than those on lists. In a big-data world in which applications may do massive amounts of processing on vast amounts of array-based data, this performance advantage can be critical. According to libraries.io, over 450 Python libraries depend on NumPy. Many popular data science libraries such as Pandas, SciPy (Scientific Python) and Keras (for deep learning) are built on or depend on NumPy.

In this chapter, we explore array's basic capabilities. Lists can have multiple dimensions. You generally process multi-dimensional lists with nested loops or list comprehensions with multiple for clauses. A strength of NumPy is "array-oriented programming," which uses functional-style programming with *internal* iteration to make array manipulations concise and straightforward, eliminating the kinds of bugs that can occur with the *external* iteration of explicitly programmed loops.

In this chapter's Intro to Data Science section, we begin our multi-section introduction to the *pandas* library that you'll use in many of the data science case study chapters. Big data applications often need more flexible collections than NumPy's arrays—collections that support mixed data types, custom indexing, missing data, data that's not structured consistently and data that needs to be manipulated into forms appropriate for the databases and data analysis packages you use. We'll introduce pandas array-like one-dimensional Series and two-dimensional DataFrames and begin demonstrating their powerful capabilities. After reading this chapter, you'll be familiar with four array-like collections—lists, arrays, Series and DataFrames. We'll add a fifth—tensors—in the "Deep Learning" chapter.

Self Check

1 *(Fill-In)* The NumPy library provides the _____ data structure, which is typically much faster than lists.
Answer: ndarray.

7.2 Creating arrays from Existing Data

The NumPy documentation recommends importing the **numpy module** as np so that you can access its members with "np.":

```
In [1]: import numpy as np
```

The numpy module provides various functions for creating arrays. Here we use the **array** function, which receives as an argument an array or other collection of elements and returns a new array containing the argument's elements. Let's pass a list:

```
In [2]: numbers = np.array([2, 3, 5, 7, 11])
```

The array function copies its argument's contents into the array. Let's look at the type of object that function array returns and display its contents:

```
In [3]: type(numbers)
Out[3]: numpy.ndarray

In [4]: numbers
Out[4]: array([ 2,  3,  5,  7, 11])
```

Note that the *type* is numpy.ndarray, but all arrays are output as "array." When outputting an array, NumPy separates each value from the next with a comma and a space and *right-aligns* all the values using the same field width. It determines the field width based on the value that occupies the *largest* number of character positions. In this case, the value 11 occupies the two character positions, so all the values are formatted in two-character fields. That's why there's a leading space between the [and 2.

Multidimensional Arguments

The array function copies its argument's dimensions. Let's create an array from a two-row-by-three-column list:

```
In [5]: np.array([[1, 2, 3], [4, 5, 6]])
Out[5]:
array([[1, 2, 3],
       [4, 5, 6]])
```

NumPy auto-formats arrays, based on their number of dimensions, aligning the columns within each row.

✓ Self Check

1 *(Fill-In)* Function array creates an array from _____.
Answer: an array or other collection of elements.

2 *(IPython Session)* Create a one-dimensional array from a list comprehension that produces the even integers from 2 through 20.
Answer:

```
In [1]: import numpy as np

In [2]: np.array([x for x in range(2, 21, 2)])
Out[2]: array([ 2,  4,  6,  8, 10, 12, 14, 16, 18, 20])
```

3 *(IPython Session)* Create a 2-by-5 array containing the even integers from 2 through 10 in the first row and the odd integers from 1 through 9 in the second row.

Answer:

```
In [3]: np.array([[2, 4, 6, 8, 10], [1, 3, 5, 7, 9]])
Out[3]:
array([[ 2, 4, 6, 8, 10],
       [ 1, 3, 5, 7, 9]])
```

7.3 array Attributes

An array object provides **attributes** that enable you to discover information about its structure and contents. In this section we'll use the following arrays:

```
In [1]: import numpy as np

In [2]: integers = np.array([[1, 2, 3], [4, 5, 6]])

In [3]: integers
Out[3]:
array([[1, 2, 3],
       [4, 5, 6]])

In [4]: floats = np.array([0.0, 0.1, 0.2, 0.3, 0.4])

In [5]: floats
Out[5]: array([ 0. , 0.1, 0.2, 0.3, 0.4])
```

NumPy does not display trailing 0s to the right of the decimal point in floating-point values.

Determining an array's Element Type

The array function determines an array's element type from its argument's elements. You can check the element type with an array's **dtype** attribute:

```
In [6]: integers.dtype
Out[6]: dtype('int64')   # int32 on some platforms

In [7]: floats.dtype
Out[7]: dtype('float64')
```

As you'll see in the next section, various array-creation functions receive a dtype keyword argument so you can specify an array's element type.

For performance reasons, NumPy is written in the C programming language and uses C's data types. By default, NumPy stores integers as the NumPy type int64 values—which correspond to 64-bit (8-byte) integers in C—and stores floating-point numbers as the NumPy type float64 values—which correspond to 64-bit (8-byte) floating-point values in C. In our examples, most commonly you'll see the types int64, float64, bool (for Boolean) and object for non-numeric data (such as strings). The complete list of supported types is at https://docs.scipy.org/doc/numpy/user/basics.types.html.

Determining an array's Dimensions

The attribute **ndim** contains an array's number of dimensions and the attribute **shape** contains a *tuple* specifying an array's dimensions:

```
In [8]: integers.ndim
Out[8]: 2

In [9]: floats.ndim
Out[9]: 1
```

```
In [10]: integers.shape
Out[10]: (2, 3)

In [11]: floats.shape
Out[11]: (5,)
```

Here, `integers` has 2 rows and 3 columns (6 elements) and `floats` is one-dimensional, so snippet `[11]` shows a one-element tuple (indicated by the comma) containing `floats`' number of elements (5).

Determining an array's Number of Elements and Element Size
You can view an array's total number of elements with the attribute **size** and the number of bytes required to store each element with **itemsize**:

```
In [12]: integers.size
Out[12]: 6

In [13]: integers.itemsize   # 4 if C compiler uses 32-bit ints
Out[13]: 8

In [14]: floats.size
Out[14]: 5

In [15]: floats.itemsize
Out[15]: 8
```

Note that `integers`' `size` is the product of the `shape` tuple's values—two rows of three elements each for a total of six elements. In each case, `itemsize` is 8 because `integers` contains `int64` values and `floats` contains `float64` values, which each occupy 8 bytes.

Iterating Through a Multidimensional array's Elements
You'll generally manipulate `array`s using concise functional-style programming techniques. However, because `array`s are *iterable*, you can use external iteration if you'd like:

```
In [16]: for row in integers:
    ...:     for column in row:
    ...:         print(column, end=' ')
    ...:     print()
    ...:
1 2 3
4 5 6
```

You can iterate through a multidimensional `array` as if it were one-dimensional by using its **flat** attribute:

```
In [17]: for i in integers.flat:
    ...:     print(i, end=' ')
    ...:
1 2 3 4 5 6
```

✓ Self Check

1. *(True/False)* By default, NumPy displays trailing 0s to the right of the decimal point in a floating-point value.
Answer: False. By default, NumPy does *not* display trailing 0s in the fractional part of a floating-point value

2 *(IPython Session)* For the two-dimensional array in the previous section's Self Check, display the number of dimensions and shape of the array.
Answer:

```
In [1]: import numpy as np

In [2]: a = np.array([[2, 4, 6, 8, 10], [1, 3, 5, 7, 9]])

In [3]: a.ndim
Out[3]: 2

In [4]: a.shape
Out[4]: (2, 5)
```

7.4 Filling arrays with Specific Values

NumPy provides functions **zeros**, **ones** and **full** for creating arrays containing 0s, 1s or a specified value, respectively. By default, zeros and ones create arrays containing float64 values. We'll show how to customize the element type momentarily. The first argument to these functions must be an integer or a tuple of integers specifying the desired dimensions. For an integer, each function returns a one-dimensional array with the specified number of elements:

```
In [1]: import numpy as np

In [2]: np.zeros(5)
Out[2]: array([ 0.,  0.,  0.,  0.,  0.])
```

For a tuple of integers, these functions return a multidimensional array with the specified dimensions. You can specify the array's element type with the zeros and ones function's dtype keyword argument:

```
In [3]: np.ones((2, 4), dtype=int)
Out[3]:
array([[1, 1, 1, 1],
       [1, 1, 1, 1]])
```

The array returned by full contains elements with the second argument's value and type:

```
In [4]: np.full((3, 5), 13)
Out[4]:
array([[13, 13, 13, 13, 13],
       [13, 13, 13, 13, 13],
       [13, 13, 13, 13, 13]])
```

7.5 Creating arrays from Ranges

NumPy provides optimized functions for creating arrays from ranges. We focus on simple evenly spaced integer and floating-point ranges, but NumPy also supports nonlinear ranges.[1]

Creating Integer Ranges with arange

Let's use NumPy's **arange** function to create integer ranges—similar to using built-in function range. In each case, arange first determines the resulting array's number of elements, allocates the memory, then stores the specified range of values in the array:

---

1. https://docs.scipy.org/doc/numpy/reference/routines.array-creation.html.

7.5 Creating arrays from Ranges

```
In [1]: import numpy as np

In [2]: np.arange(5)
Out[2]: array([0, 1, 2, 3, 4])

In [3]: np.arange(5, 10)
Out[3]: array([5, 6, 7, 8, 9])

In [4]: np.arange(10, 1, -2)
Out[4]: array([10,  8,  6,  4,  2])
```

Though you can create arrays by passing ranges as arguments, always use arange as it's optimized for arrays. Soon we'll show how to determine the execution time of various operations so you can compare their performance.

Creating Floating-Point Ranges with linspace

You can produce evenly spaced floating-point ranges with NumPy's **linspace** function. The function's first two arguments specify the starting and ending values in the range, and the ending value *is included* in the array. The optional keyword argument num specifies the number of evenly spaced values to produce—this argument's default value is 50:

```
In [5]: np.linspace(0.0, 1.0, num=5)
Out[5]: array([ 0.  ,  0.25,  0.5 ,  0.75,  1.  ])
```

Reshaping an array

You also can create an array from a range of elements, then use array method **reshape** to transform the one-dimensional array into a multidimensional array. Let's create an array containing the values from 1 through 20, then reshape it into four rows by five columns:

```
In [6]: np.arange(1, 21).reshape(4, 5)
Out[6]:
array([[ 1,  2,  3,  4,  5],
       [ 6,  7,  8,  9, 10],
       [11, 12, 13, 14, 15],
       [16, 17, 18, 19, 20]])
```

Note the *chained method calls* in the preceding snippet. First, arange produces an array containing the values 1–20. Then we call reshape on that array to get the 4-by-5 array that was displayed.

You can reshape any array, provided that the new shape has the *same* number of elements as the original. So a six-element one-dimensional array can become a 3-by-2 or 2-by-3 array, and vice versa, but attempting to reshape a 15-element array into a 4-by-4 array (16 elements) causes a ValueError.

Displaying Large arrays

When displaying an array, if there are 1000 items or more, NumPy drops the middle rows, columns or both from the output. The following snippets generate 100,000 elements. The first case shows all four rows but only the first and last three of the 25,000 columns. The notation ... represents the missing data. The second case shows the first and last three of the 100 rows, and the first and last three of the 1000 columns:

```
In [7]: np.arange(1, 100001).reshape(4, 25000)
Out[7]:
array([[    1,     2,     3, ..., 24998, 24999,  25000],
       [25001, 25002, 25003, ..., 49998, 49999,  50000],
       [50001, 50002, 50003, ..., 74998, 74999,  75000],
       [75001, 75002, 75003, ..., 99998, 99999, 100000]])

In [8]: np.arange(1, 100001).reshape(100, 1000)
Out[8]:
array([[    1,     2,     3, ...,   998,   999,   1000],
       [ 1001,  1002,  1003, ...,  1998,  1999,   2000],
       [ 2001,  2002,  2003, ...,  2998,  2999,   3000],
       ...,
       [97001, 97002, 97003, ..., 97998, 97999,  98000],
       [98001, 98002, 98003, ..., 98998, 98999,  99000],
       [99001, 99002, 99003, ..., 99998, 99999, 100000]])
```

✓ Self Check

1 *(Fill-In)* NumPy function _____ returns an ndarray containing evenly spaced floating-point values.
Answer: linspace.

2 *(IPython Session)* Use NumPy function arange to create an array of 20 even integers from 2 through 40, then reshape the result into a 4-by-5 array.
Answer:

```
In [1]: import numpy as np

In [2]: np.arange(2, 41, 2).reshape(4, 5)
Out[2]:
array([[ 2,  4,  6,  8, 10],
       [12, 14, 16, 18, 20],
       [22, 24, 26, 28, 30],
       [32, 34, 36, 38, 40]])
```

7.6 List vs. array Performance: Introducing %timeit

Most array operations execute *significantly* faster than corresponding list operations. To demonstrate, we'll use the IPython **%timeit** magic command, which times the *average* duration of operations. Note that the times displayed on your system may vary from what we show here.

Timing the Creation of a List Containing Results of 6,000,000 Die Rolls

We've demonstrated rolling a six-sided die 6,000,000 times. Here, let's use the random module's randrange function with a list comprehension to create a list of six million die rolls and time the operation using %timeit. Note that we used the line-continuation character (\) to split the statement in snippet [2] over two lines:

```
In [1]: import random

In [2]: %timeit rolls_list = \
   ...:     [random.randrange(1, 7) for i in range(0, 6_000_000)]
6.29 s ± 119 ms per loop (mean ± std. dev. of 7 runs, 1 loop each)
```

7.6 List vs. `array` Performance: Introducing `%timeit`

By default, `%timeit` executes a statement in a loop, and it runs the loop *seven* times. If you do not indicate the number of loops, `%timeit` chooses an appropriate value. In our testing, operations that on average took more than 500 milliseconds iterated only once, and operations that took fewer than 500 milliseconds iterated 10 times or more.

After executing the statement, `%timeit` displays the statement's *average* execution time, as well as the standard deviation of all the executions. On average, `%timeit` indicates that it took 6.29 seconds (s) to create the list with a standard deviation of 119 milliseconds (ms). In total, the preceding snippet took about 44 seconds to run the snippet seven times.

Timing the Creation of an `array` Containing Results of 6,000,000 Die Rolls

Now, let's use the **randint function** from the **numpy.random module** to create an `array` of 6,000,000 die rolls

```
In [3]: import numpy as np

In [4]: %timeit rolls_array = np.random.randint(1, 7, 6_000_000)
72.4 ms ± 635 µs per loop (mean ± std. dev. of 7 runs, 10 loops each)
```

On average, `%timeit` indicates that it took only 72.4 *milliseconds* with a standard deviation of 635 microseconds (µs) to create the `array`. In total, the preceding snippet took just under half a second to execute on our computer—about 1/100th of the time snippet [2] took to execute. The operation is *two orders of magnitude faster* with `array`!

60,000,000 and 600,000,000 Die Rolls

Now, let's create an `array` of 60,000,000 die rolls:

```
In [5]: %timeit rolls_array = np.random.randint(1, 7, 60_000_000)
873 ms ± 29.4 ms per loop (mean ± std. dev. of 7 runs, 1 loop each)
```

On average, it took only 873 milliseconds to create the `array`.

Finally, let's do 600,000,000 million die rolls:

```
In [6]: %timeit rolls_array = np.random.randint(1, 7, 600_000_000)
10.1 s ± 232 ms per loop (mean ± std. dev. of 7 runs, 1 loop each)
```

It took about 10 seconds to create 600,000,000 elements with NumPy vs. about 6 seconds to create only 6,000,000 elements with a list comprehension.

Based on these timing studies, you can see clearly why `array`s are preferred over lists for compute-intensive operations. In the data science case studies, we'll enter the performance-intensive worlds of big data and AI. We'll see how clever hardware, software, communications and algorithm designs combine to meet the often enormous computing challenges of today's applications.

Customizing the `%timeit` Iterations

The number of iterations within each `%timeit` loop and the number of loops are customizable with the -n and -r options. The following executes snippet [4]'s statement three times per loop and runs the loop twice:[2]

```
In [7]: %timeit -n3 -r2 rolls_array = np.random.randint(1, 7, 6_000_000)
85.5 ms ± 5.32 ms per loop (mean ± std. dev. of 2 runs, 3 loops each)
```

---

2. For most readers, using `%timeit`'s default settings should be fine.

Other IPython Magics

IPython provides dozens of magics for a variety of tasks—for a complete list, see the IPython magics documentation.[3] Here are a few helpful ones:

- **%load** to read code into IPython from a local file or URL.
- **%save** to save snippets to a file.
- **%run** to execute a .py file from IPython.
- **%precision** to change the default floating-point precision for IPython outputs.
- **%cd** to change directories without having to exit IPython first.
- **%edit** to launch an external editor—handy if you need to modify more complex snippets.
- **%history** to view a list of all snippets and commands you've executed in the current IPython session.

Self Check

1. *(IPython Session)* Use **%timeit** to compare the execution time of the following two statements. The first uses a list comprehension to create a list of the integers from 0 to 9,999,999, then totals them with the built-in **sum** function. The second statement does the same thing using an **array** and its **sum** method.

   ```
   sum([x for x in range(10_000_000)])
   np.arange(10_000_000).sum()
   ```

   Answer:

   ```
   In [1]: import numpy as np

   In [2]: %timeit sum([x for x in range(10_000_000)])
   708 ms ± 28.2 ms per loop (mean ± std. dev. of 7 runs, 1 loop each)

   In [3]: %timeit np.arange(10_000_000).sum()
   27.2 ms ± 676 µs per loop (mean ± std. dev. of 7 runs, 10 loops each)
   ```

   The statement with the list comprehension took 26 times longer to execute than the one with the **array**.

7.7 array Operators

NumPy provides many operators which enable you to write simple expressions that perform operations on entire arrays. Here, we demonstrate arithmetic between **array**s and numeric values and between **array**s of the same shape.

Arithmetic Operations with arrays and Individual Numeric Values

First, let's perform *element-wise arithmetic* with **array**s and numeric values by using arithmetic operators and augmented assignments. Element-wise operations are applied to every element, so snippet [4] multiplies every element by 2 and snippet [5] cubes every element. Each returns a *new* **array** containing the result:

---

3. http://ipython.readthedocs.io/en/stable/interactive/magics.html

7.7 array Operators

```
In [1]: import numpy as np

In [2]: numbers = np.arange(1, 6)

In [3]: numbers
Out[3]: array([1, 2, 3, 4, 5])

In [4]: numbers * 2
Out[4]: array([ 2, 4, 6, 8, 10])

In [5]: numbers ** 3
Out[5]: array([  1,   8,  27,  64, 125])

In [6]: numbers  # numbers is unchanged by the arithmetic operators
Out[6]: array([1, 2, 3, 4, 5])
```

Snippet [6] shows that the arithmetic operators did not modify numbers. Operators + and * are *commutative*, so snippet [4] could also be written as 2 * numbers.

Augmented assignments *modify* every element in the left operand.

```
In [7]: numbers += 10

In [8]: numbers
Out[8]: array([11, 12, 13, 14, 15])
```

Broadcasting

Normally, the arithmetic operations require as operands two arrays of the *same size and shape*. When one operand is a single value, called a **scalar**, NumPy performs the element-wise calculations as if the scalar were an array of the same shape as the other operand, but with the scalar value in all its elements. This is called **broadcasting**. Snippets [4], [5] and [7] each use this capability. For example, snippet [4] is equivalent to:

```
numbers * [2, 2, 2, 2, 2]
```

Broadcasting also can be applied between arrays of different sizes and shapes, enabling some concise and powerful manipulations. We'll show more examples of broadcasting later in the chapter when we introduce NumPy's universal functions.

Arithmetic Operations Between arrays

You may perform arithmetic operations and augmented assignments between arrays of the *same* shape. Let's multiply the one-dimensional arrays numbers and numbers2 (created below) that each contain five elements:

```
In [9]: numbers2 = np.linspace(1.1, 5.5, 5)

In [10]: numbers2
Out[10]: array([ 1.1, 2.2, 3.3, 4.4, 5.5])

In [11]: numbers * numbers2
Out[11]: array([ 12.1, 26.4, 42.9, 61.6, 82.5])
```

The result is a new array formed by multiplying the arrays *element-wise* in each operand—11 * 1.1, 12 * 2.2, 13 * 3.3, etc. Arithmetic between arrays of integers and floating-point numbers results in an array of floating-point numbers.

Comparing arrays

You can compare arrays with individual values and with other arrays. Comparisons are performed *element-wise*. Such comparisons produce arrays of Boolean values in which each element's True or False value indicates the comparison result:

```
In [12]: numbers
Out[12]: array([11, 12, 13, 14, 15])

In [13]: numbers >= 13
Out[13]: array([False, False,  True,  True,  True])

In [14]: numbers2
Out[14]: array([ 1.1,  2.2,  3.3,  4.4,  5.5])

In [15]: numbers2 < numbers
Out[15]: array([ True,  True,  True,  True,  True])

In [16]: numbers == numbers2
Out[16]: array([False, False, False, False, False])

In [17]: numbers == numbers
Out[17]: array([ True,  True,  True,  True,  True])
```

Snippet [13] uses broadcasting to determine whether each element of numbers is greater than or equal to 13. The remaining snippets compare the corresponding elements of each array operand.

✓ Self Check

1 *(True/False)* When one of the operands of an array operator is a scalar, NumPy uses broadcasting to perform the calculation as if the scalar were an array of the same shape as the other operand, but containing the scalar value in all its elements.
Answer: True.

2 *(IPython Session)* Create an array of the values from 1 through 5, then use broadcasting to square each value.
Answer:

```
In [1]: import numpy as np

In [2]: np.arange(1, 6) ** 2
Out[2]: array([ 1,  4,  9, 16, 25])
```

7.8 NumPy Calculation Methods

An array has various methods that perform calculations using its contents. By default, these methods ignore the array's shape and use *all* the elements in the calculations. For example, calculating the mean of an array totals all of its elements regardless of its shape, then divides by the total number of elements. You can perform these calculations on each dimension as well. For example, in a two-dimensional array, you can calculate each row's mean and each column's mean.

7.8 NumPy Calculation Methods

Consider an array representing four students' grades on three exams:

```
In [1]: import numpy as np

In [2]: grades = np.array([[87, 96, 70], [100, 87, 90],
   ...:                    [94, 77, 90], [100, 81, 82]])
   ...:

In [3]: grades
Out[3]:
array([[ 87,  96,  70],
       [100,  87,  90],
       [ 94,  77,  90],
       [100,  81,  82]])
```

We can use methods to calculate **sum**, **min**, **max**, **mean**, **std** (standard deviation) and **var** (variance)—each is a functional-style programming *reduction*:

```
In [4]: grades.sum()
Out[4]: 1054

In [5]: grades.min()
Out[5]: 70

In [6]: grades.max()
Out[6]: 100

In [7]: grades.mean()
Out[7]: 87.83333333333333

In [8]: grades.std()
Out[8]: 8.792357792739987

In [9]: grades.var()
Out[9]: 77.30555555555556
```

Calculations by Row or Column

Many calculation methods can be performed on specific array dimensions, known as the array's *axes*. These methods receive an axis keyword argument that specifies which dimension to use in the calculation, giving you a quick way to perform calculations by row or column in a two-dimensional array.

Assume that you want to calculate the average grade on each *exam*, represented by the columns of grades. Specifying axis=0 performs the calculation on all the *row* values within each column:

```
In [10]: grades.mean(axis=0)
Out[10]: array([95.25, 85.25, 83.  ])
```

So 95.25 above is the average of the first column's grades (87, 100, 94 and 100), 85.25 is the average of the second column's grades (96, 87, 77 and 81) and 83 is the average of the third column's grades (70, 90, 90 and 82). Again, NumPy does *not* display trailing 0s to the right of the decimal point in '83.'. Also note that it *does* display all element values in the same field width, which is why '83.' is followed by two spaces.

Similarly, specifying axis=1 performs the calculation on all the *column* values within each individual row. To calculate each student's average grade for all exams, we can use:

```
In [11]: grades.mean(axis=1)
Out[11]: array([84.33333333, 92.33333333, 87.        , 87.66666667])
```

This produces four averages—one each for the values in each row. So 84.33333333 is the average of row 0's grades (87, 96 and 70), and the other averages are for the remaining rows.

NumPy arrays have many more calculation methods. For the complete list, see

https://docs.scipy.org/doc/numpy/reference/arrays.ndarray.html

✓ Self Check

1 *(Fill-In)* NumPy functions _____ and _____ calculate variance and standard deviation, respectively.
Answer: var, std.

2 *(IPython Session)* Use NumPy random-number generation to create an array of twelve random grades in the range 60 through 100, then reshape the result into a 3-by-4 array. Calculate the average of all the grades, the averages of the grades in each column and the averages of the grades in each row.
Answer:

```
In [1]: import numpy as np

In [2]: grades = np.random.randint(60, 101, 12).reshape(3, 4)

In [3]: grades
Out[3]:
array([[94, 72, 76, 91],
       [65, 78, 66, 70],
       [65, 60, 63, 72]])

In [4]: grades.mean()
Out[4]: 72.66666666666667

In [5]: grades.mean(axis=0)
Out[5]: array([74.66666667, 70.        , 68.33333333, 77.66666667])

In [6]: grades.mean(axis=1)
Out[6]: array([83.25, 69.75, 65.  ])
```

7.9 Universal Functions

NumPy offers dozens of standalone **universal functions** (or **ufuncs**) that perform various element-wise operations. Each performs its task using one or two array or array-like (such as lists) arguments. Some of these functions are called when you use operators like + and * on arrays. Each returns a new array containing the results.

Let's create an array and calculate the square root of its values, using the **sqrt universal function**:

```
In [1]: import numpy as np

In [2]: numbers = np.array([1, 4, 9, 16, 25, 36])

In [3]: np.sqrt(numbers)
Out[3]: array([1., 2., 3., 4., 5., 6.])
```

Let's add two arrays with the same shape, using the **add universal function:**

```
In [4]: numbers2 = np.arange(1, 7) * 10

In [5]: numbers2
Out[5]: array([10, 20, 30, 40, 50, 60])

In [6]: np.add(numbers, numbers2)
Out[6]: array([11, 24, 39, 56, 75, 96])
```

The expression np.add(numbers, numbers2) is equivalent to:

```
numbers + numbers2
```

Broadcasting with Universal Functions

Let's use the **multiply universal function** to multiply every element of numbers2 by the scalar value 5:

```
In [7]: np.multiply(numbers2, 5)
Out[7]: array([ 50, 100, 150, 200, 250, 300])
```

The expression np.multiply(numbers2, 5) is equivalent to:

```
numbers2 * 5
```

Let's reshape numbers2 into a 2-by-3 array, then multiply its values by a one-dimensional array of three elements:

```
In [8]: numbers3 = numbers2.reshape(2, 3)

In [9]: numbers3
Out[9]:
array([[10, 20, 30],
       [40, 50, 60]])

In [10]: numbers4 = np.array([2, 4, 6])

In [11]: np.multiply(numbers3, numbers4)
Out[11]:
array([[ 20,  80, 180],
       [ 80, 200, 360]])
```

This works because numbers4 has the same length as each row of numbers3, so NumPy can apply the multiply operation by treating numbers4 as if it were the following array:

```
array([[2, 4, 6],
       [2, 4, 6]])
```

If a universal function receives two arrays of different shapes that do not support broadcasting, a ValueError occurs. You can view the broadcasting rules at:

https://docs.scipy.org/doc/numpy/user/basics.broadcasting.html

Other Universal Functions

The NumPy documentation lists universal functions in five categories—math, trigonometry, bit manipulation, comparison and floating point. The following table lists some functions from each category. You can view the complete list, their descriptions and more information about universal functions at:

https://docs.scipy.org/doc/numpy/reference/ufuncs.html

> **NumPy universal functions**
>
> *Math*—add, subtract, multiply, divide, remainder, exp, log, sqrt, power, and more.
> *Trigonometry*—sin, cos, tan, hypot, arcsin, arccos, arctan, and more.
> *Bit manipulation*—bitwise_and, bitwise_or, bitwise_xor, invert, left_shift and right_shift.
> *Comparison*—greater, greater_equal, less, less_equal, equal, not_equal, logical_and, logical_or, logical_xor, logical_not, minimum, maximum, and more.
> *Floating point*—floor, ceil, isinf, isnan, fabs, trunc, and more.

 Self Check

1 *(Fill-In)* NumPy offers dozens of standalone functions, which it calls _____.
Answer: universal functions (or ufuncs).

2 *(IPython Session)* Create an array of the values from 1 through 5, then use the power universal function and broadcasting to cube each value.
Answer:

```
In [1]: import numpy as np

In [2]: numbers = np.arange(1, 6)

In [3]: np.power(numbers, 3)
Out[3]: array([  1,   8,  27,  64, 125])
```

7.10 Indexing and Slicing

One-dimensional arrays can be indexed and sliced using the same syntax and techniques we demonstrated in the "Sequences: Lists and Tuples" chapter. Here, we focus on array-specific indexing and slicing capabilities.

Indexing with Two-Dimensional arrays

To select an element in a two-dimensional array, specify a tuple containing the element's row and column indices in square brackets (as in snippet [4]):

```
In [1]: import numpy as np

In [2]: grades = np.array([[87, 96, 70], [100, 87, 90],
   ...:                    [94, 77, 90], [100, 81, 82]])
   ...:

In [3]: grades
Out[3]:
array([[ 87,  96,  70],
       [100,  87,  90],
       [ 94,  77,  90],
       [100,  81,  82]])

In [4]: grades[0, 1]  # row 0, column 1
Out[4]: 96
```

7.10 Indexing and Slicing

Selecting a Subset of a Two-Dimensional array's Rows
To select a single row, specify only one index in square brackets:

```
In [5]: grades[1]
Out[5]: array([100, 87, 90])
```

To select multiple sequential rows, use slice notation:

```
In [6]: grades[0:2]
Out[6]:
array([[ 87, 96, 70],
       [100, 87, 90]])
```

To select multiple non-sequential rows, use a list of row indices:

```
In [7]: grades[[1, 3]]
Out[7]:
array([[100, 87, 90],
       [100, 81, 82]])
```

Selecting a Subset of a Two-Dimensional array's Columns
You can select subsets of the columns by providing a tuple specifying the row(s) and column(s) to select. Each can be a specific index, a slice or a list. Let's select only the elements in the first column:

```
In [8]: grades[:, 0]
Out[8]: array([ 87, 100, 94, 100])
```

The 0 after the comma indicates that we're selecting only column 0. The : before the comma indicates which rows within that column to select. In this case, : is a *slice* representing *all* rows. This also could be a specific row number, a slice representing a subset of the rows or a list of specific row indices to select, as in snippets [5]–[7].

You can select consecutive columns using a slice:

```
In [9]: grades[:, 1:3]
Out[9]:
array([[96, 70],
       [87, 90],
       [77, 90],
       [81, 82]])
```

or specific columns using a *list* of column indices:

```
In [10]: grades[:, [0, 2]]
Out[10]:
array([[ 87, 70],
       [100, 90],
       [ 94, 90],
       [100, 82]])
```

✓ Self Check

1. *(IPython Session)* Given the following array:

   ```
   array([[ 1, 2, 3, 4, 5],
          [ 6, 7, 8, 9, 10],
          [11, 12, 13, 14, 15]])
   ```

   write statements to perform the following tasks:

a) Select the second row.
b) Select the first and third rows.
c) Select the middle three columns.

Answer:

```
In [1]: import numpy as np

In [2]: a = np.arange(1, 16).reshape(3, 5)

In [3]: a
Out[3]:
array([[ 1,  2,  3,  4,  5],
       [ 6,  7,  8,  9, 10],
       [11, 12, 13, 14, 15]])

In [4]: a[1]
Out[4]: array([ 6,  7,  8,  9, 10])

In [5]: a[[0, 2]]
Out[5]:
array([[ 1,  2,  3,  4,  5],
       [11, 12, 13, 14, 15]])

In [6]: a[:, 1:4]
Out[6]:
array([[ 2,  3,  4],
       [ 7,  8,  9],
       [12, 13, 14]])
```

7.11 Views: Shallow Copies

The previous chapter introduced *view objects*—that is, objects that "see" the data in other objects, rather than having their own copies of the data. Views are also known as **shallow copies**. Various array methods and slicing operations produce views of an array's data.

The array method **view** returns a *new* array object with a *view* of the original array object's data. First, let's create an array and a view of that array:

```
In [1]: import numpy as np

In [2]: numbers = np.arange(1, 6)

In [3]: numbers
Out[3]: array([1, 2, 3, 4, 5])

In [4]: numbers2 = numbers.view()

In [5]: numbers2
Out[5]: array([1, 2, 3, 4, 5])
```

We can use the built-in id function to see that numbers and numbers2 are *different* objects:

```
In [6]: id(numbers)
Out[6]: 4462958592

In [7]: id(numbers2)
Out[7]: 4590846240
```

7.11 Views: Shallow Copies

To prove that numbers2 views the *same* data as numbers, let's modify an element in numbers, then display both arrays:

```
In [8]: numbers[1] *= 10
```

```
In [9]: numbers2
Out[9]: array([ 1, 20,  3,  4,  5])
```

```
In [10]: numbers
Out[10]: array([ 1, 20,  3,  4,  5])
```

Similarly, changing a value in the view also changes that value in the original array:

```
In [11]: numbers2[1] /= 10
```

```
In [12]: numbers
Out[12]: array([1, 2, 3, 4, 5])
```

```
In [13]: numbers2
Out[13]: array([1, 2, 3, 4, 5])
```

Slice Views

Slices also create views. Let's make numbers2 a slice that views only the first three elements of numbers:

```
In [14]: numbers2 = numbers[0:3]
```

```
In [15]: numbers2
Out[15]: array([1, 2, 3])
```

Again, we can confirm that numbers and numbers2 are different objects with id:

```
In [16]: id(numbers)
Out[16]: 4462958592
```

```
In [17]: id(numbers2)
Out[17]: 4590848000
```

We can confirm that numbers2 is a view of *only* the first *three* numbers elements by attempting to access numbers2[3], which produces an IndexError:

```
In [18]: numbers2[3]
-------------------------------------------------------------------
IndexError                                Traceback (most recent call last)
<ipython-input-16-582053f52daa> in <module>()
----> 1 numbers2[3]

IndexError: index 3 is out of bounds for axis 0 with size 3
```

Now, let's modify an element both arrays share, then display them. Again, we see that numbers2 is a view of numbers:

```
In [19]: numbers[1] *= 20
```

```
In [20]: numbers
Out[20]: array([1, 2, 3, 4, 5])
```

```
In [21]: numbers2
Out[21]: array([ 1, 40,  3])
```

Self Check

1 *(Fill-In)* A view is also known as a(n) _____.
Answer: shallow copy.

7.12 Deep Copies

Though views are *separate* array objects, they save memory by sharing element data from other arrays. However, when sharing *mutable* values, sometimes it's necessary to create a **deep copy** with *independent* copies of the original data. This is especially important in multi-core programming, where separate parts of your program could attempt to modify your data at the same time, possibly corrupting it.

The **array method copy** returns a new array object with a *deep copy* of the original array object's data. First, let's create an array and a deep copy of that array:

```
In [1]: import numpy as np

In [2]: numbers = np.arange(1, 6)

In [3]: numbers
Out[3]: array([1, 2, 3, 4, 5])

In [4]: numbers2 = numbers.copy()

In [5]: numbers2
Out[5]: array([1, 2, 3, 4, 5])
```

To prove that numbers2 has a separate copy of the data in numbers, let's modify an element in numbers, then display both arrays:

```
In [6]: numbers[1] *= 10

In [7]: numbers
Out[7]: array([ 1, 20,  3,  4,  5])

In [8]: numbers2
Out[8]: array([ 1,  2,  3,  4,  5])
```

As you can see, the change appears only in numbers.

Module copy—Shallow vs. Deep Copies for Other Types of Python Objects
In previous chapters, we covered *shallow copying*. In this chapter, we've covered how to *deep copy* array objects using their copy method. If you need deep copies of other types of Python objects, pass them to the **copy** module's **deepcopy** function.

 Self Check

1 *(True/False)* The array method copy returns a new array that is a view (shallow copy) of the original array.
Answer: False. The array method copy produces a *deep copy* of the original array.

2 *(True/False)* Module copy provides function deep_copy, which returns a deep copy of its argument.
Answer: False. The name of the function is deepcopy.

7.13 Reshaping and Transposing

We've used `array` method `reshape` to produce two-dimensional arrays from one-dimensional ranges. NumPy provides various other ways to reshape arrays.

reshape vs. resize

The array methods `reshape` and `resize` both enable you to change an array's dimensions. Method `reshape` returns a *view* (shallow copy) of the original array with the new dimensions. It does *not* modify the original array:

```
In [1]: import numpy as np

In [2]: grades = np.array([[87, 96, 70], [100, 87, 90]])

In [3]: grades
Out[3]:
array([[ 87,  96,  70],
       [100,  87,  90]])

In [4]: grades.reshape(1, 6)
Out[4]: array([[ 87,  96,  70, 100,  87,  90]])

In [5]: grades
Out[5]:
array([[ 87,  96,  70],
       [100,  87,  90]])
```

Method `resize` *modifies the original array's shape*:

```
In [6]: grades.resize(1, 6)

In [7]: grades
Out[7]: array([[ 87,  96,  70, 100,  87,  90]])
```

flatten vs. ravel

You can take a multidimensional array and flatten it into a single dimension with the methods **flatten** and **ravel**. Method `flatten` *deep copies* the original array's data:

```
In [8]: grades = np.array([[87, 96, 70], [100, 87, 90]])

In [9]: grades
Out[9]:
array([[ 87,  96,  70],
       [100,  87,  90]])

In [10]: flattened = grades.flatten()

In [11]: flattened
Out[11]: array([ 87,  96,  70, 100,  87,  90])

In [12]: grades
Out[12]:
array([[ 87,  96,  70],
       [100,  87,  90]])
```

To confirm that `grades` and `flattened` do *not* share the data, let's modify an element of `flattened`, then display both arrays:

```
In [13]: flattened[0] = 100

In [14]: flattened
Out[14]: array([100,  96,  70, 100,  87,  90])

In [15]: grades
Out[15]:
array([[ 87,  96,  70],
       [100,  87,  90]])
```

Method **ravel** produces a *view* of the original array, which *shares* the grades array's data:

```
In [16]: raveled = grades.ravel()

In [17]: raveled
Out[17]: array([ 87,  96,  70, 100,  87,  90])

In [18]: grades
Out[18]:
array([[ 87,  96,  70],
       [100,  87,  90]])
```

To confirm that grades and raveled *share* the same data, let's modify an element of raveled, then display both arrays:

```
In [19]: raveled[0] = 100

In [20]: raveled
Out[20]: array([100,  96,  70, 100,  87,  90])

In [21]: grades
Out[21]:
array([[100,  96,  70],
       [100,  87,  90]])
```

Transposing Rows and Columns

You can quickly **transpose** an array's rows and columns—that is "flip" the array, so the rows become the columns and the columns become the rows. The **T** attribute returns a transposed *view* (shallow copy) of the array. The original grades array represents two students' grades (the rows) on three exams (the columns). Let's transpose the rows and columns to view the data as the grades on three exams (the rows) for two students (the columns):

```
In [22]: grades.T
Out[22]:
array([[100, 100],
       [ 96,  87],
       [ 70,  90]])
```

Transposing does *not* modify the original array:

```
In [23]: grades
Out[23]:
array([[100,  96,  70],
       [100,  87,  90]])
```

7.13 Reshaping and Transposing

Horizontal and Vertical Stacking

You can combine arrays by adding more columns or more rows—known as *horizontal stacking* and *vertical stacking*. Let's create another 2-by-3 array of grades:

```
In [24]: grades2 = np.array([[94, 77, 90], [100, 81, 82]])
```

Let's assume grades2 represents three additional exam grades for the two students in the grades array. We can combine grades and grades2 with NumPy's **hstack** (horizontal stack) **function** by passing a tuple containing the arrays to combine. The extra parentheses are required because hstack expects one argument:

```
In [25]: np.hstack((grades, grades2))
Out[25]:
array([[100,  96,  70,  94,  77,  90],
       [100,  87,  90, 100,  81,  82]])
```

Next, let's assume that grades2 represents two more students' grades on three exams. In this case, we can combine grades and grades2 with NumPy's **vstack** (vertical stack) **function**:

```
In [26]: np.vstack((grades, grades2))
Out[26]:
array([[100,  96,  70],
       [100,  87,  90],
       [ 94,  77,  90],
       [100,  81,  82]])
```

✓ Self Check

1. *(IPython Session)* Given a 2-by-3 array:

   ```
   array([[1, 2, 3],
          [4, 5, 6]])
   ```

   use hstack and vstack to produce the following array:

   ```
   array([[1, 2, 3, 1, 2, 3],
          [4, 5, 6, 4, 5, 6],
          [1, 2, 3, 1, 2, 3],
          [4, 5, 6, 4, 5, 6]])
   ```

   Answer:

   ```
   In [1]: import numpy as np

   In [2]: a = np.arange(1, 7).reshape(2, 3)

   In [3]: a = np.hstack((a, a))

   In [4]: a = np.vstack((a, a))

   In [5]: a
   Out[5]:
   array([[1, 2, 3, 1, 2, 3],
          [4, 5, 6, 4, 5, 6],
          [1, 2, 3, 1, 2, 3],
          [4, 5, 6, 4, 5, 6]])
   ```

7.14 Intro to Data Science: pandas Series and DataFrames

NumPy's array is optimized for homogeneous numeric data that's accessed via integer indices. Data science presents unique demands for which more customized data structures are required. Big data applications must support mixed data types, customized indexing, missing data, data that's not structured consistently and data that needs to be manipulated into forms appropriate for the databases and data analysis packages you use.

Pandas is the most popular library for dealing with such data. It provides two key collections that you'll use in several of our Intro to Data Science sections and throughout the data science case studies—Series for one-dimensional collections and DataFrames for two-dimensional collections. You can use pandas' MultiIndex to manipulate multi-dimensional data in the context of Series and DataFrames.

Wes McKinney created pandas in 2008 while working in industry. The name pandas is derived from the term "panel data," which is data for measurements over time, such as stock prices or historical temperature readings. McKinney needed a library in which the same data structures could handle both time- and non-time-based data with support for data alignment, missing data, common database-style data manipulations, and more.[4]

NumPy and pandas are intimately related. Series and DataFrames use arrays "under the hood." Series and DataFrames are valid arguments to many NumPy operations. Similarly, arrays are valid arguments to many Series and DataFrame operations.

Pandas is a massive topic—the PDF of its documentation[5] is over 2000 pages. In this and the next chapters' Intro to Data Science sections, we present an introduction to pandas. We discuss its Series and DataFrames collections, and use them in support of data preparation. You'll see that Series and DataFrames make it easy for you to perform common tasks like selecting elements a variety of ways, filter/map/reduce operations (central to functional-style programming and big data), mathematical operations, visualization and more.

7.14.1 pandas Series

A **Series** is an enhanced one-dimensional array. Whereas arrays use only zero-based integer indices, Series support custom indexing, including even non-integer indices like strings. Series also offer additional capabilities that make them more convenient for many data-science oriented tasks. For example, Series may have missing data, and many Series operations ignore missing data by default.

Creating a Series with Default Indices
By default, a Series has integer indices numbered sequentially from 0. The following creates a Series of student grades from a list of integers:

```
In [1]: import pandas as pd

In [2]: grades = pd.Series([87, 100, 94])
```

---

4. McKinney, Wes. *Python for Data Analysis: Data Wrangling with Pandas, NumPy, and IPython*, pp. 123–165. Sebastopol, CA: OReilly Media, 2018.
5. For the latest pandas documentation, see http://pandas.pydata.org/pandas-docs/stable/.

7.14 Intro to Data Science: pandas Series and DataFrames

The initializer also may be a tuple, a dictionary, an array, another Series or a single value. We'll show a single value momentarily.

Displaying a Series
Pandas displays a Series in two-column format with the indices *left aligned* in the left column and the values *right aligned* in the right column. After listing the Series elements, pandas shows the data type (dtype) of the underlying array's elements:

```
In [3]: grades
Out[3]:
0     87
1    100
2     94
dtype: int64
```

Note how easy it is to display a Series in this format, compared to the corresponding code for displaying a list in the same two-column format.

Creating a Series with All Elements Having the Same Value
You can create a series of elements that all have the same value:

```
In [4]: pd.Series(98.6, range(3))
Out[4]:
0    98.6
1    98.6
2    98.6
dtype: float64
```

The second argument is a one-dimensional iterable object (such as a list, an array or a range) containing the Series' indices. The number of indices determines the number of elements.

Accessing a Series' Elements
You can access a Series's elements via square brackets containing an index:

```
In [5]: grades[0]
Out[5]: 87
```

Producing Descriptive Statistics for a Series
Series provides many methods for common tasks including producing various descriptive statistics. Here we show count, mean, min, max and std (standard deviation):

```
In [6]: grades.count()
Out[6]: 3

In [7]: grades.mean()
Out[7]: 93.66666666666667

In [8]: grades.min()
Out[8]: 87

In [9]: grades.max()
Out[9]: 100

In [10]: grades.std()
Out[10]: 6.506407098647712
```

Each of these is a functional-style reduction. Calling `Series` method **describe** produces all these stats and more:

```
In [11]: grades.describe()
Out[11]:
count      3.000000
mean      93.666667
std        6.506407
min       87.000000
25%       90.500000
50%       94.000000
75%       97.000000
max      100.000000
dtype: float64
```

The 25%, 50% and 75% are **quartiles**:

- 50% represents the median of the sorted values.
- 25% represents the median of the first half of the sorted values.
- 75% represents the median of the second half of the sorted values.

For the quartiles, if there are two middle elements, then their average is that quartile's median. We have only three values in our `Series`, so the 25% quartile is the average of 87 and 94, and the 75% quartile is the average of 94 and 100. Together, the **interquartile range** is the 75% quartile minus the 25% quartile, which is another measure of dispersion, like standard deviation and variance. Of course, quartiles and interquartile range are more useful in larger datasets.

Creating a Series with Custom Indices

You can specify *custom* indices with the `index` keyword argument:

```
In [12]: grades = pd.Series([87, 100, 94], index=['Wally', 'Eva', 'Sam'])

In [13]: grades
Out[13]:
Wally     87
Eva      100
Sam       94
dtype: int64
```

In this case, we used string indices, but you can use other immutable types, including integers not beginning at 0 and nonconsecutive integers. Again, notice how nicely and concisely pandas formats a `Series` for display.

Dictionary Initializers

If you initialize a `Series` with a dictionary, its keys become the `Series`' indices, and its values become the `Series`' element values:

```
In [14]: grades = pd.Series({'Wally': 87, 'Eva': 100, 'Sam': 94})

In [15]: grades
Out[15]:
Wally     87
Eva      100
Sam       94
dtype: int64
```

Accessing Elements of a Series Via Custom Indices

In a `Series` with custom indices, you can access individual elements via square brackets containing a custom index value:

```
In [16]: grades['Eva']
Out[16]: 100
```

If the custom indices are strings that could represent valid Python identifiers, pandas automatically adds them to the `Series` as attributes that you can access via a dot (.), as in:

```
In [17]: grades.Wally
Out[17]: 87
```

`Series` also has *built-in* attributes. For example, the **dtype attribute** returns the underlying array's element type:

```
In [18]: grades.dtype
Out[18]: dtype('int64')
```

and the **values attribute** returns the underlying array:

```
In [19]: grades.values
Out[19]: array([ 87, 100,  94])
```

Creating a Series of Strings

If a `Series` contains strings, you can use its **str attribute** to call string methods on the elements. First, let's create a `Series` of hardware-related strings:

```
In [20]: hardware = pd.Series(['Hammer', 'Saw', 'Wrench'])

In [21]: hardware
Out[21]:
0    Hammer
1       Saw
2    Wrench
dtype: object
```

Note that pandas also *right-aligns* string element values and that the dtype for strings is object.

Let's call string method `contains` on each element to determine whether the value of each element contains a lowercase `'a'`:

```
In [22]: hardware.str.contains('a')
Out[22]:
0     True
1     True
2    False
dtype: bool
```

Pandas returns a `Series` containing bool values indicating the `contains` method's result for each element—the element at index 2 (`'Wrench'`) does not contain an `'a'`, so its element in the resulting `Series` is `False`. Note that pandas handles the iteration internally for you—another example of functional-style programming. The `str` attribute provides many string-processing methods that are similar to those in Python's string type. For a list, see: https://pandas.pydata.org/pandas-docs/stable/api.html#string-handling.

The following uses string method `upper` to produce a *new* `Series` containing the uppercase versions of each element in `hardware`:

```
In [23]: hardware.str.upper()
Out[23]:
0     HAMMER
1        SAW
2     WRENCH
dtype: object
```

✓ Self Check

1. *(IPython Session)* Use the NumPy's random-number generation to create an array of five random integers that represent summertime temperatures in the range 60–100, then perform the following tasks:
    a) Convert the array into the Series named temperatures and display it.
    b) Determine the lowest, highest and average temperatures.
    c) Produce descriptive statistics for the Series.

Answer:

```
In [1]: import numpy as np

In [2]: import pandas as pd

In [3]: temps = np.random.randint(60, 101, 6)

In [4]: temperatures = pd.Series(temps)

In [5]: temperatures
Out[5]:
0    98
1    62
2    63
3    70
4    69
dtype: int64

In [6]: temperatures.min()
Out[6]: 62

In [7]: temperatures.max()
Out[7]: 98

In [8]: temperatures.mean()
Out[8]: 72.4

In [9]: temperatures.describe()
Out[9]:
count     5.000000
mean     72.000000
std      14.741099
min      62.000000
25%      63.000000
50%      69.000000
75%      70.000000
max      98.000000
dtype: float64
```

7.14.2 DataFrames

A **DataFrame** is an enhanced two-dimensional array. Like Series, DataFrames can have custom row and column indices, and offer additional operations and capabilities that make them more convenient for many data-science oriented tasks. DataFrames also support missing data. Each column in a DataFrame is a Series. The Series representing each column may contain different element types, as you'll soon see when we discuss loading datasets into DataFrames.

Creating a DataFrame from a Dictionary

Let's create a DataFrame from a dictionary that represents student grades on three exams:

```
In [1]: import pandas as pd

In [2]: grades_dict = {'Wally': [87, 96, 70], 'Eva': [100, 87, 90],
   ...:                'Sam': [94, 77, 90], 'Katie': [100, 81, 82],
   ...:                'Bob': [83, 65, 85]}
   ...:

In [3]: grades = pd.DataFrame(grades_dict)

In [4]: grades
Out[4]:
   Wally  Eva  Sam  Katie  Bob
0     87  100   94    100   83
1     96   87   77     81   65
2     70   90   90     82   85
```

Pandas displays DataFrames in tabular format with the indices *left aligned* in the index column and the remaining columns' values *right aligned*. The dictionary's *keys* become the column names and the *values* associated with each key become the element values in the corresponding column. Shortly, we'll show how to "flip" the rows and columns. By default, the row indices are auto-generated integers starting from 0.

Customizing a DataFrame's Indices with the index Attribute

We could have specified custom indices with the index keyword argument when we created the DataFrame, as in:

```
pd.DataFrame(grades_dict, index=['Test1', 'Test2', 'Test3'])
```

Let's use the **index** attribute to change the DataFrame's indices from sequential integers to labels:

```
In [5]: grades.index = ['Test1', 'Test2', 'Test3']

In [6]: grades
Out[6]:
       Wally  Eva  Sam  Katie  Bob
Test1     87  100   94    100   83
Test2     96   87   77     81   65
Test3     70   90   90     82   85
```

When specifying the indices, you must provide a one-dimensional collection that has the same number of elements as there are *rows* in the DataFrame; otherwise, a ValueError occurs. Series also provides an **index** attribute for changing an existing Series' indices.

Accessing a DataFrame's Columns

One benefit of pandas is that you can quickly and conveniently look at your data in many different ways, including selecting portions of the data. Let's start by getting Eva's grades by name, which displays her column as a Series:

```
In [7]: grades['Eva']
Out[7]:
Test1    100
Test2     87
Test3     90
Name: Eva, dtype: int64
```

If a DataFrame's column-name strings are valid Python identifiers, you can use them as attributes. Let's get Sam's grades with the Sam *attribute*:

```
In [8]: grades.Sam
Out[8]:
Test1    94
Test2    77
Test3    90
Name: Sam, dtype: int64
```

Selecting Rows via the loc and iloc Attributes

Though DataFrames support indexing capabilities with [], the pandas documentation recommends using the attributes loc, iloc, at and iat, which are optimized to access DataFrames and also provide additional capabilities beyond what you can do only with []. Also, the documentation states that indexing with [] *often* produces a copy of the data, which is a logic error if you attempt to assign new values to the DataFrame by assigning to the result of the [] operation.

You can access a row by its label via the DataFrame's **loc** attribute. The following lists all the grades in the row 'Test1':

```
In [9]: grades.loc['Test1']
Out[9]:
Wally     87
Eva      100
Sam       94
Katie    100
Bob       83
Name: Test1, dtype: int64
```

You also can access rows by integer zero-based indices using the **iloc** attribute (the i in iloc means that it's used with integer indices). The following lists all the grades in the second row:

```
In [10]: grades.iloc[1]
Out[10]:
Wally    96
Eva      87
Sam      77
Katie    81
Bob      65
Name: Test2, dtype: int64
```

Selecting Rows via Slices and Lists with the loc and iloc Attributes

The index can be a *slice*. When using slices containing labels with loc, the range specified *includes* the high index ('Test3'):

```
In [11]: grades.loc['Test1':'Test3']
Out[11]:
       Wally  Eva  Sam  Katie  Bob
Test1     87  100   94    100   83
Test2     96   87   77     81   65
Test3     70   90   90     82   85
```

When using slices containing integer indices with iloc, the range you specify *excludes* the high index (2):

```
In [12]: grades.iloc[0:2]
Out[12]:
       Wally  Eva  Sam  Katie  Bob
Test1     87  100   94    100   83
Test2     96   87   77     81   65
```

To select *specific rows*, use a *list* rather than slice notation with loc or iloc:

```
In [13]: grades.loc[['Test1', 'Test3']]
Out[13]:
       Wally  Eva  Sam  Katie  Bob
Test1     87  100   94    100   83
Test3     70   90   90     82   85

In [14]: grades.iloc[[0, 2]]
Out[14]:
       Wally  Eva  Sam  Katie  Bob
Test1     87  100   94    100   83
Test3     70   90   90     82   85
```

Selecting Subsets of the Rows and Columns

So far, we've selected only *entire* rows. You can focus on small subsets of a DataFrame by selecting rows *and* columns using two slices, two lists or a combination of slices and lists.

Suppose you want to view only Eva's and Katie's grades on Test1 and Test2. We can do that by using loc with a slice for the two consecutive rows and a list for the two non-consecutive columns:

```
In [15]: grades.loc['Test1':'Test2', ['Eva', 'Katie']]
Out[15]:
       Eva  Katie
Test1  100    100
Test2   87     81
```

The slice 'Test1':'Test2' selects the rows for Test1 and Test2. The list ['Eva', 'Katie'] selects only the corresponding grades from those two columns.

Let's use iloc with a list and a slice to select the first and third tests and the first three columns for those tests:

```
In [16]: grades.iloc[[0, 2], 0:3]
Out[16]:
       Wally  Eva  Sam
Test1     87  100   94
Test3     70   90   90
```

Boolean Indexing
One of pandas' more powerful selection capabilities is **Boolean indexing**. For example, let's select all the A grades—that is, those that are greater than or equal to 90:

```
In [17]: grades[grades >= 90]
Out[17]:
         Wally    Eva    Sam  Katie   Bob
Test1      NaN  100.0   94.0  100.0   NaN
Test2     96.0    NaN    NaN    NaN   NaN
Test3      NaN   90.0   90.0    NaN   NaN
```

Pandas checks every grade to determine whether its value is greater than or equal to 90 and, if so, includes it in the new `DataFrame`. Grades for which the condition is `False` are represented as **NaN** (**not a number**) in the new `DataFrame`. NaN is pandas' notation for missing values.

Let's select all the B grades in the range 80–89:

```
In [18]: grades[(grades >= 80) & (grades < 90)]
Out[18]:
         Wally   Eva   Sam  Katie   Bob
Test1     87.0   NaN   NaN    NaN  83.0
Test2      NaN  87.0   NaN   81.0   NaN
Test3      NaN   NaN   NaN   82.0  85.0
```

Pandas Boolean indices combine multiple conditions with the Python operator & (bitwise AND), *not* the and Boolean operator. For or conditions, use | (bitwise OR). NumPy also supports Boolean indexing for `arrays`, but always returns a one-dimensional array containing only the values that satisfy the condition.

Accessing a Specific DataFrame Cell by Row and Column
You can use a `DataFrame`'s **at** and **iat** attributes to get a single value from a `DataFrame`. Like `loc` and `iloc`, at uses labels and iat uses integer indices. In each case, the row and column indices must be separated by a comma. Let's select Eva's Test2 grade (87) and Wally's Test3 grade (70):

```
In [19]: grades.at['Test2', 'Eva']
Out[19]: 87

In [20]: grades.iat[2, 0]
Out[20]: 70
```

You also can assign new values to specific elements. Let's change Eva's Test2 grade to 100 using at, then change it back to 87 using iat:

```
In [21]: grades.at['Test2', 'Eva'] = 100

In [22]: grades.at['Test2', 'Eva']
Out[22]: 100

In [23]: grades.iat[1, 2] = 87

In [24]: grades.iat[1, 2]
Out[24]: 87.0
```

Descriptive Statistics

Both Series and DataFrames have a **describe method** that calculates basic descriptive statistics for the data and returns them as a DataFrame. In a DataFrame, the statistics are calculated by column (again, soon you'll see how to flip rows and columns):

```
In [25]: grades.describe()
Out[25]:
            Wally        Eva        Sam      Katie        Bob
count    3.000000   3.000000   3.000000   3.000000   3.000000
mean    84.333333  92.333333  87.000000  87.666667  77.666667
std     13.203535   6.806859   8.888194  10.692677  11.015141
min     70.000000  87.000000  77.000000  81.000000  65.000000
25%     78.500000  88.500000  83.500000  81.500000  74.000000
50%     87.000000  90.000000  90.000000  82.000000  83.000000
75%     91.500000  95.000000  92.000000  91.000000  84.000000
max     96.000000 100.000000  94.000000 100.000000  85.000000
```

As you can see, describe gives you a quick way to summarize your data. It nicely demonstrates the power of array-oriented programming with a clean, concise functional-style call. Pandas handles internally all the details of calculating these statistics for each column. You might be interested in seeing similar statistics on test-by-test basis so you can see how all the students performs on Tests 1, 2 and 3—we'll show how to do that shortly.

By default, pandas calculates the descriptive statistics with floating-point values and displays them with six digits of precision. You can control the precision and other default settings with pandas' **set_option function**:

```
In [26]: pd.set_option('precision', 2)

In [27]: grades.describe()
Out[27]:
       Wally     Eva    Sam   Katie    Bob
count   3.00    3.00   3.00    3.00   3.00
mean   84.33   92.33  87.00   87.67  77.67
std    13.20    6.81   8.89   10.69  11.02
min    70.00   87.00  77.00   81.00  65.00
25%    78.50   88.50  83.50   81.50  74.00
50%    87.00   90.00  90.00   82.00  83.00
75%    91.50   95.00  92.00   91.00  84.00
max    96.00  100.00  94.00  100.00  85.00
```

For student grades, the most important of these statistics is probably the mean. You can calculate that for each student simply by calling mean on the DataFrame:

```
In [28]: grades.mean()
Out[28]:
Wally    84.33
Eva      92.33
Sam      87.00
Katie    87.67
Bob      77.67
dtype: float64
```

In a moment, we'll show how to get the average of all the students' grades on each test in one line of additional code.

Transposing the DataFrame with the T Attribute

You can quickly **transpose** the rows and columns—so the rows become the columns, and the columns become the rows—by using the **T** attribute:

```
In [29]: grades.T
Out[29]:
       Test1  Test2  Test3
Wally     87     96     70
Eva      100     87     90
Sam       94     77     90
Katie    100     81     82
Bob       83     65     85
```

T returns a transposed *view* (not a copy) of the DataFrame.

Let's assume that rather than getting the summary statistics by student, you want to get them by test. Simply call describe on grades.T, as in:

```
In [30]: grades.T.describe()
Out[30]:
        Test1   Test2   Test3
count    5.00    5.00    5.00
mean    92.80   81.20   83.40
std      7.66   11.54    8.23
min     83.00   65.00   70.00
25%     87.00   77.00   82.00
50%     94.00   81.00   85.00
75%    100.00   87.00   90.00
max    100.00   96.00   90.00
```

To see the average of all the students' grades on each test, just call mean on the T attribute:

```
In [31]: grades.T.mean()
Out[31]:
Test1    92.8
Test2    81.2
Test3    83.4
dtype: float64
```

Sorting by Rows by Their Indices

You'll often sort data for easier readability. You can sort a DataFrame by its rows or columns, based on their indices or values. Let's sort the rows by their *indices* in *descending* order using **sort_index** and its keyword argument ascending=False (the default is to sort in *ascending* order). This returns a new DataFrame containing the sorted data:

```
In [32]: grades.sort_index(ascending=False)
Out[32]:
       Wally  Eva  Sam  Katie  Bob
Test3     70   90   90     82   85
Test2     96   87   77     81   65
Test1     87  100   94    100   83
```

Sorting by Column Indices

Now let's sort the columns into ascending order (left-to-right) by their column names. Passing the **axis=1** keyword argument indicates that we wish to sort the *column* indices, rather than the row indices—axis=0 (the default) sorts the *row* indices:

```
In [33]: grades.sort_index(axis=1)
Out[33]:
       Bob  Eva  Katie  Sam  Wally
Test1   83  100    100   94     87
Test2   65   87     81   77     96
Test3   85   90     82   90     70
```

Sorting by Column Values

Let's assume we want to see Test1's grades in descending order so we can see the students' names in highest-to-lowest grade order. We can call the method **sort_values** as follows:

```
In [34]: grades.sort_values(by='Test1', axis=1, ascending=False)
Out[34]:
       Eva  Katie  Sam  Wally  Bob
Test1  100    100   94     87   83
Test2   87     81   77     96   65
Test3   90     82   90     70   85
```

The by and axis keyword arguments work together to determine which values will be sorted. In this case, we sort based on the column values (axis=1) for Test1.

Of course, it might be easier to read the grades and names if they were in a column, so we can sort the transposed DataFrame instead. Here, we did not need to specify the axis keyword argument, because sort_values sorts data in a specified column by default:

```
In [35]: grades.T.sort_values(by='Test1', ascending=False)
Out[35]:
       Test1  Test2  Test3
Eva      100     87     90
Katie    100     81     82
Sam       94     77     90
Wally     87     96     70
Bob       83     65     85
```

Finally, since you're sorting only Test1's grades, you might not want to see the other tests at all. So, let's combine selection with sorting:

```
In [36]: grades.loc['Test1'].sort_values(ascending=False)
Out[36]:
Katie    100
Eva      100
Sam       94
Wally     87
Bob       83
Name: Test1, dtype: int64
```

Copy vs. In-Place Sorting

By default the sort_index and sort_values return a *copy* of the original DataFrame, which could require substantial memory in a big data application. You can sort the DataFrame *in place*, rather than *copying* the data. To do so, pass the keyword argument inplace=True to either sort_index or sort_values.

We've shown many pandas Series and DataFrame features. In the next chapter's Intro to Data Science section, we'll use Series and DataFrames for *data munging*—cleaning and preparing data for use in your database or analytics software.

Self Check

1. *(IPython Session)* Given the following dictionary;

   ```
   temps = {'Mon': [68, 89], 'Tue': [71, 93], 'Wed': [66, 82],
            'Thu': [75, 97], 'Fri': [62, 79]}
   ```

   perform the following tasks:
   a) Convert the dictionary into the DataFrame named temperatures with 'Low' and 'High' as the indices, then display the DataFrame.
   b) Use the column names to select only the columns for 'Mon' through 'Wed'.
   c) Use the row index 'Low' to select only the low temperatures for each day.
   d) Set the floating-point precision to 2, then calculate the average temperature for each day.
   e) Calculate the average low and high temperatures.

Answer:

```
In [1]: import pandas as pd

In [2]: temps = {'Mon': [68, 89], 'Tue': [71, 93], 'Wed': [66, 82],
   ...:          'Thu': [75, 97], 'Fri': [62, 79]}
   ...:

In [3]: temperatures = pd.DataFrame(temps, index=['Low', 'High'])  # (a)

In [4]: temperatures  # (a)
Out[4]:
      Mon  Tue  Wed  Thu  Fri
Low    68   71   66   75   62
High   89   93   82   97   79

In [5]: temperatures.loc[:, 'Mon':'Wed']  # (b)
Out[5]:
      Mon  Tue  Wed
Low    68   71   66
High   89   93   82

In [6]: temperatures.loc['Low']  # (c)
Out[6]:
Mon    68
Tue    71
Wed    66
Thu    75
Fri    62
Name: Low, dtype: int64

In [7]: pd.set_option('precision', 2)  # (d)

In [8]: temperatures.mean()  # (d)
Out[8]:
Mon    78.5
Tue    82.0
Wed    74.0
Thu    86.0
Fri    70.5
dtype: float64
```

```
In [9]: temperatures.mean(axis=1)   # (e)
Out[9]:
Low     68.4
High    88.0
dtype: float64
```

7.15 Wrap-Up

This chapter explored the use of NumPy's high-performance ndarrays for storing and retrieving data, and for performing common data manipulations concisely and with reduced chance of errors with functional-style programming. We refer to ndarrays simply by their synonym, arrays.

The chapter examples demonstrated how to create, initialize and refer to individual elements of one- and two-dimensional arrays. We used attributes to determine an array's size, shape and element type. We showed functions that create arrays of 0s, 1s, specific values or ranges values. We compared list and array performance with the IPython %timeit magic and saw that arrays are up to two orders of magnitude faster.

We used array operators and NumPy universal functions to perform element-wise calculations on every element of arrays that have the same shape. You also saw that NumPy uses broadcasting to perform element-wise operations between arrays and scalar values, and between arrays of different shapes. We introduced various built-in array methods for performing calculations using all elements of an array, and we showed how to perform those calculations row-by-row or column-by-column. We demonstrated various array slicing and indexing capabilities that are more powerful than those provided by Python's built-in collections. We demonstrated various ways to reshape arrays. We discussed how to shallow copy and deep copy arrays and other Python objects.

In the Intro to Data Science section, we began our multisection introduction to the popular pandas library that you'll use in many of the data science case study chapters. You learned that many big data applications need more flexible collections than NumPy's arrays, collections that support mixed data types, custom indexing, missing data, data that's not structured consistently and data that needs to be manipulated into forms appropriate for the databases and data analysis packages you use.

We showed how to create and manipulate pandas array-like one-dimensional Series and two-dimensional DataFrames. We customized Series and DataFrame indices. You saw pandas' nicely formatted outputs and customized the precision of floating-point values. We showed various ways to access and select data in Series and DataFrames. We used method describe to calculate basic descriptive statistics for Series and DataFrames. We showed how to transpose DataFrame rows and columns via the T attribute. You saw several ways to sort DataFrames using their index values, their column names, the data in their rows and the data in their columns. You're now familiar with four powerful array-like collections—lists, arrays, Series and DataFrames—and the contexts in which to use them. We'll add a fifth—tensors—in the "Deep Learning" chapter.

In the next chapter, we take a deeper look at strings, string formatting and string methods. We also introduce regular expressions, which we'll use to match patterns in text. The capabilities you'll learn will help you prepare for the "Natural Language Processing (NLP)" chapter and other key data science chapters. In the next chapter's Intro to Data Science section, we'll introduce pandas *data munging*—preparing data for use in your database or analytics software. In subsequent chapters, we'll use pandas for basic time-series analysis and introduce pandas visualization capabilities.

Exercises

Use IPython sessions for each exercise where practical. Each time you create or modify an array, Series or DataFrame, display the result.

7.1 *(Filling arrays)* Fill a 2-by-3 array with ones, a 3-by-3 array with zeros and a 2-by-5 array with 7s.

7.2 *(Broadcasting)* Use arange to create a 2-by-2 array containing the numbers 0–3. Use broadcasting to perform each of the following operations on the original array:
   a) Cube every element of the array.
   b) Add 7 to every element of the array.
   c) Multiply every element of the array by 2.

7.3 *(Element-Wise array Multiplication)* Create a 3-by-3 array containing the even integers from 2 through 18. Create a second 3-by-3 array containing the integers from 9 down to 1, then multiply the first array by the second.

7.4 *(array from List of Lists)* Create a 2-by-5 array from an argument which is a list of the two five-element lists [2, 3, 5, 7, 11] and [13, 17, 19, 23, 29].

7.5 *(Flattening arrays with flatten vs. ravel)* Create a 2-by-3 array containing the first six powers of 2 beginning with 2^0. Flatten the array first with method flatten, then with ravel. In each case, display the result then display the original array to show that it was unmodified.

7.6 *(Research: array Method astype)* Research in the NumPy documentation the array method astype, which converts an array's elements to another type. Use linspace and reshape to create a 2-by-3 array with the values 1.1, 2.2, ..., 6.6. Then use astype to convert the array to an array of integers.

7.7 *(Challenge Project: Reimplement NumPy array Output)* You saw that NumPy outputs two-dimensional arrays in a nice column-based format that right-aligns every element in a field width. The field width's size is determined by the array element value that requires the most character positions to display. To understand how powerful it is to have this formatting simply built-in, write a function that reimplements NumPy's array formatting for two-dimensional arrays using loops. Assume the array contains only positive integer values.

7.8 *(Challenge Project: Reimplement DataFrame Output)* You saw that pandas displays DataFrames in an attractive column-based format with row and column labels. The values within each column are right aligned in the same field width, which is determined by that column's widest value. To understand how powerful it is to have this formatting built-in, write a function that reimplements DataFrame formatting using loops. Assume the DataFrame contains only positive integer values and that both the row and column labels are each integer values beginning at 0.

7.9 *(Indexing and Slicing arrays)* Create an array containing the values 1–15, reshape it into a 3-by-5 array, then use indexing and slicing techniques to perform each of the following operations:
   a) Select row 2.

b) Select column 5.
c) Select rows 0 and 1.
d) Select columns 2–4.
e) Select the element that is in row 1 and column 4.
f) Select all elements from rows 1 and 2 that are in columns 0, 2 and 4.

7.10 *(Project: Two-Player, Two-Dimensional Tic-Tac-Toe)* Write a script to play two-dimensional Tic-Tac-Toe between two human players who alternate entering their moves on the same computer. Use a 3-by-3 two-dimensional `array`. Each player indicates their moves by entering a pair of numbers representing the row and column indices of the square in which they want to place their mark, either an `'X'` or an `'O'`. When the first player moves, place an `'X'` in the specified square. When the second player moves, place an `'O'` in the specified square. Each move must be to an empty square. After each move, determine whether the game has been won and whether it's a draw.

7.11 *(Challenge Project: Tic-Tac-Toe with Player Against the Computer)* Modify your script from the previous exercise so that the computer makes the moves for one of the players. Also, allow the player to specify whether he or she wants to go first or second.

7.12 *(Super Challenge Project: 3D Tic-Tac-Toe with Player Against the Computer)* Develop a script that plays three-dimensional Tic-Tac-Toe on a 4-by-4-by-4 board. [*Note:* This is an extremely challenging project! In the "Deep Learning" chapter, you'll learn techniques that will help you develop and AI-based approach to solving this problem.]

7.13 *(Research and Use Other Broadcasting Capabilities)* Research the NumPy broadcasting rules, then create your own `array`s to test the rules.

7.14 *(Horizontal and Vertical Stacking)* Create the two-dimensional arrays

```
array1 = np.array([[0, 1], [2, 3]])
array2 = np.array([[4, 5], [6, 7]])
```

a) Use vertical stacking to create the 4-by-2 array named `array3` with `array1` stacked on top of `array2`.
b) Use horizontal stacking to create the 2-by-4 array named `array4` with `array2` to the right of `array1`.
c) Use vertical stacking with two copies of `array4` to create a 4-by-4 `array5`.
d) Use horizontal stacking with two copies of `array3` to create a 4-by-4 `array6`.

7.15 *(Research and Use NumPy's **concatenate** Function)* Research NumPy function concatenate, then use it to reimplement the previous exercise.

7.16 *(Research: NumPy **tile** Function)* Research and use NumPy's `tile` function to create a checkerboard pattern of dashes and asterisks.

7.17 *(Research: NumPy **bincount** Functions)* Research and use the NumPy bincount function to count the number of occurrences of each non-negative integer in a 5-by-5 array of random integers in the range 0–99.

7.18 *(Median and Mode of an **array**)* NumPy arrays offer a mean method, but not median or mode. Write functions median and mode that use existing NumPy capabilities to determine the median (middle) and mode (most frequent) of the values in an `array`. Your functions should determine the median and mode regardless of the array's shape. Test your function on three arrays of different shapes.

7.19 *(Enhanced Median and Mode of an array)* Modify your functions from the previous exercise to allow the user to provide an `axis` keyword argument so the calculations can be performed row-by-row or column-by-column on a two-dimensional `array`.

7.20 *(Performance Analysis)* In this chapter, we used `%timeit` to compare the average execution times of generating a list of 6,000,000 random die rolls vs. generating an `array` of 6,000,000 random die rolls. Though we saw approximately two orders of magnitude performance improvement with `array`, we generated the list and the `array` using two *different* random-number generators and different techniques for building each collection. If you use the same techniques we showed to generate a one-element list and a one-element array, creating the list is slightly faster. Repeat the `%timeit` operations for one-element collections. Then do it again for 10, 100, 1000, 10,000, 100,000, and 1,000,000 elements and compare the results on your system. The table below shows the results on our system, with measurements in nanoseconds (ns), microseconds (µs), milliseconds (ms) and seconds (s).

Number of values	List average execution time	array average execution time
1	1.56 µs ± 25.2 ns	1.89 µs ± 24.4 ns
10	11.6 µs ± 59.6 ns	1.96 µs ± 27.6 ns
100	109 µs ± 1.61 µs	3 µs ± 147 ns
1000	1.09 ms ± 8.59 µs	12.3 µs ± 419 ns
10,000	11.1 ms ± 210 µs	102 µs ± 669 ns
100,000	111 ms ± 1.77 ms	1.02 ms ± 32.9 µs
1,000,000	1.1 s ± 8.47 ms	10.1 ms ± 250 µs

This analysis shows why `%timeit` is convenient for quick performance studies. However, you also need to develop performance-analysis wisdom. Many factors can affect performance—the underlying hardware, the operating system, the interpreter or compiler you're using, the other applications running on your computer at the same time, and many more. The way we thought about performance over the years is changing rapidly now with big data, data analytics and artificial intelligence. As we head into the AI portion of the book, you'll place enormous performance demands on your system, so it's always good to be thinking about performance issues.

7.21 *(Shallow vs. Deep Copy)* In this chapter, we discussed shallow vs. deep copies of arrays. Python's built-in list and dictionary types have copy methods that perform *shallow* copies. Using the following dictionary

```
dictionary = {'Sophia': [97, 88]}
```

demonstrate that a dictionary's copy method indeed performs a shallow copy. To do so, call copy to make the shallow copy, modify the list stored in the original dictionary, then display both dictionaries to see that they have the same contents.

Next, use the copy module's `deepcopy` function to create a *deep* copy of the dictionary. Modify the list stored in the original dictionary, then display both dictionaries to prove that each has its own data.

7.22 *(Pandas: Series)* Perform the following tasks with pandas `Series`:
  a) Create a `Series` from the list `[7, 11, 13, 17]`.

b) Create a Series with five elements that are all 100.0.
c) Create a Series with 20 elements that are all random numbers in the range 0 to 100. Use method describe to produce the Series' basic descriptive statistics.
d) Create a Series called temperatures of the floating-point values 98.6, 98.9, 100.2 and 97.9. Using the index keyword argument, specify the custom indices 'Julie', 'Charlie', 'Sam' and 'Andrea'.
e) Form a dictionary from the names and values in Part (d), then use it to initialize a Series.

7.23 *(Pandas: DataFrames)* Perform the following tasks with pandas DataFrames:
 a) Create a DataFrame named temperatures from a dictionary of three temperature readings each for 'Maxine', 'James' and 'Amanda'.
 b) Recreate the DataFrame temperatures in Part (a) with custom indices using the index keyword argument and a list containing 'Morning', 'Afternoon' and 'Evening'.
 c) Select from temperatures the column of temperature readings for 'Maxine'.
 d) Select from temperatures the row of 'Morning' temperature readings.
 e) Select from temperatures the rows for 'Morning' and 'Evening' temperature readings.
 f) Select from temperatures the columns of temperature readings for 'Amanda' and 'Maxine'.
 g) Select from temperatures the elements for 'Amanda' and 'Maxine' in the 'Morning' and 'Afternoon'.
 h) Use the describe method to produce temperatures' descriptive statistics.
 i) Transpose temperatures.
 j) Sort temperatures so that its column names are in alphabetical order.

7.24 *(AI Project: Introducing Heuristic Programming with the Knight's Tour)* An interesting puzzler for chess buffs is the Knight's Tour problem, originally proposed by the mathematician Euler. Can the knight piece move around an empty chessboard and touch each of the 64 squares once and only once? We study this intriguing problem in depth here.
 The knight makes only L-shaped moves (two spaces in one direction and one space in a perpendicular direction). Thus, as shown in the figure below, from a square near the middle of an empty chessboard, the knight (labeled K) can make eight different moves (numbered 0 through 7).

a) Draw an eight-by-eight chessboard on a sheet of paper, and attempt a Knight's Tour by hand. Put a 1 in the starting square, a 2 in the second square, a 3 in the third, and so on. Before starting the tour, estimate how far you think you'll get, remembering that a full tour consists of 64 moves. How far did you get? Was this close to your estimate?

b) Now let's develop a script that will move the knight around a chessboard represented by an eight-by-eight two-dimensional array named board. Initialize each square to zero. We describe each of the eight possible moves in terms of its horizontal and vertical components. For example, a move of type 0, as shown in the preceding figure, consists of moving two squares horizontally to the right and one square vertically upward. A move of type 2 consists of moving one square horizontally to the left and two squares vertically upward. Horizontal moves to the left and vertical moves upward are indicated with negative numbers. The eight moves may be described by two one-dimensional arrays, horizontal and vertical, as follows:

```
horizontal[0] = 2      vertical[0] = -1
horizontal[1] = 1      vertical[1] = -2
horizontal[2] = -1     vertical[2] = -2
horizontal[3] = -2     vertical[3] = -1
horizontal[4] = -2     vertical[4] = 1
horizontal[5] = -1     vertical[5] = 2
horizontal[6] = 1      vertical[6] = 2
horizontal[7] = 2      vertical[7] = 1
```

Let the variables current_row and current_column indicate the row and column, respectively, of the knight's current position. To make a move of type move_number (a value 0–7), your script should use the statements

```
current_row += vertical[move_number]
current_column += horizontal[move_number]
```

Write a script to move the knight around the chessboard. Keep a counter that varies from 1 to 64. Record the latest count in each square the knight moves to. Test each potential move to see if the knight has already visited that square. Test every potential move to ensure that the knight does not land off the chessboard. Run the application. How many moves did the knight make?

c) After attempting to write and run a Knight's Tour script, you've probably developed some valuable insights. We'll use these insights to develop a *heuristic* (i.e., a common-sense rule) for moving the knight. Heuristics do not guarantee success, but a carefully developed heuristic greatly improves the chance of success. You may have observed that the outer squares are more troublesome than the squares nearer the center of the board. In fact, the most troublesome or inaccessible squares are the four corners.

Intuition may suggest that you should attempt to move the knight to the most troublesome squares first and leave open those that are easiest to get to so that when the board gets congested near the end of the tour, there will be a greater chance of success.

We could develop an "accessibility heuristic" by classifying each of the squares according to how accessible it is and always moving the knight (using the knight's L-shaped moves) to the most inaccessible square. We fill two-dimensional array `accessibility` with numbers indicating from how many squares each particular square is accessible. On a blank chessboard, each of the 16 squares nearest the center is rated as 8, each corner square is rated as 2, and the other squares have accessibility numbers of 3, 4 or 6 as follows:

```
2 3 4 4 4 4 3 2
3 4 6 6 6 6 4 3
4 6 8 8 8 8 6 4
4 6 8 8 8 8 6 4
4 6 8 8 8 8 6 4
4 6 8 8 8 8 6 4
3 4 6 6 6 6 4 3
2 3 4 4 4 4 3 2
```

Write a new version of the Knight's Tour, using the accessibility heuristic. The knight should always move to the square with the lowest accessibility number. In case of a tie, the knight may move to any of the tied squares. Therefore, the tour may begin in any of the four corners. [*Note:* As the knight moves around the chessboard, your application should reduce the accessibility numbers as more squares become occupied. In this way, at any given time during the tour, each available square's accessibility number will remain equal to precisely the number of squares from which that square may be reached.] Run this version of your script. Did you get a full tour? Modify the script to run 64 tours, one starting from each square of the chessboard. How many full tours did you get?

7.25 *(Knight's Tour Project: Brute-Force Approaches)* In Part (c) of the previous exercise, we developed a solution to the Knight's Tour problem. The approach used, called the "accessibility heuristic," generates many solutions and executes efficiently.

As computers continue to increase in power, we'll be able to solve more problems with sheer computer power and relatively unsophisticated algorithms. Let's call this approach "brute-force" problem solving.

a) Use random-number generation to enable the knight to walk around the chessboard (in its legitimate L-shaped moves) at random. Your script should run one tour and display the final chessboard. How far did the knight get?

b) Most likely, the script in Part (a) produced a relatively short tour. Now modify your script to attempt 1,000,000 tours. Use a one-dimensional `array` to keep track of the number of tours of each length. When your script finishes attempting the 1,000,000 tours, it should display this information in a neat tabular format. What was the best result?

c) Most likely, the script in Part (b) gave you some "respectable" tours, but no full tours. Now let your script run until it produces a full tour. [*Caution:* This version of the script could run for hours on a powerful computer.] Once again, keep a table of the number of tours of each length, and display this table when the first full tour is found. How many tours did your script attempt before producing a full tour? How much time did it take?

d) Compare the brute-force version of the Knight's Tour with the accessibility-heuristic version. Which required a more careful study of the problem? Which algorithm was more challenging to develop? Which required more computer power? Could we be certain (in advance) of obtaining a full tour with the accessibility-heuristic approach? Could we be certain (in advance) of obtaining a full tour with the brute-force approach? Argue the pros and cons of brute-force problem-solving in general.

7.26 *(Knight's Tour Project: Closed-Tour Test)* In the Knight's Tour, a full tour occurs when the knight makes 64 moves, touching each square of the chessboard once and only once. A closed tour occurs when the 64th move is one move away from the square in which the knight started the tour. Modify the script you wrote in Exercise 7.24 to test for a closed tour if a full tour has occurred.

Strings: A Deeper Look

8

Objectives

In this chapter, you'll:

- Understand text processing.
- Use string methods.
- Format string content.
- Concatenate and repeat strings.
- Strip whitespace from the ends of strings.
- Change characters from lowercase to uppercase and vice versa.
- Compare strings with the comparison operators.
- Search strings for substrings and replace substrings.
- Split strings into tokens.
- Concatenate strings into a single string with a specified separator between items.
- Create and use regular expressions to match patterns in strings, replace substrings and validate data.
- Use regular expression metacharacters, quantifiers, character classes and grouping.
- Understand how critical string manipulations are to natural language processing.
- Understand the data science terms data munging, data wrangling and data cleaning, and use regular expressions to munge data into preferred formats.

Strings: A Deeper Look

Outline

8.1 Introduction	**8.10** Characters and Character-Testing Methods
8.2 Formatting Strings	**8.11** Raw Strings
8.2.1 Presentation Types	**8.12** Introduction to Regular Expressions
8.2.2 Field Widths and Alignment	8.12.1 `re` Module and Function `fullmatch`
8.2.3 Numeric Formatting	8.12.2 Replacing Substrings and Splitting Strings
8.2.4 String's `format` Method	8.12.3 Other Search Functions; Accessing Matches
8.3 Concatenating and Repeating Strings	
8.4 Stripping Whitespace from Strings	**8.13** Intro to Data Science: Pandas, Regular Expressions and Data Munging
8.5 Changing Character Case	
8.6 Comparison Operators for Strings	**8.14** Wrap-Up
8.7 Searching for Substrings	Exercises
8.8 Replacing Substrings	
8.9 Splitting and Joining Strings	

8.1 Introduction

We've introduced strings, basic string formatting and several string operators and methods. You saw that strings support many of the same sequence operations as lists and tuples, and that strings, like tuples, are immutable. Now, we take a deeper look at strings and introduce regular expressions and the re module, which we'll use to match patterns[1] in text. Regular expressions are particularly important in today's data rich applications. The capabilities you learn here will help you prepare for the "Natural Language Processing (NLP)" chapter and other key data science chapters. In the NLP chapter, we'll look at other ways to have computers manipulate and even "understand" text. The table below shows many string-processing and NLP-related applications. In the Intro to Data Science section, we briefly introduce data cleaning/munging/wrangling with Pandas `Series` and `DataFrames`.

String and NLP applications		
Anagrams	Inter-language translation	Spam classification
Automated grading of written homework	Legal document preparation	Speech-to-text engines
	Monitoring social media posts	Spell checkers
Automated teaching systems	Natural language understanding	Steganography
Categorizing articles		Text editors
Chatbots	Opinion analysis	Text-to-speech engines
Compilers and interpreters	Page-composition software	Web scraping
Creative writing	Palindromes	Who authored Shakespeare's works?
Cryptography	Parts-of-speech tagging	
Document classification	Project Gutenberg free books	Word clouds
Document similarity	Reading books, articles, documentation and absorbing knowledge	Word games
Document summarization		Writing medical diagnoses from x-rays, scans, blood tests
Electronic book readers		
Fraud detection	Search engines	and many more…
Grammar checkers	Sentiment analysis	

---

1. We'll see in the data science case study chapters that searching for patterns in text is a crucial part of machine learning.

8.2 Formatting Strings

Proper text formatting makes data easier to read and understand. Here, we present many text-formatting capabilities.

8.2.1 Presentation Types

You've seen basic string formatting with f-strings. When you specify a placeholder for a value in an f-string, Python assumes the value should be displayed as a string unless you specify another type. In some cases, the type is required. For example, let's format the float value 17.489 rounded to the hundredths position:

```
In [1]: f'{17.489:.2f}'
Out[1]: '17.49'
```

Python supports precision *only* for floating-point and Decimal values. Formatting is *type dependent*—if you try to use .2f to format a string like 'hello', a ValueError occurs. So the **presentation type** f in the *format specifier* .2f is required. It indicates what type is being formatted so Python can determine whether the other formatting information is allowed for that type. Here, we show some common presentation types. You can view the complete list at

> https://docs.python.org/3/library/string.html#formatspec

Integers
The **d presentation type** formats integer values as strings:

```
In [2]: f'{10:d}'
Out[2]: '10'
```

There also are integer presentation types (b, o and x or X) that format integers using the binary, octal or hexadecimal number systems.[2]

Characters
The **c presentation type** formats an integer character code as the corresponding character:

```
In [3]: f'{65:c} {97:c}'
Out[3]: 'A a'
```

Strings
The **s presentation type** is the default. If you specify s explicitly, the value to format must be a variable that references a string, an expression that produces a string or a string literal, as in the first placeholder below. If you do not specify a presentation type, as in the second placeholder below, non-string values like the integer 7 are converted to strings:

```
In [4]: f'{"hello":s} {7}'
Out[4]: 'hello 7'
```

In this snippet, "hello" is enclosed in double quotes. Recall that you cannot place single quotes inside a single-quoted string.

---

2. See the online appendix "Number Systems" for information about the binary, octal and hexadecimal number systems.

Floating-Point and Decimal Values

You've used the f presentation type to format floating-point and Decimal values. For extremely large and small values of these types, **Exponential (scientific) notation** can be used to format the values more compactly. Let's show the difference between f and e for a large value, each with three digits of precision to the right of the decimal point:

```
In [5]: from decimal import Decimal

In [6]: f'{Decimal("10000000000000000000000000.0"):.3f}'
Out[6]: '10000000000000000000000000.000'

In [7]: f'{Decimal("10000000000000000000000000.0"):.3e}'
Out[7]: '1.000e+25'
```

For the **e presentation type** in snippet [5], the formatted value 1.000e+25 is equivalent to

$$1.000 \times 10^{25}$$

If you prefer a capital E for the exponent, use the **E presentation type** rather than e.

 Self Check

1 *(Fill-In)* Presentation types _____ and _____ format floating-point and Decimal values in scientific notation.
Answer: e, E.

2 *(Fill-In)* Presentation type _____ formats a character code as its corresponding character.
Answer: c.

3 *(IPython Session)* Use the type specifier c to display the characters that correspond to the character codes 58, 45 and 41.
Answer:
```
In [1]: print(f'{58:c}{45:c}{41:c}')
:-)
```

8.2.2 Field Widths and Alignment

Previously you used *field widths* to format text in a specified number of character positions. By default, Python *right-aligns* numbers and *left-aligns* other values such as strings—we enclose the results below in brackets ([]) so you can see how the values align in the field:

```
In [1]: f'[{27:10d}]'
Out[1]: '[        27]'

In [2]: f'[{3.5:10f}]'
Out[2]: '[  3.500000]'

In [3]: f'[{"hello":10}]'
Out[3]: '[hello     ]'
```

Snippet [2] shows that Python formats float values with six digits of precision to the right of the decimal point by default. For values that have fewer characters than the field width, the remaining character positions are filled with spaces. Values with more characters than the field width use as many character positions as they need.

8.2 Formatting Strings

Explicitly Specifying Left and Right Alignment in a Field
Recall that you can specify left and right alignment with < and >:

```
In [4]: f'[{27:<15d}]'
Out[4]: '[27             ]'

In [5]: f'[{3.5:<15f}]'
Out[5]: '[3.500000       ]'

In [6]: f'[{"hello":>15}]'
Out[6]: '[          hello]'
```

Centering a Value in a Field
In addition, you can *center* values:

```
In [7]: f'[{27:^7d}]'
Out[7]: '[  27   ]'

In [8]: f'[{3.5:^7.1f}]'
Out[8]: '[  3.5  ]'

In [9]: f'[{"hello":^7}]'
Out[9]: '[ hello ]'
```

Centering attempts to spread the remaining unoccupied character positions equally to the left and right of the formatted value. Python places the extra space to the right if an odd number of character positions remain.

✓ Self Check

1 *(True/False)* If you do not specify the alignment, all values displayed in a field are right aligned by default.
Answer: False. Only numeric values are right aligned by default.

2 *(IPython Session)* Display on separate lines the name 'Amanda' right-, center- and left-aligned in a field of 10 characters. Enclose each result in brackets so you can see the alignment results more clearly.
Answer:

```
In [1]: print(f'[{"Amanda":>10}]\n[{"Amanda":^10}]\n[{"Amanda":<10}]')
[    Amanda]
[  Amanda  ]
[Amanda    ]
```

8.2.3 Numeric Formatting
There are a variety of numeric formatting capabilities.

Formatting Positive Numbers with Signs
Sometimes it's desirable to force the sign on a positive number:

```
In [1]: f'[{27:+10d}]'
Out[1]: '[       +27]'
```

The + before the field width specifies that a positive number should be preceded by a +. A negative number always starts with a -. To fill the remaining characters of the field with 0s rather than spaces, place a 0 before the field width (and *after* the + if there is one):

```
In [2]: f'[{27:+010d}]'
Out[2]: '[+000000027]'
```

Using a Space Where a + Sign Would Appear in a Positive Value
A space indicates that positive numbers should show a space character in the sign position. This is useful for aligning positive and negative values for display purposes:

```
In [3]: print(f'{27:d}\n{27: d}\n{-27: d}')
27
 27
-27
```

Note that the two numbers with a space in their format specifiers align. If a field width is specified, the space should appear *before* the field width.

Grouping Digits
You can format numbers with **thousands separators** by using a **comma** (,), as follows:

```
In [4]: f'{12345678:,d}'
Out[4]: '12,345,678'

In [5]: f'{123456.78:,.2f}'
Out[5]: '123,456.78'
```

✓ Self Check

1 *(Fill-In)* To display all numeric values with their sign, use a(n) _____ in the format specifier; to display a space rather than a sign for positive values use a(n) _____ instead.
Answer: +, space character.

2 *(IPython Session)* Print the values 10240.473 and -3210.9521, each preceded by its sign, in 10-character fields with thousands separators, their decimal points aligned vertically and two digits of precision.
Answer:

```
In [1]: print(f'{10240.473:+10,.2f}\n{-3210.9521:+10,.2f}')
+10,240.47
 -3,210.95
```

8.2.4 String's format Method

Python's f-strings were added to the language in version 3.6. Before that, formatting was performed with the string method **format**. In fact, f-string formatting is based on the format method's capabilities. We show you the format method here because you'll encounter it in code written prior to Python 3.6. You'll often see the format method in the Python documentation and in the many Python books and articles written before f-strings were introduced. However, we recommend using the newer f-string formatting that we've presented to this point.

You call method format on a *format string* containing curly brace ({}) *placeholders*, possibly with format specifiers. You pass to the method the values to be formatted. Let's format the float value 17.489 rounded to the hundredths position:

```
In [1]: '{:.2f}'.format(17.489)
Out[1]: '17.49'
```

In a placeholder, if there's a format specifier, you precede it by a colon (:), as in f-strings. The result of the `format` call is a new string containing the formatted results.

Multiple Placeholders
A format string may contain multiple placeholders, in which case the `format` method's arguments correspond to the placeholders from left to right:

```
In [2]: '{} {}'.format('Amanda', 'Cyan')
Out[2]: 'Amanda Cyan'
```

Referencing Arguments By Position Number
The format string can reference specific arguments by their position in the `format` method's argument list, starting with position 0:

```
In [3]: '{0} {0} {1}'.format('Happy', 'Birthday')
Out[3]: 'Happy Happy Birthday'
```

Note that we used the position number 0 ('Happy') twice—you can reference each argument as often as you like and in any order.

Referencing Keyword Arguments
You can reference keyword arguments by their keys in the placeholders:

```
In [4]: '{first} {last}'.format(first='Amanda', last='Gray')
Out[4]: 'Amanda Gray'

In [5]: '{last} {first}'.format(first='Amanda', last='Gray')
Out[5]: 'Gray Amanda'
```

✓ Self Check

1 *(IPython Session)* Use string method `format` to reimplement the IPython sessions in the Self Check exercises from Sections 8.2.1–8.2.3.
Answer:

```
In [1]: print('{:c}{:c}{:c}'.format(58, 45, 41))
:-)

In [2]: print('[{0:>10}]\n[{0:^10}]\n[{0:<10}]'.format('Amanda'))
[    Amanda]
[  Amanda  ]
[Amanda    ]

In [3]: print('{:+10,.2f}\n{:+10,.2f}'.format(10240.473, -3210.9521))
+10,240.47
 -3,210.95
```

Note that snippet [2] references `format`'s argument three times via its position number (0) in the argument list.

8.3 Concatenating and Repeating Strings

In earlier chapters, we used the + operator to concatenate strings and the * operator to repeat strings. You also can perform these operations with augmented assignments. Strings are immutable, so each operation assigns a new string object to the variable:

```
In [1]: s1 = 'happy'

In [2]: s2 = 'birthday'

In [3]: s1 += ' ' + s2

In [4]: s1
Out[4]: 'happy birthday'

In [5]: symbol = '>'

In [6]: symbol *= 5

In [7]: symbol
Out[7]: '>>>>>'
```

Self Check

1. *(IPython Session)* Use the += operator to concatenate your first and last name. Then use the *= operator to create a bar of asterisks with the same number of characters as your full name and display the bar above and below your name.
Answer:

```
In [1]: name = 'Pam'

In [2]: name += ' Black'

In [3]: bar = '*'

In [4]: bar *= len(name)

In [5]: print(f'{bar}\n{name}\n{bar}')
*********
Pam Black
*********
```

8.4 Stripping Whitespace from Strings

There are several string methods for removing whitespace from the ends of a string. Each returns a new string leaving the original unmodified. Strings are immutable, so each method that appears to modify a string returns a new one.

Removing Leading and Trailing Whitespace
Let's use string method **strip** to remove the leading and trailing whitespace from a string:

```
In [1]: sentence = '\t  \n  This is a test string. \t\t \n'

In [2]: sentence.strip()
Out[2]: 'This is a test string.'
```

Removing Leading Whitespace
Method **lstrip** removes only leading whitespace:

```
In [3]: sentence.lstrip()
Out[3]: 'This is a test string. \t\t \n'
```

Removing Trailing Whitespace
Method **rstrip** removes only trailing whitespace:

```
In [4]: sentence.rstrip()
Out[4]: '\t  \n   This is a test string.'
```

As the outputs demonstrate, these methods remove all kinds of whitespace, including spaces, newlines and tabs.

Self Check

1 *(IPython Session)* Use the methods in this section to strip the whitespace from the following string, which has five spaces at the beginning and end of the string:

```
name = '     Margo Magenta     '
```

Answer:

```
In [1]: name = '     Margo Magenta     '

In [2]: name.strip()
Out[2]: 'Margo Magenta'

In [3]: name.lstrip()
Out[3]: 'Margo Magenta     '

In [4]: name.rstrip()
Out[4]: '     Margo Magenta'
```

8.5 Changing Character Case

In earlier chapters, you used string methods **lower** and **upper** to convert strings to all lowercase or all uppercase letters. You also can change a string's capitalization with methods **capitalize** and **title**.

Capitalizing Only a String's First Character
Method **capitalize** copies the original string and returns a new string with only the first letter capitalized (this is sometimes called *sentence capitalization*):

```
In [1]: 'happy birthday'.capitalize()
Out[1]: 'Happy birthday'
```

Capitalizing the First Character of Every Word in a String
Method **title** copies the original string and returns a new string with only the first character of each word capitalized (this is sometimes called *book-title capitalization*):

```
In [2]: 'strings: a deeper look'.title()
Out[2]: 'Strings: A Deeper Look'
```

Self Check

1 *(IPython Session)* Demonstrate the results of calling **capitalize** and **title** on the string 'happy new year'.

Answer:

```
In [1]: test_string = 'happy new year'

In [2]: test_string.capitalize()
Out[2]: 'Happy new year'

In [3]: test_string.title()
Out[3]: 'Happy New Year'
```

8.6 Comparison Operators for Strings

Strings may be compared with the comparison operators. Recall that strings are compared based on their underlying integer numeric values. So uppercase letters compare as less than lowercase letters because uppercase letters have lower integer values. For example, 'A' is 65 and 'a' is 97. You've seen that you can check character codes with ord:

```
In [1]: print(f'A: {ord("A")}; a: {ord("a")}')
A: 65; a: 97
```

Let's compare the strings 'Orange' and 'orange' using the comparison operators:

```
In [2]: 'Orange' == 'orange'
Out[2]: False

In [3]: 'Orange' != 'orange'
Out[3]: True

In [4]: 'Orange' < 'orange'
Out[4]: True

In [5]: 'Orange' <= 'orange'
Out[5]: True

In [6]: 'Orange' > 'orange'
Out[6]: False

In [7]: 'Orange' >= 'orange'
Out[7]: False
```

8.7 Searching for Substrings

You can search in a string for one or more adjacent characters—known as a *substring*—to count the number of occurrences, determine whether a string contains a substring, or determine the index at which a substring resides in a string. Each method shown in this section compares characters lexicographically using their underlying numeric values.

Counting Occurrences
String method **count** returns the number of times its argument occurs in the string on which the method is called:

```
In [1]: sentence = 'to be or not to be that is the question'

In [2]: sentence.count('to')
Out[2]: 2
```

If you specify as the second argument a *start_index*, count searches only the slice *string*[*start_index*:]—that is, from *start_index* through end of the string:

```
In [3]: sentence.count('to', 12)
Out[3]: 1
```

If you specify as the second and third arguments the *start_index* and *end_index*, count searches only the slice *string[start_index:end_index]*—that is, from *start_index* up to, but not including, *end_index*:

```
In [4]: sentence.count('that', 12, 25)
Out[4]: 1
```

Like count, each of the other string methods presented in this section has *start_index* and *end_index* arguments for searching only a slice of the original string.

Locating a Substring in a String

String method **index** searches for a substring within a string and returns the first index at which the substring is found; otherwise, a ValueError occurs:

```
In [5]: sentence.index('be')
Out[5]: 3
```

String method **rindex** performs the same operation as index, but searches from the end of the string and returns the *last* index at which the substring is found; otherwise, a ValueError occurs:

```
In [6]: sentence.rindex('be')
Out[6]: 16
```

String methods **find** and **rfind** perform the same tasks as index and rindex but, if the substring is not found, return -1 rather than causing a ValueError.

Determining Whether a String Contains a Substring

If you need to know only whether a string contains a substring, use operator in or not in:

```
In [7]: 'that' in sentence
Out[7]: True

In [8]: 'THAT' in sentence
Out[8]: False

In [9]: 'THAT' not in sentence
Out[9]: True
```

Locating a Substring at the Beginning or End of a String

String methods **startswith** and **endswith** return True if the string starts with or ends with a specified substring:

```
In [10]: sentence.startswith('to')
Out[10]: True

In [11]: sentence.startswith('be')
Out[11]: False

In [12]: sentence.endswith('question')
Out[12]: True

In [13]: sentence.endswith('quest')
Out[13]: False
```

294 Strings: A Deeper Look

Self Check

1 *(Fill-In)* Method _____ returns the number of times a given substring occurs in a string.
Answer: count.

2 *(True/False)* String method find causes a ValueError if it does not find the specified substring.
Answer: False. String method find returns -1 in this case. String method index causes a ValueError.

3 *(IPython Session)* Create a loop that locates and displays every word that starts with 't' in the string 'to be or not to be that is the question'.
Answer:

```
In [1]: for word in 'to be or not to be that is the question'.split():
   ...:     if word.startswith('t'):
   ...:         print(word, end=' ')
   ...:
to to that the
```

8.8 Replacing Substrings

A common text manipulation is to locate a substring and replace its value. Method **replace** takes two substrings. It searches a string for the substring in its first argument and replaces *each* occurrence with the substring in its second argument. The method returns a new string containing the results. Let's replace tab characters with commas:

```
In [1]: values = '1\t2\t3\t4\t5'

In [2]: values.replace('\t', ',')
Out[2]: '1,2,3,4,5'
```

Method replace can receive an optional third argument specifying the maximum number of replacements to perform.

Self Check

1 *(IPython Session)* Replace the spaces in the string '1 2 3 4 5' with ' --> '.
Answer:

```
In [1]: '1 2 3 4 5'.replace(' ', ' --> ')
Out[1]: '1 --> 2 --> 3 --> 4 --> 5'
```

8.9 Splitting and Joining Strings

When you read a sentence, your brain breaks it into individual words, or **tokens**, each of which conveys meaning. Interpreters like IPython tokenize statements, breaking them into individual components such as keywords, identifiers, operators and other elements of a programming language. Tokens typically are separated by whitespace characters such as blank, tab and newline, though other characters may be used—the separators are known as **delimiters**.

Splitting Strings

We showed previously that string method `split` with *no* arguments tokenizes a string by breaking it into substrings at each whitespace character, then returns a list of tokens. To tokenize a string at a custom delimiter (such as each comma-and-space pair), specify the delimiter string (such as, `', '`) that `split` uses to tokenize the string:

```
In [1]: letters = 'A, B, C, D'

In [2]: letters.split(', ')
Out[2]: ['A', 'B', 'C', 'D']
```

If you provide an integer as the second argument, it specifies the maximum number of splits. The last token is the remainder of the string after the maximum number of splits:

```
In [3]: letters.split(', ', 2)
Out[3]: ['A', 'B', 'C, D']
```

There is also an `rsplit` method that performs the same task as `split` but processes the maximum number of splits from the end of the string toward the beginning.

Joining Strings

String method `join` concatenates the strings in its argument, which must be an iterable containing only string values; otherwise, a `TypeError` occurs. The separator between the concatenated items is the string on which you call `join`. The following code creates strings containing comma-separated lists of values:

```
In [4]: letters_list = ['A', 'B', 'C', 'D']

In [5]: ','.join(letters_list)
Out[5]: 'A,B,C,D'
```

The next snippet joins the results of a list comprehension that creates a list of strings:

```
In [6]: ','.join([str(i) for i in range(10)])
Out[6]: '0,1,2,3,4,5,6,7,8,9'
```

In the "Files and Exceptions" chapter, you'll see how to work with files that contain comma-separated values. These are known as **CSV files** and are a common format for storing data that can be loaded by spreadsheet applications like Microsoft Excel or Google Sheets. In the data science case study chapters, you'll see that many key libraries, such as NumPy, Pandas and Seaborn, provide built-in capabilities for working with CSV data.

String Methods `partition` and `rpartition`

String method **`partition`** splits a string into a tuple of three strings based on the method's *separator* argument. The three strings are

- the part of the original string before the separator,
- the separator itself, and
- the part of the string after the separator.

This might be useful for splitting more complex strings. Consider a string representing a student's name and grades:

```
'Amanda: 89, 97, 92'
```

Let's split the original string into the student's name, the separator ': ' and a string representing the list of grades:

```
In [7]: 'Amanda: 89, 97, 92'.partition(': ')
Out[7]: ('Amanda', ': ', '89, 97, 92')
```

To search for the separator from the end of the string instead, use method **rpartition** to split. For example, consider the following URL string:

```
'http://www.deitel.com/books/PyCDS/table_of_contents.html'
```

Let's use rpartition split 'table_of_contents.html' from the rest of the URL:

```
In [8]: url = 'http://www.deitel.com/books/PyCDS/table_of_contents.html'

In [9]: rest_of_url, separator, document = url.rpartition('/')

In [10]: document
Out[10]: 'table_of_contents.html'

In [11]: rest_of_url
Out[11]: 'http://www.deitel.com/books/PyCDS'
```

String Method splitlines

In the "Files and Exceptions" chapter, you'll read text from a file. If you read large amounts of text into a string, you might want to split the string into a list of lines based on newline characters. Method **splitlines** returns a list of new strings representing the lines of text split at each newline character in the original string. Recall that Python stores multiline strings with embedded \n characters to represent the line breaks, as shown in snippet [13]:

```
In [12]: lines = """This is line 1
   ...: This is line2
   ...: This is line3"""

In [13]: lines
Out[13]: 'This is line 1\nThis is line2\nThis is line3'

In [14]: lines.splitlines()
Out[14]: ['This is line 1', 'This is line2', 'This is line3']
```

Passing True to splitlines keeps the newlines at the end of each string:

```
In [15]: lines.splitlines(True)
Out[15]: ['This is line 1\n', 'This is line2\n', 'This is line3']
```

✓ Self Check

1. *(Fill-In)* Tokens are separated from one another by _____.
Answer: delimiters.

2. *(IPython Session)* Use split and join in one statement to reformat the string

    'Pamela White'

into the string

    'White, Pamela'

Answer:

```
In [1]: ', '.join(reversed('Pamela White'.split()))
Out[1]: 'White, Pamela'
```

3 *(IPython Session)* Use partition and rpartition to extract from the URL string

```
'http://www.deitel.com/books/PyCDS/table_of_contents.html'
```

the substrings 'www.deitel.com' and 'books/PyCDS'.

Answer:

```
In [2]: url = 'http://www.deitel.com/books/PyCDS/table_of_contents.html'

In [3]: protocol, separator, rest_of_url = url.partition('://')

In [4]: host, separator, document_with_path = rest_of_url.partition('/')

In [5]: host
Out[5]: 'www.deitel.com'

In [6]: path, separator, document = document_with_path.rpartition('/')

In [7]: path
Out[7]: 'books/PyCDS'
```

8.10 Characters and Character-Testing Methods

Characters (digits, letters and symbols such as $, @, % and *) are the fundamental building blocks of programs. Every program is composed of characters that, when grouped meaningfully, represent instructions and data that the interpreter uses to perform tasks. Many programming languages have separate string and character types. In Python, a character is simply a one-character string.

Python provides string methods for testing whether a string matches certain characteristics. For example, string method **isdigit** returns True if the string on which you call the method contains only the digit characters (0–9). You might use this when validating user input that must contain only digits:

```
In [1]: '-27'.isdigit()
Out[1]: False

In [2]: '27'.isdigit()
Out[2]: True
```

and the string method isalnum returns True if the string on which you call the method is alphanumeric—that is, it contains only digits and letters:

```
In [3]: 'A9876'.isalnum()
Out[3]: True

In [4]: '123 Main Street'.isalnum()
Out[4]: False
```

The table below shows many of the character-testing methods. Each method returns `False` if the condition described is not satisfied:

String Method	Description
`isalnum()`	Returns `True` if the string contains only *alphanumeric* characters (i.e., digits and letters).
`isalpha()`	Returns `True` if the string contains only *alphabetic* characters (i.e., letters).
`isdecimal()`	Returns `True` if the string contains only *decimal integer* characters (that is, base 10 integers) and does not contain a + or – sign.
`isdigit()`	Returns `True` if the string contains only digits (e.g., `'0'`, `'1'`, `'2'`).
`isidentifier()`	Returns `True` if the string represents a valid *identifier*.
`islower()`	Returns `True` if all alphabetic characters in the string are *lowercase* characters (e.g., `'a'`, `'b'`, `'c'`).
`isnumeric()`	Returns `True` if the characters in the string represent a *numeric value* without a + or – sign and without a decimal point.
`isspace()`	Returns `True` if the string contains only *whitespace* characters.
`istitle()`	Returns `True` if the first character of each word in the string is the only *uppercase* character in the word.
`isupper()`	Returns `True` if all alphabetic characters in the string are *uppercase* characters (e.g., `'A'`, `'B'`, `'C'`).

✓ Self Check

1 *(Fill-In)* Method _____ returns `True` if a string contains only letters and numbers.
Answer: `isalnum`.

2 *(Fill-In)* Method _____ returns `True` if a string contains only letters.
Answer: `isalpha`.

8.11 Raw Strings

Recall that backslash characters in strings introduce *escape sequences*—like \n for newline and \t for tab. So, if you wish to include a backslash in a string, you must use two backslash characters \\. This makes some strings difficult to read. For example, Microsoft Windows uses backslashes to separate folder names when specifying a file's location. To represent a file's location on Windows, you might write:

```
In [1]: file_path = 'C:\\MyFolder\\MySubFolder\\MyFile.txt'

In [2]: file_path
Out[2]: 'C:\\MyFolder\\MySubFolder\\MyFile.txt'
```

For such cases, **raw strings**—preceded by the character r—are more convenient. They treat each backslash as a regular character, rather than the beginning of an escape sequence:

```
In [3]: file_path = r'C:\MyFolder\MySubFolder\MyFile.txt'

In [4]: file_path
Out[4]: 'C:\\MyFolder\\MySubFolder\\MyFile.txt'
```

Python converts the raw string to a regular string that still uses the two backslash characters in its internal representation, as shown in the last snippet. Raw strings can make your code more readable, particularly when using the regular expressions that we discuss in the next section. Regular expressions often contain many backslash characters.

 Self Check

1. *(Fill-In)* The raw string `r'\\Hi!\\'` represents the regular string _____.
Answer: `'\\\\Hi!\\\\'`.

8.12 Introduction to Regular Expressions

Sometimes you'll need to recognize *patterns* in text, like phone numbers, e-mail addresses, ZIP Codes, web page addresses, Social Security numbers and more. A **regular expression** string describes a *search pattern* for *matching* characters in other strings.

Regular expressions can help you extract data from unstructured text, such as social media posts. They're also important for ensuring that data is in the correct format before you attempt to process it.[3]

Validating Data

Before working with text data, you'll often use regular expressions to *validate the data*. For example, you can check that:

- A U.S. ZIP Code consists of five digits (such as 02215) or five digits followed by a hyphen and four more digits (such as 02215-4775).
- A string last name contains only letters, spaces, apostrophes and hyphens.
- An e-mail address contains only the allowed characters in the allowed order.
- A U.S. Social Security number contains three digits, a hyphen, two digits, a hyphen and four digits, and adheres to other rules about the specific numbers that can be used in each group of digits.

You'll rarely need to create your own regular expressions for common items like these. Websites like

- `https://regex101.com`
- `http://www.regexlib.com`
- `https://www.regular-expressions.info`

and others offer repositories of existing regular expressions that you can copy and use. Many sites like these also provide interfaces in which you can test regular expressions to determine whether they'll meet your needs. We ask you to do this in the exercises.

---

3. The topic of regular expressions might feel more challenging than most other Python features you've used. After mastering this subject, you'll often write more concise code than with conventional string-processing techniques, speeding the code-development process. You'll also deal with "fringe" cases you might not ordinarily think about, possibly avoiding subtle bugs.

Other Uses of Regular Expressions
In addition to validating data, regular expressions often are used to:
- Extract data from text (sometimes known as *scraping*)—For example, locating all URLs in a web page. [You might prefer tools like BeautifulSoup, XPath and lxml.]
- Clean data—For example, removing data that's not required, removing duplicate data, handling incomplete data, fixing typos, ensuring consistent data formats, dealing with outliers and more.
- Transform data into other formats—For example, reformatting data that was collected as tab-separated or space-separated values into comma-separated values (CSV) for an application that requires data to be in CSV format.

8.12.1 re Module and Function fullmatch
To use regular expressions, import the Python Standard Library's **re** module:

```
In [1]: import re
```

One of the simplest regular expression functions is **fullmatch**, which checks whether the *entire* string in its second argument matches the pattern in its first argument.

Matching Literal Characters
Let's begin by matching *literal characters*—that is, characters that match themselves:

```
In [2]: pattern = '02215'

In [3]: 'Match' if re.fullmatch(pattern, '02215') else 'No match'
Out[3]: 'Match'

In [4]: 'Match' if re.fullmatch(pattern, '51220') else 'No match'
Out[4]: 'No match'
```

The function's first argument is the regular expression pattern to match. Any string can be a regular expression. The variable pattern's value, '02215', contains only *literal digits* that match *themselves* in the specified order. The second argument is the string that should entirely match the pattern.

If the second argument matches the pattern in the first argument, fullmatch returns an object containing the matching text, which evaluates to True. We'll say more about this object later. In snippet [4], even though the second argument contains the *same digits* as the regular expression, they're in a *different* order. So there's no match, and fullmatch returns None, which evaluates to False.

Metacharacters, Character Classes and Quantifiers
Regular expressions typically contain various special symbols called **metacharacters**, which are shown in the table below:

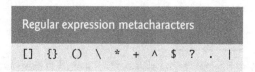

The \ metacharacter begins each of the predefined **character classes**, each matching a specific set of characters. Let's validate a five-digit ZIP Code:

```
In [5]: 'Valid' if re.fullmatch(r'\d{5}', '02215') else 'Invalid'
Out[5]: 'Valid'

In [6]: 'Valid' if re.fullmatch(r'\d{5}', '9876') else 'Invalid'
Out[6]: 'Invalid'
```

In the regular expression \d{5}, **\d** is a character class representing a digit (0–9). A character class is a *regular expression escape sequence* that matches *one* character. To match more than one, follow the character class with a **quantifier**. The quantifier {5} repeats \d five times, as if we had written \d\d\d\d\d, to match five consecutive digits. In snippet [6], fullmatch returns None because '9876' contains only four consecutive digit characters.

Other Predefined Character Classes

The table below shows some common predefined character classes and the groups of characters they match. To match any metacharacter as its *literal* value, precede it by a backslash (\). For example, \\ matches a backslash (\) and \$ matches a dollar sign ($).

Character class	Matches
\d	Any digit (0–9).
\D	Any character that is *not* a digit.
\s	Any whitespace character (such as spaces, tabs and newlines).
\S	Any character that is *not* a whitespace character.
\w	Any **word character** (also called an **alphanumeric character**)—that is, any uppercase or lowercase letter, any digit or an underscore
\W	Any character that is *not* a word character.

Custom Character Classes

Square brackets, [], define a **custom character class** that matches a *single* character. For example, [aeiou] matches a lowercase vowel, [A-Z] matches an uppercase letter, [a-z] matches a lowercase letter and [a-zA-Z] matches any lowercase or uppercase letter.

Let's validate a simple first name with no spaces or punctuation. We'll ensure that it begins with an uppercase letter (A–Z) followed by any number of lowercase letters (a–z):

```
In [7]: 'Valid' if re.fullmatch('[A-Z][a-z]*', 'Wally') else 'Invalid'
Out[7]: 'Valid'

In [8]: 'Valid' if re.fullmatch('[A-Z][a-z]*', 'eva') else 'Invalid'
Out[8]: 'Invalid'
```

A first name might contain many letters. The * **quantifier** matches *zero or more occurrences* of the subexpression to its left (in this case, [a-z]). So [A-Z][a-z]* matches an uppercase letter followed by *zero or more* lowercase letters, such as 'Amanda', 'Bo' or even 'E'.

When a custom character class starts with a **caret** (^), the class matches any character that's *not* specified. So [^a-z] matches any character that's *not* a lowercase letter:

```
In [9]: 'Match' if re.fullmatch('[^a-z]', 'A') else 'No match'
Out[9]: 'Match'

In [10]: 'Match' if re.fullmatch('[^a-z]', 'a') else 'No match'
Out[10]: 'No match'
```

Metacharacters in a custom character class are treated as literal characters—that is, the characters themselves. So [*+$] matches a *single* *, + or $ character:

```
In [11]: 'Match' if re.fullmatch('[*+$]', '*') else 'No match'
Out[11]: 'Match'

In [12]: 'Match' if re.fullmatch('[*+$]', '!') else 'No match'
Out[12]: 'No match'
```

* vs. + Quantifier

If you want to require *at least one* lowercase letter in a first name, you can replace the * quantifier in snippet [7] with +, which matches *at least one occurrence* of a subexpression:

```
In [13]: 'Valid' if re.fullmatch('[A-Z][a-z]+', 'Wally') else 'Invalid'
Out[13]: 'Valid'

In [14]: 'Valid' if re.fullmatch('[A-Z][a-z]+', 'E') else 'Invalid'
Out[14]: 'Invalid'
```

Both * and + are **greedy**—they match as many characters as possible. So the regular expression [A-Z][a-z]+ matches 'Al', 'Eva', 'Samantha', 'Benjamin' and any other words that begin with a capital letter followed at least one lowercase letter.

Other Quantifiers

The **?** **quantifier** matches *zero or one occurrences* of a subexpression:

```
In [15]: 'Match' if re.fullmatch('labell?ed', 'labelled') else 'No match'
Out[15]: 'Match'

In [16]: 'Match' if re.fullmatch('labell?ed', 'labeled') else 'No match'
Out[16]: 'Match'

In [17]: 'Match' if re.fullmatch('labell?ed', 'labellled') else 'No match'
Out[17]: 'No match'
```

The regular expression labell?ed matches labelled (the U.K. English spelling) and labeled (the U.S. English spelling), but not the misspelled word labellled. In each snippet above, the first five literal characters in the regular expression (label) match the first five characters of the second arguments. Then l? indicates that there can be *zero or one more* l characters before the remaining literal ed characters.

You can match *at least n occurrences* of a subexpression with the **{n,}** **quantifier**. The following regular expression matches strings containing *at least* three digits:

```
In [18]: 'Match' if re.fullmatch(r'\d{3,}', '123') else 'No match'
Out[18]: 'Match'

In [19]: 'Match' if re.fullmatch(r'\d{3,}', '1234567890') else 'No match'
Out[19]: 'Match'

In [20]: 'Match' if re.fullmatch(r'\d{3,}', '12') else 'No match'
Out[20]: 'No match'
```

You can match *between n and m (inclusive) occurrences* of a subexpression with the **{n,m}** quantifier. The following regular expression matches strings containing 3 to 6 digits:

```
In [21]: 'Match' if re.fullmatch(r'\d{3,6}', '123') else 'No match'
Out[21]: 'Match'

In [22]: 'Match' if re.fullmatch(r'\d{3,6}', '123456') else 'No match'
Out[22]: 'Match'

In [23]: 'Match' if re.fullmatch(r'\d{3,6}', '1234567') else 'No match'
Out[23]: 'No match'

In [24]: 'Match' if re.fullmatch(r'\d{3,6}', '12') else 'No match'
Out[24]: 'No match'
```

Self Check

1 *(True/False)* Any string can be a regular expression.
Answer: True.

2 *(True/False)* The ? quantifier matches *exactly one* occurrence of a subexpression.
Answer: False. The ? quantifier matches *zero or one* occurrences of a subexpression.

3 *(True/False)* The character class [^0-9] matches any digit.
Answer: False. The character class [^0-9] matches anything that is *not* a digit.

4 *(IPython Session)* Create and test a regular expression that matches a street address consisting of a number with one or more digits followed by two words of one or more characters each. The tokens should be separated by one space each, as in 123 Main Street.
Answer:

```
In [1]: import re

In [2]: street = r'\d+ [A-Z][a-z]* [A-Z][a-z]*'

In [3]: 'Match' if re.fullmatch(street, '123 Main Street') else 'No match'
Out[3]: 'Match'

In [4]: 'Match' if re.fullmatch(street, 'Main Street') else 'No match'
Out[4]: 'No match'
```

8.12.2 Replacing Substrings and Splitting Strings

The re module provides function sub for replacing patterns in a string, and function split for breaking a string into pieces, based on patterns.

Function sub—Replacing Patterns

By default, the re module's **sub function** replaces *all* occurrences of a pattern with the replacement text you specify. Let's convert a tab-delimited string to comma-delimited:

```
In [1]: import re

In [2]: re.sub(r'\t', ', ', '1\t2\t3\t4')
Out[2]: '1, 2, 3, 4'
```

The sub function receives three required arguments:

- the *pattern to match* (the tab character '\t')
- the *replacement text* (', ') and
- the *string to be searched* ('1\t2\t3\t4')

and returns a new string. The keyword argument count can be used to specify the maximum number of replacements:

```
In [3]: re.sub(r'\t', ', ', '1\t2\t3\t4', count=2)
Out[3]: '1, 2, 3\t4'
```

Function split
The **split** function *tokenizes* a string, using a regular expression to specify the *delimiter*, and returns a list of strings. Let's tokenize a string by splitting it at any comma that's followed by 0 or more whitespace characters—\s is the whitespace character class and * indicates *zero or more* occurrences of the preceding subexpression:

```
In [4]: re.split(r',\s*', '1,   2,   3,4,    5,6,7,8')
Out[4]: ['1', '2', '3', '4', '5', '6', '7', '8']
```

Use the keyword argument maxsplit to specify the maximum number of splits:

```
In [5]: re.split(r',\s*', '1,   2,   3,4,    5,6,7,8', maxsplit=3)
Out[5]: ['1', '2', '3', '4,    5,6,7,8']
```

In this case, after the 3 splits, the fourth string contains the rest of the original string.

Self Check

1 *(IPython Session)* Replace each occurrence of one or more adjacent tab characters in the following string with a comma and a space:

'A\tB\t\tC\t\t\tD'

Answer:

```
In [1]: import re

In [2]: re.sub(r'\t+', ', ', 'A\tB\t\tC\t\t\tD')
Out[2]: 'A, B, C, D'
```

2 *(IPython Session)* Use a regular expression and the split function to split the following string at *one or more* adjacent $ characters.

'123$Main$$Street'

Answer:

```
In [3]: re.split('\$+', '123$Main$$Street')
Out[3]: ['123', 'Main', 'Street']
```

8.12.3 Other Search Functions; Accessing Matches

Earlier we used the fullmatch function to determine whether an *entire* string matched a regular expression. There are several other searching functions. Here, we discuss the search, match, findall and finditer functions, and show how to access the matching substrings.

Function search—Finding the First Match Anywhere in a String
Function **search** looks in a string for the *first* occurrence of a substring that matches a regular expression and returns a **match object** (of type **SRE_Match**) that contains the matching substring. The match object's **group** method returns that substring:

```
In [1]: import re

In [2]: result = re.search('Python', 'Python is fun')

In [3]: result.group() if result else 'not found'
Out[3]: 'Python'
```

Function search returns None if the string does *not* contain the pattern:

```
In [4]: result2 = re.search('fun!', 'Python is fun')

In [5]: result2.group() if result2 else 'not found'
Out[5]: 'not found'
```

You can search for a match only at the *beginning* of a string with function **match**.

Ignoring Case with the Optional flags Keyword Argument

Many re module functions receive an optional flags keyword argument that changes how regular expressions are matched. For example, matches are *case sensitive* by default, but by using the re module's IGNORECASE constant, you can perform a *case-insensitive* search:

```
In [6]: result3 = re.search('Sam', 'SAM WHITE', flags=re.IGNORECASE)

In [7]: result3.group() if result3 else 'not found'
Out[7]: 'SAM'
```

Here, 'SAM' matches the pattern 'Sam' because both have the same letters, even though 'SAM' contains only uppercase letters.

Metacharacters That Restrict Matches to the Beginning or End of a String

The ^ metacharacter at the beginning of a regular expression (and not inside square brackets) is an anchor indicating that the expression matches only the *beginning* of a string:

```
In [8]: result = re.search('^Python', 'Python is fun')

In [9]: result.group() if result else 'not found'
Out[9]: 'Python'

In [10]: result = re.search('^fun', 'Python is fun')

In [11]: result.group() if result else 'not found'
Out[11]: 'not found'
```

Similarly, the $ metacharacter at the end of a regular expression is an anchor indicating that the expression matches only the *end* of a string:

```
In [12]: result = re.search('Python$', 'Python is fun')

In [13]: result.group() if result else 'not found'
Out[13]: 'not found'

In [14]: result = re.search('fun$', 'Python is fun')

In [15]: result.group() if result else 'not found'
Out[15]: 'fun'
```

Function findall and finditer—Finding All Matches in a String

Function **findall** finds *every* matching substring in a string and returns a list of the matching substrings. Let's extract all the U.S. phone numbers from a string. For simplicity we'll assume that U.S. phone numbers have the form ###-###-####:

```
In [16]: contact = 'Wally White, Home: 555-555-1234, Work: 555-555-4321'

In [17]: re.findall(r'\d{3}-\d{3}-\d{4}', contact)
Out[17]: ['555-555-1234', '555-555-4321']
```

Function **finditer** works like findall, but returns a lazy *iterable* of match objects. For large numbers of matches, using finditer can save memory because it returns one match at a time, whereas findall returns all the matches at once:

```
In [18]: for phone in re.finditer(r'\d{3}-\d{3}-\d{4}', contact):
   ...:     print(phone.group())
   ...:
555-555-1234
555-555-4321
```

Capturing Substrings in a Match

You can use **parentheses metacharacters**—(and)—to capture substrings in a match. For example, let's capture as separate substrings the name and e-mail address in the string text:

```
In [19]: text = 'Charlie Cyan, e-mail: demo1@deitel.com'

In [20]: pattern = r'([A-Z][a-z]+ [A-Z][a-z]+), e-mail: (\w+@\w+\.\w{3})'

In [21]: result = re.search(pattern, text)
```

The regular expression specifies two substrings to capture, each denoted by the metacharacters (and). These metacharacters do *not* affect whether the pattern is found in the string text—the match function returns a match object *only* if the *entire* pattern is found in the string text.

Let's consider the regular expression:

- '([A-Z][a-z]+ [A-Z][a-z]+)' matches two words separated by a space. Each word must have an initial capital letter.
- ', e-mail: ' contains literal characters that match themselves.
- (\w+@\w+\.\w{3}) matches a *simple* e-mail address consisting of one or more alphanumeric characters (\w+), the @ character, one or more alphanumeric characters (\w+), a dot (\.) and three alphanumeric characters (\w{3}). We preceded the dot with \ because a dot (.) is a regular expression metacharacter that matches one character.

The match object's **groups** method returns a tuple of the captured substrings:

```
In [22]: result.groups()
Out[22]: ('Charlie Cyan', 'demo1@deitel.com')
```

The match object's group method returns the *entire* match as a single string:

```
In [23]: result.group()
Out[23]: 'Charlie Cyan, e-mail: demo1@deitel.com'
```

You can access each captured substring by passing an integer to the group method. The captured substrings are *numbered from 1* (unlike list indices, which start at 0):

```
In [24]: result.group(1)
Out[24]: 'Charlie Cyan'

In [25]: result.group(2)
Out[25]: 'demo1@deitel.com'
```

Self Check

1 *(Fill-In)* Function _____ finds in a string the first substring that matches a regular expression.
Answer: search.

2 *(IPython Session)* Assume you have a string representing an addition problem such as

```
'10 + 5'
```

Use a regular expression to break the string into three groups representing the two operands and the operator, then display the groups.
Answer:

```
In [1]: import re

In [2]: result = re.search(r'(\d+) ([-+*/]) (\d+)', '10 + 5')

In [3]: result.groups()
Out[3]: ('10', '+', '5')

In [4]: result.group(1)
Out[4]: '10'

In [5]: result.group(2)
Out[5]: '+'

In [6]: result.group(3)
Out[6]: '5'
```

8.13 Intro to Data Science: Pandas, Regular Expressions and Data Munging

Data does not always come in forms ready for analysis. It could, for example, be in the wrong format, incorrect or even missing. Industry experience has shown that data scientists can spend as much as 75% of their time preparing data before they begin their studies. Preparing data for analysis is called **data munging** or **data wrangling**. These are synonyms—from this point forward, we'll say data munging.

Two of the most important steps in data munging are *data cleaning* and *transforming data* into the optimal formats for your database systems and analytics software. Some common data cleaning examples are:

- deleting observations with missing values,
- substituting reasonable values for missing values,
- deleting observations with bad values,
- substituting reasonable values for bad values,
- tossing outliers (although sometimes you'll want to keep them),
- duplicate elimination (although sometimes duplicates are valid),
- dealing with inconsistent data,
- and more.

You're probably already thinking that data cleaning is a difficult and messy process where you could easily make bad decisions that would negatively impact your results. This is correct. When you get to the data science case studies in the later chapters, you'll see that data science is more of an **empirical science**, like medicine, and less of a theoretical science, like theoretical physics. Empirical sciences base their conclusions on observations and experience. For example, many medicines that effectively solve medical problems today were developed by observing the effects that early versions of these medicines had on lab animals and eventually humans, and gradually refining ingredients and dosages. The actions data scientists take can vary per project, be based on the quality and nature of the data and be affected by evolving organization and professional standards.

Some common data transformations include:

- removing unnecessary data and *features* (we'll say more about features in the data science case studies),
- combining related features,
- sampling data to obtain a representative subset (we'll see in the data science case studies that *random sampling* is particularly effective for this and we'll say why),
- standardizing data formats,
- grouping data,
- and more.

It's always wise to hold onto your original data. We'll show simple examples of cleaning and transforming data in the context of Pandas `Series` and `DataFrames`.

Cleaning Your Data

Bad data values and missing values can significantly impact data analysis. Some data scientists advise against any attempts to insert "reasonable values." Instead, they advocate clearly marking missing data and leaving it up to the data analytics package to handle the issue. Others offer strong cautions.[4]

Let's consider a hospital that records patients' temperatures (and probably other vital signs) four times per day. Assume that the data consists of a name and four `float` values, such as

    ['Brown, Sue', 98.6, 98.4, 98.7, 0.0]

The preceding patient's first three recorded temperatures are 99.7, 98.4 and 98.7. The last temperature was missing and recorded as 0.0, perhaps because the sensor malfunctioned. The average of the first three values is 98.57, which is close to normal. However, if you calculate the average temperature *including* the missing value for which 0.0 was sub-

---

4. This footnote was abstracted from a comment sent to us July 20, 2018 by one of the book's academic reviewers, Dr. Alison Sanchez of the University of San Diego School of Business. She commented: "Be cautious when mentioning 'substituting reasonable values' for missing or bad values.' A stern warning: 'Substituting' values that increase statistical significance or give more 'reasonable' or 'better' results is not permitted. 'Substituting' data should not turn into 'fudging' data. The first rule students should learn is not to eliminate or change values that contradict their hypotheses. 'Substituting reasonable values' does not mean students should feel free to change values to get the results they want."

stituted, the average is only 73.93, clearly a questionable result. Certainly, doctors would not want to take drastic remedial action on this patient—it's crucial to "get the data right."

One common way to clean the data is to substitute a *reasonable* value for the missing temperature, such as the average of the patient's other readings. Had we done that above, then the patient's average temperature would remain 98.57—a much more likely average temperature, based on the other readings.

Data Validation

Let's begin by creating a Series of five-digit ZIP Codes from a dictionary of city-name/five-digit-ZIP-Code key–value pairs. We intentionally entered an invalid ZIP Code for Miami:

```
In [1]: import pandas as pd

In [2]: zips = pd.Series({'Boston': '02215', 'Miami': '3310'})

In [3]: zips
Out[3]:
Boston     02215
Miami       3310
dtype: object
```

Though zips looks like a two-dimensional array, it's actually one-dimensional. The "second column" represents the Series' ZIP Code *values* (from the dictionary's values), and the "first column" represents their *indices* (from the dictionary's keys).

We can use regular expressions with Pandas to validate data. The **str attribute** of a Series provides string-processing and various regular expression methods. Let's use the str attribute's **match method** to check whether each ZIP Code is valid:

```
In [4]: zips.str.match(r'\d{5}')
Out[4]:
Boston     True
Miami     False
dtype: bool
```

Method match applies the regular expression \d{5} to *each* Series element, attempting to ensure that the element is comprised of exactly five digits. You do not need to loop explicitly through all the ZIP Codes—match does this for you. This is another example of functional-style programming with internal rather than external iteration. The method returns a new Series containing True for each valid element. In this case, the ZIP Code for Miami did *not* match, so its element is False.

There are several ways to deal with invalid data. One is to catch it at its source and interact with the source to correct the value. That's not always possible. For example, the data could be coming from high-speed sensors in the Internet of Things. In that case, we would not be able to correct it at the source, so we could apply data cleaning techniques. In the case of the bad Miami ZIP Code of 3310, we might look for Miami ZIP Codes beginning with 3310. There are two—33101 and 33109—and we could pick one of those.

Sometimes, rather than matching an *entire* value to a pattern, you'll want to know whether a value contains a *substring* that matches the pattern. In this case, use method **contains** instead of match. Let's create a Series of strings, each containing a U.S. city, state and ZIP Code, then determine whether each string contains a substring matching the pattern ' [A-Z]{2} ' (a space, followed by two uppercase letters, followed by a space):

```
In [5]: cities = pd.Series(['Boston, MA 02215', 'Miami, FL 33101'])

In [6]: cities
Out[6]:
0    Boston, MA 02215
1       Miami, FL 33101
dtype: object

In [7]: cities.str.contains(r' [A-Z]{2} ')
Out[7]:
0    True
1    True
dtype: bool

In [8]: cities.str.match(r' [A-Z]{2} ')
Out[8]:
0    False
1    False
dtype: bool
```

We did not specify the index values, so the Series uses zero-based indexes by default (snippet [6]). Snippet [7] uses contains to show that both Series elements contain substrings that match ' [A-Z]{2} '. Snippet [8] uses match to show that neither element's value matches that pattern in its entirety, because each has other characters in its complete value.

Reformatting Your Data

We've discussed data cleaning. Now let's consider munging data into a different format. As a simple example, assume that an application requires U.S. phone numbers in the format ###-###-####, with hyphens separating each group of digits. The phone numbers have been provided to us as 10-digit strings without hyphens. Let's create the DataFrame:

```
In [9]: contacts = [['Mike Green', 'demo1@deitel.com', '5555555555'],
   ...:             ['Sue Brown', 'demo2@deitel.com', '5555551234']]
   ...:

In [10]: contactsdf = pd.DataFrame(contacts,
   ...:                            columns=['Name', 'Email', 'Phone'])
   ...:

In [11]: contactsdf
Out[11]:
         Name             Email       Phone
0   Mike Green  demo1@deitel.com  5555555555
1    Sue Brown  demo2@deitel.com  5555551234
```

In this DataFrame, we specified column indices via the columns keyword argument but did *not* specify row indices, so the rows are indexed from 0. Also, the output shows the column values right aligned by default. This differs from Python formatting in which numbers in a field are *right aligned* by default but non-numeric values are *left aligned* by default.

Now, let's munge the data with a little more functional-style programming. We can *map* the phone numbers to the proper format by calling the Series method **map** on the DataFrame's 'Phone' column. Method map's argument is a *function* that receives a value and returns the *mapped* value. The function get_formatted_phone maps 10 consecutive digits into the format ###-###-####:

8.13 Intro to Data Science: Pandas, Regular Expressions and Data Munging

```
In [12]: import re
In [13]: def get_formatted_phone(value):
    ...:     result = re.fullmatch(r'(\d{3})(\d{3})(\d{4})', value)
    ...:     return '-'.join(result.groups()) if result else value
    ...:
    ...:
```

The regular expression in the block's first statement matches *only* 10 consecutive digits. It captures substrings containing the first three digits, the next three digits and the last four digits. The `return` statement operates as follows:

- If `result` is `None`, we simply return `value` unmodified.
- Otherwise, we call `result.groups()` to get a tuple containing the captured substrings and pass that tuple to string method `join` to concatenate the elements, separating each from the next with `'-'` to form the mapped phone number.

`Series` method `map` returns a new `Series` containing the results of calling its function argument for each value in the column. Snippet [15] displays the result, including the column's name and type:

```
In [14]: formatted_phone = contactsdf['Phone'].map(get_formatted_phone)

In [15]: formatted_phone
0    555-555-5555
1    555-555-1234
Name: Phone, dtype: object
```

Once you've confirmed that the data is in the correct format, you can update it in the original `DataFrame` by assigning the new `Series` to the `'Phone'` column:

```
In [16]: contactsdf['Phone'] = formatted_phone

In [17]: contactsdf
Out[17]:
         Name            Email         Phone
0  Mike Green  demo1@deitel.com  555-555-5555
1   Sue Brown  demo2@deitel.com  555-555-1234
```

We'll continue our pandas discussion in the next chapter's Intro to Data Science section, and we'll use pandas in several later chapters.

✓ Self Check

1 *(Fill-In)* Preparing data for analysis is called _____ or _____. A subset of this process is data cleaning.
Answer: data munging, data wrangling.

2 *(IPython Session)* Let's assume that an application requires U.S. phone numbers in the format (###) ###-####. Modify the `get_formatted_phone` function in snippet [13] to return the phone number in this new format. Then recreate the `DataFrame` from snippets [9] and [10] and use the updated `get_formatted_phone` function to munge the data.
Answer:

```
In [1]: import pandas as pd

In [2]: import re
```

```
In [3]: contacts = [['Mike Green', 'demo1@deitel.com', '5555555555'],
   ...:             ['Sue Brown', 'demo2@deitel.com', '5555551234']]
   ...:

In [4]: contactsdf = pd.DataFrame(contacts,
   ...:                           columns=['Name', 'Email', 'Phone'])
   ...:

In [5]: def get_formatted_phone(value):
   ...:     result = re.fullmatch(r'(\d{3})(\d{3})(\d{4})', value)
   ...:     if result:
   ...:         part1, part2, part3 = result.groups()
   ...:         return '(' + part1 + ') ' + part2 + '-' + part3
   ...:     else:
   ...:         return value
   ...:

In [6]: contactsdf['Phone'] = contactsdf['Phone'].map(get_formatted_phone)

In [7]: contactsdf
Out[7]:
         Name              Email            Phone
0  Mike Green   demo1@deitel.com   (555) 555-5555
1   Sue Brown   demo2@deitel.com   (555) 555-1234
```

8.14 Wrap-Up

In this chapter, we presented various string formatting and processing capabilities. You formatted data in f-strings and with the string method `format`. We showed the augmented assignments for concatenating and repeating strings. You used string methods to remove whitespace from the beginning and end of strings and to change their case. We discussed additional methods for splitting strings and for joining iterables of strings. We introduced various character-testing methods.

We showed raw strings that treat backslashes (\) as literal characters rather than the beginning of escape sequences. These were particularly useful for defining regular expressions, which often contain many backslashes.

Next, we introduced the powerful pattern-matching capabilities of regular expressions with functions from the `re` module. We used the `fullmatch` function to ensure that an entire string matched a pattern, which is useful for validating data. We showed how to use the `replace` function to search for and replace substrings. We used the `split` function to tokenize strings based on delimiters that match a regular expression pattern. Then we showed various ways to search for patterns in strings and to access the resulting matches.

In the Intro to Data Science section, we introduced the synonyms data munging and data wrangling and showed an sample data munging operation, namely and transforming data. We continued our discussion of Panda's `Series` and `DataFrames` by using regular expressions to validate and munge data.

In the next chapter, we'll continue using various string-processing capabilities as we introduce reading text from files and writing text to files. We'll introduce the `csv` module for manipulating comma-separated value (CSV) files. We'll also introduce exception handling so we can process exceptions as they occur, rather than displaying a traceback.

Exercises

Use IPython sessions for each exercise where practical.

8.1 *(Check Protection)* Although electronic deposit has become extremely popular, payroll and accounts payable applications often print checks. A serious problem is the intentional alteration of a check amount by someone who plans to cash a check fraudulently. To prevent a dollar amount from being altered, some computerized check-writing systems employ a technique called *check protection*. Checks designed for printing by computer typically contain a fixed number of spaces for the printed amount. Suppose a paycheck contains eight blank spaces in which the computer is supposed to print the amount of a weekly paycheck. If the amount is large, then all eight of the spaces will be filled:

```
1,230.60 (check amount)
--------
01234567 (position numbers)
```

On the other hand, if the amount is smaller, then several of the spaces would ordinarily be left blank. For example,

```
  399.87
--------
01234567
```

contains two blank spaces. If a check is printed with blank spaces, it's easier for someone to alter the amount. Check-writing systems often insert *leading asterisks* to prevent alteration and protect the amount as follows:

```
**399.87
--------
01234567
```

Write a script that inputs a dollar amount, then prints the amount in check-protected format in a field of 10 characters with leading asterisks if necessary. [*Hint*: In a format string that explicitly specifies alignment with <, ^ or >, you can precede the alignment specifier with the fill character of your choice.]

8.2 *(Random Sentences)* Write a script that uses random-number generation to compose sentences. Use four arrays of strings called `article`, `noun`, `verb` and `preposition`. Create a sentence by selecting a word at random from each array in the following order: `article`, `noun`, `verb`, `preposition`, `article` and `noun`. As each word is picked, concatenate it to the previous words in the sentence. Spaces should separate the words. When the final sentence is output, it should start with a capital letter and end with a period. The script should generate and display 20 sentences.

8.3 *(Pig Latin)* Write a script that encodes English-language phrases into a form of coded language called pig Latin. There are many different ways to form pig Latin phrases. For simplicity, use the following algorithm:

To form a pig Latin phrase from an English-language phrase, tokenize the phrase into words with string method `split`. To translate each English word into a pig Latin word, place the first letter of the English word at the end of the word and add the letters "ay." Thus, the word "jump" becomes "umpjay," the word "the" becomes "hetay," and the word "computer" becomes "omputercay." If the word starts with a vowel, just add "ay." Blanks between words remain as blanks. Assume the following: The English phrase consists of words separated by blanks, there are no punctuation marks and all words have two or more letters. Enable the user to enter a sentence, then display the sentence in pig Latin.

8.4 *(Reversing a Sentence)* Write a script that reads a line of text as a string, tokenizes the string with the `split` method and outputs the tokens in reverse order. Use space characters as delimiters.

8.5 *(Tokenizing and Comparing Strings)* Write a script that reads a line of text, tokenizes the line using space characters as delimiters and outputs only those words beginning with the letter `'b'`.

8.6 *(Tokenizing and Comparing Strings)* Write a script that reads a line of text, tokenizes it using space characters as delimiters and outputs only those words ending with the letters `'ed'`.

8.7 *(Converting Integers to Characters)* Use the `c` presentation type to display a table of the character codes in the range 0 to 255 and their corresponding characters.

8.8 *(Converting Integers to Emojis)* Modify the previous exercise to display 10 emojis beginning with the smiley face, which has the value `0x1F600`:[5]

The value `0x1F600` is a hexadecimal (base 16) integer. See the online appendix "Number Systems" for information on the hexadecimal number system. You can find emoji codes by searching online for "Unicode full emoji list." The Unicode website precedes each character code with `"U+"` (representing Unicode). Replace `"U+"` with `"0x"` to properly format the code as a Python hexadecimal integer.

8.9 *(Creating Three-Letter Strings from a Five-Letter Word)* Write a script that reads a five-letter word from the user and produces every possible three-letter string, based on the word's letters. For example, the three-letter words produced from the word "bathe" include "ate," "bat," "bet," "tab," "hat," "the" and "tea." *Challenge*: Investigate the functions from the `itertools` module, then use an appropriate function to automate this task.

8.10 *(Project: Simple Sentiment Analysis)* Search online for lists of positive sentiment words and negative sentiment words. Create a script that inputs text, then determines whether that text is positive or negative, based on the total number of positive words and the total number of negative words. Test your script by searching for Twitter tweets on a topic of your choosing, then entering the text for several tweets. In the data science case study chapters, we'll take a deeper look at sentiment analysis.

8.11 *(Project: Evaluate Word Problems)* Write a script that enables the user to enter mathematical word problems like "two times three" and "seven minus five", then use string processing to break apart the string into the numbers and the operation and return the result. So "two times three" would return 6 and "seven minus five" would return 2. To keep things simple, assume the user enters only the words for the numbers 0 through 9 and only the operations `'plus'`, `'minus'`, `'times'` and `'divided by'`.

---

5. The look-and-feel of emojis varies across systems. The emoji shown here is from macOS. Also, depending on your system's fonts the emoji symbols might not display correctly.

8.12 *(Project: Scrambled Text)* Use string-processing capabilities to keep the first and last letter of a word and scramble the remaining letters in between the first and last. Search online for "University of Cambridge scrambled text" for an intriguing paper on the readability of texts consisting of such scrambled words. Investigate the random module's shuffle function to help you implement this exercise's solution.

Regular Expression Exercises

8.13 *(Regular Expressions: Condense Spaces to a Single Space)* Check whether a sentence contains more than one space between words. If so, remove the extra spaces and display the results. For example, 'Hello   World' should become 'Hello World'.

8.14 *(Regular Expressions: Capturing Substrings)* Reimplement Exercises 8.5 and 8.6 using regular expressions that capture the matching substrings, then display them.

8.15 *(Regular Expressions: Counting Characters and Words)* Use regular expressions and the findall function to count the number of digits, non-digit characters, whitespace characters and words in a string.

8.16 *(Regular Expressions: Locating URLs)* Use a regular expression to search through a string and to locate all valid URLs. For this exercise, assume that a valid URL has the form http://www.*domain_name*.*extension*, where *extension* must be two or more characters.

8.17 *(Regular Expressions: Matching Numeric Values)* Write a regular expression that searches a string and matches a valid number. A number can have any number of digits, but it can have only digits and a decimal point and possibly a leading sign. The decimal point is optional, but if it appears in the number, there must be only one, and it must have digits on its left and its right. There should be whitespace or a beginning or end-of-line character on either side of a valid number.

8.18 *(Regular Expression: Password Format Validator)* Search online for secure password recommendations, then research existing regular expressions that validate secure passwords. Two examples of password requirements are:

- Passwords must contain at least five words, each separated by a hyphen, a space, a period, a comma or an underscore.

- Passwords must have a minimum of 8 characters and contain at least one each from uppercase characters, lowercase characters, digits and punctuation characters (such as characters in '!@#$%<^>&*?').

Write regular expressions for each of the two requirements above, then use them to test sample passwords.

8.19 *(Regular Expressions: Testing Regular Expressions Online)* Before using any regular expression in your code, you should thoroughly test it to ensure that it meets your needs. Use a regular expression website like regex101.com to explore and test existing regular expressions, then write your own regular expression tester.

8.20 *(Regular Expressions: Munging Dates)* Dates are stored and displayed in several common formats. Three common formats are

```
042555
04/25/1955
April 25, 1955
```

Use regular expressions to search a string containing dates, find substrings that match these formats and munge them into the other formats. The original string should have one date in each format, so there will be a total of six transformations.

8.21 *(Project: Metric Conversions)* Write a script that assists the user with some common metric-to-English conversions. Your script should allow the user to specify the names of the units as strings (i.e., centimeters, liters, grams, and so on for the metric system and inches, quarts, pounds, and so on for the English system) and should respond to simple questions, such as

```
'How many inches are in 2 meters?'
'How many liters are in 10 quarts?'
```

Your script should recognize invalid conversions. For example, the following question is not meaningful, because 'feet' is a unit of length and 'kilograms' is a unit of mass:

```
'How many feet are in 5 kilograms?'
```

Assume that all questions are in the form shown above. Use regular expressions to capture the important substrings, such as 'inches', '2' and 'meters' in the first sample question above. Recall that functions int and float can convert strings to numbers.

More Challenging String-Manipulation Exercises

The preceding exercises are keyed to the text and designed to test your understanding of fundamental string manipulation and regular expression concepts. This section includes a collection of intermediate and advanced string-manipulation exercises. You should find these problems challenging, yet entertaining. The problems vary considerably in difficulty. Some require an hour or two of coding. Others are useful for lab assignments that might require two or three weeks of study and implementation. Some are challenging term projects. In the "Natural Language Processing (NLP)" chapter, you'll learn other text-processing techniques that will enable you to approach some of these exercises from a machine learning perspective.

8.22 *(Project: Cooking with Healthier Ingredients)* In the "Dictionaries and Sets" chapter's exercises, you created a dictionary that mapped ingredients to lists of their possible substitutions. Use that dictionary in a script that helps users choose healthier ingredients when cooking. The script should read a recipe from the user and suggest healthier replacements for some of the ingredients. For simplicity, your script should assume the recipe has no abbreviations for measures such as teaspoons, cups, and tablespoons, and uses numerical digits for quantities (e.g., 1 egg, 2 cups) rather than spelling them out (one egg, two cups). Your program should display a warning such as, "Always consult your healthcare professional before making significant changes to your diet." Your program should take into consideration that replacements are not always one-for-one. For example, each whole egg in a recipe can be replaced with two egg whites.

8.23 *(Project: Spam Scanner)* Spam (or junk e-mail) costs U.S. organizations billions of dollars a year in spam-prevention software, equipment, network resources, bandwidth, and lost productivity. Research online some of the most common spam e-mail messages and words, and check your junk e-mail folder. Create a list of 30 words and phrases commonly found in spam messages. Write an application in which the user enters an e-mail message. Then, scan the message for each of the 30 keywords or phrases. For each occurrence of one of these within the message, add a point to the message's "spam score." Next, rate the likelihood that the message is spam, based on the number of points it received. In the data science case study chapters, you'll be able to attack this problem in a more sophisticated way.

8.24 *(Research: Inter-Language Translation)* This exercise will help you explore one of the most challenging problems in natural language processing and artificial intelligence. The Internet brings us all together in ways that make inter-language translation particularly important. As authors, we frequently receive messages from non-English speaking readers worldwide. Not long ago, we'd write back asking them to write to us in English so we could understand.

With advances in machine learning, artificial intelligence and natural language processing, services like Google Translate (100+ languages) and Bing Microsoft Translator (60+ languages) can translate between languages instantly. In fact, the translations are so good that when non-English speakers write to us in English, we often ask them to write back in their native language, then we translate their message online.

There are many challenges in natural language translation. To get a sense of this, use online translation services to perform the following tasks:
    a) Start with a sentence in English. A popular sentence in machine translation lore is from the Bible's Matthew 26:41, "The spirit is willing, but the flesh is weak."
    b) Translate that sentence to another language, like Japanese.
    c) Translate the Japanese text back to English.

Do you get the original sentence? Often, translating from one language to another and back gives the original sentence or something close. Try chaining multiple language translations together. For instance, we took the phrase in Part (a) above and translated it from English to Chinese Traditional to Japanese to Arabic and back to English. The result was, "The soul is very happy, but the flesh is very crisp." Send us your favorite translations!

8.25 *(Project: State of the Union Speeches)* All U.S. Presidents' State of the Union speeches are available online. Copy and paste one into a large multiline string, then display statistics, including the total word count, the total character count, the average word length, the average sentence length, a word distribution of all words, a word distribution of words ending in 'ly' and the top 10 longest words. In the "Natural Language Processing (NLP)" chapter, you'll find lots of more sophisticated techniques for analyzing and comparing such texts.

8.26 *(Research: Grammarly)* Copy and paste State of the Union speeches into the free version of Grammarly or similar software. Compare the reading grade levels for speeches from several presidents.

Files and Exceptions

9

Objectives

In this chapter, you'll:

- Understand the notions of files and persistent data.
- Read, write and update files.
- Read and write CSV files, a common format for machine-learning datasets.
- Serialize objects into the JSON data-interchange format—commonly used to transmit over the Internet—and deserialize JSON into objects.
- Use the `with` statement to ensure that resources are properly released, avoiding "resource leaks."
- Use the `try` statement to delimit code in which exceptions may occur and handle those exceptions with associated `except` clauses.
- Use the `try` statement's `else` clause to execute code when no exceptions occur in the `try` suite.
- Use the `try` statement's `finally` clause to execute code regardless of whether an exception occurs in the `try`.
- `raise` exceptions to indicate runtime problems.
- Understand the traceback of functions and methods that led to an exception.
- Use pandas to load into a `DataFrame` and process the Titanic Disaster CSV dataset.

Outline

9.1 Introduction
9.2 Files
9.3 Text-File Processing
  9.3.1 Writing to a Text File: Introducing the `with` Statement
  9.3.2 Reading Data from a Text File
9.4 Updating Text Files
9.5 Serialization with JSON
9.6 Focus on Security: `pickle` Serialization and Deserialization
9.7 Additional Notes Regarding Files
9.8 Handling Exceptions
  9.8.1 Division by Zero and Invalid Input
  9.8.2 `try` Statements
  9.8.3 Catching Multiple Exceptions in One `except` Clause
  9.8.4 What Exceptions Does a Function or Method Raise?
  9.8.5 What Code Should Be Placed in a `try` Suite?
9.9 `finally` Clause
9.10 Explicitly Raising an Exception
9.11 (Optional) Stack Unwinding and Tracebacks
9.12 Intro to Data Science: Working with CSV Files
  9.12.1 Python Standard Library Module `csv`
  9.12.2 Reading CSV Files into Pandas `DataFrames`
  9.12.3 Reading the Titanic Disaster Dataset
  9.12.4 Simple Data Analysis with the Titanic Disaster Dataset
  9.12.5 Passenger Age Histogram
9.13 Wrap-Up

Exercises

9.1 Introduction

Variables, lists, tuples, dictionaries, sets, arrays, pandas `Series` and pandas `DataFrames` offer only *temporary* data storage. The data is lost when a local variable "goes out of scope" or when the program terminates. **Files** provide long-term retention of typically large amounts of data, even after the program that created the data terminates, so data maintained in files is **persistent**. Computers store files on **secondary storage devices**, including solid-state drives, hard disks and more. In this chapter, we explain how Python programs create, update and process data files.

We consider text files in several popular formats—plain text, JSON (JavaScript Object Notation) and CSV (comma-separated values). We'll use JSON to serialize and deserialize objects to facilitate saving those objects to secondary storage and transmitting them over the Internet. Be sure to read this chapter's Intro to Data Science section in which we'll use both the Python Standard Library's `csv` module and pandas to load and manipulate CSV data. In particular, we'll look at the CSV version of the Titanic disaster dataset. We'll use many popular datasets in upcoming data-science case-study chapters on machine learning, deep learning and more.

As part of our continuing emphasis on Python security, we'll discuss the security vulnerabilities of serializing and deserializing data with the Python Standard Library's `pickle` module. We recommend JSON serialization in preference to `pickle`.

We also introduce **exception handling**. An exception indicates an execution-time problem. You've seen exceptions of types `ZeroDivisionError`, `NameError`, `ValueError`, `StatisticsError`, `TypeError`, `IndexError`, `KeyError` and `RuntimeError`. We'll show how to deal with exceptions as they occur by using `try` statements and associated `except` clauses to *handle* exceptions. We'll also discuss the `try` statement's `else` and `finally` clauses. The features presented here help you write *robust, fault-tolerant* programs that can deal with problems and continue executing or *terminate gracefully*.

Programs typically request and release resources (such as files) during program execution. Often, these are in limited supply or can be used only by one program at a time. We show how to guarantee that after a program uses a resource, it's released for use by other programs, even if an exception has occurred. You'll use the `with` statement for this purpose.

9.2 Files

Python views a **text file** as a sequence of characters and a **binary file** (for images, videos and more) as a sequence of bytes. As in lists and arrays, the first character in a text file and byte in a binary file is located at position 0, so in a file of n characters or bytes, the highest position number is $n - 1$. The diagram below shows a conceptual view of a file:

For each file you **open**, Python creates a **file object** that you'll use to interact with the file.

End of File

Every operating system provides a mechanism to denote the end of a file. Some represent it with an **end-of-file marker** (as in the preceding figure), while others might maintain a count of the total characters or bytes in the file. Programming languages generally hide these operating-system details from you.

Standard File Objects

When a Python program begins execution, it creates three **standard file objects**:

- **sys.stdin**—the **standard input file object**
- **sys.stdout**—the **standard output file object**, and
- **sys.stderr**—the **standard error file object**.

Though these are considered file objects, they do not read from or write to files by default. The `input` function implicitly uses `sys.stdin` to get user input from the keyboard. Function `print` implicitly outputs to `sys.stdout`, which appears in the command line. Python implicitly outputs program errors and tracebacks to `sys.stderr`, which also appears in the command line. You must import the `sys` module if you need to refer to these objects explicitly in your code, but this is rare.

9.3 Text-File Processing

In this section, we'll write a simple text file that might be used by an accounts-receivable system to track the money owed by a company's clients. We'll then read that text file to confirm that it contains the data. For each client, we'll store the client's account number, last name and account balance owed to the company. Together, these data fields represent a client **record**. Python imposes no structure on a file, so notions such as records do not exist natively in Python. Programmers must structure files to meet their applications' requirements. We'll create and maintain this file in order by account number. In this sense, the account number may be thought of as a **record key**. For this chapter, we assume that you launch IPython from the `ch09` examples folder.

9.3.1 Writing to a Text File: Introducing the `with` Statement

Let's create an `accounts.txt` file and write five client records to the file. Generally, records in text files are stored one per line, so we end each record with a newline character:

```
In [1]: with open('accounts.txt', mode='w') as accounts:
   ...:     accounts.write('100 Jones 24.98\n')
   ...:     accounts.write('200 Doe 345.67\n')
   ...:     accounts.write('300 White 0.00\n')
   ...:     accounts.write('400 Stone -42.16\n')
   ...:     accounts.write('500 Rich 224.62\n')
   ...:
```

You can also write to a file with `print` (which automatically outputs a \n), as in

```
print('100 Jones 24.98', file=accounts)
```

The `with` Statement

Many applications *acquire* resources, such as files, network connections, database connections and more. You should *release* resources as soon as they're no longer needed. This practice ensures that other applications can use the resources. Python's **`with` statement**:

- acquires a resource (in this case, the file object for `accounts.txt`) and assigns its corresponding object to a variable (`accounts` in this example),
- allows the application to use the resource via that variable, and
- calls the resource object's `close` method to release the resource when program control reaches the end of the `with` statement's suite.

Built-In Function open

The built-in **open function** opens the file `accounts.txt` and associates it with a file object. The `mode` argument specifies the **file-open mode**, indicating whether to open a file for reading from the file, for writing to the file or both. The mode `'w'` opens the file for *writing*, creating the file if it does not exist. If you do not specify a path to the file, Python creates it in the current folder (ch09). Be careful—opening a file for writing *deletes* all the existing data in the file. By convention, the **.txt file extension** indicates a plain text file.

Writing to the File

The `with` statement assigns the object returned by `open` to the variable `accounts` in the **as clause**. In the `with` statement's suite, we use the variable `accounts` to interact with the file. In this case, we call the file object's **`write` method** five times to write five records to the file, each as a separate line of text ending in a newline. At the end of the `with` statement's suite, the `with` statement *implicitly* calls the file object's **`close`** method to close the file.

Contents of accounts.txt File

After executing the previous snippet, your ch09 directory contains the file `accounts.txt` with the following contents, which you can view by opening the file in a text editor:

```
100 Jones 24.98
200 Doe 345.67
300 White 0.00
400 Stone -42.16
500 Rich 224.62
```

In the next section, you'll read the file and display its contents.

Self Check

1 *(Fill-In)* The _____ implicitly releases resources when its suite finishes executing.
Answer: with.

2 *(True/False)* It's good practice to keep resources open until your program terminates.
Answer: False. It's good practice to close resources as soon as the program no longer needs them.

3 *(IPython Session)* Create a grades.txt file and write to it the following three records consisting of student IDs, last names and letter grades:

```
1 Red A
2 Green B
3 White A
```

Answer:

```
In [1]: with open('grades.txt', mode='w') as grades:
   ...:     grades.write('1 Red A\n')
   ...:     grades.write('2 Green B\n')
   ...:     grades.write('3 White A\n')
   ...:
```

After the preceding snippet executed, we used a text editor to view the grades.txt file:

```
1 Red A
2 Green B
3 White A
```

9.3.2 Reading Data from a Text File

We just created the text file accounts.txt and wrote data to it. Now let's read that data from the file sequentially from beginning to end. The following session reads records from the file accounts.txt and displays the contents of each record in columns with the Account and Name columns *left aligned* and the Balance column *right aligned*, so the decimal points align vertically:

```
In [1]: with open('accounts.txt', mode='r') as accounts:
   ...:     print(f'{"Account":<10}{"Name":<10}{"Balance":>10}')
   ...:     for record in accounts:
   ...:         account, name, balance = record.split()
   ...:         print(f'{account:<10}{name:<10}{balance:>10}')
   ...:
Account   Name         Balance
100       Jones          24.98
200       Doe           345.67
300       White           0.00
400       Stone         -42.16
500       Rich          224.62
```

If the contents of a file should not be modified, open the file for reading only—another example of the principle of least privilege. This prevents the program from accidentally modifying the file. You open a file for reading by passing the `'r'` file-open mode as function open's second argument. If you do not specify the folder in which to store the file, open assumes the file is in the current folder.

Iterating through a file object, as shown in the preceding for statement, reads one line at a time from the file and returns it as a string. For each record (that is, line) in the file, string method split returns tokens in the line as a list, which we unpack into the variables account, name and balance.[1] The last statement in the for statement's suite displays these variables in columns using field widths.

File Method readlines
The file object's **readlines** method also can be used to read an *entire* text file. The method returns each line as a string in a list of strings. For small files, this works well, but iterating over the lines in a file object, as shown above, can be more efficient.[2] Calling readlines for a large file can be a time-consuming operation, which must complete before you can begin using the list of strings. Using the file object in a for statement enables your program to process each text line as it's read.

Seeking to a Specific File Position
While reading through a file, the system maintains a **file-position pointer** representing the location of the next character to read. Sometimes it's necessary to process a file sequentially from the beginning *several times* during a program's execution. Each time, you must reposition the file-position pointer to the beginning of the file, which you can do either by closing and reopening the file, or by calling the file object's **seek** method, as in

> *file_object*.seek(0)

The latter approach is faster.

✓ Self Check

1 *(Fill-In)* A file object's _____ method can be used to reposition the file-position pointer.
Answer: seek.

2 *(True/False)* By default, iterating through a file object with a for statement reads one line at a time from the file and returns it as a string.
Answer: True.

3 *(IPython Session)* Read the file grades.txt that you created in the previous section's Self Check and display it in columns with the column heads 'ID', 'Name' and 'Grade'.
Answer:

```
In [1]: with open('grades.txt', 'r') as grades:
   ...:     print(f'{"ID":<4}{"Name":<7}{"Grade"}')
   ...:     for record in grades:
   ...:         student_id, name, grade = record.split()
   ...:         print(f'{student_id:<4}{name:<7}{grade}')
   ...:
ID  Name   Grade
1   Red    A
2   Green  B
3   White  A
```

---

1. When splitting strings on spaces (the default), split automatically discards the newline character.
2. https://docs.python.org/3/tutorial/inputoutput.html#methods-of-file-objects.

9.4 Updating Text Files

Formatted data written to a text file cannot be modified without the risk of destroying other data. If the name `'White'` needs to be changed to `'Williams'` in accounts.txt, the old name cannot simply be overwritten. The original record for White is stored as

```
300 White 0.00
```

If you overwrite the name `'White'` with the name `'Williams'`, the record becomes

```
300 Williams00
```

The new last name contains three more characters than the original one, so the characters beyond the second "i" in `'Williams'` overwrite other characters in the line. The problem is that in the formatted input–output model, records and their fields can vary in size. For example, 7, 14, –117, 2074 and 27383 are all integers and are stored in the same number of "raw data" bytes internally (typically 4 or 8 bytes in today's systems). However, when these integers are output as formatted text, they become different-sized fields. For example, 7 is one character, 14 is two characters and 27383 is five characters.

To make the preceding name change, we can:

- copy the records before 300 White 0.00 into a temporary file,
- write the updated and correctly formatted record for account 300 to this file,
- copy the records after 300 White 0.00 to the temporary file,
- delete the old file and
- rename the temporary file to use the original file's name.

This can be cumbersome because it requires processing *every* record in the file, even if you need to update only one record. Updating a file as described above is more efficient when an application needs to update many records in one pass of the file.[3]

Updating accounts.txt

Let's use a `with` statement to update the accounts.txt file to change account 300's name from `'White'` to `'Williams'` as described above:

```
In [1]: accounts = open('accounts.txt', 'r')

In [2]: temp_file = open('temp_file.txt', 'w')

In [3]: with accounts, temp_file:
   ...:     for record in accounts:
   ...:         account, name, balance = record.split()
   ...:         if account != '300':
   ...:             temp_file.write(record)
   ...:         else:
   ...:             new_record = ' '.join([account, 'Williams', balance])
   ...:             temp_file.write(new_record + '\n')
   ...:
```

For readability, we opened the file objects (snippets [1] and [2]), then specified their variable names in the first line of snippet [3]. This `with` statement manages two resource objects,

---

3. In the chapter, "Big Data: Hadoop, Spark, NoSQL and IoT," you'll see that database systems solve this "update in place" problem efficiently.

specified in a comma-separated list after with. The for statement unpacks each record into account, name and balance. If the account is not '300', we write record (which contains a newline) to temp_file. Otherwise, we assemble the new record containing 'Williams' in place of 'White' and write it to the file. After snippet [3], temp_file.txt contains:

```
100 Jones 24.98
200 Doe 345.67
300 Williams 0.00
400 Stone -42.16
500 Rich 224.62
```

os Module File-Processing Functions

At this point, we have the old accounts.txt file and the new temp_file.txt. To complete the update, let's delete the old accounts.txt file, then rename temp_file.txt as accounts.txt. The **os module**[4] provides functions for interacting with the operating system, including several that manipulate your system's files and directories. Now that we've created the temporary file, let's use the **remove function**[5] to delete the original file:

```
In [4]: import os

In [5]: os.remove('accounts.txt')
```

Next, let's use the **rename function** to rename the temporary file as 'accounts.txt':

```
In [6]: os.rename('temp_file.txt', 'accounts.txt')
```

Self Check

1 *(Fill-In)* The os module's _____ and _____ functions delete a file and specify a new name for a file, respectively.
Answer: remove, rename.

2 *(True/False)* Formatted data in a text file can be updated in place because records and their fields are fixed in size.
Answer: False. Such data cannot be modified without the risk of destroying other data in the file, because records and their fields can vary in size.

3 *(IPython Session)* In the accounts.txt file, update the last name 'Doe' to 'Smith'.
Answer:

```
In [1]: accounts = open('accounts.txt', 'r')

In [2]: temp_file = open('temp_file.txt', 'w')

In [3]: with accounts, temp_file:
   ...:     for record in accounts:
   ...:         account, name, balance = record.split()
   ...:         if name != 'Doe':
   ...:             temp_file.write(record)
   ...:         else:
   ...:             new_record = ' '.join([account, 'Smith', balance])
   ...:             temp_file.write(new_record + '\n')
   ...:
```

---

4. https://docs.python.org/3/library/os.html.
5. Use remove with caution—it does not warn you that you're *permanently* deleting the file.

```
In [4]: import os

In [5]: os.remove('accounts.txt')

In [6]: os.rename('temp_file.txt', 'accounts.txt')
```

9.5 Serialization with JSON

Many libraries we'll use to interact with cloud-based services such as Twitter, IBM Watson and others communicate with your applications via JSON objects. **JSON (JavaScript Object Notation)** is a text-based, human-and-computer-readable, data-interchange format used to represent objects (such as dictionaries, lists and more) as collections of name–value pairs. JSON can even represent objects of custom classes like those you'll build in the next chapter.

JSON has become the preferred data format for transmitting objects across platforms. This is especially true for invoking cloud-based web services, which are functions and methods that you call over the Internet. You'll become proficient at working with JSON data. In the "Big Data: Hadoop, Spark, NoSQL and IoT" chapter, we'll store JSON tweet objects that we obtain from Twitter in MongoDB, a popular NoSQL database.

JSON Data Format

JSON objects are similar to Python dictionaries. Each JSON object contains a comma-separated list of *property names* and *values*, in curly braces. For example, the following key–value pairs might represent a client record:

```
{"account": 100, "name": "Jones", "balance": 24.98}
```

JSON also supports arrays which, like Python lists, are comma-separated values in square brackets. For example, the following is an acceptable JSON array of numbers:

```
[100, 200, 300]
```

Values in JSON objects and arrays can be:

- strings in *double quotes* (like `"Jones"`),
- numbers (like 100 or 24.98),
- JSON Boolean values (represented as `true` or `false` in JSON),
- `null` (to represent no value, like `None` in Python),
- arrays (like [100, 200, 300]), and
- other JSON objects.

Python Standard Library Module `json`

The **`json` module** enables you to convert objects to JSON (JavaScript Object Notation) text format. This is known as **serializing** the data. Consider the following dictionary, which contains one key–value pair consisting of the key `'accounts'` with its associated value being a list of dictionaries representing two accounts. Each account dictionary contains three key–value pairs for the account number, name and balance:

```
In [1]: accounts_dict = {'accounts': [
   ...:     {'account': 100, 'name': 'Jones', 'balance': 24.98},
   ...:     {'account': 200, 'name': 'Doe', 'balance': 345.67}]}
```

Serializing an Object to JSON
Let's write that object in JSON format to a file:

```
In [2]: import json

In [3]: with open('accounts.json', 'w') as accounts:
   ...:     json.dump(accounts_dict, accounts)
   ...:
```

Snippet [3] opens the file accounts.json and uses the json module's **dump function** to serialize the dictionary accounts_dict into the file. The resulting file contains the following text, which we reformatted slightly for readability:

```
{"accounts":
    [{"account": 100, "name": "Jones", "balance": 24.98},
     {"account": 200, "name": "Doe", "balance": 345.67}]}
```

Note that JSON delimits strings with *double-quote characters*.

Deserializing the JSON Text
The json module's **load function** reads the entire JSON contents of its file object argument and converts the JSON into a Python object. This is known as **deserializing** the data. Let's reconstruct the original Python object from this JSON text:

```
In [4]: with open('accounts.json', 'r') as accounts:
   ...:     accounts_json = json.load(accounts)
   ...:
   ...:
```

We can now interact with the loaded object. For example, we can display the dictionary:

```
In [5]: accounts_json
Out[5]:
{'accounts': [{'account': 100, 'name': 'Jones', 'balance': 24.98},
    {'account': 200, 'name': 'Doe', 'balance': 345.67}]}
```

As you'd expect, you can access the dictionary's contents. Let's get the list of dictionaries associated with the 'accounts' key:

```
In [6]: accounts_json['accounts']
Out[6]:
[{'account': 100, 'name': 'Jones', 'balance': 24.98},
 {'account': 200, 'name': 'Doe', 'balance': 345.67}]
```

Now, let's get the individual account dictionaries:

```
In [7]: accounts_json['accounts'][0]
Out[7]: {'account': 100, 'name': 'Jones', 'balance': 24.98}

In [8]: accounts_json['accounts'][1]
Out[8]: {'account': 200, 'name': 'Doe', 'balance': 345.67}
```

Though we did not do so here, you can modify the dictionary as well. For example, you could add accounts to or remove accounts from the list, then write the dictionary back into the JSON file.

Displaying the JSON Text
The json module's **dumps function** (dumps is short for "dump string") returns a Python string representation of an object in JSON format. Using dumps with load, you can read

the JSON from the file and display it in a nicely indented format—sometimes called "pretty printing" the JSON. When the dumps function call includes the indent keyword argument, the string contains newline characters and indentation for pretty printing—you also can use indent with the dump function when writing to a file:

```
In [9]: with open('accounts.json', 'r') as accounts:
   ...:     print(json.dumps(json.load(accounts), indent=4))
   ...:
{
    "accounts": [
        {
            "account": 100,
            "name": "Jones",
            "balance": 24.98
        },
        {
            "account": 200,
            "name": "Doe",
            "balance": 345.67
        }
    ]
}
```

✓ Self Check

1 *(Fill-In)* Converting objects to JSON text format is known as _____, and reconstructing the original Python object from the JSON text is known as _____.
Answer: serialization, deserialization.

2 *(True/False)* JSON is both a human-readable *and* computer-readable format that makes it convenient to send and receive objects across the Internet.
Answer: True.

3 *(IPython Session)* Create a JSON file named grades.json and write into it the following dictionary:

```
grades_dict = {'gradebook':
    [{'student_id': 1, 'name': 'Red', 'grade': 'A'},
     {'student_id': 2, 'name': 'Green', 'grade': 'B'},
     {'student_id': 3, 'name': 'White', 'grade': 'A'}]}
```

Then, read the file and display its pretty-printed JSON.
Answer:

```
In [1]: import json

In [2]: grades_dict = {'gradebook':
   ...:     [{'student_id': 1, 'name': 'Red', 'grade': 'A'},
   ...:      {'student_id': 2, 'name': 'Green', 'grade': 'B'},
   ...:      {'student_id': 3, 'name': 'White', 'grade': 'A'}]}
   ...:

In [3]: with open('grades.json', 'w') as grades:
   ...:     json.dump(grades_dict, grades)
   ...:
```

```
In [4]: with open('grades.json', 'r') as grades:
   ...:     print(json.dumps(json.load(grades), indent=4))
   ...:
{
    "gradebook": [
        {
            "student_id": 1,
            "name": "Red",
            "grade": "A"
        },
        {
            "student_id": 2,
            "name": "Green",
            "grade": "B"
        },
        {
            "student_id": 3,
            "name": "White",
            "grade": "A"
        }
    ]
}
```

9.6 Focus on Security: `pickle` Serialization and Deserialization

The Python Standard Library's `pickle` module can serialize objects into in a Python-specific data format. **Caution: The Python documentation provides the following warnings about `pickle`:**

- "Pickle files can be hacked. If you receive a raw pickle file over the network, don't trust it! It could have malicious code in it, that would run arbitrary Python when you try to de-pickle it. However, if you are doing your own pickle writing and reading, you're safe (provided no one else has access to the pickle file, of course.)"[6]
- "Pickle is a protocol which allows the serialization of arbitrarily complex Python objects. As such, it is specific to Python and cannot be used to communicate with applications written in other languages. It is also insecure by default: deserializing pickle data coming from an untrusted source can execute arbitrary code, if the data was crafted by a skilled attacker."[7]

We do not recommend using `pickle`, but it's been used for many years, so you're likely to encounter it in **legacy code**—old code that's often no longer supported. For this reason, we've included an end-of-chapter `pickle` exercise, which explains how to use it.

9.7 Additional Notes Regarding Files

The following table summarizes the various file-open modes for text files, including the modes for reading and writing we've introduced. The *writing* and *appending* modes create the file if it does not exist. The *reading* modes raise a `FileNotFoundError` if the file does

---

6. https://wiki.python.org/moin/UsingPickle.
7. https://docs.python.org/3/tutorial/inputoutput.html#reading-and-writing-files.

not exist. Each text-file mode has a corresponding binary-file mode specified with b, as in 'rb' or 'wb+'. You'd use these modes, for example, if you were reading or writing binary files, such as images, audio, video, compressed ZIP files and many other popular custom file formats.

Mode	Description
'r'	Open a text file for reading. This is the default if you do not specify the file-open mode when you call open.
'w'	Open a text file for writing. Existing file contents are *deleted*.
'a'	Open a text file for appending at the end, creating the file if it does not exist. New data is written at the end of the file.
'r+'	Open a text file reading and writing.
'w+'	Open a text file reading and writing. Existing file contents are *deleted*.
'a+'	Open a text file reading and appending at the end. New data is written at the end of the file. If the file does not exist, it is created.

Other File Object Methods

Here are a few more useful file-object methods.

- For a text file, the **read** method returns a string containing the number of characters specified by the method's integer argument. For a binary file, the method returns the specified number of bytes. If no argument is specified, the method returns the entire contents of the file.
- The **readline** method returns one line of text as a string, including the newline character if there is one. This method returns an empty string when it encounters the end of the file.
- The **writelines** method receives a list of strings and writes its contents to a file.

The classes that Python uses to create file objects are defined in the Python Standard Library's **io** module (https://docs.python.org/3/library/io.html).

 Self Check

1 *(Fill-In)* The classes that Python uses to create file objects are defined in the Python Standard Library's _____ module.
Answer: io.

2 *(True/False)* The read method always returns the entire contents of a file.
Answer: False. You may specify an argument indicating the number of characters (or bytes for a binary file) to read from the file.

9.8 Handling Exceptions

Various types of exceptions can occur when you work with files, including:

- A **FileNotFoundError** occurs if you attempt to open a non-existent file for reading with the 'r' or 'r+' modes.

- A **PermissionsError** occurs if you attempt an operation for which you do not have permission. This might occur if you try to open a file that your account is not allowed to access or create a file in a folder where your account does not have permission to write, such as where your computer's operating system is stored.
- A **ValueError** (with the error message `'I/O operation on closed file.'`) occurs when you attempt to write to a file that has already been closed.

9.8.1 Division by Zero and Invalid Input

Let's revisit two exceptions that you saw earlier in the book.

Division By Zero
Recall that attempting to divide by 0 results in a ZeroDivisionError:

```
In [1]: 10 / 0
-------------------------------------------------------------------------
ZeroDivisionError                        Traceback (most recent call last)
<ipython-input-1-a243dfbf119d> in <module>()
----> 1 10 / 0

ZeroDivisionError: division by zero

In [2]:
```

In this case, the interpreter is said to **raise an exception** of type ZeroDivisionError. When an exception is raised in IPython, it:

- terminates the snippet,
- displays the exception's traceback, then
- shows the next In [] prompt so you can input the next snippet.

If an exception occurs in a script, IPython terminates the script and displays the exception's traceback.

Invalid Input
Recall that the int function raises a ValueError if you attempt to convert to an integer a string (like 'hello') that does not represent a number:

```
In [2]: value = int(input('Enter an integer: '))
Enter an integer: hello
-------------------------------------------------------------------------
ValueError                               Traceback (most recent call last)
<ipython-input-2-b521605464d6> in <module>()
----> 1 value = int(input('Enter an integer: '))

ValueError: invalid literal for int() with base 10: 'hello'

In [3]:
```

9.8.2 try Statements

Now let's see how to *handle* these exceptions so that you can enable code to continue processing. Consider the following script and sample execution. Its loop attempts to read two

9.8 Handling Exceptions

integers from the user, then display the first number divided by the second. The script uses exception handling to catch and handle (i.e., deal with) any ZeroDivisionErrors and ValueErrors that arise—in this case, allowing the user to re-enter the input.

```
1   # dividebyzero.py
2   """Simple exception handling example."""
3
4   while True:
5       # attempt to convert and divide values
6       try:
7           number1 = int(input('Enter numerator: '))
8           number2 = int(input('Enter denominator: '))
9           result = number1 / number2
10      except ValueError:  # tried to convert non-numeric value to int
11          print('You must enter two integers\n')
12      except ZeroDivisionError:  # denominator was 0
13          print('Attempted to divide by zero\n')
14      else:  # executes only if no exceptions occur
15          print(f'{number1:.3f} / {number2:.3f} = {result:.3f}')
16          break  # terminate the loop
```

```
Enter numerator: 100
Enter denominator: 0
Attempted to divide by zero

Enter numerator: 100
Enter denominator: hello
You must enter two integers

Enter numerator: 100
Enter denominator: 7
100.000 / 7.000 = 14.286
```

try Clause
Python uses **try statements** (like lines 6–16) to enable exception handling. The try statement's **try clause** (lines 6–9) begins with keyword try, followed by a colon (:) and a suite of statements that *might* raise exceptions.

except Clause
A try clause may be followed by one or more **except clauses** (lines 10–11 and 12–13) that immediately follow the try clause's suite. These also are known as *exception handlers*. Each except clause specifies the type of exception it handles. In this example, each exception handler just displays a message indicating the problem that occurred.

else Clause
After the last except clause, an optional **else clause** (lines 14–16) specifies code that should execute only if the code in the try suite did not raise exceptions. If no exceptions occur in this example's try suite, line 15 displays the division result and line 16 terminates the loop.

Flow of Control for a ZeroDivisionError

Now let's consider this example's flow of control, based on the first three lines of the sample output:

- First, the user enters 100 for the numerator in response to line 7 in the try suite.
- Next, the user enters 0 for the denominator in response to line 8 in the try suite.
- At this point, we have two integer values, so line 9 attempts to divide 100 by 0, causing Python to raise a ZeroDivisionError. The point in the program at which an exception occurs is often referred to as the **raise point**.

When an exception occurs in a try suite, it terminates immediately. If there are any except handlers following the try suite, program control transfers to the first one. If there are no except handlers, a process called *stack unwinding* occurs, which we discuss later in the chapter.

In this example, there *are* except handlers, so the interpreter searches for the *first one* that matches the type of the raised exception:

- The except clause at lines 10–11 handles ValueErrors. This does not match the type ZeroDivisionError, so that except clause's suite does not execute and program control transfers to the next except handler.
- The except clause at lines 12–13 handles ZeroDivisionErrors. This *is* a match, so that except clause's suite executes, displaying "Attempted to divide by zero".

When an except clause successfully handles the exception, program execution resumes with the finally clause (if there is one), then with the next statement after the try statement. In this example, we reach the end of the loop, so execution resumes with the next loop iteration. Note that after an exception is handled, program control does *not* return to the raise point. Rather, control resumes after the try statement. We'll discuss the finally clause shortly.

Flow of Control for a ValueError

Now let's consider the flow of control, based on the next three lines of the sample output:

- First, the user enters 100 for the numerator in response to line 7 in the try suite.
- Next, the user enters hello for the denominator in response to line 8 in the try suite. The input is not a valid integer, so the int function raises a ValueError.

The exception terminates the try suite and program control transfers to the first except handler. In this case, the except clause at lines 10–11 is a match, so its suite executes, displaying "You must enter two integers". Then, program execution resumes with the next statement after the try statement. Again, that's the end of the loop, so execution resumes with the next loop iteration.

Flow of Control for a Successful Division

Now let's consider the flow of control, based on the last three lines of the sample output:

- First, the user enters 100 for the numerator in response to line 7 in the try suite.
- Next, the user enters 7 for the denominator in response to line 8 in the try suite.
- At this point, we have two valid integer values and the denominator is not 0, so line 9 successfully divides 100 by 7.

When no exceptions occur in the try suite, program execution resumes with the else clause (if there is one); otherwise, program execution resumes with the next statement after the try statement. In this example's else clause, we display the division result, then terminate the loop, and the program terminates.

Self Check

1 *(Fill-In)* The statement that raises an exception is sometimes called the _____ of the exception.
Answer: raise point.

2 *(True/False)* In Python, it's possible to return to the raise point of an exception via keyword return.
Answer: False. Program control continues from the first statement after the try statement in which the exception was handled.

3 *(IPython Session)* Before executing the IPython session, determine what the following function displays if you call it with the value 10.7 then the value 'Python'?

```
def try_it(value)
    try:
        x = int(value)
    except ValueError:
        print(f'{value} could not be converted to an integer')
    else:
        print(f'int({value}) is {x}')
```

Answer:

```
In [1]: def try_it(value):
   ...:     try:
   ...:         x = int(value)
   ...:     except ValueError:
   ...:         print(f'{value} could not be converted to an integer')
   ...:     else:
   ...:         print(f'int({value}) is {x}')
   ...:

In [2]: try_it(10.7)
int(10.7) is 10

In [3]: try_it('Python')
Python could not be converted to an integer
```

9.8.3 Catching Multiple Exceptions in One except Clause

It's relatively common for a try clause to be followed by several except clauses to handle various types of exceptions. If several except suites are identical, you can catch those exception types by specifying them as a tuple in a *single* except handler, as in:

> except (*type1*, *type2*, ...) as *variable_name*:

The as clause is optional. Typically, programs do not need to reference the caught exception object directly. If you do, you can use the variable in the as clause to reference the exception object in the except suite.

9.8.4 What Exceptions Does a Function or Method Raise?

Exceptions may surface via statements in a `try` suite, via functions or methods called directly or indirectly from a `try` suite, or via the Python interpreter as it executes the code (for example, `ZeroDivisionErrors`).

Before using any function or method, read its online API documentation, which specifies what exceptions are thrown (if any) by the function or method and indicates reasons why such exceptions may occur. Next, read the online API documentation for each exception type to see potential reasons why such an exception occurs.

9.8.5 What Code Should Be Placed in a `try` Suite?

Place in a `try` suite a significant logical section of a program in which several statements can raise exceptions, rather than wrapping a separate `try` statement around every statement that raises an exception. However, for proper exception-handling granularity, each `try` statement should enclose a section of code small enough that, when an exception occurs, the specific context is known and the `except` handlers can process the exception properly. If many statements in a `try` suite raise the same exception types, multiple `try` statements may be required to determine each exception's context.

9.9 `finally` Clause

Operating systems typically can prevent more than one program from manipulating a file at once. When a program finishes processing a file, the program should close it to release the resource. This enables other programs to use the file (if they're allowed to access it). Closing the file helps prevent a **resource leak** in which the file resource is not available to other programs because a program using the file never closes it.

The `finally` Clause of the `try` Statement

A `try` statement may have a `finally` clause as its last clause after any `except` clauses or `else` clause. The `finally` clause is guaranteed to execute, *regardless* of whether its `try` suite executes successfully or an exception occurs.[8] In other languages that have `finally`, this makes the `finally` suite an ideal location to place resource-deallocation code for resources acquired in the corresponding `try` suite. In Python, we prefer the `with` statement for this purpose and place other kinds of "clean up" code in the `finally` suite.

Example

The following IPython session demonstrates that the `finally` clause always executes, regardless of whether an exception occurs in the corresponding `try` suite. First, let's consider a `try` statement in which no exceptions occur in the `try` suite:

---

8. The only reason a `finally` suite will not execute if program control enters the corresponding `try` suite is if the application terminates first, for example by calling the `sys` module's `exit` function. In this case, the operating system would "clean up" any resources that the program did not release.

9.9 finally Clause

```
In [1]: try:
   ...:      print('try suite with no exceptions raised')
   ...: except:
   ...:      print('this will not execute')
   ...: else:
   ...:      print('else executes because no exceptions in the try suite')
   ...: finally:
   ...:      print('finally always executes')
   ...:
try suite with no exceptions raised
else executes because no exceptions in the try suite
finally always executes

In [2]:
```

The preceding try suite displays a message but does not raise any exceptions. When program control successfully reaches the end of the try suite, the except clause is skipped, the else clause executes and the finally clause displays a message showing that it always executes. When the finally clause terminates, program control continues with the next statement after the try statement. In an IPython session, the next In [] prompt appears.

Now let's consider a try statement in which an exception occurs in the try suite:

```
In [2]: try:
   ...:      print('try suite that raises an exception')
   ...:      int('hello')
   ...:      print('this will not execute')
   ...: except ValueError:
   ...:      print('a ValueError occurred')
   ...: else:
   ...:      print('else will not execute because an exception occurred')
   ...: finally:
   ...:      print('finally always executes')
   ...:
try suite that raises an exception
a ValueError occurred
finally always executes

In [3]:
```

This try suite begins by displaying a message. The second statement attempts to convert the string 'hello' to an integer, which causes the int function to raise a ValueError. The try suite immediately terminates, skipping its last print statement. The except clause catches the ValueError exception and displays a message. The else clause does not execute because an exception occurred. Then, the finally clause displays a message showing that it always executes. When the finally clause terminates, program control continues with the next statement after the try statement. In an IPython session, the next In [] prompt appears.

Combining with Statements and try…except Statements
Most resources that require explicit release, such as files, network connections and database connections, have potential exceptions associated with processing those resources. For example, a program that processes a file might raise IOErrors. For this reason, *robust* file-processing code normally appears in a try suite containing a with statement to guarantee

that the resource gets released. The code is in a try suite, so you can catch in except handlers any exceptions that occur and you do not need a finally clause because the with statement handles resource deallocation.

To demonstrate this, first let's assume you're asking the user to supply the name of a file and they provide that name incorrectly, such as gradez.txt rather than the file we created earlier grades.txt. In this case, the open call raises a FileNotFoundError by attempting to open a non-existent file:

```
In [3]: open('gradez.txt')
-------------------------------------------------------------------------
FileNotFoundError                         Traceback (most recent call last)
<ipython-input-3-b7f41b2d5969> in <module>()
----> 1 open('gradez.txt')

FileNotFoundError: [Errno 2] No such file or directory: 'gradez.txt'
```

To catch exceptions like FileNotFoundError that occur when you try to open a file for reading, wrap the with statement in a try suite, as in:

```
In [4]: try:
   ...:     with open('gradez.txt', 'r') as accounts:
   ...:         print(f'{"ID":<3}{"Name":<7}{"Grade"}')
   ...:         for record in accounts:
   ...:             student_id, name, grade = record.split()
   ...:             print(f'{student_id:<3}{name:<7}{grade}')
   ...: except FileNotFoundError:
   ...:     print('The file name you specified does not exist')
   ...:
The file name you specified does not exist
```

✓ Self Check

1. *(True/False)* If a finally clause appears in a function, that finally clause is guaranteed to execute when the function executes, regardless of whether the function raises an exception.
Answer: False. The finally clause will execute only if program control enters the corresponding try suite.

2. *(Fill-In)* Closing a file helps prevent a(n) _____ in which the file resource is not available to other programs because a program using the file never closes it.
Answer: resource leak.

3. *(IPython Session)* Before executing the IPython session, determine what the following function displays if you call it with the value 10.7, then the value 'Python'?

```
def try_it(value)
    try:
        x = int(value)
    except ValueError:
        print(f'{value} could not be converted to an integer')
    else:
        print(f'int({value}) is {x}')
    finally:
        print('finally executed')
```

Answer:

```
In [1]: def try_it(value):
   ...:     try:
   ...:         x = int(value)
   ...:     except ValueError:
   ...:         print(f'{value} could not be converted to an integer')
   ...:     else:
   ...:         print(f'int({value}) is {int(value)}')
   ...:     finally:
   ...:         print('finally executed')
   ...:

In [2]: try_it(10.7)
int(10.7) is 10
finally executed

In [3]: try_it('Python')
Python could not be converted to an integer
finally executed
```

9.10 Explicitly Raising an Exception

You've seen various exceptions raised by your Python code. Sometimes you might need to write functions that raise exceptions to inform callers of errors that occur. The `raise` statement explicitly raises an exception. The simplest form of the `raise` statement is

> raise *ExceptionClassName*

The `raise` statement creates an object of the specified exception class. Optionally, the exception class name may be followed by parentheses containing arguments to initialize the exception object—typically to provide a custom error message string. Code that raises an exception first should release any resources acquired before the exception occurred. In the next section, we'll show an example of raising an exception.

In most cases, when you need to raise an exception, it's recommended that you use one of Python's many built-in exception types[9] listed at:

> https://docs.python.org/3/library/exceptions.html

 Self Check

1 *(Fill-In)* Use the _____ statement to indicate that a problem occurred at execution time.
Answer: `raise`.

9.11 (Optional) Stack Unwinding and Tracebacks

Each exception object stores information indicating the precise series of function calls that led to the exception. This is helpful when debugging your code. Consider the following function definitions—`function1` calls `function2` and `function2` raises an Exception:

---

9. You may be tempted to create custom exception classes that are specific to your application. We'll say more about custom exceptions in the next chapter.

```
In [1]: def function1():
   ...:     function2()
   ...:
   
In [2]: def function2():
   ...:     raise Exception('An exception occurred')
   ...:
```

Calling `function1` results in the following traceback. For emphasis, we placed in bold the parts of the traceback indicating the lines of code that led to the exception:

```
In [3]: function1()
---------------------------------------------------------------------------
Exception                                 Traceback (most recent call last)
<ipython-input-3-c0b3cafe2087> in <module>()
----> 1 function1()

<ipython-input-1-a9f4faeeeb0c> in function1()
      1 def function1():
----> 2     function2()
      3

<ipython-input-2-c65e19d6b45b> in function2()
      1 def function2():
----> 2     raise Exception('An exception occurred')

Exception: An exception occurred
```

Traceback Details

The traceback shows the type of exception that occurred (`Exception`) followed by the complete function call stack that led to the raise point. The stack's bottom function call is listed *first* and the top is *last*, so the interpreter displays the following text as a reminder:

```
Traceback (most recent call last)
```

In this traceback, the following text indicates the bottom of the function-call stack—the `function1` call in snippet [3] (indicated by `ipython-input-3`):

```
<ipython-input-3-c0b3cafe2087> in <module>()
----> 1 function1()
```

Next, we see that `function1` called `function2` from line 2 in snippet [1]:

```
<ipython-input-1-a9f4faeeeb0c> in function1()
      1 def function1():
----> 2     function2()
      3
```

Finally, we see the *raise point*—in this case, line 2 in snippet [2] raised the exception:

```
<ipython-input-2-c65e19d6b45b> in function2()
      1 def function2():
----> 2     raise Exception('An exception occurred')
```

Stack Unwinding

In our previous exception-handling examples, the raise point occurred in a `try` suite, and the exception was handled in one of the `try` statement's corresponding `except` handlers.

When an exception is *not* caught in a given function, **stack unwinding** occurs. Let's consider stack unwinding in the context of this example:

- In function2, the raise statement raises an exception. This is not in a try suite, so function2 terminates, its stack frame is removed from the function-call stack, and control returns to the statement in function1 that called function2.
- In function1, the statement that called function2 is not in a try suite, so function1 terminates, its stack frame is removed from the function-call stack, and control returns to the statement that called function1—snippet [3] in the IPython session.
- The call in snippet [3] call is not in a try suite, so that function call terminates. Because the exception was not caught (known as an **uncaught exception**), IPython displays the traceback, then awaits your next input. If this occurred in a typical script, the script would terminate.[10]

Tip for Reading Tracebacks
You'll often call functions and methods that belong to libraries of code you did not write. Sometimes those functions and methods raise exceptions. When reading a traceback, start from the end of the traceback and read the error message first. Then, read upward through the traceback, looking for the first line that indicates code you wrote in your program. Typically, this is the location in your code that led to the exception.

Exceptions in finally Suites
Raising an exception in a finally suite can lead to subtle, hard-to-find problems. If an exception occurs and is not processed by the time the finally suite executes, stack unwinding occurs. If the finally suite raises a *new* exception that the suite does not catch, the first exception is *lost*, and the *new* exception is passed to the next enclosing try statement. For this reason, a finally suite should always enclose in a try statement any code that may raise an exception, so that the exceptions will be processed within that suite.

Self Check

1 *(Fill-In)* An uncaught exception in a function causes _____. The function's stack frame is removed from the function-call stack.
Answer: stack unwinding.

2 *(True/False)* Exceptions always are handled in the function that raises the exception.
Answer: False. Although it is possible to handle an exception in the function that raises it, normally an exception is handled by a calling function on the function-call stack.

3 *(True/False)* Exceptions can be raised only by code in try statements.
Answer: False. Exceptions can be raised by any code, regardless of whether the code is wrapped in a try statement.

---

10. In more advanced applications that use threads, an uncaught exception terminates only the thread in which the exception occurs, not necessarily the entire application.

9.12 Intro to Data Science: Working with CSV Files

Throughout this book, you'll work with many datasets as you learn data-science concepts. CSV (**comma-separated values**) is a particularly popular file format. In this section, we'll demonstrate CSV file processing with a Python Standard Library module and pandas.

9.12.1 Python Standard Library Module `csv`

The **csv module**[11] provides functions for working with CSV files. Many other Python libraries also have built-in CSV support.

Writing to a CSV File

Let's create an `accounts.csv` file using CSV format. The `csv` module's documentation recommends opening CSV files with the additional keyword argument `newline=''` to ensure that newlines are processed properly:

```
In [1]: import csv

In [2]: with open('accounts.csv', mode='w', newline='') as accounts:
   ...:     writer = csv.writer(accounts)
   ...:     writer.writerow([100, 'Jones', 24.98])
   ...:     writer.writerow([200, 'Doe', 345.67])
   ...:     writer.writerow([300, 'White', 0.00])
   ...:     writer.writerow([400, 'Stone', -42.16])
   ...:     writer.writerow([500, 'Rich', 224.62])
   ...:
```

The **.csv file extension** indicates a CSV-format file. The `csv` module's **writer function** returns an object that writes CSV data to the specified file object. Each call to the writer's **writerow method** receives an iterable to store in the file. Here we're using lists. By default, `writerow` delimits values with commas, but you can specify custom delimiters.[12] After the preceding snippet, `accounts.csv` contains:

```
100,Jones,24.98
200,Doe,345.67
300,White,0.00
400,Stone,-42.16
500,Rich,224.62
```

CSV files generally do not contain spaces after commas, but some people use them to enhance readability. The `writerow` calls above can be replaced with one **writerows** call that outputs a comma-separated list of iterables representing the records.

If you write data that contains commas within a given string, `writerow` encloses that string in double quotes. For example, consider the following Python list:

```
[100, 'Jones, Sue', 24.98]
```

The single-quoted string `'Jones, Sue'` contains a comma separating the last name and first name. In this case, `writerow` would output the record as

```
100,"Jones, Sue",24.98
```

The quotes around `"Jones, Sue"` indicate that this is a *single* value. Programs reading this from a CSV file would break the record into *three* pieces—100, `'Jones, Sue'` and 24.98.

---

11. https://docs.python.org/3/library/csv.html.
12. https://docs.python.org/3/library/csv.html#csv-fmt-params.

Reading from a CSV File

Now let's read the CSV data from the file. The following snippet reads records from the file accounts.csv and displays the contents of each record, producing the same output we showed earlier:

```
In [3]: with open('accounts.csv', 'r', newline='') as accounts:
   ...:     print(f'{"Account":<10}{"Name":<10}{"Balance":>10}')
   ...:     reader = csv.reader(accounts)
   ...:     for record in reader:
   ...:         account, name, balance = record
   ...:         print(f'{account:<10}{name:<10}{balance:>10}')
   ...:
Account   Name         Balance
100       Jones          24.98
200       Doe           345.67
300       White           0.0
400       Stone         -42.16
500       Rich          224.62
```

The csv module's **reader function** returns an object that reads CSV-format data from the specified file object. Just as you can iterate through a file object, you can iterate through the reader object one record of comma-delimited values at a time. The preceding for statement returns each record as a list of values, which we unpack into the variables account, name and balance, then display.

Caution: Commas in CSV Data Fields

Be careful when working with strings containing embedded commas, such as the name 'Jones, Sue'. If you accidentally enter this as the two strings 'Jones' and 'Sue', then writerow would, of course, create a CSV record with *four* fields, not *three*. Programs that read CSV files typically expect every record to have the *same* number of fields; otherwise, problems occur. For example, consider the following two lists:

```
[100, 'Jones', 'Sue', 24.98]
[200, 'Doe'   , 345.67]
```

The first list contains *four* values and the second contains only *three*. If these two records were written into the CSV file, then read into a program using the previous snippet, the following statement would fail when we attempt to unpack the four-field record into only three variables:

```
account, name, balance = record
```

Caution: Missing Commas and Extra Commas in CSV Files

Be careful when preparing and processing CSV files. For example, suppose your file is composed of records, each with *four* comma-separated int values, such as:

```
100,85,77,9
```

If you accidentally omit one of these commas, as in:

```
100,8577,9
```

then the record has only *three* fields, one with the invalid value 8577.

If you put two adjacent commas where only one is expected, as in:

```
100,85,,77,9
```

then you have *five* fields rather than *four*, and one of the fields erroneously would be *empty*. Each of these comma-related errors could confuse programs trying to process the record.

Self Check

1. *(Fill-In)* The csv module provides capabilities for writing and reading files in _____ (CSV) format.
Answer: comma-separated values.

2. *(True/False)* The csv module's reader function returns an object that reads from the specified file object CSV-format data.
Answer: True.

3. *(IPython Session)* Create a text file named grades.csv and write to it the following three records consisting of student IDs, last names and letter grades:

    ```
    1,Red,A
    2,Green,B
    3,White,A
    ```

    Then, read the file grades.csv and display it in columns with the column heads 'ID', 'Name' and 'Grade'.
Answer:

```
In [1]: import csv

In [2]: with open('grades.csv', mode='w', newline='') as grades:
   ...:     writer = csv.writer(grades)
   ...:     writer.writerow([1, 'Red', 'A'])
   ...:     writer.writerow([2, 'Green', 'B'])
   ...:     writer.writerow([3, 'White', 'A'])
   ...:

In [3]: with open('grades.csv', 'r', newline='') as grades:
   ...:     print(f'{"ID":<4}{"Name":<7}{"Grade"}')
   ...:     reader = csv.reader(grades)
   ...:     for record in reader:
   ...:         student_id, name, grade = record
   ...:         print(f'{student_id:<4}{name:<7}{grade}')
   ...:
ID  Name   Grade
1   Red    A
2   Green  B
3   White  A
```

9.12.2 Reading CSV Files into Pandas DataFrames

In the Intro to Data Science sections in the previous two chapters, we introduced many Pandas fundamentals. Here, we demonstrate pandas' ability to load files in CSV format, then perform some basic data-analysis tasks.

Datasets
In the data-science case studies, we'll use various free and open datasets to demonstrate machine learning and natural language processing concepts. There's an enormous variety of free datasets available online. The popular **Rdatasets repository** provides links to over

1100 free datasets in comma-separated values (CSV) format. These were originally provided with the R programming language for people learning about and developing statistical software, though they are not specific to R. They are now available on GitHub at:

> https://vincentarelbundock.github.io/Rdatasets/datasets.html

This repository is so popular that there's a **pydataset module** specifically for accessing Rdatasets. For instructions on installing pydataset and accessing datasets with it, see:

> https://github.com/iamaziz/PyDataset

Another large source of datasets is:

> https://github.com/awesomedata/awesome-public-datasets

A commonly used machine-learning dataset for beginners is the **Titanic disaster dataset**, which lists all the passengers and whether they survived when the ship *Titanic* struck an iceberg and sank April 14–15, 1912. We'll use it here to show how to load a dataset, view some of its data and display some descriptive statistics. We'll dig deeper into a variety of popular datasets in the data-science chapters later in the book.

Working with Locally Stored CSV Files

You can load a CSV dataset into a DataFrame with the pandas function **read_csv**. The following loads and displays the CSV file accounts.csv that you created earlier in this chapter:

```
In [1]: import pandas as pd

In [2]: df = pd.read_csv('accounts.csv',
   ...:                  names=['account', 'name', 'balance'])
   ...:

In [3]: df
Out[3]:
   account   name  balance
0      100  Jones    24.98
1      200    Doe   345.67
2      300  White     0.00
3      400  Stone   -42.16
4      500   Rich   224.62
```

The names keyword argument specifies the DataFrame's column names. If you do not supply this argument, read_csv assumes that the CSV file's first row is a comma-delimited list of column names.

To save a DataFrame to a file using CSV format, call DataFrame method **to_csv**:

```
In [4]: df.to_csv('accounts_from_dataframe.csv', index=False)
```

The index=False keyword argument indicates that the row names (0–4 at the left of the DataFrame's output in snippet [3]) are not written to the file. The resulting file contains the column names as the first row:

```
account,name,balance
100,Jones,24.98
200,Doe,345.67
300,White,0.0
400,Stone,-42.16
500,Rich,224.62
```

9.12.3 Reading the Titanic Disaster Dataset

The Titanic disaster dataset is one of the most popular machine-learning datasets. In the "Machine Learning" chapter, an exercise asks you to use this dataset to "predict" whether passengers would live or die, based *only* on attributes like gender, age and passenger class. The dataset is available in many formats, including CSV.

Loading the Titanic Dataset via a URL

If you have a URL representing a CSV dataset, you can load it into a DataFrame with read_csv. Let's load the Titanic Disaster dataset directly from GitHub:

```
In [1]: import pandas as pd

In [2]: titanic = pd.read_csv('https://vincentarelbundock.github.io/' +
   ...:     'Rdatasets/csv/carData/TitanicSurvival.csv')
   ...:
```

Viewing Some of the Rows in the Titanic Dataset

This dataset contains over 1300 rows, each representing one passenger. According to Wikipedia, there were approximately 1317 passengers and 815 of them died.[13] For large datasets, displaying the DataFrame shows only the first 30 rows, followed by "..." and the last 30 rows. To save space, let's view the first five and last five rows with DataFrame methods **head** and **tail**. Both methods return five rows by default, but you can specify the number of rows to display as an argument:

```
In [3]: pd.set_option('precision', 2)   # format for floating-point values

In [4]: titanic.head()
Out[4]:
                      Unnamed: 0  survived     sex    age  passengerClass
0       Allen, Miss. Elisabeth Walton       yes  female  29.00             1st
1      Allison, Master. Hudson Trevor       yes    male   0.92             1st
2        Allison, Miss. Helen Loraine        no  female   2.00             1st
3      Allison, Mr. Hudson Joshua Crei       no    male  30.00             1st
4      Allison, Mrs. Hudson J C (Bessi       no  female  25.00             1st

In [5]: titanic.tail()
Out[5]:
                      Unnamed: 0  survived     sex    age  passengerClass
1304            Zabour, Miss. Hileni         no  female  14.50             3rd
1305           Zabour, Miss. Thamine         no  female    NaN             3rd
1306      Zakarian, Mr. Mapriededer         no    male  26.50             3rd
1307            Zakarian, Mr. Ortin         no    male  27.00             3rd
1308            Zimmerman, Mr. Leo         no    male  29.00             3rd
```

Note that pandas adjusts each column's width, based on the widest value in the column or based on the column name, whichever is wider. Also, note the value in the age column of row 1305 is NaN (not a number), indicating a missing value in the dataset.

Customizing the Column Names

The first column in this dataset has a strange name ('Unnamed: 0'). We can clean that up by setting the column names. Let's change 'Unnamed: 0' to 'name' and let's shorten 'passengerClass' to 'class':

---

13. https://en.wikipedia.org/wiki/Passengers_of_the_RMS_Titanic.

```
In [6]: titanic.columns = ['name', 'survived', 'sex', 'age', 'class']

In [7]: titanic.head()
Out[7]:
                          name survived     sex    age class
0     Allen, Miss. Elisabeth Walton      yes  female  29.00   1st
1     Allison, Master. Hudson Trevor     yes    male   0.92   1st
2      Allison, Miss. Helen Loraine       no  female   2.00   1st
3     Allison, Mr. Hudson Joshua Crei     no    male  30.00   1st
4     Allison, Mrs. Hudson J C (Bessi     no  female  25.00   1st
```

9.12.4 Simple Data Analysis with the Titanic Disaster Dataset

Now, you can use pandas to perform some simple analysis. For example, let's look at some descriptive statistics. When you call `describe` on a `DataFrame` containing both numeric and non-numeric columns, `describe` calculates these statistics *only for the numeric columns*—in this case, just the age column:

```
In [8]: titanic.describe()
Out[8]:
               age
count    1046.00
mean       29.88
std        14.41
min         0.17
25%        21.00
50%        28.00
75%        39.00
max        80.00
```

Note the discrepancy in the `count` (1046) vs. the dataset's number of rows (1309—the last row's index was 1308 when we called `tail`). Only 1046 (the `count` above) of the records contained an age value. The rest were *missing* and marked as NaN, as in row 1305. When performing calculations, Pandas *ignores missing data (NaN) by default*. For the 1046 people with valid ages, the average (`mean`) age was 29.88 years old. The youngest passenger (`min`) was just over two months old (0.17 * 12 is 2.04), and the oldest (`max`) was 80. The median age was 28 (indicated by the 50% quartile). The 25% quartile is the median age in the first half of the passengers (sorted by age), and the 75% quartile is the median of the second half of passengers.

Let's say you want to determine some statistics about people who survived. We can compare the survived column to `'yes'` to get a new `Series` containing True/False values, then use `describe` to summarize the results:

```
In [9]: (titanic.survived == 'yes').describe()
Out[9]:
count      1309
unique        2
top       False
freq        809
Name: survived, dtype: object
```

For non-numeric data, `describe` displays different descriptive statistics:

- count is the total number of items in the result.

- **unique** is the number of unique values (2) in the result—True (survived) and False (died).
- **top** is the most frequently occurring value in the result.
- **freq** is the number of occurrences of the **top** value.

9.12.5 Passenger Age Histogram

Visualization is a nice way to get to know your data. Pandas has many built-in visualization capabilities that are implemented with Matplotlib. To use them, first enable Matplotlib support in IPython:

```
In [10]: %matplotlib
```

A histogram visualizes the distribution of numerical data over a range of values. A DataFrame's **hist** method automatically analyzes each numerical column's data and produces a corresponding histogram. To view histograms of each numerical data column, call hist on your DataFrame:

```
In [11]: histogram = titanic.hist()
```

The Titanic dataset contains only one numerical data column, so the diagram below shows one histogram for the age distribution. For datasets with multiple numerical columns (as we'll see in the exercises), hist creates a separate histogram for each numerical column.

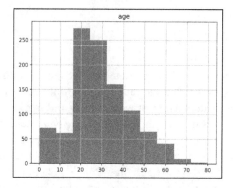

✓ Self Check

1. *(Fill-In)* Pandas function _____ loads a CSV dataset from a URL or the local file system into a DataFrame.
Answer: read_csv.

2. *(IPython Session)* Load the grades.csv file you created in the Section 9.12.1's Self Check into a DataFrame, then display it.
Answer:

```
In [12]: pd.read_csv('grades.csv', names=['ID', 'Name', 'Grade'])
Out[12]:
   ID   Name  Grade
0   1    Red      A
1   2  Green      B
2   3  White      A
```

9.13 Wrap-Up

In this chapter, we introduced text-file processing and exception handling. Files are used to store data persistently. We discussed file objects and mentioned that Python views a file as a sequence of characters or bytes. We also mentioned the standard file objects that are automatically created for you when a Python program begins executing.

We showed how to create, read, write and update text files. We considered several popular file formats—plain text, JSON (JavaScript Object Notation) and CSV (comma-separated values). We used the built-in open function and the with statement to open a file, write to or read from the file and automatically close the file to prevent resource leaks when the with statement terminates. We used the Python Standard Library's json module to serialize objects into JSON format and store them in a file, load JSON objects from a file, deserialize them into Python objects and pretty-print a JSON object for readability.

We discussed how exceptions indicate execution-time problems and listed the various exceptions you've already seen. We showed how to deal with exceptions by wrapping code in try statements that provide except clauses to handle specific types of exceptions that may occur in the try suite, making your programs more robust and fault-tolerant.

We discussed the try statement's finally clause for executing code if program flow entered the corresponding try suite. You can use either the with statement or a try statement's finally clause for this purpose—we prefer the with statement.

In the Intro to Data Science section, we used both the Python Standard Library's csv module and capabilities of the pandas library to load, manipulate and store CSV data. Finally, we loaded the Titanic disaster dataset into a pandas DataFrame, changed some column names for readability, displayed the head and tail of the dataset, and performed simple analysis of the data. In the next chapter, we'll discuss Python's object-oriented programming capabilities.

Exercises

9.1 *(Class Average: Writing Grades to a Plain Text File)* Figure 3.2 presented a class-average script in which you could enter any number of grades followed by a sentinel value, then calculate the class average. Another approach would be to read the grades from a file. In an IPython session, write code that enables you to store any number of grades into a grades.txt plain text file.

9.2 *(Class Average: Reading Grades from a Plain Text File)* In an IPython session, write code that reads the grades from the grades.txt file you created in the previous exercise. Display the individual grades and their total, count and average.

9.3 *(Class Average: Writing Student Records to a CSV File)* An instructor teaches a class in which each student takes three exams. The instructor would like to store this information in a file named grades.csv for later use. Write code that enables an instructor to enter each student's first name and last name as strings and the student's three exam grades as integers. Use the csv module to write each record into the grades.csv file. Each record should be a single line of text in the following CSV format:

firstname, lastname, exam1grade, exam2grade, exam3grade

9.4 *(Class Average: Reading Student Records from a CSV File)* Use the csv module to read the grades.csv file from the previous exercise. Display the data in tabular format.

9.5 *(Class Average: Creating a Grade Report from a CSV File)* Modify your solution to the preceding exercise to create a grade report that displays each student's average to the right of that student's row and the class average for each exam below that exam's column.

9.6 *(Class Average: Writing a Gradebook Dictionary to a JSON File)* Reimplement Exercise 9.3 using the `json` module to write the student information to the file in JSON format. For this exercise, create a dictionary of student data in the following format:

gradebook_dict = {'students': [*student1dictionary*, *student2dictionary*, ...]}

Each dictionary in the list represents one student and contains the keys `'first_name'`, `'last_name'`, `'exam1'`, `'exam2'` and `'exam3'`, which map to the values representing each student's first name (string), last name (string) and three exam scores (integers). Output the gradebook_dict in JSON format to the file grades.json.

9.7 *(Class Average: Reading a Gradebook Dictionary from a JSON File)* Reimplement Exercise 9.4 using the `json` module to read the grades.json file created in the previous exercise. Display the data in tabular format, including an additional column showing each student's average to the right of that student's three exam grades and an additional row showing the class average on each exam below that exam's column.

9.8 *(`pickle` Object Serialization and Deserialization)* We mentioned that we prefer to use JSON for object serialization due to the Python documentation's stern security warnings about `pickle`. However, `pickle` has been used to serialize objects for many years, so you're likely to encounter it in Python legacy code. According to the documentation, "If you are doing your own `pickle` writing and reading, you're safe (provided no one else has access to the `pickle` file, of course.)"[14] Reimplement your solutions to Exercises 9.6–9.7 using the `pickle` module's `dump` function to serialize the dictionary into a file and its `load` function to deserialize the object. Pickle is a *binary* format, so this exercise requires binary files. Use the file-open mode `'wb'` to open the binary file for writing and `'rb'` to open the binary file for reading. Function `dump` receives as arguments an object to serialize and a file object in which to write the serialized object. Function `load` receives the file object containing the serialized data and returns the original object. The Python documentation suggests the `pickle` file extension .p.

9.9 *(Telephone-Number Word Generator)* In Exercise 5.12, you created a telephone-number word-generator program. Modify that program to write its output to a text file.

9.10 *(Project: Analyzing a Book from Project Gutenberg)* A great source of plain text files is the collection of over 57,000 free e-books at Project Gutenberg:

https://www.gutenberg.org

These books are out of copyright in the United States. For information about Project Gutenberg's Terms of Use and copyright in other countries, see:

https://www.gutenberg.org/wiki/Gutenberg:Terms_of_Use

Download the text-file version of *Pride and Prejudice* from Project Gutenberg

https://www.gutenberg.org/ebooks/1342

Create a script that reads *Pride and Prejudice* from a text file. Produce statistics about the book, including the total word count, the total character count, the average word length,

---

14. https://wiki.python.org/moin/UsingPickle.

the average sentence length, a word distribution containing frequency counts of all words, and the top 10 longest words. In the "Natural Language Processing (NLP)" chapter, you'll find lots of more sophisticated techniques for analyzing and comparing such texts.

Each Project Gutenberg e-book begins and ends with some additional text, such as licensing information, which is not part of the e-book itself. You may want to remove that text from your copy of the book before analyzing its text.

9.11 *(Project: Visualizing Word Frequencies with a Word Cloud)* A word cloud visualizes words, displaying more frequently occurring words in larger fonts. In this exercise, you'll create a word cloud that visualizes the top 200 words in *Pride and Prejudice*. You'll use the open-source `wordcloud` module's[15] `WordCloud` class to generate a word cloud with just a few lines of code.

To install `wordcloud`, open your Anaconda Prompt (Windows), Terminal (macOS/Linux) or shell (Linux) and enter the command:

```
conda install -c conda-forge wordcloud
```

You create and configure a `WordCloud` object as follows:

```
from wordcloud import WordCloud
wordcloud = WordCloud(colormap='prism', background_color='white')
```

Using the techniques from the previous exercise, create a `frequencies` dictionary containing the frequencies of the top-200 words in *Pride and Prejudice*. Then execute the following statements to generate a rectangular word cloud and save its image to a file on disk:

```
wordcloud = wordcloud.fit_words(frequencies)
wordcloud = wordcloud.to_file('PrideAndPrejudice.png')
```

You can then double-click the `PrideAndPrejudice.png` image file on your system to view it. In the "Natural Language Processing" chapter, we'll show you how to place your word clouds into shapes. For example, we placed our *Romeo and Juliet* word cloud into a heart.

9.12 *(Project: State-of-the-Union Speeches)* Text files of all U.S. Presidents' State-of-the-Union speeches are available online. Download one of these speeches. Write a script that reads the speech from the file, then displays statistics about the speech, including the total word count, the total character count, the average word length, the average sentence length, a word distribution of the words frequencies, and the top 10 longest words. In the "Natural Language Processing (NLP)" chapter, you'll find lots of more sophisticated techniques for analyzing and comparing such texts.

9.13 *(Project: Building a Basic Sentiment Analyzer)* We'll do lots of sentiment analysis in the data-science chapters. For example, we'll look at large numbers of tweets from Twitter on various topics, determining whether people had positive or negative opinions about those topics. We'll see that many software packages have built-in sentiment-analysis capabilities. In this exercise, you'll build a basic sentiment analyzer. A basic way to do this is to search online for files of positive words (like happy, pleasant, ...) and files of negative words (like sad, angry, ...). Then, search through a text to see how many positive words and how many negative words it contains. Based on those counts, rate the text as positive, negative or neutral.

---

15. https://github.com/amueller/word_cloud.

9.14 *(Project: Basic Similarity Detection via Average Sentence Length and Average Word Length)* Who actually wrote William Shakespeare's works? Some researchers believe that Sir Francis Bacon may have authored some or all of these works. Download one of Shakespeare's works and one of Bacon's works from Project Gutenberg. For each, calculate the average sentence length and average word length. Are these close? Compute other statistics as well.

9.15 *(Project: Working with CSV Datasets Using the csv Module)* In the Intro to Data Science section, we loaded the Titanic disaster dataset into a pandas DataFrame, then used DataFrame capabilities to perform some simple analysis of that data. For this exercise, use the csv module to read the Titanic disaster dataset, then manually count the records that contain a value for the age column. Those that do not will have the value 'NA'. For only those records that have an age value, calculate the average age. For this exercise, investigate and use the csv module's DictReader class.

9.16 *(Working with the diamonds.csv Dataset in Pandas)* In this book's data-science chapters, you'll work extensively with datasets, many in CSV format. You'll frequently use pandas to load datasets and prepare their data for use in machine-learning studies. Datasets are available for almost anything you'd want to study. There are numerous dataset repositories from which you can download datasets in CSV and other formats. In this chapter, we mentioned:

> https://vincentarelbundock.github.io/Rdatasets/datasets.html

and

> https://github.com/awesomedata/awesome-public-datasets

The Kaggle competition site:[16]

> https://www.kaggle.com/datasets?filetype=csv

has approximately 11,000 datasets with over 7500 in CSV format. The U.S. government's data.gov site:

> https://catalog.data.gov/dataset?res_format=CSV&_res_format_limit=0

has over 300,000 datasets with approximately 19,000 in CSV format.

In this exercise, you'll use the diamonds dataset to perform tasks similar to those you saw in the Intro to Data Science section. This dataset is available as diamonds.csv from various sources, including the Kaggle and Rdatasets sites listed above. The dataset contains information on 53,940 diamonds, including each diamond's carats, cut, color, clarity, depth, table (flat top surface), price and *x*, *y* and *z* measurements. The Kaggle site's web page for this dataset describes each column's content.[17]

Perform the following tasks to study and analyze the diamonds dataset:
  a) Download diamonds.csv from one of the dataset repositories.
  b) Load the dataset into a pandas DataFrame with the following statement, which uses the first column of each record as the row index:

      df = pd.read_csv('diamonds.csv', index_col=0)

  c) Display the first seven rows of the DataFrame.

---

16. To download data from Kaggle, you must register for a free account. This is true of various other dataset repository sites as well.
17. https://www.kaggle.com/shivam2503/diamonds.

d) Display the last seven rows of the DataFrame.
e) Use the DataFrame method describe (which looks only at the numerical columns) to calculate the descriptive statistics for the numerical columns—carat, depth, table, price, x, y and z.
f) Use Series method describe to calculate the descriptive statistics for the categorical data (text) columns—cut, color and clarity.
g) What are the unique category values (use the Series method unique)?
h) Pandas has many built-in graphing capabilities. Execute the %matplotlib magic to enable Matplotlib support in IPython. Then, to view histograms of each numerical data column, call your DataFrame's hist method. The following figure shows the results for the DataFrame's seven numerical columns:

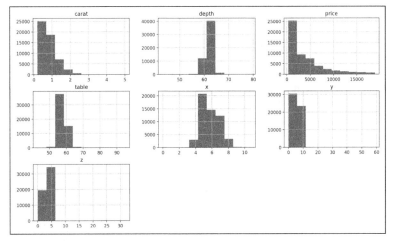

9.17 *(Working with the Iris.csv Dataset in Pandas)* Another popular dataset for machine-learning novices is the Iris dataset, which contains 150 records of information about three Iris plant species. Like this diamonds dataset, the Iris dataset is available from various online sources, including Kaggle. Investigate the Iris dataset's columns,[18] then perform the following tasks to study and analyze the dataset:
a) Download Iris.csv from one of the dataset repositories.
b) Load the dataset into a pandas DataFrame with the following statement, which uses the first column of each record as the row index:

   df = pd.read_csv('Iris.csv', index_col=0)

c) Display the DataFrame's head.
d) Display the DataFrame's tail.
e) Use the DataFrame method describe to calculate the descriptive statistics for the numerical data columns—SepalLengthCm, SepalWidthCm, PetalLengthCm and PetalWidthCm.
f) Pandas has many built-in graphing capabilities. Execute the %matplotlib magic to enable Matplotlib support in IPython. Then, to view histograms of each numerical data column, call your DataFrame's hist method.

---

18. https://www.kaggle.com/uciml/iris/home.

9.18 *(Project: Anscombe's Quartet CSV)* Locate a CSV file online containing the data for Anscombe's Quartet. Load the data into a pandas `DataFrame`. Investigate pandas built-in scatter-plot capability for plotting *x-y* coordinate pairs and use it to plot the *x-y* coordinate pairs in Anscombe's Quartet.

9.19 *(Challenging Project: A Crossword-Puzzle Generator)* Most people have worked crossword puzzles, but few have ever attempted to generate one by hand. Generating a crossword puzzle is suggested here as a string-manipulation and file-processing project requiring substantial sophistication and effort.

There are many issues you must resolve to get even the most straightforward crossword-puzzle-generator application working. For example, how do you represent the grid of squares of a crossword puzzle inside the computer? Consider using a two-dimensional list where each element is one square. Some of those elements will be "black" and some will be "white." Some of the "white" cells will include a number that corresponds to a number in your across and down clues.

You need a source of words (i.e., a computerized dictionary) that can be directly referenced by the script. In what form should these words be stored to facilitate the complex manipulations required by the application? Consider using a Python dictionary for this purpose.

You'll want to generate the clues portion of the puzzle, in which the word definitions for each across and down word are printed. Merely printing a version of the blank puzzle itself is not a simple problem, especially if you'd like the black-squared regions to be symmetric as they often are in published crossword puzzles.

9.20 *(Challenging Project: A Spell Checker)* Many apps you use daily have built-in spell checkers. In the "Natural Language Processing," "Machine Learning" and "Deep Learning" chapters, you'll learn techniques that can be used to build sophisticated spell checkers. In this project, you'll take a simpler, more mechanical approach. You'll need a computerized dictionary as a source of words.

Why do we type so many words incorrectly? In some cases, it's because we do not know the correct spelling, so we guess. In some cases, it's because we transpose two letters (e.g., "defualt" instead of "default"). Sometimes we accidentally double-type a letter (e.g., "hanndy" instead of "handy"). Sometimes we type a nearby key instead of the one we intended (e.g., "biryhday" instead of "birthday"), and so on.

Design and implement a spell-checker application. Create a text file that has some words spelled correctly and some misspelled. Your script should look up each word in the dictionary. Your script should point out each incorrect word and suggest some correct alternatives that might have been what was intended.

For example, you can try all possible single transpositions of adjacent letters to discover that the word "defualt" is a direct match for "default." Of course, this implies that your application will check all other single transpositions, such as "edfault," "dfeault," "deaflut," "defalut" and "defautl." When you find a new transposition that matches a word in the dictionary, print it in a message, such as

```
Did you mean "default"?
```

Implement other tests, such as replacing each double letter with a single letter, and any other tests you can develop to improve the value of your spell checker.

Object-Oriented Programming

Objectives

In this chapter, you'll:

- Create custom classes and objects of those classes.
- Understand the benefits of crafting valuable classes.
- Control access to attributes.
- Appreciate the value of object orientation.
- Use Python special methods __repr__, __str__ and __format__ to get an object's string representations.
- Use Python special methods to overload (redefine) operators to use them with objects of new classes.
- Inherit methods, properties and attributes from existing classes into new classes, then customize those classes.
- Understand the inheritance notions of base classes (superclasses) and derived classes (subclasses).
- Understand duck typing and polymorphism that enable "programming in the general."
- Understand class **object** from which all classes inherit fundamental capabilities.
- Compare composition and inheritance.
- Build test cases into docstrings and run these tests with **doctest**,
- Understand namespaces and how they affect scope.

Outline

10.1 Introduction
10.2 Custom Class Account
  10.2.1 Test-Driving Class Account
  10.2.2 Account Class Definition
  10.2.3 Composition: Object References as Members of Classes
10.3 Controlling Access to Attributes
10.4 Properties for Data Access
  10.4.1 Test-Driving Class Time
  10.4.2 Class Time Definition
  10.4.3 Class Time Definition Design Notes
10.5 Simulating "Private" Attributes
10.6 Case Study: Card Shuffling and Dealing Simulation
  10.6.1 Test-Driving Classes Card and DeckOfCards
  10.6.2 Class Card—Introducing Class Attributes
  10.6.3 Class DeckOfCards
  10.6.4 Displaying Card Images with Matplotlib
10.7 Inheritance: Base Classes and Subclasses
10.8 Building an Inheritance Hierarchy; Introducing Polymorphism
  10.8.1 Base Class CommissionEmployee
  10.8.2 Subclass SalariedCommissionEmployee
  10.8.3 Processing Commission-Employees and Salaried-CommissionEmployees Polymorphically
  10.8.4 A Note About Object-Based and Object-Oriented Programming
10.9 Duck Typing and Polymorphism
10.10 Operator Overloading
  10.10.1 Test-Driving Class Complex
  10.10.2 Class Complex Definition
10.11 Exception Class Hierarchy and Custom Exceptions
10.12 Named Tuples
10.13 A Brief Intro to Python 3.7's New Data Classes
  10.13.1 Creating a Card Data Class
  10.13.2 Using the Card Data Class
  10.13.3 Data Class Advantages over Named Tuples
  10.13.4 Data Class Advantages over Traditional Classes
10.14 Unit Testing with Docstrings and doctest
10.15 Namespaces and Scopes
10.16 Intro to Data Science: Time Series and Simple Linear Regression
10.17 Wrap-Up
Exercises

10.1 Introduction

Section 1.5 introduced the basic terminology and concepts of object-oriented programming. Everything in Python is an object, so you've been using objects constantly throughout this book. Just as houses are built from blueprints, objects are built from classes—one of the core technologies of object-oriented programming. Building a new object from even a large class is simple—you typically write one statement.

Crafting Valuable Classes
You've already used lots of classes created by other people. In this chapter you'll learn how to create your own *custom* classes. You'll focus on "crafting valuable classes" that help you meet the requirements of the applications you'll build. You'll use object-oriented programming with its core technologies of classes, objects, inheritance and polymorphism. Software applications are becoming larger and more richly functional. Object-oriented programming makes it easier for you to design, implement, test, debug and update such edge-of-the-practice applications. Read Sections 10.1 through 10.9 for a code-intensive introduction to these technologies. Most people can skip Sections 10.10 through 10.15, which provide additional perspectives on these technologies and present additional related features.

Class Libraries and Object-Based Programming

The vast majority of object-oriented programming you'll do in Python is **object-based programming** in which you primarily create and use objects of *existing* classes. You've been doing this throughout the book with built-in types like int, float, str, list, tuple, dict and set, with Python Standard Library types like Decimal, and with NumPy arrays, Matplotlib Figures and Axes, and pandas Series and DataFrames.

To take maximum advantage of Python you must familiarize yourself with lots of pre-existing classes. Over the years, the Python open-source community has crafted an enormous number of valuable classes and packaged them into class libraries. This makes it easy for you to reuse existing classes rather than "reinventing the wheel." Widely used open-source library classes are more likely to be thoroughly tested, bug free, performance tuned and portable across a wide range of devices, operating systems and Python versions. You'll find abundant Python libraries on the Internet at sites like GitHub, BitBucket, SourceForge and more—most easily installed with conda or pip. This is a key reason for Python's popularity. The vast majority of the classes you'll need are likely to be freely available in open-source libraries.

Creating Your Own Custom Classes

Classes are new data types. Each Python Standard Library class and third-party library class is a custom type built by someone else. In this chapter, you'll develop application-specific classes, like CommissionEmployee, Time, Card, DeckOfCards and more. The dozens of chapter exercises challenge you to create additional classes for a wide variety of applications.

Most applications you'll build for your own use will commonly use either no custom classes or just a few. If you become part of a development team in industry, you may work on applications that contain hundreds, or even thousands, of classes. You can contribute your custom classes to the Python open-source community, but you are not obligated to do so. Organizations often have policies and procedures related to open-sourcing code.

Inheritance

Perhaps most exciting is the notion that new classes can be formed through inheritance and composition from classes in abundant class libraries. Eventually, software will be constructed predominantly from **standardized, reusable components** just as hardware is constructed from interchangeable parts today. This will help meet the challenges of developing ever more powerful software.

When creating a new class, instead of writing all new code, you can designate that the new class is to be formed initially by **inheriting** the attributes (variables) and methods (the class version of functions) of a previously defined **base class** (also called a **superclass**). The new class is called a **derived class** (or **subclass**). After inheriting, you then customize the derived class to meet the specific needs of your application. To minimize the customization effort, you should always try to inherit from the base class that's closest to your needs. To do that effectively, you should familiarize yourself with the class libraries that are geared to the kinds of applications you'll be building.

Polymorphism

We explain and demonstrate **polymorphism**, which enables you to conveniently program "in the general" rather than "in the specific." You simply send the *same* method call to objects possibly of many *different* types. Each object responds by "doing the right thing." So the same method call takes on "many forms," hence the term "poly-morphism." We'll explain how to implement polymorphism through inheritance and a Python feature called duck typing. We'll explain both and show examples of each.

An Entertaining Case Study: Card-Shuffling-and-Dealing Simulation
You've already used a random-numbers-based die-rolling simulation and used those techniques to implement the popular dice game craps. Here, we present a card-shuffling-and-dealing simulation, which you can use to implement your favorite card games. You'll use Matplotlib with attractive public-domain card images to display the full deck of cards both before and after the deck is shuffled. In the exercises, you can implement the popular card games blackjack and solitaire, and evaluate a five-card poker hand.

Data Classes
Python 3.7's new *data classes* help you build classes faster by using a more concise notation and by autogenerating portions of the classes. The Python community's early reaction to data classes has been positive. As with any major new feature, it may take time before it's widely used. We present class development with both the older and newer technologies.

Other Concepts Introduced in This Chapter
Other concepts you'll learn include:

- How to specify that certain identifiers should be used only inside a class and not be accessible to clients of the class.
- Special methods for creating string representations of your classes' objects and specifying how objects of your classes work with Python's built-in operators (a process called *operator overloading*).
- An introduction to the Python exception class hierarchy and creating custom exception classes.
- Testing code with the Python Standard Library's `doctest` module.
- How Python uses namespaces to determine the scopes of identifiers.

10.2 Custom Class Account

Let's begin with a bank Account class that holds an account holder's name and balance. An actual bank account class would likely include lots of other information, such as address, birth date, telephone number, account number and more. The Account class accepts deposits that increase the balance and withdrawals that decrease the balance.

10.2.1 Test-Driving Class Account

Each new class you create becomes a new *data type* that can be used to create objects. This is one reason why Python is said to be an **extensible language**. Before we look at class Account's definition, let's demonstrate its capabilities.

Importing Classes Account and Decimal
To use the new Account class, launch your IPython session from the ch10 examples folder, then import class Account:

```
In [1]: from account import Account
```

Class `Account` maintains and manipulates the account balance as a `Decimal`, so we also import class `Decimal`:

```
In [2]: from decimal import Decimal
```

Create an Account Object with a Constructor Expression
To create a `Decimal` object, we can write:

```
value = Decimal('12.34')
```

This is known as a **constructor expression** because it builds and initializes an object of the class, similar to the way a house is constructed from a blueprint then painted with the buyer's preferred colors. Constructor expressions create new objects and initialize their data using argument(s) specified in parentheses. The parentheses following the class name are required, even if there are no arguments.

Let's use a constructor expression to create an `Account` object and initialize it with an account holder's name (a string) and balance (a `Decimal`):

```
In [3]: account1 = Account('John Green', Decimal('50.00'))
```

Getting an Account's Name and Balance
Let's access the `Account` object's name and balance attributes:

```
In [4]: account1.name
Out[4]: 'John Green'

In [5]: account1.balance
Out[5]: Decimal('50.00')
```

Depositing Money into an Account
An `Account`'s deposit method receives a positive dollar amount and adds it to the balance:

```
In [6]: account1.deposit(Decimal('25.53'))

In [7]: account1.balance
Out[7]: Decimal('75.53')
```

Account Methods Perform Validation
Class `Account`'s methods validate their arguments. For example, if a deposit amount is negative, deposit raises a `ValueError`:

```
In [8]: account1.deposit(Decimal('-123.45'))
-------------------------------------------------------------------------
ValueError                                Traceback (most recent call last)
<ipython-input-8-27dc468365a7> in <module>()
----> 1 account1.deposit(Decimal('-123.45'))

~/Documents/examples/ch10/account.py in deposit(self, amount)
     21         # if amount is less than 0.00, raise an exception
     22         if amount < Decimal('0.00'):
---> 23             raise ValueError('Deposit amount must be positive.')
     24
     25         self.balance += amount

ValueError: Deposit amount must be positive.
```

Self Check

1 *(Fill-In)* Each new class you create becomes a new data type that can be used to create objects. This is one reason why Python is said to be a(n) _____ language.
Answer: extensible.

2 *(Fill-In)* A(n) _____ expression creates and initializes an object of a class.
Answer: constructor.

10.2.2 Account Class Definition

Now, let's look at Account's class definition, which is located in the file account.py.

Defining a Class

A class definition begins with the keyword `class` (line 5) followed by the class's name and a colon (:). This line is called the **class header**. The *Style Guide for Python Code* recommends that you begin each word in a multi-word class name with an uppercase letter (for example, `CommissionEmployee`). Every statement in a class's suite is indented.

```
1   # account.py
2   """Account class definition."""
3   from decimal import Decimal
4
5   class Account:
6       """Account class for maintaining a bank account balance."""
7
```

Each class typically provides a descriptive docstring (line 6). When provided, it must appear in the line or lines immediately following the class header. To view any class's docstring in IPython, type the class name and a question mark, then press *Enter*:

```
In [9]: Account?
Init signature:  Account(name, balance)
Docstring:       Account class for maintaining a bank account balance.
Init docstring:  Initialize an Account object.
File:            ~/Documents/examples/ch10/account.py
Type:            type
```

The identifier Account is both the class name and the name used in a constructor expression to create an Account object and invoke the class's __init__ method. For this reason, IPython's help mechanism shows both the class's docstring ("Docstring:") and the __init__ method's docstring ("Init docstring:").

Initializing Account Objects: Method __init__

The constructor expression in snippet [3] from the preceding section:

```
account1 = Account('John Green', Decimal('50.00'))
```

creates a new object, then initializes its data by calling the class's __init__ method. Each new class you create can provide an __init__ method that specifies how to initialize an object's data attributes. Returning a value other than None from __init__ results in a TypeError. Recall that None is returned by any function or method that does not contain a return statement. Class Account's __init__ method (lines 8–16) initializes an Account object's name and balance attributes if the balance is valid:

```
 8      def __init__(self, name, balance):
 9          """Initialize an Account object."""
10
11          # if balance is less than 0.00, raise an exception
12          if balance < Decimal('0.00'):
13              raise ValueError('Initial balance must be >= to 0.00.')
14
15          self.name = name
16          self.balance = balance
17
```

When you call a method for a specific object, Python implicitly passes a reference to that object as the method's first argument. For this reason, all methods of a class must specify at least one parameter. By convention most Python programmers call a method's first parameter self. A class's methods must use that reference (self) to access the object's attributes and other methods. Class Account's __init__ method also specifies parameters for the name and balance.

The if statement *validates* the balance parameter. If balance is less than 0.00, __init__ raises a ValueError, which terminates the __init__ method. Otherwise, the method creates and initializes the new Account object's name and balance attributes.

When an object of class Account is created, it does not yet have any attributes. They're added *dynamically* via assignments of the form:

> self.*attribute_name* = *value*

Python classes may define many **special methods**, like __init__, each identified by leading and trailing double-underscores (__) in the method name. Python class **object**, which we'll discuss later in this chapter, defines the special methods that are available for *all* Python objects.

Method deposit
The Account class's deposit method adds a positive amount to the account's balance attribute. If the amount argument is less than 0.00, the method raises a ValueError, indicating that only positive deposit amounts are allowed. If the amount is valid, line 25 adds it to the object's balance attribute.

```
18      def deposit(self, amount):
19          """Deposit money to the account."""
20
21          # if amount is less than 0.00, raise an exception
22          if amount < Decimal('0.00'):
23              raise ValueError('amount must be positive.')
24
25          self.balance += amount
```

10.2.3 Composition: Object References as Members of Classes
An Account *has a* name, and an Account *has a* balance. Recall that "everything in Python is an object." This means that an object's attributes are references to objects of other classes. For example, an Account object's name attribute is a reference to a string object and an Account object's balance attribute is a reference to a Decimal object. Embedding references to objects of other types is a form of software reusability known as **composition**

and is sometimes referred to as the **"has a"** relationship. An end-of-chapter exercise asks you to implement composition with `Circle` and `Point` classes—a `Circle` "has a" `Point` that represents the `Circle`'s center location. Later in this chapter, we'll discuss *inheritance*, which establishes *"is a"* relationships.

Self Check

1 *(Fill-In)* A class's _____ method is called by a constructor expression to initialize a new object of the class.
Answer: `__init__`.

2 *(True/False)* A class's `__init__` method returns an object of the class.
Answer: False. A class's `__init__` method initializes an object of the class and implicitly returns `None`.

3 *(IPython Session)* Add a `withdraw` method to class `Account`. If the withdrawal amount is greater than the `balance`, raise a `ValueError`, indicating that the withdrawal amount must be less than or equal to the `balance`. If the withdrawal amount is less than 0.00, raise a `ValueError` indicating that the withdrawal amount must be positive. If the withdrawal amount is valid, subtract it from the `balance` attribute. Create an `Account` object, then test method `withdraw` first with a valid withdrawal amount, then with a withdrawal amount greater than the `balance` and finally with a negative withdrawal amount.
Answer: The new method in class `Account` is:

```
def withdraw(self, amount):
    """Withdraw money from the account."""

    # if amount is greater than balance, raise an exception
    if amount > self.balance:
        raise ValueError('amount must be <= to balance.')
    elif amount < Decimal('0.00'):
        raise ValueError('amount must be positive.')

    self.balance -= amount
```

Testing method `withdraw`:

```
In [1]: from account import Account

In [2]: from decimal import Decimal

In [3]: account1 = Account('John Green', Decimal('50.00'))

In [4]: account1.withdraw(Decimal('20.00'))

In [5]: account1.balance
Out[5]: Decimal('30.00')

In [6]: account1.withdraw(Decimal('100.00'))
-------------------------------------------------------------------------
ValueError                                Traceback (most recent call last)
<ipython-input-6-61bb6aa89aa4> in <module>()
----> 1 account1.withdraw(Decimal('100.00'))
```

```
~/Documents/examples/ch10/snippets_py/account.py in withdraw(self,
amount)
     30             # if amount is greater than balance, raise an exception
     31             if amount > self.balance:
---> 32                 raise ValueError('amount must be <= to balance.')
     33             elif amount < Decimal('0.00'):
     34                 raise ValueError('amount must be positive.')

ValueError: amount must be <= to balance.

In [7]: account1.withdraw(Decimal('-10.00'))
---------------------------------------------------------------------------
ValueError                                Traceback (most recent call last)
<ipython-input-7-ab50927d9727> in <module>()
----> 1 account1.withdraw(Decimal('-10.00'))

~/Documents/examples/ch10/snippets_py/account.py in withdraw(self,
amount)
     32                 raise ValueError('amount must be <= to balance.')
     33             elif amount < Decimal('0.00'):
---> 34                 raise ValueError('amount must be positive.')
     35
     36             self.balance -= amount

ValueError: amount must be positive.
```

10.3 Controlling Access to Attributes

Class Account's methods validate their arguments to ensure that the balance is *always* valid—that is, always greater than or equal to 0.00. In the previous example, we used the attributes name and balance only to *get* the values of those attributes. It turns out that we also can use those attributes to *modify* their values. Consider the Account object in the following IPython session:

```
In [1]: from account import Account

In [2]: from decimal import Decimal

In [3]: account1 = Account('John Green', Decimal('50.00'))

In [4]: account1.balance
Out[4]: Decimal('50.00')
```

Initially, account1 contains a valid balance. Now, let's set the balance attribute to an *invalid* negative value, then display the balance:

```
In [5]: account1.balance = Decimal('-1000.00')

In [6]: account1.balance
Out[6]: Decimal('-1000.00')
```

Snippet [6]'s output shows that account1's balance is now negative. As you can see, unlike methods, data attributes cannot validate the values you assign to them.

Encapsulation

A class's **client code** is any code that uses objects of the class. Most object-oriented programming languages enable you to **encapsulate** (or *hide*) an object's data from the client code. Such data in these languages is said to be *private data*.

Leading Underscore (_) Naming Convention

Python does *not* have private data. Instead, you use *naming conventions* to design classes that encourage correct use. By convention, Python programmers know that any attribute name beginning with an underscore (_) is for a class's *internal use only*. Client code should use the class's methods and—as you'll see in the next section—the class's properties to interact with each object's internal-use data attributes. Attributes whose identifiers do *not* begin with an underscore (_) are considered *publicly accessible* for use in client code. In the next section, we'll define a Time class and use these naming conventions. However, even when we use these conventions, attributes are always accessible.

✓ Self Check

1 *(True/False)* Like most object-oriented programming languages, Python provides capabilities for encapsulating an object's data attributes so client code cannot access the data directly.
Answer: False. In Python, all data attributes are accessible. You use attribute naming conventions to indicate that attributes should not be accessed directly from client code.

10.4 Properties for Data Access

Let's develop a Time class that stores the time in 24-hour clock format with hours in the range 0–23, and minutes and seconds each in the range 0–59. For this class, we'll provide *properties*, which look like data attributes to client-code programmers, but control the manner in which they get and modify an object's data. This assumes that other programmers follow Python conventions to correctly use objects of your class.

10.4.1 Test-Driving Class Time

Before we look at class Time's definition, let's demonstrate its capabilities. First, ensure that you're in the ch10 folder, then import class Time from timewithproperties.py:

```
In [1]: from timewithproperties import Time
```

Creating a Time Object

Next, let's create a Time object. Class Time's __init__ method has hour, minute and second parameters, each with a default argument value of 0. Here, we specify the hour and minute—second defaults to 0:

```
In [2]: wake_up = Time(hour=6, minute=30)
```

Displaying a Time Object

Class Time defines two methods that produce string representations of Time object. When you evaluate a variable in IPython as in snippet [3], IPython calls the object's __repr__ special method to produce a string representation of the object. Our __repr__ implementation creates a string in the following format:

```
In [3]: wake_up
Out[3]: Time(hour=6, minute=30, second=0)
```

We'll also provide the __str__ special method, which is called when an object is converted to a string, such as when you output the object with print.[1] Our __str__ implementation creates a string in 12-hour clock format:

```
In [4]: print(wake_up)
6:30:00 AM
```

Getting an Attribute Via a Property

Class time provides hour, minute and second **properties**, which provide the convenience of data attributes for getting and modifying an object's data. However, as you'll see, properties are implemented as methods, so they may contain additional logic, such as specifying the format in which to return a data attribute's value or validating a new value before using it to modify a data attribute. Here, we get the wake_up object's hour value:

```
In [5]: wake_up.hour
Out[5]: 6
```

Though this snippet appears to simply get an hour data attribute's value, it's actually a call to an hour *method* that returns the value of a data attribute (which we named _hour, as you'll see in the next section).

Setting the Time

You can set a new time with the Time object's set_time method. Like method __init__, method set_time provides hour, minute and second parameters, each with a default of 0:

```
In [6]: wake_up.set_time(hour=7, minute=45)

In [7]: wake_up
Out[7]: Time(hour=7, minute=45, second=0)
```

Setting an Attribute via a Property

Class Time also supports setting the hour, minute and second values individually via its properties. Let's change the hour value to 6:

```
In [8]: wake_up.hour = 6

In [9]: wake_up
Out[9]: Time(hour=6, minute=45, second=0)
```

Though snippet [8] appears to simply assign a value to a data attribute, it's actually a call to an hour method that takes 6 as an argument. The method validates the value, then assigns it to a corresponding data attribute (which we named _hour, as you'll see in the next section).

Attempting to Set an Invalid Value

To prove that class Time's properties *validate* the values you assign to them, let's try to assign an invalid value to the hour property, which results in a ValueError:

---

1. If a class does not provide __str__ and an object of the class is converted to a string, the class's __repr__ method is called instead.

```
In [10]: wake_up.hour = 100
```
```
-------------------------------------------------------------------
ValueError                                Traceback (most recent call last)
<ipython-input-10-1fce0716ef14> in <module>()
----> 1 wake_up.hour = 100

~/Documents/examples/ch10/timewithproperties.py in hour(self, hour)
     20         """Set the hour."""
     21         if not (0 <= hour < 24):
---> 22             raise ValueError(f'Hour ({hour}) must be 0-23')
     23
     24         self._hour = hour

ValueError: Hour (100) must be 0-23
```

Self Check

1 *(Fill-In)* The print function implicitly invokes special method _____.
Answer: __str__.

2 *(Fill-In)* IPython calls an object's special method _____ to produce a string representation of the object
Answer: __repr__.

3 *(True/False)* Properties are used like methods.
Answer: False. Properties are used like data attributes, but (as we'll see in the next section) are implemented as methods.

10.4.2 Class Time Definition

Now that we've seen class Time in action, let's look at its definition.

Class Time: __init__ Method with Default Parameter Values

Class Time's __init__ method specifies hour, minute and second parameters, each with a default argument of 0. Similar to class Account's __init__ method, recall that the self parameter is a reference to the Time object being initialized. The statements containing self.hour, self.minute and self.second *appear* to create hour, minute and second attributes for the new Time object (self). However, these statements actually call methods that implement the class's hour, minute and second *properties* (lines 13–50). Those methods then create attributes named _hour, _minute and _second that are meant for use only inside the class:

```
 1   # timewithproperties.py
 2   """Class Time with read-write properties."""
 3
 4   class Time:
 5       """Class Time with read-write properties."""
 6
 7       def __init__(self, hour=0, minute=0, second=0):
 8           """Initialize each attribute."""
 9           self.hour = hour      # 0-23
10           self.minute = minute  # 0-59
11           self.second = second  # 0-59
12
```

Class Time: hour Read-Write Property

Lines 13–24 define a *publicly accessible* **read-write property** named hour that manipulates a data attribute named _hour. The single-leading-underscore (_) naming convention indicates that client code should not access _hour directly. As you saw in the previous section's snippets [5] and [8], properties look like data attributes to programmers working with Time objects. However, notice that properties are implemented as *methods*. Each property defines a *getter* method which *gets* (that is, returns) a data attribute's value and can *optionally* define a *setter* method which *sets* a data attribute's value:

```
13      @property
14      def hour(self):
15          """Return the hour."""
16          return self._hour
17
18      @hour.setter
19      def hour(self, hour):
20          """Set the hour."""
21          if not (0 <= hour < 24):
22              raise ValueError(f'Hour ({hour}) must be 0-23')
23
24          self._hour = hour
25
```

The **@property decorator** precedes the property's *getter* method, which receives only a self parameter. Behind the scenes, a decorator adds code to the decorated function—in this case to make the hour function work with attribute syntax. The *getter* method's name is the property name. This *getter* method returns the _hour data attribute's value. The following client-code expression invokes the *getter* method:

    wake_up.hour

You also can use the *getter* method inside the class, as you'll see shortly.

A decorator of the form *@property_name*.setter (in this case, @hour.setter) precedes the property's *setter* method. The method receives two parameters—self and a parameter (hour) representing the value being assigned to the property. If the hour parameter's value is *valid*, this method assigns it to the self object's _hour attribute; otherwise, the method raises a ValueError. The following client-code expression invokes the *setter* by assigning a value to the property:

    wake_up.hour = 8

We also invoked this setter inside the class at line 9 of __init__:

    self.hour = hour

Using the *setter* enabled us to *validate* __init__'s hour argument *before* creating and initializing the object's _hour attribute, which occurs the *first* time the hour property's *setter* executes as a result of line 9. A **read-write property** has both a *getter* and a *setter*. A **read-only property** has only a *getter*.

Class Time: minute and second Read-Write Properties

Lines 26–37 and 39–50 define read-write minute and second properties. Each property's setter ensures that its second argument is in the range 0–59 (the valid range of values for minutes and seconds):

```
26      @property
27      def minute(self):
28          """Return the minute."""
29          return self._minute
30
31      @minute.setter
32      def minute(self, minute):
33          """Set the minute."""
34          if not (0 <= minute < 60):
35              raise ValueError(f'Minute ({minute}) must be 0-59')
36
37          self._minute = minute
38
39      @property
40      def second(self):
41          """Return the second."""
42          return self._second
43
44      @second.setter
45      def second(self, second):
46          """Set the second."""
47          if not (0 <= second < 60):
48              raise ValueError(f'Second ({second}) must be 0-59')
49
50          self._second = second
51
```

Class Time: Method set_time

We provide method set_time as a convenient way to change *all three* attributes with a *single* method call. Lines 54–56 invoke the *setters* for the hour, minute and second properties:

```
52      def set_time(self, hour=0, minute=0, second=0):
53          """Set values of hour, minute, and second."""
54          self.hour = hour
55          self.minute = minute
56          self.second = second
57
```

Class Time: Special Method __repr__

When you pass an object to built-in function repr—which happens implicitly when you evaluate a variable in an IPython session—the corresponding class's **__repr__** special method is called to get a string representation of the object:

```
58      def __repr__(self):
59          """Return Time string for repr()."""
60          return (f'Time(hour={self.hour}, minute={self.minute}, ' +
61                  f'second={self.second})')
62
```

The Python documentation indicates that __repr__ returns the "official" string representation of the object. Typically this string looks like a constructor expression that creates and initializes the object,[2] as in:

    'Time(hour=6, minute=30, second=0)'

---

2. https://docs.python.org/3/reference/datamodel.html.

10.4 Properties for Data Access

This is similar to the constructor expression in the previous section's snippet [2]. Python has a built-in function **eval** that could receive the preceding string as an argument and use it to create and initialize a Time object containing values specified in the string.

Class Time: Special Method __str__

For our class Time we also define the **__str__** special method. This method is called implicitly when you convert an object to a string with the built-in function str, such as when you print an object or call str explicitly. Our implementation of __str__ creates a string in 12-hour clock format, such as '7:59:59 AM' or '12:30:45 PM':

```
63    def __str__(self):
64        """Print Time in 12-hour clock format."""
65        return (('12' if self.hour in (0, 12) else str(self.hour % 12)) +
66                f':{self.minute:0>2}:{self.second:0>2}' +
67                (' AM' if self.hour < 12 else ' PM'))
```

✓ Self Check

1 *(Fill-In)* The print function implicitly invokes special method _____.
Answer: __str__.

2 *(Fill-In)* A(n) _____ property has both a getter and setter. If only a getter is provided, the property is a(n) _____ property, meaning that you only can get the property's value.
Answer: read-write, read-only.

3 *(IPython Session)* Add to class Time a read-write property time in which the *getter* returns a tuple containing the values of the hour, minute and second properties, and the *setter* receives a tuple containing hour, minute and second values and uses them to set the time. Create a Time object and test the new property.
Answer: The new read-write property definition is shown below:

```
@property
def time(self):
    """Return hour, minute and second as a tuple."""
    return (self.hour, self.minute, self.second)

@time.setter
def time(self, time_tuple):
    """Set time from a tuple containing hour, minute and second."""
    self.set_time(time_tuple[0], time_tuple[1], time_tuple[2])
```

```
In [1]: from timewithproperties import Time

In [2]: t = Time()

In [3]: t
Out[3]: Time(hour=0, minute=0, second=0)

In [4]: t.time = (12, 30, 45)

In [5]: t
Out[5]: Time(hour=12, minute=30, second=45)

In [6]: t.time
Out[6]: (12, 30, 45)
```

Note that the `self.set_time` call in the time property's `setter` method may be expressed more concisely as

```
self.set_time(*time_tuple)
```

The expression `*time_tuple` uses the **unary * operator** to *unpack* the `time_tuple`'s values, then passes them as individual arguments. In the preceding IPython session, the `setter` would receive the tuple (12, 30, 45), then unpack the tuple and call `self.set_time` as follows:

```
self.set_time(12, 30, 45)
```

10.4.3 Class Time Definition Design Notes

Let's consider some class-design issues in the context of our `Time` class.

Interface of a Class

Class `Time`'s properties and methods define the class's **public interface**—that is, the set of properties and methods programmers should use to interact with objects of the class.

Attributes Are Always Accessible

Though we provided a well-defined interface, Python does *not* prevent you from directly manipulating the data attributes `_hour`, `_minute` and `_second`, as in:

```
In [1]: from timewithproperties import Time

In [2]: wake_up = Time(hour=7, minute=45, second=30)

In [3]: wake_up._hour
Out[3]: 7

In [4]: wake_up._hour = 100

In [5]: wake_up
Out[5]: Time(hour=100, minute=45, second=30)
```

After snippet [4], the `wake_up` object contains *invalid* data. Unlike many other object-oriented programming languages, such as C++, Java and C#, data attributes in Python cannot be hidden from client code. The Python tutorial says, "**nothing in Python makes it possible to enforce data hiding—it is all based upon convention.**"[3]

Internal Data Representation

We chose to represent the time as three integer values for hours, minutes and seconds. It would be perfectly reasonable to represent the time internally as the number of seconds since midnight. Though we'd have to reimplement the properties hour, minute and second, programmers could use the *same* interface and get the *same* results without being aware of these changes. An exercise at the end of this chapter asks you to make this change and show that client code using `Time` objects does not need to change.

Evolving a Class's Implementation Details

When you design a class, carefully consider the class's interface before making that class available to other programmers. Ideally, you'll design the interface such that existing code

---

3. https://docs.python.org/3/tutorial/classes.html#random-remarks.

will not break if you update the class's implementation details—that is, the internal data representation or how its method bodies are implemented.

If Python programmers follow convention and do not access attributes that begin with leading underscores, then class designers can evolve class implementation details without breaking client code.

Properties
It may seem that providing properties with both *setters* and *getters* has no benefit over accessing the data attributes directly, but there are subtle differences. A *getter* seems to allow clients to read the data at will, but the *getter* can control the formatting of the data. A *setter* can scrutinize attempts to modify the value of a data attribute to prevent the data from being set to an invalid value.

Utility Methods
Not all methods need to serve as part of a class's interface. Some serve as **utility methods** used only *inside* the class and are not intended to be part of the class's public interface used by client code. Such methods should be named with a single leading underscore. In other object-oriented languages like C++, Java and C#, such methods typically are implemented as private methods.

Module datetime
In professional Python development, rather than building your own classes to represent times and dates, you'll typically use the Python Standard Library's datetime module capabilities. You can learn more about the datetime module at:

https://docs.python.org/3/library/datetime.html

An exercise at the end of the chapter has you manipulate dates and times with this module.

Self Check

1 *(Fill-In)* A class's _____ is the set of public properties and methods programmers should use to interact with objects of the class.
Answer: interface.

2 *(Fill-In)* A class's _____ methods are used only inside the class and are not intended to be used by client code.
Answer: utility.

10.5 Simulating "Private" Attributes

In programming languages such as C++, Java and C#, classes state explicitly which class members are *publicly accessible*. Class members that may not be accessed outside a class definition are **private** and visible only within the class that defines them. Python programmers often use "private" attributes for data or utility methods that are essential to a class's inner workings but are not part of the class's public interface.

As you've seen, Python objects' attributes are *always* accessible. However, Python has a naming convention for "private" attributes. Suppose we want to create an object of class Time and to *prevent* the following assignment statement:

```
wake_up._hour = 100
```

that would set the hour to an invalid value. Rather than _hour, we can name the attribute __hour with *two* leading underscores. This convention indicates that __hour is "private" and should not be accessible to the class's clients. To help prevent clients from accessing "private" attributes, Python *renames* them by preceding the attribute name with _*ClassName*, as in _Time__hour. This is called **name mangling**. If you try assign to __hour, as in

```
wake_up.__hour = 100
```

Python raises an AttributeError, indicating that the class does not have an __hour attribute. We'll show this momentarily.

IPython Auto-Completion Shows Only "Public" Attributes
In addition, IPython does not show attributes with one or two leading underscores when you try to auto-complete an expression like

```
wake_up.
```

by pressing *Tab*. Only attributes that are part of the wake_up object's "public" interface are displayed in the IPython auto-completion list.

Demonstrating "Private" Attributes
To demonstrate name mangling, consider class PrivateClass with one "public" data attribute public_data and one "private" data attribute __private_data:

```
 1  # private.py
 2  """Class with public and private attributes."""
 3
 4  class PrivateClass:
 5      """Class with public and private attributes."""
 6
 7      def __init__(self):
 8          """Initialize the public and private attributes."""
 9          self.public_data = "public"   # public attribute
10          self.__private_data = "private"  # private attribute
```

Let's create an object of class PrivateData to demonstrate these data attributes:

```
In [1]: from private import PrivateClass

In [2]: my_object = PrivateClass()
```

Snippet [3] shows that we can access the public_data attribute directly:

```
In [3]: my_object.public_data
Out[3]: 'public'
```

However, when we attempt to access __private_data directly in snippet [4], we get an AttributeError indicating that the class does not have an attribute by that name:

```
In [4]: my_object.__private_data
-------------------------------------------------------------------------
AttributeError                            Traceback (most recent call last)
<ipython-input-4-d896bfdf2053> in <module>()
----> 1 my_object.__private_data

AttributeError: 'PrivateClass' object has no attribute '__private_data'
```

10.6 Case Study: Card Shuffling and Dealing Simulation

This occurs because python changed the attribute's name. Unfortunately, as you'll see in one of this section's Self Check exercises, __private_data is still indirectly accessible.

 Self Check

1 *(Fill-In)* Python mangles attribute names that begin with _____ underscore(s).
Answer: two.

2 *(True/False)* An attribute that begins with a single underscore is a private attribute.
Answer: False. An attribute that begins with a single underscore simply conveys the convention that a client of the class should not access the attribute directly, but it *does* allow access. Again, "**nothing in Python makes it possible to enforce data hiding—it is all based upon convention.**"[4]

3 *(IPython Session)* Even with double-underscore (__) naming, we can still access and modify __private_data, because we know that Python renames attributes simply by prefixing their names with '_*ClassName*'. Demonstrate this for class PrivateData's data attribute __private_data.
Answer:

```
In [5]: my_object._PrivateClass__private_data
Out[5]: 'private'

In [6]: my_object._PrivateClass__private_data = 'modified'

In [7]: my_object._PrivateClass__private_data
Out[7]: 'modified'
```

10.6 Case Study: Card Shuffling and Dealing Simulation

Our next example presents two custom classes that you can use to shuffle and deal a deck of cards. Class Card represents a playing card that has a face ('Ace', '2', '3', ..., 'Jack', 'Queen', 'King') and a suit ('Hearts', 'Diamonds', 'Clubs', 'Spades'). Class DeckOfCards represents a deck of 52 playing cards as a list of Card objects. First, we'll test-drive these classes in an IPython session to demonstrate card shuffling and dealing capabilities and displaying the cards as text. Then we'll look at the class definitions. Finally, we'll use another IPython session to display the 52 cards as images using Matplotlib. We'll show you where to get nice-looking public-domain card images.

10.6.1 Test-Driving Classes Card and DeckOfCards

Before we look at classes Card and DeckOfCards, let's use an IPython session to demonstrate their capabilities.

Creating, Shuffling and Dealing the Cards
First, import class DeckOfCards from deck.py and create an object of the class:

```
In [1]: from deck import DeckOfCards

In [2]: deck_of_cards = DeckOfCards()
```

---

4. https://docs.python.org/3/tutorial/classes.html#random-remarks.

DeckOfCards method __init__ creates the 52 Card objects in order by suit and by face within each suit. You can see this by printing the deck_of_cards object, which calls the DeckOfCards class's __str__ method to get the deck's string representation. Read each row left-to-right to confirm that all the cards are displayed in order from each suit (Hearts, Diamonds, Clubs and Spades):

```
In [3]: print(deck_of_cards)
Ace of Hearts     2 of Hearts       3 of Hearts       4 of Hearts
5 of Hearts       6 of Hearts       7 of Hearts       8 of Hearts
9 of Hearts       10 of Hearts      Jack of Hearts    Queen of Hearts
King of Hearts    Ace of Diamonds   2 of Diamonds     3 of Diamonds
4 of Diamonds     5 of Diamonds     6 of Diamonds     7 of Diamonds
8 of Diamonds     9 of Diamonds     10 of Diamonds    Jack of Diamonds
Queen of Diamonds King of Diamonds  Ace of Clubs      2 of Clubs
3 of Clubs        4 of Clubs        5 of Clubs        6 of Clubs
7 of Clubs        8 of Clubs        9 of Clubs        10 of Clubs
Jack of Clubs     Queen of Clubs    King of Clubs     Ace of Spades
2 of Spades       3 of Spades       4 of Spades       5 of Spades
6 of Spades       7 of Spades       8 of Spades       9 of Spades
10 of Spades      Jack of Spades    Queen of Spades   King of Spades
```

Next, let's shuffle the deck and print the deck_of_cards object again. We did not specify a seed for reproducibility, so each time you shuffle, you'll get different results:

```
In [4]: deck_of_cards.shuffle()
```

```
In [5]: print(deck_of_cards)
King of Hearts    Queen of Clubs    Queen of Diamonds 10 of Clubs
5 of Hearts       7 of Hearts       4 of Hearts       2 of Hearts
5 of Clubs        8 of Diamonds     3 of Hearts       10 of Hearts
8 of Spades       5 of Spades       Queen of Spades   Ace of Clubs
8 of Clubs        7 of Spades       Jack of Diamonds  10 of Spades
4 of Diamonds     8 of Hearts       6 of Spades       King of Spades
9 of Hearts       4 of Spades       6 of Clubs        King of Clubs
3 of Spades       9 of Diamonds     3 of Clubs        Ace of Spades
Ace of Hearts     3 of Diamonds     2 of Diamonds     6 of Hearts
King of Diamonds  Jack of Spades    Jack of Clubs     2 of Spades
5 of Diamonds     4 of Clubs        Queen of Hearts   9 of Clubs
10 of Diamonds    2 of Clubs        Ace of Diamonds   7 of Diamonds
9 of Spades       Jack of Hearts    6 of Diamonds     7 of Clubs
```

Dealing Cards

We can deal one Card at a time by calling method deal_card. IPython calls the returned Card object's __repr__ method to produce the string output shown in the Out[] prompt:

```
In [6]: deck_of_cards.deal_card()
Out[6]: Card(face='King', suit='Hearts')
```

Class Card's Other Features

To demonstrate class Card's __str__ method, let's deal another card and pass it to the built-in str function:

```
In [7]: card = deck_of_cards.deal_card()
```

```
In [8]: str(card)
Out[8]: 'Queen of Clubs'
```

Each Card has a corresponding image file name, which you can get via the `image_name` read-only property. We'll use this soon when we display the Cards as images:

```
In [9]: card.image_name
Out[9]: 'Queen_of_Clubs.png'
```

10.6.2 Class Card—Introducing Class Attributes

Each Card object contains three string properties representing that Card's face, suit and image_name (a file name containing a corresponding image). As you saw in the preceding section's IPython session, class Card also provides methods for initializing a Card and for getting various string representations.

Class Attributes FACES and SUITS

Each object of a class has its own copies of the class's data attributes. For example, each Account object has its own name and balance. Sometimes, an attribute should be shared by *all* objects of a class. A **class attribute** (also called a **class variable**) represents *class-wide* information. It belongs to the *class*, not to a specific object of that class. Class Card defines two class attributes (lines 5–7):

- FACES is a list of the card face names.
- SUITS is a list of the card suit names.

```
1  # card.py
2  """Card class that represents a playing card and its image file name."""
3
4  class Card:
5      FACES = ['Ace', '2', '3', '4', '5', '6',
6               '7', '8', '9', '10', 'Jack', 'Queen', 'King']
7      SUITS = ['Hearts', 'Diamonds', 'Clubs', 'Spades']
8
```

You define a class attribute by assigning a value to it inside the class's definition, but not inside any of the class's methods or properties (in which case, they'd be local variables). FACES and SUITS are *constants* that are not meant to be modified. Recall that the *Style Guide for Python Code* recommends naming your constants with all capital letters.[5]

We'll use elements of these lists to initialize each Card we create. However, we do not need a separate copy of each list in every Card object. Class attributes can be accessed through any object of the class, but are typically accessed through the class's name (as in, Card.FACES or Card.SUITS). Class attributes exist as soon as you import their class's definition.

Card Method __init__

When you create a Card object, method __init__ defines the object's _face and _suit data attributes:

```
9      def __init__(self, face, suit):
10         """Initialize a Card with a face and suit."""
11         self._face = face
12         self._suit = suit
13
```

---

5. Recall that Python does not have true constants, so FACES and SUITS are still modifiable.

Read-Only Properties `face`, `suit` and `image_name`

Once a `Card` is created, its `face`, `suit` and `image_name` do not change, so we implement these as read-only properties (lines 14–17, 19–22 and 24–27). Properties `face` and `suit` return the corresponding data attributes `_face` and `_suit`. A property is not required to have a corresponding data attribute. To demonstrate this, the `Card` property `image_name`'s value is *created dynamically* by getting the `Card` object's string representation with `str(self)`, replacing any spaces with underscores and appending the `'.png'` filename extension. So, `'Ace of Spades'` becomes `'Ace_of_Spades.png'`. We'll use this file name to load a PNG-format image representing the `Card`. PNG (Portable Network Graphics) is a popular image format for web-based images.

```
14      @property
15      def face(self):
16          """Return the Card's self._face value."""
17          return self._face
18
19      @property
20      def suit(self):
21          """Return the Card's self._suit value."""
22          return self._suit
23
24      @property
25      def image_name(self):
26          """Return the Card's image file name."""
27          return str(self).replace(' ', '_') + '.png'
28
```

Methods That Return String Representations of a Card

Class `Card` provides three special methods that return string representations. As in class `Time`, method `__repr__` returns a string representation that looks like a constructor expression for creating and initializing a `Card` object:

```
29      def __repr__(self):
30          """Return string representation for repr()."""
31          return f"Card(face='{self.face}', suit='{self.suit}')"
32
```

Method `__str__` returns a string of the format *'face of suit'*, such as `'Ace of Hearts'`:

```
33      def __str__(self):
34          """Return string representation for str()."""
35          return f'{self.face} of {self.suit}'
36
```

When the preceding section's IPython session printed the entire deck, you saw that the `Card`s were displayed in four left-aligned columns. As you'll see in the `__str__` method of class `DeckOfCards`, we use f-strings to format the `Card`s in fields of 19 characters each. Class `Card`'s special method **`__format__`** is called when a `Card` object is *formatted* as a string, such as in an f-string:

```
37      def __format__(self, format):
38          """Return formatted string representation for str()."""
39          return f'{str(self):{format}}'
```

This method's second argument is the format string used to format the object. To use the format parameter's value as the format specifier, enclose the parameter name in braces to the *right* of the colon. In this case, we're formatting the Card object's string representation returned by str(self). We'll discuss __format__ again when we present the __str__ method in class DeckOfCards.

10.6.3 Class DeckOfCards

Class DeckOfCards has a class attribute NUMBER_OF_CARDS, representing the number of Cards in a deck, and creates two data attributes:

- _current_card keeps track of which Card will be dealt next (0–51) and
- _deck (line 12) is a list of 52 Card objects.

Method __init__

DeckOfCards method __init__ initializes a _deck of Cards. The for statement fills the list _deck by appending new Card objects, each initialized with two strings—one from the list Card.FACES and one from Card.SUITS. The calculation count % 13 *always* results in a value from 0 to 12 (the 13 indices of Card.FACES), and the calculation count // 13 *always* results in a value from 0 to 3 (the four indices of Card.SUITS). When the _deck list is initialized, it contains the Cards with faces 'Ace' through 'King' in order for all the Hearts, then the Diamonds, then the Clubs, then the Spades.

```
1  # deck.py
2  """Deck class represents a deck of Cards."""
3  import random
4  from card import Card
5
6  class DeckOfCards:
7      NUMBER_OF_CARDS = 52  # constant number of Cards
8
9      def __init__(self):
10         """Initialize the deck."""
11         self._current_card = 0
12         self._deck = []
13
14         for count in range(DeckOfCards.NUMBER_OF_CARDS):
15             self._deck.append(Card(Card.FACES[count % 13],
16                 Card.SUITS[count // 13]))
17
```

Method shuffle

Method shuffle resets _current_card to 0, then shuffles the Cards in _deck using the random module's shuffle function:

```
18     def shuffle(self):
19         """Shuffle deck."""
20         self._current_card = 0
21         random.shuffle(self._deck)
22
```

Method deal_card

Method `deal_card` deals one `Card` from `_deck`. Recall that `_current_card` indicates the index (0–51) of the next `Card` to be dealt (that is, the `Card` at the top of the deck). Line 26 tries to get the `_deck` element at index `_current_card`. If successful, the method increments `_current_card` by 1, then returns the `Card` being dealt; otherwise, the method returns `None` to indicate there are no more `Cards` to deal.

```
23      def deal_card(self):
24          """Return one Card."""
25          try:
26              card = self._deck[self._current_card]
27              self._current_card += 1
28              return card
29          except:
30              return None
31
```

Method __str__

Class `DeckOfCards` also defines special method `__str__` to get a string representation of the deck in four columns with each `Card` left aligned in a field of 19 characters. When line 37 formats a given `Card`, its `__format__` special method is called with format specifier `'<19'` as the method's `format` argument. Method `__format__` then uses `'<19'` to create the `Card`'s formatted string representation.

```
32      def __str__(self):
33          """Return a string representation of the current _deck."""
34          s = ''
35
36          for index, card in enumerate(self._deck):
37              s += f'{self._deck[index]:<19}'
38              if (index + 1) % 4 == 0:
39                  s += '\n'
40
41          return s
```

10.6.4 Displaying Card Images with Matplotlib

So far, we've displayed `Cards` as text. Now, let's display `Card` images. For this demonstration, we downloaded public-domain[6] card images from Wikimedia Commons:

> https://commons.wikimedia.org/wiki/
>   Category:SVG_English_pattern_playing_cards

These are located in the `ch10` examples folder's `card_images` subfolder. First, let's create a `DeckOfCards`:

```
In [1]: from deck import DeckOfCards

In [2]: deck_of_cards = DeckOfCards()
```

---

6. https://creativecommons.org/publicdomain/zero/1.0/deed.en.

10.6 Case Study: Card Shuffling and Dealing Simulation

Enable Matplotlib in IPython
Next, enable Matplotlib support in IPython by using the %matplotlib magic:

```
In [3]: %matplotlib
Using matplotlib backend: Qt5Agg
```

Create the Base Path for Each Image
Before displaying each image, we must load it from the card_images folder. We'll use the pathlib module's **Path class** to construct the full path to each image on our system. Snippet [5] creates a Path object for the current folder (the ch10 examples folder), which is represented by '.', then uses Path method **joinpath** to append the subfolder containing the card images:

```
In [4]: from pathlib import Path
```

```
In [5]: path = Path('.').joinpath('card_images')
```

Import the Matplotlib Features
Next, let's import the Matplotlib modules we'll need to display the images. We'll use a function from **matplotlib.image** to load the images:

```
In [6]: import matplotlib.pyplot as plt
```

```
In [7]: import matplotlib.image as mpimg
```

Create the Figure and Axes Objects
The following snippet uses Matplotlib function **subplots** to create a Figure object in which we'll display the images as 52 *subplots* with four rows (nrows) and 13 columns (ncols). The function returns a tuple containing the Figure and an array of the subplots' Axes objects. We unpack these into variables figure and axes_list:

```
In [8]: figure, axes_list = plt.subplots(nrows=4, ncols=13)
```

When you execute this statement in IPython, the Matplotlib window appears immediately with 52 empty subplots.

Configure the Axes Objects and Display the Images
Next, we iterate through all the Axes objects in axes_list. Recall that ravel provides a one-dimensional view of a multidimensional array. For each Axes object, we perform the following tasks:

- We're not plotting data, so we do not need axis lines and labels for each image. The first two statements in the loop hide the *x*- and *y*-axes.
- The third statement deals a Card and gets its image_name.
- The fourth statement uses Path method joinpath to append the image_name to the Path, then calls Path method **resolve** to determine the full path to the image on our system. We pass the resulting Path object to the built-in str function to get the string representation of the image's location. Then, we pass that string to the matplotlib.image module's **imread function**, which loads the image.
- The last statement calls Axes method **imshow** to display the current image in the current subplot.

```
In [9]: for axes in axes_list.ravel():
   ...:     axes.get_xaxis().set_visible(False)
   ...:     axes.get_yaxis().set_visible(False)
   ...:     image_name = deck_of_cards.deal_card().image_name
   ...:     img = mpimg.imread(str(path.joinpath(image_name).resolve()))
   ...:     axes.imshow(img)
   ...:
```

Maximize the Image Sizes

At this point, all the images are displayed. To make the cards as large as possible, you can maximize the window, then call the Matplotlib `Figure` object's **`tight_layout`** method. This removes most of the extra white space in the window:

```
In [10]: figure.tight_layout()
```

The following image shows the contents of the resulting window:

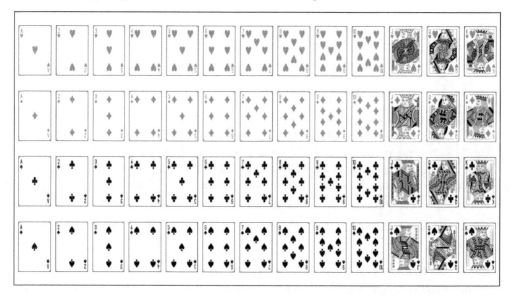

Shuffle and Re-Deal the Deck

To see the images shuffled, call method `shuffle`, then re-execute snippet [9]'s code:

```
In [11]: deck_of_cards.shuffle()

In [12]: for axes in axes_list.ravel():
    ...:     axes.get_xaxis().set_visible(False)
    ...:     axes.get_yaxis().set_visible(False)
    ...:     image_name = deck_of_cards.deal_card().image_name
    ...:     img = mpimg.imread(str(path.joinpath(image_name).resolve()))
    ...:     axes.imshow(img)
    ...:
```

10.6 Case Study: Card Shuffling and Dealing Simulation

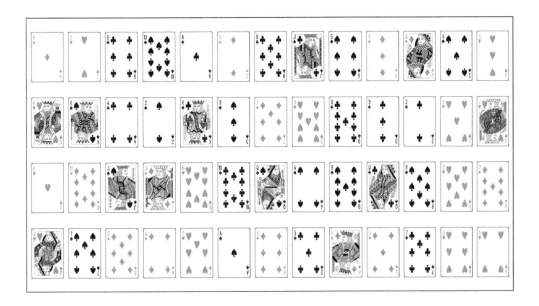

✓ Self Check

1 *(Fill-In)* Matplotlib function _____ returns a tuple containing a `Figure` and an array of `Axes` objects.
Answer: `subplots`.

2 *(True/False)* `Path` method `appendpath` appends to a `Path` object.
Answer: False. `Path` method `joinpath` appends to a `Path` object

3 *(Fill-In)* The `Path` object `Path('.')` represents _____.
Answer: the current folder from which the code was executed.

4 *(IPython Session)* Continue this section's session by reshuffling the cards, then creating a new `Figure` containing two rows of five cards each—these might represent two five-card poker hands.

```
In [13]: deck_of_cards.shuffle()

In [14]: figure, axes_list = plt.subplots(nrows=2, ncols=5)

In [15]: for axes in axes_list.ravel():
    ...:     axes.get_xaxis().set_visible(False)
    ...:     axes.get_yaxis().set_visible(False)
    ...:     image_name = deck_of_cards.deal_card().image_name
    ...:     img = mpimg.imread(str(path.joinpath(image_name).resolve()))
    ...:     axes.imshow(img)
    ...:

In [16]: figure.tight_layout()
```

Answer:

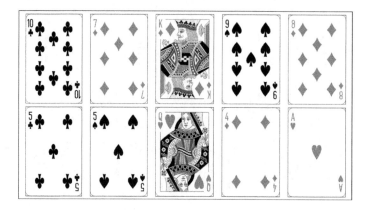

10.7 Inheritance: Base Classes and Subclasses

Often, an object of one class *is an* object of another class as well. For example, a `CarLoan` *is a* `Loan` as are `HomeImprovementLoans` and `MortgageLoans`. Class `CarLoan` can be said to inherit from class `Loan`. In this context, class `Loan` is a base class, and class `CarLoan` is a subclass. A `CarLoan` *is a* specific type of `Loan`, but it's incorrect to claim that every `Loan` *is a* `CarLoan`—the `Loan` could be any type of loan. The following table lists several simple examples of base classes and subclasses—base classes tend to be "more general" and subclasses "more specific":

Base class	Subclasses
Student	GraduateStudent, UndergraduateStudent
Shape	Circle, Triangle, Rectangle, Sphere, Cube
Loan	CarLoan, HomeImprovementLoan, MortgageLoan
Employee	Faculty, Staff
BankAccount	CheckingAccount, SavingsAccount

Because every subclass object *is an* object of its base class, and one base class can have many subclasses, the set of objects represented by a base class is often larger than the set of objects represented by any of its subclasses. For example, the base class `Vehicle` represents *all* vehicles, including cars, trucks, boats, bicycles and so on. By contrast, subclass `Car` represents a smaller, more specific subset of vehicles.

CommunityMember Inheritance Hierarchy
Inheritance relationships form tree-like *hierarchical* structures. A base class exists in a hierarchical relationship with its subclasses. Let's develop a sample class hierarchy (shown in the following diagram), also called an **inheritance hierarchy**. A university community has thousands of members, including employees, students and alumni. Employees are either faculty or staff members. Faculty members are either administrators (e.g., deans and department chairpersons) or teachers. The hierarchy could contain many other classes. For example, stu-

dents can be graduate or undergraduate students. Undergraduate students can be freshmen, sophomores, juniors or seniors. With **single inheritance**, a class is derived from *one* base class. With **multiple inheritance**, a subclass inherits from *two or more* base classes. Single inheritance is straightforward. Multiple inheritance is beyond the scope of this book. Before you use it, search online for the "diamond problem in Python multiple inheritance."

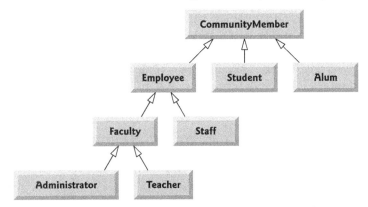

Each arrow in the hierarchy represents an *is-a* relationship. As we follow the arrows upward in this class hierarchy, we can state, for example, that "an Employee *is a* CommunityMember" and "a Teacher *is a* Faculty member." CommunityMember is the direct base class of Employee, Student and Alum and is an indirect base class of all the other classes in the diagram. Starting from the bottom, you can follow the arrows and apply the *is-a* relationship up to the topmost superclass. For example, Administrator *is a* Faculty member, *is an* Employee, *is a* CommunityMember and, of course, ultimately *is an* object.

Shape Inheritance Hierarchy
Now consider the Shape inheritance hierarchy in the following class diagram, which begins with base class Shape, followed by subclasses TwoDimensionalShape and ThreeDimensionalShape. Each Shape is either a TwoDimensionalShape or a ThreeDimensionalShape. The third level of this hierarchy contains *specific* types of TwoDimensionalShapes and ThreeDimensionalShapes. Again, we can follow the arrows from the bottom of the diagram to the topmost base class in this class hierarchy to identify several *is-a* relationships. For example, a Triangle *is a* TwoDimensionalShape and *is a* Shape, while a Sphere *is a* ThreeDimensionalShape and *is a* Shape. This hierarchy could contain many other classes. For example, ellipses and trapezoids also are TwoDimensionalShapes, and cones and cylinders also are ThreeDimensionalShapes.

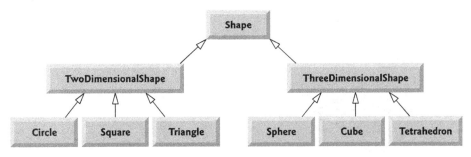

"is a" vs. "has a"

Inheritance produces **"is-a" relationships** in which an object of a subclass type may also be treated as an object of the base-class type. You've also seen "has-a" (composition) relationships in which a class has references to one or more objects of other classes as members.

 Self Check

1 *(Fill-In)* A base class exists in a(n) _____ relationship with its subclasses.
Answer: hierarchical.

2 *(Fill-In)* In this section's Shape class hierarchy, TwoDimensionalShape is a(n) _____ of Shape and a(n) _____ of Circle, Square and Triangle.
Answer: subclass, base class.

10.8 Building an Inheritance Hierarchy; Introducing Polymorphism

Let's use a hierarchy containing types of employees in a company's payroll app to discuss the relationship between a base class and its subclass. All employees of the company have a lot in common, but *commission employees* (who will be represented as objects of a base class) are paid a percentage of their sales, while *salaried commission employees* (who will be represented as objects of a subclass) receive a percentage of their sales *plus* a base salary.

First, we present *base class* CommissionEmployee. Next, we create a *subclass* SalariedCommissionEmployee that inherits from class CommissionEmployee. Then, we use an IPython session to create a SalariedCommissionEmployee object and demonstrate that it has all the capabilities of the base class *and* the subclass, but calculates its earnings differently.

10.8.1 Base Class CommissionEmployee

Consider class CommissionEmployee, which provides the following features:

- Method __init__ (lines 8–15), which creates the data attributes _first_name, _last_name and _ssn (Social Security number), and uses the setters of properties gross_sales and commission_rate to create their corresponding data attributes.

- Read-only properties first_name (lines 17–19), last_name (lines 21–23) and ssn (line 25–27), which return the corresponding data attributes.

- Read-write properties gross_sales (lines 29–39) and commission_rate (lines 41–52), in which the setters perform data validation.

- Method earnings (lines 54–56), which calculates and returns a CommissionEmployee's earnings.

- Method __repr__ (lines 58–64), which returns a string representation of a CommissionEmployee.

```
1   # commmissionemployee.py
2   """CommissionEmployee base class."""
3   from decimal import Decimal
4
```

10.8 Building an Inheritance Hierarchy; Introducing Polymorphism

```python
 5  class CommissionEmployee:
 6      """An employee who gets paid commission based on gross sales."""
 7
 8      def __init__(self, first_name, last_name, ssn,
 9                   gross_sales, commission_rate):
10          """Initialize CommissionEmployee's attributes."""
11          self._first_name = first_name
12          self._last_name = last_name
13          self._ssn = ssn
14          self.gross_sales = gross_sales  # validate via property
15          self.commission_rate = commission_rate  # validate via property
16
17      @property
18      def first_name(self):
19          return self._first_name
20
21      @property
22      def last_name(self):
23          return self._last_name
24
25      @property
26      def ssn(self):
27          return self._ssn
28
29      @property
30      def gross_sales(self):
31          return self._gross_sales
32
33      @gross_sales.setter
34      def gross_sales(self, sales):
35          """Set gross sales or raise ValueError if invalid."""
36          if sales < Decimal('0.00'):
37              raise ValueError('Gross sales must be >= to 0')
38
39          self._gross_sales = sales
40
41      @property
42      def commission_rate(self):
43          return self._commission_rate
44
45      @commission_rate.setter
46      def commission_rate(self, rate):
47          """Set commission rate or raise ValueError if invalid."""
48          if not (Decimal('0.0') < rate < Decimal('1.0')):
49              raise ValueError(
50                  'Interest rate must be greater than 0 and less than 1')
51
52          self._commission_rate = rate
53
54      def earnings(self):
55          """Calculate earnings."""
56          return self.gross_sales * self.commission_rate
57
```

```
58      def __repr__(self):
59          """Return string representation for repr()."""
60          return ('CommissionEmployee: ' +
61              f'{self.first_name} {self.last_name}\n' +
62              f'social security number: {self.ssn}\n' +
63              f'gross sales: {self.gross_sales:.2f}\n' +
64              f'commission rate: {self.commission_rate:.2f}')
```

Properties first_name, last_name and ssn are read-only. We chose not to validate them, though we could have. For example, we could validate the first and last names—perhaps by ensuring that they're of a reasonable length. We could validate the Social Security number to ensure that it contains nine digits, with or without dashes (for example, to ensure that it's in the format ###-##-#### or #########, where each # is a digit).

All Classes Inherit Directly or Indirectly from Class object

You use inheritance to create new classes from existing ones. In fact, *every* Python class inherits from an existing class. When you do not explicitly specify the base class for a new class, Python assumes that the class inherits directly from class object. The Python class hierarchy begins with class object, the direct or indirect base class of *every* class. So, class CommissionEmployee's header could have been written as

```
class CommissionEmployee(object):
```

The parentheses after CommissionEmployee indicate inheritance and may contain a single class for single inheritance or a comma-separated list of base classes for multiple inheritance. Once again, multiple inheritance is beyond the scope of this book.

Class CommissionEmployee inherits all the methods of class object. Class object does not have any data attributes. Two of the many methods inherited from object are __repr__ and __str__. So *every* class has these methods that return string representations of the objects on which they're called. When a base-class method implementation is inappropriate for a derived class, that method can be **overridden** (i.e., redefined) in the derived class with an appropriate implementation. Method __repr__ (lines 58–64) overrides the default implementation inherited into class CommissionEmployee from class object.[7]

Testing Class CommissionEmployee

Let's quickly test some of CommissionEmployee's features. First, create and display a CommissionEmployee:

```
In [1]: from commissionemployee import CommissionEmployee

In [2]: from decimal import Decimal

In [3]: c = CommissionEmployee('Sue', 'Jones', '333-33-3333',
   ...:     Decimal('10000.00'), Decimal('0.06'))
   ...:

In [4]: c
Out[4]:
CommissionEmployee: Sue Jones
social security number: 333-33-3333
gross sales: 10000.00
commission rate: 0.06
```

---

7. See https://docs.python.org/3/reference/datamodel.html for object's overridable methods.

10.8 Building an Inheritance Hierarchy; Introducing Polymorphism

Next, let's calculate and display the CommissionEmployee's earnings:

```
In [5]: print(f'{c.earnings():,.2f}')
600.00
```

Finally, let's change the CommissionEmployee's gross sales and commission rate, then recalculate the earnings:

```
In [6]: c.gross_sales = Decimal('20000.00')

In [7]: c.commission_rate = Decimal('0.1')

In [8]: print(f'{c.earnings():,.2f}')
2,000.00
```

Self Check

1 *(Fill-In)* When a base-class method implementation is inappropriate for a derived class, that method can be _____ (i.e., redefined) in the derived class with an appropriate implementation.
Answer: overridden.

2 *(What Does This Code Do?)* In this section's IPython session, explain in detail what snippet [6] does:

```
c.gross_sales = Decimal('20000.00')
```

Answer: This statement creates a Decimal object and assigns it to a CommissionEmployee's gross_sales property, invoking the property's setter. The setter checks whether the new value is less than Decimal('0.00'). If so, the setter raises a ValueError, indicating that the value must be greater than or equal to 0; otherwise, the setter assigns the new value to the CommissionEmployee's _gross_sales attribute.

10.8.2 Subclass SalariedCommissionEmployee

With single inheritance, the subclass starts essentially the same as the base class. The real strength of inheritance comes from the ability to define in the subclass additions, replacements or refinements for the features inherited from the base class.

Many of a SalariedCommissionEmployee's capabilities are similar, if not identical, to those of class CommissionEmployee. Both types of employees have first name, last name, Social Security number, gross sales and commission rate data attributes, and properties and methods to manipulate that data. To create class SalariedCommissionEmployee *without* using inheritance, we could have *copied* class CommissionEmployee's code and *pasted* it into class SalariedCommissionEmployee. Then we could have modified the new class to include a base salary data attribute, and the properties and methods that manipulate the base salary, including a new earnings method. This *copy-and-paste approach* is often error-prone. Worse yet, it can spread many physical copies of the same code (including errors) throughout a system, making your code less maintainable. Inheritance enables us to "absorb" the features of a class *without* duplicating code. Let's see how.

Declaring Class SalariedCommissionEmployee

We now declare the subclass SalariedCommissionEmployee, which *inherits* most of its capabilities from class CommissionEmployee (line 6). A SalariedCommissionEmployee *is* a CommissionEmployee (because inheritance passes on the capabilities of class CommissionEmployee), but class SalariedCommissionEmployee also has the following features:

- Method __init__ (lines 10–15), which initializes all the data inherited from class CommissionEmployee (we'll say more about this momentarily), then uses the base_salary property's setter to create a _base_salary data attribute.
- Read-write property base_salary (lines 17–27), in which the setter performs data validation.
- A customized version of method earnings (lines 29–31).
- A customized version of method __repr__ (lines 33–36).

```
1   # salariedcommissionemployee.py
2   """SalariedCommissionEmployee derived from CommissionEmployee."""
3   from commissionemployee import CommissionEmployee
4   from decimal import Decimal
5
6   class SalariedCommissionEmployee(CommissionEmployee):
7       """An employee who gets paid a salary plus
8       commission based on gross sales."""
9
10      def __init__(self, first_name, last_name, ssn,
11                   gross_sales, commission_rate, base_salary):
12          """Initialize SalariedCommissionEmployee's attributes."""
13          super().__init__(first_name, last_name, ssn,
14                           gross_sales, commission_rate)
15          self.base_salary = base_salary  # validate via property
16
17      @property
18      def base_salary(self):
19          return self._base_salary
20
21      @base_salary.setter
22      def base_salary(self, salary):
23          """Set base salary or raise ValueError if invalid."""
24          if salary < Decimal('0.00'):
25              raise ValueError('Base salary must be >= to 0')
26
27          self._base_salary = salary
28
29      def earnings(self):
30          """Calculate earnings."""
31          return super().earnings() + self.base_salary
32
33      def __repr__(self):
34          """Return string representation for repr()."""
35          return ('Salaried' + super().__repr__() +
36              f'\nbase salary: {self.base_salary:.2f}')
```

Inheriting from Class CommissionEmployee

To inherit from a class, you must first import its definition (line 3). Line 6

```
class SalariedCommissionEmployee(CommissionEmployee):
```

10.8 Building an Inheritance Hierarchy; Introducing Polymorphism

specifies that class `SalariedCommissionEmployee` *inherits* from `CommissionEmployee`. Though you do not see class `CommissionEmployee`'s data attributes, properties and methods in class `SalariedCommissionEmployee`, they're nevertheless part of the new class, as you'll soon see.

Method __init__ and Built-In Function super
Each subclass __init__ must explicitly call its base class's __init__ to initialize the data attributes inherited from the base class. This call should be the first statement in the subclass's __init__ method. SalariedCommissionEmployee's __init__ method explicitly calls class CommissionEmployee's __init__ method (lines 13–14) to initialize the base-class portion of a SalariedCommissionEmployee object (that is, the five inherited data attributes from class CommissionEmployee). The notation super().__init__ uses the built-in function **super** to locate and call the base class's __init__ method, passing the five arguments that initialize the inherited data attributes.

Overriding Method earnings
Class SalariedCommissionEmployee's earnings method (lines 29–31) overrides class CommissionEmployee's earnings method (Section 10.8.1, lines 54–56) to calculate the earnings of a SalariedCommissionEmployee. The new version obtains the portion of the earnings based on *commission alone* by calling CommissionEmployee's earnings method with the expression super().earnings() (line 31). SalariedCommissionEmployee's earnings method then adds the base_salary to this value to calculate the total earnings. By having SalariedCommissionEmployee's earnings method invoke CommissionEmployee's earnings method to calculate part of a SalariedCommissionEmployee's earnings, we avoid duplicating the code and reduce code-maintenance problems.

Overriding Method __repr__
SalariedCommissionEmployee's __repr__ method (lines 33–36) overrides class CommissionEmployee's __repr__ method (Section 10.8.1, lines 58–64) to return a String representation that's appropriate for a SalariedCommissionEmployee. The subclass creates part of the string representation by concatenating 'Salaried' and the string returned by super().__repr__(), which calls CommissionEmployee's __repr__ method. The overridden method then concatenates the base salary information and returns the resulting string.

Testing Class SalariedCommissionEmployee
Let's test class SalariedCommissionEmployee to show that it indeed inherited capabilities from class CommissionEmployee. First, let's create a SalariedCommissionEmployee and print all of its properties:

```
In [9]: from salariedcommissionemployee import SalariedCommissionEmployee

In [10]: s = SalariedCommissionEmployee('Bob', 'Lewis', '444-44-4444',
   ...:         Decimal('5000.00'), Decimal('0.04'), Decimal('300.00'))
   ...:

In [11]: print(s.first_name, s.last_name, s.ssn, s.gross_sales,
   ...:         s.commission_rate, s.base_salary)
Bob Lewis 444-44-4444 5000.00 0.04 300.00
```

Notice that the SalariedCommissionEmployee object has *all* of the properties of classes CommissionEmployee *and* SalariedCommissionEmployee.

Next, let's calculate and display the SalariedCommissionEmployee's earnings. Because we call method earnings on a SalariedCommissionEmployee object, the *subclass version* of the method executes:

```
In [12]: print(f'{s.earnings():,.2f}')
500.00
```

Now, let's modify the gross_sales, commission_rate and base_salary properties, then display the updated data via the SalariedCommissionEmployee's __repr__ method:

```
In [13]: s.gross_sales = Decimal('10000.00')

In [14]: s.commission_rate = Decimal('0.05')

In [15]: s.base_salary = Decimal('1000.00')

In [16]: print(s)
SalariedCommissionEmployee: Bob Lewis
social security number: 444-44-4444
gross sales: 10000.00
commission rate: 0.05
base salary: 1000.00
```

Again, because this method is called on a SalariedCommissionEmployee object, the *subclass version* of the method executes. Finally, let's calculate and display the SalariedCommissionEmployee's updated earnings:

```
In [17]: print(f'{s.earnings():,.2f}')
1,500.00
```

Testing the "is a" Relationship
Python provides two built-in functions—**issubclass** and **isinstance**—for testing "is a" relationships. Function issubclass determines whether one class is derived from another:

```
In [18]: issubclass(SalariedCommissionEmployee, CommissionEmployee)
Out[18]: True
```

Function isinstance determines whether an object has an "is a" relationship with a specific type. Because SalariedCommissionEmployee inherits from CommissionEmployee, both of the following snippets return True, confirming the "is a" relationship

```
In [19]: isinstance(s, CommissionEmployee)
Out[19]: True

In [20]: isinstance(s, SalariedCommissionEmployee)
Out[20]: True
```

✓ Self Check

1 *(Fill-In)* Function _____ determines whether an object has an "is a" relationship with a specific type.
Answer: isinstance.

2 *(Fill-In)* Function _____ determines whether one class is derived from another.
Answer: issubclass.

3 *(What Does This Code Do?)* Explain in detail what the following statement from class SalariedCommissionEmployee's earnings method does:

```
return super().earnings() + self.base_salary
```

Answer: This statement calculates a SalariedCommissionEmployee's earnings by using the built-in function super to invoke the base class CommissionEmployee's version of method earnings then adding to the result the base_salary.

10.8.3 Processing CommissionEmployees and SalariedCommissionEmployees Polymorphically

With inheritance, every object of a subclass also may be treated as an object of that subclass's base class. We can take advantage of this "subclass-object-is-a-base-class-object" relationship to perform some interesting manipulations. For example, we can place objects related through inheritance into a list, then iterate through the list and treat each element as a base-class object. This allows a variety of objects to be processed in a *general* way. Let's demonstrate this by placing the CommissionEmployee and SalariedCommissionEmployee objects in a list, then for each element displaying its string representation and earnings:

```
In [21]: employees = [c, s]

In [22]: for employee in employees:
    ...:     print(employee)
    ...:     print(f'{employee.earnings():,.2f}\n')
    ...:
CommissionEmployee: Sue Jones
social security number: 333-33-3333
gross sales: 20000.00
commission rate: 0.10
2,000.00

SalariedCommissionEmployee: Bob Lewis
social security number: 444-44-4444
gross sales: 10000.00
commission rate: 0.05
base salary: 1000.00
1,500.00
```

As you can see, the correct string representation and earnings are displayed for each employee. This is called *polymorphism*—a key capability of object-oriented programming (OOP).

✓ Self Check

1 *(Fill-In)* _____ enables us to take advantage of the "subclass-object-is-a-base-class-object" relationship to process objects in a general way.
Answer: Polymorphism.

10.8.4 A Note About Object-Based and Object-Oriented Programming

Inheritance with method overriding is a powerful way to build software components that are *like* existing components but need to be customized to your application's unique needs. In the Python open-source world, there are a huge number of well-developed class libraries for which your programming style is:

- know what libraries are available,
- know what classes are available,
- make objects of existing classes, and
- send them messages (that is, call their methods).

This style of programming called *object-based programming (OBP)*. When you do composition with objects of known classes, you're still doing object-based programming. Adding inheritance with overriding to customize methods to the unique needs of your applications and possibly process objects polymorphically is called *object-oriented programming (OOP)*. If you do composition with objects of inherited classes, that's also object-oriented programming.

10.9 Duck Typing and Polymorphism

Most other object-oriented programming languages require inheritance-based "is a" relationships to achieve polymorphic behavior. Python is more flexible. It uses a concept called **duck typing**, which the Python documentation describes as:

> *A programming style which does not look at an object's type to determine if it has the right interface; instead, the method or attribute is simply called or used ("If it looks like a duck and quacks like a duck, it must be a duck.").*[8]

So, when processing an object at execution time, its type does not matter. As long as the object has the data attribute, property or method (with the appropriate parameters) you wish to access, the code will work.

Let's reconsider the loop at the end of Section 10.8.3, which processes a list of employees:

```
for employee in employees:
    print(employee)
    print(f'{employee.earnings():,.2f}\n')
```

In Python, this loop works properly as long as `employees` contains only objects that:

- can be displayed with `print` (that is, they have a string representation) and
- have an `earnings` method which can be called with no arguments.

All classes inherit from `object` directly or indirectly, so they *all* inherit the default methods for obtaining string representations that `print` can display. If a class has an `earnings` method that can be called with no arguments, we can include objects of that class in the list `employees`, even if the object's class does not have an "is a" relationship with class `CommissionEmployee`. To demonstrate this, consider class `WellPaidDuck`:

```
In [1]: class WellPaidDuck:
   ...:     def __repr__(self):
   ...:         return 'I am a well-paid duck'
   ...:     def earnings(self):
   ...:         return Decimal('1_000_000.00')
   ...:
```

---

8. https://docs.python.org/3/glossary.html#term-duck-typing.

WellPaidDuck objects, which clearly are not meant to be employees, will work with the preceding loop. To prove this, let's create objects of our classes CommissionEmployee, SalariedCommissionEmployee and WellPaidDuck and place them in a list:

```
In [2]: from decimal import Decimal

In [3]: from commissionemployee import CommissionEmployee

In [4]: from salariedcommissionemployee import SalariedCommissionEmployee

In [5]: c = CommissionEmployee('Sue', 'Jones', '333-33-3333',
   ...:                        Decimal('10000.00'), Decimal('0.06'))
   ...:

In [6]: s = SalariedCommissionEmployee('Bob', 'Lewis', '444-44-4444',
   ...:       Decimal('5000.00'), Decimal('0.04'), Decimal('300.00'))
   ...:

In [7]: d = WellPaidDuck()

In [8]: employees = [c, s, d]
```

Now, let's process the list using the loop from Section 10.8.3. As you can see in the output, Python is able to use duck typing to *polymorphically* process all three objects in the list:

```
In [9]: for employee in employees:
   ...:     print(employee)
   ...:     print(f'{employee.earnings():,.2f}\n')
   ...:
CommissionEmployee: Sue Jones
social security number: 333-33-3333
gross sales: 10000.00
commission rate: 0.06
600.00

SalariedCommissionEmployee: Bob Lewis
social security number: 444-44-4444
gross sales: 5000.00
commission rate: 0.04
base salary: 300.00
500.00

I am a well-paid duck
1,000,000.00
```

10.10 Operator Overloading

You've seen that you can interact with objects by accessing their attributes and properties and by calling their methods. Method-call notation can be cumbersome for certain kinds of operations, such as arithmetic. In these cases, it would be more convenient to use Python's rich set of built-in operators.

This section shows how to use **operator overloading** to define how Python's operators should handle objects of your own types. You've already used operator overloading frequently across wide ranges of types. For example, you've used:

- the + operator for adding numeric values, concatenating lists, concatenating strings and adding a value to every element in a NumPy array.
- the [] operator for accessing elements in lists, tuples, strings and arrays and for accessing the value for a specific key in a dictionary.
- the * operator for multiplying numeric values, repeating a sequence and multiplying every element in a NumPy array by a specific value.

You can overload most operators. For every overloadable operator, class `object` defines a special method, such as __add__ for the addition (+) operator or __mul__ for the multiplication (*) operator. Overriding these methods enables you to define how a given operator works for objects of your custom class. For a complete list of special methods, see

> https://docs.python.org/3/reference/datamodel.html#special-method-names

Operator Overloading Restrictions

There are some restrictions on operator overloading:

- The precedence of an operator cannot be changed by overloading. However, parentheses can be used to force evaluation order in an expression.
- The left-to-right or right-to-left grouping of an operator cannot be changed by overloading.
- The "arity" of an operator—that is, whether it's a unary or binary operator—cannot be changed.
- You cannot create new operators—only existing operators can be overloaded.
- The meaning of how an operator works on objects of built-in types cannot be changed. You cannot, for example, change + so that it subtracts two integers.
- Operator overloading works only with objects of custom classes or with a mixture of an object of a custom class and an object of a built-in type.

Complex Numbers

To demonstrate operator overloading, we'll define a class named `Complex` that represents complex numbers. Complex numbers, like –3 + 4i and 6.2 – 11.73i, have the form

> *realPart* + *imaginaryPart* * i

where i is $\sqrt{-1}$. Like `ints`, `floats` and `Decimals`, complex numbers are arithmetic types. In this section, we'll create a class `Complex` that overloads just the + addition operator and the += augmented assignment, so we can add `Complex` objects using Python's mathematical notations.[9]

10.10.1 Test-Driving Class Complex

First, let's use class `Complex` to demonstrate its capabilities. We'll discuss the class's details in the next section. Import class `Complex` from `complexnumber.py`:

```
In [1]: from complexnumber import Complex
```

---

9. Note that Python has built-in support for complex values. In the exercises, we'll ask you to explore using these built-in capabilities.

Next, create and display a couple of `Complex` objects. Snippets [3] and [5] implicitly call the `Complex` class's __repr__ method to get a string representation of each object:

```
In [2]: x = Complex(real=2, imaginary=4)

In [3]: x
Out[3]: (2 + 4i)

In [4]: y = Complex(real=5, imaginary=-1)

In [5]: y
Out[5]: (5 - 1i)
```

We chose the __repr__ string format shown in snippets [3] and [5] to mimic the __repr__ strings produced by Python's built-in `complex` type.[10]

Now, let's use the + operator to add the `Complex` objects x and y. This expression adds the real parts of the two operands (2 and 5) and the imaginary parts of the two operands (4i and -1i), then returns a new `Complex` object containing the result:

```
In [6]: x + y
Out[6]: (7 + 3i)
```

The + operator does not modify either of its operands:

```
In [7]: x
Out[7]: (2 + 4i)

In [8]: y
Out[8]: (5 - 1i)
```

Finally, let's use the += operator to add y to x and store the result in x. The += operator *modifies* its left operand but not its right operand:

```
In [9]: x += y

In [10]: x
Out[10]: (7 + 3i)

In [11]: y
Out[11]: (5 - 1i)
```

10.10.2 Class Complex Definition

Now that we've seen class `Complex` in action, let's look at its definition to see how those capabilities were implemented.

Method __init__
The class's __init__ method receives parameters to initialize the `real` and `imaginary` data attributes:

---

10. Python uses j rather than i for $\sqrt{-1}$. For example, 3+4j (with no spaces around the operator) creates a complex object with `real` and `imag` attributes. The __repr__ string for this complex value is '(3+4j)'.

```
 1  # complexnumber.py
 2  """Complex class with overloaded operators."""
 3
 4  class Complex:
 5      """Complex class that represents a complex number
 6      with real and imaginary parts."""
 7
 8      def __init__(self, real, imaginary):
 9          """Initialize Complex class's attributes."""
10          self.real = real
11          self.imaginary = imaginary
12
```

Overloaded + Operator

The following overridden special method **__add__** defines how to overload the + operator for use with two `Complex` objects:

```
13      def __add__(self, right):
14          """Overrides the + operator."""
15          return Complex(self.real + right.real,
16              self.imaginary + right.imaginary)
17
```

Methods that overload binary operators must provide two parameters—the *first* (`self`) is the *left* operand and the *second* (`right`) is the *right* operand. Class `Complex`'s __add__ method takes two `Complex` objects as arguments and returns a new `Complex` object containing the sum of the operands' `real` parts and the sum of the operands' `imaginary` parts.

We do *not* modify the contents of either of the original operands. This matches our intuitive sense of how this operator should behave. Adding two numbers does not modify either of the original values.

Overloaded += Augmented Assignment

Lines 18–22 overload special method **__iadd__** to define how the += operator adds two `Complex` objects:

```
18      def __iadd__(self, right):
19          """Overrides the += operator."""
20          self.real += right.real
21          self.imaginary += right.imaginary
22          return self
23
```

Augmented assignments modify their left operands, so method __iadd__ modifies the `self` object, which represents the left operand, then returns `self`.

Method __repr__

Lines 24–28 return the string representation of a `Complex` number.

```
24      def __repr__(self):
25          """Return string representation for repr()."""
26          return (f'({self.real} ' +
27              ('+' if self.imaginary >= 0 else '-') +
28              f' {abs(self.imaginary)}i)')
```

Self Check

1 *(Fill-In)* Suppose a and b are integer variables and a program calculates a + b. Now suppose c and d are string variables and a program performs the concatenation c + d. The two + operators here are clearly being used for different purposes. This is an example of _____.
Answer: operator overloading.

2 *(True/False)* Python allows you to create new operators to overload and to change how existing operators work for built-in types.
Answer: False. Python prohibits you from creating new operators, and operator overloading cannot change how an operator works with built-in types.

3 *(IPython Session)* Modify class Complex to support operators - and -=, then test these operators.
Answer: The new method definitions (located in complexnumber2.py) are:

```
def __sub__(self, right):
    """Overrides the - operator."""
    return Complex(self.real - right.real,
                   self.imaginary - right.imaginary)

def __isub__(self, right):
    """Overrides the -= operator."""
    self.real -= right.real
    self.imaginary -= right.imaginary
    return self
```

```
In [1]: from complexnumber2 import Complex

In [2]: x = Complex(real=2, imaginary=4)

In [3]: y = Complex(real=5, imaginary=-1)

In [4]: x - y
Out[4]: (-3 + 5i)

In [5]: x -= y

In [6]: x
Out[6]: (-3 + 5i)

In [7]: y
Out[7]: (5 - 1i)
```

10.11 Exception Class Hierarchy and Custom Exceptions

In the previous chapter, we introduced exception handling. Every exception is an object of a class in Python's exception class hierarchy[11] or an object of a class that inherits from one of those classes. Exception classes inherit directly or indirectly from base class BaseException and are defined in module **exceptions**.

---

11. https://docs.python.org/3/library/exceptions.html.

Python defines four primary `BaseException` subclasses—`SystemExit`, `KeyboardInterrupt`, `GeneratorExit` and `Exception`:

- `SystemExit` terminates program execution (or terminates an interactive session) and when uncaught does not produce a traceback like other exception types.
- `KeyboardInterrupt` exceptions occur when the user types the interrupt command—*Ctrl* + *C* (or *control* + *C*) on most systems.
- `GeneratorExit` exceptions occur when a generator closes—normally when a generator finishes producing values or when its `close` method is called explicitly.
- `Exception` is the base class for most common exceptions you'll encounter. You've seen exceptions of the `Exception` subclasses `ZeroDivisionError`, `NameError`, `ValueError`, `StatisticsError`, `TypeError`, `IndexError`, `KeyError`, `RuntimeError` and `AttributeError`. Often, `StandardErrors` can be caught and handled, so the program can continue running.

Catching Base-Class Exceptions

One of the benefits of the exception class hierarchy is that an `except` handler can catch exceptions of a particular type or can use a base-class type to catch those base-class exceptions and all related subclass exceptions. For example, an `except` handler that specifies the base class `Exception` can catch objects of *any* subclass of `Exception`. Placing an `except` handler that catches type `Exception` before other `except` handlers is a logic error, because all exceptions would be caught before other exception handlers could be reached. Thus, subsequent exception handlers are unreachable.

Custom Exception Classes

When you raise an exception from your code, you should generally use one of the existing exception classes from the Python Standard Library. However, using the inheritance techniques you learned earlier in this chapter, you can create your own custom exception classes that derive directly or indirectly from class `Exception`. Generally, that's discouraged, especially among novice programmers. Before creating custom exception classes, look for an appropriate existing exception class in the Python exception hierarchy. Define new exception classes only if you need to catch and handle the exceptions differently from other existing exception types. That should be rare.

✓ Self Check

1. *(Fill-In)* Most exceptions you'll encounter inherit from base class _____ and are defined in module _____.
Answer: Exception, exceptions.

2. *(True/False)* When you raise an exception from your code, you should generally use a new exception class.
Answer: False. When you raise an exception from your code, you should generally use one of the existing exception classes from the Python Standard Library.

10.12 Named Tuples

You've used tuples to aggregate several data attributes into a single object. The Python Standard Library's **collections** module also provides **named tuples** that enable you to reference a tuple's members by name rather than by index number.

Let's create a simple named tuple that might be used to represent a card in a deck of cards. First, import function namedtuple:

```
In [1]: from collections import namedtuple
```

Function **namedtuple** creates a subclass of the built-in tuple type. The function's first argument is your new type's name and the second is a list of strings representing the identifiers you'll use to reference the new type's members:

```
In [2]: Card = namedtuple('Card', ['face', 'suit'])
```

We now have a new tuple type named Card that we can use anywhere a tuple can be used. Let's create a Card object, access its members by name and display its string representation:

```
In [3]: card = Card(face='Ace', suit='Spades')

In [4]: card.face
Out[4]: 'Ace'

In [5]: card.suit
Out[5]: 'Spades'

In [6]: card
Out[6]: Card(face='Ace', suit='Spades')
```

Other Named Tuple Features

Each named tuple type has additional methods. The type's **_make class method** (that is, a method called on the *class*) receives an iterable of values and returns an object of the named tuple type:

```
In [7]: values = ['Queen', 'Hearts']

In [8]: card = Card._make(values)

In [9]: card
Out[9]: Card(face='Queen', suit='Hearts')
```

This could be useful, for example, if you have a named tuple type representing records in a CSV file. As you read and tokenize CSV records, you could convert them into named tuple objects.

For a given object of a named tuple type, you can get an **OrderedDict** dictionary representation of the object's member names and values. An OrderedDict remembers the order in which its key–value pairs were inserted in the dictionary:

```
In [10]: card._asdict()
Out[10]: OrderedDict([('face', 'Queen'), ('suit', 'Hearts')])
```

For additional named tuple features see:

> https://docs.python.org/3/library/
>   collections.html#collections.namedtuple

Self Check

1. *(Fill-In)* The Python Standard Library's collections module's _____ function creates a custom tuple type that enables you to reference the tuple's members by name rather than by index number.
Answer: namedtuple.

2. *(IPython Session)* Create a namedtuple called Time with members named hour, minute and second. Then, create a Time object, access its members and display its string representation.
Answer:

```
In [1]: from collections import namedtuple

In [2]: Time = namedtuple('Time', ['hour', 'minute', 'second'])

In [3]: t = Time(13, 30, 45)

In [4]: print(t.hour, t.minute, t.second)
13 30 45

In [5]: t
Out[5]: Time(hour=13, minute=30, second=45)
```

10.13 A Brief Intro to Python 3.7's New Data Classes

Though named tuples allow you to reference their members by name, they're still just tuples, not classes. For some of the benefits of named tuples, plus the capabilities that traditional Python classes provide, you can use Python 3.7's new **data classes**[12] from the Python Standard Library's **dataclasses module**.

Data classes are among Python 3.7's most important new features. They help you build classes *faster* by using more *concise* notation and by *autogenerating* "boilerplate" code that's common in most classes. They could become the preferred way to define many Python classes. In this section, we'll present data-class fundamentals. At the end of the section, we'll provide links to more information.

Data Classes Autogenerate Code

Most classes you'll define provide an __init__ method to create and initialize an object's attributes and a __repr__ method to specify an object's custom string representation. If a class has many data attributes, creating these methods can be tedious.

Data classes *autogenerate* the data attributes and the __init__ and __repr__ methods for you. This can be particularly useful for classes that primarily aggregate related data items. For example, in an application that processes CSV records, you might want a class that represents each record's fields as data attributes in an object. You'll see in an exercise that data classes can be generated *dynamically* from a list of field names.

Data classes also autogenerate method __eq__, which overloads the == operator. Any class that has an __eq__ method also implicitly supports !=. *All* classes inherit class object's default __ne__ (not equals) method implementation, which returns the opposite of __eq__ (or NotImplemented if the class does not define __eq__). Data classes do *not* automatically generate methods for the <, <=, > and >= comparison operators, but they can.

---

12. https://www.python.org/dev/peps/pep-0557/.

10.13.1 Creating a Card Data Class

Let's reimplement class Card from Section 10.6.2 as a data class. The new class is defined in carddataclass.py. As you'll see, defining a data class requires some new syntax. In the subsequent subsections, we'll use our new Card data class in class DeckOfCards to show that it's interchangeable with the original Card class, then discuss some of the benefits of data classes over named tuples and traditional Python classes.

Importing from the dataclasses and typing Modules

The Python Standard Library's dataclasses module defines decorators and functions for implementing data classes. We'll use the **@dataclass decorator** (imported at line 4) to specify that a new class is a data class and causes various code to be written for you. Recall that our original Card class defined *class variables* FACES and SUITS, which are lists of the strings used to initialize Cards. We use ClassVar and List from the Python Standard Library's **typing module** (imported at line 5) to indicate that FACES and SUITS are *class variables* that refer to *lists*. We'll say more about these momentarily:

```
1   # carddataclass.py
2   """Card data class with class attributes, data attributes,
3   autogenerated methods and explicitly defined methods."""
4   from dataclasses import dataclass
5   from typing import ClassVar, List
6
```

Using the @dataclass Decorator

To specify that a class is a *data class*, precede its definition with the @dataclass decorator:[13]

```
7   @dataclass
8   class Card:
```

Optionally, the @dataclass decorator may specify parentheses containing arguments that help the data class determine what autogenerated methods to include. For example, the decorator @dataclass(order=True) would cause the data class to autogenerate overloaded comparison operator methods for <, <=, > and >=. This might be useful, for example, if you need to sort your data-class objects.

Variable Annotations: Class Attributes

Unlike regular classes, data classes declare both class attributes and data attributes *inside* the class, but *outside* the class's methods. In a regular class, only *class attributes* are declared this way, and data attributes typically are created in __init__. Data classes require additional information, or *hints*, to distinguish class attributes from data attributes, which also affects the autogenerated methods' implementation details.

Lines 9–11 define and initialize the *class attributes* FACES and SUITS:

```
9       FACES: ClassVar[List[str]] = ['Ace', '2', '3', '4', '5', '6', '7',
10                                    '8', '9', '10', 'Jack', 'Queen', 'King']
11      SUITS: ClassVar[List[str]] = ['Hearts', 'Diamonds', 'Clubs', 'Spades']
12
```

---

13. https://docs.python.org/3/library/dataclasses.html#module-level-decorators-classes-and-functions.

In lines 9 and 11, The notation

    `: ClassVar[List[str]]`

is a **variable annotation**[14,15] (sometimes called a *type hint*) specifying that `FACES` is a class attribute (`ClassVar`) which refers to a *list* of strings (`List[str]`). `SUITS` also is a class attribute which refers to a list of strings.

Class variables are initialized in their definitions and are specific to the *class*, not individual *objects* of the class. Methods __init__, __repr__ and __eq__, however, are for use with *objects* of the class. When a data class generates these methods, it inspects all the variable annotations and includes only the *data attributes* in the method implementations.

Variable Annotations: Data Attributes

Normally, we create an object's data attributes in the class's __init__ method (or methods called by __init__) via assignments of the form `self.attribute_name = value`. Because a data class *autogenerates* its __init__ method, we need another way to specify data attributes in a data class's definition. We cannot simply place their names inside the class, which generates a `NameError`, as in:

```
In [1]: from dataclasses import dataclass

In [2]: @dataclass
   ...: class Demo:
   ...:     x   # attempting to create a data attribute x
   ...:
-------------------------------------------------------------------------
NameError                                 Traceback (most recent call last)
<ipython-input-2-79ffe37b1ba2> in <module>()
----> 1 @dataclass
      2 class Demo:
      3     x   # attempting to create a data attribute x
      4

<ipython-input-2-79ffe37b1ba2> in Demo()
      1 @dataclass
      2 class Demo:
----> 3     x   # attempting to create a data attribute x
      4

NameError: name 'x' is not defined
```

Like class attributes, each data attribute must be declared with a variable annotation. Lines 13–14 define the data attributes `face` and `suit`. The variable annotation ": str" indicates that each should refer to string objects:

```
13      face: str
14      suit: str
```

---

14. https://www.python.org/dev/peps/pep-0526/.
15. Variable annotations are a recent language feature and are optional for regular classes. You will not see them in most legacy Python code.

10.13 A Brief Intro to Python 3.7's New Data Classes

Defining a Property and Other Methods

Data classes are classes, so they may contain properties and methods and participate in class hierarchies. For this Card data class, we defined the same read-only image_name property and custom special methods __str__ and __format__ as in our original Card class earlier in the chapter:

```
15      @property
16      def image_name(self):
17          """Return the Card's image file name."""
18          return str(self).replace(' ', '_') + '.png'
19
20      def __str__(self):
21          """Return string representation for str()."""
22          return f'{self.face} of {self.suit}'
23
24      def __format__(self, format):
25          """Return formatted string representation."""
26          return f'{str(self):{format}}'
```

Variable Annotation Notes

You can specify variable annotations using built-in type names (like str, int and float), class types or types defined by the typing module (such as ClassVar and List shown earlier). Even with type annotations, Python is still a *dynamically typed language*. So, type annotations are *not* enforced at execution time. So, even though a Card's face is meant to be a string, you can assign any type of object to face, as you'll do in a Self Check exercise.

 Self Check

1 *(Fill-In)* Data classes require _____ that specify each class attribute's or data attribute's data type.
Answer: variable annotations.

2 *(Fill-In)* The _____ decorator specifies that a new class is a data class.
Answer: @dataclass.

3 *(True/False)* The Python Standard Library's annotations module defines the variable annotations that are required in data class definitions.
Answer: False. The typing module defines the variable annotations that are required in data-class definitions.

4 *(True/False)* Data classes have auto-generated <, <=, > and >= operators, by default.
Answer: False. The == and != operators are autogenerated by default. The <, <=, > and >= operators are autogenerated only if the @dataclass decorator specifies the keyword argument order=True.

10.13.2 Using the Card Data Class

Let's demonstrate the new Card data class. First, create a Card:

```
In [1]: from carddataclass import Card

In [2]: c1 = Card(Card.FACES[0], Card.SUITS[3])
```

Next, let's use Card's autogenerated __repr__ method to display the Card:

```
In [3]: c1
Out[3]: Card(face='Ace', suit='Spades')
```

Our custom __str__ method, which print calls when passing it a Card object, returns a string of the form '*face* of *suit*':

```
In [4]: print(c1)
Ace of Spades
```

Let's access our data class's attributes and read-only property:

```
In [5]: c1.face
Out[5]: 'Ace'

In [6]: c1.suit
Out[6]: 'Spades'

In [7]: c1.image_name
Out[7]: 'Ace_of_Spades.png'
```

Next, let's demonstrate that Card objects can be compared via the *autogenerated* == operator and inherited != operator. First, create two additional Card objects—one identical to the first and one different:

```
In [8]: c2 = Card(Card.FACES[0], Card.SUITS[3])

In [9]: c2
Out[9]: Card(face='Ace', suit='Spades')

In [10]: c3 = Card(Card.FACES[0], Card.SUITS[0])

In [11]: c3
Out[11]: Card(face='Ace', suit='Hearts')
```

Now, compare the objects using == and !=:

```
In [12]: c1 == c2
Out[12]: True

In [13]: c1 == c3
Out[13]: False

In [14]: c1 != c3
Out[14]: True
```

Our Card data class is interchangeable with the Card class developed earlier in this chapter. To demonstrate this, we created the deck2.py file containing a copy of class DeckOfCards from earlier in the chapter and imported the Card data class into the file. The following snippets import class DeckOfCards, create an object of the class and print it. Recall that print implicitly calls the DeckOfCards __str__ method, which formats each Card in a field of 19 characters, resulting in a call to each Card's __format__ method. Read each row left-to-right to confirm that all the Cards are displayed in order from each suit (Hearts, Diamonds, Clubs and Spades):

```
In [15]: from deck2 import DeckOfCards  # uses Card data class

In [16]: deck_of_cards = DeckOfCards()
```

```
In [17]: print(deck_of_cards)
Ace of Hearts      2 of Hearts        3 of Hearts        4 of Hearts
5 of Hearts        6 of Hearts        7 of Hearts        8 of Hearts
9 of Hearts        10 of Hearts       Jack of Hearts     Queen of Hearts
King of Hearts     Ace of Diamonds    2 of Diamonds      3 of Diamonds
4 of Diamonds      5 of Diamonds      6 of Diamonds      7 of Diamonds
8 of Diamonds      9 of Diamonds      10 of Diamonds     Jack of Diamonds
Queen of Diamonds  King of Diamonds   Ace of Clubs       2 of Clubs
3 of Clubs         4 of Clubs         5 of Clubs         6 of Clubs
7 of Clubs         8 of Clubs         9 of Clubs         10 of Clubs
Jack of Clubs      Queen of Clubs     King of Clubs      Ace of Spades
2 of Spades        3 of Spades        4 of Spades        5 of Spades
6 of Spades        7 of Spades        8 of Spades        9 of Spades
10 of Spades       Jack of Spades     Queen of Spades    King of Spades
```

Self Check

1. *(IPython Session)* Python is a dynamically typed language, so variable annotations are not enforced on objects of data classes. To prove this, create a Card object, then assign the integer 100 to its face attribute and display the Card. Display the face attribute's type before and after the assignment
Answer:

```
In [1]: from carddataclass import Card

In [2]: c = Card('Ace', 'Spades')

In [3]: c
Out[3]: Card(face='Ace', suit='Spades')

In [4]: type(c.face)
Out[4]: str

In [5]: c.face = 100

In [6]: c
Out[6]: Card(face=100, suit='Spades')

In [7]: type(c.face)
Out[7]: int
```

10.13.3 Data Class Advantages over Named Tuples

Data classes offer several advantages over named tuples[16]:

- Although each named tuple technically represents a different type, a named tuple *is a* tuple and *all* tuples can be compared to one another. So, objects of *different* named tuple types could compare as equal if they have the same number of members and the same values for those members. Comparing objects of different data classes *always* returns False, as does comparing a data class object to a tuple object.

---
16. https://www.python.org/dev/peps/pep-0526/.

- If you have code that unpacks a tuple, adding more members to that tuple breaks the unpacking code. Data class objects cannot be unpacked. So you can add more data attributes to a data class without breaking existing code.
- A data class can be a base class or a subclass in an inheritance hierarchy.

10.13.4 Data Class Advantages over Traditional Classes

Data classes also offer various advantages over the traditional Python classes you saw earlier in this chapter:

- A data class autogenerates `__init__`, `__repr__` and `__eq__`, saving you time.
- A data class can autogenerate the special methods that overload the <, <=, > and >= comparison operators.
- When you change data attributes defined in a data class, then use it in a script or interactive session, the autogenerated code updates automatically. So, you have less code to maintain and debug.
- The required variable annotations for class attributes and data attributes enable you to take advantage of static code analysis tools. So, you might be able to eliminate additional errors before they can occur at execution time.
- Some static code analysis tools and IDEs can inspect variable annotations and issue warnings if your code uses the wrong type. This can help you locate logic errors in your code *before* you execute it. In an end-of-chapter exercise, we ask you to use the static code analysis tool MyPy to demonstrate such warnings.

More Information

Data classes have additional capabilities, such as creating "frozen" instances which do not allow you to assign values to a data class object's attributes after the object is created. For a complete list of data class benefits and capabilities, see

> https://www.python.org/dev/peps/pep-0557/

and

> https://docs.python.org/3/library/dataclasses.html

We'll ask you to experiment with additional data class features in this chapter's exercises.

10.14 Unit Testing with Docstrings and doctest

A key aspect of software development is testing your code to ensure that it works correctly. Even with extensive testing, however, your code may still contain bugs. According to the famous Dutch computer scientist Edsger Dijkstra, "Testing shows the presence, not the absence of bugs."[17]

Module doctest and the `testmod` Function

The Python Standard Library provides the **doctest module** to help you test your code and conveniently retest it after you make modifications. When you execute the doctest mod-

---

17. J. N. Buxton and B. Randell, eds, *Software Engineering Techniques*, April 1970, p. 16. Report on a conference sponsored by the NATO Science Committee, Rome, Italy, 27–31 October 1969

10.14 Unit Testing with Docstrings and doctest

ule's **testmod function**, it inspects your functions', methods' and classes' docstrings looking for sample Python statements preceded by >>>, each followed on the next line by the given statement's expected output (if any).[18] The testmod function then executes those statements and confirms that they produce the expected output. If they do not, testmod reports errors indicating which tests failed so you can locate and fix the problems in your code. Each test you define in a docstring typically tests a specific *unit of code*, such as a function, a method or a class. Such tests are called **unit tests**.

Modified Account Class

The file accountdoctest.py contains the class Account from this chapter's first example. We modified the __init__ method's docstring to include four tests which can be used to ensure that the method works correctly:

- The test in line 11 creates a sample Account object named account1. This statement does not produce any output.
- The test in line 12 shows what the value of account1's name attribute should be if line 11 executed successfully. The sample output is shown in line 13.
- The test in line 14 shows what the value of account1's balance attribute should be if line 11 executed successfully. The sample output is shown in line 15.
- The test in line 18 creates an Account object with an invalid initial balance. The sample output shows that a ValueError exception should occur in this case. For exceptions, the doctest module's documentation recommends showing just the first and last lines of the traceback.[19]

You can intersperse your tests with descriptive text, such as line 17.

```
1   # accountdoctest.py
2   """Account class definition."""
3   from decimal import Decimal
4
5   class Account:
6       """Account class for demonstrating doctest."""
7
8       def __init__(self, name, balance):
9           """Initialize an Account object.
10
11          >>> account1 = Account('John Green', Decimal('50.00'))
12          >>> account1.name
13          'John Green'
14          >>> account1.balance
15          Decimal('50.00')
16
17          The balance argument must be greater than or equal to 0.
18          >>> account2 = Account('John Green', Decimal('-50.00'))
19          Traceback (most recent call last):
20              ...
21          ValueError: Initial balance must be >= to 0.00.
22          """
```

---

18. The notation >>> mimics the standard **python** interpreter's input prompts.
19. https://docs.python.org/3/library/doctest.html?highlight=doctest#module-doctest.

```
23
24              # if balance is less than 0.00, raise an exception
25              if balance < Decimal('0.00'):
26                  raise ValueError('Initial balance must be >= to 0.00.')
27
28              self.name = name
29              self.balance = balance
30
31          def deposit(self, amount):
32              """Deposit money to the account."""
33
34              # if amount is less than 0.00, raise an exception
35              if amount < Decimal('0.00'):
36                  raise ValueError('amount must be positive.')
37
38              self.balance += amount
39
40      if __name__ == '__main__':
41          import doctest
42          doctest.testmod(verbose=True)
```

Module __main__

When you load any module, Python assigns a string containing the module's name to a global attribute of the module called __name__. When you execute a Python source file (such as accountdoctest.py) as a *script*, Python uses the string '__main__' as the module's name. You can use __name__ in an if statement like lines 40–42 to specify code that should execute only if the source file is executed as a *script*. In this example, line 41 imports the doctest module and line 42 calls the module's testmod function to execute the docstring unit tests.

Running Tests

Run the file accountdoctest.py as a script to execute the tests. By default, if you call testmod with no arguments, it does not show test results for *successful* tests. In that case, if you get no output, all the tests executed successfully. In this example, line 42 calls testmod with the keyword argument verbose=True. This tells testmod to produce verbose output showing *every* test's results:

```
Trying:
    account1 = Account('John Green', Decimal('50.00'))
Expecting nothing
ok
Trying:
    account1.name
Expecting:
    'John Green'
ok
Trying:
    account1.balance
Expecting:
    Decimal('50.00')
ok
Trying:
    account2 = Account('John Green', Decimal('-50.00'))
```

```
        Expecting:
            Traceback (most recent call last):
            ...
            ValueError: Initial balance must be >= to 0.00.
        ok
        3 items had no tests:
            __main__
            __main__.Account
            __main__.Account.deposit
        1 items passed all tests:
            4 tests in __main__.Account.__init__
        4 tests in 4 items.
        4 passed and 0 failed.
        Test passed.
```

In verbose mode, testmod shows for each test what it's "Trying" to do and what it's "Expecting" as a result, followed by "ok" if the test is successful. After completing the tests in verbose mode, testmod shows a summary of the results.

To demonstrate a *failed* test, "comment out" lines 25–26 in accountdoctest.py by preceding each with a #, then run accountdoctest.py as a script. To save space, we show just the portions of the doctest output indicating the failed test:

```
        ...
        **********************************************************************
        File "accountdoctest.py", line 18, in __main__.Account.__init__
        Failed example:
            account2 = Account('John Green', Decimal('-50.00'))
        Expected:
            Traceback (most recent call last):
            ...
            ValueError: Initial balance must be >= to 0.00.
        Got nothing
        **********************************************************************
        1 items had failures:
            1 of   4 in __main__.Account.__init__
        4 tests in 4 items.
        3 passed and 1 failed.
        ***Test Failed*** 1 failures.
```

In this case, we see that line 18's test failed. The testmod function was *expecting* a traceback indicating that a ValueError was raised due to the invalid initial balance. That exception did *not* occur, so the test failed. As the programmer responsible for defining this class, this failing test would be an indication that something is wrong with the validation code in your __init__ method.

IPython %doctest_mode Magic

A convenient way to create doctests for existing code is to use an IPython interactive session to test your code, then copy and paste that session into a docstring. IPython's In [] and Out[] prompts are not compatible with doctest, so IPython provides the magic **%doctest_mode** to display prompts in the correct doctest format. The magic toggles between the two prompt styles. The first time you execute %doctest_mode, IPython switches to >>> prompts for input and no output prompts. The second time you execute %doctest_mode, IPython switches back to In [] and Out[] prompts.

Self Check

1 *(Fill-In)* When you execute a Python source file as a script, Python creates a global attribute __name__ and assigns it the string _____.
Answer: '__main__'.

2 *(True/False)* When you execute the `doctest` module's `testmod` function, it inspects your code and automatically creates tests for you.
Answer: False. When you execute the `doctest` module's `testmod` function, it inspects your code's function, method and class docstrings looking for sample Python statements preceded by >>>, each followed on the next line by the given statement's expected output (if any).

3 *(IPython Session)* Add tests to the `deposit` method's docstring, then execute the tests. Your test should create an `Account` object, deposit a valid amount into it, then attempt to deposit an invalid negative amount, which raises a `ValueError`.
Answer: The updated docstring for method deposit is shown below, followed by the verbose `doctest` results:

```
"""Deposit money to the account.

>>> account1 = Account('John Green', Decimal('50.00'))
>>> account1.deposit(Decimal('10.55'))
>>> account1.balance
Decimal('60.55')

>>> account1.deposit(Decimal('-100.00'))
Traceback (most recent call last):
    ...
ValueError: amount must be positive.
"""
```

```
Trying:
    account1 = Account('John Green', Decimal('50.00'))
Expecting nothing
ok
Trying:
    account1.name
Expecting:
    'John Green'
ok
Trying:
    account1.balance
Expecting:
    Decimal('50.00')
ok
Trying:
    account2 = Account('John Green', Decimal('-50.00'))
Expecting:
    Traceback (most recent call last):
        ...
    ValueError: Initial balance must be >= to 0.00.
ok
Trying:
    account1 = Account('John Green', Decimal('50.00'))
```

```
    Expecting nothing
    ok
    Trying:
        account1.deposit(Decimal('10.55'))
    Expecting nothing
    ok
    Trying:
        account1.balance
    Expecting:
        Decimal('60.55')
    ok
    Trying:
        account1.deposit(Decimal('-100.00'))
    Expecting:
        Traceback (most recent call last):
            ...
        ValueError: amount must be positive.
    ok
    2 items had no tests:
        __main__
        __main__.Account
    2 items passed all tests:
       4 tests in __main__.Account.__init__
       4 tests in __main__.Account.deposit
    8 tests in 4 items.
    8 passed and 0 failed.
    Test passed.
```

10.15 Namespaces and Scopes

In the "Functions" chapter, we showed that each identifier has a scope that determines where you can use it in your program, and we introduced the local and global scopes. Here we continue our discussion of scopes with an introduction to namespaces.

Scopes are determined by **namespaces**, which associate identifiers with objects and are implemented "under the hood" as dictionaries. All namespaces are independent of one another. So, the same identifier may appear in multiple namespaces. There are three primary namespaces—local, global and built-in.

Local Namespace

Each function and method has a **local namespace** that associates local identifiers (such as, parameters and local variables) with objects. The local namespace exists from the moment the function or method is called until it terminates and is accessible *only* to that function or method. In a function's or method's suite, *assigning* to a variable that does not exist creates a local variable and adds it to the local namespace. Identifiers in the local namespace are *in scope* from the point at which you define them until the function or method terminates.

Global Namespace

Each module has a **global namespace** that associates a module's global identifiers (such as global variables, function names and class names) with objects. Python creates a module's global namespace when it loads the module. A module's global namespace exists and its identifiers are *in scope* to the code within that module until the program (or interactive session) terminates. An IPython session has its own global namespace for all the identifiers you create in that session.

Each module's global namespace also has an identifier called **__name__** containing the module's name, such as `'math'` for the math module or `'random'` for the random module. As you saw in the previous section's doctest example, __name__ contains `'__main__'` for a .py file that you run as a script.

Built-In Namespace

The **built-in namespace** associates identifiers for Python's built-in functions (such as input and range) and types (such as int, float and str) with objects that define those functions and types. Python creates the built-in namespace when the interpreter starts executing. The built-in namespace's identifiers remain *in scope* for all code until the program (or interactive session) terminates.[20]

Finding Identifiers in Namespaces

When you use an identifier, Python searches for that identifier in the currently accessible namespaces, proceeding from *local* to *global* to *built-in*. To help you understand the namespace search order, consider the following IPython session:

```
In [1]: z = 'global z'

In [2]: def print_variables():
   ...:     y = 'local y in print_variables'
   ...:     print(y)
   ...:     print(z)
   ...:

In [3]: print_variables()
local y in print_variables
global z
```

The identifiers you define in an IPython session are placed in the session's *global* namespace. When snippet [3] calls print_variables, Python searches the *local*, *global* and *built-in* namespaces as follows:

- Snippet [3] is not in a function or method, so the session's *global* namespace and the *built-in* namespace are currently accessible. Python first searches the session's *global* namespace, which contains print_variables. So print_variables is *in scope* and Python uses the corresponding object to call print_variables.

- As print_variables begins executing, Python creates the function's *local* namespace. When function print_variables defines the local variable y, Python adds y to the function's *local* namespace. The variable y is now *in scope* until the function finishes executing.

---

20. This assumes you do not shadow the built-in functions or types by redefining their identifiers in a local or global namespace. We discussed shadowing in the "Functions" chapter.

- Next, `print_variables` calls the *built-in* function `print`, passing y as the argument. To execute this call, Python must resolve the identifiers y and `print`. The identifier y is defined in the *local* namespace, so it's *in scope* and Python will use the corresponding object (the string `'local y in print_variables'`) as `print`'s argument. To call the function, Python must find `print`'s corresponding object. First, it looks in the *local* namespace, which does *not* define `print`. Next, it looks in the session's *global* namespace, which does *not* define `print`. Finally, it looks in the *built-in* namespace, which *does* define `print`. So, `print` is *in scope* and Python uses the corresponding object to call `print`.

- Next, `print_variables` calls the *built-in* function `print` again with the argument z, which is *not* defined in the *local* namespace. So, Python looks in the *global* namespace. The argument z *is* defined in the *global* namespace, so z is *in scope* and Python will use the corresponding object (the string `'global z'`) as `print`'s argument. Again, Python finds the identifier `print` in the *built-in* namespace and uses the corresponding object to call `print`.

- At this point, we reach the end of the `print_variables` function's suite, so the function terminates and its *local* namespace no longer exists, meaning the local variable y is now undefined.

To prove that y is undefined, let's try to display y:

```
In [4]: y
-------------------------------------------------------------------------
NameError                                 Traceback (most recent call last)
<ipython-input-4-9063a9f0e032> in <module>()
----> 1 y

NameError: name 'y' is not defined
```

In this case, there's no *local* namespace, so Python searches for y in the session's *global* namespace. The identifier y is *not* defined there, so Python searches for y in the *built-in* namespace. Again, Python does not find y. There are no more namespaces to search, so Python raises a `NameError`, indicating that y is not defined.

The identifiers `print_variables` and z still exist in the session's *global* namespace, so we can continue using them. For example, let's evaluate z to see its value:

```
In [5]: z
Out[5]: 'global z'
```

Nested Functions

One namespace we did not cover in the preceding discussion is the **enclosing namespace**. Python allows you to define **nested functions** inside other functions or methods. For example, if a function or method performs the same task several times, you might define a nested function to avoid repeating code in the enclosing function. When you access an identifier inside a nested function, Python searches the nested function's *local* namespace first, then the *enclosing* function's namespace, then the *global* namespace and finally the *built-in* namespace. This is sometimes referred to as the **LEGB (local, enclosing, global, built-in) rule**. In an exercise, we ask you to create a nested function to demonstrate this namespace search order.

Class Namespace
A class has a namespace in which its class attributes are stored. When you access a class attribute, Python looks for that attribute first in the class's namespace, then in the base class's namespace, and so on, until either it finds the attribute or it reaches class `object`. If the attribute is not found, a `NameError` occurs.

Object Namespace
Each object has its own namespace containing the object's methods and data attributes. The class's `__init__` method starts with an empty object (`self`) and adds each attribute to the object's namespace. Once you define an attribute in an object's namespace, clients using the object may access the attribute's value.

Self Check

1 *(Fill-In)* A function's _____ namespace stores information about identifiers created in the function, such as its parameters and local variables.
Answer: local.

2 *(True/False)* When a function attempts to get an attribute's value, Python searches the local namespace, then the global namespace, then the built-in namespace until it finds the attribute; otherwise, a `NameError` occurs.
Answer: True.

10.16 Intro to Data Science: Time Series and Simple Linear Regression

We've looked at sequences, such as lists, tuples and arrays. In this section, we'll discuss **time series**, which are sequences of values (called **observations**) associated with points in time. Some examples are daily closing stock prices, hourly temperature readings, the changing positions of a plane in flight, annual crop yields and quarterly company profits. Perhaps the ultimate time series is the stream of time-stamped tweets coming from Twitter users worldwide. In the "Data Mining Twitter" chapter, we'll study Twitter data in depth.

In this section, we'll use a technique called simple linear regression to make predictions from time series data. We'll use the 1895 through 2018 January average high temperatures in New York City to predict future average January high temperatures and to estimate the average January high temperatures for years preceding 1895.

In the "Machine Learning" chapter, we'll revisit this example using the scikit-learn library. In the "Deep Learning" chapter, we'll use *recurrent neural networks (RNNs)* to analyze time series.

In later chapters, we'll see that time series are popular in financial applications and with the Internet of Things (IoT), which we'll discuss in the "Big Data: Hadoop, Spark, NoSQL and IoT" chapter.

In this section, we'll display graphs with Seaborn and pandas, which both use Matplotlib, so launch IPython with Matplotlib support:

```
ipython --matplotlib
```

10.16 Intro to Data Science: Time Series and Simple Linear Regression

Time Series
The data we'll use is a time series in which the observations are *ordered* by year. **Univariate time series** have *one* observation per time, such as the average of the January high temperatures in New York City for a particular year. **Multivariate time series** have *two or more* observations per time, such as temperature, humidity and barometric pressure readings in a weather application. Here, we'll analyze a univariate time series.

Two tasks often performed with time series are:

- **Time series analysis**, which looks at existing time series data for patterns, helping data analysts understand the data. A common analysis task is to look for **seasonality** in the data. For example, in New York City, the monthly average high temperature varies significantly based on the seasons (winter, spring, summer or fall).

- **Time series forecasting**, which uses past data to predict the future.

We'll perform time series forecasting in this section.

Simple Linear Regression
Using a technique called **simple linear regression**, we'll make predictions by finding a linear relationship between the months (January of each year) and New York City's average January high temperatures. Given a collection of values representing an **independent variable** (the month/year combination) and a **dependent variable** (the average high temperature for that month/year), simple linear regression describes the relationship between these variables with a straight line, known as the **regression line**.

Linear Relationships
To understand the general concept of a linear relationship, consider Fahrenheit and Celsius temperatures. Given a Fahrenheit temperature, we can calculate the corresponding Celsius temperature using the following formula:

```
c = 5 / 9 * (f - 32)
```

In this formula, f (the Fahrenheit temperature) is the *independent variable*, and c (the Celsius temperature) is the *dependent variable*—each value of c *depends on* the value of f used in the calculation.

Plotting Fahrenheit temperatures and their corresponding Celsius temperatures produces a straight line. To show this, let's first create a lambda for the preceding formula and use it to calculate the Celsius equivalents of the Fahrenheit temperatures 0–100 in 10-degree increments. We store each Fahrenheit/Celsius pair as a tuple in temps:

```
In [1]: c = lambda f: 5 / 9 * (f - 32)

In [2]: temps = [(f, c(f)) for f in range(0, 101, 10)]
```

Next, let's place the data in a DataFrame, then use its **plot** method to display the linear relationship between the Fahrenheit and Celsius temperatures. The plot method's style keyword argument controls the data's appearance. The period in the string '.-' indicates that each point should appear as a dot, and the dash indicates that lines should connect the dots. We manually set the *y*-axis label to 'Celsius' because the plot method shows 'Celsius' only in the graph's upper-left corner legend, by default.

```
In [3]: import pandas as pd

In [4]: temps_df = pd.DataFrame(temps, columns=['Fahrenheit', 'Celsius'])

In [5]: axes = temps_df.plot(x='Fahrenheit', y='Celsius', style='.-')

In [6]: y_label = axes.set_ylabel('Celsius')
```

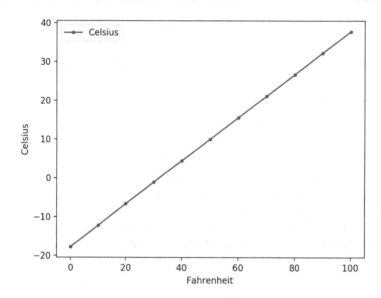

Components of the Simple Linear Regression Equation
The points along any straight line (in two dimensions) like those shown in the preceding graph can be calculated with the equation:

$$y = mx + b$$

where

- m is the line's **slope**,
- b is the line's **intercept** with the y-axis (at $x = 0$),
- x is the independent variable (the date in this example), and
- y is the dependent variable (the temperature in this example).

In simple linear regression, y is the *predicted value* for a given x.

Function linregress from the SciPy's stats Module
Simple linear regression determines the slope (m) and intercept (b) of a straight line that best fits your data. Consider the following diagram, which shows a few of the time-series data points we'll process in this section and a corresponding regression line. We added vertical lines to indicate each data point's distance from the regression line:

10.16 Intro to Data Science: Time Series and Simple Linear Regression 417

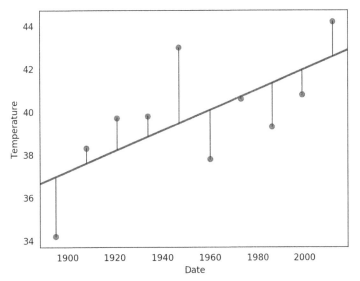

The simple linear regression algorithm iteratively adjusts the slope and intercept and, for each adjustment, calculates the square of each point's distance from the line. The "best fit" occurs when the slope and intercept values minimize the sum of those squared distances. This is known as an **ordinary least squares** calculation.[21]

The **SciPy (Scientific Python) library** is widely used for engineering, science and math in Python. This library's `linregress` function (from the `scipy.stats` module) performs simple linear regression for you. After calling `linregress`, you'll plug the resulting slope and intercept into the $y = mx + b$ equation to make predictions.

Pandas
In the three previous Intro to Data Science sections, you used pandas to work with data. You'll continue using pandas throughout the rest of the book. In this example, we'll load the data for New York City's 1895–2018 average January high temperatures from a CSV file into a `DataFrame`. We'll then format the data for use in this example.

Seaborn Visualization
We'll use Seaborn to plot the `DataFrame`'s data with a regression line that shows the average high-temperature trend over the period 1895–2018.

Getting Weather Data from NOAA
Let's get the data for our study. The National Oceanic and Atmospheric Administration (NOAA)[22] offers lots of public historical data including time series for average high temperatures in specific cities over various time intervals.

We obtained the January average high temperatures for New York City from 1895 through 2018 from NOAA's "Climate at a Glance" time series at:

https://www.ncdc.noaa.gov/cag/

---

21. https://en.wikipedia.org/wiki/Ordinary_least_squares.
22. http://www.noaa.gov.

On that web page, you can select temperature, precipitation and other data for the entire U.S., regions within the U.S., states, cities and more. Once you've set the area and time frame, click **Plot** to display a diagram and view a table of the selected data. At the top of that table are links for downloading the data in several formats including CSV, which we discussed in the "Files and Exceptions" chapter. NOAA's maximum date range available at the time of this writing was 1895–2018. For your convenience, we provided the data in the ch10 examples folder in the file ave_hi_nyc_jan_1895-2018.csv. If you download the data on your own, delete the rows above the line containing "Date,Value,Anomaly".

This data contains three columns per observation:

- Date—A value of the form 'YYYYMM' (such as '201801'). MM is always 01 because we downloaded data for only January of each year.
- Value—A floating-point Fahrenheit temperature.
- Anomaly—The difference between the value for the given date and average values for all dates. We do not use the Anomaly value in this example, so we'll ignore it.

Loading the Average High Temperatures into a DataFrame
Let's load and display the New York City data from ave_hi_nyc_jan_1895-2018.csv:

```
In [7]: nyc = pd.read_csv('ave_hi_nyc_jan_1895-2018.csv')
```

We can look at the DataFrame's head and tail to get a sense of the data:

```
In [8]: nyc.head()
Out[8]:
      Date  Value  Anomaly
0   189501   34.2     -3.2
1   189601   34.7     -2.7
2   189701   35.5     -1.9
3   189801   39.6      2.2
4   189901   36.4     -1.0

In [9]: nyc.tail()
Out[9]:
        Date  Value  Anomaly
119   201401   35.5     -1.9
120   201501   36.1     -1.3
121   201601   40.8      3.4
122   201701   42.8      5.4
123   201801   38.7      1.3
```

Cleaning the Data
We'll soon use Seaborn to graph the Date-Value pairs and a regression line. When plotting data from a DataFrame, Seaborn labels a graph's axes using the DataFrame's column names. For readability, let's rename the 'Value' column as 'Temperature':

```
In [10]: nyc.columns = ['Date', 'Temperature', 'Anomaly']

In [11]: nyc.head(3)
Out[11]:
      Date  Temperature  Anomaly
0   189501         34.2     -3.2
1   189601         34.7     -2.7
2   189701         35.5     -1.9
```

Seaborn labels the tick marks on the *x*-axis with Date values. Since this example processes only January temperatures, the *x*-axis labels will be more readable if they do not contain 01 (for January), we'll remove it from each Date. First, let's check the column's type:

```
In [12]: nyc.Date.dtype
Out[12]: dtype('int64')
```

The values are integers, so we can divide by 100 to truncate the last two digits. Recall that each column in a DataFrame is a Series. Calling Series method floordiv performs *integer division* on every element of the Series:

```
In [13]: nyc.Date = nyc.Date.floordiv(100)

In [14]: nyc.head(3)
Out[14]:
   Date  Temperature  Anomaly
0  1895         34.2     -3.2
1  1896         34.7     -2.7
2  1897         35.5     -1.9
```

Calculating Basic Descriptive Statistics for the Dataset

For some quick statistics on the dataset's temperatures, call describe on the Temperature column. We can see that there are 124 observations, the mean value of the observations is 37.60, and the lowest and highest observations are 26.10 and 47.60 degrees, respectively:

```
In [15]: pd.set_option('precision', 2)

In [16]: nyc.Temperature.describe()
Out[16]:
count    124.00
mean      37.60
std        4.54
min       26.10
25%       34.58
50%       37.60
75%       40.60
max       47.60
Name: Temperature, dtype: float64
```

Forecasting Future January Average High Temperatures

The **SciPy (Scientific Python) library** is widely used for engineering, science and math in Python. Its **stats module** provides function **linregress**, which calculates a regression line's *slope* and *intercept* for a given set of data points:

```
In [17]: from scipy import stats

In [18]: linear_regression = stats.linregress(x=nyc.Date,
    ...:                                     y=nyc.Temperature)
    ...:
```

Function linregress receives two one-dimensional arrays[23] of the same length representing the data points' *x*- and *y*-coordinates. The keyword arguments x and y represent the independent and dependent variables, respectively. The object returned by linregress contains the regression line's slope and intercept:

---

23. These arguments also can be one-dimensional array-like objects, such as lists or pandas Series.

```
In [19]: linear_regression.slope
Out[19]: 0.00014771361132966167

In [20]: linear_regression.intercept
Out[20]: 8.694845520062952
```

We can use these values with the simple linear regression equation for a straight line, $y = mx + b$, to predict the average January temperature in New York City for a given year. Let's predict the average Fahrenheit temperature for January of 2019. In the following calculation, `linear_regression.slope` is m, 2019 is x (the date value for which you'd like to predict the temperature), and `linear_regression.intercept` is b:

```
In [21]: linear_regression.slope * 2019 + linear_regression.intercept
Out[21]: 38.51837136113298
```

We also can approximate what the average temperature might have been in the years before 1895. For example, let's approximate the average temperature for January of 1890:

```
In [22]: linear_regression.slope * 1890 + linear_regression.intercept
Out[22]: 36.612865774980335
```

For this example, we had data for 1895–2018. You should expect that the further you go outside this range, the less reliable the predictions will be.

Plotting the Average High Temperatures and a Regression Line

Next, let's use Seaborn's **regplot function** to plot each data point with the dates on the *x*-axis and the temperatures on the *y*-axis. The regplot function creates the **scatter plot** or **scattergram** below in which the scattered blue dots represent the Temperatures for the given Dates, and the straight line displayed through the points is the regression line:

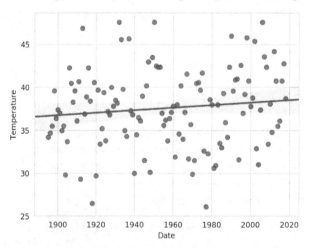

First, close the prior Matplotlib window if you have not done so already—otherwise, regplot will use the existing window that already contains a graph. Function regplot's x and y keyword arguments are one-dimensional arrays[24] of the same length representing the *x*-*y* coordinate pairs to plot. Recall that pandas automatically creates attributes for each column name if the name can be a valid Python identifier:[25]

---

24. These arguments also can be one-dimensional array-like objects, such as lists or pandas Series.

10.16 Intro to Data Science: Time Series and Simple Linear Regression

```
In [23]: import seaborn as sns

In [24]: sns.set_style('whitegrid')

In [25]: axes = sns.regplot(x=nyc.Date, y=nyc.Temperature)
```

The regression line's slope (lower at the left and higher at the right) indicates a warming trend over the last 124 years. In this graph, the y-axis represents a 21.5-degree temperature range between the minimum of 26.1 and the maximum of 47.6, so the data appears to be spread significantly above and below the regression line, making it difficult to see the linear relationship. This is a common issue in data analytics visualizations. When you have axes that reflect different kinds of data (dates and temperatures in this case), how do you reasonably determine their respective scales? In the preceding graph, this is purely an issue of the graph's height—Seaborn and Matplotlib *auto-scale* the axes, based on the data's range of values. We can scale the y-axis range of values to emphasize the linear relationship. Here, we scaled the y-axis from a 21.5-degree range to a 60-degree range (from 10 to 70 degrees):

```
In [26]: axes.set_ylim(10, 70)
Out[26]: (10, 70)
```

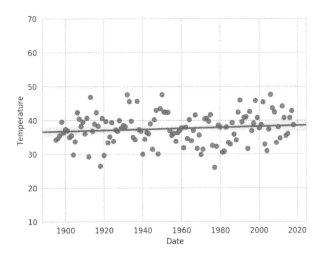

Getting Time Series Datasets
Here are some popular sites where you can find time series to use in your studies:

Time-series dataset sources
https://data.gov/   This is the U.S. government's open data portal. Searching for "time series" yields over 7200 time-series datasets.

---

25. For readers with more statistics background, the shaded area surrounding the regression line is the 95% *confidence interval* for the regression line (https://en.wikipedia.org/wiki/Simple_linear_regression#Confidence_intervals). To draw the diagram without a confidence interval, add the keyword argument ci=None to the regplot function's argument list.

> **Time-series dataset sources**
>
> https://www.ncdc.noaa.gov/cag/
> The National Oceanic and Atmospheric Administration (NOAA) Climate at a Glance portal provides both global and U.S. weather-related time series.
>
> https://www.esrl.noaa.gov/psd/data/timeseries/
> NOAA's Earth System Research Laboratory (ESRL) portal provides monthly and seasonal climate-related time series.
>
> https://www.quandl.com/search
> Quandl provides hundreds of free financial-related time series, as well as fee-based time series.
>
> https://datamarket.com/data/list/?q=provider:tsdl
> The Time Series Data Library (TSDL) provides links to hundreds of time series datasets across many industries.
>
> http://archive.ics.uci.edu/ml/datasets.html
> The University of California Irvine (UCI) Machine Learning Repository contains dozens of time-series datasets for a variety of topics.
>
> http://inforumweb.umd.edu/econdata/econdata.html
> The University of Maryland's EconData service provides links to thousands of economic time series from various U.S. government agencies.

Self Check

1 *(Fill-In)* Time series _____ looks at existing time series data for patterns, helping data analysts understand the data. Time series _____ uses data from the past to predict the future.
Answer: analysis, forecasting.

2 *(True/False)* In the formula, c = 5 / 9 * (f - 32), f (the Fahrenheit temperature) is the independent variable and c (the Celsius temperature) is the dependent variable.
Answer: True.

3 *(IPython Session)* Assuming that this linear trend continues, based on the slope and intercept values calculated in this section's interactive session, in what year might the average January temperature in New York City reach 40 degrees Fahrenheit.
Answer:

```
In [27]: year = 2019

In [28]: slope = linear_regression.slope

In [29]: intercept = linear_regression.intercept

In [30]: temperature = slope * year + intercept

In [31]: while temperature < 40.0:
   ...:     year += 1
   ...:     temperature = slope * year + intercept
   ...:

In [32]: year
Out[32]: 2120
```

10.17 Wrap-Up

In this chapter, we discussed the details of crafting valuable classes. You saw how to define a class, create objects of the class, access an object's attributes and call its methods. You define the special method __init__ to create and initialize a new object's data attributes.

We discussed controlling access to attributes and using properties. We showed that all object attributes may be accessed directly by a client. We discussed identifiers with single leading underscores (_), which indicate attributes that are not meant to be accessed by client code. We showed how to implement "private" attributes via the double-leading-underscore (__) naming convention, which tells Python to mangle an attribute's name.

We implemented a card shuffling and dealing simulation consisting of a Card class and a DeckOfCards class that maintained a list of Cards, and displayed the deck both as strings and as card images using Matplotlib. We introduced special methods __repr__, __str__ and __format__ for creating string representations of objects.

Next, we looked at Python's capabilities for creating base classes and subclasses. We showed how to create a subclass that inherits many of its capabilities from its superclass, then adds more capabilities, possibly by overriding the base class's methods. We created a list containing both base class and subclass objects to demonstrate Python's polymorphic programming capabilities.

We introduced operator overloading for defining how Python's built-in operators work with objects of custom class types. You saw that overloaded operator methods are implemented by overriding various special methods that all classes inherit from class object. We discussed the Python exception class hierarchy and creating custom exception classes.

We showed how to create a named tuple that enables you to access tuple elements via attribute names rather than index numbers. Next, we introduced Python 3.7's new data classes, which can autogenerate various boilerplate code commonly provided in class definitions, such as the __init__, __repr__ and __eq__ special methods.

You saw how to write unit tests for your code in docstrings, then execute those tests conveniently via the doctest module's testmod function. Finally, we discussed the various namespaces that Python uses to determine the scopes of identifiers. In the next chapter, we'll introduce the computer science concepts of recursion, searching and sorting and Big O.

Exercises

10.1 *(What's Wrong with This Code?)* What is wrong with the code in the following IPython session?

```
In [1]: try:
   ...:     raise RuntimeError()
   ...: except Exception:
   ...:     print('An Exception occurred')
   ...: except RuntimeError:
   ...:     print('A RuntimeError occurred')
   ...:
An Exception occurred
```

10.2 *(Account Class with Read-Only Properties)* Modify Section 10.2.2's Account class to provide read-only properties for the name and balance. Rename the class attributes with single leading underscores. Re-execute Section 10.2.2's IPython session to test your updated class. To show that name and balance are read-only, try to assign new values to them.

10.3 *(Time Class Enhancement)* Modify Section 10.4.2's Time class to provide a read-only property universal_str that returns a string representation of a Time in 24-hour clock format with two digits each for the hour, minute and second, as in '22:30:00' (for 10:30 PM) or '06:30:00' (for 6:30 AM). Test your new read-only property.

10.4 *(Modifying the Internal Data Representation of a Class)* Section 10.4.2's Time class represents the time as three integer values. Modify the class to store the time as the total number of seconds since midnight. Replace the _hour, _minute and _second attributes with one _total_seconds attribute. Modify the bodies of the hour, minute and second properties' methods to get and set _total_seconds. Re-execute Section 10.4's IPython session using the modified Time class to show that the updated class Time is interchangeable with the original one.

10.5 *(Duck Typing)* Recall that with duck typing, objects of unrelated classes can respond to the same method calls if they implement those methods. In Section 10.8, you created a list containing a CommissionEmployee and a SalariedCommissionEmployee. Then, you iterated through it, displaying each employee's string representation and earnings. Create a class SalariedEmployee for an employee that gets paid a fixed weekly salary. Do *not* inherit from CommissionEmployee or SalariedCommissionEmployee. In class SalariedEmployee, override method __repr__ and provide an earnings method. Demonstrate duck typing by creating an object of your class, adding it to the list at the end of Section 10.8, then executing the loop to show that it properly processes objects of all three classes.

10.6 *(Composition: A Circle "Has a" Point at Its Center)* A circle has a point at its center. Create a class Point that represents an *(x-y)* coordinate pair and provides x and y read-write properties for the attributes _x and _y. Include __init__ and __repr__ methods, and a move method that receives *x*- and *y*-coordinate values and sets the Point's new location. Create a class Circle that has as its attributes _radius and _point (a Point that represents the Circle's center location). Include __init__ and __repr__ methods, and a move method that receives *x*- and *y*-coordinate values and sets a new location for the Circle by calling the composed Point object's move method. Test your Circle class by creating a Circle object, displaying its string representation, moving the Circle and displaying its string representation again.

10.7 *(Manipulating Dates and Times with Module **datetime**)* The Python Standard Library's **datetime** module contains a **datetime** class for manipulating dates and times. The class provides various overloaded operators. Research class datetime's capabilities, then perform the following tasks:
  a) Get the current date and time and store it in variable x.
  b) Repeat Part (a) and store the result in variable y.
  c) Display each datetime object.
  d) Display each datetime object's data attributes individually.
  e) Use the comparison operators to compare the two datetime objects.
  f) Calculate the difference between y and x.

10.8 *(Converting Data Class Objects to Tuples and Dictionaries)* In some cases, you might want to treat data class objects as tuples or dictionaries. The dataclasses module provides functions **astuple** and **asdict** for this purpose. Research these functions, then create an object of this chapter's Card data class and use these functions to convert the Card to a tuple and a dictionary. Display the results.

10.9 *(Square Class)* Write a class that implements a Square shape. The class should contain a side property. Provide an __init__ method that takes the side length as an argument. Also, provide the following read-only properties:
   a) perimeter returns 4 × side.
   b) area returns side × side.
   c) diagonal returns the square root of the expression (2 × side^2).

The perimeter, area and diagonal should not have corresponding data attributes; rather, they should use side in calculations that return the desired values. Create a Square object and display its side, perimeter, area and diagonal properties' values.

10.10 *(Invoice Class)* Create a class called Invoice that a hardware store might use to represent an invoice for an item sold at the store. An Invoice should include four pieces of information as data attributes—a part number (a string), a part description (a string), a quantity of the item being purchased (an int) and a price per item (a Decimal). Your class should have an __init__ method that initializes the four data attributes. Provide a property for each data attribute. The quantity and price per item should each be non-negative—use validation in the properties for these data attributes to ensure that they remain valid. Provide a calculate_invoice method that returns the invoice amount (that is, multiplies the quantity by the price per item). Demonstrate class Invoice's capabilities.

10.11 *(Class Fraction)* The Python Standard Library module fractions provides a Fraction class that stores the numerator and denominator of a fraction, such as:

$$\frac{2}{4}$$

Research Fraction's capabilities, then demonstrate:
   a) Adding two Fractions.
   b) Subtracting two Fractions.
   c) Multiplying two Fractions.
   d) Dividing two Fractions.
   e) Printing Fractions in the form a/b, where a is the numerator and b is the denominator.
   f) Converting Fractions to floating-point numbers with built-in function float.

10.12 *(Built-in Type complex)* Python supports complex numbers with the built-in type complex. Research complex's capabilities, then demonstrate:
   a) Adding two complex numbers.
   b) Subtracting two complex numbers.
   c) Printing complex numbers.
   d) Getting the real and imaginary parts of complex numbers.

10.13 *(doctest)* Create a script containing the following maximum function:

```
def maximum(value1, value2, value3):
    """Return the maximum of three values."""
    max_value = value1
    
    if value2 > max_value:
        max_value = value2
    
    if value3 > max_value:
        max_value = value3
    
    return max_value
```

Modify the function's docstring to define tests for calling function maximum with three ints, three floats and three strings. For each type, provide three tests—one with the largest value as the first argument, one with the largest value as the second argument, one with the largest value as the third argument. Use doctest to run your tests and confirm that all execute correctly. Next, modify the maximum function to use < operators rather than > operators. Run your tests again to see which tests fail.

10.14 *(Creating an Account Data Class Dynamically)* The dataclasses module's make_dataclass function creates a data class *dynamically* from a list of strings that represent the data class's attributes. Research function make_dataclass, then use it to generate an Account class from the following list of strings:

    ['account', 'name', 'balance']

Create objects of the new Account class, then display their string representations and compare the objects with the == and != operators.

10.15 *(Immutable Data Class Objects)* Built-in types int, float, str and tuple are immutable. Data classes can simulate immutability by designating that objects of the class should be "frozen" after they're created. Client code cannot assign values to the attributes of a frozen object. Research "frozen" data classes, then reimplement this chapter's Complex class as a "frozen" data class. Show that you cannot modify a Complex object after you create it.

10.16 *(Account Inheritance Hierarchy)* Create an inheritance hierarchy that a bank might use to represent customer bank accounts. All customers at this bank can deposit money into their accounts and withdraw money from their accounts. More specific types of accounts also exist. Savings accounts, for instance, earn interest on the money they hold. Checking accounts, on the other hand, don't earn interest and charge a fee per transaction.

Start with class Account from this chapter and create two subclasses SavingsAccount and CheckingAccount. A SavingsAccount should also include a data attribute indicating the interest rate. A SavingsAccount's calculate_interest method should return the Decimal result of multiplying the interest rate by the account balance. SavingsAccount should inherit methods deposit and withdraw without redefining them.

A CheckingAccount should include a Decimal data attribute that represents the fee charged per transaction. Class CheckingAccount should override methods deposit and withdraw so that they subtract the fee from the account balance whenever either transaction is performed successfully. CheckingAccount's versions of these methods should invoke the base-class Account versions to update the account balance. CheckingAccount's withdraw method should charge a fee only if money is withdrawn (that is, the withdrawal amount does not exceed the account balance).

Create objects of each class and tests their methods. Add interest to the SavingsAccount object by invoking its calculate_interest method, then passing the returned interest amount to the object's deposit method.

10.17 *(Nested Functions and Namespaces)* Section 10.15 discussed namespaces and how Python uses them to determine which identifiers are in scope. We also mentioned the LEGB (local, enclosing, global, built-in) rule for the order in which Python searches for identifiers in namespaces. For each of the print function calls in the following IPython session, list the namespaces that Python searches for print's argument:

    In [1]: z = 'global z'

```
In [2]: def print_variables():
   ...:     y = 'local y in print_variables'
   ...:     print(y)
   ...:     print(z)
   ...:     def nested_function():
   ...:         x = 'x in nested function'
   ...:         print(x)
   ...:         print(y)
   ...:         print(z)
   ...:     nested_function()
   ...:

In [3]: print_variables()
local y in print_variables
global z
x in nested function
local y in print_variables
global z
```

10.18 *(Intro to Data Science: Time Series)* Reimplement the Intro to Data Science section's study using the Los Angeles Average January High Temperatures for 1985 through 2018, which can be found in the file ave_hi_la_jan_1895-2018.csv located in the ch10 examples folder. How does the Los Angeles temperature trend compare to that of New York City?

10.19 *(Project: Static Code Analysis with Prospector and MyPy)* In Exercise 3.24, you used the prospector static code analysis tool to check your code for common errors and suggested improvements. The prospector tool includes support for checking variable annotations with the MyPy static code analysis tool. Research MyPy online. Write a script that creates objects of this chapter's Card data class. In the script, assign *integers* to a Card's face and suit string attributes. Then, use MyPy to analyze the script and see the warning messages that MyPy produces. For instructions on using MyPy with prospector, see

> https://github.com/PyCQA/prospector/blob/master/docs/
>   supported_tools.rst

10.20 *(Project: Solitaire)* Using classes Card and DeckOfCards from this chapter's examples, implement your favorite solitaire card game.

10.21 *(Project: Blackjack)* Using the DeckOfCards class from this chapter, create a simple Blackjack game. The rules of the game are as follows:

- Two cards each are dealt to the dealer and the player. The player's cards are dealt face up. Only one of the dealer's cards is dealt face up.
- Each card has a value. A card numbered 2 through 10 is worth its face value. Jacks, queens and kings each count as 10. Aces can count as 1 or 11—whichever value is more beneficial to the player (as we'll soon see).
- If the sum of the player's first two cards is 21 (that is, the player was dealt a card valued at 10 and an ace, which counts as 11 in this situation), the player has "blackjack" and immediately wins the game—if the dealer does not also have blackjack, which would result in a "push" (or tie).
- Otherwise, the player can begin taking additional cards one at a time. These cards are dealt face up, and the player decides when to stop taking cards. If the player "busts" (that is, the sum of the player's cards exceeds 21), the game is over and the

player loses. When the player is satisfied with the current set of cards, the player "stands" (that is, stops taking cards), and the dealer's hidden card is revealed.

- If the dealer's total is 16 or less, the dealer must take another card; otherwise, the dealer must stand. The dealer must continue taking cards until the sum of the cards is greater than or equal to 17. If the dealer exceeds 21, the player wins. Otherwise, the hand with the higher point total wins. If the dealer and the player have the same point total, the game is a "push," and no one wins.

An ace's value for a dealer depends on the dealer's other card(s) and the casino's house rules. A dealer typically must hit for totals of 16 or less and must stand for 17 or more. For a "soft 17"—a total of 17 with one ace counted as 11—some casinos require the dealer to hit and some require the dealer to stand (we require the dealer to stand). Such a hand is known as a "soft 17" because taking another card cannot bust the hand.

Enable a player to interact with the game using the keyboard—'H' means hit (take another card and 'S' means stand (do not take another card). Display the dealer's and player's hands as card images using Matplotlib, as we did in this chapter.

10.22 *(Project: Card Class with Overloaded Comparison Operators)* Modify class Card to support the comparison operators, so you can determine whether one Card is less than, equal to or greater than another. Investigate the functools module's total_ordering decorator. If your class is preceded by @total_ordering and defines methods __eq__ and __lt__ (for the < operator), the remaining comparison methods for <=, > and >= are autogenerated.

10.23 *(Project: Poker)* Exercises 5.25–5.26 asked you to create functions for comparing poker hands. Develop equivalent features for use with this chapter's DeckOfCards class. Develop a new class called Hand that represents a five-card poker hand. Use operator overloading to enable two Hands to be compared with the comparison operators. Use your new capabilities in a simple poker game script.

10.24 *(Project: PyDealer Library)* We demonstrated basic card shuffling and dealing in this chapter, but many card games require significant additional capabilities. As is often the case in Python, libraries already exist that can help you build more substantial card games. One such library is PyDealer. Research this library's extensive capabilities, then use it to implement your favorite card game.

10.25 *(Project: Enumerations)* Many programming languages provide a language element called an enumeration for creating sets of named constants. Often, these are used to make code more readable. The Python Standard Library's enum module enables you to emulate this concept by creating subclasses of the Enum base class. Investigate the enum module's capabilities, then create subclasses of Enum that represent card faces and card suits. Modify class Card to use these to represent the face and suit as enum constants rather than as strings.

10.26 *(Software Engineering with Abstract Classes and Abstract Methods)* When we think of a class, we assume that programs use it to create objects. Sometimes, it's useful to declare classes for which you *never* instantiate objects, because in some way they are *incomplete*. As you'll see, such classes can help you engineer effective inheritance hierarchies.

Concrete Classes—Consider Section 10.7's Shape hierarchy. If Circle, Square and Triangle objects all have draw methods, its reasonable to expect that calling draw on a Circle will display a Circle, calling draw on a Square will display a Square and calling draw on a Triangle will display a Triangle. Objects of each class know all the details of the specific shapes to draw. Classes that provide (or inherit) implementations of *every* method they define and that can be used to create objects are called **concrete classes**.

Abstract Classes—Now, let's consider class TwoDimensionalShape in the Shape hierarchy's second level. If we were to create a TwoDimensionalShape object and call its draw method, class TwoDimensionalShape knows that all two-dimensional shapes are *drawable*, but it does *not* know what *specific* two-dimensional shape to draw—there are many! So it does not make sense for TwoDimensionalShape to fully implement a draw method. A method that is defined in a given class, but for which you cannot provide an implementation is called an **abstract method**. Any class with an abstract method has a "hole"—the incomplete method implementation—and is called an **abstract class**. TypeErrors occur when you try to create objects of abstract classes. In the Shape hierarchy, classes Shape, TwoDimensionalShape and ThreeDimensionalShape all are abstract classes. They all know that shapes should be *drawable*, but do not know *what specific shape to draw*. Abstract base classes are *too general* to create real objects.

Inheriting a Common Design—An abstract class's purpose is to provide a *base class* from which subclasses can *inherit a common design*, such as a specific set of attributes and methods. So, such classes often are called **abstract base classes**. In the Shape hierarchy, subclasses inherit from the abstract base class Shape the notion of what it means to be a Shape—that is, common properties, such as location and color, and common behaviors, such as draw, move and resize.

Polymorphic Employee Payroll System—Now, let's develop an Employee class hierarchy that begins with an abstract class, then use polymorphism to perform payroll calculations for objects of two concrete subclasses. Consider the following problem statement:

> *A company pays its employees weekly. The employees are of two types. Salaried employees are paid a fixed weekly salary regardless of the number of hours worked. Hourly employees are paid by the hour and receive overtime pay (1.5 times their hourly salary rate) for all hours worked in excess of 40 hours. The company wants to implement an app that performs its payroll calculations polymorphically.*

Employee Hierarchy Class Diagram—The following diagram shows the Employee hierarchy. Abstract class Employee represents the *general concept* of an employee. Subclasses SalariedEmployee and HourlyEmployee inherit from Employee. Employee is *italicized* by convention to indicate that it's an abstract class. Concrete class names are *not* italicized:

Abstract Base Class Employee—The Python Standard Library's **abc (abstract base class) module** helps you define abstract classes by inheriting from the module's **ABC class**. Your abstract base class Employee class should declare the methods and properties that all employees should have. Each employee, regardless of the way his or her earnings are calculated, has a first name, a last name and a Social Security number. Also, every employee should have an earnings method, but the *specific* calculation depends on the employee's type, so you'll make earnings an abstract method that the subclasses must override. Your Employee class should contain:

- An __init__ method that initializes the first name, last name and Social Security number data attributes.
- Read-only properties for the first name, last name and Social Security number data attributes.
- An abstract method earnings preceded by the abc module's **@abstractmethod** decorator. Concrete subclasses *must* implement this method. The Python documentation says you should raise a NotImplementedError in abstract methods.[26]
- A __repr__ method that returns a string containing the first name, last name and Social Security number of the employee.

Concrete Subclass SalariedEmployee—This Employee subclass should override earnings to return a SalariedEmployee's weekly salary. The class also should include:

- An __init__ method that initializes the first name, last name, Social Security number and weekly salary data attributes. The first three of these should be initialized by calling base class Employee's __init__ method.
- A read-write weekly_salary property in which the setter ensures that the property is always non-negative.
- A __repr__ method that returns a string starting with 'SalariedEmployee:' and followed by all the information about a SalariedEmployee. This overridden method should call Employee's version.

Concrete Subclass HourlyEmployee—This Employee subclass should override earnings to return an HourlyEmployee's earnings, based on the hours worked and wage per hour. The class also should include:

- An __init__ method to initialize the first name, last name, Social Security number, hours and wages data attributes. The first name, last name and Social Security number should be initialized by calling base class Employee's __init__ method.
- Read-write hours and wages properties in which the setters ensure that the hours are in range (0–168) and wage per hour is always non-negative.
- A __repr__ method that returns a string starting with 'HourlyEmployee:' and followed by all the information about a HourlyEmployee. This overridden method should call Employee's version.

Testing Your Classes—In an IPython session, test your hierarchy:

- Import the classes Employee, SalariedEmployee and HourlyEmployee.
- Attempt to create an Employee object to see the TypeError that occurs and prove that you cannot create an object of an abstract class.
- Assign objects of the concrete classes SalariedEmployee and HourlyEmployee to variables, then display each employee's string representation and earnings.
- Place the objects into a list, then iterate through the list and polymorphically process each object, displaying its string representation and earnings.

---

26. https://docs.python.org/3.7/library/exceptions.html#NotImplementedError.

Computer Science Thinking: Recursion, Searching, Sorting and Big O

Objectives

In this chapter you'll:

- Learn the concept of recursion.
- Write and use recursive functions.
- Determine the base case and recursion step in a recursive algorithm.
- Learn how the system handles recursive function calls.
- Compare recursion and iteration, including when it's best to use each approach.
- Search for a given value in an array using linear search and binary search.
- Sort arrays using the simple iterative selection and insertion sort algorithms.
- Sort arrays using the more complex but higher-performance recursive merge sort algorithm.
- Use Big O notation to compare the efficiencies of searching and sorting algorithms.
- Use Seaborn and Matplotlib to build an animated selection sort algorithm visualization.
- In the exercises, implement additional sorting algorithms and animated visualizations, and determine the Big O of additional algorithms.

Outline

11.1 Introduction
11.2 Factorials
11.3 Recursive Factorial Example
11.4 Recursive Fibonacci Series Example
11.5 Recursion vs. Iteration
11.6 Searching and Sorting
11.7 Linear Search
11.8 Efficiency of Algorithms: Big O
11.9 Binary Search
  11.9.1 Binary Search Implementation
  11.9.2 Big O of the Binary Search
11.10 Sorting Algorithms
11.11 Selection Sort
  11.11.1 Selection Sort Implementation
  11.11.2 Utility Function `print_pass`
  11.11.3 Big O of the Selection Sort
11.12 Insertion Sort
  11.12.1 Insertion Sort Implementation
  11.12.2 Big O of the Insertion Sort
11.13 Merge Sort
  11.13.1 Merge Sort Implementation
  11.13.2 Big O of the Merge Sort
11.14 Big O Summary for This Chapter's Searching and Sorting Algorithms
11.15 Visualizing Algorithms
  11.15.1 Generator Functions
  11.15.2 Implementing the Selection Sort Animation
11.16 Wrap-Up
Exercises

11.1 Introduction

In this chapter, we concentrate on some key aspects of computer science thinking that go beyond programming fundamentals. Our focus on performance issues is a nice way to transition into the data science chapters, where we'll use AI and big data techniques that can place extraordinary performance demands on a system's resources.

We begin with a treatment of recursion. **Recursive functions** (or methods) call themselves, either directly or indirectly through other functions (or methods). Recursion can often help you solve problems more naturally when an iterative solution is not apparent. We'll show examples and compare the recursive programming style to the iterative style we've used to this point. We'll indicate where each might be preferable.

Next, we'll look at the crucial topics of searching and sorting arrays and other sequences. These are fascinating problems, because no matter what algorithms you use, the final result is the same. So you'll want to choose algorithms that perform "the best"—most likely, the ones that run the fastest or use the least memory. For big data applications, you'll also want to choose algorithms that are easy to *parallelize*. That will enable you to put lots of processors to work simultaneously—much as Google does, for example, when answering your search queries quickly.

This chapter focuses on the intimate relationship between algorithm design and performance. You'll see that the simplest and most apparent algorithms often perform poorly and that developing more sophisticated algorithms can lead to superior performance. We introduce **Big O notation**, which concisely classifies algorithms by how hard they have to work to get the job done. Big O helps you compare the efficiency of algorithms.

In an optional section at the end of this chapter, we develop an animated visualization of the selection sort so you can see it "in action." This is a great technique for understanding how algorithms work. It can often help you develop better performing algorithms.

The chapter includes a rich selection of recursion, searching and sorting exercises. You'll attempt some of the classic problems in recursion, implement alternative searching and sorting algorithms, and build animated visualizations of some of these to better understand the deep ties between algorithm design and performance.

11.2 Factorials

Let's write a program to perform a famous mathematical calculation. Consider the *factorial* of a positive integer n, which is written $n!$ and pronounced "n factorial." This is the product

$$n \cdot (n-1) \cdot (n-2) \cdot \ldots \cdot 1$$

with 1! equal to 1 and 0! defined to be 1. For example, 5! is the product $5 \cdot 4 \cdot 3 \cdot 2 \cdot 1$, which is equal to 120.

Iterative Factorial Approach
You can calculate 5! *iteratively* with a for statement, as in:

```
In [1]: factorial = 1

In [2]: for number in range(5, 0, -1):
   ...:     factorial *= number
   ...:

In [3]: factorial
Out[3]: 120
```

11.3 Recursive Factorial Example

Recursive problem-solving approaches have several elements in common. When you call a recursive function to solve a problem, it's actually capable of solving only the *simplest case(s)*, or **base case**(s). If you call the function with a *base case*, it immediately returns a result. If you call the function with a more complex problem, it typically divides the problem into two pieces—one that the function knows how to do and one that it does not know how to do. To make recursion feasible, this latter piece must be a slightly simpler or smaller version of the original problem. Because this new problem resembles the original problem, the function calls a fresh *copy* of itself to work on the smaller problem—this is referred to as a **recursive call** and is also called the **recursion step**. This concept of separating the problem into two smaller portions is a form of the *divide-and-conquer* approach introduced earlier in the book.

The recursion step executes while the original function call is still active (i.e., it has not finished executing). It can result in many more recursive calls as the function divides each new subproblem into two conceptual pieces. For the recursion to eventually terminate, each time the function calls itself with a simpler version of the original problem, the sequence of smaller and smaller problems must *converge on a base case*. When the function recognizes the base case, it returns a result to the previous copy of the function. A sequence of returns ensues until the original function call returns the final result to the caller.

Recursive Factorial Approach
You can arrive at a recursive factorial representation by observing that $n!$ can be written as:

$$n! = n \cdot (n-1)!$$

For example, 5! is equal to $5 \cdot 4!$, as in:

$$5! = 5 \cdot 4 \cdot 3 \cdot 2 \cdot 1$$
$$5! = 5 \cdot (4 \cdot 3 \cdot 2 \cdot 1)$$
$$5! = 5 \cdot (4!)$$

Visualizing Recursion

The evaluation of 5! would proceed as shown below. The left column shows how the succession of recursive calls proceeds until 1! (the base case) is evaluated to be 1, which terminates the recursion. The right column shows from bottom to top the values returned from each recursive call to its caller until the final value is calculated and returned.

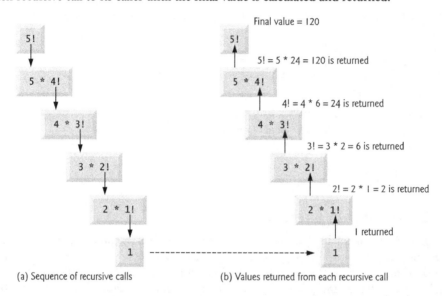

(a) Sequence of recursive calls   (b) Values returned from each recursive call

Implementing a Recursive Factorial Function

The following session uses recursion to calculate and display the factorials of the integers 0 through 10:

```
In [1]: def factorial(number):
   ...:     """Return factorial of number."""
   ...:     if number <= 1:
   ...:         return 1
   ...:     return number * factorial(number - 1)   # recursive call
   ...:

In [2]: for i in range(11):
   ...:     print(f'{i}! = {factorial(i)}')
   ...:
0! = 1
1! = 1
2! = 2
3! = 6
4! = 24
5! = 120
6! = 720
7! = 5040
8! = 40320
9! = 362880
10! = 3628800
```

Snippet [1]'s recursive function factorial first determines whether the *terminating condition* number <= 1 is True. If this condition *is* True (the base case), factorial returns

1 and no further recursion is necessary. If number is greater than 1, the second return statement expresses the problem as the product of number and a *recursive call* to factorial that evaluates factorial(number - 1). This is a slightly smaller problem than the original calculation, factorial(number). Note that function factorial must receive a *nonnegative* argument. We do not test for this case.

The loop in snippet [2] calls the factorial function for the values from 0 through 10. The output shows that factorial values grow quickly. *Python does not limit the size of an integer*, unlike many other programming languages.

Indirect Recursion

A recursive function may call another function, which may, in turn, make a call back to the recursive function. This is known as an **indirect recursive call** or **indirect recursion**. For example, function A calls function B, which makes a call back to function A. This is still recursion because the second call to function A is made while the first call to function A is active. That is, the first call to function A has not yet finished executing (because it is waiting on function B to return a result to it) and has not returned to function A's original caller.

Stack Overflow and Infinite Recursion

Of course, the amount of memory in a computer is finite, so only a certain amount of memory can be used to store activation records on the function-call stack. If more recursive function calls occur than can have their activation records stored on the stack, a fatal error known as **stack overflow** occurs.[1] This typically is the result of **infinite recursion**, which can be caused by omitting the base case or writing the recursion step incorrectly so that it does not converge on the base case. This error is analogous to the problem of an *infinite loop* in an *iterative* (nonrecursive) solution.

Recursion and the Function-Call Stack

In the "Functions" chapter, we introduced the *stack* data structure in the context of understanding how Python performs function calls. We discussed both the *function-call stack* and *stack frames*. That discussion also applies to recursive function calls. Each recursive function call gets its own stack frame on the function-call stack. When a given call completes, the system pops the function's stack frame from the stack and control returns to the caller, possibly another copy of the same function.

Self Check

1 *(Fill-In)* A(n) _____ case is needed to successfully terminate recursion.
Answer: base.

2 *(True/False)* A function calling itself indirectly is not an example of recursion.
Answer: False. A function calling itself in this manner is an example of indirect recursion.

3 *(True/False)* When a recursive function is called to solve a problem, it's capable of solving only the simplest case(s), or base case(s)—anything else requires a recursive call.
Answer: True.

---

1. This is how the website stackoverflow.com got its name. This is an excellent site for getting answers to your programming questions.

4 *(True/False)* To make recursion feasible, the recursion step in a recursive solution must resemble the original problem, but be a slightly larger version of it.
Answer: False. To make recursion feasible, the recursion step in a recursive solution must resemble the original problem, but be a slightly *smaller* or *simpler* version of it.

5 *(IPython Session)* Most other programming languages store integers in a fixed amount of space. So their built-in integer types can represent only a limited range of integer values. For example, Java's `int` type can represent only values in the range −2,147,483,648 to +2,147,483,647. Python allows integers to become *arbitrarily* large. Continue this section's IPython session and execute the function call `factorial(50)` to demonstrate that Python supports much larger integers.
Answer:

```
In [3]: factorial(50)
Out[3]: 30414093201713378043612608166064768844377641568960512000000000000
```

11.4 Recursive Fibonacci Series Example

The **Fibonacci series**,

> 0, 1, 1, 2, 3, 5, 8, 13, 21, …

begins with 0 and 1 and has the property that each subsequent Fibonacci number is the sum of the previous two. This series occurs in nature and describes a form of spiral.

The ratio of successive Fibonacci numbers converges on a constant value of 1.618…, a number that has been called the **golden ratio** or the **golden mean**. Humans tend to find the golden mean aesthetically pleasing. Architects often design windows, rooms and buildings whose length and width are in the ratio of the golden mean. Postcards often are designed with a golden-mean length-to-width ratio.

The Fibonacci series may be defined recursively as follows:

> fibonacci(0) is defined to be 0
> fibonacci(1) is defined to be 1
> fibonacci(n) = fibonacci(n − 1) + fibonacci(n − 2)

There are *two base cases* for the Fibonacci calculation:

- `fibonacci(0)` is 0, and
- `fibonacci(1)` is 1.

Function fibonacci

Let's define function `fibonacci`, which calculates the nth Fibonacci number recursively:

```
In [1]: def fibonacci(n):
   ...:     if n in (0, 1):    # base cases
   ...:         return n
   ...:     else:
   ...:         return fibonacci(n - 1) + fibonacci(n - 2)
   ...:
```

The initial call to function `fibonacci` is *not* a recursive call, but all subsequent calls to `fibonacci` performed from function `fibonacci`'s block *are* recursive because at that point the calls are initiated by the function itself. Each time you call `fibonacci`, it immediately tests for the *base cases*—n equal to 0 or n equal to 1, which we test simply by checking whether n is in the tuple (0, 1). If so, fibonacci returns n, because fibonacci(0) is 0 and fibonacci(1) is 1. Interestingly, if n is greater than 1, the recursion step generates *two* recursive calls, each for a slightly smaller problem than the original call to fibonacci.

Testing Function fibonacci
The following for loop tests fibonacci, displaying the Fibonacci values of 0–40. We omitted the outputs for the Fibonacci values of 7–37 to save space:

```
In [2]: for n in range(41):
   ...:     print(f'Fibonacci({n}) = {fibonacci(n)}')
   ...:
Fibonacci(0) = 0
Fibonacci(1) = 1
Fibonacci(2) = 1
Fibonacci(3) = 2
Fibonacci(4) = 3
Fibonacci(5) = 5
Fibonacci(6) = 8
...
Fibonacci(38) = 39088169
Fibonacci(39) = 63245986
Fibonacci(40) = 102334155
```

You'll notice that the speed of the calculation slows substantially as you get near the end of the loop. The variable n indicates which Fibonacci number to calculate in each iteration of the loop.

Analyzing the Calls to Function fibonacci
The following diagram shows how function fibonacci evaluates fibonacci(3). At the bottom of the diagram, we're left with the values 1, 0 and 1—the results of evaluating the *base cases*. The first two return values (from left to right), 1 and 0, are returned as the values for the calls fibonacci(1) and fibonacci(0). The sum 1 plus 0 is returned as the value of fibonacci(2). This is added to the result (1) of the rightmost call to fibonacci(1), producing the value 2. This final value is then returned as the value of fibonacci(3).

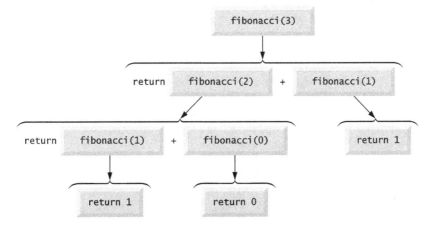

Complexity Issues

A word of caution is in order about recursive programs like the one we use here to generate Fibonacci numbers. Each invocation of the `fibonacci` function that does not match one of the *base cases* (0 or 1) results in *two more recursive calls* to the `fibonacci` function (snippet [1]). Hence, this set of recursive calls rapidly gets out of hand. Calculating the Fibonacci value of 20 with the recursive implementation requires 21,891 calls to the `fibonacci` function; calculating the Fibonacci value of 30 requires 2,692,537 calls!

As you try to calculate larger Fibonacci values, you'll notice that each consecutive Fibonacci number you calculate results in a substantial increase in calculation time and in the number of calls to the `fibonacci` function. For example, the Fibonacci value of 31 requires 4,356,617 calls, and the Fibonacci value of 32 requires 7,049,155 calls! As you can see, the number of calls to `fibonacci` increases quickly—1,664,080 additional calls between Fibonacci values of 30 and 31 and 2,692,538 additional calls between Fibonacci values of 31 and 32! The difference in the number of calls made between Fibonacci values of 31 and 32 is more than 1.5 times the difference in the number of calls for Fibonacci values between 30 and 31. Problems of this nature can humble even the world's most powerful computers.

In the field of complexity theory, computer scientists study how hard algorithms work to complete their tasks—that is, how many operations do they perform. Complexity issues are discussed in detail in the upper-level computer science curriculum course generally called "Algorithms." We introduce various complexity issues later in this chapter. In general, you should avoid Fibonacci-style recursive programs, because they result in an exponential "explosion" of function calls.

✓ Self Check

1 *(Fill-In)* The ratio of successive Fibonacci numbers converges on a constant value of 1.618…, a number that has been called the _____ or the _____.
Answer: golden ratio, golden mean.

2 *(True/False)* In the field of complexity theory, computer scientists study how hard algorithms work to complete their tasks.
Answer: True.

3 *(IPython Session)* Continuing this section's IPython session, create a function named `iterative_fibonacci` that uses looping rather than recursion to calculate Fibonacci numbers. Use both the iterative and recursive versions to calculate the 32nd, 33rd and 34th Fibonacci numbers. Time the calls with `%timeit` to see the difference in computation time.
Answer:

```
In [3]: def iterative_fibonacci(n):
   ...:     result = 0
   ...:     temp = 1
   ...:     for j in range(0, n):
   ...:         temp, result = result, result + temp
   ...:     return result
   ...:
   ...:
```

```
In [4]: %timeit fibonacci(32)
960 ms ± 14.3 ms per loop (mean ± std. dev. of 7 runs, 1 loop each)

In [5]: %timeit iterative_fibonacci(32)
1.72 µs ± 80.8 ns per loop (mean ± std. dev. of 7 runs, 1000000 loops each)

In [6]: %timeit fibonacci(33)
1.54 s ± 12.1 ms per loop (mean ± std. dev. of 7 runs, 1 loop each)

In [7]: %timeit iterative_fibonacci(33)
1.71 µs ± 20.9 ns per loop (mean ± std. dev. of 7 runs, 1000000 loops each)

In [8]: %timeit fibonacci(34)
2.81 s ± 212 ms per loop (mean ± std. dev. of 7 runs, 1 loop each)

In [9]: %timeit iterative_fibonacci(34)
2.05 µs ± 165 ns per loop (mean ± std. dev. of 7 runs, 100000 loops each)
```

11.5 Recursion vs. Iteration

We've studied functions factorial and fibonacci, which can be implemented either recursively or iteratively. In this section, we compare the two approaches and discuss why you might choose one approach over the other in a particular situation.

Both iteration and recursion are *based on a control statement*: Iteration uses an iteration statement (e.g., for or while), whereas recursion uses a selection statement (e.g., if or if…else or if…elif…else):

- Both iteration and recursion involve *iteration*: Iteration explicitly uses an iteration statement, whereas recursion achieves iteration through repeated function calls.

- Iteration and recursion each involve a *termination test*: Iteration terminates when the loop-continuation condition fails, whereas recursion terminates when a base case is reached.

- Iteration with counter-controlled iteration and recursion both *gradually approach termination*: Iteration keeps modifying a counter until the counter assumes a value that makes the loop-continuation condition fail, whereas recursion keeps producing smaller versions of the original problem until the base case is reached.

- Both iteration and recursion *can occur infinitely*: An infinite loop occurs with iteration if the loop-continuation test never becomes false, whereas infinite recursion occurs if the recursion step does not reduce the problem each time in a manner that converges on the base case, or if the base case is mistakenly not tested.

Negatives of Recursion

Recursion has many *negatives*. It repeatedly invokes the mechanism, and consequently the *overhead, of function calls*. This overhead can be *expensive* in terms of both processor time and memory space. Each recursive call causes another *copy* of the function (actually, only the function's variables, stored in the stack frame) to be created—this set of copies *can consume considerable memory space*. Iteration avoids these repeated function calls and extra memory assignments. However, for some algorithms that are easily expressed and understood with recursion, iterative solutions are not readily apparent.

Self Check

1 *(Fill-In)* Recursion terminates when _____.
Answer: a base case is reached.

2 *(True/False)* Iteration and recursion can occur infinitely.
Answer: True.

11.6 Searching and Sorting

Searching data involves determining whether a value (referred to as the **search key**) is present in the data and, if so, finding its location. Two popular search algorithms are the simple linear search and the faster but more complex binary search. **Sorting** places data in ascending or descending order, based on one or more **sort keys**. Your cell phone contacts list is sorted alphabetically, bank accounts are sorted by account number, employee payroll records are sorted by Social Security number, and so on. This chapter introduces two *simple* sorting algorithms, the selection sort and the insertion sort, along with the more efficient but *more complex* merge sort. The table below summarizes the searching and sorting algorithms, functions and methods discussed in the examples and exercises of this book.

Chapter	Algorithm	Location
Searching Algorithms, Functions and Methods:		
5	List method index	Section 5.9
8	String methods count, index and rindex.	Section 8.7
	re module functions search, match, findall and finditer	Section 8.12
	Linear search	Section 11.7
15	Binary search	Section 11.9
	Recursive binary search	Exercise 11.18
Sorting Algorithms, Functions and Methods:		
5	List method sort	Section 5.8
	Built-in function sorted	Section 5.8
	Built-in function sorted with a key	Exercise 5.15
7	DataFrame methods sort_index and sort_values	Section 7.14
15	Selection sort	Section 11.11
	Insertion sort	Section 11.12
	Recursive merge sort	Section 11.13
	Bucket sort	Exercise 11.17
	Recursive quicksort	Exercise 11.19

The techniques in this chapter are provided to introduce students to the concepts behind searching and sorting algorithms—upper-level computer science courses typically discuss additional algorithms.

11.7 Linear Search

Looking up a phone number, finding a website via a search engine and checking the definition of a word in a dictionary all involve searching large amounts of data. This section and Section 11.9 discuss two common search algorithms—one that's easy to program yet relatively inefficient (linear search) and one that's extremely efficient but more complex to program (binary search).

Linear Search Algorithm

The **linear search algorithm** searches each element in an array sequentially. If the search key does not match an element in the array, the algorithm informs the user that the search key is not present. If the search key is in the array, the algorithm tests each element until it finds one that matches the search key and returns the index of that element.

As an example, consider an array containing the following values:

    35 73 90 65 23 86 43 81 34 58

and a program that's searching for 86. The linear search algorithm first checks whether 35 matches the search key. It does not, so the algorithm checks whether 73 matches the search key. The program continues moving through the array sequentially, testing 90, then 65, then 23. When the program tests 86, which matches the search key, the program returns the index 5, which is the location of 86 in the array. If, after checking every array element, the program determines that the search key does not match any element in the array, it returns a sentinel value (e.g., -1).

Linear Search Implementation

Let's define a function `linear_search` for performing linear searches of an array of integers. The function receives as parameters the array to search (`data`) and the `search_key`. The `for` loop iterates through `data`'s elements and compares each with `search_key`. If the values are equal, `linear_search` returns the `index` of the matching element. If there are *duplicate* values in the array, the linear search returns the index of the *first* element that matches the search key. If the loop ends without finding the value, the function returns -1.

```
In [1]: def linear_search(data, search_key):
   ...:     for index, value in enumerate(data)::
   ...:         if value == search_key:
   ...:             return index
   ...:     return -1
   ...:
   ...:
```

To test the function, let's create an array of 10 random integers in the range 10–90:

```
In [2]: import numpy as np

In [3]: np.random.seed(11)

In [4]: values = np.random.randint(10, 91, 10)

In [5]: values
Out[5]: array([35, 73, 90, 65, 23, 86, 43, 81, 34, 58])
```

The following snippets call `linear_search` with the values 23 (found at index 4), 61 (not found) and 34 (found at index 8):

```
In [6]: linear_search(values, 23)
Out[6]: 4

In [7]: linear_search(values, 61)
Out[7]: -1
```

```
In [8]: linear_search(values, 34)
Out[8]: 8
```

Self Check

1 *(Fill-In)* The _____ algorithm searches each element in an array sequentially.
Answer: linear search.

2 *(True/False)* If an array contains duplicate values, the linear search finds the last matching value.
Answer: False. If there are duplicate values, linear search finds the first matching value.

11.8 Efficiency of Algorithms: Big O

Searching algorithms all accomplish the *same* goal—finding an element (or elements) matching a given search key if such an element exists. There are, however, several things that differentiate search algorithms from one another. *The major difference is the amount of effort they require to complete the search.* One way to describe this effort is with **Big O notation**, which indicates how hard an algorithm may have to work to solve a problem. For searching and sorting algorithms, this depends mainly on how many data elements there are. In this chapter, we use Big O to describe the worst-case run times for various searching and sorting algorithms.

O(1) Algorithms

Suppose an algorithm is designed to test whether the first element of an array is equal to the second. If the array has 10 elements, this algorithm requires one comparison. If the array has 1000 elements, it still requires one comparison. The algorithm is completely independent of the number of elements in the array. This algorithm is said to have a **constant run time**, which is represented in Big O notation as $O(1)$ and pronounced as "order one." An algorithm that's $O(1)$ does not necessarily require only one comparison. $O(1)$ means that the number of comparisons is *constant*—it does not grow as the size of the array increases. An algorithm that tests whether the first element of an array is equal to any of the next three elements is still $O(1)$ even though it requires three comparisons.

O(n) Algorithms

An algorithm that tests whether the first array element is equal to *any* of the other array elements requires at most $n - 1$ comparisons, where n is the number of array elements. If the array has 10 elements, this algorithm requires up to nine comparisons. If the array has 1000 elements, it requires up to 999 comparisons. As n grows larger, the n part of the expression $n - 1$ "dominates," and subtracting 1 becomes inconsequential. Big O is designed to highlight these dominant terms and ignore terms that become unimportant as n grows. So, an algorithm that requires a total of $n - 1$ comparisons (such as the one we described earlier) is said to be $O(n)$. An $O(n)$ algorithm is said to have a **linear run time**. $O(n)$ is often pronounced "on the order of n" or simply "order n."

O(n²) Algorithms

Now suppose you have an algorithm that tests whether *any* element of an array is duplicated elsewhere in the array. The first element must be compared with every other element in the array. The second element must be compared with every other element except the first (it was already compared to the first). The third element must be compared with every other element except the first two. In the end, this algorithm makes $(n - 1) + (n - 2) + \ldots + 2 + 1$ or $n^2/2 - n/2$ comparisons. As n increases, the n^2 term dominates, and the n term becomes inconsequential. Again, Big O notation highlights the n^2 term, ignoring $n/2$.

Big O is concerned with how an algorithm's run time grows in relation to the number of items processed. Suppose an algorithm requires n^2 comparisons. With four elements, the algorithm requires 16 comparisons; with eight elements, 64 comparisons. With this algorithm, *doubling* the number of elements *quadruples* the number of comparisons. Consider a similar algorithm requiring $n^2/2$ comparisons. With four elements, the algorithm requires eight comparisons; with eight elements, 32 comparisons. Again, *doubling* the number of elements *quadruples* the number of comparisons. Both of these algorithms grow as the square of n, so Big O ignores the constant, and both algorithms are considered to be $O(n^2)$, which is referred to as **quadratic run time** and pronounced "on the order of *n*-squared" or more simply "order *n*-squared."

When n is small, $O(n^2)$ algorithms (on today's computers) will not noticeably affect performance, but as n grows, you'll start to notice performance degradation. An $O(n^2)$ algorithm running on a million-element array would require a trillion "operations" (where each could execute several machine instructions). We tested one of this chapter's $O(n^2)$ algorithms on a 100,000-element array using a current desktop computer, and it ran for many minutes. A billion-element array (not unusual in today's big-data applications) would require a quintillion operations, which on that same desktop computer would take approximately 13.3 years to complete! As you'll see, $O(n^2)$ algorithms, unfortunately, are easy to write. You'll also see algorithms with more favorable Big O measures. These more efficient algorithms often take a bit more cleverness and work to create, but their superior performance can be well worth the extra effort, especially as n gets large and as the algorithms are integrated into larger programs.

Big O of the Linear Search

The linear search algorithm runs in $O(n)$ time. The worst case in this algorithm is that every element must be checked to determine whether the search item exists in the array. If the size of the array is *doubled*, the number of comparisons that the algorithm must perform is also *doubled*. Linear search can provide outstanding performance if the element matching the search key happens to be at or near the front of the array. However, we seek algorithms that perform well, on average, across *all* searches, including those where the element matching the search key is near the end of the array.

Linear search is easy to program, but it can be slow compared to other search algorithms. If a program needs to perform many searches on large arrays, it's better to implement a more efficient algorithm, such as the binary search, which we present next.

✓ **Self Check**

1 *(Fill-In)* _____ notation indicates how hard an algorithm may have to work to solve a problem.
Answer: Big O.

2   *(True/False)* An $O(n)$ algorithm is referred to as having a quadratic run time.
Answer: False. An $O(n)$ algorithm is referred to as having a linear run time. $O(n^2)$ algorithms have quadratic run time.

3   *(Discussion)* When might you choose to write an $O(n^2)$ algorithm?
Answer: When n is small and you do not have the time (or need) to invest in developing a better-performing algorithm.

11.9 Binary Search

The **binary search algorithm** is more efficient than linear search, but it requires a sorted array. The first iteration of this algorithm tests the *middle* element in the array. If this matches the search key, the algorithm ends. Assuming the array is sorted in *ascending* order, then if the search key is *less than* the middle element, it cannot match any element in the second half of the array and the algorithm continues with only the first half of the array (i.e., the first element up to, but not including, the middle element). If the search key is *greater than* the middle element, it cannot match any element in the first half of the array and the algorithm continues with only the second half (i.e., the element *after* the middle element through the last element). Each iteration tests the middle value of the remaining portion of the array. If the search key does not match the element, the algorithm eliminates half of the remaining elements. The algorithm ends by either finding an element that matches the search key or reducing the subarray to zero size.

Example

As an example consider the sorted 15-element array

   2   3   5   10   27   30   34   51   56   65   77   81   82   93   99

and a search key of 65. A program implementing the binary search algorithm would first check whether 51 is the search key (because 51 is the *middle* element of the array). The search key (65) is *larger* than 51, so 51 is ignored along with the first half of the array (all elements smaller than 51), leaving

   56   65   77   81   82   93   99

Next, the algorithm checks whether 81 (the middle element of the remainder of the array) matches the search key. The search key (65) is *smaller* than 81, so 81 is discarded along with the elements larger than 81. After just two tests, the algorithm has narrowed the number of values to check to only three (56, 65 and 77). It then checks 65 (which indeed matches the search key) and returns the index of the array element containing 65. This algorithm required just three comparisons to determine whether the search key matched an element of the array. Using a linear search algorithm would have required 10 comparisons. [*Note:* In this example, we used an array with 15 elements so that there's an obvious middle element. With an even number of elements, the middle of the array lies between two elements. We implement the algorithm to choose the higher of those two elements.]

✓ Self Check

1   *(True/False)* The linear search and binary search algorithms require arrays to be sorted.
Answer: False. Only the binary search requires a sorted array.

2 *(Fill-In)* With binary search, the smallest number of comparisons that would be needed to find a matching element in a 1,000,001-element array is _____.
Answer: One. This happens if on the first comparison the key matches the middle array element.

11.9.1 Binary Search Implementation

The file binarysearch.py contains the following definitions:

- Function binary_search searches an array for a specified key.
- Function remaining_elements displays the remaining elements in the array being searched, to visualize how the algorithm works.
- Function main tests function binary_search.

Each is discussed below.

Function binary_search

Lines 5–30 define function binary_search, which receives as parameters the array to search (data) and the search key (key).

```
1   # binarysearch.py
2   """Use binary search to locate an item in an array."""
3   import numpy as np
4
5   def binary_search(data, key):
6       """Perform binary search of data looking for key."""
7       low = 0   # low end of search area
8       high = len(data) - 1  # high end of search area
9       middle = (low + high + 1) // 2  # middle element index
10      location = -1  # return value -1 if not found
11
12      # loop to search for element
13      while low <= high and location == -1:
14          # print remaining elements of array
15          print(remaining_elements(data, low, high))
16
17          print('   ' * middle, end='')  # output spaces for alignment
18          print(' * ')  # indicate current middle
19
20          # if the element is found at the middle
21          if key == data[middle]:
22              location = middle  # location is the current middle
23          elif key < data[middle]:  # middle element is too high
24              high = middle - 1  # eliminate the higher half
25          else:  # middle element is too low
26              low = middle + 1  # eliminate the lower half
27
28          middle = (low + high + 1) // 2  # recalculate the middle
29
30      return location  # return location of search key
31
```

Lines 7–9 calculate the low end index, high end index and middle index of the portion of the array that the program is currently searching. Initially, the low end is 0, the high end is the length of the array minus 1 and the `middle` is the average of these two values. Line 10 initializes the `location` of the element to -1. This is the value that `binary_search` returns if it does not find the key. Lines 13–28 loop as long as `low` is less than or equal to `high` (the search is not complete) and `location` is equal to -1 (the key has not yet been found). Line 21 tests whether the value in the `middle` element is equal to the key. If so, line 22 assigns `middle` to `location`, the loop terminates and the function returns `location` to the caller. Each iteration of the loop tests a single value (line 21) and *eliminates half of the remaining values in the array* (lines 23–24 or 25–26) if the value is not the key.

Function `remaining_elements`

During each loop iteration in `binary_search`, we call function `remaining_elements` (line 15) to show the portion of the array being searched, then `binary_search` displays an asterisk under the `middle` element (lines 17–18). Line 34 in `remaining_elements` first repeats a three-space string `low` times for alignment purposes. The remainder of the line appends to that a string representation of the array values from the index `low` up to `high + 1`. The built-in function `str` converts its argument (an integer element of the array) to a string

```
32    def remaining_elements(data, low, high):
33        """Display remaining elements of the binary search."""
34        return '   ' * low + ' '.join(str(s) for s in data[low:high + 1])
35
```

Function `main`

The `main` function (lines 36–53) is called if you run the file `binarysearch.py` as a script. Line 38 creates a 15-element array of random values in the range 10–90, and line 39 sorts the values into ascending order. Recall that the binary search algorithm works only on a *sorted* array. The output shows that when the user instructs the program to search for 23, the program first tests the middle element, which is 52 (as indicated by *) in our sample execution. The search key is less than 52, so the program eliminates the second half of the array and tests the middle element from the first half. The search key is smaller than 35, so the program eliminates the second half of the array, leaving only three elements. Finally, the program checks 23 (which matches the search key) and returns the index 1.

```
36    def main():
37        # create and display array of random values
38        data = np.random.randint(10, 91, 15)
39        data.sort()
40        print(data, '\n')
41
42        search_key = int(input('Enter an integer value (-1 to quit): '))
43
44        # repeatedly input an integer; -1 terminates the program
45        while search_key != -1:
46            location = binary_search(data, search_key)  # perform search
47
48            if location == -1:  # not found
49                print(f'{search_key} was not found\n')
50            else:
51                print(f'{search_key} found in position {location}\n')
52
53            search_key = int(input('Enter an integer value (-1 to quit): '))
```

```
54
55   # call main if this file is executed as a script
56   if __name__ == '__main__':
57       main()
```

```
[16 23 31 35 36 46 48 52 54 57 63 76 83 89 90]

Enter an integer value (-1 to quit): 23
16 23 31 35 36 46 48 52 54 57 63 76 83 89 90
             *
16 23 31 35 36 46 48
         *
16 23 31
   *
23 found in position 1

Enter an integer value (-1 to quit): 83
16 23 31 35 36 46 48 52 54 57 63 76 83 89 90
             *
                     54 57 63 76 83 89 90
                              *
                                    83 89 90
                                     *
                                    83
                                     *
83 found in position 12

Enter an integer value (-1 to quit): 60
16 23 31 35 36 46 48 52 54 57 63 76 83 89 90
             *
                     54 57 63 76 83 89 90
                              *
                     54 57 63
                          *
                              63
                               *
60 was not found

Enter an integer value (-1 to quit): -1
```

11.9.2 Big O of the Binary Search

In the worst-case scenario, searching a *sorted* array of 1023 elements takes *only 10 comparisons* when using a binary search. Repeatedly dividing 1023 by 2 (because after each comparison we can eliminate half the array) and rounding down (because we also remove the middle element) yields the values 511, 255, 127, 63, 31, 15, 7, 3, 1 and 0. The number 1023 (which is $2^{10} - 1$) is divided by 2 only 10 times to get the value 0, which indicates that there are no more elements to test.

Dividing by 2 is equivalent to one comparison in the binary search algorithm. Thus, an array of 1,048,575 ($2^{20} - 1$) elements takes a *maximum of 20 comparisons* to find the key, and an array of about one billion elements takes a *maximum of 30 comparisons* to find the key. This is a tremendous performance improvement over the linear search. For a one-billion-element array, this is a difference between an average of 500 million comparisons for the linear search and a maximum of only 30 comparisons for the binary search!

The maximum number of comparisons needed for the binary search of any sorted array is the exponent of the first power of 2 greater than the number of elements in the array, which is represented as $\log_2 n$. From a Big O perspective, all logarithms grow at roughly the same rate, so in big O notation the base can be omitted. This results in a big O of $O(\log n)$ for a binary search, which is also known as **logarithmic run time** and pronounced as "order log n." This assumes the array is sorted, which could take time. We'll discuss sorting next.

11.10 Sorting Algorithms

Sorting data (i.e., placing the data in a particular order, such as ascending or descending) is one of the most important computing applications. Virtually every organization must sort some data, and often massive amounts of it. Sorting data is an intriguing, computer-intensive problem that has attracted intense research efforts.

An important item to understand about sorting is that the end result—the sorted array—will be the *same* no matter which algorithm you use to sort the array. The choice of algorithm affects only the run time and memory use of the program. The rest of this chapter introduces three common sorting algorithms. The first two—*selection sort* and *insertion sort*—are relatively simple to program but *inefficient*. The last algorithm—*merge sort*—is *much faster* than selection sort and insertion sort but *harder to program*. We focus on sorting arrays of primitive-type data, namely `int`s.

11.11 Selection Sort

Selection sort is a simple, but inefficient, sorting algorithm. If you're sorting in increasing order, its first iteration selects the *smallest* element in the array and swaps it with the first element. The second iteration selects the *second-smallest* item (which is the smallest item of the remaining elements) and swaps it with the second element. The algorithm continues until the last iteration selects the *second-largest* element and swaps it with the second-to-last index, leaving the largest element in the last index. After the *i*th iteration, the smallest *i* items of the array will be sorted into increasing order in the first *i* elements of the array.

As an example, consider the array

    34   56   14   20   77   51   93   30   15   52

A program that implements selection sort first determines the smallest element (14) of this array, which is contained in index 2. The program swaps 14 with 34, resulting in

    14   56   34   20   77   51   93   30   15   52

The program then determines the smallest value of the remaining elements (all elements except 14), which is 15, contained in index 8. The program swaps 15 with 56, resulting in

    14   15   34   20   77   51   93   30   56   52

On the third iteration, the program determines the next smallest value (20) and swaps it with 34.

    14   15   20   34   77   51   93   30   56   52

The process continues until the array is fully sorted.

    14   15   20   30   34   51   52   56   77   93

11.11.1 Selection Sort Implementation

The file `selectionsort.py` defines:

- Function `selection_sort`, which implements the selection sort algorithm, and
- Function `main` to test the `selection_sort` function.

The following shows a sample output of the program. The -- notation below a given number indicates that after the algorithm's specified pass, the number above the -- is in its final sorted position in the array, and the * indicates which value was swapped out of the rightmost --'s position.

```
Unsorted array: [34 56 14 20 77 51 93 30 15 52]
after pass 1: 14  56  34* 20  77  51  93  30  15  52
              --
after pass 2: 14  15  34  20  77  51  93  30  56* 52
              --  --
after pass 3: 14  15  20  34* 77  51  93  30  56  52
              --  --  --
after pass 4: 14  15  20  30  77  51  93  34* 56  52
              --  --  --  --
after pass 5: 14  15  20  30  34  51  93  77* 56  52
              --  --  --  --  --
after pass 6: 14  15  20  30  34  51* 93  77  56  52
              --  --  --  --  --  --
after pass 7: 14  15  20  30  34  51  52  77  56  93*
              --  --  --  --  --  --  --
after pass 8: 14  15  20  30  34  51  52  56  77* 93
              --  --  --  --  --  --  --  --
after pass 9: 14  15  20  30  34  51  52  56  77* 93
              --  --  --  --  --  --  --  --  --

Sorted array: [14 15 20 30 34 51 52 56 77 93]
```

Function `selection_sort`

Lines 6–19 define the `selection_sort` function that implements the algorithm. Lines 9–19 loop `len(data) - 1` times. The variable `smallest` stores the index of the smallest element in the remaining array. Line 10 initializes `smallest` to the current index `index1`. Lines 13–15 loop over the remaining elements in the array. For each, line 14 compares its value to the value of the smallest element. If the current element is smaller than the smallest element, line 15 assigns the current element's index to `smallest`. When this loop finishes, `smallest` will contain the index of the smallest element in the remaining array. Line 18 uses tuple packing and unpacking to swap the smallest remaining element into the next ordered spot in the array.

```
1  # selectionsort.py
2  """Sorting an array with selection sort."""
3  import numpy as np
4  from ch11utilities import print_pass
5
```

```python
 6  def selection_sort(data):
 7      """Sort array using selection sort."""
 8      # loop over len(data) - 1 elements
 9      for index1 in range(len(data) - 1):
10          smallest = index1  # first index of remaining array
11
12          # loop to find index of smallest element
13          for index2 in range(index1 + 1, len(data)):
14              if data[index2] < data[smallest]:
15                  smallest = index2
16
17          # swap smallest element into position
18          data[smallest], data[index1] = data[index1], data[smallest]
19          print_pass(data, index1 + 1, smallest)
20
```

Function main

In the main function, line 22 creates an array of ten integers. Line 24 calls selection_sort to sort the array's elements into ascending order.

```python
21  def main():
22      data = np.array([35, 73, 90, 65, 23, 86, 43, 81, 34, 58])
23      print(f'Unsorted array: {data}\n')
24      selection_sort(data)
25      print(f'\nSorted array: {data}\n')
26
27  # call main if this file is executed as a script
28  if __name__ == '__main__':
29      main()
30
```

11.11.2 Utility Function print_pass

Line 19 in the selection_sort function displays the array at the end of the current pass by calling function print_pass from the file ch11utilities.py. Function print_pass performs the following tasks:

- Lines 7–8 create and display a label containing the pass number for the beginning of each pass's output.
- Lines 11–12 create and display a string containing the elements from the beginning of the array up to position index, separated by two spaces each. The built-in function str converts its argument (an integer element of the array) to a string.
- Line 14 indicates the swap element position by displaying the element at that index followed by an asterisk (*).
- Line 17 creates and displays a string containing the rest of the array's elements.
- Line 20 displays dashes under the sorted portion of the array to help visualize the sort.

On each pass, the element next to the asterisk and the element above the rightmost set of dashes were swapped. We'll use this function again when we implement the insertion sort algorithm in the next section.

```
 1  # ch11utilities.py
 2  """Utility function for printing a pass of the
 3  insertion_sort and selection_sort algorithms"""
 4
 5  def print_pass(data, pass_number, index):
 6      """Print a pass of the algorithm."""
 7      label = f'after pass {pass_number}: '
 8      print(label, end='')
 9
10      # output elements up to selected item
11      print('  '.join(str(d) for d in data[:index]),
12            end='  ' if index != 0 else '')
13
14      print(f'{data[index]}* ', end='')  # indicate swap with *
15
16      # output rest of elements
17      print('  '.join(str(d) for d in data[index + 1:len(data)]))
18
19      # underline elements that are sorted after this pass_number
20      print(f'{" " * len(label)}{"-- " * pass_number}')
```

11.11.3 Big O of the Selection Sort

The selection sort algorithm runs in $O(n^2)$ time. Function `selection_sort` uses nested `for` loops. The outer one (lines 9–19) iterates over the first $n-1$ elements in the array, swapping the smallest remaining item into its sorted position. The inner loop (lines 13–15) iterates over each item in the remaining array, searching for the smallest element. This loop executes $n-1$ times during the first iteration of the outer loop, $n-2$ times during the second iteration, then $n-3, \ldots, 3, 2, 1$. This inner loop will iterate a total of $n(n-1)/2$ or $(n^2-n)/2$. In Big O notation, smaller terms drop out, and constants are ignored, leaving a Big O of $O(n^2)$. Note that the algorithm iterates the same number of times regardless of whether the array's elements are randomly ordered, partially ordered or already sorted.

Self Check

1 *(Fill-In)* A selection sort application would take approximately _____ times as long to run on a 128-million-element array as on a 32-million-element array.
Answer: 16, because an $O(n^2)$ algorithm takes 16 times as long to sort four times as much information.

11.12 Insertion Sort

Insertion sort is another *simple, but inefficient*, sorting algorithm. The first iteration of this algorithm takes the *second element* in the array and, if it's *less than* the *first element, swaps it with the first element*. The second iteration looks at the third element and inserts it into the correct position with respect to the first two, so all three elements are in order. At the *i*th iteration of this algorithm, the first *i* elements in the original array will be sorted.

Consider as an example the following array, which is identical to the one used in the discussion of selection sort.

```
34  56  14  20  77  51  93  30  15  52
```

A program that implements the insertion sort algorithm will first look at the first two elements of the array, 34 and 56. These are already in order, so the program continues. If they were out of order, the program would swap them.

In the next iteration, the program looks at the third value, 14. This value is less than 56, so the program stores 14 in a temporary variable and moves 56 one element to the right. The program then checks and determines that 14 is less than 34, so it moves 34 one element to the right. The program has now reached the beginning of the array, so it places 14 in element 0. The array now is

14   34   56   20   77   51   93   30   15   52

In the next iteration, the program stores 20 in a temporary variable. Then it compares 20 to 56 and moves 56 one element to the right because it's larger than 20. The program then compares 20 to 34, moving 34 right one element. When the program compares 20 to 14, it observes that 20 is larger than 14 and places 20 in element 1. The array now is

14   20   34   56   77   51   93   30   15   52

Using this algorithm, at the *i*th iteration, the first *i* elements of the original array are sorted, but they may not be in their final locations—smaller values may be located later in the array.

11.12.1 Insertion Sort Implementation

The file `insertionsort.py` defines:

- Function `insertion_sort`, which implements the insertion sort algorithm, and
- Function `main` to test the `insertion_sort` function.

Function `main` (lines 22–27) is identical to `main` in Section 11.11.1 except that line 26 calls function `insertion_sort` to sort the array's elements into ascending order.

The following is a sample output of the program. The -- notation below a given number indicates the values that have been sorted so far after the algorithm's specified pass.

```
Unsorted array: [34 56 14 20 77 51 93 30 15 52]

after pass 1:  34   56*  14   20   77   51   93   30   15   52
                    --
after pass 2:  14*  34   56   20   77   51   93   30   15   52
               --   --
after pass 3:  14   20*  34   56   77   51   93   30   15   52
               --   --   --
after pass 4:  14   20   34   56   77*  51   93   30   15   52
               --   --   --   --
after pass 5:  14   20   34   51*  56   77   93   30   15   52
               --   --   --   --
after pass 6:  14   20   34   51   56   77   93*  30   15   52
               --   --   --   --   --   --
after pass 7:  14   20   30*  34   51   56   77   93   15   52
               --   --   --   --   --   --   --
after pass 8:  14   15*  20   30   34   51   56   77   93   52
               --   --   --   --   --   --   --   --
after pass 9:  14   15   20   30   34   51   52*  56   77   93
               --   --   --   --   --   --   --   --   --

Sorted array: [14 15 20 30 34 51 52 56 77 93]
```

11.12.1 Function insertion_sort

Function insertion_sort

Lines 6–20 declare the `insertion_sort` function. Lines 9–20 loop over the elements at indices 1 up to `len(data)` items in the array. In each iteration, line 10 declares and initializes variable `insert`, which holds the value of the element that will be inserted into the sorted portion of the array. Line 11 declares and initializes the variable `move_item`, which keeps track of where to insert the element. Lines 14–17 loop to locate the correct position where the element should be inserted. The loop will terminate either when the program reaches the front of the array or when it reaches an element that's less than the value to be inserted. Line 19 moves an element to the right in the array, and line 17 decrements the position at which to insert the next element. After the loop ends, line 19 inserts the element into place.

```python
 1  # insertionsort.py
 2  """Sorting an array with insertion sort."""
 3  import numpy as np
 4  from ch11utilities import print_pass
 5
 6  def insertion_sort(data):
 7      """Sort an array using insertion sort."""
 8      # loop over len(data) - 1 elements
 9      for next in range(1, len(data)):
10          insert = data[next]   # value to insert
11          move_item = next      # location to place element
12
13          # search for place to put current element
14          while move_item > 0 and data[move_item - 1] > insert:
15              # shift element right one slot
16              data[move_item] = data[move_item - 1]
17              move_item -= 1
18
19          data[move_item] = insert   # place inserted element
20          print_pass(data, next, move_item)   # output pass of algorithm
21
22  def main():
23      data = np.array([35, 73, 90, 65, 23, 86, 43, 81, 34, 58])
24      print(f'Unsorted array: {data}\n')
25      insertion_sort(data)
26      print(f'\nSorted array: {data}\n')
27
28  # call main if this file is executed as a script
29  if __name__ == '__main__':
30      main()
```

11.12.2 Big O of the Insertion Sort

The insertion sort algorithm also runs in $O(n^2)$ time. Like selection sort, the implementation of insertion sort contains two loops. The `for` loop iterates `len(data) - 1` times, inserting an element into the appropriate position among the elements sorted so far. For the purposes of this application, `len(data) - 1` is equivalent to $n - 1$ (as `len(data)` is the size of the array). The `while` loop (lines 14–17) iterates over the preceding elements in the array. In the worst case, this `while` loop will require $n - 1$ comparisons. Each individual loop runs in $O(n)$ time. In Big O notation, nested loops mean that you must *multiply* the number of comparisons. For each iteration of an outer loop, there will be a certain number of iterations of the inner loop. In this algorithm, for each $O(n)$ iterations of the outer loop, there will be $O(n)$ iterations of the inner loop. Multiplying these values results in a Big O of $O(n^2)$.

Self Check

1 *(True/False)* Like the selection sort algorithm, the insertion sort algorithm has linear run time.
Answer: False. Both algorithms have quadratic run time.

2 *(True/False)* Each iteration of the selection sort algorithm inserts one value into sorted order among the values that have been sorted so far.
Answer: True.

11.13 Merge Sort

Merge sort is an *efficient* sorting algorithm but is conceptually *more complex* than selection sort and insertion sort. The merge sort algorithm sorts an array by *splitting* it into two equal-sized subarrays, *sorting* each subarray, then *merging* them into one larger array. With an odd number of elements, the algorithm creates the two subarrays such that one has one more element than the other.

The implementation of merge sort in this example is recursive. The base case is an array with one element, which is, of course, sorted, so the merge sort immediately returns in this case. The recursion step splits the array into two approximately equal pieces, recursively sorts them, then merges the two sorted arrays into one larger, sorted array.

Suppose the algorithm has already merged smaller arrays to create sorted arrays array1:

    14 20 34 56 77

and array2:

    15 30 51 52 93

Merge sort combines these two arrays into one larger, sorted array. The smallest element in array1 is 14 (located in index 0 of array1). The smallest element in array2 is 15 (located in index 0 of array2). To determine the smallest element in the larger array, the algorithm compares 14 and 15. The value from array1 is smaller, so 14 becomes the first element in the merged array. The algorithm continues by comparing 20 (the second element in array1) to 15 (the first element in array2). The value from array2 is smaller, so 15 becomes the second element in the larger array. The algorithm continues by comparing 20 to 30, with 20 becoming the third element in the array, and so on.

11.13.1 Merge Sort Implementation

The file mergesort.py defines:
- Function merge_sort to initiate the sorting.
- Function sort_array implements the recursive merge sort algorithm—this is called by function mergeSort.

- Function merge merges two sorted subarrays into a single sorted subarray.
- Function subarray_string gets a subarray's string representation for output purposes to help visualize the sort.
- Function main tests function merge_sort.

Function main (lines 69–73) is identical to main in the previous sorting examples, except that line 72 calls function merge_sort to sort the array elements.

The following sample output visualizes the splits and merges performed by merge sort, showing the progress of the sort at each step of the algorithm. It's well worth your time to step through these outputs to fully understand this elegant and *fast* sorting algorithm.

```
Unsorted array: [34 56 14 20 77 51 93 30 15 52]
split:    34 56 14 20 77 51 93 30 15 52
          34 56 14 20 77
                         51 93 30 15 52

split:    34 56 14 20 77
          34 56 14
                   20 77

split:    34 56 14
          34 56
                14

split:    34 56
          34
             56

merge:    34
             56
          34 56

merge:    34 56
                14
          14 34 56

split:              20 77
                    20
                       77

merge:              20
                       77
                    20 77

merge:    14 34 56
                    20 77
          14 20 34 56 77

split:                   51 93 30 15 52
                         51 93 30
                                  15 52
```

```
split:              51 93 30
                    51 93
                          30

split:              51 93
                    51
                       93

merge:              51
                       93
                    51 93

merge:              51 93
                          30
                    30 51 93

split:                    15 52
                          15
                             52

merge:                    15
                             52
                          15 52

merge:              30 51 93
                          15 52
                    15 30 51 52 93

merge:    14 20 34 56 77
                    15 30 51 52 93
          14 15 20 30 34 51 52 56 77 93

Sorted array: [14 15 20 30 34 51 52 56 77 93]
```

Function merge_sort

Lines 6–7 define the merge_sort function. Line 7 calls function sort_array to initiate the recursive algorithm, passing 0 and len(data) - 1 as the low and high indices of the array to be sorted. These values tell function sort_array to operate on the entire array.

```
1   # mergesort.py
2   """Sorting an array with merge sort."""
3   import numpy as np
4
5   # calls recursive sort_array method to begin merge sorting
6   def merge_sort(data):
7       sort_array(data, 0, len(data) - 1)
8
```

Recursive Function sort_array

Function sort_array (lines 9–26) performs the recursive merge sort algorithm. Line 12 tests the base case. If the size of the array is 1, the array is already sorted, so the function returns immediately. If the size of the array is greater than 1, the function splits the array in two, recursively calls function sort_array to sort the two subarrays, then merges them.

Line 22 recursively calls function `sort_array` on the first half of the array, and line 23 recursively calls function `sort_array` on the second half. When these two function calls return, each half of the array has been sorted. Line 26 calls function `merge` (lines 29–61) with the indices for the two halves of the array to combine the two sorted arrays into one larger sorted array.

```
 9  def sort_array(data, low, high):
10      """Split data, sort subarrays and merge them into sorted array."""
11      # test base case size of array equals 1
12      if (high - low) >= 1:  # if not base case
13          middle1 = (low + high) // 2  # calculate middle of array
14          middle2 = middle1 + 1  # calculate next element over
15
16          # output split step
17          print(f'split:   {subarray_string(data, low, high)}')
18          print(f'         {subarray_string(data, low, middle1)}')
19          print(f'         {subarray_string(data, middle2, high)}\n')
20
21          # split array in half then sort each half (recursive calls)
22          sort_array(data, low, middle1)  # first half of array
23          sort_array(data, middle2, high)  # second half of array
24
25          # merge two sorted arrays after split calls return
26          merge(data, low, middle1, middle2, high)
27
```

Function merge

Lines 29–61 define function `merge`. Lines 40–50 in merge loop until the end of either subarray is reached. Line 43 tests which element at the beginning of the arrays is smaller. If the element in the left array is smaller or equal, line 44 places it in position in the combined array. If the element in the right array is smaller, line 48 places it in position in the combined array. When the `while` loop completes, one entire subarray has been placed in the combined array, but the other subarray still contains data. Line 53 tests whether the left array has reached the end. If so, line 54 uses slices to fill the appropriate elements of the `combined` array with the elements of `data` that represent the right array. If the left array has not reached the end, then the right array must have reached the end, and line 56 uses slices to fill the appropriate elements of the `combined` array with the elements of `data` that represent the left array. Finally, line 58 copies the `combined` array into the original array that `data` references.

```
28  # merge two sorted subarrays into one sorted subarray
29  def merge(data, left, middle1, middle2, right):
30      left_index = left  # index into left subarray
31      right_index = middle2  # index into right subarray
32      combined_index = left  # index into temporary working array
33      merged = [0] * len(data)  # working array
34
35      # output two subarrays before merging
36      print(f'merge:   {subarray_string(data, left, middle1)}')
37      print(f'         {subarray_string(data, middle2, right)}')
38
```

```
39          # merge arrays until reaching end of either
40          while left_index <= middle1 and right_index <= right:
41              # place smaller of two current elements into result
42              # and move to next space in arrays
43              if data[left_index] <= data[right_index]:
44                  merged[combined_index] = data[left_index]
45                  combined_index += 1
46                  left_index += 1
47              else:
48                  merged[combined_index] = data[right_index]
49                  combined_index += 1
50                  right_index += 1
51
52          # if left array is empty
53          if left_index == middle2:  # if True, copy in rest of right array
54              merged[combined_index:right + 1] = data[right_index:right + 1]
55          else:  # right array is empty, copy in rest of left array
56              merged[combined_index:right + 1] = data[left_index:middle1 + 1]
57
58          data[left:right + 1] = merged[left:right + 1]  # copy back to data
59
60          # output merged array
61          print(f'          {subarray_string(data, left, right)}\n')
62
```

Function `subarray_string`

Throughout the algorithm, we display portions of the array to show the split and merge operations. Each time we call function `subarray_string` to create and display a string containing the appropriate subarray's items. Line 65 creates a string of spaces that ensures the first subarray element aligns properly. Line 66 joins the appropriate elements of data separated by a space and appends that string to the one created in line 65. Line 67 returns the result.

```
63  # method to output certain values in array
64  def subarray_string(data, low, high):
65      temp = '   ' * low  # spaces for alignment
66      temp += ' '.join(str(item) for item in data[low:high + 1])
67      return temp
68
```

Function `main`

The `main` function creates the array to sort and calls `merge_sort` to sort the data:

```
69  def main():
70      data = np.array([35, 73, 90, 65, 23, 86, 43, 81, 34, 58])
71      print(f'Unsorted array: {data}\n')
72      merge_sort(data)
73      print(f'\nSorted array: {data}\n')
74
75  # call main if this file is executed as a script
76  if __name__ == '__main__':
77      main()
```

11.13.2 Big O of the Merge Sort

Merge sort is *far more efficient* than insertion or selection sort. Consider the first (non-recursive) call to sort_array. This results in two recursive calls to sort_array with subarrays each approximately half the original array's size, and a single call to merge, which requires, at worst, $n - 1$ comparisons to fill the original array, which is $O(n)$. (Recall that each array element can be chosen by comparing one element from each subarray.) The two calls to sort_array result in four more recursive sort_array calls, each with a subarray approximately a quarter of the original array's size, along with two calls to merge that each require, at worst, $n/2 - 1$ comparisons, for a total number of comparisons of $O(n)$. This process continues, each sort_array call generating two additional sort_array calls and a merge call until the algorithm has *split* the array into one-element subarrays. At each level, $O(n)$ comparisons are required to *merge* the subarrays. Each level splits the arrays in half, so doubling the array size requires one more level. Quadrupling the array size requires two more levels. This pattern is logarithmic and results in $\log_2 n$ levels. This results in a total efficiency of $O(n \log n)$ which, of course, is much faster than the $O(n^2)$ sorts we studied.

Self Check

1 *(Fill-In)* The efficiency of merge sort is _____.
Answer: $O(n \log n)$.

11.14 Big O Summary for This Chapter's Searching and Sorting Algorithms

The following table summarizes the searching and sorting algorithms covered in this chapter with the Big O for each.

Algorithm	Location	Big O
Searching Algorithms:		
Linear search	Section 11.7	$O(n)$
Binary search	Section 11.9	$O(\log n)$
Recursive binary search	Exercise 11.18	$O(\log n)$
Sorting Algorithms:		
Selection sort	Section 11.11	$O(n^2)$
Insertion sort	Section 11.12	$O(n^2)$
Merge sort	Section 11.13	$O(n \log n)$

The following table lists the Big O values we've covered in this chapter along with a number of values for n to highlight the differences in the growth rates.

n =	$O(\log n)$	$O(n)$	$O(n \log n)$	$O(n^2)$
1	0	1	0	1
2	1	2	2	4
3	1	3	3	9
4	1	4	4	16
5	1	5	5	25
10	1	10	10	100
100	2	100	200	10,000
1000	3	1000	3000	10^6
1,000,000	6	1,000,000	6,000,000	10^{12}
1,000,000,000	9	1,000,000,000	9,000,000,000	10^{18}

11.15 Visualizing Algorithms

Animated visualizations can help you understand how algorithms work. In this section, you'll animate the selection sort algorithm using:

- the Seaborn bar plot capabilities from the "Sequences: Lists and Tuples" chapter's Intro to Data Science section,
- the Matplotlib animation techniques from the "Dictionaries and Sets" chapter's Intro to Data Science section where we introduced `FuncAnimations`, and
- the `yield` and `yield from` statements, which we'll introduce here.

We've already presented the selection sort algorithm and the technologies above. Here, we'll focus on how we modified this chapter's earlier selection sort example to implement the animation.

Pysine: Playing Musical Notes

To enhance the animation, we'll use the **Pysine** module to play musical notes representing the bar's magnitudes. Pysine uses **MIDI (Musical Instrument Digital Interface)** to generate musical notes based on sound frequencies specified in hertz.

To install Pysine, open your Anaconda Prompt (Windows[2]), Terminal (macOS/Linux) or shell (Linux), then execute the following command:

```
pip install pysine
```

Executing the Animation

You can execute the animation as follows:

```
ipython selectionsortanimation.py 10
```

to sort the values in the range 1–10. Consider the following sample screen captures:

---

2. Windows users might need to run the Anaconda Prompt as an administrator for proper software installation privileges. To do so, right-click **Anaconda Prompt** in the start menu and select **More > Run as administrator**.

a) After 1-4 sorted, 8 (purple) is the current smallest element.

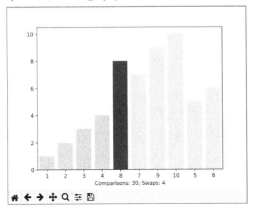

b) 7 (red) is being compared with 8 (purple).

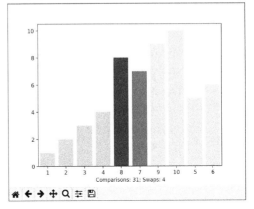

c) 5 and 8 before the swap.

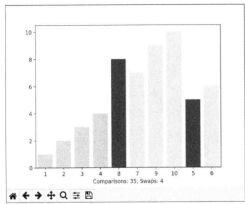

d) 5 and 8 after the swap.

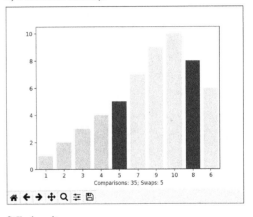

e) Emphasizing the final sorted elements in dark green.

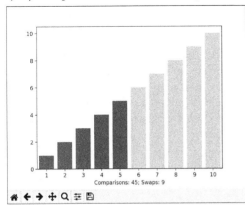

f) Final result.

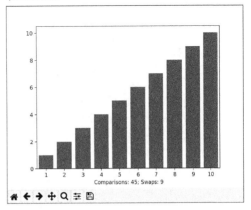

The colored bars in each have the following meanings:

- Gray bars, as in Parts (a)–(d), are not yet sorted and are not currently being compared.

- When locating the smallest element to swap into position, the animation displays one purple and one red bar, as in Part (b). During a pass of the unsorted elements to find the smallest remaining value, the purple bar represents the value at index `smallest`. The red bar is the other remaining element currently being compared. The red bar moves once for each comparison. The purple bar moves only if a smaller element is found during the current pass.
- The animation displays in purple both bars participating in a swap, as in Parts (c) and (d). Part (c) shows the bars before the swap and Part (d) shows them after.
- A light green bar indicates an element that has been placed into its final position by the sort.
- The dark green bars, as in Parts (e) and (f), are displayed one at a time when the sort completes to emphasize the final sorted order.

11.15.1 Generator Functions

In the "Sequences: Lists and Tuples" chapter, we introduced generator expressions. You saw that they're similar to list comprehensions, but create generator objects that produce values *on demand* using lazy evaluation.

The `selection_sort` function that we'll present in the next subsection implements the selection sort algorithm but, as you'll see, is now interspersed with:

- statements that keep track of the information we display in the animation,
- `yield` and `yield from` statements that provide values for the `FuncAnimation` to pass to the `update` function, and
- `play_sound` calls to play musical notes.

The `yield` and `yield from` statements, make `selection_sort` a **generator function**. Like generator expressions, generators functions are *lazy* and return values *on demand*. The `FuncAnimation` we'll implement obtains values on demand from the generator and passes them to `update` to animate the algorithm.

`yield` Statements

A generator function uses the **`yield` keyword** to return the next generated item, then its execution suspends until the program requests another item. When the Python interpreter encounters a generator function call, it creates an iterable **generator object** that keeps track of the next value to generate. We'll use generator functions in the next chapter to help create an animated visualization of an algorithm that sorts array values into ascending order.

Let's create a generator function that iterates through a sequence of values and returns the square of each value:

```
In [1]: def square_generator(values):
   ...:     for value in values:
   ...:         yield value ** 2
   ...:
```

Snippet [2] calls `square_generator` to create the generator object, which will not return any values until you access its values:

```
In [2]: squares = square_generator(numbers)
```

One way to access the values it to iterate over the generator object:

```
In [3]: for number in squares:
   ...:     print(number, end=' ')
   ...:
100 400 900
```

You must create a new generator object each time you wish to iterate through the generator's values again. Let's recreate the generator and access its values one at a time with built in function **next**, which receives an iterable argument and returns its next item:

```
In [4]: squares = square_generator(numbers)

In [5]: next(squares)
Out[5]: 100

In [6]: next(squares)
Out[6]: 400

In [7]: next(squares)
Out[7]: 900

In [8]: next(squares)
-------------------------------------------------------------------------
StopIteration                             Traceback (most recent call last)
<ipython-input-8-e7cf8d24b3b2> in <module>()
----> 1 next(squares)

StopIteration:
```

When there are no more items to process, the generator raises a StopIteration exception, which is how a for statement knows when to stop iterating through any iterable object. We'll discuss yield from when we encounter it in the selection sort animation's code.

11.15.2 Implementing the Selection Sort Animation

Now, let's implement the animation.

import Statements
Lines 3–8 import the modules we use. The file ch11soundutilities.py from which we import play_sound (line 8) is discussed after the example.

```
1  # selectionsortanimation.py
2  """Animated selection sort visualization."""
3  from matplotlib import animation
4  import matplotlib.pyplot as plt
5  import numpy as np
6  import seaborn as sns
7  import sys
8  from ch11soundutilities import play_sound
9
```

update Function That Displays Each Animation Frame
As in the "Dictionaries and Sets" chapter's Intro to Data Science section, we define an update function that the Matplotlib FuncAnimation calls once per animation frame to redraw the animation's graphical elements.

```
10  def update(frame_data):
11      """Display bars representing the current state."""
12      # unpack info for graph update
13      data, colors, swaps, comparisons = frame_data
14      plt.cla()  # clear old contents of current Figure
15
16      # create barplot and set its xlabel
17      bar_positions = np.arange(len(data))
18      axes = sns.barplot(bar_positions, data, palette=colors)  # new bars
19      axes.set(xlabel=f'Comparisons: {comparisons}; Swaps: {swaps}',
20              xticklabels=data)
21
```

The update method's frame_data parameter receives a tuple containing information for updating the bar plot. Line 13 unpacks frame_data into:

- data—the array being sorted,
- colors—an array of color names containing the specific color for each bar,
- swaps—an integer representing the number of swaps performed so far, and
- comparisons—an integer representing the number of comparisons performed so far.

We show the comparisons and swaps values below the *x*-axis so you can see them update as the sort executes.

Rather than letting Seaborn choose colors for the bars, line 18 specifies the colors array as the barplot's palette. When barplot creates each bar for the values in data, it uses the color at the corresponding index in colors. This enables us to emphasize with color the values sorted so far, the values the algorithm currently is comparing and the values the algorithm is swapping. Lines 19–20 display the comparisons and swaps performed so far as the *x*-axis label and set the data array's values as the *x*-axis tick labels below the bars. As the sort animation proceeds, you'll see both the bars and their corresponding tick labels change positions.

flash_bars Function That Flashes the Bars About to Be Swapped

We emphasize when the algorithm swaps two element values by calling flash_bars before and after a swap to flash the corresponding bars. The function receives the indices of the values about to be swapped, as well as the data array, colors array and the swaps and comparisons values. The last two parameters are used only in the yield statements. Each such statement throughout this example returns a tuple of the values the FuncAnimation passes to the update function's frame_data parameter.

```
22  def flash_bars(index1, index2, data, colors, swaps, comparisons):
23      """Flash the bars about to be swapped and play their notes."""
24      for x in range(0, 2):
25          colors[index1], colors[index2] = 'white', 'white'
26          yield (data, colors, swaps, comparisons)
27          colors[index1], colors[index2] = 'purple', 'purple'
28          yield (data, colors, swaps, comparisons)
29          play_sound(data[index1], seconds=0.05)
30          play_sound(data[index2], seconds=0.05)
31
```

11.15 Visualizing Algorithms

The loop (lines 24–30) iterates twice and performs the following tasks:

- To create the flashing effect, we first "erase" the bars by drawing them in white. So line 25 sets the colors array's elements at index1 and index2 to 'white'.
- Line 26 yields data, colors, swaps and comparisons so that the FuncAnimation can pass them to function update for use in drawing the next animation frame.
- Lines 27 and 28 repeat the two previous steps but redisplay the bars in purple to complete the flash effect.
- Lines 29 and 30 call play_sound with the data value at each index to play for 0.05 seconds a note corresponding to the bar's magnitude.

When the play_sound calls occur before the swap, the notes play in decreasing pitch to indicate that the values are out of order. The larger the bar, the higher its pitch and vice versa. When the calls occur after the swap, the notes play in increasing pitch to indicate that the values are now in order.

selection_sort Generator Function

Before the algorithm begins, lines 35–36 initialize the swaps and comparisons counters. Line 37 creates the colors array specifying that all the bars should be 'lightgray' initially. We'll modify this array's elements throughout the algorithm to emphasize bars in different ways. Line 40 yields data, colors, swaps and comparisons to the FuncAnimation. Each time values are yielded, the FuncAnimation passes them to the update function. This first yield causes the animation to display the bars in their initial unsorted order.

```
32  def selection_sort(data):
33      """Sort data using the selection sort algorithm and
34      yields values that update uses to visualize the algorithm."""
35      swaps = 0
36      comparisons = 0
37      colors = ['lightgray'] * len(data)  # list of bar colors
38
39      # display initial bars representing shuffled values
40      yield (data, colors, swaps, comparisons)
41
42      # loop over len(data) - 1 elements
43      for index1 in range(0, len(data) - 1):
44          smallest = index1
45
46          # loop to find index of smallest element's index
47          for index2 in range(index1 + 1, len(data)):
48              comparisons += 1
49              colors[smallest] = 'purple'
50              colors[index2] = 'red'
51              yield (data, colors, swaps, comparisons)
52              play_sound(data[index2], seconds=0.05)
53
54              # compare elements at positions index and smallest
55              if data[index2] < data[smallest]:
56                  colors[smallest] = 'lightgray'
57                  smallest = index2
58                  colors[smallest] = 'purple'
59                  yield (data, colors, swaps, comparisons)
```

```
60              else:
61                  colors[index2] = 'lightgray'
62                  yield (data, colors, swaps, comparisons)
63
64          # ensure that last bar is not purple
65          colors[-1] = 'lightgray'
66
67          # flash the bars about to be swapped
68          yield from flash_bars(index1, smallest, data, colors,
69                                swaps, comparisons)
70
71          # swap the elements at positions index1 and smallest
72          swaps += 1
73          data[smallest], data[index1] = data[index1], data[smallest]
74
75          # flash the bars that were just swapped
76          yield from flash_bars(index1, smallest, data, colors,
77                                swaps, comparisons)
78
79          # indicate that bar index1 is now in its final spot
80          colors[index1] = 'lightgreen'
81          yield (data, colors, swaps, comparisons)
82
83      # indicate that last bar is now in its final spot
84      colors[-1] = 'lightgreen'
85      yield (data, colors, swaps, comparisons)
86      play_sound(data[-1], seconds=0.05)
87
88      # play each bar's note once and color it darker green
89      for index in range(len(data)):
90          colors[index] = 'green'
91          yield (data, colors, swaps, comparisons)
92          play_sound(data[index], seconds=0.05)
93
```

The nested loops in lines 43–81 implement the selection sort algorithm. Here, we focus on the animation enhancements:

- Every iteration of the nested loop adds one to comparisons (line 48) to keep track of the total number of comparisons performed.
- The bar at the index smallest is always colored purple (line 49) during a pass of the algorithm. This bar's position frequently changes as the nested loop executes.
- The bar at the index index2 is always colored red (line 50) during a pass of the algorithm's nested loop. The red bar represents the value currently being compared with the smallest value during the current pass.
- Line 51 yields data, colors, swaps and comparisons during each iteration of the nested loop, so we can see the purple and red bars indicating the values that are being compared.
- Line 52 plays a musical note corresponding to the red bar's magnitude. We'll soon overview how we choose the note frequencies.
- If lines 55–62 find a smaller value, we color the original smallest bar light gray and the new smallest bar purple. Otherwise, we color the bar at index2 light

gray. These steps ensure that only one bar is red and one is purple during the comparisons. In either case, we `yield` the values for the next animation frame.

- Line 65 ensures that the last bar does not remain purple at the end of a pass.
- Each iteration of the outer loop ends with a swap. Lines 68–77 flash the corresponding bars before the swap, increment the `swaps` counter, perform the swap and flash the corresponding bars again. When one generator function (in this case `selection_sort`) needs to `yield` the results of another, chained generator functions are required. In this case, you must call the latter function (`flash_bars`), in a `yield from` statement (lines 68–69 and 76–77). This ensures that the values yielded by `flash_bars` "pass through" to the `FuncAnimation` for the next update call.
- Line 80 colors the element at `index1` light green to indicate that it is now in its final sorted position, and line 81 `yield`s to draw the next animation frame.

When the algorithm's outer loop completes, the last element of `data` is in its final position, so lines 84–86 color it light green, `yield` to draw the next animation frame and play that bar's note. To complete the animation, lines 89–92 convert each bar one at a time to a darker green, `yield` to draw the next animation frame and play that bar's musical note. The `FuncAnimation` terminates when the source of its animation-frame data has no more values. In this example, that occurs when the `selection_sort` function terminates.

main Function That Launches the Animation

The `main` function configures and starts the animation. If you specify the number of values to sort as a command-line argument, line 95 uses that value; otherwise, it uses 10 as the default. Lines 98–99 create and shuffle the array.

```
94  def main():
95      number_of_values = int(sys.argv[1]) if len(sys.argv) == 2 else 10
96
97      figure = plt.figure('Selection Sort')  # Figure to display barplot
98      numbers = np.arange(1, number_of_values + 1)  # create array
99      np.random.shuffle(numbers)  # shuffle the array
100
101     # start the animation
102     anim = animation.FuncAnimation(figure, update, repeat=False,
103         frames=selection_sort(numbers), interval=50)
104
105     plt.show()  # display the Figure
106
107 # call main if this file is executed as a script
108 if __name__ == '__main__':
109     main()
```

Lines 102–103 create the `FuncAnimation`. Recall that the `frames` keyword argument receives a value that specifies how many animation frames to execute. In this example, the `frames` keyword argument receives a call to the generator function `selection_sort`. So the number of animation frames depends on when the generator "runs out" of values. The values yielded by `selection_sort` are the source of the `update` function's `frame_data`. When `selection_sort` yields a new tuple of values, `selection_sort` pauses execution (keeping track of where it left off) and the `FuncAnimation` passes the tuple to `update`, which displays the next animation frame. When update finishes executing, selection_sort continues executing the algorithm from the point at which it paused.

Sound Utility Functions

The file ch11soundutilities.py contains two constants that we use to calculate musical note frequencies programmatically and three functions that play MIDI notes:

- TWELFTH_ROOT_2 represents the 12th root of 2, and A3 represents the frequency of the note A in the third octave on a piano. Positive values of i in line 11's calculation produce notes with higher frequencies than A3 and negative values produce notes with lower frequencies. Each consecutive value of i represents the next key on the piano, so if i is 12, the note would be A4—that is, A in the next higher octave. Similarly, if i is -12, the note would be A2.

- Function play_sound (lines 8–11) receives two arguments specifying the number of steps (that is, keys on the piano) above A3 for which to produce a note and the length of time in seconds to play that note. Line 11 calls the pysine module's **sine function** to play a note for the specified frequency and duration.

- The functions play_found_sound and play_not_found_sound are not used in this example. We provided them for you to use in your solution to Exercise 11.23, in which you'll create an animated binary search visualization. Each function plays specific notes to indicate whether a search key is found or not found.

```
 1  # ch11soundutilities.py
 2  """Functions to play sounds."""
 3  from pysine import sine
 4
 5  TWELFTH_ROOT_2 = 1.059463094359  # 12th root of 2
 6  A3 = 220  # hertz frequency for musical note A from third octave
 7
 8  def play_sound(i, seconds=0.1):
 9      """Play a note representing a bar's magnitude. Calculation
10      based on https://pages.mtu.edu/~suits/NoteFreqCalcs.html."""
11      sine(frequency=(A3 * TWELFTH_ROOT_2 ** i), duration=seconds)
12
13  def play_found_sound(seconds=0.1):
14      """Play sequence of notes indicating a found item."""
15      sine(frequency=523.25, duration=seconds)  # C5
16      sine(frequency=698.46, duration=seconds)  # F5
17      sine(frequency=783.99, duration=seconds)  # G5
18
19  def play_not_found_sound(seconds=0.3):
20      """Play a note indicating an item was not found."""
21      sine(frequency=220, duration=seconds)  # A3
```

11.16 Wrap-Up

In this chapter, which concludes our introduction to Python programming, you immersed yourself in some computer science thinking that goes beyond programming fundamentals. You created recursive functions—i.e., functions that call themselves. These functions (or methods) typically divide a problem into two conceptual pieces—the base case and the recursion step. The latter is a slightly smaller version of the original problem and is performed by a recursive function call. You saw some popular recursion examples, including calculating factorials and Fibonacci numbers. We compared recursive and iterative problem-solving approaches.

We also introduced searching and sorting. We discussed two searching algorithms—linear search and binary search—and three sorting algorithms—selection sort, insertion sort and recursive merge sort. We introduced Big O notation, which helps you express the efficiency of algorithms, making it easy to compare different algorithms for solving the same problem.

We presented an animated visualization of the selection sort using Seaborn and Matplotlib. We used the Matplotlib `FuncAnimation` class to drive the animation. We recast our `selection_sort` function as a generator function to provide values for display in the animation.

In the next part of the book, we present a series of implementation case studies that use a mix of AI and big-data technologies. We explore natural language processing, data mining Twitter, IBM Watson and cognitive computing, supervised and unsupervised machine learning, and deep learning with convolutional neural networks and recurrent neural networks. We discuss big-data software and hardware infrastructure, including NoSQL databases, Hadoop and Spark with a major emphasis on performance. You're about to learn some really cool stuff!

Exercises

11.1 What does the following code do?

```
In [1]: def mystery(a, b):
   ...:     if b == 1:
   ...:         return a
   ...:     else:
   ...:         return a + mystery(a, b - 1)
   ...:

In [2]: mystery(2, 10)
Out[2]: ?????
```

11.2 Find the logic error(s) in the following recursive function, and explain how to correct it (them). This function should find the sum of the values from 0 to n.

```
In [3]: def sum(n):
   ...:     if n == 0:
   ...:         return 0
   ...:     else:
   ...:         return n + sum(n)
   ...:
```

11.3 What does the following code do?

```
In [4]: def mystery(a_array, size):
   ...:     if size == 1:
   ...:         return a_array[0]
   ...:     else:
   ...:         return a_array[size - 1] + mystery(a_array, size - 1)
   ...:
```

```
In [5]: import numpy as np

In [6]: numbers = np.arange(1, 11)

In [7]: mystery(numbers, len(numbers))
Out[7]: ?????
```

11.4   In Section 11.3, we presented a recursive `factorial` function. What happens if you remove the `if` statement from the `factorial` function, then call the function?

11.5   *(Recursive **power** Function)* Write a recursive function `power(base, exponent)` that, when called, returns

$$base^{exponent}$$

For example, `power(3,4)` = 3 * 3 * 3 * 3. Assume that `exponent` is an integer greater than or equal to 1. *Hint:* The recursion step should use the relationship

$$base^{exponent} = base \cdot base^{exponent-1}$$

and the terminating condition occurs when `exponent` is equal to 1, because

$$base^1 = base$$

Incorporate this function into a program that enables the user to enter the `base` and `exponent`.

11.6   *(Recursive Fibonacci Modification)* Modify Section 11.4's recursive `fibonacci` function to keep track of the total number of function calls. Display the number of calls for `fibonacci(10)`, `fibonacci(20)` and `fibonacci(30)`.

11.7   *(Improving Recursive Fibonacci Performance: Memoization)* Research the performance enhancement technique called memoization. Modify Section 11.4's recursive `fibonacci` function to incorporate memoization. Compare the performance of both versions for `fibonacci(10)`, `fibonacci(20)` and `fibonacci(30)`.

11.8   *(Visualizing Recursion)* It's interesting to watch recursion "in action." Modify the recursive `factorial` function presented in this chapter to print its local variable and recursive-call parameter. For each recursive call, display the outputs on a separate line and add a level of indentation. Make the outputs clear, interesting and meaningful. Your goal here is to design and implement an output format that makes it easier to understand recursion.

11.9   *(Greatest Common Divisor)* The greatest common divisor of integers x and y is the largest integer that evenly divides into both x and y. Write and test a recursive function `gcd` that returns the greatest common divisor of x and y. The gcd of x and y is defined recursively as follows: If y is equal to 0, then `gcd(x, y)` is x; otherwise, `gcd(x, y)` is `gcd(y, x % y)`, where % is the remainder operator.

11.10   *(Palindromes)* A palindrome is a string that is spelled the same way forward and backward. Some examples of palindromes are "radar," "able was i ere i saw elba" and (if spaces are ignored) "a man a plan a canal panama." Write a recursive `test_palindrome` function that returns `True` if the string stored in an array is a palindrome and `False` otherwise. The function should ignore spaces and punctuation in the string.

11.11   *(Eight Queens)* A puzzler for chess buffs is the Eight Queens problem, which asks: Is it possible to place eight queens on an empty chessboard so that no queen is "attacking" any other (i.e., no two queens are in the same row, in the same column or along the same diagonal)? For example, if a queen is placed in the upper-left corner of the board, no other queens could be placed in any of the marked squares shown in the following figure. Solve the problem recursively. [*Hint:* Your solution should begin with the first column and look for a location in that column where a queen can be placed—initially, place the queen in the first row. The solution should then recursively search the remaining columns. In the first few columns, there will be several locations where a queen may be placed. Take the first available location. If a column is reached with no possible location for a queen, the program should return to the previous column, and move the queen in that column to a new row. This continuous backing up and trying new alternatives is an example of recursive backtracking.]

11.12 *(Towers of Hanoi)* In this chapter, you studied functions that can be easily implemented both recursively and iteratively. In this exercise, we present a problem whose recursive solution demonstrates the elegance of recursion, and whose iterative solution may not be as apparent.

The **Towers of Hanoi** is one of the most famous classic problems every budding computer scientist must grapple with. Legend has it that in a temple in the Far East, priests are attempting to move a stack of golden disks from one diamond peg to another. The following diagram shows the pegs with four disks on peg 1.

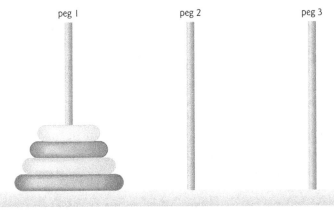

The initial stack has 64 disks threaded onto one peg and arranged from bottom to top by decreasing size. The priests are attempting to move the stack from one peg to another under the constraints that exactly one disk is moved at a time and at no time may a larger disk be placed above a smaller disk. Three pegs are provided, one being used for temporarily holding disks. Supposedly, the world will end when the priests complete their task, so there is little incentive for us to facilitate their efforts.

Let's assume that the priests are attempting to move the disks from peg 1 to peg 3. We wish to develop an algorithm that prints the precise sequence of peg-to-peg disk transfers.

If we were to approach this problem with conventional functions, we would rapidly find ourselves hopelessly knotted up in managing the disks. Instead, attacking this problem with recursion in mind allows the steps to be simple. Moving n disks can be viewed in terms of moving only $n-1$ disks (hence, the recursion), as follows:

  a) Move $n-1$ disks from peg 1 to peg 2, using peg 3 as a temporary holding area.
  b) Move the last disk (the largest) from peg 1 to peg 3.
  c) Move the $n-1$ disks from peg 2 to peg 3, using peg 1 as a temporary holding area.

The process ends when the last task involves moving $n = 1$ disk (i.e., the base case). This task is accomplished by simply moving the disk, without the need for a temporary holding area. Write a program to solve the Towers of Hanoi problem. Use a recursive function with four parameters:

  a) The number of disks to be moved.
  b) The peg on which these disks are initially threaded.
  c) The peg to which this stack of disks is to be moved.
  d) The peg to be used as a temporary holding area.

Display the precise instructions for moving the disks from the starting peg to the destination peg. To move a stack of three disks from peg 1 to peg 3, the program displays the following moves:

  1 → 3 (This means move one disk from peg 1 to peg 3.)
  1 → 2
  3 → 2
  1 → 3
  2 → 1
  2 → 3
  1 → 3

11.13 *(Logarithmic Portion of Big O)* What key aspect of both the binary search and the merge sort accounts for the logarithmic portion of their respective Big Os?

11.14 *(Comparing Insertion Sort to Merge Sort)* In what sense is the insertion sort superior to the merge sort? In what sense is the merge sort superior to the insertion sort?

11.15 *(Merge Sort: Sorting Subarrays)* In the text, we say that after the merge sort splits the array into two subarrays, it then sorts these two subarrays and merges them. Why might someone be puzzled by our statement that "it then sorts these two subarrays"?

11.16 *(Timing Sorting Algorithms)* Remove the output statements from the functions `selection_sort`, `insertion_sort` and `merge_sort` defined in this chapter, then import each example's source-code file into IPython. Create a 100,000-element array of random integers named `data1` and make two additional copies of the array (`data2` and `data3`) by calling method `copy` on the original array. Next, use `%timeit` as follows to compare the performance of each sorting algorithm:

```
%timeit -n 1 -r 1 selectionsort.selection_sort(data1)
%timeit -n 1 -r 1 insertionsort.insertion_sort(data2)
%timeit -n 1 -r 1 mergesort.merge_sort(data3)
```

Do the `selection_sort` and `insertion_sort` take approximately the same amount of time? Is `merge_sort` much faster?

11.17 *(Bucket Sort)* A bucket sort begins with a one-dimensional array of positive integers to be sorted and a two-dimensional array of integers with rows indexed from 0 to 9 and columns indexed from 0 to $n - 1$, where n is the number of values to be sorted. Each row of the two-dimensional array is referred to as a *bucket*. Write a class named `BucketSort` containing a function called `sort` that operates as follows:

  a) Place each value of the one-dimensional array into a row of the bucket array, based on the value's "ones" (rightmost) digit. For example, 97 is placed in row 7, 3 is placed in row 3, and 100 is placed in row 0. This procedure is called a *distribution pass*.
  b) Loop through the bucket array row by row, and copy the values back to the original array. This procedure is called a *gathering pass*. The new order of the preceding values in the one-dimensional array is 100, 3 and 97.
  c) Repeat this process for each subsequent digit position (tens, hundreds, thousands, etc.). On the second (tens digit) pass, 100 is placed in row 0, 3 is placed in row 0 (because 3 has no tens digit), and 97 is placed in row 9. After the gathering pass, the order of the values in the one-dimensional array is 100, 3 and 97. On the third (hundreds digit) pass, 100 is placed in row 1, 3 is placed in row 0, and 97 is placed in row 0 (after the 3). After this last gathering pass, the original array is in sorted order.

  The two-dimensional array of buckets is 10 times the length of the integer array being sorted. This sorting technique provides better performance than a selection and insertion sorts but requires much more memory—the selection and insertion sorts require space for only one additional element of data. This comparison is an example of a space/time trade-off: The bucket sort uses more memory than the selection and insertion sorts, but performs better. This version of the bucket sort requires copying all the data back to the original array on each pass. Another possibility is to create a second two-dimensional bucket array and repeatedly swap the data between the two bucket arrays.

11.18 *(Recursive Binary Search)* Modify this chapter's `binary_search` function to perform a recursive binary search of the array. The function should receive the search key, starting index and ending index as arguments. If the search key is found, return its index in the array. If the search key is not found, return -1.

11.19 *(Quicksort)* The recursive sorting technique called quicksort uses the following basic algorithm for a one-dimensional array of values:

  a) *Partitioning Step*: Take the first element of the unsorted array and determine its final location in the sorted array (i.e., all values to the left of the element in the array are less than the element, and all values to the right of the element in the array are greater than the element—we show how to do this below). We now have one element in its proper location and two unsorted subarrays.
  b) *Recursive Step*: Perform *Step 1* on each unsorted subarray. Each time *Step 1* is performed on a subarray, another element is placed in its final location of the sorted array, and two unsorted subarrays are created. When a subarray consists

of one element, that element is in its final location (because a one-element array is already sorted).

The basic algorithm seems simple enough, but how do we determine the final position of the first element of each subarray? As an example, consider the following set of values (the element in bold is the partitioning element—it will be placed in its final location in the sorted array):

37 2 6 4 89 8 10 12 68 45

Starting from the rightmost element of the array, compare each element with 37 until an element less than 37 is found; then swap 37 and that element. The first element less than 37 is 12, so 37 and 12 are swapped. The new array is

12 2 6 4 89 8 10 **37** 68 45

Element 12 is in italics to indicate that it was just swapped with 37.

Starting from the left of the array, but beginning with the element after 12, compare each element with 37 until an element greater than 37 is found—then swap 37 and that element. The first element greater than 37 is 89, so 37 and 89 are swapped. The new array is

12 2 6 4 **37** 8 10 *89* 68 45

Starting from the right, but beginning with the element before 89, compare each element with 37 until an element less than 37 is found—then swap 37 and that element. The first element less than 37 is 10, so 37 and 10 are swapped. The new array is

12 2 6 4 *10* 8 **37** 89 68 45

Starting from the left, but beginning with the element after 10, compare each element with 37 until an element greater than 37 is found—then swap 37 and that element. There are no more elements greater than 37, so when we compare 37 with itself, we know that 37 has been placed in its final location in the sorted array. Every value to the left of 37 is smaller than it, and every value to the right of 37 is larger than it.

Once the partition has been applied on the previous array, there are two unsorted subarrays. The subarray with values less than 37 contains 12, 2, 6, 4, 10 and 8. The subarray with values greater than 37 contains 89, 68 and 45. The sort continues recursively, with both subarrays being partitioned in the same manner as the original array.

Based on the preceding discussion, write recursive function `quick_sort_helper` to sort a one-dimensional integer array. The function should receive as arguments a starting index and an ending index on the original array being sorted. Call this function from a `quick_sort` function that receives just the original array to sort.

11.20 *(Determining Big O of Various Algorithms)* Determine the Big O for each of the following. You may need to research some of these items online:
   a) Get or set an item by index in a Python list.
   b) Insert a new value in order in a Python sorted list.
   c) Shell short an array.
   d) Bubble sort an array.

e) Towers of Hanoi for *n* disks. [*Hint:* For *n*=1, 2, 3, 4, 5 or 6, the number of operations is 1, 3, 7, 15, 31 or 63, respectively.] You'll see that the Big O for the Towers of Hanoi—$O(2^n)$—is far worse than that of the $O(n^2)$ sorts in this chapter. By the way, for 64 disks, the number of operations is 18,446,744,073,709,551,615. If the priests could move one disk per second, it would take 584,942,417,355 years to move the stack of 64 disks!

f) Find all the permutations (unique arrangements) of *n* unique items. [*Hint:* For the digits 1, 2 and 3 there are six permutations—123, 132, 213, 231, 312 and 321.] For the digits 1, 2, 3 and 4 there are 24 permutations. For the digits 1, 2, 3, 4 and 5 there are 120 permutations. You'll see that this Big O is far worse even than that of the Towers of Hanoi.

11.21 *(Project: Quicksort Animation)* Look at the `QuickSort.mp4` video file provided with this chapter's examples. Using the techniques you learned in the selection sort animation, modify your solution to Exercise 11.19 to display an animation of the quicksort algorithm in action.

11.22 *(Project: Merge Sort Animation)* Using the techniques you learned in the selection sort animation, modify the merge sort presented in this chapter to display an animation of the algorithm in action.

11.23 *(Project: Binary Search Animation)* Using the techniques you learned in the selection sort animation, modify the binary search presented in this chapter to display an animation of the algorithm in action. The `ch11soundutilities.py` files includes functions `play_found_sound` and `play_not_found_sound` for use in this exercise.

11.24 *(Challenge Project: Animated Towers of Hanoi)* The following website

> https://svn.python.org/projects/stackless/trunk/Demo/tkinter/guido/hanoi.py

has an animated implementation of the Towers of Hanoi using the `tkinter` module. Study the code, then modify it to run faster. [*Note:* The code has a typo—the `import` statement for `tkinter` has a capital T, which should be lowercase.]

11.25 *(Project: Recursive Directory Searching)* To better understand the concept of recursion, let's look at an example that's quite familiar to computer users—the recursive definition of a file-system directory on a computer. A computer normally stores related files in a directory (also called a folder). A directory can be empty, can contain files and/or can contain other directories, usually referred to as subdirectories. Each of these subdirectories, in turn, may also contain both files and directories. If we want to list each file in a directory (including all the files in the directory's subdirectories), we need to create a function that first lists the initial directory's files, then makes recursive calls to list the files in each of that directory's subdirectories. The base case occurs when a directory is reached that does not contain any subdirectories. At this point, all the files in the original directory have been listed, and no further recursion is necessary. Write a `print_directory` function that recursively walks through the files and subdirectories of a directory specified as an argument. The output from each recursive call should indent the names of the files and directories it prints one additional "level" so you can see the file and directory structure. Precede each file or directory name with F (for file) or D (for directory).

Natural Language Processing (NLP)

12

Objectives

In this chapter you'll:

- Perform natural language processing (NLP) tasks, which are fundamental to many of the forthcoming data science case study chapters.
- Run lots of NLP demos.
- Use the TextBlob, NLTK, Textatistic and spaCy NLP libraries and their pretrained models to perform various NLP tasks.
- Tokenize text into words and sentences.
- Use parts-of-speech tagging.
- Use sentiment analysis to determine whether text is positive, negative or neutral.
- Detect the language of text and translate between languages using TextBlob's Google Translate support.
- Get word roots via stemming and lemmatization.
- Use TextBlob's spell checking and correction capabilities.
- Get word definitions, synonyms and antonyms.
- Remove stop words from text.
- Create word clouds.
- Determine text readability with Textatistic.
- Use the spaCy library for named entity recognition and similarity detection.

Outline

12.1 Introduction
12.2 TextBlob
    12.2.1 Create a TextBlob
    12.2.2 Tokenizing Text into Sentences and Words
    12.2.3 Parts-of-Speech Tagging
    12.2.4 Extracting Noun Phrases
    12.2.5 Sentiment Analysis with TextBlob's Default Sentiment Analyzer
    12.2.6 Sentiment Analysis with the `NaiveBayesAnalyzer`
    12.2.7 Language Detection and Translation
    12.2.8 Inflection: Pluralization and Singularization
    12.2.9 Spell Checking and Correction
    12.2.10 Normalization: Stemming and Lemmatization
    12.2.11 Word Frequencies
    12.2.12 Getting Definitions, Synonyms and Antonyms from WordNet
    12.2.13 Deleting Stop Words
    12.2.14 n-grams
12.3 Visualizing Word Frequencies with Bar Charts and Word Clouds
    12.3.1 Visualizing Word Frequencies with Pandas
    12.3.2 Visualizing Word Frequencies with Word Clouds
12.4 Readability Assessment with Textatistic
12.5 Named Entity Recognition with spaCy
12.6 Similarity Detection with spaCy
12.7 Other NLP Libraries and Tools
12.8 Machine Learning and Deep Learning Natural Language Applications
12.9 Natural Language Datasets
12.10 Wrap-Up
Exercises

12.1 Introduction

Your alarm wakes you, and you hit the "Alarm Off" button. You reach for your smartphone and read your text messages and check the latest news clips. You listen to TV hosts interviewing celebrities. You speak to family, friends and colleagues and listen to their responses. You have a hearing-impaired friend with whom you communicate via sign language and who enjoys close-captioned video programs. You have a blind colleague who reads braille, listens to books being read by a computerized book reader and listens to a screen reader speak about what's on his computer screen. You read emails, distinguishing junk from important communications and send email. You read novels or works of nonfiction. You drive, observing road signs like "Stop," "Speed Limit 35" and "Road Under Construction." You give your car verbal commands, like "call home," "play classical music" or ask questions like, "Where's the nearest gas station?" You teach a child how to speak and read. You send a sympathy card to a friend. You study from textbooks. You read newspapers and magazines. You take notes during a class or meeting. You learn a foreign language to prepare for a semester abroad. You receive a client email in Spanish and run it through a free translation program. You respond in English knowing that your client can easily translate your email back to Spanish. You are uncertain about the language of an email, but language detection software instantly figures that out for you and translates the email to English.

These are examples of **natural language** communications in text, voice, video, sign language, braille and other forms with languages like English, Spanish, French, Russian, Chinese, Japanese and hundreds more. In this chapter, you'll master many natural language processing (NLP) capabilities through a series of hands-on demos, IPython sessions, Self-Check exercises and a broad range of end-of-chapter exercises and projects. You'll use many of these NLP capabilities in the upcoming data science case study chapters.

Natural language processing is performed on text collections, composed of Tweets, Facebook posts, conversations, movie reviews, Shakespeare's plays, historic documents, news items, meeting logs, and so much more. A text collection is known as a **corpus**, the plural of which is **corpora**.

Natural language lacks mathematical precision. Nuances of meaning make natural language understanding difficult. A text's meaning can be influenced by its context and the reader's "world view." Search engines, for example, can get to "know you" through your prior searches. The upside is better search results. The downside could be invasion of privacy.

12.2 TextBlob[1]

TextBlob is an object-oriented NLP text-processing library that is built on the **NLTK** and **pattern** NLP libraries and simplifies many of their capabilities. Some of the NLP tasks TextBlob can perform include:

- **Tokenization**—splitting text into pieces called **tokens**, which are meaningful units, such as words and numbers.
- **Parts-of-speech (POS) tagging**—identifying each word's part of speech, such as noun, verb, adjective, etc.
- **Noun phrase extraction**—locating groups of words that represent nouns, such as "red brick factory."[2]
- **Sentiment analysis**—determining whether text has positive, neutral or negative sentiment.
- **Inter-language translation** and **language detection** powered by Google Translate.
- **Inflection**[3]—pluralizing and singularizing words. There are other aspects of inflection that are not part of TextBlob.
- **Spell checking** and **spelling correction**.
- **Stemming**—reducing words to their stems by removing prefixes or suffixes. For example, the stem of "varieties" is "varieti."
- **Lemmatization**—like stemming, but produces real words based on the original words' context. For example, the lemmatized form of "varieties" is "variety."
- **Word frequencies**—determining how often each word appears in a corpus.
- **WordNet integration** for finding word definitions, synonyms and antonyms.
- **Stop word elimination**—removing common words, such as a, an, the, I, we, you and more to analyze the important words in a corpus.
- **n-grams**—producing sets of consecutive words in a corpus for use in identifying words that frequently appear adjacent to one another.

---

1. https://textblob.readthedocs.io/en/latest/.
2. The phrase "red brick factory" illustrates why natural language is such a difficult subject. Is a "red brick factory" a factory that makes red bricks? Is it a red factory that makes bricks of any color? Is it a factory built of red bricks that makes products of any type? In today's music world, it could even be the name of a rock band or the name of a game on your smartphone.
3. https://en.wikipedia.org/wiki/Inflection.

Many of these capabilities are used as part of more complex NLP tasks. In this section, we'll perform these NLP tasks using TextBlob and NLTK.

Installing the TextBlob Module
To install TextBlob, open your Anaconda Prompt (Windows), Terminal (macOS/Linux) or shell (Linux), then execute the following command:

```
conda install -c conda-forge textblob
```

Windows users might need to run the Anaconda Prompt as an Administrator for proper software installation privileges. To do so, right-click Anaconda Prompt in the start menu and select **More > Run as administrator**.

Once installation completes, execute the following command to download the NLTK corpora used by TextBlob:

```
ipython -m textblob.download_corpora
```

These include:

- The Brown Corpus (created at Brown University[4]) for parts-of-speech tagging.
- Punkt for English sentence tokenization.
- WordNet for word definitions, synonyms and antonyms.
- Averaged Perceptron Tagger for parts-of-speech tagging.
- conll2000 for breaking text into components, like nouns, verbs, noun phrases and more—known as **chunking** the text. The name conll2000 is from the conference that created the chunking data—Conference on Computational Natural Language Learning.
- Movie Reviews for sentiment analysis.

Project Gutenberg
A great source of text for analysis is the free e-books at Project Gutenberg:

> https://www.gutenberg.org

The site contains over 57,000 e-books in various formats, including plain text files. These are out of copyright in the United States. For information about Project Gutenberg's Terms of Use and copyright in other countries, see:

> https://www.gutenberg.org/wiki/Gutenberg:Terms_of_Use

In some of this section's examples, we use the plain-text e-book file for Shakespeare's *Romeo and Juliet*, which you can find at:

> https://www.gutenberg.org/ebooks/1513

Project Gutenberg does not allow programmatic access to its e-books. You're required to copy the books for that purpose.[5] To download *Romeo and Juliet* as a plain-text e-book, right click the **Plain Text UTF-8** link on the book's web page, then select **Save Link As...** (Chrome/FireFox), **Download Linked File As...** (Safari) or **Save target as** (Microsoft Edge) option to save the book to your system. Save it as RomeoAndJuliet.txt in the ch12 exam-

---

4. https://en.wikipedia.org/wiki/Brown_Corpus.
5. https://www.gutenberg.org/wiki/Gutenberg:Information_About_Robot_Access_to_our_Pages.

ples folder to ensure that our code examples will work correctly. For analysis purposes, we removed the Project Gutenberg text before "THE TRAGEDY OF ROMEO AND JULIET", as well as the Project Guttenberg information at the end of the file starting with:

```
End of the Project Gutenberg EBook of Romeo and Juliet,
by William Shakespeare
```

Self Check

1 *(Fill-In)* TextBlob is an object-oriented NLP text-processing library built on the _____ and _____ NLP libraries, and simplifies accessing their capabilities.
Answer: NLTK, pattern.

12.2.1 Create a TextBlob

TextBlob[6] is the fundamental class for NLP with the **textblob module**. Let's create a TextBlob containing two sentences:

```
In [1]: from textblob import TextBlob

In [2]: text = 'Today is a beautiful day. Tomorrow looks like bad weather.'

In [3]: blob = TextBlob(text)

In [4]: blob
Out[4]: TextBlob("Today is a beautiful day. Tomorrow looks like bad
weather.")
```

TextBlobs—and, as you'll see shortly, Sentences and Words—support string methods and can be compared with strings. They also provide methods for various NLP tasks. Sentences, Words and TextBlobs inherit from **BaseBlob**, so they have many common methods and properties.

[*Note:* We use snippet [3]'s TextBlob in several of the following Self Checks and subsections, in which we continue the previous interactive session.]

Self Check

1 *(Fill-In)* _____ is the fundamental class for NLP with the textblob module.
Answer: TextBlob.

2 *(True/False)* TextBlobs support string methods and can be compared with strings using the comparison operators.
Answer: True.

3 *(IPython Session)* Create a TextBlob named exercise_blob containing 'This is a TextBlob'.
Answer:

```
In [5]: exercise_blob = TextBlob('This is a TextBlob')

In [6]: exercise_blob
Out[6]: TextBlob("This is a TextBlob")
```

---

6. http://textblob.readthedocs.io/en/latest/api_reference.html#textblob.blob.TextBlob.

12.2.2 Tokenizing Text into Sentences and Words

Natural language processing often requires tokenizing text before performing other NLP tasks. TextBlob provides convenient properties for accessing the sentences and words in TextBlobs. Let's use the **sentence property** to get a list of **Sentence** objects:

```
In [7]: blob.sentences
Out[7]:
[Sentence("Today is a beautiful day."),
 Sentence("Tomorrow looks like bad weather.")]
```

The **words property** returns a `WordList` object containing a list of `Word` objects, representing each word in the `TextBlob` with the punctuation removed:

```
In [8]: blob.words
Out[8]: WordList(['Today', 'is', 'a', 'beautiful', 'day', 'Tomorrow',
'looks', 'like', 'bad', 'weather'])
```

Self Check

1 *(IPython Session)* Create a `TextBlob` with two sentences, then tokenize it into Sentences and Words, displaying all the tokens.
Answer:

```
In [9]: ex = TextBlob('My old computer is slow. My new one is fast.')

In [10]: ex.sentences
Out[10]: [Sentence("My old computer is slow."), Sentence("My new one is fast.")]

In [11]: ex.words
Out[11]: WordList(['My', 'old', 'computer', 'is', 'slow', 'My', 'new',
'one', 'is', 'fast'])
```

12.2.3 Parts-of-Speech Tagging

Parts-of-speech (POS) **tagging** is the process of evaluating words based on their context to determine each word's part of speech. There are eight primary English parts of speech—nouns, pronouns, verbs, adjectives, adverbs, prepositions, conjunctions and interjections (words that express emotion and that are typically followed by punctuation, like "Yes!" or "Ha!"). Within each category there are many subcategories.

Some words have multiple meanings. For example, the words "set" and "run" have hundreds of meanings each! If you look at the dictionary.com definitions of the word "run," you'll see that it can be a verb, a noun, an adjective or a part of a verb phrase. An important use of POS tagging is determining a word's meaning among its possibly many meanings. This is important for helping computers "understand" natural language.

The **tags property** returns a list of tuples, each containing a word and a string representing its part-of-speech tag:

```
In [12]: blob
Out[12]: TextBlob("Today is a beautiful day. Tomorrow looks like bad weather.")
```

```
In [13]: blob.tags
Out[13]:
[('Today', 'NN'),
 ('is', 'VBZ'),
 ('a', 'DT'),
 ('beautiful', 'JJ'),
 ('day', 'NN'),
 ('Tomorrow', 'NNP'),
 ('looks', 'VBZ'),
 ('like', 'IN'),
 ('bad', 'JJ'),
 ('weather', 'NN')]
```

By default, TextBlob uses a PatternTagger to determine parts-of-speech. This class uses the parts-of-speech tagging capabilities of the *pattern library*:

> https://www.clips.uantwerpen.be/pattern

You can view the library's 63 parts-of-speech tags at

> https://www.clips.uantwerpen.be/pages/MBSP-tags

In the preceding snippet's output:

- Today, day and weather are tagged as NN—a singular noun or mass noun.
- is and looks are tagged as VBZ—a third person singular present verb.
- a is tagged as DT—a determiner.[7]
- beautiful and bad are tagged as JJ—an adjective.
- Tomorrow is tagged as NNP—a proper singular noun.
- like is tagged as IN—a subordinating conjunction or preposition.

✓ Self Check

1 *(Fill-In)* _____ is the process of evaluating words based on their context to determine each word's part of speech
Answer: Parts-of-speech (POS) tagging.

2 *(IPython Session)* Display the parts-of-speech tags for the sentence, 'My dog is cute'.
Answer:

```
In [14]: TextBlob('My dog is cute').tags
Out[14]: [('My', 'PRP$'), ('dog', 'NN'), ('is', 'VBZ'), ('cute', 'JJ')]
```

In the preceding output, the POS tag PRP$ indicates a possessive pronoun.

12.2.4 Extracting Noun Phrases

Let's say you're preparing to purchase a water ski so you're researching them online. You might search for "best water ski." In this case, "water ski" is a noun phrase. If the search engine does not parse the noun phrase properly, you probably will not get the best search results. Go online and try searching for "best water," "best ski" and "best water ski" and see what you get.

---

7. https://en.wikipedia.org/wiki/Determiner.

A TextBlob's **noun_phrases property** returns a WordList object containing a list of Word objects—one for each noun phrase in the text:

```
In [15]: blob
Out[15]: TextBlob("Today is a beautiful day. Tomorrow looks like bad
weather.")

In [16]: blob.noun_phrases
Out[16]: WordList(['beautiful day', 'tomorrow', 'bad weather'])
```

Note that a Word representing a noun phrase can contain multiple words. A WordList is an extension of Python's built-in list type. WordLists provide additional methods for stemming, lemmatizing, singularizing and pluralizing.

Self Check

1. *(IPython Session)* Show the noun phrase(s) in the sentence, 'The red brick factory is for sale'.
Answer:

```
In [17]: TextBlob('The red brick factory is for sale').noun_phrases
Out[17]: WordList(['red brick factory'])
```

12.2.5 Sentiment Analysis with TextBlob's Default Sentiment Analyzer

One of the most common and valuable NLP tasks is **sentiment analysis**, which determines whether text is positive, neutral or negative. For instance, companies might use this to determine whether people are speaking positively or negatively online about their products. Consider the positive word "good" and the negative word "bad." Just because a sentence contains "good" or "bad" does not mean the sentence's sentiment necessarily is positive or negative. For example, the sentence

```
The food is not good.
```

clearly has negative sentiment. Similarly, the sentence

```
The movie was not bad.
```

clearly has positive sentiment, though perhaps not as positive as something like

```
The movie was excellent!
```

Sentiment analysis is a complex machine-learning problem. However, libraries like TextBlob have pretrained machine learning models for performing sentiment analysis.

Getting the Sentiment of a TextBlob

A TextBlob's **sentiment property** returns a **Sentiment** object indicating whether the text is positive or negative and whether it's objective or subjective:

```
In [18]: blob
Out[18]: TextBlob("Today is a beautiful day. Tomorrow looks like bad
weather.")

In [19]: blob.sentiment
Out[19]: Sentiment(polarity=0.07500000000000007,
subjectivity=0.8333333333333333)
```

12.2 TextBlob

In the preceding output, the polarity indicates sentiment with a value from -1.0 (negative) to 1.0 (positive) with 0.0 being neutral. The subjectivity is a value from 0.0 (objective) to 1.0 (subjective). Based on the values for our TextBlob, the overall sentiment is close to neutral, and the text is mostly subjective.

Getting the polarity and subjectivity from the Sentiment Object

The values displayed above probably provide more precision that you need in most cases. This can detract from numeric output's readability. The IPython magic **%precision** allows you to specify the default precision for *standalone* float objects and float objects in *built-in types* like lists, dictionaries and tuples. Let's use the magic to *round* the polarity and subjectivity values to three digits to the right of the decimal point:

```
In [20]: %precision 3
Out[20]: '%.3f'

In [21]: blob.sentiment.polarity
Out[21]: 0.075

In [22]: blob.sentiment.subjectivity
Out[22]: 0.833
```

Getting the Sentiment of a Sentence

You also can get the sentiment at the individual sentence level. Let's use the sentence property to get a list of **Sentence**[8] objects, then iterate through them and display each Sentence's sentiment property:

```
In [23]: for sentence in blob.sentences:
    ...:     print(sentence.sentiment)
    ...:
Sentiment(polarity=0.85, subjectivity=1.0)
Sentiment(polarity=-0.6999999999999998, subjectivity=0.6666666666666666)
```

This might explain why the entire TextBlob's sentiment is close to 0.0 (neutral)—one sentence is positive (0.85) and the other negative (-0.6999999999999998).

Self Check

1. *(IPython Session)* Import Sentence from the TextBlob module then make Sentence objects to check the sentiment of the three sentences used in this section's introduction.
Answer: Snippet [25]'s output shows that the sentence's sentiment is somewhat negative (due to "not good"). Snippet [26]'s output shows that the sentence's sentiment is somewhat positive (due to "not bad"). Snippet [27]'s output shows that the sentence's sentiment is totally positive (due to "excellent"). The outputs indicate that all three sentences are subjective, with the last being perfectly positive and subjective.

```
In [24]: from textblob import Sentence

In [25]: Sentence('The food is not good.').sentiment
Out[25]: Sentiment(polarity=-0.35, subjectivity=0.6000000000000001)
```

---

8. http://textblob.readthedocs.io/en/latest/api_reference.html#textblob.blob.Sentence.

```
In [26]: Sentence('The movie was not bad.').sentiment
Out[26]: Sentiment(polarity=0.3499999999999999,
subjectivity=0.6666666666666666)

In [27]: Sentence('The movie was excellent!').sentiment
Out[27]: Sentiment(polarity=1.0, subjectivity=1.0)
```

12.2.6 Sentiment Analysis with the NaiveBayesAnalyzer

By default, a TextBlob and the Sentences and Words you get from it determine sentiment using a PatternAnalyzer, which uses the same sentiment analysis techniques as in the Pattern library. The TextBlob library also comes with a **NaiveBayesAnalyzer**[9] (module **textblob.sentiments**), which was trained on a database of movie reviews. Naive Bayes[10] is a commonly used machine learning text-classification algorithm. The following uses the analyzer keyword argument to specify a TextBlob's sentiment analyzer. Recall from earlier in this ongoing IPython session that text contains 'Today is a beautiful day. Tomorrow looks like bad weather.':

```
In [28]: from textblob.sentiments import NaiveBayesAnalyzer

In [29]: blob = TextBlob(text, analyzer=NaiveBayesAnalyzer())

In [30]: blob
Out[30]: TextBlob("Today is a beautiful day. Tomorrow looks like bad
weather.")
```

Let's use the TextBlob's sentiment property to display the text's sentiment using the NaiveBayesAnalyzer:

```
In [31]: blob.sentiment
Out[31]: Sentiment(classification='neg', p_pos=0.47662917962091056,
p_neg=0.5233708203790892)
```

In this case, the overall sentiment is classified as negative (classification='neg'). The Sentiment object's p_pos indicates that the TextBlob is 47.66% positive, and its p_neg indicates that the TextBlob is 52.34% negative. Since the overall sentiment is just slightly more negative we'd probably view this TextBlob's sentiment as neutral overall.

Now, let's get the sentiment of each Sentence:

```
In [32]: for sentence in blob.sentences:
    ...:     print(sentence.sentiment)
    ...:
Sentiment(classification='pos', p_pos=0.8117563121751951,
p_neg=0.18824368782480477)
Sentiment(classification='neg', p_pos=0.174363226578349,
p_neg=0.8256367734216521)
```

Notice that rather than polarity and subjectivity, the Sentiment objects we get from the NaiveBayesAnalyzer contain a *classification*—'pos' (positive) or 'neg' (negative)—and p_pos (percentage positive) and p_neg (percentage negative) values from 0.0 to 1.0. Once again, we see that the first sentence is positive and the second is negative.

---

9. https://textblob.readthedocs.io/en/latest/api_reference.html#module-textblob.en.sentiments.
10. https://en.wikipedia.org/wiki/Naive_Bayes_classifier.

Self Check

1. *(IPython Session)* Check the sentiment of the sentence `'The movie was excellent!'` using the `NaiveBayesAnalyzer`.
Answer:

```
In [33]: text = ('The movie was excellent!')

In [34]: exblob = TextBlob(text, analyzer=NaiveBayesAnalyzer())

In [35]: exblob.sentiment
Out[35]: Sentiment(classification='pos', p_pos=0.7318278242290406,
p_neg=0.26817217577095936)
```

12.2.7 Language Detection and Translation

Inter-language translation is a challenging problem in natural language processing and artificial intelligence. With advances in machine learning, artificial intelligence and natural language processing, services like Google Translate (100+ languages) and Microsoft Bing Translator (60+ languages) can translate between languages instantly.

Inter-language translation also is great for people traveling to foreign countries. They can use translation apps to translate menus, road signs and more. There are even efforts at live speech translation so that you'll be able to converse in real time with people who do not know your natural language.[11,12] Some smartphones, can now work together with in-ear headphones to provide near-live translation of many languages.[13,14,15] In the "IBM Watson and Cognitive Computing" chapter, we develop a script that does near real-time inter-language translation among languages supported by Watson.

The TextBlob library uses Google Translate to detect a text's language and translate TextBlobs, Sentences and Words into other languages.[16] Let's use **detect_language method** to detect the language of the text we're manipulating (`'en'` is English):

```
In [36]: blob
Out[36]: TextBlob("Today is a beautiful day. Tomorrow looks like bad
weather.")

In [37]: blob.detect_language()
Out[37]: 'en'
```

Next, let's use the **translate method** to translate the text to Spanish (`'es'`) then detect the language on the result. The to keyword argument specifies the target language.

```
In [38]: spanish = blob.translate(to='es')

In [39]: spanish
Out[39]: TextBlob("Hoy es un hermoso dia. Mañana parece mal tiempo.")
```

---

11. https://www.skype.com/en/features/skype-translator/.
12. https://www.microsoft.com/en-us/translator/business/live/.
13. https://www.telegraph.co.uk/technology/2017/10/04/googles-new-headphones-can-translate-foreign-languages-real/.
14. https://store.google.com/us/product/google_pixel_buds?hl=en-US.
15. http://www.chicagotribune.com/bluesky/originals/ct-bsi-google-pixel-buds-review-20171115-story.html.
16. These features require an Internet connection.

```
In [40]: spanish.detect_language()
Out[40]: 'es'
```

Next, let's translate our TextBlob to simplified Chinese (specified as 'zh' or 'zh-CN') then detect the language on the result:

```
In [41]: chinese = blob.translate(to='zh')

In [42]: chinese
Out[42]: TextBlob("今天是美好的一天。明天看起来像恶劣的天气。")

In [43]: chinese.detect_language()
Out[43]: 'zh-CN'
```

Method detect_language's output always shows simplified Chinese as 'zh-CN', even though the translate function can receive simplified Chinese as 'zh' or 'zh-CN'.

In each of the preceding cases, Google Translate automatically detects the source language. You can specify a source language explicitly by passing the from_lang keyword argument to the translate method, as in

```
chinese = blob.translate(from_lang='en', to='zh')
```

Google Translate uses iso-639-1[17] language codes listed at

https://en.wikipedia.org/wiki/List_of_ISO_639-1_codes

For the supported languages, you'd use these codes as the values of the from_lang and to keyword arguments. Google Translate's list of supported languages is at:

https://cloud.google.com/translate/docs/languages

Calling translate without arguments translates from the detected source language to English:

```
In [44]: spanish.translate()
Out[44]: TextBlob("Today is a beautiful day. Tomorrow seems like bad
weather.")

In [45]: chinese.translate()
Out[45]: TextBlob("Today is a beautiful day. Tomorrow looks like bad
weather.")
```

Note the slight difference in the English results.

✓ Self Check

1 *(IPython Session)* Translate 'Today is a beautiful day.' into French, then detect the language.
Answer:

```
In [46]: blob = TextBlob('Today is a beautiful day.')

In [47]: french = blob.translate(to='fr')

In [48]: french
Out[48]: TextBlob("Aujourd'hui est un beau jour.")
```

---

17. ISO is the International Organization for Standardization (https://www.iso.org/).

12.2.8 Inflection: Pluralization and Singularization

```
In [49]: french.detect_language()
Out[49]: 'fr'
```

Inflections are different forms of the same words, such as singular and plural (like "person" and "people") and different verb tenses (like "run" and "ran"). When you're calculating word frequencies, you might first want to convert all inflected words to the same form for more accurate word frequencies. Words and WordLists each support converting words to their singular or plural forms. Let's pluralize and singularize a couple of Word objects:

```
In [1]: from textblob import Word

In [2]: index = Word('index')

In [3]: index.pluralize()
Out[3]: 'indices'

In [4]: cacti = Word('cacti')

In [5]: cacti.singularize()
Out[5]: 'cactus'
```

Pluralizing and singularizing are sophisticated tasks which, as you can see above, are not as simple as adding or removing an "s" or "es" at the end of a word.

You can do the same with a WordList:

```
In [6]: from textblob import TextBlob

In [7]: animals = TextBlob('dog cat fish bird').words

In [8]: animals.pluralize()
Out[8]: WordList(['dogs', 'cats', 'fish', 'birds'])
```

Note that the word "fish" is the same in both its singular and plural forms.

✓ Self Check

1 *(IPython Session)* Singularize the word `'children'` and pluralize `'focus'`.
Answer:

```
In [1]: from textblob import Word

In [2]: Word('children').singularize()
Out[2]: 'child'

In [3]: Word('focus').pluralize()
Out[3]: 'foci'
```

12.2.9 Spell Checking and Correction

For natural language processing tasks, it's important that the text be free of spelling errors. Software packages for writing and editing text, like Microsoft Word, Google Docs and others automatically check your spelling as you type and typically display a red line under misspelled words. Other tools enable you to manually invoke a spelling checker.

You can check a Word's spelling with its **spellcheck** method, which returns a list of tuples containing possible correct spellings and a confidence value. Let's assume we meant to type the word "they" but we misspelled it as "theyr." The spell checking results show two possible corrections with the word 'they' having the highest confidence value:

```
In [1]: from textblob import Word

In [2]: word = Word('theyr')

In [3]: %precision 2
Out[3]: '%.2f'

In [4]: word.spellcheck()
Out[4]: [('they', 0.57), ('their', 0.43)]
```

Note that the word with the highest confidence value might *not* be the correct word for the given context.

TextBlobs, Sentences and Words all have a **correct method** that you can call to correct spelling. Calling correct on a Word returns the correctly spelled word that has the highest confidence value (as returned by spellcheck):

```
In [5]: word.correct()  # chooses word with the highest confidence value
Out[5]: 'they'
```

Calling correct on a TextBlob or Sentence checks the spelling of each word. For each incorrect word, correct replaces it with the correctly spelled one that has the highest confidence value:

```
In [6]: from textblob import Word

In [7]: sentence = TextBlob('Ths sentense has missplled wrds.')

In [8]: sentence.correct()
Out[8]: TextBlob("The sentence has misspelled words.")
```

✓ Self Check

1 *(True/False)* You can check a Word's spelling with its correct method, which returns a list of tuples containing possible correct spellings and a confidence value.
Answer: False. You can check a Word's spelling with its spellcheck method, which returns a list of tuples containing potential correct spellings and a confidence value.

2 *(IPython Session)* Correct the spelling in 'I canot beleive I misspeled thees werds'.
Answer:

```
In [1]: from textblob import TextBlob

In [2]: sentence = TextBlob('I canot beleive I misspeled thees werds')

In [3]: sentence.correct()
Out[3]: TextBlob("I cannot believe I misspelled these words")
```

12.2.10 Normalization: Stemming and Lemmatization

Stemming removes a prefix or suffix from a word leaving only a stem, which may or may not be a real word. **Lemmatization** is similar, but factors in the word's part of speech and meaning and results in a real word.

Stemming and lemmatization are **normalization** operations, in which you prepare words for analysis. For example, before calculating statistics on words in a body of text, you might convert all words to lowercase so that capitalized and lowercase words are not treated differently. Sometimes, you might want to use a word's root to represent the word's many forms. For example, in a given application, you might want to treat all of the following words as "program": program, programs, programmer, programming and programmed (and perhaps U.K. English spellings, like programmes as well).

Words and WordLists each support stemming and lemmatization via the methods **stem** and **lemmatize**. Let's use both on a Word:

```
In [1]: from textblob import Word

In [2]: word = Word('varieties')

In [3]: word.stem()
Out[3]: 'varieti'

In [4]: word.lemmatize()
Out[4]: 'variety'
```

✓ Self Check

1 *(True/False)* Stemming is similar to lemmatization, but factors in the word's part of speech and meaning and results in a real word.
Answer: False. Lemmatization is similar to stemming, but factors in the word's part of speech and meaning and results in a real word.

2 *(IPython Session)* Stem and lemmatize the word 'strawberries'.
Answer:

```
In [1]: from textblob import Word

In [2]: word = Word('strawberries')

In [3]: word.stem()
Out[3]: 'strawberri'

In [4]: word.lemmatize()
Out[4]: 'strawberry'
```

12.2.11 Word Frequencies

Various techniques for detecting similarity between documents rely on word frequencies. As you'll see here, TextBlob automatically counts word frequencies. First, let's load the e-book for Shakespeare's Romeo and Juliet into a TextBlob. To do so, we'll use the **Path** class from the Python Standard Library's **pathlib module**:

```
In [1]: from pathlib import Path

In [2]: from textblob import TextBlob

In [3]: blob = TextBlob(Path('RomeoAndJuliet.txt').read_text())
```

Use the file `RomeoAndJuliet.txt`[18] that you downloaded earlier. We assume here that you started your IPython session from that folder. When you read a file with `Path`'s **read_text method**, it closes the file immediately after it finishes reading the file.

You can access the word frequencies through the `TextBlob`'s **word_counts dictionary**. Let's get the counts of several words in the play:

```
In [4]: blob.word_counts['juliet']
Out[4]: 190

In [5]: blob.word_counts['romeo']
Out[5]: 315

In [6]: blob.word_counts['thou']
Out[6]: 278
```

If you already have tokenized a `TextBlob` into a `WordList`, you can count specific words in the list via the **count method**:

```
In [7]: blob.words.count('joy')
Out[7]: 14

In [8]: blob.noun_phrases.count('lady capulet')
Out[8]: 46
```

Self Check

1 *(True/False)* You can access the word frequencies through the `TextBlob`'s `counts` dictionary.
Answer: False. You can access the word frequencies through the `word_counts` dictionary.

2 *(IPython Session)* Using the `TextBlob` from this section's IPython session, determine how many times the stop words "a," "an" and "the" appear in *Romeo and Juliet*.
Answer:

```
In [9]: blob.word_counts['a']
Out[9]: 483

In [10]: blob.word_counts['an']
Out[10]: 71

In [11]: blob.word_counts['the']
Out[11]: 688
```

12.2.12 Getting Definitions, Synonyms and Antonyms from WordNet

WordNet[19] is a word database created by Princeton University. The TextBlob library uses the NLTK library's WordNet interface, enabling you to look up word definitions, and get synonyms and antonyms. For more information, check out the NLTK WordNet interface documentation at:

> https://www.nltk.org/api/nltk.corpus.reader.html#module-nltk.corpus.reader.wordnet

---

18. Each Project Gutenberg e-book includes additional text, such as their licensing information, that's not part of the e-book itself. For this example, we used a text editor to remove that text from our copy of the e-book.
19. https://wordnet.princeton.edu/.

Getting Definitions

First, let's create a Word:

```
In [1]: from textblob import Word

In [2]: happy = Word('happy')
```

The Word class's **definitions property** returns a list of all the word's definitions in the WordNet database:

```
In [3]: happy.definitions
Out[3]:
['enjoying or showing or marked by joy or pleasure',
 'marked by good fortune',
 'eagerly disposed to act or to be of service',
 'well expressed and to the point']
```

The database does not necessarily contain every dictionary definition of a given word. There's also a **define method** that enables you to pass a part of speech as an argument so you can get definitions matching only that part of speech.

Getting Synonyms

You can get a Word's **synsets**—that is, its sets of synonyms—via the **synsets property**. The result is a list of Synset objects:

```
In [4]: happy.synsets
Out[4]:
[Synset('happy.a.01'),
 Synset('felicitous.s.02'),
 Synset('glad.s.02'),
 Synset('happy.s.04')]
```

Each Synset represents a group of synonyms. In the notation happy.a.01:

- happy is the original Word's lemmatized form (in this case, it's the same).
- a is the part of speech, which can be a for adjective, n for noun, v for verb, r for adverb or s for adjective satellite. Many adjective synsets in WordNet have satellite synsets that represent similar adjectives.
- 01 is a 0-based index number. Many words have multiple meanings, and this is the index number of the corresponding meaning in the WordNet database.

There's also a **get_synsets method** that enables you to pass a part of speech as an argument so you can get Synsets matching only that part of speech.

You can iterate through the synsets list to find the original word's synonyms. Each Synset has a **lemmas method** that returns a list of Lemma objects representing the synonyms. A Lemma's name method returns the synonymous word as a string. In the following code, for each Synset in the synsets list, the nested for loop iterates through that Synset's Lemmas (if any). Then we add the synonym to the set named synonyms. We used a set collection because it automatically eliminates any duplicates we add to it:

```
In [5]: synonyms = set()

In [6]: for synset in happy.synsets:
   ...:     for lemma in synset.lemmas():
   ...:         synonyms.add(lemma.name())
   ...:
```

```
In [7]: synonyms
Out[7]: {'felicitous', 'glad', 'happy', 'well-chosen'}
```

Getting Antonyms

If the word represented by a Lemma has antonyms in the WordNet database, invoking the Lemma's antonyms method returns a list of Lemmas representing the antonyms (or an empty list if there are no antonyms in the database). In snippet [4] you saw there were four Synsets for 'happy'. First, let's get the Lemmas for the Synset at index 0 of the synsets list:

```
In [8]: lemmas = happy.synsets[0].lemmas()

In [9]: lemmas
Out[9]: [Lemma('happy.a.01.happy')]
```

In this case, lemmas returned a list of one Lemma element. We can now check whether the database has any corresponding antonyms for that Lemma:

```
In [10]: lemmas[0].antonyms()
Out[10]: [Lemma('unhappy.a.01.unhappy')]
```

The result is list of Lemmas representing the antonym(s). Here, we see that the one antonym for 'happy' in the database is 'unhappy'.

✓ Self Check

1 *(Fill-In)* A(n) _____ represents synonyms of a given word.
Answer: Synset.

2 *(IPython Session)* Display the synsets and definitions for the word "boat."
Answer:

```
In [1]: from textblob import Word

In [2]: word = Word('boat')

In [3]: word.synsets
Out[3]: [Synset('boat.n.01'), Synset('gravy_boat.n.01'),
Synset('boat.v.01')]

In [4]: word.definitions
Out[4]:
['a small vessel for travel on water',
 'a dish (often boat-shaped) for serving gravy or sauce',
 'ride in a boat on water']
```

In this case, there were three Synsets, and the definitions property displayed the corresponding definitions.

12.2.13 Deleting Stop Words

Stop words are common words in text that are often removed from text before analyzing it because they typically do not provide useful information. The following table shows NLTK's list of English stop words, which is returned by the NLTK stopwords module's words function[20] (which we'll use momentarily):

NLTK's English stop words list

```
['a', 'about', 'above', 'after', 'again', 'against', 'ain', 'all', 'am', 'an', 'and',
'any', 'are', 'aren', "aren't", 'as', 'at', 'be', 'because', 'been', 'before', 'being',
'below', 'between', 'both', 'but', 'by', 'can', 'couldn', "couldn't", 'd', 'did', 'didn',
"didn't", 'do', 'does', 'doesn', "doesn't", 'doing', 'don', "don't", 'down', 'during',
'each', 'few', 'for', 'from', 'further', 'had', 'hadn', "hadn't", 'has', 'hasn', "hasn't",
'have', 'haven', "haven't", 'having', 'he', 'her', 'here', 'hers', 'herself', 'him', 'him-
self', 'his', 'how', 'i', 'if', 'in', 'into', 'is', 'isn', "isn't", 'it', "it's", 'its',
'itself', 'just', 'll', 'm', 'ma', 'me', 'mightn', "mightn't", 'more', 'most', 'mustn',
"mustn't", 'my', 'myself', 'needn', "needn't", 'no', 'nor', 'not', 'now', 'o', 'of', 'off',
'on', 'once', 'only', 'or', 'other', 'our', 'ours', 'ourselves', 'out', 'over', 'own',
're', 's', 'same', 'shan', "shan't", 'she', "she's", 'should', "should've", 'shouldn',
"shouldn't", 'so', 'some', 'such', 't', 'than', 'that', "that'll", 'the', 'their',
'theirs', 'them', 'themselves', 'then', 'there', 'these', 'they', 'this', 'those',
'through', 'to', 'too', 'under', 'until', 'up', 've', 'very', 'was', 'wasn', "wasn't",
'we', 'were', 'weren', "weren't", 'what', 'when', 'where', 'which', 'while', 'who', 'whom',
'why', 'will', 'with', 'won', "won't", 'wouldn', "wouldn't", 'y', 'you', "you'd", "you'll",
"you're", "you've", 'your', 'yours', 'yourself', 'yourselves']
```

The NLTK library has lists of stop words for several other natural languages as well. Before using NLTK's stop-words lists, you must download them, which you do with the nltk module's **download function**:

```
In [1]: import nltk

In [2]: nltk.download('stopwords')
[nltk_data] Downloading package stopwords to
[nltk_data]     C:\Users\PaulDeitel\AppData\Roaming\nltk_data...
[nltk_data]   Unzipping corpora\stopwords.zip.
Out[2]: True
```

For this example, we'll load the 'english' stop words list. First import stopwords from the nltk.corpus module, then use stopwords method words to load the 'english' stop words list:

```
In [3]: from nltk.corpus import stopwords
```

```
In [4]: stops = stopwords.words('english')
```

Next, let's create a TextBlob from which we'll remove stop words:

```
In [5]: from textblob import TextBlob
```

```
In [6]: blob = TextBlob('Today is a beautiful day.')
```

Finally, to remove the stop words, let's use the TextBlob's words in a list comprehension that adds each word to the resulting list only if the word is not in stops:

```
In [7]: [word for word in blob.words if word not in stops]
Out[7]: ['Today', 'beautiful', 'day']
```

---

20. https://www.nltk.org/book/ch02.html.

Self Check

1 *(Fill-In)* _____ are common words in text that are often removed from text before analyzing it.
Answer: Stop words.

2 *(IPython Session)* Eliminate stop words from a `TextBlob` containing the sentence `'TextBlob is easy to use.'`
Answer:

```
In [1]: from nltk.corpus import stopwords

In [2]: stops = stopwords.words('english')

In [3]: from textblob import TextBlob

In [4]: blob = TextBlob('TextBlob is easy to use.')

In [5]: [word for word in blob.words if word not in stops]
Out[5]: ['TextBlob', 'easy', 'use']
```

12.2.14 n-grams

An **n-gram**[21] is a sequence of *n* text items, such as letters in words or words in a sentence. In natural language processing, *n*-grams can be used to identify letters or words that frequently appear adjacent to one another. For text-based user input, this can help predict the next letter or word a user will type—such as when completing items in IPython with tab-completion or when entering a message to a friend in your favorite smartphone messaging app. For speech-to-text, *n*-grams might be used to improve the quality of the transcription. N-grams are a form of *co-occurrence* in which words or letters appear near each other in a body of text.

`TextBlob`'s **ngrams method** produces a list of `WordList` *n*-grams of length three by default—known as *trigrams*. You can pass the keyword argument n to produce *n*-grams of any desired length. The output shows that the first trigram contains the first three words in the sentence (`'Today'`, `'is'` and `'a'`). Then, `ngrams` creates a trigram starting with the second word (`'is'`, `'a'` and `'beautiful'`) and so on until it creates a trigram containing the last three words in the `TextBlob`:

```
In [1]: from textblob import TextBlob

In [2]: text = 'Today is a beautiful day. Tomorrow looks like bad weather.'

In [3]: blob = TextBlob(text)

In [4]: blob.ngrams()
Out[4]:
[WordList(['Today', 'is', 'a']),
 WordList(['is', 'a', 'beautiful']),
 WordList(['a', 'beautiful', 'day']),
 WordList(['beautiful', 'day', 'Tomorrow']),
 WordList(['day', 'Tomorrow', 'looks']),
 WordList(['Tomorrow', 'looks', 'like']),
 WordList(['looks', 'like', 'bad']),
 WordList(['like', 'bad', 'weather'])]
```

---

21. https://en.wikipedia.org/wiki/N-gram.

The following produces *n*-grams consisting of five words:

```
In [5]: blob.ngrams(n=5)
Out[5]:
[WordList(['Today', 'is', 'a', 'beautiful', 'day']),
 WordList(['is', 'a', 'beautiful', 'day', 'Tomorrow']),
 WordList(['a', 'beautiful', 'day', 'Tomorrow', 'looks']),
 WordList(['beautiful', 'day', 'Tomorrow', 'looks', 'like']),
 WordList(['day', 'Tomorrow', 'looks', 'like', 'bad']),
 WordList(['Tomorrow', 'looks', 'like', 'bad', 'weather'])]
```

✓ Self Check

1 *(Fill-In)* N-grams are a form of _____ in which words appear near each other in a body of text.
Answer: co-occurrence.

2 *(IPython Session)* Produce n-grams consisting of three words each for 'TextBlob is easy to use.'
Answer:

```
In [1]: from textblob import TextBlob

In [2]: blob = TextBlob('TextBlob is easy to use.')

In [3]: blob.ngrams()
Out[3]:
[WordList(['TextBlob', 'is', 'easy']),
 WordList(['is', 'easy', 'to']),
 WordList(['easy', 'to', 'use'])]
```

12.3 Visualizing Word Frequencies with Bar Charts and Word Clouds

Earlier, we obtained frequencies for a few words in *Romeo and Juliet*. Sometimes frequency visualizations enhance your corpus analyses. There's often more than one way to visualize data, and sometimes one is superior to others. For example, you might be interested in word frequencies relative to one another, or you may just be interested in relative uses of words in a corpus. In this section, we'll look at two ways to visualize word frequencies:

- A bar chart that *quantitatively* visualizes the top 20 words in *Romeo and Juliet* as bars representing each word and its frequency.
- A **word cloud** that *qualitatively* visualizes more frequently occurring words in bigger fonts and less frequently occurring words in smaller fonts.

12.3.1 Visualizing Word Frequencies with Pandas

Let's visualize *Romeo and Juliet*'s top 20 words that are *not* stop words. To do this, we'll use features from TextBlob, NLTK and pandas. Pandas visualization capabilities are based on Matplotlib, so launch IPython with the following command for this session:

```
ipython --matplotlib
```

Loading the Data

First, let's load *Romeo and Juliet*. Launch IPython from the ch12 examples folder before executing the following code so you can access the e-book file `RomeoAndJuliet.txt` that you downloaded earlier in the chapter:

```
In [1]: from pathlib import Path

In [2]: from textblob import TextBlob

In [3]: blob = TextBlob(Path('RomeoAndJuliet.txt').read_text())
```

Next, load the NLTK stopwords:

```
In [4]: from nltk.corpus import stopwords

In [5]: stop_words = stopwords.words('english')
```

Getting the Word Frequencies

To visualize the top 20 words, we need each word and its frequency. Let's call the `blob.word_counts` dictionary's `items` method to get a list of word-frequency tuples:

```
In [6]: items = blob.word_counts.items()
```

Eliminating the Stop Words

Next, let's use a list comprehension to eliminate any tuples containing stop words:

```
In [7]: items = [item for item in items if item[0] not in stop_words]
```

The expression `item[0]` gets the word from each tuple so we can check whether it's in `stop_words`.

Sorting the Words by Frequency

To determine the top 20 words, let's sort the tuples in `items` in descending order by frequency. We can use built-in function `sorted` with a key argument to sort the tuples by the frequency element in each tuple. To specify the tuple element to sort by, use the **itemgetter** function from the Python Standard Library's **operator module**:

```
In [8]: from operator import itemgetter

In [9]: sorted_items = sorted(items, key=itemgetter(1), reverse=True)
```

As `sorted` orders `items`' elements, it accesses the element at index 1 in each tuple via the expression `itemgetter(1)`. The `reverse=True` keyword argument indicates that the tuples should be sorted in *descending* order.

Getting the Top 20 Words

Next, we use a slice to get the top 20 words from `sorted_items`. When TextBlob tokenizes a corpus, it splits all contractions at their apostrophes and counts the total number of apostrophes as one of the "words." *Romeo and Juliet* has many contractions. If you display `sorted_items[0]`, you'll see that they are the most frequently occurring "word"

12.3 Visualizing Word Frequencies with Bar Charts and Word Clouds

with 867 of them.[22] We want to display only words, so we ignore element 0 and get a slice containing elements 1 through 20 of `sorted_items`:

```
In [10]: top20 = sorted_items[1:21]
```

Convert top20 to a DataFrame

Next, let's convert the `top20` list of tuples to a pandas `DataFrame` so we can visualize it conveniently:

```
In [11]: import pandas as pd

In [12]: df = pd.DataFrame(top20, columns=['word', 'count'])

In [13]: df
Out[13]:
        word  count
0      romeo    315
1       thou    278
2     juliet    190
3        thy    170
4     capulet    163
5      nurse    149
6       love    148
7       thee    138
8       lady    117
9      shall    110
10     friar    105
11      come     94
12   mercutio     88
13   lawrence     82
14      good     80
15   benvolio     79
16     tybalt     79
17     enter     75
18        go     75
19     night     73
```

Visualizing the DataFrame

To visualize the data, we'll use the **bar method** of the `DataFrame`'s **plot property**. The arguments indicate which column's data should be displayed along the *x*- and *y*-axes, and that we do not want to display a legend on the graph:

```
In [14]: axes = df.plot.bar(x='word', y='count', legend=False)
```

The bar method creates and displays a Matplotlib bar chart.

When you look at the initial bar chart that appears, you'll notice that some of the words are truncated. To fix that, use Matplotlib's `gcf` (get current figure) function to get the Matplotlib figure that pandas displayed, then call the figure's `tight_layout` method. This compresses the bar chart to ensure all its components fit:

```
In [15]: import matplotlib.pyplot as plt

In [16]: plt.gcf().tight_layout()
```

---

22. In some locales this does not happen and element 0 is indeed 'romeo'.

The final graph is shown below:

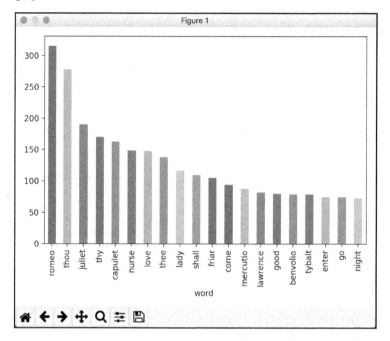

12.3.2 Visualizing Word Frequencies with Word Clouds

Next, we'll build a word cloud that visualizes the top 200 words in *Romeo and Juliet*. You can use the open source **wordcloud** module's[23] **WordCloud** class to generate word clouds with just a few lines of code. By default, **wordcloud** creates rectangular word clouds, but as you'll see the library can create word clouds with arbitrary shapes.

Installing the wordcloud Module
To install **wordcloud**, open your Anaconda Prompt (Windows), Terminal (macOS/Linux) or shell (Linux) and enter the command:

```
conda install -c conda-forge wordcloud
```

Windows users might need to run the Anaconda Prompt as an Administrator for proper software installation privileges. To do so, right-click Anaconda Prompt in the start menu and select **More > Run as administrator**.

Loading the Text
First, let's load *Romeo and Juliet*. Launch IPython from the ch12 examples folder before executing the following code so you can access the e-book file RomeoAndJuliet.txt you downloaded earlier:

```
In [1]: from pathlib import Path

In [2]: text = Path('RomeoAndJuliet.txt').read_text()
```

---

23. https://github.com/amueller/word_cloud.

12.3 Visualizing Word Frequencies with Bar Charts and Word Clouds

Loading the Mask Image that Specifies the Word Cloud's Shape

To create a word cloud of a given shape, you can initialize a `WordCloud` object with an image known as a *mask*. The `WordCloud` fills non-white areas of the mask image with text. We'll use a heart shape in this example, provided as `mask_heart.png` in the ch12 examples folder. More complex masks require more time to create the word cloud.

Let's load the mask image by using the **imread function** from the `imageio` module that comes with Anaconda:

```
In [3]: import imageio

In [4]: mask_image = imageio.imread('mask_heart.png')
```

This function returns the image as a NumPy array, which is required by `WordCloud`.

Configuring the WordCloud Object

Next, let's create and configure the `WordCloud` object:

```
In [5]: from wordcloud import WordCloud

In [6]: wordcloud = WordCloud(colormap='prism', mask=mask_image,
   ...:     background_color='white')
   ...:
```

The default `WordCloud` width and height in pixels is 400x200, unless you specify `width` and `height` keyword arguments or a mask image. For a mask image, the `WordCloud` size is the image's size. `WordCloud` uses Matplotlib under the hood. `WordCloud` assigns random colors from a color map. You can supply the `colormap` keyword argument and use one of Matplotlib's named color maps. For a list of color map names and their colors, see:

https://matplotlib.org/examples/color/colormaps_reference.html

The mask keyword argument specifies the `mask_image` we loaded previously. By default, the word is drawn on a black background, but we customized this with the `background_color` keyword argument by specifying a `'white'` background. For a complete list of `WordCloud`'s keyword arguments, see

http://amueller.github.io/word_cloud/generated/wordcloud.WordCloud.html

Generating the Word Cloud

`WordCloud`'s **generate method** receives the text to use in the word cloud as an argument and creates the word cloud, which it returns as a `WordCloud` object:

```
In [7]: wordcloud = wordcloud.generate(text)
```

Before creating the word cloud, `generate` first removes stop words from the text argument using the `wordcloud` module's built-in stop-words list. Then `generate` calculates the word frequencies for the remaining words. The method uses a maximum of 200 words in the word cloud by default, but you can customize this with the `max_words` keyword argument.

Saving the Word Cloud as an Image File

Finally, we use `WordCloud`'s **to_file** method to save the word cloud image into the specified file:

```
In [8]: wordcloud = wordcloud.to_file('RomeoAndJulietHeart.png')
```

You can now go to the `ch12` examples folder and double-click the `RomeoAndJuliet.png` image file on your system to view it—your version might have the words in different positions and different colors:

Generating a Word Cloud from a Dictionary

If you already have a dictionary of key–value pairs representing word counts, you can pass it to `WordCloud`'s `fit_words` method. This method assumes you've already removed the stop words.

Displaying the Image with Matplotlib

If you'd like to display the image on the screen, you can use the IPython magic

    %matplotlib

to enable interactive Matplotlib support in IPython, then execute the following statements:

    import matplotlib.pyplot as plt
    plt.imshow(wordcloud)

✓ Self Check

1 *(Fill-In)* A(n) _____ is a graphic that shows more frequently occurring words in bigger fonts and less frequently occurring words in smaller fonts.
Answer: word cloud.

2 *(IPython Session)* We provided oval, circle and star masks for you to try when creating your own word clouds. Continue this section's IPython session and generate another *Romeo and Juliet* word cloud using the `mask_star.png` image.

12.4 Readability Assessment with Textatistic

Answer:

```
In [9]: mask_image2 = imageio.imread('mask_star.png')

In [10]: wordcloud2 = WordCloud(width=1000, height=1000,
    ...:     colormap='prism', mask=mask_image2, background_color='white')
    ...:

In [11]: wordcloud2 = wordcloud2.generate(text)

In [12]: wordcloud2 = wordcloud2.to_file('RomeoAndJulietStar.png')
```

12.4 Readability Assessment with Textatistic

An interesting use of natural language processing is assessing text **readability**, which is affected by the vocabulary used, sentence structure, sentence length, topic and more. While writing this book, we used the paid tool Grammarly to help tune the writing and ensure the text's readability for a wide audience.

In this section, we'll use the **Textatistic library**[24] to assess readability.[25] There are many formulas used in natural language processing to calculate readability. Textatistic uses five popular readability formulas—Flesch Reading Ease, Flesch-Kincaid, Gunning Fog, Simple Measure of Gobbledygook (SMOG) and Dale-Chall.

Install Textatistic

To install Textatistic, open your Anaconda Prompt (Windows), Terminal (macOS/Linux) or shell (Linux), then execute the following command:

    pip install textatistic

Windows users might need to run the Anaconda Prompt as an Administrator for proper software installation privileges. To do so, right-click Anaconda Prompt in the start menu and select **More > Run as administrator.**

---

24. https://github.com/erinhengel/Textatistic.
25. Some other Python readability assessment libraries include readability-score, textstat, readability and pylinguistics.

Calculating Statistics and Readability Scores

First, let's load *Romeo and Juliet* into the `text` variable:

```
In [1]: from pathlib import Path

In [2]: text = Path('RomeoAndJuliet.txt').read_text()
```

Calculating statistics and readability scores requires a **Textatistic** object that's initialized with the text you want to assess:

```
In [3]: from textatistic import Textatistic

In [4]: readability = Textatistic(text)
```

Textatistic method **dict** returns a dictionary containing various statistics and the readability scores[26]:

```
In [5]: %precision 3
Out[5]: '%.3f'

In [6]: readability.dict()
Out[6]:
{'char_count': 115141,
 'word_count': 26120,
 'sent_count': 3218,
 'sybl_count': 30166,
 'notdalechall_count': 5823,
 'polysyblword_count': 549,
 'flesch_score': 100.892,
 'fleschkincaid_score': 1.203,
 'gunningfog_score': 4.087,
 'smog_score': 5.489,
 'dalechall_score': 7.559}
```

Each of the values in the dictionary is also accessible via a `Textatistic` property of the same name as the keys shown in the preceding output. The statistics produced include:

- `char_count`—The number of characters in the text.
- `word_count`—The number of words in the text.
- `sent_count`—The number of sentences in the text.
- `sybl_count`—The number of syllables in the text.
- `notdalechall_count`—A count of the words that are not on the Dale-Chall list, which is a list of words understood by 80% of 5th graders.[27] The higher this number is compared to the total word count, the less readable the text is considered to be.
- `polysyblword_count`—The number of words with three or more syllables.
- `flesch_score`—The Flesch Reading Ease score, which can be mapped to a grade level. Scores over 90 are considered readable by 5th graders. Scores under 30 require a college degree. Ranges in between correspond to the other grade levels.

---

26. Each Project Gutenberg e-book includes additional text, such as their licensing information, that's not part of the e-book itself. For this example, we used a text editor to remove that text from our copy of the e-book.
27. http://www.readabilityformulas.com/articles/dale-chall-readability-word-list.php.

- **fleschkincaid_score**—The Flesch-Kincaid score, which corresponds to a specific grade level.
- **gunningfog_score**—The Gunning Fog index value, which corresponds to a specific grade level.
- **smog_score**—The Simple Measure of Gobbledygook (SMOG), which corresponds to the years of education required to understand text. This measure is considered particularly effective for healthcare materials.[28]
- **dalechall_score**—The Dale-Chall score, which can be mapped to grade levels from 4 and below to college graduate (grade 16) and above. This score considered to be most reliable for a broad range of text types.[29,30]

You can learn about each of these readability scores produced here and several others at

> https://en.wikipedia.org/wiki/Readability

The Textatistic documentation also shows the readability formulas used:

> http://www.erinhengel.com/software/textatistic/

 Self Check

1 *(Fill-In)* _____ indicates how easy is it for readers to understand text.
Answer: Readability.

2 *(IPython Session)* Using the results in this section's IPython session, calculate the average numbers of words per sentence, characters per word and syllables per word.
Answer:

```
In [7]: readability.word_count / readability.sent_count   # sentence length
Out[7]: 8.117

In [8]: readability.char_count / readability.word_count   # word length
Out[8]: 4.408

In [9]: readability.sybl_count / readability.word_count   # syllables
Out[9]: 1.155
```

12.5 Named Entity Recognition with spaCy

NLP can determine what a text is about. A key aspect of this is **named entity recognition**, which attempts to locate and categorize items like dates, times, quantities, places, people, things, organizations and more. In this section, we'll use the named entity recognition capabilities in the **spaCy NLP library**[31,32] to analyze text.

---

28. https://en.wikipedia.org/wiki/SMOG.
29. https://en.wikipedia.org/wiki/Readability#The_Dale%E2%80%93Chall_formula.
30. http://www.readabilityformulas.com/articles/how-do-i-decide-which-readability-formula-to-use.php.
31. https://spacy.io/.
32. You may also want to check out Textacy (https://github.com/chartbeat-labs/textacy)—an NLP library built on spaCy that supports additional NLP tasks.

Install spaCy

To install spaCy, open your Anaconda Prompt (Windows), Terminal (macOS/Linux) or shell (Linux), then execute the following command:

```
conda install -c conda-forge spacy
```

Windows users might need to run the Anaconda Prompt as an Administrator for proper software installation privileges. To do so, right-click Anaconda Prompt in the start menu and select **More > Run as administrator.**

Once the install completes, you also need to execute the following command, so spaCy can download additional components it needs for processing English (**en**) text:

```
python -m spacy download en
```

Loading the Language Model

The first step in using spaCy is to load the language model representing the natural language of the text you're analyzing. To do this, you'll call the spacy module's **load** function. Let's load the English model that we downloaded above:

```
In [1]: import spacy

In [2]: nlp = spacy.load('en')
```

The spaCy documentation recommends the variable name **nlp**.

Creating a spaCy Doc

Next, you use the **nlp** object to create a spaCy **Doc** object[33] representing the document to process. Here we used a sentence from the introduction to the World Wide Web in many of our books:

```
In [3]: document = nlp('In 1994, Tim Berners-Lee founded the ' +
   ...:     'World Wide Web Consortium (W3C), devoted to ' +
   ...:     'developing web technologies')
   ...:
```

Getting the Named Entities

The **Doc** object's **ents property** returns a tuple of **Span** objects representing the named entities found in the Doc. Each Span has many properties.[34] Let's iterate through the Spans and display the **text** and **label_** properties:

```
In [4]: for entity in document.ents:
   ...:     print(f'{entity.text}: {entity.label_}')
   ...:
1994: DATE
Tim Berners-Lee: PERSON
the World Wide Web Consortium: ORG
```

Each Span's **text property** returns the entity as a string, and the **label_ property** returns a string indicating the entity's kind. Here, spaCy found three entities representing a DATE (1994), a PERSON (Tim Berners-Lee) and an ORG (organization; the World Wide Web Consortium). To learn more about spaCy, see its Quickstart guide at

https://spacy.io/usage/models#section-quickstart

---

33. https://spacy.io/api/doc.
34. https://spacy.io/api/span.

Self Check

1. *(True/False)* Named entity recognition attempts to locate only people's names in text.
Answer: False. Named entity recognition attempts to locate and categorize items like dates, times, quantities, places, people, things, organizations and more.

2. *(IPython Session)* Display the named entities in

    ```
    'Paul J. Deitel is CEO of Deitel & Associates, Inc.'
    ```

Answer:

```
In [1]: import spacy

In [2]: nlp = spacy.load('en')

In [3]: document = nlp('Paul J. Deitel is CEO of ' +
   ...:                'Deitel & Associates, Inc.')
   ...:

In [4]: for entity in document.ents:
   ...:     print(f'{entity.text}: {entity.label_}')
   ...:
Paul J. Deitel: PERSON
Deitel & Associates, Inc.: ORG
```

12.6 Similarity Detection with spaCy

Similarity detection is the process of analyzing documents to determine how alike they are. One possible similarity detection technique is word frequency counting. For example, some people believe that the works of William Shakespeare actually might have been written by Sir Francis Bacon, Christopher Marlowe or others.[35] Comparing the word frequencies of their works with those of Shakespeare can reveal writing-style similarities.

Various machine-learning techniques we'll discuss in later chapters can be used to study document similarity. However, as is often the case in Python, there are libraries such as spaCy and Gensim that can do this for you. Here, we'll use spaCy's similarity detection features to compare Doc objects representing Shakespeare's *Romeo and Juliet* with Christopher Marlowe's *Edward the Second*. You can download *Edward the Second* from Project Gutenberg as we did for *Romeo and Juliet* earlier in the chapter.[36]

Loading the Language Model and Creating a spaCy Doc

As in the preceding section, we first load the English model:

```
In [1]: import spacy

In [2]: nlp = spacy.load('en')
```

---

35. https://en.wikipedia.org/wiki/Shakespeare_authorship_question.
36. Each Project Gutenberg e-book includes additional text, such as their licensing information, that's not part of the e-book itself. For this example, we used a text editor to remove that text from our copies of the e-books.

Creating the spaCy Docs

Next, we create two Doc objects—one for *Romeo and Juliet* and one for *Edward the Second*:

```
In [3]: from pathlib import Path

In [4]: document1 = nlp(Path('RomeoAndJuliet.txt').read_text())

In [5]: document2 = nlp(Path('EdwardTheSecond.txt').read_text())
```

Comparing the Books' Similarity

Finally, we use the Doc class's **similarity** method to get a value from 0.0 (not similar) to 1.0 (identical) indicating how similar the documents are:

```
In [6]: document1.similarity(document2)
Out[6]: 0.9349950179100041
```

As you can see, spaCy believes these two documents have significant similarities. For comparison purposes, we also created a Doc representing a current news story and compared it with *Romeo and Juliet*. As expected, spaCy returned a low value indicating little similarity between those documents. Try copying a current news article into a text file, then performing the similarity comparison yourself.

Self Check

1 *(Fill-In)* _____ is the process of analyzing documents to determine how similar they are.
Answer: Similarity detection.

2 *(IPython Session)* You'd expect that if you compare Shakespeare's works to one another that you'd get a high similarity value, especially among plays of the same genre. *Romeo and Juliet* is one of Shakespeare's tragedies. Three other Shakespearian tragedies are *Hamlet*, *Macbeth* and *King Lear*. Download these three plays from Project Gutenberg, then compare each one for similarity with *Romeo and Juliet* using the spaCy code presented in this section.
Answer:

```
In [1]: import spacy

In [2]: nlp = spacy.load('en')

In [3]: from pathlib import Path

In [4]: document1 = nlp(Path('RomeoAndJuliet.txt').read_text())

In [5]: document2 = nlp(Path('Hamlet.txt').read_text())

In [6]: document1.similarity(document2)
Out[6]: 0.9653729533870296

In [7]: document3 = nlp(Path('Macbeth.txt').read_text())

In [8]: document1.similarity(document3)
Out[8]: 0.9601267484020871

In [9]: document4 = nlp(Path('KingLear.txt').read_text())

In [10]: document1.similarity(document4)
Out[10]: 0.9966456936385792
```

12.7 Other NLP Libraries and Tools

We've shown you various NLP libraries, but it's always a good idea to investigate the range of options available to you so you can leverage the best tools for your tasks. Below are some additional mostly free and open source NLP libraries and APIs:

- Gensim—Similarity detection and topic modeling.
- Google Cloud Natural Language API—Cloud-based API for NLP tasks such as named entity recognition, sentiment analysis, parts-of-speech analysis and visualization, determining content categories and more.
- Microsoft Linguistic Analysis API.
- Bing sentiment analysis—Microsoft's Bing search engine now uses sentiment in its search results. At the time of this writing, sentiment analysis in search results is available only in the United States.
- PyTorch NLP—Deep learning library for NLP.
- Stanford CoreNLP—Extensive NLP library written in Java, which also provides a Python wrapper. Includes corefererence resolution, which finds all references to the same thing.
- Apache OpenNLP—Another Java-based NLP library for common tasks, including coreference resolution. Python wrappers are available.
- PyNLPl (pineapple)—Python NLP library, includes basic and more sophisticated NLP capabilities.
- SnowNLP—Python library that simplifies Chinese text processing.
- KoNLPy—Korean language NLP.
- `stop-words`—Python library with stop words for many languages. We used NLTK's stop words lists in this chapter.
- `TextRazor`—A paid cloud-based NLP API that provides a free tier.

12.8 Machine Learning and Deep Learning Natural Language Applications

There are many natural language applications that require machine learning and deep learning techniques. We'll discuss some of the following in our machine learning and deep learning chapters:

- Answering natural language questions—For example, our publisher Pearson Education, has a partnership with IBM Watson that uses Watson as a virtual tutor. Students ask Watson natural language questions and get answers.
- Summarizing documents—analyzing documents and producing short summaries (also called abstracts) that can, for example, be included with search results and can help you decide what to read.
- Speech synthesis (speech-to-text) and speech recognition (text-to-speech)—We use these in our "IBM Watson" chapter, along with inter-language text-to-text translation, to develop a near real-time inter-language voice-to-voice translator.

- Collaborative filtering—used to implement recommender systems ("if you liked this movie, you might also like…").
- Text classification—for example, classifying news articles by categories, such as world news, national news, local news, sports, business, entertainment, etc.
- Topic modeling—finding the topics discussed in documents.
- Sarcasm detection—often used with sentiment analysis.
- Text simplification—making text more concise and easier to read.
- Speech to sign language and vice versa—to enable a conversation with a hearing-impaired person.
- Lip reader technology—for people who can't speak, convert lip movement to text or speech to enable conversation.
- Closed captioning—adding text captions to video.

12.9 Natural Language Datasets

There's a tremendous number of text data sources available to you for working with Natural language processing:

- Wikipedia—some or all of Wikipedia (https://meta.wikimedia.org/wiki/Datasets).
- IMDB (Internet Movie Database)—various movie and TV datasets are available.
- UCI's text datasets—many datasets, including the Spambase dataset.
- Project Gutenberg—50,000+ free e-books that are out-of-copyright in the U.S.
- Jeopardy! dataset—200,000+ questions from the Jeopardy! TV show. A milestone in AI occurred in 2011 when IBM Watson famously beat two of the world's best Jeopardy! players.
- Natural language processing datasets: https://machinelearningmastery.com/datasets-natural-language-processing/.
- NLTK data: https://www.nltk.org/data.html.
- Sentiment labeled sentences data set (from sources including IMDB.com, amazon.com, yelp.com.)
- Registry of Open Data on AWS—a searchable directory of datasets hosted on Amazon Web Services (https://registry.opendata.aws).
- Amazon Customer Reviews Dataset—130+ million product reviews (https://registry.opendata.aws/amazon-reviews/).
- Pitt.edu corpora (http://mpqa.cs.pitt.edu/corpora/).

12.10 Wrap-Up

In this chapter, you performed a broad range of natural language processing (NLP) tasks using several NLP libraries. You learned that NLP is performed on text collections known as corpora. We discussed nuances of meaning that make natural language understanding difficult.

We focused on the TextBlob NLP library, which is built on the NLTK and pattern libraries, but easier to use. You created `TextBlob`s and tokenized them into `Sentence`s and `Word`s. You determined the part of speech for each word in a `TextBlob`, and you extracted noun phrases.

We demonstrated how to evaluate the positive or negative sentiment of `TextBlob`s and `Sentence`s with the TextBlob library's default sentiment analyzer and with the `NaiveBayesAnalyzer`. You learned how to use the TextBlob library's integration with Google Translate to detect the language of text and perform inter-language translation.

We showed various other NLP tasks, including singularization and pluralization, spell checking and correction, normalization with stemming and lemmatization, and getting word frequencies. You obtained word definitions, synonyms and antonyms from WordNet. You also used NLTK's stop words list to eliminate stop words from text, and you created n-grams containing groups of consecutive words.

We showed how to visualize word frequencies quantitatively as a bar chart using pandas' built-in plotting capabilities. Then, we used the `wordcloud` library to visualize word frequencies qualitatively as word clouds. You performed readability assessments using the Textatistic library. Finally, you used spaCy to locate named entities and to perform similarity detection among documents. In the next chapter, you'll continue using natural language processing as we introduce data mining tweets using the Twitter APIs.

Exercises

12.1 *(Web Scraping with the Requests and Beautiful Soup Libraries)* Web pages are excellent sources of text to use in NLP tasks. In the following IPython session, you'll use the **requests library** to download the www.python.org home page's content. This is called web scraping. You'll then use the **Beautiful Soup library**[37] to extract only the text from the page. Eliminate the stop words in the resulting text, then use the `wordcloud` module to create a word cloud based on the text.

```
In [1]: import requests

In [2]: response = requests.get('https://www.python.org')

In [3]: response.content  # gives back the page's HTML

In [4]: from bs4 import BeautifulSoup

In [5]: soup = BeautifulSoup(response.content, 'html5lib')

In [6]: text = soup.get_text(strip=True)  # text without tags
```

In the preceding code, snippets [1]–[3] get a web page. The **get function** receives a URL as an argument and returns the corresponding web page as a **Response** object. The Response's **content property** contains the web page's content. Snippets [4]–[6] get only the web page's text. Snippet [5] creates a BeautifulSoup object to process the text in `response.content`. BeautifulSoup method `get_text` with the keyword argument `strip=True` returns just the text of the web page without its structural information that your web browser uses to display the web page.

---
37. Its module name is bs4 for Beautiful Soup 4.

12.2 *(Tokenizing Text and Noun Phrases)* Using the text from Exercise 12.1, create a `TextBlob`, then tokenize it into `Sentences` and `Words`, and extract its noun phrases.

12.3 *(Sentiment of a News Article)* Using the techniques in Exercise 12.1, download a web page for a current news article and create a `TextBlob`. Display the sentiment for the entire `TextBlob` and for each `Sentence`.

12.4 *(Sentiment of a News Article with the `NaiveBayesAnalyzer`)* Repeat the previous exercise but use the `NaiveBayesAnalyzer` for sentiment analysis.

12.5 *(Spell Check a Project Gutenberg Book)* Download a Project Gutenberg book and create a `TextBlob`. Tokenize the `TextBlob` into `Words` and determine whether any are misspelled. If so, display the possible corrections.

12.6 *(Word Frequency Bar Chart and Word Cloud from Shakespeare's Hamlet)* Using the techniques you learned in this chapter, create a top-20 word frequency bar chart and a word cloud, based on Shakespeare's *Hamlet*. Use the `mask_oval.png` file provided in the ch12 examples folder as the mask.

12.7 *(Textatistic: Readability of News Articles)* Using the techniques in the first exercise, download from several news sites current news articles on the same topic. Perform readability assessments on them to determine which sites are the most readable. For each article, calculate the average number of words per sentence, the average number of characters per word and the average number of syllables per word.

12.8 *(spaCy: Named Entity Recognition)* Using the techniques in the first exercise, download a current news article then use the spaCy library's named entity recognition capabilities to display the named entities (people, places, organizations, etc.) in the article.

12.9 *(spaCy: Similarity Detection)* Using the techniques in the first exercise, download several news articles on the same topic and compare them for similarity.

12.10 *(spaCy: Shakespeare Similarity Detection)* Using the spaCy techniques you learned in this chapter, download a Shakespeare comedy from Project Gutenberg and compare it for similarity with *Romeo and Juliet*.

12.11 *(`textblob.utils` Utility Functions)* TextBlob's `textblob.utils` module offers several utility functions for cleaning up text, including `strip_punc` and `lowerstrip`. You call `strip_punc` with a string and the keyword argument `all=True` to remove punctuation from the string. You call `lowerstrip` with a string and the keyword argument `all=True` to get a string in all lowercase letters with whitespace and punctuation removed. Experiment with each function on *Romeo and Juliet*.

12.12 *(Research: Funny Newspaper Headlines)* To understand how tricky it is to work with natural language and its inherent ambiguity issues, research "funny newspaper headlines." List the challenges you find.

12.13 *(Try the Demos: Named Entity Recognition)* Search online for the Stanford Named Entity Tagger and the Cognitive Computation Group's Named Entity Recognition Demo. Run each with a corpus of your choice. Compare the results of both demos.

12.14 *(Try the Demo: TextRazor)* TextRazor is one of many paid commercial NLP products that offer a free tier. Search online for TextRazor Live Demo. Paste in a corpus of your choice and click **Analyze** to analyze the corpus for categories and topics, highlighting key sentences and words within them. Click the links below each piece of analyzed text for more detailed analyses. Click the **Advanced Options** link to the right of **Analyze** and **Clear** for many additional features, including the ability to analyze text in different languages.

12.15 *(Project: Readability Scores with Textatistic)* Try Textatistic with famous authors' books from Project Gutenberg.

12.16 *(Project: Who Authored the Works of Shakespeare)* Using the spaCy similarity detection code introduced in this chapter, compare Shakespeare's *Macbeth* to one major work from each of several other authors who might have written Shakespeare's works (see `https://en.wikipedia.org/wiki/Shakespeare_authorship_question`). Locate works on Project Gutenberg from a few authors listed at `https://en.wikipedia.org/wiki/List_of_Shakespeare_authorship_candidates`, then use spaCy to compare their works' similarity to *Macbeth*. Which of the authors' works are most similar to *Macbeth*?

12.17 *(Project: Similarity Detection)* One way to measure similarity between two documents is to compare frequency counts of the parts of speech used in each. Build dictionaries of the parts of speech frequencies for two Project Gutenberg books from the same author and two from different authors and compare the results.

12.18 *(Project: Text Visualization Browser)* Use `http://textvis.lnu.se/` to view hundreds of text visualizations. You can filter the visualizations by categories like the analytic tasks they perform, the kinds of visualizations, the kinds of data sources they use and more. Each visualization's summary provides a link to where you can learn more about it.

12.19 *(Project: Stanford CoreNLP)* Search online for "Stanford CoreNLP Python" to find Stanford's list of Python modules for using CoreNLP, then experiment with its features to perform tasks you learned in this chapter.

12.20 *(Project: spaCy and spacy-readability)* We used Textatistic for readability assessment in this chapter. There are many other readability libraries, such as readability-score, textstat, readability, pylinguistics and spacy-readability, which works in conjunction with spaCy. Investigate the spacy-readability module, then use it to evaluate the readability of *King Lear* (from Project Gutenberg).

12.21 *(Project: Worldwide Peace)* As you know, `TextBlob` language translation works by connecting to Google Translate. Determine the range of languages Google Translate recognizes. Write a script that translates the English word "Peace" into each of the supported languages. Display the translations in the same size text in a circular word cloud using the `mask_circle.png` file provided in the `ch12` examples folder.

12.22 *(Project: Self Tutor for Learning a New Language)* With today's natural language tools, including inter-language translation, speech-to-text and text-to-speech in various languages, you can build a self-tutor that will help you learn new languages. Find Python speech-to-text and text-to-speech libraries that can handle various languages. Write a script that allows you to indicate the language you wish to learn (use only the languages supported by Google Translate through TextBlob). The script should then allow you to say something in English, transcribe your speech to text, translate it to the selected language and use text-to-speech to speak the translated text back to you so you can hear it. Try your script with words, colors, numbers, sentences, people, places and things.

12.23 *(Project: Accessing Wikipedia with Python)* Search online for Python modules that enable you to access content from Wikipedia and similar sites. Write scripts to exercise the capabilities of those modules.

12.24 *(Project: Document Summarization with Gensim)* Document summarization involves analyzing a document and extracting content to produce a summary. For example, with today's massive flow of information, this could be useful to busy doctors studying the latest medical advances in order to provide the best care. A summary could help them decide whether a paper is worth reading. Investigate the `summarize` and `keywords` functions from the Gensim library's `gensim.summarization` module, then use them to summarize text and extract the important words. You'll need to install Gensim with

```
conda install -c conda-forge gensim
```

Assuming that `text` is a string representing a corpus, the following Gensim code summarizes the text and displays a list of keywords in the text:

```
from gensim.summarization import summarize, keywords
print(summarize(text))
print(keywords(text))
```

12.25 *(Challenge Project: Six Degrees of Separation with Wikipedia Corpus)* You may have heard of "six degrees of separation" for finding connections between any two people on the planet. The idea is that as you look at a person's connections to friends and family, then look at their friends' and family's connections, and so on, you'll often find a connection between two people within the first six levels of connection.

Research "six degrees of separation Python" online. You'll find many implementations. Execute some of them to see how they do. Next, investigate the Wikipedia APIs and Python modules for using them. Choose two famous people. Load the Wikipedia page for the first person, then use named entity recognition to locate any names in that person's Wikipedia page. Then repeat this process for the Wikipedia pages of each name you find. Continue this process six levels deep to build a graph of people connected to the original person's Wikipedia page. Along the way, check whether the other person's name appears in the graph and print the chain of people. In the "Big Data: Hadoop, Spark, NoSQL and IoT" chapter, we'll discuss the Neo4j graph database, which can be used to solve this problem.

12.26 *(Project: Synonym Chain Leading to an Antonym)* As you follow synonym chains—that is, synonyms of synonyms of synonyms, etc.—to arbitrary levels, you'll often encounter words that do not appear to be related to the original. Though rare, there actually are cases in which following a synonym chain eventually results in an antonym of the initial word. For several examples, see the paper "Websterian Synonym Chains":

```
https://digitalcommons.butler.edu/cgi/
    viewcontent.cgi?article=3342&context=wordways
```

Choose a synonym chain in the paper above. Use the WordNet features introduced in this chapter to get the first word's synonyms and antonyms. Next, for each of the words in the `Synsets`, get their synonyms, then the synonyms of those synonyms and so on. As you get the synonyms at each level, check whether any of them is one of the initial word's antonyms. If so, display the synonym chain that led to the antonym.

12.27 *(Project: Steganography)* **Steganography** hides information within other information. The term literally means "covered writing." Research online "text steganography" and "natural language steganography." Write scripts that use various steganography techniques to hide information in text and to extract that information.

Data Mining Twitter

Objectives

In this chapter, you'll:

- Understand Twitter's impact on businesses, brands, reputation, sentiment analysis, predictions and more.
- Use Tweepy, one of the most popular Python Twitter API clients for data mining Twitter.
- Use the Twitter Search API to download past tweets that meet your criteria.
- Use the Twitter Streaming API to sample the stream of live tweets as they're happening.
- See that the tweet objects returned by Twitter contain valuable information beyond the tweet text.
- Use the natural language processing techniques you learned in the last chapter to clean and preprocess tweets to prepare them for analysis.
- Perform sentiment analysis on tweets.
- Spot trends with Twitter's Trends API.
- Map tweets using folium and OpenStreetMap.
- Understand various ways to store tweets using techniques discussed throughout this book.

Outline

- 13.1 Introduction
- 13.2 Overview of the Twitter APIs
- 13.3 Creating a Twitter Account
- 13.4 Getting Twitter Credentials— Creating an App
- 13.5 What's in a Tweet?
- 13.6 Tweepy
- 13.7 Authenticating with Twitter Via Tweepy
- 13.8 Getting Information About a Twitter Account
- 13.9 Introduction to Tweepy **Cursors**: Getting an Account's Followers and Friends
  - 13.9.1 Determining an Account's Followers
  - 13.9.2 Determining Whom an Account Follows
  - 13.9.3 Getting a User's Recent Tweets
- 13.10 Searching Recent Tweets
- 13.11 Spotting Trends: Twitter Trends API
  - 13.11.1 Places with Trending Topics
  - 13.11.2 Getting a List of Trending Topics
  - 13.11.3 Create a Word Cloud from Trending Topics
- 13.12 Cleaning/Preprocessing Tweets for Analysis
- 13.13 Twitter Streaming API
  - 13.13.1 Creating a Subclass of `StreamListener`
  - 13.13.2 Initiating Stream Processing
- 13.14 Tweet Sentiment Analysis
- 13.15 Geocoding and Mapping
  - 13.15.1 Getting and Mapping the Tweets
  - 13.15.2 Utility Functions in `tweetutilities.py`
  - 13.15.3 Class `LocationListener`
- 13.16 Ways to Store Tweets
- 13.17 Twitter and Time Series
- 13.18 Wrap-Up

Exercises

13.1 Introduction

We're always trying to predict the future. Will it rain on our upcoming picnic? Will the stock market or individual securities go up or down, and when and by how much? How will people vote in the next election? What's the chance that a new oil exploration venture will strike oil and if so how much would it likely produce? Will a baseball team win more games if it changes its batting philosophy to "swing for the fences?" How much customer traffic does an airline anticipate over the next many months? And hence how should the company buy oil commodity futures to guarantee that it will have the supply it needs and hopefully at a minimal cost? What track is a hurricane likely to take and how powerful will it likely become (category 1, 2, 3, 4 or 5)? That kind of information is crucial to emergency preparedness efforts. Is a financial transaction likely to be fraudulent? Will a mortgage default? Is a disease likely to spread rapidly and, if so, to what geographic area?

Prediction is a challenging and often costly process, but the potential rewards are great. With the technologies we'll study in this and the upcoming chapters, we'll see how AI, often in concert with big data, is rapidly improving prediction capabilities.

In this chapter we concentrate on data mining Twitter, looking for the sentiment in tweets. **Data mining** is the process of searching through large collections of data, often big data, to find insights that can be valuable to individuals and organizations. The sentiment that you data mine from tweets could help predict the results of an election, the revenues a new movie is likely to generate and the success of a company's marketing campaign. It could also help companies spot weaknesses in competitors' product offerings.

You'll connect to Twitter via web services. You'll use Twitter's Search API to tap into the enormous base of past tweets. You'll use Twitter's Streaming API to sample the flood of new tweets as they happen. With the Twitter Trends API, you'll see what topics are trending. You'll find that much of what you learned in the "Natural Language Processing (NLP)" chapter will be useful in building Twitter applications.

As you've seen throughout this book, because of powerful libraries, you'll often perform significant tasks with just a few lines of code. This is why Python and its robust open-source community are appealing.

Twitter has displaced the major news organizations as the first source for newsworthy events. Most Twitter posts are public and happen in real-time as events unfold globally. People speak frankly about any subject and tweet about their personal and business lives. They comment on the social, entertainment and political scenes and whatever else comes to mind. With their mobile phones, they take and post photos and videos of events as they happen. You'll commonly hear the terms **Twitterverse** and **Twittersphere** to mean the hundreds of millions of users who have anything to do with sending, receiving and analyzing tweets.

What Is Twitter?

Twitter was founded in 2006 as a microblogging company and today is one of the most popular sites on the Internet. Its concept is simple. People write short messages called *tweets*, initially limited to 140 characters but recently increased for most languages to 280 characters. Anyone can generally choose to follow anyone else. This is different from the closed, tight communities on other social media platforms such as Facebook, LinkedIn and many others, where the "following relationships" must be reciprocal.

Twitter Statistics

Twitter has hundreds of millions of users and hundreds of millions of tweets are sent every day with many thousands sent per second.[1] Searching online for "Internet statistics" and "Twitter statistics" will help you put these numbers in perspective. Some "tweeters" have more than 100 million followers. Dedicated tweeters generally post several per day to keep their followers engaged. Tweeters with the largest followings are typically entertainers and politicians. Developers can tap into the live stream of tweets as they're happening. This has been likened to "drinking from a fire hose," because the tweets come at you so quickly.

Twitter and Big Data

Twitter has become a favorite big data source for researchers and business people worldwide. Twitter allows regular users free access to a small portion of the more recent tweets. Through special arrangements with Twitter, some third-party businesses (and Twitter itself) offer paid access to much larger portions of the all-time tweets database.

Cautions

You can't always trust everything you read on the Internet, and tweets are no exception. For example, people might use false information to try to manipulate financial markets or influence political elections. Hedge funds often trade securities based in part on the tweet streams they follow, but they're cautious. That's one of the challenges of building *business-critical* or *mission-critical* systems based on social media content.

Going forward, we use web services extensively. Internet connections can be lost, services can change and some services are not available in all countries. This is the real world of cloud-based programming. We cannot program with the same reliability as desktop apps when using web services.

---

1. http://www.internetlivestats.com/twitter-statistics/.

Self Check

1. *(Fill-In)* You connect to Twitter's APIs via _____.
Answer: web services.

2. *(True/False)* In Twitter, "following relationships" must be reciprocal.
Answer: False. This is true in most other social networks. In Twitter, you can follow people without them following you.

13.2 Overview of the Twitter APIs

Twitter's APIs are cloud-based web services, so an Internet connection is required to execute the code in this chapter. **Web services** are methods that you call in the cloud, as you'll do with the Twitter APIs in this chapter, the IBM Watson APIs in the next chapter and other APIs you'll use as computing becomes more cloud-based. Each API method has a web service **endpoint**, which is represented by a URL that's used to invoke that method over the Internet.

Twitter's APIs include many categories of functionality, some free and some paid. Most have **rate limits** that restrict the number of times you can use them in 15-minute intervals. In this chapter, you'll use the **Tweepy library** to invoke methods from the following Twitter APIs:

- **Authentication API**—Pass your Twitter credentials (discussed shortly) to Twitter so you can use the other APIs.
- **Accounts and Users API**—Access information about an account.
- **Tweets API**—Search through past tweets, access tweet streams to tap into tweets happening now and more.
- **Trends API**—Find locations of trending topics and get lists of trending topics by location.

See the extensive list of Twitter API categories, subcategories and individual methods at:

> https://developer.twitter.com/en/docs/api-reference-index.html

Rate Limits: A Word of Caution

Twitter expects developers to use its services responsibly. Each Twitter API method has a **rate limit**, which is the maximum number of requests (that is, calls) you can make during a 15-minute window. Twitter may block you from using its APIs if you continue to call a given API method after that method's rate limit has been reached.

Before using any API method, read its documentation and understand its rate limits.[2] We'll configure Tweepy to wait when it encounters rate limits. This helps prevent you from exceeding the rate-limit restrictions. Some methods list both user rate limits and app rate limits. All of this chapter's examples use *app rate limits*. User rate limits are for apps that enable individual users to log into Twitter, like third-party apps that interact with Twitter on your behalf, such as smartphone apps from other vendors.

For details on rate limiting, see

> https://developer.twitter.com/en/docs/basics/rate-limiting

---

2. Keep in mind that Twitter could change these limits in the future.

For specific rate limits on individual API methods, see

> https://developer.twitter.com/en/docs/basics/rate-limits

and each API method's documentation.

Other Restrictions
Twitter is a goldmine for data mining and they allow you to do a lot with their free services. You'll be amazed at the valuable applications you can build and how these will help you improve your personal and career endeavors. **However, if you do not follow Twitter's rules and regulations, your developer account could be terminated. You should carefully read the following and the documents they link to:**

- Terms of Service: `https://twitter.com/tos`
- Developer Agreement: `https://developer.twitter.com/en/developer-terms/agreement-and-policy.html`
- Developer Policy: `https://developer.twitter.com/en/developer-terms/policy.html`
- Other restrictions: `https://developer.twitter.com/en/developer-terms/more-on-restricted-use-cases`

You'll see later in this chapter that you can search tweets only for the last seven days and get only a limited number of tweets using the free Twitter APIs. Some books and articles say you can get around those limits by scraping tweets directly from `twitter.com`. However, the Terms of Service explicitly say that **"scraping the Services without the prior consent of Twitter is expressly prohibited."**

Self Check

1 *(Fill-In)* With the _____ API, you pass your credentials to Twitter so you can use the other APIs.
Answer: Authentication

2 *(True/False)* Twitter allows you to make as many calls as you like and as often as you like to its API methods.
Answer: False. Twitter API methods have rate limits. Twitter may block you from using its APIs if you exceed the rate limits.

13.3 Creating a Twitter Account

Twitter requires you to apply for a developer account to be able to use their APIs. Go to

> https://developer.twitter.com/en/apply-for-access

and submit your application. You'll have to register for one as part of this process if you do not already have one. You'll be asked questions about the purpose of your account (such as academic research, student, etc.). You must *carefully* read and agree to Twitter's terms to complete the application, then confirm your email address.

Twitter reviews every application. Approval is not guaranteed. At the time of this writing, *personal-use accounts* were approved immediately. For company accounts, the process was taking from a few days to several weeks, according to the Twitter developer forums.

13.4 Getting Twitter Credentials—Creating an App

Once you have a Twitter developer account, you must obtain **credentials** for interacting with the Twitter APIs. To do so, you'll create an **app**. Each app has separate credentials. To create an app, log into

> https://developer.twitter.com

and perform the following steps:

1. At the top-right of the page, click the drop-down menu for your account and select **Apps**.
2. Click **Create an app**.
3. In the **App name** field, specify your app's name. If you *send* tweets via the API, this app name will be the tweets' sender. It also will be shown to users if you create applications that require a user to log in via Twitter. We will not do either in this chapter, so a name like "*YourName* Test App" is fine for use with this chapter.
4. In the **Application description** field, enter a description for your app. When creating Twitter-based apps that will be used by other people, this would describe what your app does. For this chapter, you can use "Learning to use the Twitter API."
5. In the **Website URL** field, enter your website. When creating Twitter-based apps, this is supposed to be the website where you host your app. For learning purposes, you can use your Twitter URL: https://twitter.com/*YourUserName*, where *YourUserName* is your Twitter account screen name. For example, the URL https://twitter.com/nasa corresponds to the NASA screen name @nasa.
6. The **Tell us how this app will be used** field is a description of at least 100 characters that helps Twitter employees understand what your app does. For learning purposes, we entered "I am new to Twitter app development and am simply learning how to use the Twitter APIs for educational purposes."
7. Leave the remaining fields empty and click **Create**, then carefully review the (lengthy) developer terms and click **Create** again.

Getting Your Credentials

After you complete *Step 7* above, Twitter displays a web page for managing your app. At the top of the page are **App details**, **Keys and tokens** and **Permissions** tabs. Click the **Keys and tokens** tab to view your app's credentials. Initially, the page shows the Consumer API keys—the API key and the API secret key. Click **Create** to get an access token and access token secret. All four of these will be used to authenticate with Twitter so that you may invoke its APIs.

Storing Your Credentials

As a good practice, do not include your API keys and access tokens (or any other credentials, like usernames and passwords) directly in your source code, as that would expose them to anyone reading the code. You should store your keys in a separate file and never share that file with anyone.[3]

---

3. Good practice would be to use an encryption library such as bcrypt (https://github.com/pyca/bcrypt/) to encrypt your keys, access tokens or any other credentials you use in your code, then read them in and decrypt them only as you pass them to Twitter.

The code you'll execute in subsequent sections assumes that you place your consumer key, consumer secret, access token and access token secret values into the file keys.py shown below. You can find this file in the ch13 examples folder:

```
consumer_key='YourConsumerKey'
consumer_secret='YourConsumerSecret'
access_token='YourAccessToken'
access_token_secret='YourAccessTokenSecret'
```

Edit this file, replacing YourConsumerKey, YourConsumerSecret, YourAccessToken and YourAccessTokenSecret with your consumer key, consumer secret, access token and access token secret values. Then, save the file.

OAuth 2.0
The consumer key, consumer secret, access token and access token secret are each part of the **OAuth 2.0** authentication process[4,5]—sometimes called the *OAuth dance*—that Twitter uses to enable access to its APIs. The Tweepy library enables you to provide the consumer key, consumer secret, access token and access token secret and handles the OAuth 2.0 authentication details for you.

Self Check

1 *(Fill-In)* The consumer key, consumer secret, access token and access token secret are each part of the _____ authentication process that Twitter uses to enable access to its APIs.
Answer: OAuth 2.0.

2 *(True/False)* Once you have a Twitter developer account, you must obtain credentials to interact with APIs. To do so, you'll create an app. Each app has *separate* credentials.
Answer: True.

13.5 What's in a Tweet?

The Twitter API methods return JSON objects. **JSON (JavaScript Object Notation)** is a text-based data-interchange format used to represent objects as collections of name–value pairs. It's commonly used when invoking web services. JSON is both a human-readable and computer-readable format that makes data easy to send and receive across the Internet.

JSON objects are similar to Python dictionaries. Each JSON object contains a list of *property names* and *values*, in the following curly braced format:

{*propertyName1*: *value1*, *propertyName2*: *value2*}

As in Python, JSON lists are comma-separated values in square brackets:

[*value1*, *value2*, *value3*]

For your convenience, Tweepy handles the JSON for you behind the scenes, converting JSON to Python objects using classes defined in the Tweepy library.

---

4. https://developer.twitter.com/en/docs/basics/authentication/overview.
5. https://oauth.net/.

Key Properties of a Tweet Object

A tweet (also called a *status update*) may contain a maximum of 280 characters, but the tweet objects returned by the Twitter APIs contain many **metadata** attributes that describe aspects of the tweet, such as:

- when it was created,
- who created it,
- lists of the hashtags, urls, @-mentions and media (such as images and videos, which are specified via their URLs) included in the tweet,
- and more.

The following table lists a few key attributes of a tweet object:

Attribute	Description
created_at	The creation date and time in UTC (Coordinated Universal Time) format.
entities	Twitter extracts hashtags, urls, user_mentions (that is, @*username* mentions), media (such as images and videos), symbols and polls from tweets and places them into the entities dictionary as lists that you can access with these keys.
extended_tweet	For tweets over 140 characters, contains details such as the tweet's full_text and entities
favorite_count	Number of times other users favorited the tweet.
coordinates	The coordinates (latitude and longitude) from which the tweet was sent. This is often null (None in Python) because many users disable sending location data.
place	Users can associate a place with a tweet. If they do, this will be a place object: https://developer.twitter.com/en/docs/tweets/data-dictionary/overview/geo-objects#place-dictionary; otherwise, it'll be null (None in Python).
id	The integer ID of the tweet. Twitter recommends using id_str for portability.
id_str	The string representation of the tweet's integer ID.
lang	Language of the tweet, such as 'en' for English or 'fr' for French.
retweet_count	Number of times other users retweeted the tweet.
text	The text of the tweet. If the tweet uses the new 280-character limit and contains more than 140 characters, this property will be truncated and the truncated property will be set to *true*. This might also occur if a 140-character tweet was retweeted and became more than 140 characters as a result.
user	The User object representing the user that posted the tweet. For the User object JSON properties, see: https://developer.twitter.com/en/docs/tweets/data-dictionary/overview/user-object.

Sample Tweet JSON

Let's look at sample JSON for the following tweet from the @nasa account:

> @NoFear1075 Great question, Anthony! Throughout its seven-year mission, our Parker #SolarProbe spacecraft... https://t.co/xKd6ym8waT'

We added line numbers and reformatted some of the JSON due to wrapping. Note that some fields in Tweet JSON are not supported in every Twitter API method; such differences are explained in the online documentation for each method.

```
 1  {'created_at': 'Wed Sep 05 18:19:34 +0000 2018',
 2   'id': 1037404890354606082,
 3   'id_str': '1037404890354606082',
 4   'text': '@NoFear1075 Great question, Anthony! Throughout its seven-year
         mission, our Parker #SolarProbe spacecraft… https://t.co/xKd6ym8waT',
 5   'truncated': True,
 6   'entities': {'hashtags': [{'text': 'SolarProbe', 'indices': [84, 95]}],
 7     'symbols': [],
 8     'user_mentions': [{'screen_name': 'NoFear1075',
 9        'name': 'Anthony Perrone',
10        'id': 284339791,
11        'id_str': '284339791',
12        'indices': [0, 11]}],
13     'urls': [{'url': 'https://t.co/xKd6ym8waT',
14        'expanded_url': 'https://twitter.com/i/web/status/
           1037404890354606082',
15        'display_url': 'twitter.com/i/web/status/1…',
16        'indices': [117, 140]}]},
17   'source': '<a href="http://twitter.com" rel="nofollow">Twitter Web
     Client</a>',
18   'in_reply_to_status_id': 1037390542424956928,
19   'in_reply_to_status_id_str': '1037390542424956928',
20   'in_reply_to_user_id': 284339791,
21   'in_reply_to_user_id_str': '284339791',
22   'in_reply_to_screen_name': 'NoFear1075',
23   'user': {'id': 11348282,
24     'id_str': '11348282',
25     'name': 'NASA',
26     'screen_name': 'NASA',
27     'location': '',
28     'description': 'Explore the universe and discover our home planet with
        @NASA. We usually post in EST (UTC-5)',
29     'url': 'https://t.co/TcEE6NS8nD',
30     'entities': {'url': {'urls': [{'url': 'https://t.co/TcEE6NS8nD',
31         'expanded_url': 'http://www.nasa.gov',
32         'display_url': 'nasa.gov',
33         'indices': [0, 23]}]},
34       'description': {'urls': []}},
35     'protected': False,
36     'followers_count': 29486081,
37     'friends_count': 287,
38     'listed_count': 91928,
39     'created_at': 'Wed Dec 19 20:20:32 +0000 2007',
40     'favourites_count': 3963,
41     'time_zone': None,
42     'geo_enabled': False,
43     'verified': True,
44     'statuses_count': 53147,
45     'lang': 'en',
46     'contributors_enabled': False,
47     'is_translator': False,
48     'is_translation_enabled': False,
49     'profile_background_color': '000000',
50     'profile_background_image_url': 'http://abs.twimg.com/images/themes/
       theme1/bg.png',
```

```
51          'profile_background_image_url_https': 'https://abs.twimg.com/images/
            themes/theme1/bg.png',
52          'profile_image_url': 'http://pbs.twimg.com/profile_images/188302352/
            nasalogo_twitter_normal.jpg',
53          'profile_image_url_https': 'https://pbs.twimg.com/profile_images/
            188302352/nasalogo_twitter_normal.jpg',
54          'profile_banner_url': 'https://pbs.twimg.com/profile_banners/11348282/
            1535145490',
55          'profile_link_color': '205BA7',
56          'profile_sidebar_border_color': '000000',
57          'profile_sidebar_fill_color': 'F3F2F2',
58          'profile_text_color': '000000',
59          'profile_use_background_image': True,
60          'has_extended_profile': True,
61          'default_profile': False,
62          'default_profile_image': False,
63          'following': True,
64          'follow_request_sent': False,
65          'notifications': False,
66          'translator_type': 'regular'},
67     'geo': None,
68     'coordinates': None,
69     'place': None,
70     'contributors': None,
71     'is_quote_status': False,
72     'retweet_count': 7,
73     'favorite_count': 19,
74     'favorited': False,
75     'retweeted': False,
76     'possibly_sensitive': False,
77     'lang': 'en'}
```

Twitter JSON Object Resources

For a complete, more readable list of the tweet object attributes, see:

> https://developer.twitter.com/en/docs/tweets/data-dictionary/
>   overview/tweet-object.html

For additional details that were added when Twitter moved from a limit of 140 to 280 characters per tweet, see

> https://developer.twitter.com/en/docs/tweets/data-dictionary/
>   overview/intro-to-tweet-json.html#extendedtweet

For a general overview of all the JSON objects that Twitter APIs return, and links to the specific object details, see

> https://developer.twitter.com/en/docs/tweets/data-dictionary/
>   overview/intro-to-tweet-json

✓ Self Check

1. *(Fill-In)* Tweet objects returned by the Twitter APIs contain many _____ attributes that describe aspects of the tweet.
Answer: metadata.

2 *(True/False)* JSON is both a human-readable *and* computer-readable format that makes objects easy to send and receive across the Internet.
Answer: True.

13.6 Tweepy

We'll use the Tweepy library[6] (http://www.tweepy.org/)—one of the most popular Python libraries for interacting with the Twitter APIs. Tweepy makes it easy to access Twitter's capabilities and hides from you the details of processing the JSON objects returned by the Twitter APIs. You can view Tweepy's documentation[7] at

    http://docs.tweepy.org/en/latest/

For additional information and the Tweepy source code, visit

    https://github.com/tweepy/tweepy

Installing Tweepy

To install Tweepy, open your Anaconda Prompt (Windows), Terminal (macOS/Linux) or shell (Linux), then execute the following command:

    `pip install tweepy==3.7`

Windows users might need to run the Anaconda Prompt as an Administrator for proper software installation privileges. To do so, right-click Anaconda Prompt in the start menu and select **More > Run as administrator.**

Installing geopy

As you work with Tweepy, you'll also use functions from our `tweetutilities.py` file (provided with this chapter's example code). One of the utility functions in that file depends on the **geopy library** (https://github.com/geopy/geopy), which we'll discuss in Section 13.15. To install geopy, execute:

    `conda install -c conda-forge geopy`

13.7 Authenticating with Twitter Via Tweepy

In the next several sections, you'll invoke various cloud-based Twitter APIs via Tweepy. Here you'll begin by using Tweepy to authenticate with Twitter and create a **Tweepy API object**, which is your gateway to using the Twitter APIs over the Internet. In subsequent sections, you'll work with various Twitter APIs by invoking methods on your API object.

---

6. Other Python libraries recommended by Twitter include Birdy, python-twitter, Python Twitter Tools, TweetPony, TwitterAPI, twitter-gobject, TwitterSearch and twython. See https://developer.twitter.com/en/docs/developer-utilities/twitter-libraries.html for details.
7. The Tweepy documentation is a work in progress. At the time of this writing, Tweepy does not have documentation for their classes corresponding to the JSON objects the Twitter APIs return. Tweepy's classes use the same attribute names and structure as the JSON objects. You can determine the correct attribute names to access by looking at Twitter's JSON documentation. We'll explain any attributes we use in our code and provide footnotes with links to the Twitter JSON descriptions.

Before you can invoke any Twitter API, you must use your API key, API secret key, access token and access token secret to authenticate with Twitter.[8] Launch IPython from the ch13 examples folder, then import the **tweepy module** and the keys.py file that you modified earlier in this chapter. You can import any .py file as a module by using the file's name *without* the .py extension in an import statement:

```
In [1]: import tweepy
```

```
In [2]: import keys
```

When you import keys.py as a module, you can individually access each of the four variables defined in that file as keys.*variable_name*.

Creating and Configuring an OAuthHandler to Authenticate with Twitter
Authenticating with Twitter via Tweepy involves two steps. First, create an object of the tweepy module's **OAuthHandler** class, passing your API key and API secret key to its constructor. A **constructor** is a function that has the same name as the class (in this case, OAuthHandler) and receives the arguments used to configure the new object:

```
In [3]: auth = tweepy.OAuthHandler(keys.consumer_key,
   ...:                            keys.consumer_secret)
   ...:
```

Specify your access token and access token secret by calling the OAuthHandler object's **set_access_token method**:

```
In [4]: auth.set_access_token(keys.access_token,
   ...:                       keys.access_token_secret)
   ...:
```

Creating an API Object
Now, create the API object that you'll use to interact with Twitter:

```
In [5]: api = tweepy.API(auth, wait_on_rate_limit=True,
   ...:                  wait_on_rate_limit_notify=True)
   ...:
```

We specified three arguments in this call to the API constructor:

- auth is the OAuthHandler object containing your credentials.
- The keyword argument wait_on_rate_limit=True tells Tweepy to wait 15 minutes each time it reaches a given API method's rate limit. This ensures that you do not violate Twitter's rate-limit restrictions.
- The keyword argument wait_on_rate_limit_notify=True tells Tweepy that, if it needs to wait due to rate limits, it should notify you by displaying a message at the command line.

You're now ready to interact with Twitter via Tweepy. Note that the code examples in the next several sections are presented as a continuing IPython session, so the authorization process you went through here need not be repeated.

---

8. You may wish to create apps that enable users to log into their Twitter accounts, manage them, post tweets, read tweets from other users, search for tweets, etc. For details on user authentication see the Tweepy Authentication tutorial at http://docs.tweepy.org/en/latest/auth_tutorial.html.

Self Check

1 *(Fill-In)* Authenticating with Twitter via Tweepy involves two steps. First, create an object of the Tweepy module's _____ class, passing your API key and API secret key to its constructor.
Answer: OAuthHandler.

2 *(True/False)* The keyword argument wait_on_rate_limit_notify=True to the tweepy.API call tells Tweepy to terminate the user because of a rate-limit violation.
Answer: False. The call tells Tweepy that if it needs to wait to avoid rate-limit violations it should display a message at the command line indicating that it's waiting for the rate limit to replenish.

13.8 Getting Information About a Twitter Account

After authenticating with Twitter, you can use the Tweepy API object's **get_user** method to get a **tweepy.models.User object** containing information about a user's Twitter account. Let's get a User object for NASA's @nasa Twitter account:

```
In [6]: nasa = api.get_user('nasa')
```

The get_user method calls the Twitter API's users/show method.[9] Each Twitter method you call through Tweepy has a rate limit. You can call Twitter's users/show method up to 900 times every 15 minutes to get information on specific user accounts. As we mention other Twitter API methods, we'll provide a footnote with a link to each method's documentation in which you can view its rate limits.

The tweepy.models classes each correspond to the JSON that Twitter returns. For example, the User class corresponds to a Twitter **user object**:

> https://developer.twitter.com/en/docs/tweets/data-dictionary/
>   overview/user-object

Each tweepy.models class has a method that reads the JSON and turns it into an object of the corresponding Tweepy class.

Getting Basic Account Information

Let's access some User object properties to display information about the @nasa account:

- The **id property** is the account ID number created by Twitter when the user joined Twitter.
- The **name property** is the name associated with the user's account.
- The **screen_name property** is the user's Twitter handle (@nasa). Both the name and screen_name could be created names to protect a user's privacy.
- The **description property** is the description from the user's profile.

```
In [7]: nasa.id
Out[7]: 11348282
```

---

9. https://developer.twitter.com/en/docs/accounts-and-users/follow-search-get-users/
   api-reference/get-users-show.

```
In [8]: nasa.name
Out[8]: 'NASA'

In [9]: nasa.screen_name
Out[9]: 'NASA'

In [10]: nasa.description
Out[10]: 'Explore the universe and discover our home planet with @NASA.
We usually post in EST (UTC-5)'
```

Getting the Most Recent Status Update
The User object's **status** property returns a **tweepy.models.Status** object, which corresponds to a Twitter **tweet object**:

> https://developer.twitter.com/en/docs/tweets/data-dictionary/
> overview/tweet-object

The Status object's **text property** contains the text of the account's most recent tweet:

```
In [11]: nasa.status.text
Out[11]: 'The interaction of a high-velocity young star with the cloud of
gas and dust may have created this unusually sharp-... https://t.co/
J6uUf7MYMI'
```

The text property was originally for tweets up to 140 characters. The ... above indicates that the tweet text was *truncated*. When Twitter increased the limit to 280 characters, they added an **extended_tweet** property (demonstrated later) for accessing the text and other information from tweets between 141 and 280 characters. In this case, Twitter sets text to a truncated version of the extended_tweet's text. Also, retweeting often results in truncation because a retweet adds characters that could exceed the character limit.

Getting the Number of Followers
You can view an account's number of followers with the **followers_count** property:

```
In [12]: nasa.followers_count
Out[12]: 29453541
```

Though this number is large, there are accounts with over 100 million followers.[10]

Getting the Number of Friends
Similarly, you can view an account's number of friends (that is, the number of accounts an account follows) with the **friends_count** property:

```
In [13]: nasa.friends_count
Out[13]: 287
```

Getting Your Own Account's Information
You can use the properties in this section on your own account as well. To do so, call the Tweepy API object's **me** method, as in:

```
me = api.me()
```

This returns a User object for the account you used to authenticate with Twitter in the preceding section.

---

10. https://friendorfollow.com/twitter/most-followers/.

Self Check

1 *(Fill-In)* After authenticating with Twitter, you can use the Tweepy API object's _____ method to get a tweepy.models.User object containing information about a user's Twitter account.
Answer: get_user.

2 *(True/False)* Retweeting often results in truncation because a retweet adds characters that could exceed the character limit.
Answer: True.

3 *(IPython Session)* Use the api object to get a User object for the NASAKepler account, then display its number of followers and most recent tweet.
Answer:

```
In [14]: nasa_kepler = api.get_user('NASAKepler')

In [15]: nasa_kepler.followers_count
Out[15]: 611281

In [16]: nasa_kepler.status.text
Out[16]: 'RT @TheFantasyG: Learning that there are #MorePlanetsThanStars
means to me that there are near endless possibilities of unique
discoveries...'
```

13.9 Introduction to Tweepy Cursors: Getting an Account's Followers and Friends

When invoking Twitter API methods, you often receive as results collections of objects, such as tweets in your Twitter timeline, tweets in another account's timeline or lists of tweets that match specified search criteria. A **timeline** consists of tweets sent by that user and by that user's friends—that is, other accounts that the user follows.

Each Twitter API method's documentation discusses the maximum number of items the method can return in one call—this is known as a **page** of results. When you request more results than a given method can return, Twitter's JSON responses indicate that there are more pages to get. Tweepy's Cursor class handles these details for you. A **Cursor** invokes a specified method and checks whether Twitter indicated that there is another page of results. If so, the Cursor automatically calls the method again to get those results. This continues, subject to the method's rate limits, until there are no more results to process. If you configure the API object to wait when rate limits are reached (as we did), the Cursor will adhere to the rate limits and wait as needed between calls. The following subsections discuss Cursor fundamentals. For more details, see the Cursor tutorial at:

> http://docs.tweepy.org/en/latest/cursor_tutorial.html

13.9.1 Determining an Account's Followers

Let's use a Tweepy Cursor to invoke the API object's **followers method**, which calls the Twitter API's followers/list method[11] to obtain an account's followers. Twitter returns

---

11. https://developer.twitter.com/en/docs/accounts-and-users/follow-search-get-users/api-reference/get-followers-list.

these in groups of 20 by default, but you can request up to 200 at a time. For demonstration purposes, we'll grab 10 of NASA's followers.

Method `followers` returns `tweepy.models.User` objects containing information about each follower. Let's begin by creating a list in which we'll store the `User` objects:

```
In [17]: followers = []
```

Creating a Cursor
Next, let's create a `Cursor` object that will call the `followers` method for NASA's account, which is specified with the `screen_name` keyword argument:

```
In [18]: cursor = tweepy.Cursor(api.followers, screen_name='nasa')
```

The `Cursor`'s constructor receives as its argument the name of the method to call—`api.followers` indicates that the `Cursor` will call the `api` object's `followers` method. If the `Cursor` constructor receives any additional keyword arguments, like `screen_name`, these will be passed to the method specified in the constructor's first argument. So, this `Cursor` specifically gets followers for the @nasa Twitter account.

Getting Results
Now, we can use the `Cursor` to get some followers. The following `for` statement iterates through the results of the expression `cursor.items(10)`. The `Cursor`'s **items method** initiates the call to `api.followers` and returns the `followers` method's results. In this case, we pass 10 to the `items` method to request only 10 results:

```
In [19]: for account in cursor.items(10):
    ...:     followers.append(account.screen_name)
    ...:

In [20]: print('Followers:',
    ...:       ' '.join(sorted(followers, key=lambda s: s.lower())))
Followers: abhinavborra BHood1976 Eshwar12341 Harish90469614 heshamkisha
Highyaan2407 JiraaJaarra KimYooJ91459029 Lindsey06771483 Wendy_UAE_NL
```

The preceding snippet displays the followers in ascending order by calling the built-in `sorted` function. The function's second argument is the function used to determine how the elements of followers are sorted. In this case, we used a `lambda` that converts every follower name to lowercase letters so we can perform a case-insensitive sort.

Automatic Paging
If the number of results requested is more than can be returned by one call to `followers`, the `items` method automatically "pages" through the results by making multiple calls to `api.followers`. Recall that `followers` returns up to 20 followers at a time by default, so the preceding code needs to call `followers` only once. To get up to 200 followers at a time, we can create the `Cursor` with the `count` keyword argument, as in:

```
cursor = tweepy.Cursor(api.followers, screen_name='nasa', count=200)
```

If you do not specify an argument to the `items` method, The `Cursor` attempts to get *all* of the account's followers. For large numbers of followers, this could take a significant amount of time due to Twitter's rate limits. The Twitter API's `followers/list` method can return a maximum of 200 followers at a time and Twitter allows a maximum of 15 calls every 15 minutes. Thus, you can only get 3000 followers every 15 minutes using Twitter's free APIs. Recall that we configured the API object to automatically wait when it hits a rate limit, so if you try to get all followers and an account has more than 3000, Tweepy will automatically pause for 15 minutes after every 3000 followers and display a message. At the time of this writing, NASA has over 29.5 million followers. At 12,000 followers per hour, it would take over 100 days to get all of NASA's followers.

Note that for this example, we could have called the followers method directly, rather than using a Cursor, since we're getting only a small number of followers. We used a Cursor here to show how you'll typically call followers. In some later examples, we'll call API methods directly to get just a few results, rather than using Cursors.

Getting Follower IDs Rather Than Followers

Though you can get complete User objects for a maximum of 200 followers at a time, you can get many more Twitter ID numbers by calling the API object's **followers_ids** method. This calls the Twitter API's followers/ids method, which returns up to 5000 ID numbers at a time (again, these rate limits could change).[12] You can invoke this method up to 15 times every 15 minutes, so you can get 75,000 account ID numbers per rate-limit interval. This is particularly useful when combined with the API object's **lookup_users** method. This calls the Twitter API's users/lookup method[13] which can return up to 100 User objects at a time and can be called up to 300 times every 15 minutes. So using this combination, you could get up to 30,000 User objects per rate-limit interval.

Self Check

1 *(Fill-In)* Each Twitter API method's documentation discusses the maximum number of items the method can return in one call—this is known as a _____ of results.
Answer: page.

2 *(True/False)* Though you can get complete User objects for a maximum of 200 followers at a time, you can get many more Twitter ID numbers by calling the API object's followers_ids method.
Answer: True.

3 *(IPython Session)* Use a Cursor to get and display 10 followers of the NASAKepler account.

```
In [21]: kepler_followers = []

In [22]: cursor = tweepy.Cursor(api.followers, screen_name='NASAKepler')

In [23]: for account in cursor.items(10):
   ...:     kepler_followers.append(account.screen_name)
   ...:

In [24]: print(' '.join(kepler_followers))
cheleandre_ FranGlacierGirl Javedja88171520 Ameer90577310 c4rb0hydr8
rashadali77777 ICPN2019 usOOU5hSZ8BwnsA KHRSC1 xAquos
```

---

12. https://developer.twitter.com/en/docs/accounts-and-users/follow-search-get-users/api-reference/get-followers-ids.
13. https://developer.twitter.com/en/docs/accounts-and-users/follow-search-get-users/api-reference/get-users-lookup.

13.9.2 Determining Whom an Account Follows

The `API` object's **friends method** calls the Twitter API's `friends/list` method[14] to get a list of `User` objects representing an account's friends. Twitter returns these in groups of 20 by default, but you can request up to 200 at a time, just as we discussed for method `followers`. Twitter allows you to call the `friends/list` method up to 15 times every 15 minutes. Let's get 10 of NASA's friend accounts:

```
In [25]: friends = []

In [26]: cursor = tweepy.Cursor(api.friends, screen_name='nasa')

In [27]: for friend in cursor.items(10):
    ...:     friends.append(friend.screen_name)
    ...:

In [28]: print('Friends:',
    ...:     ' '.join(sorted(friends, key=lambda s: s.lower())))
    ...:
Friends: AFSpace Astro2fish Astro_Kimiya AstroAnnimal AstroDuke
NASA3DPrinter NASASMAP Outpost_42 POTUS44 VicGlover
```

✓ Self Check

1. *(Fill-In)* The `API` object's **friends method** calls the Twitter API's _____ method to get a list of `User` objects representing an account's friends.
Answer: friends/list.

13.9.3 Getting a User's Recent Tweets

The `API` method **user_timeline** returns tweets from the timeline of a specific account. A timeline includes that account's tweets and tweets from that account's friends. The method calls the Twitter API's `statuses/user_timeline` method[15], which returns the most recent 20 tweets, but can return up to 200 at a time. This method can return only an account's 3200 most recent tweets. Applications using this method may call it up to 1500 times every 15 minutes.

Method `user_timeline` returns `Status` objects with each one representing a tweet. Each `Status`'s user property refers to a `tweepy.models.User` object containing information about the user who sent that tweet, such as that user's `screen_name`. A `Status`'s text property contains the tweet's `text`. Let's display the `screen_name` and `text` for three tweets from @nasa:

```
In [29]: nasa_tweets = api.user_timeline(screen_name='nasa', count=3)

In [30]: for tweet in nasa_tweets:
    ...:     print(f'{tweet.user.screen_name}: {tweet.text}\n')
    ...:
NASA: Your Gut in Space: Microorganisms in the intestinal tract play an
especially important role in human health. But wh… https://t.co/
uL0sUhwn5p
```

---

14. https://developer.twitter.com/en/docs/accounts-and-users/follow-search-get-users/api-reference/get-friends-list.
15. https://developer.twitter.com/en/docs/tweets/timelines/api-reference/get-statuses-user_timeline.

NASA: We need your help! Want to see panels at @SXSW related to space exploration? There are a number of exciting panels… https://t.co/ycqMMdGKUB

NASA: "You are as good as anyone in this town, but you are no better than any of them," says retired @NASA_Langley mathem… https://t.co/nhMD4n84Nf

These tweets were truncated (as indicated by …), meaning that they probably use the newer 280-character tweet limit. We'll use the extended_tweet property shortly to access full text for such tweets.

In the preceding snippets, we chose to call the user_timeline method directly and use the count keyword argument to specify the number of tweets to retrieve. If you wish to get more than the maximum number of tweets per call (200), then you should use a Cursor to call user_timeline as demonstrated previously. Recall that a Cursor automatically pages through the results by calling the method multiple times, if necessary.

Grabbing Recent Tweets from Your Own Timeline

You can call the API method **home_timeline**, as in:

```
api.home_timeline()
```

to get tweets from *your* home timeline[16]—that is, your tweets and tweets from the people *you* follow. This method calls Twitter's statuses/home_timeline method.[17] By default, home_timeline returns the most recent 20 tweets, but can get up to 200 at a time. Again, for more than 200 tweets from your home timeline, you should use a Tweepy Cursor to call home_timeline.

 ## Self Check

1 *(Fill-In)* You can call the API method **home_timeline** to get tweets from *your* home timeline, that is, your tweets and tweets from _____.
Answer: the people you follow.

2 *(IPython Session)* Get and display two tweets from the NASAKepler account.
Answer:

```
In [31]: kepler_tweets = api.user_timeline(
    ...:     screen_name='NASAKepler', count=2)
    ...:

In [32]: for tweet in kepler_tweets:
    ...:     print(f'{tweet.user.screen_name}: {tweet.text}\n')
    ...:
```
NASAKepler: RT @TheFantasyG: Learning that there are #MorePlanetsThanStars means to me that there are near endless possibilities of unique discoveries…

NASAKepler: @KerryFoster2 @NASA Refueling Kepler is not practical since it currently sits 94 million miles from Earth. And with… https://t.co/D2P145ELON

---

16. Specifically for the account you used to authenticate with Twitter.
17. https://developer.twitter.com/en/docs/tweets/timelines/api-reference/get-statuses-home_timeline.

13.10 Searching Recent Tweets

The Tweepy API method **search** returns tweets that match a query string. According to the method's documentation, Twitter maintains its search index only for the previous seven days' tweets, and a search is not guaranteed to return all matching tweets. Method search calls Twitter's search/tweets method[18], which returns 15 tweets at a time by default, but can return up to 100.

Utility Function print_tweets from tweetutilities.py
For this section, we created a utility function print_tweets that receives the results of a call to API method search and for each tweet displays the user's screen_name and the tweet's text. If the tweet is not in English and the tweet.lang is not 'und' (undefined), we'll also translate the tweet to English using TextBlob, as you did in the "Natural Language Processing (NLP)" chapter. To use this function, import it from tweetutilities.py:

```
In [33]: from tweetutilities import print_tweets
```

Just the print_tweets function's definition from that file is shown below:

```python
def print_tweets(tweets):
    """For each Tweepy Status object in tweets, display the
    user's screen_name and tweet text. If the language is not
    English, translate the text with TextBlob."""
    for tweet in tweets:
        print(f'{tweet.screen_name}:', end=' ')

        if 'en' in tweet.lang:
            print(f'{tweet.text}\n')
        elif 'und' not in tweet.lang:  # translate to English first
            print(f'\n  ORIGINAL: {tweet.text}')
            print(f'TRANSLATED: {TextBlob(tweet.text).translate()}\n')
```

Searching for Specific Words
Let's search for three recent tweets about NASA's Mars Opportunity Rover. The search method's q keyword argument specifies the query string, which indicates what to search for and the count keyword argument specifies the number of tweets to return:

```
In [34]: tweets = api.search(q='Mars Opportunity Rover', count=3)

In [35]: print_tweets(tweets)
Jacker760: NASA set a deadline on the Mars Rover opportunity! As the dust on Mars settles the Rover will start to regain power… https://t.co/KQ7xaFgrzr

Shivak32637174: RT @Gadgets360: NASA 'Cautiously Optimistic' of Hearing Back From Opportunity Rover as Mars Dust Storm Settles https://t.co/O1iTTwRvFq

ladyanakina: NASA's Opportunity Rover Still Silent on #Mars. https://t.co/njcyP6zCm3
```

As with other methods, if you plan to request more results than can be returned by one call to search, you should use a Cursor object.

---

18. https://developer.twitter.com/en/docs/tweets/search/api-reference/get-search-tweets.

Searching with Twitter Search Operators

You can use various Twitter search operators in your query strings to refine your search results. The following table shows several Twitter search operators. Multiple operators can be combined to construct more complex queries. To see all the operators, visit

> https://twitter.com/search-home

and click the **operators** link.

Example	Finds tweets containing
python twitter	Implicit *logical and* operator—Finds tweets containing python *and* twitter.
python OR twitter	Logical OR operator—Finds tweets containing python *or* twitter *or both*.
python ?	? (question mark)—Finds tweets asking questions about python.
planets -mars	- (minus sign)—Finds tweets containing planets but not mars.
python :)	:) (happy face)—Finds *positive sentiment* tweets containing python.
python :(	:((sad face)—Finds *negative sentiment* tweets containing python.
since:2018-09-01	Finds tweets *on or after* the specified date, which must be in the form YYYY-MM-DD.
near:"New York City"	Finds tweets that were sent near "New York City".
from:nasa	Finds tweets from the account @nasa.
to:nasa	Finds tweets to the account @nasa.

Let's use the `from` and `since` operators to get three tweets from NASA since September 1, 2018—you should use a date within seven days before you execute this code:

```
In [36]: tweets = api.search(q='from:nasa since:2018-09-01', count=3)
```

```
In [37]: print_tweets(tweets)
NASA: @WYSIW Our missions detect active burning fires, track the
transport of fire smoke, provide info for fire managemen… https://t.co/
jx2iUoMIIy

NASA: Scarring of the landscape is evident in the wake of the Mendocino
Complex fire, the largest #wildfire in California… https://t.co/
Nboo5GD90m

NASA: RT @NASAglenn: To celebrate the #NASA60th anniversary, we're
exploring our history. In this image, Research Pilot Bill Swann prepares
for a…
```

Searching for a Hashtag

Tweets often contain **hashtags** that begin with # to indicate something of importance, like a trending topic. Let's get two tweets containing the hashtag #collegefootball:

```
In [38]: tweets = api.search(q='#collegefootball', count=2)
```

```
In [39]: print_tweets(tweets)
dmcreek: So much for #FAU giving #OU a game. #Oklahoma #FloridaAtlantic
#CollegeFootball #LWOS

theangrychef: It's game day folks! And our BBQ game is strong. #bbq
#atlanta #collegefootball #gameday @ Smoke Ring https://t.co/J4lkKhCQE7
```

Self Check

1 *(Fill-In)* The Tweepy API method _____ returns tweets that match a query string.
Answer: search.

2 *(True/False)* If you plan to request more results than can be returned by one call to search, you should use an API object.
Answer: False. If you plan to request more results than can be returned by one call to search, you should use a Cursor object.

3 *(IPython Session)* Search for one tweet from the nasa account containing 'astronaut'.
Answer:

```
In [40]: tweets = api.search(q='astronaut from:nasa', count=1)

In [41]: print_tweets(tweets)
NASA: Astronaut Guion "Guy" Bluford never aimed to become the first
African-American in space, but #OTD in 1983 he soared… https://t.co/
bIjl88yJdR
```

13.11 Spotting Trends: Twitter Trends API

If a topic "goes viral," you could have thousands or even millions of people tweeting about it at once. Twitter refers to these as **trending topics** and maintains lists of the trending topics worldwide. Via the Twitter Trends API, you can get lists of locations with trending topics and lists of the top 50 trending topics for each location.

13.11.1 Places with Trending Topics

The Tweepy API's **trends_available** method calls the Twitter API's trends/available[19] method to get a list of all locations for which Twitter has trending topics. Method trends_available returns a *list of dictionaries* representing these locations. When we executed this code, there were 467 locations with trending topics:

```
In [42]: trends_available = api.trends_available()

In [43]: len(trends_available)
Out[43]: 467
```

The dictionary in each list element returned by trends_available has various information, including the location's name and woeid (discussed below):

```
In [44]: trends_available[0]
Out[44]:
{'name': 'Worldwide',
 'placeType': {'code': 19, 'name': 'Supername'},
 'url': 'http://where.yahooapis.com/v1/place/1',
 'parentid': 0,
 'country': '',
 'woeid': 1,
 'countryCode': None}
```

---

19. https://developer.twitter.com/en/docs/trends/locations-with-trending-topics/api-reference/get-trends-available.

13.11 Spotting Trends: Twitter Trends API

```
In [45]: trends_available[1]
Out[45]:
{'name': 'Winnipeg',
 'placeType': {'code': 7, 'name': 'Town'},
 'url': 'http://where.yahooapis.com/v1/place/2972',
 'parentid': 23424775,
 'country': 'Canada',
 'woeid': 2972,
 'countryCode': 'CA'}
```

The Twitter Trends API's trends/place method (discussed momentarily) uses **Yahoo! Where on Earth IDs (WOEIDs)** to look up trending topics. The WOEID 1 represents *worldwide*. Other locations have unique WOEID values greater than 1. We'll use WOEID values in the next two subsections to get worldwide trending topics and trending topics for a specific city. The following table shows WOEID values for several landmarks, cities, states and continents. Note that although these are all valid WOEIDs, Twitter does not necessarily have trending topics for all these locations.

Place	WOEID	Place	WOEID
Statue of Liberty	23617050	Iguazu Falls	468785
Los Angeles, CA	2442047	United States	23424977
Washington, D.C.	2514815	North America	24865672
Paris, France	615702	Europe	24865675

You also can search for locations close to a location that you specify with latitude and longitude values. To do so, call the Tweepy API's **trends_closest method**, which invokes the Twitter API's trends/closest method.[20]

 ## Self Check

1 *(Fill-In)* If a topic "goes viral," you could have thousands or even millions of people tweeting about that topic at once. Twitter refers to these as _____ topics.
Answer: trending.

2 *(True/False)* The Twitter Trends API's trends/place method uses Yahoo! Where on Earth IDs (WOEIDs) to look up trending topics. The WOEID 1 represents worldwide.
Answer: True.

13.11.2 Getting a List of Trending Topics

The Tweepy API's **trends_place method** calls the Twitter Trends API's trends/place method[21] to get the top 50 trending topics for the location with the specified WOEID. You can get the WOEIDs from the woeid attribute in each dictionary returned by the

---

20. https://developer.twitter.com/en/docs/trends/locations-with-trending-topics/api-reference/get-trends-closest.
21. https://developer.twitter.com/en/docs/trends/trends-for-location/api-reference/get-trends-place.

`trends_available` or `trends_closest` methods discussed in the previous section, or you can find a location's Yahoo! Where on Earth ID (WOEID) by searching for a city/town, state, country, address, zip code or landmark at

http://www.woeidlookup.com

You also can look up WOEID's programmatically using Yahoo!'s web services via Python libraries like woeid[22]:

https://github.com/Ray-SunR/woeid

Worldwide Trending Topics

Let's get today's worldwide trending topics (your results will differ):

```
In [46]: world_trends = api.trends_place(id=1)
```

Method `trends_place` returns a one-element list containing a dictionary. The dictionary's `'trends'` key refers to a list of dictionaries representing each trend:

```
In [47]: trends_list = world_trends[0]['trends']
```

Each trend dictionary has `name`, `url`, `promoted_content` (indicating the tweet is an advertisement), `query` and `tweet_volume` keys (shown below). The following trend is in Spanish—`#BienvenidoSeptiembre` means "Welcome September":

```
In [48]: trends_list[0]
Out[48]:
{'name': '#BienvenidoSeptiembre',
 'url': 'http://twitter.com/search?q=%23BienvenidoSeptiembre',
 'promoted_content': None,
 'query': '%23BienvenidoSeptiembre',
 'tweet_volume': 15186}
```

For trends with more than 10,000 tweets, the `tweet_volume` is the number of tweets; otherwise, it's `None`. Let's use a list comprehension to filter the list so that it contains only trends with more than 10,000 tweets:

```
In [49]: trends_list = [t for t in trends_list if t['tweet_volume']]
```

Next, let's sort the trends in *descending* order by `tweet_volume`:

```
In [50]: from operator import itemgetter
```

```
In [51]: trends_list.sort(key=itemgetter('tweet_volume'), reverse=True)
```

Now, let's display the names of the top five trending topics:

```
In [52]: for trend in trends_list[:5]:
    ...:     print(trend['name'])
    ...:
#HBDJanaSenaniPawanKalyan
#BackToHogwarts
Khalil Mack
#ItalianGP
Alisson
```

---

22. You'll need a Yahoo! API key as described in the `woeid` module's documentation.

New York City Trending Topics

Now, let's get the top five trending topics for New York City (WOEID 2459115). The following code performs the same tasks as above, but for the different WOEID:

```
In [53]: nyc_trends = api.trends_place(id=2459115)  # New York City WOEID

In [54]: nyc_list = nyc_trends[0]['trends']

In [55]: nyc_list = [t for t in nyc_list if t['tweet_volume']]

In [56]: nyc_list.sort(key=itemgetter('tweet_volume'), reverse=True)

In [57]: for trend in nyc_list[:5]:
   ...:     print(trend['name'])
   ...:
#IDOL100M
#TuesdayThoughts
#HappyBirthdayLiam
NAFTA
#USOpen
```

✓ Self Check

1 *(Fill-In)* You also can look up WOEIDs programmatically using Yahoo!'s web services via Python libraries like _____.
Answer: woeid.

2 *(True/False)* The statement `todays_trends = api.trends_place(id=1)` gets today's U. S. trending topics.
Answer: False. Actually, it gets today's *worldwide* trending topics.

3 *(IPython Session)* Display the top 3 trending topics today in the United States.
Answer:

```
In [58]: us_trends = api.trends_place(id='23424977')

In [59]: us_list = us_trends[0]['trends']

In [60]: us_list = [t for t in us_list if t['tweet_volume']]

In [61]: us_list.sort(key=itemgetter('tweet_volume'), reverse=True)

In [62]: for trend in us_list[:3]:
   ...:     print(trend['name'])
   ...:
Cory Booker
Burt Reynolds
#ThursdayThoughts
```

13.11.3 Create a Word Cloud from Trending Topics

In the Natural Language Processing chapter, we used the WordCloud library to create word clouds. Let's use it again here, to visualize New York City's trending topics that have more than 10,000 tweets each. First, let's create a dictionary of key–value pairs consisting of the trending topic names and tweet_volumes:

```
In [63]: topics = {}
```

```
In [64]: for trend in nyc_list:
    ...:     topics[trend['name']] = trend['tweet_volume']
    ...:
```

Next, let's create a WordCloud from the topics dictionary's key–value pairs, then output the word cloud to the image file TrendingTwitter.png (shown after the code). The argument prefer_horizontal=0.5 *suggests* that 50% of the words should be horizontal, though the software may ignore that to fit the content:

```
In [65]: from wordcloud import WordCloud
```

```
In [66]: wordcloud = WordCloud(width=1600, height=900,
    ...:     prefer_horizontal=0.5, min_font_size=10, colormap='prism',
    ...:     background_color='white')
```

```
In [67]: wordcloud = wordcloud.fit_words(topics)
```

```
In [68]: wordcloud = wordcloud.to_file('TrendingTwitter.png')
```

The resulting word cloud is shown below—yours will differ based on the trending topics the day you run the code:

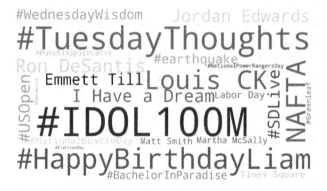

✓ Self Check

1. *(IPython Session)* Create a word cloud using the us_list list from the previous section's Self Check.
Answer:

```
In [69]: topics = {}
```

```
In [70]: for trend in us_list:
    ...:     topics[trend['name']] = trend['tweet_volume']
    ...:
```

```
In [71]: wordcloud = wordcloud.fit_words(topics)
```

```
In [72]: wordcloud = wordcloud.to_file('USTrendingTwitter.png')
```

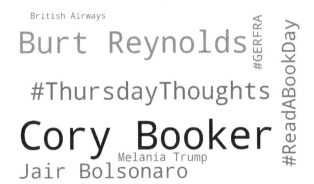

13.12 Cleaning/Preprocessing Tweets for Analysis

Data cleaning is one of the most common tasks that data scientists perform. Depending on how you intend to process tweets, you'll need to use natural language processing to normalize them by performing some or all of the data cleaning tasks in the following table. Many of these can be performed using the libraries introduced in the "Natural Language Processing (NLP)" chapter:

Tweet cleaning tasks	
Converting all text to the same case	Removing stop words
Removing # symbol from hashtags	Removing RT (retweet) and FAV (favorite)
Removing @-mentions	Removing URLs
Removing duplicates	Stemming
Removing excess whitespace	Lemmatization
Removing hashtags	Tokenization
Removing punctuation	

tweet-preprocessor Library and TextBlob Utility Functions
In this section, we'll use the **tweet-preprocessor library**

      https://github.com/s/preprocessor

to perform some basic tweet cleaning. It can automatically remove:

- URLs,
- @-mentions (like @nasa),
- hashtags (like #mars),
- Twitter reserved words (like, RT for retweet and FAV for favorite, which is similar to a "like" on other social networks),
- emojis (all or just smileys) and
- numbers

or any combination of these. The following table shows the module's constants representing each option:

Option	Option constant
@-Mentions (e.g., @nasa)	OPT.MENTION
Emoji	OPT.EMOJI
Hashtag (e.g., #mars)	OPT.HASHTAG
Number	OPT.NUMBER
Reserved Words (RT and FAV)	OPT.RESERVED
Smiley	OPT.SMILEY
URL	OPT.URL

Installing tweet-preprocessor

To install tweet-preprocessor, open your Anaconda Prompt (Windows), Terminal (macOS/Linux) or shell (Linux), then issue the following command:

```
pip install tweet-preprocessor
```

Windows users might need to run the Anaconda Prompt as an administrator for proper software installation privileges. To do so, right-click Anaconda Prompt in the start menu and select **More > Run as administrator.**

Cleaning a Tweet

Let's do some basic tweet cleaning that we'll use in a later example in this chapter. The tweet-preprocessor library's module name is preprocessor. Its documentation recommends that you import the module as follows:

```
In [1]: import preprocessor as p
```

To set the cleaning options you'd like to use call the module's **set_options** function. In this case, we'd like to remove URLs and Twitter reserved words:

```
In [2]: p.set_options(p.OPT.URL, p.OPT.RESERVED)
```

Now let's clean a sample tweet containing a reserved word (RT) and a URL:

```
In [3]: tweet_text = 'RT A sample retweet with a URL https://nasa.gov'

In [4]: p.clean(tweet_text)
Out[4]: 'A sample retweet with a URL'
```

 Self Check

1 *(True/False)* The tweet-preprocessor library can automatically remove URLs, @-mentions (like @nasa), hashtags (like #mars), Twitter reserved words (like, RT for retweet and FAV for favorite, which is similar to a "like" on other social networks), emojis (all or just smileys) and numbers, or any combination of these.
Answer: True.

13.13 Twitter Streaming API

Twitter's free Streaming API sends to your app *randomly selected* tweets dynamically as they occur—up to a maximum of one percent of the tweets per day. According to InternetLiveStats.com, there are approximately 6000 tweets per second, which is over 500 million tweets per day.[23] So the Streaming API gives you access to approximately five million tweets per day. Twitter used to allow free access to 10% of streaming tweets, but this service—called the *fire hose*—is now available only as a paid service. In this section, we'll use a class definition and an IPython session to walk through the steps for processing streaming tweets. Note that the code for receiving a tweet stream requires creating a *custom class* that *inherits* from another class. These topics are covered in Chapter 10.

13.13.1 Creating a Subclass of StreamListener

The Streaming API returns tweets as they happen that match your search criteria. Rather than connecting to Twitter on each method call, a stream uses a *persistent* connection to **push** (that is, send) tweets to your app. The rate at which those tweets arrive varies tremendously, based on your search criteria. The more popular a topic is, the more likely it is that the tweets will arrive quickly.

You create a subclass of Tweepy's **StreamListener class** to process the tweet stream. An object of this class is the *listener* that's notified when each new tweet (or other message sent by Twitter[24]) arrives. Each message Twitter sends results in a call to a StreamListener method. The following table summarizes several such methods. StreamListener already defines each method, so you redefine only the methods you need—this is known as *overriding*. For additional StreamListener methods, see:

> https://github.com/tweepy/tweepy/blob/master/tweepy/streaming.py

Method	Description
on_connect(self)	Called when you successfully connect to the Twitter stream. This is for statements that should execute only if your app is connected to the stream.
on_status(self, status)	Called when a tweet arrives—status is an object of Tweepy's Status.
on_limit(self, track)	Called when a limit notice arrives. This occurs if your search matches more tweets than Twitter can deliver based on its current streaming rate limits. In this case, the limit notice contains the number of matching tweets that could not be delivered.
on_error(self, status_code)	Called in response to error codes sent by Twitter.
on_timeout(self)	Called if the connection times out—that is, the Twitter server is not responding.
on_warning(self, notice)	Called if Twitter sends a disconnect warning to indicate that the connection might be closed. For example, Twitter maintains a queue of the tweets it's pushing to your app. If the app does not read the tweets fast enough, on_warning's notice argument will contain a warning message indicating that the connection will terminate if the queue becomes full.

---

23. http://www.internetlivestats.com/twitter-statistics/.
24. For details on the messages, see https://developer.twitter.com/en/docs/tweets/filter-realtime/guides/streaming-message-types.html.

Class TweetListener

Our `StreamListener` subclass `TweetListener` is defined in `tweetlistener.py`. We discuss the `TweetListener`'s components here. Line 6 indicates that class `TweetListener` is a subclass of `tweepy.StreamListener`. This ensures that our new class has class `StreamListener`'s default method implementations.

```
1   # tweetlistener.py
2   """tweepy.StreamListener subclass that processes tweets as they arrive."""
3   import tweepy
4   from textblob import TextBlob
5
6   class TweetListener(tweepy.StreamListener):
7       """Handles incoming Tweet stream."""
8
```

Class TweetListener: __init__ Method

The following lines define the `TweetListener` class's `__init__` method, which is called when you create a new `TweetListener` object. The `api` parameter is the Tweepy API object that `TweetListener` will use to interact with Twitter. The `limit` parameter is the total number of tweets to process—10 by default. We added this parameter to enable you to control the number of tweets to receive. As you'll soon see, we terminate the stream when that `limit` is reached. If you set `limit` to `None`, the stream will not terminate automatically. Line 11 creates an instance variable to keep track of the number of tweets processed so far, and line 12 creates a constant to store the `limit`. If you're not familiar with `__init__` and `super()` from previous chapters, line 13 ensures that the `api` object is stored properly for use by your listener object.

```
9       def __init__(self, api, limit=10):
10          """Create instance variables for tracking number of tweets."""
11          self.tweet_count = 0
12          self.TWEET_LIMIT = limit
13          super().__init__(api)  # call superclass's init
14
```

Class TweetListener: on_connect Method

Method `on_connect` is called when your app successfully connects to the Twitter stream. We override the default implementation to display a "Connection successful" message.

```
15      def on_connect(self):
16          """Called when your connection attempt is successful, enabling
17          you to perform appropriate application tasks at that point."""
18          print('Connection successful\n')
19
```

Class TweetListener: on_status Method

Method `on_status` is called by Tweepy when each tweet arrives. This method's second parameter receives a Tweepy `Status` object representing the tweet. Lines 23–26 get the tweet's text. First, we assume the tweet uses the new 280-character limit, so we attempt to access the tweet's `extended_tweet` property and get its `full_text`. An exception will occur if the tweet does not have an `extended_tweet` property. In this case, we get the `text` property instead. Lines 28–30 then display the `screen_name` of the user who sent the tweet, the `lang` (that is language) of the tweet and the `tweet_text`. If the language is not

English ('en'), lines 32–33 use a TextBlob to translate the tweet and display it in English. We increment self.tweet_count (line 36), then compare it to self.TWEET_LIMIT in the return statement. If on_status returns True, the stream remains open. When on_status returns False, Tweepy disconnects from the stream.

```
20      def on_status(self, status):
21          """Called when Twitter pushes a new tweet to you."""
22          # get the tweet text
23          try:
24              tweet_text = status.extended_tweet.full_text
25          except:
26              tweet_text = status.text
27
28          print(f'Screen name: {status.user.screen_name}:')
29          print(f'   Language: {status.lang}')
30          print(f'     Status: {tweet_text}')
31
32          if status.lang != 'en':
33              print(f' Translated: {TextBlob(tweet_text).translate()}')
34
35          print()
36          self.tweet_count += 1  # track number of tweets processed
37
38          # if TWEET_LIMIT is reached, return False to terminate streaming
39          return self.tweet_count != self.TWEET_LIMIT
```

13.13.2 Initiating Stream Processing

Let's use an IPython session to test our new TweetListener.

Authenticating

First, you must authenticate with Twitter and create a Tweepy API object:

```
In [1]: import tweepy

In [2]: import keys

In [3]: auth = tweepy.OAuthHandler(keys.consumer_key,
   ...:                           keys.consumer_secret)
   ...:

In [4]: auth.set_access_token(keys.access_token,
   ...:                       keys.access_token_secret)
   ...:

In [5]: api = tweepy.API(auth, wait_on_rate_limit=True,
   ...:                  wait_on_rate_limit_notify=True)
   ...:
```

Creating a TweetListener

Next, create an object of the TweetListener class and initialize it with the api object:

```
In [6]: from tweetlistener import TweetListener

In [7]: tweet_listener = TweetListener(api)
```

We did not specify the limit argument, so this TweetListener terminates after 10 tweets.

Creating a Stream

A Tweepy **Stream** object manages the connection to the Twitter stream and passes the messages to your `TweetListener`. The `Stream` constructor's `auth` keyword argument receives the `api` object's `auth` property, which contains the previously configured `OAuthHandler` object. The `listener` keyword argument receives your listener object:

```
In [8]: tweet_stream = tweepy.Stream(auth=api.auth,
   ...:                              listener=tweet_listener)
   ...:
```

Starting the Tweet Stream

The `Stream` object's **filter method** begins the streaming process. Let's track tweets about the NASA Mars rovers. Here, we use the `track` parameter to pass a list of search terms:

```
In [9]: tweet_stream.filter(track=['Mars Rover'], is_async=True)
```

The Streaming API will return full tweet JSON objects for tweets that match any of the terms, not just in the tweet's text, but also in @-mentions, hashtags, expanded URLs and other information that Twitter maintains in a tweet object's JSON. So, you might not see the search terms you're tracking if you look only at the tweet's text.

Asynchronous vs. Synchronous Streams

The `is_async=True` argument indicates that `filter` should initiate an **asynchronous tweet stream**. This allows your code to continue executing while your listener waits to receive tweets and is useful if you decide to terminate the stream early. When you execute an asynchronous tweet stream in IPython, you'll see the next `In []` prompt and can terminate the tweet stream by setting the `Stream` object's **running property** to `False`, as in:

```
tweet_stream.running=False
```

Without the `is_async=True` argument, `filter` initiates a **synchronous tweet stream**. In this case, IPython would display the next `In []` prompt *after* the stream terminates. Asynchronous streams are particularly handy in GUI applications so your users can continue to interact with other parts of the application while tweets arrive. The following shows a portion of the output consisting of two tweets:

```
Connection successful

Screen name: bevjoy:
   Language: en
      Status: RT @SPACEdotcom: With Mars Dust Storm Clearing, Opportunity
Rover Could Finally Wake Up https://t.co/OIRP9UyB8C https://t.co/
gTfFR3RUkG

Screen name: tourmaline1973:
   Language: en
      Status: RT @BennuBirdy: Our beloved Mars rover isn't done yet, but
she urgently needs our support! Spread the word that you want to keep
calling ou…

...
```

Other `filter` Method Parameters
Method `filter` also has parameters for refining your tweet searches by Twitter user ID numbers (to follow tweets from specific users) and by location. For details, see:

> https://developer.twitter.com/en/docs/tweets/filter-realtime/guides/
> basic-stream-parameters

Twitter Restrictions Note
Marketers, researchers and others frequently store tweets they receive from the Streaming API. If you're storing tweets, Twitter requires you to delete any message or location data for which you receive a deletion message. This will occur if a user deletes a tweet or the tweets location data after Twitter pushes that tweet to you. In each case, your listener's **on_delete** method will be called. For deletion rules and message details, see

> https://developer.twitter.com/en/docs/tweets/filter-realtime/guides/
> streaming-message-types

Self Check

1 *(Fill-In)* Rather than connecting to Twitter on each method call, a stream uses a persistent connection to _____ (that is, send) tweets to your app.
Answer: push.

2 *(True/False)* Twitter's free Streaming API sends to your app randomly selected tweets dynamically as they occur—up to a maximum of ten percent of the tweets per day.
Answer: False. Twitter's free Streaming API sends to your app randomly selected tweets dynamically as they occur—up to a maximum of *one* percent of the tweets per day.

13.14 Tweet Sentiment Analysis

In the "Natural Language Processing (NLP)" chapter, we demonstrated sentiment analysis on sentences. Many researchers and companies perform sentiment analysis on tweets. For example, political researchers might check tweet sentiment during elections season to understand how people feel about specific politicians and issues. Companies might check tweet sentiment to see what people are saying about their products and competitors' products.

In this section, we'll use the techniques introduced in the preceding section to create a script (`sentimentlistener.py`) that enables you to check the sentiment on a specific topic. The script will keep totals of all the positive, neutral and negative tweets it processes and display the results.

The script receives two command-line arguments representing the topic of the tweets you wish to receive and the number of tweets for which to check the sentiment—only those tweets that are not eliminated are counted. For viral topics, there are large numbers of retweets, which we are not counting, so it could take some time get the number of tweets you specify. You can run the script from the `ch13` folder as follows:

    ipython sentimentlistener.py football 10

which produces output like the following. Positive tweets are preceded by a +, negative tweets by a - and neutral tweets by a space:

```
- ftblNeutral: Awful game of football. So boring slow hoofball complete
waste of another 90 minutes of my life that I'll never get back #BURMUN

+ TBulmer28: I've seen 2 successful onside kicks within a 40 minute span.
I love college football

+ CMayADay12: The last normal Sunday for the next couple months. Don't
text me don't call me. I am busy. Football season is finally here?

  rpimusic: My heart legitimately hurts for Kansas football fans

+ DSCunningham30: @LeahShieldsWPSD It's awsome that u like college
football, but my favorite team is ND - GO IRISH!!!

  damanr: I'm bummed I don't know enough about football to roast
@samesfandiari properly about the Raiders

+ jamesianosborne: @TheRochaSays @WatfordFC @JackHind Haha.... just when
you think an American understands Football.... so close. Wat…

+ Tshanerbeer: @PennStateFball @PennStateOnBTN Ah yes, welcome back
college football. You've been missed.

- cougarhokie: @hokiehack @skiptyler I can verify the badness of that
football

+ Unite_Reddevils: @Pablo_di_Don Well make yourself clear it's football
not soccer we follow European football not MLS soccer

Tweet sentiment for "football"
Positive: 6
 Neutral: 2
Negative: 2
```

The script (sentimentlistener.py) is presented below. We focus only on the new capabilities in this example.

Imports
Lines 4–8 import the keys.py file and the libraries used throughout the script:

```
1  # sentimentlisener.py
2  """Script that searches for tweets that match a search string
3  and tallies the number of positive, neutral and negative tweets."""
4  import keys
5  import preprocessor as p
6  import sys
7  from textblob import TextBlob
8  import tweepy
9
```

Class SentimentListener: __init__ Method
In addition to the API object that interacts with Twitter, the __init__ method receives three additional parameters:

- sentiment_dict—a dictionary in which we'll keep track of the tweet sentiments,

- **topic**—the topic we're searching for so we can ensure that it appears in the tweet text and
- **limit**—the number of tweets to process (not including the ones we eliminate).

Each of these is stored in the current SentimentListener object (self).

```
10  class SentimentListener(tweepy.StreamListener):
11      """Handles incoming Tweet stream."""
12
13      def __init__(self, api, sentiment_dict, topic, limit=10):
14          """Configure the SentimentListener."""
15          self.sentiment_dict = sentiment_dict
16          self.tweet_count = 0
17          self.topic = topic
18          self.TWEET_LIMIT = limit
19
20          # set tweet-preprocessor to remove URLs/reserved words
21          p.set_options(p.OPT.URL, p.OPT.RESERVED)
22          super().__init__(api)  # call superclass's init
23
```

Method on_status

When a tweet is received, method on_status:

- gets the tweet's text (lines 27–30)
- skips the tweet if it's a retweet (lines 33–34)
- cleans the tweet to remove URLs and reserved words like RT and FAV (line 36)
- skips the tweet if it does not have the topic in the tweet text (lines 39–40)
- uses a TextBlob to check the tweet's sentiment and updates the sentiment_dict accordingly (lines 43–52)
- prints the tweet text (line 55) preceded by + for positive sentiment, space for neutral sentiment or - for negative sentiment and
- checks whether we've processed the specified number of tweets yet (lines 57–60).

```
24      def on_status(self, status):
25          """Called when Twitter pushes a new tweet to you."""
26          # get the tweet's text
27          try:
28              tweet_text = status.extended_tweet.full_text
29          except:
30              tweet_text = status.text
31
32          # ignore retweets
33          if tweet_text.startswith('RT'):
34              return
35
36          tweet_text = p.clean(tweet_text)  # clean the tweet
37
38          # ignore tweet if the topic is not in the tweet text
39          if self.topic.lower() not in tweet_text.lower():
40              return
```

```
41
42             # update self.sentiment_dict with the polarity
43             blob = TextBlob(tweet_text)
44             if blob.sentiment.polarity > 0:
45                 sentiment = '+'
46                 self.sentiment_dict['positive'] += 1
47             elif blob.sentiment.polarity == 0:
48                 sentiment = ' '
49                 self.sentiment_dict['neutral'] += 1
50             else:
51                 sentiment = '-'
52                 self.sentiment_dict['negative'] += 1
53
54             # display the tweet
55             print(f'{sentiment} {status.user.screen_name}: {tweet_text}\n')
56
57             self.tweet_count += 1  # track number of tweets processed
58
59             # if TWEET_LIMIT is reached, return False to terminate streaming
60             return self.tweet_count != self.TWEET_LIMIT
61
```

Main Application

The main application is defined in the function main (lines 62–87; discussed after the following code), which is called by lines 90–91 when you execute the file as a script. So sentimentlistener.py can be imported into IPython or other modules to use class SentimentListener as we did with TweetListener in the previous section:

```
62  def main():
63      # configure the OAuthHandler
64      auth = tweepy.OAuthHandler(keys.consumer_key, keys.consumer_secret)
65      auth.set_access_token(keys.access_token, keys.access_token_secret)
66
67      # get the API object
68      api = tweepy.API(auth, wait_on_rate_limit=True,
69                      wait_on_rate_limit_notify=True)
70
71      # create the StreamListener subclass object
72      search_key = sys.argv[1]
73      limit = int(sys.argv[2])  # number of tweets to tally
74      sentiment_dict = {'positive': 0, 'neutral': 0, 'negative': 0}
75      sentiment_listener = SentimentListener(api,
76          sentiment_dict, search_key, limit)
77
78      # set up Stream
79      stream = tweepy.Stream(auth=api.auth, listener=sentiment_listener)
80
81      # start filtering English tweets containing search_key
82      stream.filter(track=[search_key], languages=['en'], is_async=False)
83
84      print(f'Tweet sentiment for "{search_key}"')
85      print('Positive:', sentiment_dict['positive'])
86      print(' Neutral:', sentiment_dict['neutral'])
87      print('Negative:', sentiment_dict['negative'])
88
```

```
89    # call main if this file is executed as a script
90    if __name__ == '__main__':
91        main()
```

Lines 72–73 get the command-line arguments. Line 74 creates the `sentiment_dict` dictionary that keeps track of the tweet sentiments. Lines 75–76 create the `SentimentListener`. Line 79 creates the `Stream` object. We once again initiate the stream by calling `Stream` method `filter` (line 82). However, this example uses a synchronous stream so that lines 84–87 display the sentiment report only after the specified number of tweets (`limit`) are processed. In this call to `filter`, we also provided the keyword argument `languages`, which specifies a list of language codes. The one language code `'en'` indicates Twitter should return only English language tweets.

13.15 Geocoding and Mapping

In this section, we'll collect streaming tweets, then plot the locations of those tweets. Most tweets do not include latitude and longitude coordinates, because Twitter disables this by default for all users. Those who wish to include their precise location in tweets must opt into that feature. Though most tweets do not include precise location information, a large percentage include the user's home location information; however, even that is sometimes invalid, such as "Far Away" or a fictitious location from a user's favorite movie.

In this section, for simplicity, we'll use the `location` property of the tweet's `User` object to plot that user's location on an interactive map. The map will let you zoom in and out and drag to move the map around so you can look at different areas (known as *panning*). For each tweet, we'll display a map marker that you can click to see a popup containing the user's screen name and tweet text.

We'll ignore retweets and tweets that do not contain the search topic. For other tweets, we'll track the percentage of tweets with location information. When we get the latitude and longitude information for those locations, we'll also track the percentage of those tweets that had invalid location data.

geopy Library

We'll use the **geopy library** (https://github.com/geopy/geopy) to translate locations into latitude and longitude coordinates—known as **geocoding**—so we can place markers on a map. The library supports dozens of geocoding web services, many of which have free or lite tiers. For this example, we'll use the **OpenMapQuest geocoding service** (discussed shortly). You installed geopy in Section 13.6.

OpenMapQuest Geocoding API

We'll use the OpenMapQuest Geocoding API to convert locations, such as Boston, MA into their latitudes and longitudes, such as 42.3602534 and -71.0582912, for plotting on maps. OpenMapQuest currently allows 15,000 transactions per month on their free tier. To use the service, first sign up at

   https://developer.mapquest.com/

Once logged in, go to

   https://developer.mapquest.com/user/me/apps

and click **Create a New Key**, fill in the **App Name** field with a name of your choosing, leave the **Callback URL** empty and click **Create App** to create an API key. Next, click your app's name in the web page to see your consumer key. In the keys.py file you used earlier in the chapter, store the consumer key by replacing *YourKeyHere* in the line

```
mapquest_key = 'YourKeyHere'
```

As we did earlier in the chapter, we'll import keys.py to access this key.

Folium Library and Leaflet.js JavaScript Mapping Library
For the maps in this example, we'll use the **folium library**

```
https://github.com/python-visualization/folium
```

which uses the popular Leaflet.js JavaScript mapping library to display maps. The maps that folium produces are saved as HTML files that you can view in your web browser. To install folium, execute the following command:

```
pip install folium
```

Maps from OpenStreetMap.org
By default, Leaflet.js uses open source maps from OpenStreetMap.org. These maps are copyrighted by the OpenStreetMap.org contributors. To use these maps[25], they require the following copyright notice:

```
Map data © OpenStreetMap contributors
```

and they state:

> *You must make it clear that the data is available under the Open Database License. This can be achieved by providing a "License" or "Terms" link which links to www.openstreetmap.org/copyright or www.opendatacommons.org/licenses/odbl.*

Self Check

1 *(Fill-In)* The geopy library enables you to translate locations into latitude and longitude coordinates, known as _____, so you can plot locations on a map.
Answer: geocoding

2 *(Fill-In)* The OpenMapQuest Geocoding API converts locations, like Boston, MA into their _____ and _____ for plotting on maps.
Answer: latitudes, longitudes.

13.15.1 Getting and Mapping the Tweets

Let's interactively develop the code that plots tweet locations. We'll use utility functions from our tweetutilities.py file and class LocationListener in locationlistener.py. We'll explain the details of the utility functions and class in the subsequent sections.

---

25. https://wiki.osmfoundation.org/wiki/Licence/Licence_and_Legal_FAQ.

13.15 Geocoding and Mapping 553

Get the API Object
As in the other streaming examples, let's authenticate with Twitter and get the Tweepy API object. In this case, we do this via the get_API utility function in tweetutilities.py:

```
In [1]: from tweetutilities import get_API

In [2]: api = get_API()
```

Collections Required By LocationListener
Our LocationListener class requires two collections: A list (tweets) to store the tweets we collect and a dictionary (counts) to track the total number of tweets we collect and the number that have location data:

```
In [3]: tweets = []

In [4]: counts = {'total_tweets': 0, 'locations': 0}
```

Creating the LocationListener
For this example, the LocationListener will collect 50 tweets about 'football':

```
In [5]: from locationlistener import LocationListener

In [6]: location_listener = LocationListener(api, counts_dict=counts,
   ...:     tweets_list=tweets, topic='football', limit=50)
   ...:
```

The LocationListener will use our utility function get_tweet_content to extract the screen name, tweet text and location from each tweet, place that data in a dictionary.

Configure and Start the Stream of Tweets
Next, let's set up our Stream to look for English language 'football' tweets:

```
In [7]: import tweepy

In [8]: stream = tweepy.Stream(auth=api.auth, listener=location_listener)

In [9]: stream.filter(track=['football'], languages=['en'], is_async=False)
```

Now wait to receive the tweets. Though we do not show them here (to save space), the LocationListener displays each tweet's screen name and text so you can see the live stream. If you're not receiving any (perhaps because it is not football season), you might want to type *Ctrl* + *C* to terminate the previous snippet then try again with a different search term.

Displaying the Location Statistics
When the next In [] prompt displays, we can check how many tweets we processed, how many had locations and the percentage that had locations:

```
In [10]: counts['total_tweets']
Out[10]: 63

In [11]: counts['locations']
Out[11]: 50
```

```
In [12]: print(f'{counts["locations"] / counts["total_tweets"]:.1%}')
79.4%
```
In this particular execution, 79.4% of the tweets contained location data.

Geocoding the Locations
Now, let's use our `get_geocodes` utility function from `tweetutilities.py` to geocode the location of each tweet stored in the list tweets:
```
In [13]: from tweetutilities import get_geocodes

In [14]: bad_locations = get_geocodes(tweets)
Getting coordinates for tweet locations...
OpenMapQuest service timed out. Waiting.
OpenMapQuest service timed out. Waiting.
Done geocoding
```
Sometimes the OpenMapQuest geocoding service times out, meaning that it cannot handle your request immediately and you need to try again. In that case, our function `get_geocodes` displays a message, waits for a short time, then retries the geocoding request.

As you'll soon see, for each tweet with a *valid* location, the `get_geocodes` function adds to the tweet's dictionary in the tweets list two new keys—`'latitude'` and `'longitude'`. For the corresponding values, the function uses the tweet's coordinates that OpenMapQuest returns.

Displaying the Bad Location Statistics
When the next `In []` prompt displays, we can check the percentage of tweets that had invalid location data:
```
In [15]: bad_locations
Out[15]: 7

In [16]: print(f'{bad_locations / counts["locations"]:.1%}')
14.0%
```
In this case, of the 50 tweets with location data, 7 (14%) had invalid locations.

Cleaning the Data
Before we plot the tweet locations on a map, let's use a pandas `DataFrame` to clean the data. When you create a `DataFrame` from the tweets list, it will contain the value `NaN` for the `'latitude'` and `'longitude'` of any tweet that did not have a valid location. We can remove any such rows by calling the `DataFrame`'s **dropna method**:
```
In [17]: import pandas as pd

In [18]: df = pd.DataFrame(tweets)

In [19]: df = df.dropna()
```

Creating a Map with Folium
Now, let's create a folium **Map** on which we'll plot the tweet locations:

```
In [20]: import folium

In [21]: usmap = folium.Map(location=[39.8283, -98.5795],
    ...:                   tiles='Stamen Terrain',
    ...:                   zoom_start=5, detect_retina=True)
    ...:
```

The `location` keyword argument specifies a sequence containing latitude and longitude coordinates for the map's center point. The values above are the geographic center of the continental United States (http://bit.ly/CenterOfTheUS). It's possible that some of the tweets we plot will be outside the U.S. In this case, you will not see them initially when you open the map. You can zoom in and out using the + and - buttons at the top-left of the map, or you can pan the map by dragging it with the mouse to see anywhere in the world.

The `zoom_start` keyword argument specifies the map's initial zoom level, lower values show more of the world and higher values show less. On our system, 5 displays the entire continental United States. The `detect_retina` keyword argument enables folium to detect high-resolution screens. When it does, it requests higher-resolution maps from OpenStreetMap.org and changes the zoom level accordingly.

Creating Popup Markers for the Tweet Locations
Next, let's iterate through the `DataFrame` and add to the `Map` folium `Popup` objects containing each tweet's text. In this case, we'll use method **itertuples** to create tuples from each row of the `DataFrame`. Each tuple will contain a property for each `DataFrame` column:

```
In [22]: for t in df.itertuples():
    ...:     text = ': '.join([t.screen_name, t.text])
    ...:     popup = folium.Popup(text, parse_html=True)
    ...:     marker = folium.Marker((t.latitude, t.longitude),
    ...:                            popup=popup)
    ...:     marker.add_to(usmap)
    ...:
```

First, we create a string (`text`) containing the user's `screen_name` and tweet `text` separated by a colon. This will be displayed on the map if you click the corresponding marker. The second statement creates a folium **Popup** to display the text. The third statement creates a folium **Marker** object using a tuple to specify the `Marker`'s latitude and longitude. The popup keyword argument associates the tweet's `Popup` object with the new `Marker`. Finally, the last statement calls the `Marker`'s **add_to method** to specify the `Map` that will display the `Marker`.

Saving the Map
The last step is to call the `Map`'s **save** method to store the map in an HTML file, which you can then double click to open in your web browser:

```
In [23]: usmap.save('tweet_map.html')
```

The resulting map follows. The `Markers` on your map will differ:

556 Data Mining Twitter

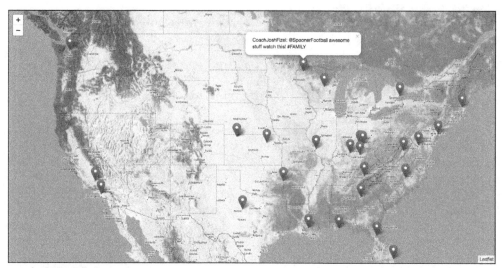

Map data © OpenStreetMap contributors.
The data is available under the Open Database License www.openstreetmap.org/copyright.

✓ Self Check

1 *(Fill-In)* The folium classes _____ and _____ enable you to mark locations on a map and add text that displays when the user clicks a marked location.
Answer: Marker, Popup.

2 *(Fill-In)* Pandas `DataFrame` method _____ creates an iterator for accessing the rows of a `DataFrame` as tuples.
Answer: itertuples.

13.15.2 Utility Functions in tweetutilities.py

Here we present the utility functions `get_tweet_content` and `get_geo_codes` used in the preceding section's IPython session. In each case, the line numbers start from 1 for discussion purposes. These are both defined in tweetutilities.py, which is included in the ch13 examples folder.

get_tweet_content Utility Function

Function `get_tweet_content` receives a `Status` object (tweet) and creates a dictionary containing the tweet's `screen_name` (line 4), `text` (lines 7–10) and `location` (lines 12–13). The location is included only if the `location` keyword argument is `True`. For the tweet's text, we try to use the `full_text` property of an `extended_tweet`. If it's not available, we use the `text` property:

```
1  def get_tweet_content(tweet, location=False):
2      """Return dictionary with data from tweet (a Status object)."""
3      fields = {}
4      fields['screen_name'] = tweet.user.screen_name
5
```

13.15 Geocoding and Mapping

```
6        # get the tweet's text
7        try:
8            fields['text'] = tweet.extended_tweet.full_text
9        except:
10           fields['text'] = tweet.text
11
12       if location:
13           fields['location'] = tweet.user.location
14
15       return fields
```

get_geocodes Utility Function

Function `get_geocodes` receives a list of dictionaries containing tweets and geocodes their locations. If geocoding is successful for a given tweet, the function adds the latitude and longitude to the corresponding tweet's dictionary in `tweet_list`. This code requires class **OpenMapQuest** from the geopy module, which we import into the file tweetutilities.py as follows:

```
from geopy import OpenMapQuest
```

```
1   def get_geocodes(tweet_list):
2       """Get the latitude and longitude for each tweet's location.
3       Returns the number of tweets with invalid location data."""
4       print('Getting coordinates for tweet locations...')
5       geo = OpenMapQuest(api_key=keys.mapquest_key)  # geocoder
6       bad_locations = 0
7
8       for tweet in tweet_list:
9           processed = False
10          delay = .1  # used if OpenMapQuest times out to delay next call
11          while not processed:
12              try:  # get coordinates for tweet['location']
13                  geo_location = geo.geocode(tweet['location'])
14                  processed = True
15              except:  # timed out, so wait before trying again
16                  print('OpenMapQuest service timed out. Waiting.')
17                  time.sleep(delay)
18                  delay += .1
19
20          if geo_location:
21              tweet['latitude'] = geo_location.latitude
22              tweet['longitude'] = geo_location.longitude
23          else:
24              bad_locations += 1  # tweet['location'] was invalid
25
26      print('Done geocoding')
27      return bad_locations
```

The function operates as follows:

- Line 5 creates the `OpenMapQuest` object we'll use to geocode locations. The api_key keyword argument is loaded from the keys.py file you edited earlier.

- Line 6 initializes `bad_locations` which we use to keep track of the number of invalid locations in the tweet objects we collected.

- In the loop, lines 9–18 attempt to geocode the current tweet's location. As we mentioned, sometimes the OpenMapQuest geocoding service will time out, meaning that it's temporarily unavailable. This can happen if you make too many requests too quickly. For this reason, the `while` loop continues executing as long as processed is `False`. In each iteration, this loop calls the `OpenMapQuest` object's **geocode method** with the tweet's location string as an argument. If successful, processed is set to `True` and the loop terminates. Otherwise, lines 16–18 display a time-out message, tell the loop to wait for `delay` seconds and increase the delay in case we get another time out. Line 17 calls the Python Standard Library `time` module's `sleep` method to pause the code execution.

- After the `while` loop terminates, lines 20–24 check whether location data was returned and, if so, add it to the tweet's dictionary. Otherwise, line 24 increments the `bad_locations` counter.

- Finally, the function prints a message that it's done geocoding and returns the `bad_locations` value.

Self Check

1. *(IPython Session)* Use an OpenMapQuest geocoding object to get the latitude and Longitude for Chicago, IL.
Answer:

```
In [1]: import keys

In [2]: from geopy import OpenMapQuest

In [3]: geo = OpenMapQuest(api_key=keys.mapquest_key)

In [4]: geo.geocode('Chicago, IL')
Out[4]: Location(Chicago, Cook County, Illinois, United States of America, (41.8755546, -87.6244212, 0.0))
```

13.15.3 Class LocationListener

Class `LocationListener` performs many of the same tasks we demonstrated in the prior streaming examples, so we'll focus on just a few lines in this class:

```
1  # locationlistener.py
2  """Receives tweets matching a search string and stores a list of
3  dictionaries containing each tweet's screen_name/text/location."""
4  import tweepy
5  from tweetutilities import get_tweet_content
6
7  class LocationListener(tweepy.StreamListener):
8      """Handles incoming Tweet stream to get location data."""
9
10     def __init__(self, api, counts_dict, tweets_list, topic, limit=10):
11         """Configure the LocationListener."""
12         self.tweets_list = tweets_list
13         self.counts_dict = counts_dict
14         self.topic = topic
15         self.TWEET_LIMIT = limit
16         super().__init__(api)  # call superclass's init
```

```python
17
18      def on_status(self, status):
19          """Called when Twitter pushes a new tweet to you."""
20          # get each tweet's screen_name, text and location
21          tweet_data = get_tweet_content(status, location=True)
22
23          # ignore retweets and tweets that do not contain the topic
24          if (tweet_data['text'].startswith('RT') or
25              self.topic.lower() not in tweet_data['text'].lower()):
26              return
27
28          self.counts_dict['total_tweets'] += 1  # original tweet
29
30          # ignore tweets with no location
31          if not status.user.location:
32              return
33
34          self.counts_dict['locations'] += 1  # tweet with location
35          self.tweets_list.append(tweet_data)  # store the tweet
36          print(f'{status.user.screen_name}: {tweet_data["text"]}\n')
37
38          # if TWEET_LIMIT is reached, return False to terminate streaming
39          return self.counts_dict['locations'] != self.TWEET_LIMIT
```

In this case, the __init__ method receives a counts dictionary that we use to keep track of the total number of tweets processed and a tweet_list in which we store the dictionaries returned by the get_tweet_content utility function.

Method on_status:

- Calls get_tweet_content to get the screen name, text and location of each tweet.

- Ignores the tweet if it is a retweet or if the text does not include the topic we're searching for—we'll use only original tweets containing the search string.

- Adds 1 to the value of the 'total_tweets' key in the counts dictionary to track the number of original tweets we process.

- Ignores tweets that have no location data.

- Adds 1 to the value of the 'locations' key in the counts dictionary to indicate that we found a tweet with a location.

- Appends to the tweets_list the tweet_data dictionary that get_tweet_content returned.

- Displays the tweet's screen name and tweet text so you can see that the app is making progress.

- Checks whether the TWEET_LIMIT has been reached and, if so, returns False to terminate the stream.

13.16 Ways to Store Tweets

For analysis, you'll commonly store tweets in:

- CSV files—A file format that we introduced in the "Files and Exceptions" chapter.

- pandas `DataFrames` in memory—CSV files can be loaded easily into `DataFrames` for cleaning and manipulation.
- SQL databases—Such as MySQL, a free and open source relational database management system (RDBMS).
- NoSQL databases—Twitter returns tweets as JSON documents, so the natural way to store them is in a NoSQL JSON document database, such as MongoDB. Tweepy generally hides the JSON from the developer. If you'd like to manipulate the JSON directly, use the techniques we present in the "Big Data: Hadoop, Spark, NoSQL and IoT Databases" chapter, where we'll look at the PyMongo library.

13.17 Twitter and Time Series

A time series is a sequence of values with timestamps. Some examples are daily closing stock prices, daily high temperatures at a given location, monthly U.S. job-creation numbers, quarterly earnings for a given company and more. Tweets are natural for time-series analysis because they're time stamped. In the "Machine Learning" chapter, we'll use a technique called simple linear regression to make predictions with time series. We'll take another look at time series in the "Deep Learning" chapter when we study recurrent neural networks.

13.18 Wrap-Up

In this chapter, we explored data mining Twitter, perhaps the most open and accessible of all the social media sites, and one of the most commonly used big-data sources. You created a Twitter developer account and connected to Twitter using your account credentials. We discussed Twitter's rate limits and some additional rules, and the importance of conforming to them.

We looked at the JSON representation of a tweet. We used Tweepy—one of the most widely used Twitter API clients—to authenticate with Twitter and access its APIs. We saw that tweets returned by the Twitter APIs contain much metadata in addition to a tweet's text. We determined an account's followers and whom an account follows, and looked at a user's recent tweets.

We used Tweepy `Cursors` to conveniently request successive pages of results from various Twitter APIs. We used Twitter's Search API to download past tweets that met specified criteria. We used Twitter's Streaming API to tap into the flow of live tweets as they happened. We used the Twitter Trends API to determine trending topics for various locations and created a word cloud from trending topics.

We used the tweet-preprocessor library to clean and preprocess tweets to prepare them for analysis, and performed sentiment analysis on tweets. We used the folium library to create a map of tweet locations and interacted with it to see the tweets at particular locations. We enumerated common ways to store tweets and noted that tweets are a natural form of time series data. In the next chapter, we'll study IBM's Watson and its cognitive computing capabilities.

Exercises

13.1 *(Percentage of English Tweets)* Twitter is truly an international social network. Use the Twitter search API to look at 10,000 tweets. Look at each tweet's `lang` property. Count and display the number of tweets in each language.

13.2 *(Percentage of Retweets)* Look at 10,000 tweets and determine the percentage of tweets that begin with Twitter's reserved word `RT` (for retweet).

13.3 *(Percentage of Extended Tweets)* Look at 10,000 tweets and determine what percentage of them are extended tweets.

13.4 *(Basic Account Information)* Get the ID, name, screen name and description of a Twitter account of interest to you.

13.5 *(User Timeline)* Get the last 10 tweets from an account of interest to you.

13.6 *(Sentiment Analysis)* When searching for tweets, you can include :) or :(to look for positive and negative tweets, respectively. Perform searches for 10 positive tweets and 10 negative tweets, then use `TextBlob` sentiment analysis to confirm that each is positive or negative.

13.7 *(Condensing Tweet Objects)* You've already seen a complete JSON representation of a typical tweet. That's about 9000 characters of information for the new 280-character tweet text limit. When you work with Tweepy it forms a large `Status` object. For most applications you'll need a relatively small number of that object's properties. Write a script that will extract only a small subset of a tweet's common properties and place those in a CSV file.

13.8 *(Trends Bar Chart Using Pandas)* Use the pandas plotting you learned in the "Natural Language Processing (NLP)" chapter to create a bar chart showing the tweet counts for Twitter's trending topics in a city of your choice.

13.9 *(Trending Topics Word Cloud)* Use the Twitter Trends API to determine the locations for which Twitter has trending topics. Pick one of the locations and display its trending-topics list.

13.10 *(Tweet Mapping Modification)* In this chapter's tweet mapping example, for simplicity, we used the `location` property of a `Status` object to grab the user's location. Another level of location is to check the tweet object's `coordinates` property to see if it contains latitude and longitude information. This field is included in a small percentage of tweets. Update your code to look only at tweets with `coordinates` and use those to plot the map. You might need to look through a large number of tweets before you have enough information to make the map worthwhile. Count the number of tweets you find and divide by the total number of tweets you received to determine the percentage of tweets that included latitude and longitude information directly.

13.11 *(Project: Mapping Only Tweets Inside the Continental U.S.)* Look at geopy's supported geocoding APIs in its online documentation. Locate one that supports reverse geocoding in which you provide the coordinates to the geocoder object's `reverse` method and it returns the location. Display and study the JSON properties in the result. Next, modify this chapter's mapping example to use this capability. Check each tweet's location and plot on a map only those tweets inside the continental U.S.

13.12 *(Project: Twitter Geo API)* Use the Twitter Geo API's `reverse_geocode` method to locate up to 20 places near the latitude and longitude 47.6205, -122.3493 (the Seattle Space Needle, built for the 1962 World's Fair).

13.13 *(Project: Twitter Geo API)* Use the Twitter Geo API's `search` method to locate places near the Eiffel Tower. This method can receive latitude and longitude, a place name or an IP address.

13.14 *(Project: Twitter Geo API)* The results returned by the `reverse_geocode` and `search` methods in the two previous exercises include place IDs. Use the Twitter Geo API's `place_id` method to get the information for each of the places returned.

13.15 *(Project: Heat Maps with Folium)* In this chapter, you used the folium library to create an interactive map showing tweet locations. Investigate creating heat maps with folium. Build a folium heat map showing the tweeting activity on a given subject throughout the United States.

13.16 *(Project: Live Translating the Flow of Tweets to English)* Twitter is a global network. Use Twitter and the language translation services you learned in the "Natural Language Processing (NLP)" chapter to data mine tweets for a Spanish-speaking city. In particular, get the trending topics list then stream 10 tweets on that city's top trending topic. Use `TextBlob` to translate the tweets to English.

13.17 *(Project: Data Mining Foreign Language Tweets)* Add this capability into one of your existing examples. Enhance your application with the language-translation services you'll learn in the next chapter, "IBM Watson and Cognitive Computing."

13.18 *(Project: Tweet Cleaner/Preprocessor)* Section 13.12 discussed cleaning and preprocessing tweets and demonstrated basic cleaning with the tweet-preprocessor library. Use the search API to get 100 tweets on a topic of your choice. Preprocess the tweets using all of tweet-preprocessor's features. Then, investigate and use `TextBlob`'s `lowerstrip` utility function to remove all punctuation and convert the text to lowercase letters. Display the original and cleaned version of each tweet.

13.19 *(Project: Data Mining Facebook)* Now that you're familiar with data mining Twitter, research data mining Facebook and implement several examples like those here in this chapter. Develop some examples of data mining with capabilities unique to Facebook.

13.20 *(Project: Data Mining LinkedIn)* Now that you're familiar with data mining Twitter, research data mining LinkedIn and implement several examples like those here in this chapter. Develop some examples of data mining with capabilities unique to the LinkedIn social network, especially those for professional people.

13.21 *(Project: Predicting the Stock Market with Twitter)* Many articles and research papers have been published on predicting the stock market with Twitter. Some of the approaches are quite mathematical. Choose a few public companies listed on the major stock exchanges. Use sentiment analysis with tweets mentioning these companies. Based on the strength of the sentiment values, determine what recommendations you would have made for buying and selling the securities of these companies. Would these trades have been profitable? If you're successful with stocks, you may want to apply a similar approach to the bond and commodities markets.

13.22 *(Project: Hedge Funds Use Twitter to Predict the Securities Markets)* Some hedge funds employ powerful computer equipment and sophisticated software to predict the securities markets. They must distinguish between correct information about companies and their products, and fake information from people who are trying to influence stock prices. Research the kinds of things this software should find. Implement a system for detecting fake information.

13.23 *(Project: Predicting Movie Revenues)* Research "Using Twitter to Predict How Well New Movies Will Do at the Box Office." Try to do this only with the techniques you've learned so far in this book. You may want to refine your effort with techniques you'll learn in the forthcoming "Machine Learning" and "Deep Learning" chapters. You can use similar techniques to predict the success of stage plays, TV programs and products of all kinds. The quality of these kinds of predictions will surely improve with time. Eventually, it's reasonable to expect that the product design process will be influenced by what is learned from years of prediction efforts.

13.24 *(Project: Generating the Social Graph)* Because you can look at whom a Twitter account follows and who follows that account, you can build "social graphs" showing the relationships among Twitter accounts. Study the NetworkX tool. Write a script that uses NetworkX to draw the social graph of a small "sub-community" in Twitter.

13.25 *(Project: Using Twitter to Predict Elections)* Research online "Predicting Elections with Twitter." Develop and test your approach on local, statewide and/or national elections. Try refining your approach after you study the "Machine Learning" and "Deep Learning" chapters.

13.26 *(Project: Predicting a User's Gender on Twitter)* A person's gender often is valuable to marketers. Try determining gender from tweet text by using the techniques you've learned so far. Later, try using the techniques you'll learn in the "Machine Learning" and "Deep Learning" chapters. Always check Twitters latest rules and regulations to be sure you're not compromising a user's privacy or other rights.

13.27 *(Project: Using Twitter to Predict If a User Is Conservative or Liberal)* This kind of information is valuable to people who run political campaigns. Try doing this with the techniques you've learned so far. Then try using the techniques you'll learn in the "Machine Learning" and "Deep Learning" chapters. Always check Twitter's latest rules and regulations to be sure you're not compromising a user's privacy or other rights.

13.28 *(Project: Using Twitter Find Job Opportunities)* Many companies encourage their employees to tweet regularly about ongoing development efforts and job opportunities. Analyze the tweet streams of a possibly large number of companies in your field companies to determine if the specific projects they're doing interests you.

13.29 *(Project: Using Twitter to Examine Tweets By Congressional District)* Investigate the site govtrack.us, which includes the statement, "You are encouraged to reuse any material on this site." Analyze the trending topics in key cities in several congressional districts of interest to you. Try to determine from the tweets the relative percentages of Democrats, Republicans and Independents in each district. Research the term "gerrymandering," which is often used in a negative context, to see how politicians have used changes in these percentages over time for political advantage. Find instances of where gerrymandering has been used in a positive context.

13.30 *(Project: Accessing the YouTube API)* In this chapter, you used web services to access Twitter through its APIs. The hugely popular YouTube website serves up billions of videos per day. Look for Python libraries that conveniently access the YouTube APIs, then use them to integrate YouTube videos into one of your Twitter applications. You might, for example, display YouTube videos for trending topics.

13.31 *(Project: Tracking Natural Disasters with Twitter and Spatial Data)* Research spatial data, then use Twitter and spatial data to implement a system for tracking natural disasters like hurricanes, earthquakes and tornadoes.

13.32 *(Project: Twitter Sentiment Analysis with Emoticons)* Emoticons *scream* emotions, making them useful for sentiment analysis. Identify common emoticons as positive, negative or neutral, then look for them in tweets and use them to classify the sentiment of those tweets.

13.33 *(Project: Tweet Normalization—Expanding Common Abbreviations)* Search for common social media abbreviations and expansions. Add expanding common abbreviations to your tweet preprocessing script. Find tools that do these expansions. Some of the tools are likely to be domain specific.

13.34 *(Project: Tweet Normalization—Shortening "Stretched Words")* Shorten "stretched words" like "sooooooo" to "so." Make a list of stretched words commonly used in social media.

13.35 *(Project: Sentiment Analysis of Streaming Tweets)* Stream tweets during an event and note how sentiment changes throughout the event.

13.36 *(Project: Finding Positive and Negative Sentiment Words)* There are lots of free and open source sentiment datasets online, such as IMDB (the Internet Movie Database) and others. Many of these have labeled descriptions of movies, airline service, and more, with sentiment tags, such as positive, negative and neutral. Analyze one or more of these datasets. Find the most common words used in the positive sentiment descriptions and the most common words in the negative sentiment descriptions. Then, search through tweets looking for these positive and negative words. Based on the matches, decide whether the tweets have positive or negative sentiment. Compare your sentiment results to what `TextBlob` returns for each tweet.

13.37 *(For the Entrepreneur)* Check out `business.twitter.com`. Research Twitter business applications. Developer a Twitter-based business application.

13.38 *(Uber Visualization Video)* In this chapter, we visualized tweets on a map. To learn more about visualizing live data, watch the following visualization video to see how Uber is using visualization to optimize their business:

> https://www.youtube.com/watch?v=nLy3OQYsXWA

IBM Watson and Cognitive Computing

Objectives

In this chapter, you'll:

- Learn Watson's range of services and use their Lite tier to become familiar with them at no charge.
- Try lots of demos of Watson services.
- Understand what cognitive computing is and how you can incorporate it into your applications.
- Register for an IBM Cloud account and get credentials to use various services.
- Install the Watson Developer Cloud Python SDK to interact with Watson services.
- Develop a traveler's companion language translator app by using Python to weave together a mashup of the Watson Speech to Text, Language Translator and Text to Speech services.
- Check out additional resources, such as IBM Watson Redbooks that will help you jump start your custom Watson application development.

Outline

14.1 Introduction: IBM Watson and Cognitive Computing	14.6 Case Study: Traveler's Companion Translation App
14.2 IBM Cloud Account and Cloud Console	14.6.1 Before You Run the App
	14.6.2 Test-Driving the App
14.3 Watson Services	14.6.3 `SimpleLanguageTranslator.py` Script Walkthrough
14.4 Additional Services and Tools	14.7 Watson Resources
14.5 Watson Developer Cloud Python SDK	14.8 Wrap-Up
	Exercises

14.1 Introduction: IBM Watson and Cognitive Computing

In Chapter 1, we discussed some key IBM artificial-intelligence accomplishments, including beating the two best human Jeopardy players in a $1 million match. Watson won the competition and IBM donated the prize money to charity. Watson simultaneously executed hundreds of language-analysis algorithms to locate correct answers in 200 million pages of content (including all of Wikipedia) requiring four terabytes of storage.[1,2] IBM researchers trained Watson using machine-learning and reinforcement-learning techniques (which we discuss in upcoming chapters).[3]

Early in our research for this book, we recognized the rapidly growing importance of Watson, so we placed Google Alerts on Watson and related topics. Through those alerts and the newsletters and blogs we follow, we accumulated 900+ current Watson-related articles, documentation pieces and videos. We investigated many competitive services and found Watson's "no credit card required" policy and free *Lite tier* services[4] to be among friendliest to people who'd like to experiment with Watson's services at no charge.

IBM Watson is a cloud-based cognitive-computing platform being employed across a wide range of real-world scenarios. Cognitive-computing systems simulate the pattern-recognition and decision-making capabilities of the human brain to "learn" as they consume more data.[5,6,7] We overview Watson's broad range of web services and provide a hands-on Watson treatment, demonstrating many Watson capabilities. The table on the next page shows just a few of the ways in which organizations are using Watson.

Watson offers an intriguing set of capabilities that you can incorporate into your applications. In this chapter, you'll set up an IBM Cloud account[8] and use the *Lite tier* and IBM's Watson demos to experiment with various web services, such as natural language translation, speech-to-text, text-to-speech, natural language understanding, chatbots, analyzing text for tone and visual object recognition in images and video. We'll briefly overview some additional Watson services and tools.

---

1. https://www.techrepublic.com/article/ibm-watson-the-inside-story-of-how-the-jeopardy-winning-supercomputer-was-born-and-what-it-wants-to-do-next/.
2. https://en.wikipedia.org/wiki/Watson_(computer).
3. https://www.aaai.org/Magazine/Watson/watson.php, *AI Magazine*, Fall 2010.
4. Always check the latest terms on IBM's website as the terms and services may change.
5. http://whatis.techtarget.com/definition/cognitive-computing.
6. https://en.wikipedia.org/wiki/Cognitive_computing.
7. https://www.forbes.com/sites/bernardmarr/2016/03/23/what-everyone-should-know-about-cognitive-computing.
8. IBM Cloud previously was called Bluemix. You'll still see "`bluemix`" in many of this chapter's URLs.

14.1 Introduction: IBM Watson and Cognitive Computing

Watson use cases

ad targeting	fraud prevention	personal assistants
artificial intelligence	game playing	predictive maintenance
augmented intelligence	genetics	product recommendations
augmented reality	healthcare	robots and drones
chatbots	image processing	self-driving cars
closed captioning	IoT (Internet of Things)	sentiment and mood analysis
cognitive computing	language translation	smart homes
conversational interfaces	machine learning	sports
crime prevention	malware detection	supply-chain management
customer support	medical diagnosis and treatment	threat detection
detecting cyberbullying	medical imaging	virtual reality
drug development	music	voice analysis
education	natural language processing	weather forecasting
facial recognition	natural language understanding	workplace safety
finance	object recognition	

You'll install the Watson Developer Cloud Python Software Development Kit (SDK) for programmatic access to Watson services from your Python code. Then, in our hands-on implementation case study, you'll develop a traveler's companion translation app by quickly and conveniently *mashing up* several Watson services. The app enables English-only and Spanish-only speakers to communicate with one another verbally, despite the language barrier. You'll transcribe English and Spanish audio recordings to text, translate the text to the other language, then synthesize and play English and Spanish audio from the translated text. The chapter concludes with one of the richest exercise/project sets in the book, enabling you to develop Watson-based solutions to a broad range of interesting problems. Watson services can be used in many of this book's other data science chapters, so we'll include Watson exercises and projects in later chapters, too.

Watson is a dynamic and evolving set of capabilities. During the time we worked on this book, new services were added and existing services were updated and/or removed multiple times. The descriptions of the Watson services and the steps we present were accurate as of the time of this writing. We'll post updates as necessary on the book's web page at www.deitel.com.

✓ Self Check

1 *(Fill-In)* IBM researchers trained Watson using _____-learning and reinforcement-learning techniques (which we discuss in upcoming chapters).
Answer: machine.

2 *(True/False)* IBM Watson is a desktop-based cognitive-computing platform being employed across a wide range of real-world scenarios.
Answer: False. IBM Watson is *cloud*-based, not desktop-based.

3 *(True/False)* The Watson Developer Cloud Python Software Development Kit (SDK) enables you to programmatically access Watson services from your Python code.
Answer: True.

14.2 IBM Cloud Account and Cloud Console

You'll need a free IBM Cloud account to access Watson's Lite tier services. Each service's description web page lists the service's tiered offerings and what you get with each tier. Though the Lite tier services limit your use, they typically offer what you'll need to familiarize yourself with Watson features and begin using them to develop apps. The limits are subject to change, so rather than list them here, we point you to each service's web page. IBM increased the limits significantly on some services while we were writing this book. Paid tiers are available for use in commercial-grade applications.

To get a free IBM Cloud account, follow the instructions at:

> https://console.bluemix.net/docs/services/watson/index.html#about

You'll receive an e-mail. Follow its instructions to confirm your account. Then you can log in to the IBM Cloud console. Once there, you can go to the **Watson dashboard** at:

> https://console.bluemix.net/developer/watson/dashboard

where you can:

- Browse the Watson services.
- Link to the services you've already registered to use.
- Look at the developer resources, including the Watson documentation, SDKs and various resources for learning Watson.
- View the apps you've created with Watson.

Later, you'll register for and get your credentials to use various Watson services. You can view and manage your list of services and your credentials in the **IBM Cloud dashboard** at:

> https://console.bluemix.net/dashboard/apps

You can also click **Existing Services** in the Watson dashboard to get to this list.

✓ Self Check

1. *(Fill-In)* Accessing Watson's Lite tier services requires a free _____.
Answer: IBM Cloud account.

14.3 Watson Services

This section overviews many of Watson's services and provides links to the details for each. Be sure to run the demos to see the services in action. For links to each Watson service's documentation and API reference, visit:

> https://console.bluemix.net/developer/watson/documentation

We provide footnotes with links to each service's details. When you're ready to use a particular service, click the **Create** button on its details page to set up your credentials.

Watson Assistant
The **Watson Assistant service**[9] helps you build chatbots and virtual assistants that enable users to interact via natural language text. IBM provides a web interface that you can use

---

9. https://console.bluemix.net/catalog/services/watson-assistant-formerly-conversation.

to *train* the Watson Assistant service for specific scenarios associated with your app. For example, a weather chatbot could be trained to respond to questions like, "What is the weather forecast for New York City?" In a customer service scenario, you could create chatbots that answer customer questions and route customers to the correct department, if necessary. Try the demo at the following site to see some sample interactions:

> https://www.ibm.com/watson/services/conversation/demo/index.html#demo

Visual Recognition
The **Visual Recognition service**[10] enables apps to locate and understand information in images and video, including colors, objects, faces, text, food and inappropriate content. IBM provides predefined models (used in the service's demo), or you can train and use your own (as you'll do in the "Deep Learning" chapter). Try the following demo with the images provided and upload some of your own:

> https://watson-visual-recognition-duo-dev.ng.bluemix.net/

Speech to Text
The **Speech to Text service**,[11] which we'll use in building this chapter's app, converts speech audio files to text transcriptions of the audio. You can give the service keywords to "listen" for, and it tells you whether it found them, what the likelihood of a match was and where the match occurred in the audio. The service can distinguish among multiple speakers. You could use this service to help implement voice-controlled apps, transcribe live audio and more. Try the following demo with its sample audio clips or upload your own:

> https://speech-to-text-demo.ng.bluemix.net/

Text to Speech
The **Text to Speech service**,[12] which we'll also use in building this chapter's app, enables you to synthesize speech from text. You can use **Speech Synthesis Markup Language** (**SSML**) to embed instructions in the text for control over voice inflection, cadence, pitch and more. Currently, this service supports English (U.S. and U.K.), French, German, Italian, Spanish, Portuguese and Japanese. Try the following demo with its plain sample text, its sample text that includes SSML and text that you provide:

> https://text-to-speech-demo.ng.bluemix.net/

Language Translator
The **Language Translator service**,[13] which we'll also use in building in this chapter's app, has two key components:

- translating text between languages and
- identifying text as being written in one of over 60 languages.

Translation is supported to and from English and many languages, as well as between other languages. Try translating text into various languages with the following demo:

> https://language-translator-demo.ng.bluemix.net/

---

10. https://console.bluemix.net/catalog/services/visual-recognition.
11. https://console.bluemix.net/catalog/services/speech-to-text.
12. https://console.bluemix.net/catalog/services/text-to-speech.
13. https://console.bluemix.net/catalog/services/language-translator.

Natural Language Understanding

The **Natural Language Understanding service**[14] analyzes text and produces information including the text's overall sentiment and emotion and keywords ranked by their relevance. Among other things, the service can identify

- people, places, job titles, organizations, companies and quantities.
- categories and concepts like sports, government and politics.
- parts of speech like subjects and verbs.

You also can train the service for industry- and application-specific domains with Watson Knowledge Studio (discussed shortly). Try the following demo with its sample text, with text that you paste in or by providing a link to an article or document online:

> https://natural-language-understanding-demo.ng.bluemix.net/

Discovery

The **Watson Discovery service**[15] shares many features with the Natural Language Understanding service but also enables enterprises to store and manage documents. So, for example, organizations can use Watson Discovery to store all their text documents and be able to use natural language understanding across the entire collection. Try this service's demo, which enables you to search recent news articles for companies:

> https://discovery-news-demo.ng.bluemix.net/

Personality Insights

The **Personality Insights service**[16] analyzes text for personality traits. According to the service description, it can help you "gain insight into how and why people think, act, and feel the way they do. This service applies linguistic analytics and personality theory to infer attributes from a person's unstructured text." This information could be used to target product advertising at the people most likely to purchase those products. Try the following demo with tweets from various Twitter accounts or documents built into the demo, with text documents that you paste into the demo or with your own Twitter account:

> https://personality-insights-livedemo.ng.bluemix.net/

Tone Analyzer

The **Tone Analyzer service**[17] analyzes text for its tone in three categories:

- emotions—anger, disgust, fear, joy, sadness.
- social propensities—openness, conscientiousness, extroversion, agreeableness and emotional range.
- language style—analytical, confident, tentative.

Try the following demo with sample tweets, a sample product review, a sample e-mail or text you provide. You'll see the tone analyses at both the document and sentence levels:

> https://tone-analyzer-demo.ng.bluemix.net/

---

14. https://console.bluemix.net/catalog/services/natural-language-understanding.
15. https://console.bluemix.net/catalog/services/discovery.
16. https://console.bluemix.net/catalog/services/personality-insights.
17. https://console.bluemix.net/catalog/services/tone-analyzer.

Natural Language Classifier

You *train* the **Natural Language Classifier service**[18] with sentences and phrases that are specific to your application and classify each sentence or phrase. For example, you might classify "I need help with your product" as "tech support" and "My bill is incorrect" as "billing." Once you've trained your classifier, the service can receive sentences and phrases, then use Watson's cognitive computing capabilities and your classifier to return the best matching classifications and their match probabilities. You might then use the returned classifications and probabilities to determine the next steps in your app. For example, in a customer service app where someone is calling in with a question about a particular product, you might use Speech to Text to convert a question into text, use the Natural Language Classifier service to classify the text, then route the call to the appropriate person or department. This service *does not offer a Lite tier*. In the following demo, enter a question about the weather—the service will respond by indicating whether your question was about the temperature or the weather conditions:

```
https://natural-language-classifier-demo.ng.bluemix.net/
```

Synchronous and Asynchronous Capabilities

Many of the APIs we discuss throughout the book are **synchronous**—when you call a function or method, the program *waits* for the function or method to return before moving on to the next task. **Asynchronous** programs can start a task, continue doing other things, then be *notified* when the original task completes and returns its results. Many Watson services offer both synchronous and asynchronous APIs.

The Speech to Text demo is a good example of asynchronous APIs. The demo processes sample audio of two people speaking. As the service transcribes the audio, it returns intermediate transcription results, even if it has not yet been able to distinguish among the speakers. The demo displays these intermediate results in parallel with the service's continued work. Sometimes the demo displays "Detecting speakers" while the service figures out who is speaking. Eventually, the service sends updated transcription results for distinguishing among the speakers, and the demo then replaces the prior transcription results.

With today's multi-core computers and multi-computer clusters, the asynchronous APIs can help you improve program performance. However, programming with them can be more complicated than programming with synchronous APIs. When we discuss installing the Watson Developer Cloud Python SDK, we provide a link to the SDK's code examples on GitHub, where you can see examples that use synchronous and asynchronous versions of several services. Each service's API reference provides the complete details.

Self Check

1 *(Fill-In)* You can use _____ to embed instructions in the text for control over voice inflection, cadence, pitch and more.
Answer: Speech Synthesis Markup Language (SSML).

2 *(Fill-In)* The _____ service analyzes text and produces information including the text's overall sentiment and emotion and keywords ranked by their relevance.
Answer: Natural Language Understanding.

---

18. `https://console.bluemix.net/catalog/services/natural-language-classifier`.

3. *(True/False)* Synchronous programs can start a task, continue doing other things, then be notified when the original task completes and returns its results.
Answer: False. Asynchronous programs can start a task, continue doing other things, then be notified when the original task completes and returns its results.

14.4 Additional Services and Tools

In this section, we overview several Watson advanced services and tools.

Watson Studio

Watson Studio[19] is the new Watson interface for creating and managing your Watson projects and for collaborating with your team members on those projects. You can add data, prepare your data for analysis, create Jupyter Notebooks for interacting with your data, create and train models and work with Watson's deep-learning capabilities. Watson Studio offers a single-user Lite tier. Once you've set up your Watson Studio Lite access by clicking **Create** on the service's details web page

> https://console.bluemix.net/catalog/services/data-science-experience

you can access Watson Studio at

> https://dataplatform.cloud.ibm.com/

Watson Studio contains preconfigured projects.[20] Click **Create a project** to view them:

- Standard—"Work with any type of asset. Add services for analytical assets as you need them."
- Data Science—"Analyze data to discover insights and share your findings with others."
- Visual Recognition—"Tag and classify visual content using the Watson Visual Recognition service."
- Deep Learning—"Build neural networks and deploy deep learning models."
- Modeler—"Build modeler flows to train SPSS models or design deep neural networks."
- Business Analytics—"Create visual dashboards from your data to gain insights faster."
- Data Engineering—"Combine, cleanse, analyze, and shape data using Data Refinery."
- Streams Flow—"Ingest and analyze streaming data using the Streaming Analytics service."

Knowledge Studio

Various Watson services work with *predefined* models, but also allow you to provide custom models that are trained for specific industries or applications. Watson's **Knowledge**

---

19. https://console.bluemix.net/catalog/services/data-science-experience.
20. https://dataplatform.cloud.ibm.com/.

Studio[21] helps you build custom models. It allows enterprise teams to work together to create and train new models, which can then be deployed for use by Watson services.

Machine Learning
The **Watson Machine Learning service**[22] enables you to add predictive capabilities to your apps via popular machine-learning frameworks, including Tensorflow, Keras, scikit-learn and others. You'll use scikit-learn and Keras in the next two chapters.

Knowledge Catalog
The **Watson Knowledge Catalog**[23,24] is an advanced enterprise-level tool for securely managing, finding and sharing your organization's data. The tool offers:

- Central access to an enterprise's local and cloud-based data and machine learning models.
- Watson Studio support so users can find and access data, then easily use it in machine-learning projects.
- Security policies that ensure only the people who should have access to specific data actually do.
- Support for over 100 data cleaning and wrangling operations.
- And more.

Cognos Analytics
The IBM **Cognos Analytics**[25] service, which has a 30-day free trial, uses AI and machine learning to discover and visualize information in your data, without any programming on your part. It also provides a natural-language interface that enables you to ask questions which Cognos Analytics answers based on the knowledge it gathers from your data.

Self Check

1 *(Fill-In)* Watson's _____ helps you build custom models.
Answer: Knowledge Studio.

2 *(Fill-In)* The Watson Machine Learning service enables you to add _____ capabilities to your apps via popular machine-learning frameworks, including Tensorflow, Keras, scikit-learn and others.
Answer: predictive.

14.5 Watson Developer Cloud Python SDK

In this section, you'll install the modules required for the next section's full-implementation Watson case study. For your coding convenience, IBM provides the **Watson Developer Cloud Python SDK** (software development kit). Its `watson_developer_cloud` module

---

21. https://console.bluemix.net/catalog/services/knowledge-studio.
22. https://console.bluemix.net/catalog/services/machine-learning.
23. https://medium.com/ibm-watson/introducing-ibm-watson-knowledge-catalog-cf42c13032c1.
24. https://dataplatform.cloud.ibm.com/docs/content/catalog/overview-wkc.html.
25. https://www.ibm.com/products/cognos-analytics.

contains classes that you'll use to interact with Watson services. You'll create objects for each service you need, then interact with the service by calling the object's methods.

To install the SDK[26] open an Anaconda Prompt (Windows; open as Administrator), Terminal (macOS/Linux) or shell (Linux), then execute the following command[27]:

```
pip install --upgrade watson-developer-cloud
```

Modules We'll Need for Audio Recording and Playback
You'll also need two additional modules for audio recording (PyAudio) and playback (PyDub). To install these, use the following commands[28]:

```
pip install pyaudio
pip install pydub
```

SDK Examples
On GitHub, IBM provides sample code demonstrating how to access Watson services using the Watson Developer Cloud Python SDK's classes. You can find the examples at:

```
https://github.com/watson-developer-cloud/python-sdk/tree/master/examples
```

Self Check

1. *(True/False)* The Watson Developer Cloud Python SDK's watson_developer_cloud module contains classes for interacting with each of the Watson services.
Answer: True.

14.6 Case Study: Traveler's Companion Translation App

Suppose you're traveling in a Spanish-speaking country, but you do not speak Spanish, and you need to communicate with someone who does not speak English. You could use a translation app to speak in English, and the app could translate that, then speak it in Spanish. The Spanish-speaking person could then respond, and the app could translate that and speak it to you in English.

Here, you'll use three powerful IBM Watson services to implement such a traveler's companion translation app,[29] enabling people who speak different languages to converse in near real time. Combining services like this is known as creating a **mashup**. This app also uses simple file-processing capabilities that we introduced in the "Files and Exceptions" chapter.

Self Check

1. *(Fill-In)* Combining services is known as creating a(n) _____.
Answer: mashup.

---

26. For detailed installation instructions and troubleshooting tips, see https://github.com/watson-developer-cloud/python-sdk/blob/develop/README.md.
27. Windows users might need to install Microsoft's C++ build tools from https://visualstudio.microsoft.com/visual-cpp-build-tools/, then install the watson-developer-cloud module.
28. Mac users might need to first execute conda install -c conda-forge portaudio.
29. These services could change in the future. If they do, we'll post updates on the book's web page at http://www.deitel.com/books/IntroToPython.

14.6.1 Before You Run the App

You'll build this app using the Lite (free) tiers of several IBM Watson services. Before executing the app, make sure that you've registered for an IBM Cloud account, as we discussed earlier in the chapter, so you can get credentials for each of the three services the app uses. Once you have your credentials (described below), you'll insert them in our keys.py file (located in the ch14 examples folder) that we import into the example. Never share your credentials.

As you configure the services below, each service's credentials page also shows you the service's URL. These are the default URLs used by the Watson Developer Cloud Python SDK, so you do not need to copy them. In Section 14.6.3, we present the SimpleLanguageTranslator.py script and a detailed walkthrough of the code.

Registering for the Speech to Text Service

This app uses the Watson Speech to Text service to transcribe English and Spanish audio files to English and Spanish text, respectively. To interact with the service, you must get a username and password. To do so:

1. *Create a Service Instance:* Go to https://console.bluemix.net/catalog/services/speech-to-text and click the **Create** button on the bottom of the page. This auto-generates an API key for you and takes you to a tutorial for working with the Speech to Text service.

2. *Get Your Service Credentials:* To see your API key, click **Manage** at the top-left of the page. To the right of **Credentials**, click **Show credentials**, then copy the **API Key**, and paste it into the variable speech_to_text_key's string in the keys.py file provided in this chapter's ch14 examples folder.

Registering for the Text to Speech Service

In this app, you'll use the Watson Text to Speech service to synthesize speech from text. This service also requires you to get a username and password. To do so:

1. *Create a Service Instance:* Go to https://console.bluemix.net/catalog/services/text-to-speech and click the **Create** button on the bottom of the page. This auto-generates an API key for you and takes you to a tutorial for working with the Text to Speech service.

2. *Get Your Service Credentials:* To see your API key, click **Manage** at the top-left of the page. To the right of **Credentials**, click **Show credentials**, then copy the **API Key** and paste it into the variable text_to_speech_key's string in the keys.py file provided in this chapter's ch14 examples folder.

Registering for the Language Translator Service

In this app, you'll use the Watson Language Translator service to pass text to Watson and receive back the text translated into another language. This service requires you to get an API key. To do so:

1. *Create a Service Instance:* Go to https://console.bluemix.net/catalog/services/language-translator and click the **Create** button on the bottom of the page. This auto-generates an API key for you and takes you to a page to manage your instance of the service.

2. *Get Your Service Credentials:* To the right of **Credentials**, click **Show credentials**, then copy the **API Key** and paste it into the variable `translate_key`'s string in the `keys.py` file provided in this chapter's `ch14` examples folder.

Retrieving Your Credentials
To view your credentials at any time, click the appropriate service instance at:

> https://console.bluemix.net/dashboard/apps

✓ Self Check

1 *(Fill-In)* Once you have an IBM Cloud account, you can get your _____ for interacting with Watson services.
Answer: credentials.

14.6.2 Test-Driving the App

Once you've added your credentials to the script, open an Anaconda Prompt (Windows), a Terminal (macOS/Linux) or a shell (Linux). Run the script[30] by executing the following command from the `ch14` examples folder:

```
ipython SimpleLanguageTranslator.py
```

Processing the Question
The app performs 10 steps, which we point out via comments in the code. When the app begins executing:

Step 1 prompts for and records a question. First, the app displays:

```
Press Enter then ask your question in English
```

and waits for you to press *Enter*. When you do, the app displays:

```
Recording 5 seconds of audio
```

Speak your question. We said, "Where is the closest bathroom?" After five seconds, the app displays:

```
Recording complete
```

Step 2 interacts with Watson's Speech to Text service to transcribe your audio to text and displays the result:

```
English: where is the closest bathroom
```

Step 3 then uses Watson's Language Translator service to translate the English text to Spanish and displays the translated text returned by Watson:

```
Spanish: ¿Dónde está el baño más cercano?
```

Step 4 passes this Spanish text to Watson's Text to Speech service to convert the text to an audio file.

Step 5 plays the resulting Spanish audio file.

---

30. The `pydub.playback` module we use in this app issues a warning when you run our script. The warning has to do with module features we don't use and can be ignored. To eliminate this warning, you can install `ffmpeg` for Windows, macOS or Linux from https://www.ffmpeg.org.

Processing the Response

At this point, we're ready to process the Spanish speaker's response.
Step 6 displays:

```
Press Enter then speak the Spanish answer
```

and waits for you to press *Enter*. When you do, the app displays:

```
Recording 5 seconds of audio
```

and the Spanish speaker records a response. We do not speak Spanish, so we used Watson's Text to Speech service to *prerecord* Watson saying the Spanish response "El baño más cercano está en el restaurante," then played that audio loud enough for our computer's microphone to record it. We provided this prerecorded audio for you as `SpokenResponse.wav` in the ch14 folder. If you use this file, play it quickly after pressing *Enter* above as the app records for only 5 seconds.[31] To ensure that the audio loads and plays quickly, you might want to play it once before you press *Enter* to begin recording. After five seconds, the app displays:

```
Recording complete
```

Step 7 interacts with Watson's Speech to Text service to transcribe the Spanish audio to text and displays the result:

```
Spanish response: el baño más cercano está en el restaurante
```

Step 8 then uses Watson's Language Translator service to translate the Spanish text to English and displays the result:

```
English response: The nearest bathroom is in the restaurant
```

Step 9 passes the English text to Watson's Text to Speech service to convert the text to an audio file.
Step 10 then plays the resulting English audio.

✓ Self Check

1. *(Fill-In)* Watson's Text to Speech service converts text to _____.
Answer: audio.

14.6.3 SimpleLanguageTranslator.py Script Walkthrough

In this section, we present the `SimpleLanguageTranslator.py` script's source code, which we've divided into small consecutively numbered pieces. Let's use a top-down approach as we did in the "Control Statements and Program Development" chapter. Here's the top:

> Create a translator app that enables English and Spanish speakers to communicate.

The first refinement is:

> Translate a question spoken in English into Spanish speech.
> Translate the answer spoken in Spanish into English speech.

We can break the first line of the second refinement into five steps:

---

31. For simplicity, we set the app to record five seconds of audio. You can control the duration with the variable `SECONDS` in function `record_audio`. It's possible to create a recorder that begins recording once it detects sound and stops recording after a period of silence, but the code is more complicated.

Step 1: Prompt for then record English speech into an audio file.
Step 2: Transcribe the English speech to English text.
Step 3: Translate the English text into Spanish text.
Step 4: Synthesize the Spanish text into Spanish speech and save it into an audio file.
Step 5: Play the Spanish audio file.

We can break the second line of the second refinement into five steps:

Step 6: Prompt for then record Spanish speech into an audio file.
Step 7: Transcribe the Spanish speech to Spanish text.
Step 8: Translate the Spanish text into English text.
Step 9: Synthesize the English text into English speech and save it into an audio file.
Step 10: Play the English audio.

This top-down development makes the benefits of the divide-and-conquer approach clear, focusing our attention on small pieces of a more significant problem.

In this section's script, we implement the 10 steps specified in the second refinement. **Steps 2** and **7** use the Watson Speech to Text service, **Steps 3** and **8** use the Watson Language Translator service, and **Steps 4** and **9** use the Watson Text to Speech service.

Importing Watson SDK Classes

Lines 4–6 import classes from the `watson_developer_cloud` module that was installed with the Watson Developer Cloud Python SDK. Each of these classes uses the Watson credentials you obtained earlier to interact with a corresponding Watson service:

- Class `SpeechToTextV1`[32] enables you to pass an audio file to the Watson Speech to Text service and receive a JSON[33] document containing the text transcription.

- Class `LanguageTranslatorV3` enables you to pass text to the Watson Language Translator service and receive a JSON document containing the translated text.

- Class `TextToSpeechV1` enables you to pass text to the Watson Text to Speech service and receive audio of the text spoken in a specified language.

```
1   # SimpleLanguageTranslator.py
2   """Use IBM Watson Speech to Text, Language Translator and Text to Speech
3      APIs to enable English and Spanish speakers to communicate."""
4   from watson_developer_cloud import SpeechToTextV1
5   from watson_developer_cloud import LanguageTranslatorV3
6   from watson_developer_cloud import TextToSpeechV1
```

Other Imported Modules

Line 7 imports the `keys.py` file containing your Watson credentials. Lines 8–11 import modules that support this app's audio-processing capabilities:

- The `pyaudio` module enables us to record audio from the microphone.

---

32. The `V1` in the class name indicates the service's version number. As IBM revises its services, it adds new classes to the `watson_developer_cloud` module, rather than modifying the existing classes. This ensures that existing apps do not break when the services are updated. The Speech to Text and Text to Speech services are each Version 1 (`V1`) and the Language Translator service is Version 3 (`V3`) at the time of this writing.
33. We introduced JSON in the previous chapter, "Data Mining Twitter."

14.6 Case Study: Traveler's Companion Translation App

- pydub and pydub.playback modules enable us to load and play audio files.
- The Python Standard Library's wave module enables us to save WAV (Waveform Audio File Format) files. WAV is a popular audio format originally developed by Microsoft and IBM. This app uses the wave module to save the recorded audio to a .wav file that we send to Watson's Speech to Text service for transcription.

```
 7  import keys  # contains your API keys for accessing Watson services
 8  import pyaudio  # used to record from mic
 9  import pydub  # used to load a WAV file
10  import pydub.playback  # used to play a WAV file
11  import wave  # used to save a WAV file
12
```

Main Program: Function run_translator
Let's look at the main part of the program defined in function run_translator (lines 13–54), which calls the functions defined later in the script. For discussion purposes, we broke run_translator into the 10 steps it performs. In **Step 1** (lines 15–17), we prompt in English for the user to press *Enter*, then speak a question. Function record_audio then records audio for five seconds and stores it in the file english.wav:

```
13  def run_translator():
14      """Calls the functions that interact with Watson services."""
15      # Step 1: Prompt for then record English speech into an audio file
16      input('Press Enter then ask your question in English')
17      record_audio('english.wav')
18
```

In **Step 2**, we call function speech_to_text, passing the file english.wav for transcription and telling the Speech to Text service to transcribe the text using its *predefined* model 'en-US_BroadbandModel'.[34] We then display the transcribed text:

```
19      # Step 2: Transcribe the English speech to English text
20      english = speech_to_text(
21          file_name='english.wav', model_id='en-US_BroadbandModel')
22      print('English:', english)
23
```

In **Step 3**, we call function translate, passing the transcribed text from **Step 2** as the text to translate. Here we tell the Language Translator service to translate the text using its *predefined* model 'en-es' to translate from English (en) to Spanish (es). We then display the Spanish translation:

```
24      # Step 3: Translate the English text into Spanish text
25      spanish = translate(text_to_translate=english, model='en-es')
26      print('Spanish:', spanish)
27
```

---

[34]. For most languages, the Watson Speech to Text service supports *broadband* and *narrowband* models. Each has to do with the audio quality. For audio captured at 16 kHZ and higher, IBM recommends using the broadband models. In this app, we capture the audio at 44.1 kHZ.

In **Step 4**, we call function `text_to_speech`, passing the Spanish text from **Step 3** for the Text to Speech service to speak using its voice `'es-US_SofiaVoice'`. We also specify the file in which the audio should be saved:

```
28      # Step 4: Synthesize the Spanish text into Spanish speech
29      text_to_speech(text_to_speak=spanish, voice_to_use='es-US_SofiaVoice',
30          file_name='spanish.wav')
31
```

In **Step 5**, we call function `play_audio` to play the file `'spanish.wav'`, which contains the Spanish audio for the text we translated in **Step 3**.

```
32      # Step 5: Play the Spanish audio file
33      play_audio(file_name='spanish.wav')
34
```

Finally, **Steps 6–10** repeat what we did in **Steps 1–5**, but for Spanish speech to English speech:

- **Step 6** records the *Spanish* audio.
- **Step 7** transcribes the Spanish audio to Spanish text using the Speech to Text service's predefined model `'es-ES_BroadbandModel'`.
- **Step 8** translates the Spanish text to English text using the Language Translator Service's `'es-en'` (Spanish-to-English) model.
- **Step 9** creates the English audio using the Text to Speech Service's voice `'en-US_AllisonVoice'`.
- **Step 10** plays the English audio.

```
35      # Step 6: Prompt for then record Spanish speech into an audio file
36      input('Press Enter then speak the Spanish answer')
37      record_audio('spanishresponse.wav')
38
39      # Step 7: Transcribe the Spanish speech to Spanish text
40      spanish = speech_to_text(
41          file_name='spanishresponse.wav', model_id='es-ES_BroadbandModel')
42      print('Spanish response:', spanish)
43
44      # Step 8: Translate the Spanish text into English text
45      english = translate(text_to_translate=spanish, model='es-en')
46      print('English response:', english)
47
48      # Step 9: Synthesize the English text into English speech
49      text_to_speech(text_to_speak=english,
50          voice_to_use='en-US_AllisonVoice',
51          file_name='englishresponse.wav')
52
53      # Step 10: Play the English audio
54      play_audio(file_name='englishresponse.wav')
55
```

Now let's implement the functions we call from *Steps 1* through *10*.

14.6 Case Study: Traveler's Companion Translation App   **581**

Function speech_to_text

To access Watson's Speech to Text service, function `speech_to_text` (lines 56–87) creates a `SpeechToTextV1` object named `stt` (short for speech-to-text), passing as the argument the API key you set up earlier. The `with` statement (lines 62–65) opens the audio file specified by the `file_name` parameter and assigns the resulting file object to `audio_file`. The open mode `'rb'` indicates that we'll read (r) binary data (b)—audio files are stored as bytes in binary format. Next, lines 64–65 use the `SpeechToTextV1` object's **recognize** method to invoke the Speech to Text service. The method receives three keyword arguments:

- `audio` is the file (`audio_file`) to pass to the Speech to Text service.
- `content_type` is the media type of the file's contents—`'audio/wav'` indicates that this is an audio file stored in WAV format.[35]
- `model` indicates which spoken language model the service will use to recognize the speech and transcribe it to text. This app uses predefined models—either `'en-US_BroadbandModel'` (for English) or `'es-ES_BroadbandModel'` (for Spanish).

```
56  def speech_to_text(file_name, model_id):
57      """Use Watson Speech to Text to convert audio file to text."""
58      # create Watson Speech to Text client
59      stt = SpeechToTextV1(iam_apikey=keys.speech_to_text_key)
60
61      # open the audio file
62      with open(file_name, 'rb') as audio_file:
63          # pass the file to Watson for transcription
64          result = stt.recognize(audio=audio_file,
65              content_type='audio/wav', model=model_id).get_result()
66
67      # Get the 'results' list. This may contain intermediate and final
68      # results, depending on method recognize's arguments. We asked
69      # for only final results, so this list contains one element.
70      results_list = result['results']
71
72      # Get the final speech recognition result--the list's only element.
73      speech_recognition_result = results_list[0]
74
75      # Get the 'alternatives' list. This may contain multiple alternative
76      # transcriptions, depending on method recognize's arguments. We did
77      # not ask for alternatives, so this list contains one element.
78      alternatives_list = speech_recognition_result['alternatives']
79
80      # Get the only alternative transcription from alternatives_list.
81      first_alternative = alternatives_list[0]
82
83      # Get the 'transcript' key's value, which contains the audio's
84      # text transcription.
85      transcript = first_alternative['transcript']
86
87      return transcript  # return the audio's text transcription
88
```

---

35. Media types were formerly known as MIME (Multipurpose Internet Mail Extensions) types—a standard that specifies data formats, which programs can use to interpret data correctly.

The `recognize` method returns a `DetailedResponse` object. Its `getResult` method returns a JSON document containing the transcribed text, which we store in `result`. The JSON will look similar to the following but depends on the question you ask:

```
{
  "results": [                                              Line 70
    {                                                       Line 73
      "alternatives": [                                     Line 78
        {                                                   Line 81
          "confidence": 0.983,
          "transcript": "where is the closest bathroom "   Line 85
        }
      ],
      "final": true
    }
  ],
  "result_index": 0
}
```

The JSON contains *nested* dictionaries and lists. To simplify navigating this data structure, lines 70–85 use separate small statements to "pick off" one piece at a time until we get the transcribed text—`"where is the closest bathroom "`, which we then return. The boxes around portions of the JSON and the line numbers in each box correspond to the statements in lines 70–85. The statements operate as follows:

- Line 70 assigns to `results_list` the list associated with the key `'results'`:

    `results_list = result['results']`

    Depending on the arguments you pass to method `recognize`, this list may contain intermediate and final results. Intermediate results might be useful, for example, if you were transcribing live audio, such as a newscast. We asked for only final results, so this list contains one element.[36]

- Line 73 assigns to `speech_recognition_result` the final speech-recognition result—the only element in `results_list`:

    `speech_recognition_result = results_list[0]`

- Line 78

    `alternatives_list = speech_recognition_result['alternatives']`

    assigns to `alternatives_list` the list associated with the key `'alternatives'`. This list may contain multiple alternative transcriptions, depending on method `recognize`'s arguments. The arguments we passed result in a one-element list.

- Line 81 assigns to `first_alternative` the only element in `alternatives_list`:

    `first_alternative = alternatives_list[0]`

- Line 85 assigns to `transcript` the `'transcript'` key's value, which contains the audio's text transcription:

    `transcript = first_alternative['transcript']`

- Finally, line 87 returns the audio's text transcription.

---

36. For method recognize's arguments and JSON response details, see https://www.ibm.com/watson/developercloud/speech-to-text/api/v1/python.html?python#recognize-sessionless.

14.6 Case Study: Traveler's Companion Translation App

Lines 70–85 could be replaced with the denser statement

```
return result['results'][0]['alternatives'][0]['transcript']
```

but we prefer the separate simpler statements.

Function translate

To access the Watson Language Translator service, function translate (lines 89–111) first creates a LanguageTranslatorV3 object named language_translator, passing as arguments the service version ('2018-05-31'[37]), the API Key you set up earlier and the service's URL. Lines 93–94 use the LanguageTranslatorV3 object's **translate** method to invoke the Language Translator service, passing two keyword arguments:

- text is the string to translate to another language.
- model_id is the predefined model that the Language Translator service will use to understand the original text and translate it into the appropriate language. In this app, model will be one of IBM's *predefined* translation models—'en-es' (for English to Spanish) or 'es-en' (for Spanish to English).

```
89  def translate(text_to_translate, model):
90      """Use Watson Language Translator to translate English to Spanish
91         (en-es) or Spanish to English (es-en) as specified by model."""
92      # create Watson Translator client
93      language_translator = LanguageTranslatorV3(version='2018-05-31',
94          iam_apikey=keys.translate_key)
95
96      # perform the translation
97      translated_text = language_translator.translate(
98          text=text_to_translate, model_id=model).get_result()
99
100     # Get 'translations' list. If method translate's text argument has
101     # multiple strings, the list will have multiple entries. We passed
102     # one string, so the list contains only one element.
103     translations_list = translated_text['translations']
104
105     # get translations_list's only element
106     first_translation = translations_list[0]
107
108     # get 'translation' key's value, which is the translated text
109     translation = first_translation['translation']
110
111     return translation  # return the translated string
112
```

The method returns a DetailedResponse. That object's getResult method returns a JSON document, like:

---

[37]. According to the Language Translator service's API reference, '2018-05-31' is the current version string at the time of this writing. IBM changes the version string only if they make API changes that are not backward compatible. Even when they do, the service will respond to your calls using the API version you specify in the version string. For more details, see https://www.ibm.com/watson/developercloud/language-translator/api/v3/python.html?python#versioning.

```
{
    "translations": [                                              Line 103
    {                                                              Line 106
        "translation": "¿Dónde está el baño más cercano? "  Line 109
    }
    ],
    "word_count": 5,
    "character_count": 30
}
```

The JSON you get as a response depends on the question you asked and, again, contains nested dictionaries and lists. Lines 103–109 use small statements to pick off the translated text "¿Dónde está el baño más cercano? ". The boxes around portions of the JSON and the line numbers in each box correspond to the statements in lines 103–109. The statements operate as follows:

- Line 103 gets the 'translations' list:

    translations_list = translated_text['translations']

    If method translate's text argument has multiple strings, the list will have multiple entries. We passed only one string, so the list contains only one element.

- Line 106 gets translations_list's only element:

    first_translation = translations_list[0]

- Line 109 gets the 'translation' key's value, which is the translated text:

    translation = first_translation['translation']

- Line 111 returns the translated string.

Lines 103–109 could be replaced with the more concise statement

    return translated_text['translations'][0]['translation']

but again, we prefer the separate simpler statements.

Function text_to_speech

To access the Watson Text to Speech service, function text_to_speech (lines 113–122) creates a TextToSpeechV1 object named tts (short for text-to-speech), passing as the argument the API key you set up earlier. The with statement opens the file specified by file_name and associates the file with the name audio_file. The mode 'wb' opens the file for writing (w) in binary (b) format. We'll write into that file the contents of the audio returned by the Speech to Text service.

```
113  def text_to_speech(text_to_speak, voice_to_use, file_name):
114      """Use Watson Text to Speech to convert text to specified voice
115         and save to a WAV file."""
116      # create Text to Speech client
117      tts = TextToSpeechV1(iam_apikey=keys.text_to_speech_key)
118
119      # open file and write the synthesized audio content into the file
120      with open(file_name, 'wb') as audio_file:
121          audio_file.write(tts.synthesize(text_to_speak,
122              accept='audio/wav', voice=voice_to_use).get_result().content)
123
```

14.6 Case Study: Traveler's Companion Translation App

Lines 121–122 call two methods. First, we invoke the Speech to Text service by calling the TextToSpeechV1 object's **synthesize method**, passing three arguments:

- text_to_speak is the string to speak.
- the keyword argument accept is the media type indicating the audio format the Speech to Text service should return—again, 'audio/wav' indicates an audio file in WAV format.
- the keyword argument voice is one of the Speech to Text service's predefined voices. In this app, we'll use 'en-US_AllisonVoice' to speak English text and 'es-US_SofiaVoice' to speak Spanish text. Watson provides many male and female voices across various languages.[38]

Watson's DetailedResponse contains the spoken text audio file, accessible via get_result. We access the returned file's content attribute to get the bytes of the audio and pass them to the audio_file object's write method to output the bytes to a .wav file.

Function record_audio

The pyaudio module enables you to record audio from the microphone. The function record_audio (lines 124–154) defines several constants (lines 126–130) used to configure the stream of audio information coming from your computer's microphone. We used the settings from the pyaudio module's online documentation:

- FRAME_RATE—44100 frames-per-second represents 44.1 kHz, which is common for CD-quality audio.
- CHUNK—1024 is the number of frames streamed into the program at a time.
- FORMAT—pyaudio.paInt16 is the size of each frame (in this case, 16-bit or 2-byte integers).
- CHANNELS—2 is the number of samples per frame.
- SECONDS—5 is the number of seconds for which we'll record audio in this app.

```
124  def record_audio(file_name):
125      """Use pyaudio to record 5 seconds of audio to a WAV file."""
126      FRAME_RATE = 44100  # number of frames per second
127      CHUNK = 1024  # number of frames read at a time
128      FORMAT = pyaudio.paInt16  # each frame is a 16-bit (2-byte) integer
129      CHANNELS = 2  # 2 samples per frame
130      SECONDS = 5  # total recording time
131
132      recorder = pyaudio.PyAudio()  # opens/closes audio streams
133
134      # configure and open audio stream for recording (input=True)
135      audio_stream = recorder.open(format=FORMAT, channels=CHANNELS,
136          rate=FRAME_RATE, input=True, frames_per_buffer=CHUNK)
137      audio_frames = []  # stores raw bytes of mic input
138      print('Recording 5 seconds of audio')
139
```

---

38. For a complete list, see https://www.ibm.com/watson/developercloud/text-to-speech/api/v1/python.html?python#get-voice. Try experimenting with other voices.

```
140        # read 5 seconds of audio in CHUNK-sized pieces
141        for i in range(0, int(FRAME_RATE * SECONDS / CHUNK)):
142            audio_frames.append(audio_stream.read(CHUNK))
143
144        print('Recording complete')
145        audio_stream.stop_stream()  # stop recording
146        audio_stream.close()
147        recorder.terminate()  # release underlying resources used by PyAudio
148
149        # save audio_frames to a WAV file
150        with wave.open(file_name, 'wb') as output_file:
151            output_file.setnchannels(CHANNELS)
152            output_file.setsampwidth(recorder.get_sample_size(FORMAT))
153            output_file.setframerate(FRAME_RATE)
154            output_file.writeframes(b''.join(audio_frames))
155
```

Line 132 creates the **PyAudio** object from which we'll obtain the input stream to record audio from the microphone. Lines 135–136 use the PyAudio object's **open method** to open the input stream, using the constants FORMAT, CHANNELS, FRAME_RATE and CHUNK to configure the stream. Setting the input keyword argument to True indicates that the stream will be used to *receive* audio input. The open method returns a pyaudio **Stream** object for interacting with the stream.

Lines 141–142 use the Stream object's **read method** to get 1024 (that is, CHUNK) frames at a time from the input stream, which we then append to the audio_frames list. To determine the total number of loop iterations required to produce 5 seconds of audio using CHUNK frames at a time, we multiply the FRAME_RATE by SECONDS, then divide the result by CHUNK. Once reading is complete, line 145 calls the Stream object's **stop_stream** method to terminate recording, line 146 calls the Stream object's **close** method to close the Stream, and line 147 calls the PyAudio object's **terminate** method to release the underlying audio resources that were being used to manage the audio stream.

The with statement in lines 150–154 uses the wave module's open function to open the WAV file specified by file_name for writing in binary format ('wb'). Lines 151–153 configure the WAV file's number of channels, sample width (obtained from the PyAudio object's **get_sample_size** method) and frame rate. Then line 154 writes the audio content to the file. The expression b''.join(audio_frames) concatenates all the frames' bytes into a **byte string**. Prepending a string with b indicates that it's a string of bytes rather than a string of characters.

Function play_audio

To play the audio files returned by Watson's Text to Speech service, we use features of the pydub and pydub.playback modules. First, from the pydub module, line 158 uses the **AudioSegment** class's **from_wav method** to load a WAV file. The method returns a new AudioSegment object representing the audio file. To play the AudioSegment, line 159 calls the pydub.playback module's **play function**, passing the AudioSegment as an argument.

```
156    def play_audio(file_name):
157        """Use the pydub module (pip install pydub) to play a WAV file."""
158        sound = pydub.AudioSegment.from_wav(file_name)
159        pydub.playback.play(sound)
160
```

Executing the run_translator Function
We call the `run_translator` function when you execute `SimpleLanguageTranslator.py` as a script:

```
161  if __name__ == '__main__':
162      run_translator()
```

Hopefully, the fact that we took a divide-and-conquer approach on this substantial case study script made it manageable. Many of the steps matched up nicely with some key Watson services, enabling us to quickly create a powerful mashup application.

✓ Self Check

1. *(True/False)* Class `SpeechToTextV1` enables you to pass an audio file to the Watson Speech to Text service and receive an XML document containing the text transcription.
Answer: False. Watson returns a JSON document, not an XML document.

2. *(Fill-In)* The Language Translator service's _____ model translates from English to Spanish.
Answer: `'en-es'`.

3. *(Fill-In)* The _____ type `'audio/wav'` indicates that data is audio in WAV format.
Answer: media.

4. *(True/False)* 44100 frames-per-second is common for BluRay quality audio.
Answer: False. This is a common frame rate for CD quality sound.

5. *(Code Explanation)* In lines 121–122 of the script, what's the `content` attribute?
Answer: The `content` attribute represents the bytes of the audio file received from the Text to Speech service. We write those bytes to an audio file.

6. *(Code Explanation)* In lines 97–98 of the script, what's the purpose of the keyword argument `model_id`?
Answer: The `model_id` specifies the model that the Language Translator service uses to understand the original text and translate it into the appropriate language.

14.7 Watson Resources

IBM provides a wide range of developer resources to help you familiarize yourself with their services and begin using them to build applications.

Watson Services Documentation
The Watson Services documentation is at:

> https://console.bluemix.net/developer/watson/documentation

For each service, there are documentation and API reference links. Each service's documentation typically includes some or all of the following:

- a getting started tutorial.
- a video overview of the service.
- a link to a service demo.
- links to more specific how-to and tutorial documents.

- sample apps.
- additional resources, such as more advanced tutorials, videos, blog posts and more.

Each service's API reference shows all the details of interacting with the service using any of several languages, including Python. Click the **Python** tab to see the Python-specific documentation and corresponding code samples for the Watson Developer Cloud Python SDK. The API reference explains all the options for invoking a given service, the kinds of responses it can return, sample responses, and more.

Watson SDKs
We used the Watson Developer Cloud Python SDK to develop this chapter's script. There are SDKs for many other languages and platforms. The complete list is located at:

> https://console.bluemix.net/developer/watson/sdks-and-tools

Learning Resources
On the Learning Resources page

> https://console.bluemix.net/developer/watson/learning-resources

you'll find links to:

- Blog posts on Watson features and how Watson and AI are being used in industry.
- Watson's GitHub repository (developer tools, SDKs and sample code).
- The Watson YouTube channel (discussed below).
- Code patterns, which IBM refers to as "roadmaps for solving complex programming challenges." Some are implemented in Python, but you may still find the other code patterns helpful in designing and implementing your Python apps.

Watson Videos
The Watson YouTube channel

> https://www.youtube.com/user/IBMWatsonSolutions/

contains hundreds of videos showing you how to use all aspects of Watson. There are also spotlight videos showing how Watson is being used.

IBM Redbooks
The following IBM Redbooks publications cover IBM Cloud and Watson services in detail, helping you develop your Watson skills.

- Essentials of Application Development on IBM Cloud: http://www.redbooks.ibm.com/abstracts/sg248374.html
- Building Cognitive Applications with IBM Watson Services: Volume 1 **Getting Started**: http://www.redbooks.ibm.com/abstracts/sg248387.html
- Building Cognitive Applications with IBM Watson Services: Volume 2 **Conversation** (now called Watson Assistant): http://www.redbooks.ibm.com/abstracts/sg248394.html
- Building Cognitive Applications with IBM Watson Services: Volume 3 **Visual Recognition**: http://www.redbooks.ibm.com/abstracts/sg248393.html

- Building Cognitive Applications with IBM Watson Services: Volume 4 **Natural Language Classifier**: http://www.redbooks.ibm.com/abstracts/sg248391.html
- Building Cognitive Applications with IBM Watson Services: Volume 5 **Language Translator**: http://www.redbooks.ibm.com/abstracts/sg248392.html
- Building Cognitive Applications with IBM Watson Services: Volume 6 **Speech to Text and Text to Speech**:
http://www.redbooks.ibm.com/abstracts/sg248388.html
- Building Cognitive Applications with IBM Watson Services: Volume 7 **Natural Language Understanding**:
http://www.redbooks.ibm.com/abstracts/sg248398.html

Self Check

1 *(Fill-In)* IBM provides dozens of _____, which IBM refers to as "roadmaps for solving complex programming challenges."
Answer: code patterns.

14.8 Wrap-Up

In this chapter, we introduced IBM's Watson cognitive-computing platform and overviewed its broad range of services. You saw that Watson offers intriguing capabilities that you can integrate into in your applications. IBM encourages learning and experimentation via its free Lite tiers. To take advantage of that, you set up an IBM Cloud account. You tried Watson demos to experiment with various services, such as natural language translation, speech-to-text, text-to-speech, natural language understanding, chatbots, analyzing text for tone, and visual object recognition in images and video.

You installed the Watson Developer Cloud Python SDK for programmatic access to Watson services from your Python code. In the traveler's companion translation app, we mashed up several Watson services to enable English-only and Spanish-only speakers to communicate easily with one another verbally. We transcribed English and Spanish audio recordings to text, translated the text to the other language, then synthesized English and Spanish audio from the translated text. Finally, we discussed various Watson resources, including documentation, blogs, the Watson GitHub repository, the Watson YouTube channel, code patterns implemented in Python (and other languages) and IBM Redbooks.

Exercises

14.1 *(Try It: Watson Speech to Text)* Use the microphone on your computer to record yourself speaking a paragraph of text. Upload that audio to the Watson Speech to Text demo: https://speech-to-text-demo.ng.bluemix.net/. Check the transcription results to see whether there are any words Watson has trouble understanding.

14.2 *(Try It: Watson Speech to Text—Detecting Separate Speakers)* With a friend's permission and using the microphone on your computer, record a conversation between you and a friend, then upload that audio file to the Watson Speech to Text demo at https://speech-to-text-demo.ng.bluemix.net/. Enable the option to detect *multiple* speakers. As the demo transcribes your voices to text, check whether Watson accurately distinguishes between your voices and transcribes the text accordingly.

14.3 *(Visual Object Recognition)* Investigate the Visual Recognition service and use its demo to locate various items in your photos and your friends' photos.

14.4 *(Language Translator App Enhancement)* In our Traveler's Assistant Translator app's *Steps 1* and *6*, we displayed only English text prompting the user to press *Enter* and record. Display the instructions in both English and Spanish.

14.5 *(Language Translator App Enhancement)* The Text to Speech service supports multiple voices for some languages. For example, there are four English voices and four Spanish voices. Experiment with the different voices. For the names of the voices, see

```
https://www.ibm.com/watson/developercloud/text-to-speech/api/v1/
    python.html?python#get-voice
```

14.6 *(Language Translator App Enhancement)* Our Traveler's Assistant Translator app supports only English and Spanish. Investigate the languages Watson currently supports in common for the Speech to Text, Language Translator and Text to Speech services. Pick one and convert our app to use that language rather than Spanish.

14.7 *(United Nations Dilemma: Inter-Language Translation)* Inter-language translation is one of the most challenging artificial intelligence and natural language processing problems. Literally hundreds of languages are spoken at the United Nations. As of this writing, the Watson Language Translator service will allow you to translate an English sentence to Spanish, then the Spanish to French, then the French to German, then the German to Italian, then the Italian back to English. You may be surprised with how the final result differs from the original. For a list of all the inter-language translations that Watson allows, see

```
https://console.bluemix.net/catalog/services/language-translator
```

Use the Watson Language Translator service to build a Python application that performs the preceding series of translations, showing the text in each language along the way and the final English result. This will help you appreciate the challenge of having people from many countries understand one another.

14.8 *(Python Pizza Parlor)* Use Watson Text to Speech and Speech to Text services to communicate verbally with a person ordering a pizza. Your app should welcome the person and ask them what size pizza they'd like (small or large). Then ask the person if they'd like pepperoni (yes or no). Then ask if they'd like mushrooms (yes or no). The user responds by speaking each answer. After processing the user's responses, the app should summarize the order verbally and thank the customer for their order. For an extra challenge, consider researching and using the Watson Assistant service to build a chatbot to solve this problem.

14.9 *(Language Translator: Language Identification)* Investigate the Language Translator service's ability to detect the language of text. Then write an app that will send text strings in a variety of languages to the Language Translator service and see if it identifies the source languages correctly. See `https://console.bluemix.net/catalog/services/language-translator` for a list of the dozens of supported languages.

14.10 *(Watson Internet of Things Platform)* Watson also provides the Watson Internet of Things (IoT) Platform for analyzing live data streams from devices in the Internet of Things, such as temperature sensors, motion sensors and more. To get a sense of a live data stream, you can follow the instructions at

```
https://discover-iot.eu-gb.mybluemix.net/#/play
```

to connect your smartphone to the demo, then watch on your computer and phone screens as live sensor data displays. On your computer screen, a phone image moves dynamically to show the your phone's orientation as you move and rotate it in your hand.

14.11 *(Pig Latin Translator App)* Research the rules for translating English-language words into pig Latin. Read a sentence from the user. Then, encode the sentence into pig Latin, display the pig Latin text and use speech synthesis to speak `'The sentence` *insertOriginalSentenceHere* `in pig Latin is` *insertPigLatinSentenceHere*`'` (replace the italicized English text with the corresponding pig Latin text). For simplicity, assume that the English sentence consists of words separated by blanks, there are no punctuation marks and all words have two or more letters.

14.12 *(Random Story Writer App)* Write a script that uses random-number generation to create, display and speak sentences. Use four arrays of strings called `article`, `noun`, `verb` and `preposition`. Create a sentence by selecting a word at random from each array in the following order: `article`, `noun`, `verb`, `preposition`, `article`, `noun`. As each word is picked, concatenate it to the previous words in the sentence. Spaces should separate the words. When a sentence is displayed, it should start with a capital letter and end with a period. Allow the script to produce a short story consisting of several sentences. Use Text to Speech to read the story aloud to the user.

14.13 *(Eyesight Tester App)* You've probably had your eyesight tested. In an eye exam, you're asked to cover one eye, then read out loud the letters from an eyesight chart called a Snellen chart. The letters are arranged in 11 rows and include only the letters C, D, E, F, L, N, O, P, T, Z. The first row has one letter in a huge font. As you move down the page, the number of letters in each row increases and the font size decreases, ending with a row of 11 letters in a tiny font. Your ability to read the letters accurately measures your visual acuity. Create an eyesight testing chart similar to the Snellen chart used by medical professionals (`https://en.wikipedia.org/wiki/Snellen_chart`).

The app should prompt the user to say each letter. Then use speech synthesis to determine if the user said the correct letter. At the end of the test, display—and speak—`'Your vision is 20/20'` or whatever the appropriate value is for the user's visual acuity.

14.14 *(Project: Speech Synthesis Markup Language)* Investigate SSML (Speech Synthesis Markup Language), then use it to mark up a paragraph of text to see how the SSML you specify affects Watson's voices. Experiment with inflection, cadence, pitch and more. Try out your text with various voices in the Watson Text to Speech demo at:

`https://text-to-speech-demo.ng.bluemix.net/`

You can learn more about SSML at `https://www.w3.org/TR/speech-synthesis/`.

14.15 *(Project: Text to Speech and SSML—Singing Happy Birthday)* Use Watson Text to Speech and SSML to have Watson sing Happy Birthday. Let users enter their names.

14.16 *(Enhanced Tortoise and Hare)* Add speech-synthesis capabilities to your solution to the simulation of the tortoise-and-hare race in Exercise 4.12. Use speech to call the race as it proceeds, dropping in phrases like `'On your mark. Get set. Go!'`, `"And they're off!"`, `'The tortoise takes the lead!'`, `'The hare is taking a snooze'`, etc. At the end of the race, announce the winner.

14.17 *(Project: Enhanced Tortoise and Hare with SSML)* Use SSML in your solution to Exercise 14.16 to make the speech sound like a sportscaster announcing a race on TV.

14.18 *(Challenge Project: Language Translator App—Supporting Any Length Audio)* Our Traveler's Assistant Translator app allows each speaker to record for five seconds. Investigate how to use PyAudio (which is not a Watson capability) to detect when someone starts speaking and stops speaking so you can record audio of any length. *Caution:* The code for doing this is complex.

14.19 *(Project: Building a Chatbot with Watson Assistant)* Investigate the Watson Assistant service. Next, go to `https://console.bluemix.net/developer/watson/dashboard` and try **Build a chatbot**. After you click **Create**, follow the steps provided to build your chatbot. Be sure to follow the **Getting started tutorial** at

> `https://console.bluemix.net/docs/services/conversation/getting-started.html#getting-started-tutorial`

14.20 *(For the Entrepreneur: Bot Applications)* Research common bot applications. Indicate how they can improve things like call center operations. For example, you would eliminate time spent on the phone waiting for a human to become available. The bot can ask if the caller is satisfied with the answer, then the caller can hang up or the bot can route the caller to a human. Bots can accumulate massive expertise over time. If you're entrepreneurial, you could develop sophisticated bots for organizations to purchase. Opportunities abound in fields, such as health care, answering Social Security and Medicare questions, helping travelers plan itineraries, and many more.

14.21 *(Project: Metric Conversion App)* Write an app that uses speech recognition and speech synthesis to assist users with metric conversions. Allow the user to specify the names of the units (e.g., centimeters, liters, grams, for the metric system and inches, quarts, pounds, for the English system) and should respond to simple questions, such as

> `'How many inches are in 2 meters?'`
> `'How many liters are in 10 quarts?'`

Your program should recognize invalid conversions. For example, the question

> `'How many feet are in 5 kilograms?'`

is not a meaningful question because `'feet'` is a unit of length whereas `'kilograms'` is a unit of mass. Use speech synthesis to speak the result and display the result in text. If the question is invalid, the app should speak, `'That is an invalid conversion.'` and display the same message as text.

14.22 *(Accessibility Challenge Project: Voice-Driven Text Editor)* The speech synthesis and recognition technologies you learned in this chapter are particularly useful for implementing apps for people who cannot use their hands. Create a simple text editor app that allows the user to speak some text, then edit the text verbally. Provide basic editing features such as insert and delete via voice commands.

14.23 *(Project: Watson Sentiment Analysis)* In the "Natural Language Processing" chapter and in the Data Mining Twitter chapter, we performed sentiment analysis. Run some of the tweets whose sentiment you analyzed in the Twitter chapter through the Watson Natural Language Understanding service's sentiment analysis capability. Compare the analysis results.

Machine Learning: Classification, Regression and Clustering

Objectives

In this chapter you'll:

- Use scikit-learn with popular datasets to perform machine learning studies.
- Use Seaborn and Matplotlib to visualize and explore data.
- Perform supervised machine learning with k-nearest neighbors classification and linear regression.
- Perform multi-classification with Digits dataset.
- Divide a dataset into training, test and validation sets.
- Tune model hyperparameters with k-fold cross-validation.
- Measure model performance.
- Display a confusion matrix showing classification prediction hits and misses.
- Perform multiple linear regression with the California Housing dataset.
- Perform dimensionality reduction with PCA and t-SNE on the Iris and Digits datasets to prepare them for two-dimensional visualizations.
- Perform unsupervised machine learning with k-means clustering and the Iris dataset.

Outline

15.1 Introduction to Machine Learning
  15.1.1 Scikit-Learn
  15.1.2 Types of Machine Learning
  15.1.3 Datasets Bundled with Scikit-Learn
  15.1.4 Steps in a Typical Data Science Study
15.2 Case Study: Classification with k-Nearest Neighbors and the Digits Dataset, Part 1
  15.2.1 k-Nearest Neighbors Algorithm
  15.2.2 Loading the Dataset
  15.2.3 Visualizing the Data
  15.2.4 Splitting the Data for Training and Testing
  15.2.5 Creating the Model
  15.2.6 Training the Model
  15.2.7 Predicting Digit Classes
15.3 Case Study: Classification with k-Nearest Neighbors and the Digits Dataset, Part 2
  15.3.1 Metrics for Model Accuracy
  15.3.2 K-Fold Cross-Validation
  15.3.3 Running Multiple Models to Find the Best One
  15.3.4 Hyperparameter Tuning
15.4 Case Study: Time Series and Simple Linear Regression
15.5 Case Study: Multiple Linear Regression with the California Housing Dataset
  15.5.1 Loading the Dataset
  15.5.2 Exploring the Data with Pandas
  15.5.3 Visualizing the Features
  15.5.4 Splitting the Data for Training and Testing
  15.5.5 Training the Model
  15.5.6 Testing the Model
  15.5.7 Visualizing the Expected vs. Predicted Prices
  15.5.8 Regression Model Metrics
  15.5.9 Choosing the Best Model
15.6 Case Study: Unsupervised Machine Learning, Part 1—Dimensionality Reduction
15.7 Case Study: Unsupervised Machine Learning, Part 2—k-Means Clustering
  15.7.1 Loading the Iris Dataset
  15.7.2 Exploring the Iris Dataset: Descriptive Statistics with Pandas
  15.7.3 Visualizing the Dataset with a Seaborn `pairplot`
  15.7.4 Using a `KMeans` Estimator
  15.7.5 Dimensionality Reduction with Principal Component Analysis
  15.7.6 Choosing the Best Clustering Estimator
15.8 Wrap-Up
Exercises

15.1 Introduction to Machine Learning

In this chapter and the next, we'll present machine learning—one of the most exciting and promising subfields of artificial intelligence. You'll see how to quickly solve challenging and intriguing problems that novices and most experienced programmers probably would not have attempted just a few years ago. Machine learning is a big, complex topic that raises lots of subtle issues. Our goal here is to give you a friendly, hands-on introduction to a few of the simpler machine-learning techniques.

What Is Machine Learning?
Can we really make our machines (that is, our computers) learn? In this and the next chapter, we'll show exactly how that magic happens. What's the "secret sauce" of this new application-development style? It's data and lots of it. Rather than programming expertise into our applications, we program them to learn from data.

We'll present many Python-based code examples that build working machine-learning models then use them to make remarkably accurate predictions. The chapter is loaded with exercises and projects that will give you the opportunity to broaden and deepen your machine-learning expertise.

Prediction

Wouldn't it be fantastic if you could improve weather forecasting to save lives, minimize injuries and property damage? What if we could improve cancer diagnoses and treatment regimens to save lives, or improve business forecasts to maximize profits and secure people's jobs? What about detecting fraudulent credit-card purchases and insurance claims? How about predicting customer "churn," what prices houses are likely to sell for, ticket sales of new movies, and anticipated revenue of new products and services? How about predicting the best strategies for coaches and players to use to win more games and championships? All of these kinds of predictions are happening today with machine learning.

Machine Learning Applications

Here's a table of some popular machine-learning applications:

Machine learning applications		
Anomaly detection	Detecting objects in scenes	Recommender systems ("people who bought this product also bought...")
Chatbots	Detecting patterns in data	
Classifying emails as spam or not spam	Diagnostic medicine	
	Facial recognition	Self-Driving cars (more generally, autonomous vehicles)
Classifying news articles as sports, financial, politics, etc.	Insurance fraud detection	Sentiment analysis (like classifying movie reviews as positive, negative or neutral)
	Intrusion detection in computer networks	
Computer vision and image classification	Handwriting recognition	
Credit-card fraud detection	Marketing: Divide customers into clusters	Spam filtering
Customer churn prediction		Time series predictions like stock-price forecasting and weather forecasting
Data compression	Natural language translation (English to Spanish, French to Japanese, etc.)	
Data exploration		Voice recognition
Data mining social media (like Facebook, Twitter, LinkedIn)	Predict mortgage loan defaults	

15.1.1 Scikit-Learn

We'll use the popular *scikit-learn machine learning library*. Scikit-learn, also called sklearn, conveniently packages the most effective machine-learning algorithms as *estimators*. Each is encapsulated, so you don't see the intricate details and heavy mathematics of how these algorithms work. You should feel comfortable with this—you drive your car without knowing the intricate details of how engines, transmissions, braking systems and steering systems work. Think about this the next time you step into an elevator and select your destination floor, or turn on your television and select the program you'd like to watch. Do you really understand the internal workings of your smart phone's hardware and software?

With scikit-learn and a small amount of Python code, you'll create powerful models quickly for analyzing data, extracting insights from the data and most importantly making predictions. You'll use scikit-learn to *train* each model on a subset of your data, then *test* each model on the rest to see how well your model works. Once your models are trained, you'll put them to work making predictions based on data they have not seen. You'll often be amazed at the results. All of a sudden your computer that you've used mostly on rote chores will take on characteristics of intelligence.

Scikit-learn has tools that automate training and testing your models. Although you can specify parameters to customize the models and possibly improve their performance, in this chapter, we'll typically use the models' *default parameters*, yet still obtain impressive results. It gets even better. In the exercises, you'll investigate auto-sklearn which automates many of the tasks you perform with scikit-learn.

Which Scikit-Learn Estimator Should You Choose for Your Project
It's difficult to know in advance which model(s) will perform best on your data, so you typically try many models and pick the one that performs best. As you'll see, scikit-learn makes this convenient for you. A popular approach is to run many models and pick the best one(s). How do we evaluate which model performed best?

You'll want to experiment with lots of different models on different kinds of datasets. You'll rarely get to know the details of the complex mathematical algorithms in the sklearn estimators, but with experience, you'll become familiar with which algorithms may be best for particular types of datasets and problems. Even with that experience, it's unlikely that you'll be able to intuit the best model for each new dataset. So scikit-learn makes it easy for you to "try 'em all." It takes at most a few lines of code for you to create and use each model. The models report their performance so you can compare the results and pick the model(s) with the best performance.

15.1.2 Types of Machine Learning

We'll study the two main types of machine learning—*supervised machine learning*, which works with *labeled data*, and *unsupervised machine learning*, which works with *unlabeled data*.

If, for example, you're developing a computer vision application to recognize dogs and cats, you'll train your model on lots of dog photos labeled "dog" and cat photos labeled "cat." If your model is effective, when you put it to work processing unlabeled photos it will recognize dogs and cats it has never seen before. The more photos you train with, the greater the chance that your model will accurately predict which new photos are dogs and which are cats. In this era of big data and massive, economical computer power, you should be able to build some pretty accurate models with the techniques you're about to learn.

How can looking at unlabeled data be useful? Online booksellers sell lots of books. They record enormous amounts of (unlabeled) book purchase transaction data. They noticed early on that people who bought certain books were likely to purchase other books on the same or similar topics. That led to their *recommendation systems*. Now, when you browse a bookseller site for a particular book, you're likely to see recommendations like, "people who bought this book also bought these other books." Recommendation systems are big business today, helping to maximize product sales of all kinds.

Supervised Machine Learning
Supervised machine learning falls into two categories—*classification* and *regression*. You train machine-learning models on datasets that consist of rows and columns. Each row represents a data *sample*. Each column represents a *feature* of that sample. In supervised machine learning, each sample has an associated label called a *target* (like "dog" or "cat"). This is the value you're trying to predict for new data that you present to your models.

Datasets

You'll work with some "toy" datasets, each with a small number of samples with a limited number of features. You'll also work with several richly featured real-world datasets, one containing a few thousand samples and one containing tens of thousands of samples. In the world of big data, datasets commonly have, millions and billions of samples, or even more.

There's an enormous number of free and open datasets available for data science studies. Libraries like scikit-learn package up popular datasets for you to experiment with and provide mechanisms for loading datasets from various repositories (such as openml.org). Governments, businesses and other organizations worldwide offer datasets on a vast range of subjects. Between the text examples and the exercises and projects, you'll work with many popular free datasets, using a variety of machine learning techniques.

Classification

We'll use one of the simplest classification algorithms, *k-nearest neighbors*, to analyze the Digits dataset bundled with scikit-learn. Classification algorithms predict the discrete classes (categories) to which samples belong. Binary classification uses two classes, such as "spam" or "not spam" in an email classification application. Multi-classification uses more than two classes, such as the 10 classes, 0 through 9, in the Digits dataset. A classification scheme looking at movie descriptions might try to classify them as "action," "adventure," "fantasy," "romance," "history" and the like.

Regression

Regression models predict a *continuous output*, such as the predicted temperature output in the weather *time series* analysis from Chapter 10's Intro to Data Science section. In this chapter, we'll revisit that *simple linear regression* example, this time implementing it using scikit-learn's LinearRegression estimator. Next, we use a LinearRegression estimator to perform *multiple linear regression* with the California Housing dataset that's bundled with scikit-learn. We'll predict the median house value of a U. S. census block of homes, considering eight features per block, such as the average number of rooms, median house age, average number of bedrooms and median income. The LinearRegression estimator, by default, uses *all* the numerical features in a dataset to make more sophisticated predictions than you can with a single-feature simple linear regression.

Unsupervised Machine Learning

Next, we'll introduce unsupervised machine learning with *clustering* algorithms. We'll use *dimensionality reduction* (with scikit-learn's TSNE estimator) to *compress* the Digits dataset's 64 features down to two for visualization purposes. This will enable us to see how nicely the Digits data "cluster up." This dataset contains handwritten digits like those the post office's computers must recognize to route each letter to its designated zip code. This is a challenging computer-vision problem, given that each person's handwriting is unique. Yet, we'll build this clustering model with just a few lines of code and achieve impressive results. And we'll do this without having to understand the inner workings of the clustering algorithm. This is the beauty of object-based programming. We'll see this kind of convenient object-based programming again in the next chapter, where we'll build powerful deep learning models using the open source Keras library.

K-Means Clustering and the Iris Dataset

We'll present the simplest unsupervised machine-learning algorithm, *k-means clustering*, and use it on the Iris dataset that's also bundled with scikit-learn. We'll use dimensionality reduction (with scikit-learn's PCA estimator) to compress the Iris dataset's four features to two for visualization purposes. We'll show the clustering of the three *Iris* species in the dataset and graph each cluster's *centroid*, which is the cluster's center point. Finally, we'll run multiple clustering estimators to compare their ability to divide the Iris dataset's samples effectively into three clusters.

You normally specify the desired number of clusters, *k*. K-means works through the data trying to divide it into that many clusters. As with many machine learning algorithms, k-means is *iterative* and gradually zeros in on the clusters to match the number you specify.

K-means clustering can find similarities in unlabeled data. This can ultimately help with assigning labels to that data so that supervised learning estimators can then process it. Given that it's tedious and error-prone for humans to have to assign labels to unlabeled data, and given that the vast majority of the world's data is unlabeled, unsupervised machine learning is an important tool.

Big Data and Big Computer Processing Power

The amount of data that's available today is already enormous and continues to grow exponentially. The data produced in the world in the last few years equals the amount produced up to that point since the dawn of civilization. We commonly talk about big data, but "big" may not be a strong enough term to describe truly how huge data is getting.

People used to say "I'm drowning in data and I don't know what to do with it." With machine learning, we now say, "Flood me with big data so I can use machine-learning technology to extract insights and make predictions from it."

This is occurring at a time when computing power is *exploding* and computer memory and secondary storage are *exploding* in capacity while costs dramatically decline. All of this enables us to think differently about the solution approaches. We now can program computers to *learn* from data, and lots of it. It's now all about predicting from data.

15.1.3 Datasets Bundled with Scikit-Learn

The following table lists scikit-learn's bundled datasets.[1] It also provides capabilities for loading datasets from other sources, such as the 20,000+ datasets available at openml.org.

Datasets bundled with scikit-learn	
"Toy" datasets	*Real-world datasets*
Boston house prices	Olivetti faces
Iris plants	20 newsgroups text
Diabetes	Labeled Faces in the Wild face recognition
Optical recognition of handwritten digits	Forest cover types
Linnerrud	RCV1
Wine recognition	Kddcup 99
Breast cancer Wisconsin (diagnostic)	California Housing

---

1. http://scikit-learn.org/stable/datasets/index.html

15.1.4 Steps in a Typical Data Science Study

We'll perform the steps of a typical machine-learning case study, including:

- loading the dataset
- exploring the data with pandas and visualizations
- transforming your data (converting non-numeric data to numeric data because scikit-learn requires numeric data; in the chapter, we use datasets that are "ready to go," but we'll discuss the issue again in the "Deep Learning" chapter)
- splitting the data for training and testing
- creating the model
- training and testing the model
- tuning the model and evaluating its accuracy
- making predictions on live data that the model hasn't seen before.

In the "Array-Oriented Programming with NumPy" and "Strings: A Deeper Look" chapters' Intro to Data Science sections, we discussed using pandas to deal with missing and erroneous values. These are important steps in cleaning your data before using it for machine learning.

Self Check

1 *(Fill-In)* Machine learning falls into two main categories—_____ machine learning, which works with labeled data and _____ machine learning, which works with unlabeled data
Answer: supervised, unsupervised.

2 *(True/False)* With machine learning, rather than programming expertise into our applications, we program them to learn from data.
Answer: True.

15.2 Case Study: Classification with k-Nearest Neighbors and the Digits Dataset, Part 1

To process mail efficiently and route each letter to the correct destination, postal service computers must be able to scan handwritten names, addresses and zip codes and recognize the letters and digits. As you'll see in this chapter, powerful libraries like scikit-learn enable even novice programmers to make such machine-learning problems manageable. In the next chapter, we'll use even more powerful computer-vision capabilities as we study the deep learning technology of convolutional neural networks.

Classification Problems
In this section, we'll look at **classification** in supervised machine learning, which attempts to predict the distinct class[2] to which a sample belongs. For example, if you have images

---

2. Note that the term "class" in this case means "category," not the Python concept of a class.

of dogs and images of cats, you can classify each image as a "dog" or a "cat." This is a **binary classification problem** because there are *two* classes.

We'll use the **Digits dataset**[3] bundled with scikit-learn, which consists of 8-by-8 pixel images representing 1797 hand-written digits (0 through 9). Our goal is to predict which digit an image represents. Since there are 10 possible digits (the classes), this is a **multi-classification problem**. You train a classification model using **labeled data**—we know in advance each digit's class. In this case study, we'll use one of the simplest machine-learning classification algorithms, *k-nearest neighbors (k-NN)*, to recognize handwritten digits.

The following low-resolution digit visualization of a 5 was produced with Matplotlib from one digit's 8-by-8 pixel raw data. We'll show how to display images like this with Matplotlib momentarily:

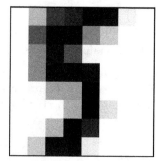

Researchers created the images in this dataset from the MNIST database's tens of thousands of 32-by-32 pixel images that were produced in the early 1990s. At today's high-definition camera and scanner resolutions, such images can be captured with much higher resolutions.

Our Approach

We'll cover this case study over two sections. In this section, we'll begin with the basic steps of a machine learning case study:

- Decide the data from which to train a model.
- Load and explore the data.
- Split the data for training and testing.
- Select and build the model.
- Train the model.
- Make predictions.

As you'll see, in scikit-learn each of these steps requires at most a few lines of code. In the next section, we'll

- Evaluate the results.
- Tune the model.
- Run several classification models to choose the best one(s).

---

3. http://scikit-learn.org/stable/datasets/index.html#optical-recognition-of-handwritten-digits-dataset.

15.2 Classification with k-Nearest Neighbors and the Digits Dataset, Part 1

We'll visualize the data using Matplotlib and Seaborn, so launch IPython with Matplotlib support:

```
ipython --matplotlib
```

✓ Self Check

1. *(Fill-In)* _____ classification divides samples into two distinct classes, and _____-classification divides samples into many distinct classes.
Answer: Binary, multi.

15.2.1 k-Nearest Neighbors Algorithm

Scikit-learn supports many **classification algorithms**, including the simplest—**k-nearest neighbors (k-NN)**. This algorithm attempts to predict a test sample's class by looking at the *k* training samples that are nearest (in distance) to the test sample. For example, consider the following diagram in which the blue, purple, green and red dots represent four sample classes. For this discussion, we'll use the color names as the class names:

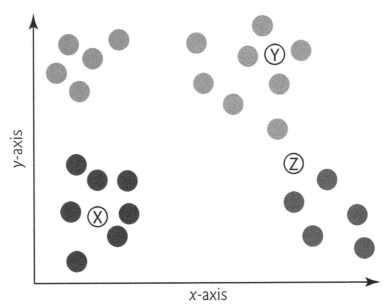

We want to predict the classes to which the new samples **X**, **Y** and **Z** belong. Let's assume we'd like to make these predictions using each sample's *three* nearest neighbors—*three* is *k* in the k-nearest neighbors algorithm:

- Sample **X**'s three nearest neighbors are all purple dots, so we'd predict that **X**'s class is purple.
- Sample **Y**'s three nearest neighbors are all green dots, so we'd predict that **Y**'s class is green.
- For **Z**, the choice is not as clear, because it appears *between* the green *and* red dots. Of the three nearest neighbors, one is green and two are red. In the k-nearest neighbors algorithm, the class with the most "votes" wins. So, based on two red votes to

one green vote, we'd predict that **Z**'s class is red. Picking an odd *k* value in the kNN algorithm avoids ties by ensuring there's never an equal number of votes.

Hyperparameters and Hyperparameter Tuning
In machine learning, a **model** implements a machine-learning algorithm. In scikit-learn, models are called **estimators**. There are two parameter types in machine learning:
- those the estimator calculates as it learns from the data you provide and
- those you specify in advance when you create the scikit-learn estimator object that represents the model.

The parameters specified in advance are called **hyperparameters**.
  In the k-nearest neighbors algorithm, *k* is a hyperparameter. For simplicity, we'll use scikit-learn's *default* hyperparameter values. In real-world machine-learning studies, you'll want to experiment with different values of *k* to produce the best possible models for your studies. This process is called **hyperparameter tuning**. Later we'll use hyperparameter tuning to choose the value of *k* that enables the k-nearest neighbors algorithm to make the best predictions for the Digits dataset. Scikit-learn also has *automated* hyperparameter tuning capabilities that you'll explore in the exercises.

Self Check

1 *(True/False)* In machine learning, a model implements a machine-learning algorithm. In scikit-learn, models are called estimators.
Answer: True.

2 *(Fill-In)* The process of choosing the best value of *k* for the k-nearest neighbors algorithm is called _____.
Answer: hyperparameter tuning.

15.2.2 Loading the Dataset

The `load_digits` function from the `sklearn.datasets` module returns a scikit-learn **Bunch** object containing the digits data and information *about* the Digits dataset (called metadata):

```
In [1]: from sklearn.datasets import load_digits

In [2]: digits = load_digits()
```

Bunch is a subclass of `dict` that has additional attributes for interacting with the dataset.

Displaying the Description
The Digits dataset bundled with scikit-learn is a subset of the UCI (University of California Irvine) ML hand-written digits dataset at:

> http://archive.ics.uci.edu/ml/datasets/
>   Optical+Recognition+of+Handwritten+Digits

The original UCI dataset contains 5620 samples—3823 for training and 1797 for testing. The version of the dataset bundled with scikit-learn contains only the *1797 testing samples*. A Bunch's **DESCR** attribute contains a description of the dataset. According to the Digits dataset's description[4], each sample has 64 features (as specified by `Number of Attributes`) that represent an 8-by-8 image with pixel values in the range 0–16 (specified by Attribute Information). This dataset has *no missing values* (as specified by Missing Attribute Values). The 64 features may seem like a lot, but real-world datasets can sometimes have hundreds, thousands or even millions of features.

```
In [3]: print(digits.DESCR)
.. _digits_dataset:

Optical recognition of handwritten digits dataset
--------------------------------------------------

**Data Set Characteristics:**

    :Number of Instances: 5620
    :Number of Attributes: 64
    :Attribute Information: 8x8 image of integer pixels in the range
    0..16.
    :Missing Attribute Values: None
    :Creator: E. Alpaydin (alpaydin '@' boun.edu.tr)
    :Date: July; 1998

This is a copy of the test set of the UCI ML hand-written digits datasets
http://archive.ics.uci.edu/ml/datasets/
    Optical+Recognition+of+Handwritten+Digits

The data set contains images of hand-written digits: 10 classes where
each class refers to a digit.

Preprocessing programs made available by NIST were used to extract
normalized bitmaps of handwritten digits from a preprinted form. From a
total of 43 people, 30 contributed to the training set and different 13
to the test set. 32x32 bitmaps are divided into nonoverlapping blocks of
4x4 and the number of on pixels are counted in each block. This generates
an input matrix of 8x8 where each element is an integer in the range
0..16. This reduces dimensionality and gives invariance to small
distortions.

For info on NIST preprocessing routines, see M. D. Garris, J. L. Blue, G.
T. Candela, D. L. Dimmick, J. Geist, P. J. Grother, S. A. Janet, and C.
L. Wilson, NIST Form-Based Handprint Recognition System, NISTIR 5469,
1994.

.. topic:: References

  - C. Kaynak (1995) Methods of Combining Multiple Classifiers and Their
    Applications to Handwritten Digit Recognition, MSc Thesis, Institute
    of Graduate Studies in Science and Engineering, Bogazici University.
  - E. Alpaydin, C. Kaynak (1998) Cascading Classifiers, Kybernetika.
  - Ken Tang and Ponnuthurai N. Suganthan and Xi Yao and A. Kai Qin.
    Linear dimensionality reduction using relevance weighted LDA. School
    of Electrical and Electronic Engineering Nanyang Technological
    University. 2005.
```

---

4. We highlighted some key information in bold.

- Claudio Gentile. A New Approximate Maximal Margin Classification Algorithm. NIPS. 2000.

Checking the Sample and Target Sizes
The Bunch object's **data** and **target** attributes are NumPy arrays:

- The data array contains the 1797 samples (the digit images), each with 64 features, having values in the range 0–16, representing *pixel intensities*. With Matplotlib, we'll visualize these intensities in grayscale shades from white (0) to black (16):

- The target array contains the images' labels—that is, the classes indicating which digit each image represents. The array is called target because, when you make predictions, you're aiming to "hit the target" values. To see labels of samples throughout the dataset, let's display the target values of every 100th sample:

```
In [4]: digits.target[::100]
Out[4]: array([0, 4, 1, 7, 4, 8, 2, 2, 4, 4, 1, 9, 7, 3, 2, 1, 2, 5])
```

We can confirm the number of samples and features (per sample) by looking at the data array's shape attribute, which shows that there are 1797 rows (samples) and 64 columns (features):

```
In [5]: digits.data.shape
Out[5]: (1797, 64)
```

You can confirm that the number of target values matches the number of samples by looking at the target array's shape:

```
In [6]: digits.target.shape
Out[6]: (1797,)
```

A Sample Digit Image
Each image is two-dimensional—it has a width and a height in pixels. The Bunch object returned by load_digits contains an images attribute—an array in which each element is a two-dimensional 8-by-8 array representing a digit image's pixel intensities. Though the original dataset represents each pixel as an integer value from 0–16, scikit-learn stores these values as *floating-point* values (NumPy type float64). For example, here's the two-dimensional array representing the sample image at index 13:

```
In [7]: digits.images[13]
Out[7]:
array([[ 0.,  2.,  9., 15., 14.,  9.,  3.,  0.],
       [ 0.,  4., 13.,  8.,  9., 16.,  8.,  0.],
       [ 0.,  0.,  0.,  6., 14., 15.,  3.,  0.],
       [ 0.,  0.,  0., 11., 14.,  2.,  0.,  0.],
       [ 0.,  0.,  0.,  2., 15., 11.,  0.,  0.],
       [ 0.,  0.,  0.,  0.,  2., 15.,  4.,  0.],
       [ 0.,  1.,  5.,  6., 13., 16.,  6.,  0.],
       [ 0.,  2., 12., 12., 13., 11.,  0.,  0.]])
```

and here's the image represented by this two-dimensional array—we'll soon show the code for displaying this image:

Preparing the Data for Use with Scikit-Learn

Scikit-learn's machine-learning algorithms require samples to be stored in a *two-dimensional array* of *floating-point values* (or two-dimensional *array-like* collection, such as a list of lists or a pandas DataFrame):

- Each row represents one *sample*.
- Each column in a given row represents one *feature* for that sample.

To represent every sample as one row, multi-dimensional data like the two-dimensional image array shown in snippet [7] must be *flattened* into a one-dimensional array.

If you were working with a data containing **categorical features** (typically represented as strings, such as 'spam' or 'not-spam'), you'd also have to *preprocess* those features into numerical values—known as one-hot encoding, which we cover in the next chapter. Scikit-learn's **sklearn.preprocessing** module provides capabilities for converting categorical data to numeric data. The Digits dataset has no categorical features.

For your convenience, the load_digits function returns the preprocessed data ready for machine learning. The Digits dataset is numerical, so load_digits simply flattens each image's two-dimensional array into a one-dimensional array. For example, the 8-by-8 array digits.images[13] shown in snippet [7] corresponds to the 1-by-64 array digits.data[13] shown below:

```
In [8]: digits.data[13]
Out[8]:
array([ 0.,  2.,  9., 15., 14.,  9.,  3.,  0.,  0.,  4., 13.,  8.,  9.,
       16.,  8.,  0.,  0.,  0.,  0.,  6., 14., 15.,  3.,  0.,  0.,  0.,
        0., 11., 14.,  2.,  0.,  0.,  0.,  0.,  0.,  2., 15., 11.,  0.,
        0.,  0.,  0.,  0.,  0.,  0.,  2., 15.,  4.,  0.,  0.,  1.,  5.,  6.,
       13., 16.,  6.,  0.,  0.,  2., 12., 12., 13., 11.,  0.,  0.])
```

In this one-dimensional array, the first eight elements are the two-dimensional array's row 0, the next eight elements are the two-dimensional array's row 1, and so on.

✓ Self Check

1 *(Fill-In)* A Bunch object's _____ and _____ attributes are NumPy arrays containing the dataset's samples and labels, respectively.
Answer: data, target.

2 *(True/False)* A scikit-learn Bunch object contains only a dataset's data.
Answer: False. A scikit-learn Bunch object contains a dataset's data and information about the dataset (called metadata), available through the DESCR attribute.

3 *(IPython Session)* For sample number 22 in the Digits dataset, display the 8-by-8 image data and numeric value of the digit the image represents.
Answer:

```
In [9]: digits.images[22]
Out[9]:
array([[ 0.,  0.,  8., 16.,  5.,  0.,  0.,  0.],
       [ 0.,  1., 13., 11., 16.,  0.,  0.,  0.],
       [ 0.,  0., 10.,  0., 13.,  3.,  0.,  0.],
       [ 0.,  0.,  3.,  1., 16.,  1.,  0.,  0.],
       [ 0.,  0.,  0.,  9., 12.,  0.,  0.,  0.],
       [ 0.,  0.,  3., 15.,  5.,  0.,  0.,  0.],
       [ 0.,  0., 14., 15.,  8.,  8.,  3.,  0.],
       [ 0.,  0.,  7., 12., 12., 12., 13.,  1.]])

In [10]: digits.target[22]
Out[10]: 2
```

15.2.3 Visualizing the Data

You should always familiarize yourself with your data. This process is called **data exploration**. For the digit images, you can get a sense of what they look like by displaying them with the Matplotlib `implot` function. The following image shows the dataset's first 24 images. To see how difficult a problem handwritten digit recognition is, consider the *variations* among the images of the 3s in the first, third and fourth rows, and look at the images of the 2s in the first, third and fourth rows.

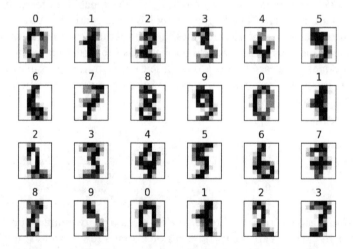

Creating the Diagram
Let's look at the code that displayed these 24 digits. The following call to function `subplots` creates a 6-by-4 inch `Figure` (specified by the `figsize(6, 4)` keyword argument) containing 24 subplots arranged in 4 rows (`nrows=4`) and 6 columns (`ncols=6`). Each subplot has its own `Axes` object, which we'll use to display one digit image:

```
In [11]: import matplotlib.pyplot as plt

In [12]: figure, axes = plt.subplots(nrows=4, ncols=6, figsize=(6, 4))
```

15.2 Classification with k-Nearest Neighbors and the Digits Dataset, Part 1

Function `subplots` returns the `Axes` objects in a two-dimensional NumPy array. Initially, the `Figure` appears as shown below with labels (which we'll remove) on every subplot's *x*- and *y*-axes:

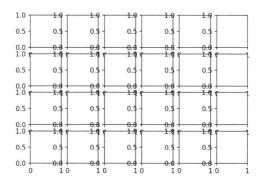

Displaying Each Image and Removing the Axes Labels

Next, use a `for` statement with the built-in `zip` function to iterate in parallel through the 24 Axes objects, the first 24 images in `digits.images` and the first 24 values in `digits.target`:

```
In [13]: for item in zip(axes.ravel(), digits.images, digits.target):
    ...:     axes, image, target = item
    ...:     axes.imshow(image, cmap=plt.cm.gray_r)
    ...:     axes.set_xticks([])  # remove x-axis tick marks
    ...:     axes.set_yticks([])  # remove y-axis tick marks
    ...:     axes.set_title(target)
    ...: plt.tight_layout()
    ...:
    ...:
```

Recall that NumPy array method `ravel` creates a *one-dimensional view* of a multidimensional array. Also, recall that each tuple `zip` produces contains elements from the same index in each of `zip`'s arguments and that argument with the fewest elements determines how many tuples `zip` returns.

Each iteration of the loop:

- Unpacks one tuple from the `zipped` items into three variables representing the Axes object, image and target value.

- Calls the Axes object's `imshow` method to display one image. The keyword argument `cmap=plt.cm.gray_r` determines the colors displayed in the image. The value `plt.cm.gray_r` is a **color map**—a group of colors that are typically chosen to work well together. This particular color map enables us to display the image's pixels in grayscale, with 0 as white, 16 as black and the values in between as gradually darkening shades of gray. For Matplotlib's color map names see https://matplotlib.org/examples/color/colormaps_reference.html. Each can be accessed through the `plt.cm` object or via a string, like `'gray_r'`.

- Calls the Axes object's `set_xticks` and `set_yticks` methods with empty lists to indicate that the *x*- and *y*-axes should not have tick marks.

- Calls the Axes object's `set_title` method to display the `target` value above the image—this shows the actual value that the image represents.

608 Machine Learning: Classification, Regression and Clustering

After the loop, we call `tight_layout` to remove the extra whitespace at the Figure's top, right, bottom and left, so the rows and columns of digit images can fill more of the Figure.

 Self Check

1 *(Fill-In)* The process of familiarizing yourself with your data is called _____ .
Answer: data exploration.

2 *(IPython Session)* Display the image for sample number 22 of the Digits dataset.
Answer:
```
In [14]: axes = plt.subplot()

In [15]: image = plt.imshow(digits.images[22], cmap=plt.cm.gray_r)

In [16]: xticks = axes.set_xticks([])

In [17]: yticks = axes.set_yticks([])
```

15.2.4 Splitting the Data for Training and Testing

You typically train a machine-learning model with a subset of a dataset. Typically, the more data you have for training, the better you can train the model. It's important to set aside a portion of your data for testing, so you can evaluate a model's performance using data that the model has not yet seen. Once you're confident that the model is performing well, you can use it to make predictions using new data it hasn't seen.

We first break the data into a **training set** and a **testing set** to prepare to train and test the model. The function `train_test_split` from the `sklearn.model_selection` module *shuffles* the data to randomize it, then splits the samples in the `data` array and the target values in the `target` array into training and testing sets. This helps ensure that the training and testing sets have similar characteristics. The shuffling and splitting is performed conveniently for you by a `ShuffleSplit` object from the `sklearn.model_selection` module. Function `train_test_split` returns a tuple of four elements in which the first two are the *samples* split into training and testing sets, and the last two are the corresponding *target values* split into training and testing sets. By convention, uppercase X is used to represent the samples, and lowercase y is used to represent the target values:

```
In [18]: from sklearn.model_selection import train_test_split

In [19]: X_train, X_test, y_train, y_test = train_test_split(
    ...:     digits.data, digits.target, random_state=11)
    ...:
```

We assume the data has **balanced classes**—that is, the samples are divided evenly among the classes. This is the case for each of scikit-learn's bundled classification datasets. Unbalanced classes could lead to incorrect results.

15.2 Classification with k-Nearest Neighbors and the Digits Dataset, Part 1

In the "Functions" chapter, you saw how to *seed* a random-number generator for *reproducibility*. In machine-learning studies, this helps others confirm your results by working with the *same* randomly selected data. Function train_test_split provides the keyword argument random_state for *reproducibility*. When you run the code in the future with the *same* seed value, train_test_split will select the *same* data for the training set and the *same* data for the testing set. We chose the seed value (11) arbitrarily.

Training and Testing Set Sizes

Looking at X_train's and X_test's shapes, you can see that, *by default*, train_test_split reserves 75% of the data for training and 25% for testing:

```
In [20]: X_train.shape
Out[20]: (1347, 64)

In [21]: X_test.shape
Out[21]: (450, 64)
```

To specify *different* splits, you can set the sizes of the testing and training sets with the train_test_split function's keyword arguments test_size and train_size. Use floating-point values from 0.0 through 1.0 to specify the percentages of the data to use for each. You can use integer values to set the precise numbers of samples. If you specify one of these keyword arguments, the other is inferred. For example, the statement

```
X_train, X_test, y_train, y_test = train_test_split(
    digits.data, digits.target, random_state=11, test_size=0.20)
```

specifies that 20% of the data is for testing, so train_size is inferred to be 0.80.

✓ Self Check

1 *(True/False)* You should typically use all of a dataset's data to train a model.
Answer: False. It's important to set aside a portion of your data for testing, so you can evaluate a model's performance using data that the model has not yet seen.

2 *(Discussion)* For the Digits dataset, what numbers of samples would the following statement reserve for training and testing purposes?

```
X_train, X_test, y_train, y_test = train_test_split(
    digits.data, digits.target, test_size=0.40)
```

Answer: 1078 and 719.

15.2.5 Creating the Model

The **KNeighborsClassifier** estimator (module **sklearn.neighbors**) implements the k-nearest neighbors algorithm. First, we create the KNeighborsClassifier estimator object:

```
In [22]: from sklearn.neighbors import KNeighborsClassifier

In [23]: knn = KNeighborsClassifier()
```

To create an estimator, you simply create an object. The internal details of how this object implements the k-nearest neighbors algorithm are hidden in the object. You'll simply call its methods. This is the essence of Python *object-based programming*.

15.2.6 Training the Model

Next, we invoke the `KNeighborsClassifier` object's **fit method**, which loads the sample training set (X_train) and target training set (y_train) into the estimator:

```
In [24]: knn.fit(X=X_train, y=y_train)
Out[24]:
KNeighborsClassifier(algorithm='auto', leaf_size=30, metric='minkowski',
        metric_params=None, n_jobs=None, n_neighbors=5, p=2,
        weights='uniform')
```

For most, scikit-learn estimators, the `fit` method loads the data into the estimator then uses that data to perform complex calculations behind the scenes that learn from the data and train the model. The `KNeighborsClassifier`'s `fit` method just loads the data into the estimator, because k-NN actually has no initial learning process. The estimator is said to be *lazy* because its work is performed only when you use it to make predictions. In this and the next chapter, you'll use lots of models that have significant training phases. In the real-world machine-learning applications, it can sometimes take minutes, hours, days or even months to train your models. We'll see in the next chapter, "Deep Learning," that special-purpose, high-performance hardware called GPUs and TPUs can significantly reduce model training time.

As shown in snippet [24]'s output, the `fit` method returns the estimator, so IPython displays its string representation, which includes the estimator's *default* settings. The `n_neighbors` value corresponds to *k* in the k-nearest neighbors algorithm. By default, a `KNeighborsClassifier` looks at the five nearest neighbors to make its predictions. For simplicity, we generally use the default estimator settings. For `KNeighborsClassifier`, these are described at:

> http://scikit-learn.org/stable/modules/generated/
>   sklearn.neighbors.KNeighborsClassifier.html

Many of these settings are beyond the scope of this book. In Part 2 of this case study, we'll discuss how to choose the best value for `n_neighbors`.

 Self Check

1 *(Fill-In)* The `KNeighborsClassifier` is said to be _____ because its work is performed only when you use it to make predictions.
Answer: lazy.

2 *(True/False)* Each scikit-learn estimator's `fit` method simply loads a dataset.
Answer: False. For most, scikit-learn estimators, the `fit` method loads the data into the estimator then uses that data to perform complex calculations behind the scenes that learn from the data and train the model.

15.2.7 Predicting Digit Classes

Now that we've loaded the data into the `KNeighborsClassifier`, we can use it with the test samples to make predictions. Calling the estimator's **predict method** with X_test as an argument returns an array containing the predicted class of each test image:

```
In [25]: predicted = knn.predict(X=X_test)

In [26]: expected = y_test
```

15.2 Classification with k-Nearest Neighbors and the Digits Dataset, Part 1

Let's look at the predicted digits vs. expected digits for the first 20 test samples:

```
In [27]: predicted[:20]
Out[27]: array([0, 4, 9, 9, 3, 1, 4, 1, 5, 0, 4, 9, 4, 1, 5, 3, 3, 8, 5, 6])

In [28]: expected[:20]
Out[28]: array([0, 4, 9, 9, 3, 1, 4, 1, 5, 0, 4, 9, 4, 1, 5, 3, 3, 8, 3, 6])
```

As you can see, in the first 20 elements, only the predicted and expected arrays' values at index 18 do not match. We expected a 3, but the model predicted a 5.

Let's use a list comprehension to locate *all* the incorrect predictions for the *entire* test set—that is, the cases in which the predicted and expected values do *not* match:

```
In [29]: wrong = [(p, e) for (p, e) in zip(predicted, expected) if p != e]

In [30]: wrong
Out[30]:
[(5, 3),
 (8, 9),
 (4, 9),
 (7, 3),
 (7, 4),
 (2, 8),
 (9, 8),
 (3, 8),
 (3, 8),
 (1, 8)]
```

The list comprehension uses zip to create tuples containing the corresponding elements in predicted and expected. We include a tuple in the result only if its p (the predicted value) and e (the expected value) differ—that is, the predicted value was incorrect. In this example, the estimator incorrectly predicted only 10 of the 450 test samples. So the prediction accuracy of this estimator is an impressive 97.78%, even though we used only the estimator's default parameters.

✓ Self Check

1 *(IPython Session)* Using the predicted and expected arrays, calculate and display the prediction accuracy percentage.
Answer:

```
In [31]: print(f'{(len(expected) - len(wrong)) / len(expected):.2%}')
97.78%
```

2 *(IPython Session)* Rewrite the list comprehension in snippet [29] using a for loop. Which coding style do you prefer?
Answer:

```
In [32]: wrong = []

In [33]: for p, e in zip(predicted, expected):
   ...:     if p != e:
   ...:         wrong.append((p, e))

In [34]: wrong
Out[34]:
[(5, 3),
 (8, 9),
```

```
          (4, 9),
          (7, 3),
          (7, 4),
          (2, 8),
          (9, 8),
          (3, 8),
          (3, 8),
          (1, 8)]
```

15.3 Case Study: Classification with k-Nearest Neighbors and the Digits Dataset, Part 2

In this section, we continue the digit classification case study. We'll:

- evaluate the k-NN classification estimator's accuracy,
- execute multiple estimators and can compare their results so you can choose the best one(s), and
- show how to tune k-NN's hyperparameter k to get the best performance out of a `KNeighborsClassifier`.

15.3.1 Metrics for Model Accuracy

Once you've trained and tested a model, you'll want to measure its accuracy. Here, we'll look at two ways of doing this—a classification estimator's `score` method and a *confusion matrix*.

Estimator Method `score`

Each estimator has a **`score`** method that returns an indication of how well the estimator performs for the test data you pass as arguments. For classification estimators, this method returns the *prediction accuracy* for the test data:

```
In [35]: print(f'{knn.score(X_test, y_test):.2%}')
97.78%
```

The `kNeighborsClassifier`'s with its default k (that is, n_neighbors=5) achieved 97.78% prediction accuracy. Shortly, we'll perform hyperparameter tuning to try to determine the optimal value for k, hoping that we get even better accuracy.

Confusion Matrix

Another way to check a classification estimator's accuracy is via a **confusion matrix**, which shows the correct and incorrect predicted values (also known as the *hits* and *misses*) for a given class. Simply call the function `confusion_matrix` from the `sklearn.metrics` module, passing the expected classes and the predicted classes as arguments, as in:

```
In [36]: from sklearn.metrics import confusion_matrix

In [37]: confusion = confusion_matrix(y_true=expected, y_pred=predicted)
```

The y_true keyword argument specifies the test samples' actual classes. People looked at the dataset's images and labeled them with specific classes (the digit values). The y_pred keyword argument specifies the predicted digits for those test images.

15.3 Classification with k-Nearest Neighbors and the Digits Dataset, Part 2

Below is the confusion matrix produced by the preceding call. The correct predictions are shown on the diagonal from top-left to bottom-right. This is called the **principal diagonal**. The nonzero values that are not on the principal diagonal indicate incorrect predictions:

```
In [38]: confusion
Out[38]:
array([[45,  0,  0,  0,  0,  0,  0,  0,  0,  0],
       [ 0, 45,  0,  0,  0,  0,  0,  0,  0,  0],
       [ 0,  0, 54,  0,  0,  0,  0,  0,  0,  0],
       [ 0,  0,  0, 42,  0,  1,  0,  1,  0,  0],
       [ 0,  0,  0,  0, 49,  0,  0,  1,  0,  0],
       [ 0,  0,  0,  0,  0, 38,  0,  0,  0,  0],
       [ 0,  0,  0,  0,  0,  0, 42,  0,  0,  0],
       [ 0,  0,  0,  0,  0,  0,  0, 45,  0,  0],
       [ 0,  1,  1,  2,  0,  0,  0,  0, 39,  1],
       [ 0,  0,  0,  0,  1,  0,  0,  0,  1, 41]])
```

Each row represents one distinct class—that is, one of the digits 0–9. The columns within a row specify how many of the test samples were classified into each distinct class. For example, row 0:

```
[45,  0,  0,  0,  0,  0,  0,  0,  0,  0]
```

represents the digit 0 class. The columns represent the ten possible target classes 0 through 9. Because we're working with digits, the classes (0–9) and the row and column index numbers (0–9) happen to match. According to row 0, 45 test samples were classified as the digit 0, and *none* of the test samples were misclassified as any of the digits 1 through 9. So 100% of the 0s were correctly predicted.

On the other hand, consider row 8 which represents the results for the digit 8:

```
[ 0,  1,  1,  2,  0,  0,  0,  0, 39,  1]
```

- The 1 at column index 1 indicates that one 8 was *incorrectly* classified as a 1.
- The 1 at column index 2 indicates that one 8 was *incorrectly* classified as a 2.
- The 2 at column index 3 indicates that two 8s were *incorrectly* classified as 3s.
- The 39 at column index 8 indicates that 39 8s were *correctly* classified as 8s.
- The 1 at column index 9 indicates that one 8 was *incorrectly* classified as a 9.

So the algorithm correctly predicted 88.63% (39 of 44) of the 8s. Earlier we saw that the overall prediction accuracy of this estimator was 97.78%. The lower prediction accuracy for 8s indicates that they're apparently harder to recognize than the other digits.

Classification Report

The sklearn.metrics module also provides function **classification_report**, which produces a table of **classification metrics**[5] based on the expected and predicted values:

```
In [39]: from sklearn.metrics import classification_report

In [40]: names = [str(digit) for digit in digits.target_names]
```

---

5. http://scikit-learn.org/stable/modules/model_evaluation.html#precision-recall-and-f-measures.

```
In [41]: print(classification_report(expected, predicted,
   ...:     target_names=names))
   ...:
              precision    recall  f1-score   support

           0       1.00      1.00      1.00        45
           1       0.98      1.00      0.99        45
           2       0.98      1.00      0.99        54
           3       0.95      0.95      0.95        44
           4       0.98      0.98      0.98        50
           5       0.97      1.00      0.99        38
           6       1.00      1.00      1.00        42
           7       0.96      1.00      0.98        45
           8       0.97      0.89      0.93        44
           9       0.98      0.95      0.96        43

   micro avg       0.98      0.98      0.98       450
   macro avg       0.98      0.98      0.98       450
weighted avg       0.98      0.98      0.98       450
```

In the report:

- **precision** is the total number of correct predictions for a given digit divided by the total number of predictions for that digit. You can confirm the precision by looking at each column in the confusion matrix. For example, if you look at column index 7, you'll see 1s in rows 3 and 4, indicating that one 3 and one 4 were incorrectly classified as 7s and a 45 in row 7 indicating the 45 images were correctly classified as 7s. So the *precision* for the digit 7 is 45/47 or 0.96.

- **recall** is the total number of correct predictions for a given digit divided by the total number of samples that should have been predicted as that digit. You can confirm the recall by looking at each row in the confusion matrix. For example, if you look at row index 8, you'll see three 1s and a 2 indicating that some 8s were incorrectly classified as other digits and a 39 indicating that 39 images were correctly classified. So the *recall* for the digit 8 is 39/44 or 0.89.

- **f1-score**—This is the average of the *precision* and the *recall*.

- **support**—The number of samples with a given expected value. For example, 50 samples were labeled as 4s, and 38 samples were labeled as 5s.

For details on the averages displayed at the bottom of the report, see:

> http://scikit-learn.org/stable/modules/generated/
> sklearn.metrics.classification_report.html

Visualizing the Confusion Matrix

A **heat map** displays values as colors, often with values of higher magnitude displayed as more intense colors. Seaborn's graphing functions work with two-dimensional data. When using a pandas `DataFrame` as the data source, Seaborn automatically labels its visualizations using the column names and row indices. Let's convert the confusion matrix into a `DataFrame`, then graph it:

```
In [42]: import pandas as pd
```

15.3 Classification with k-Nearest Neighbors and the Digits Dataset, Part 2

```
In [43]: confusion_df = pd.DataFrame(confusion, index=range(10),
    ...:     columns=range(10))
    ...:

In [44]: import seaborn as sns

In [45]: axes = sns.heatmap(confusion_df, annot=True,
    ...:                    cmap='nipy_spectral_r')
    ...:
```

The Seaborn function **heatmap** creates a heat map from the specified DataFrame. The keyword argument annot=True (short for "annotation") displays a color bar to the right of the diagram, showing how the values correspond to the heat map's colors. The cmap='nipy_spectral_r' keyword argument specifies which color map to use. We used the nipy_spectral_r color map with the colors shown in the heat map's color bar. When you display a confusion matrix as a heat map, the principal diagonal and the incorrect predictions stand out nicely.

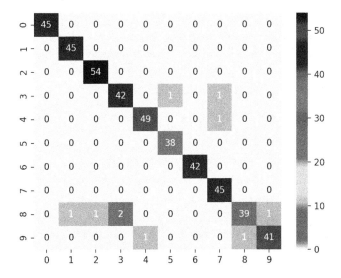

✓ Self Check

1 *(Fill-In)* A Seaborn _____ displays values as colors, often with values of higher magnitude displayed as more intense colors.
Answer: heat map.

2 *(True/False)* In a classification report, the precision specifies the total number of correct predictions for a class divided by the total number of samples for that class.
Answer: True.

3 *(Discussion)* Explain row 3 of the confusion matrix presented in this section:

[0, 0, 0, 42, 0, 1, 0, 1, 0, 0]

Answer: The number 42 in column index 3 indicates that 42 3s were correctly predicted as 3s. The number 1 at column indices 5 and 7 indicates that one 3 was incorrectly classified as a 5 and one was incorrectly classified as a 7.

15.3.2 K-Fold Cross-Validation

K-fold cross-validation enables you to use all of your data for *both* training *and* testing, to get a better sense of how well your model will make predictions for new data by repeatedly training and testing the model with different portions of the dataset. K-fold cross-validation splits the dataset into k equal-size **folds** (this k is unrelated to k in the k-nearest neighbors algorithm). You then repeatedly train your model with $k-1$ folds and test the model with the remaining fold. For example, consider using $k = 10$ with folds numbered 1 through 10. With 10 folds, we'd do 10 successive training and testing cycles:

- First, we'd train with folds 1–9, then test with fold 10.
- Next, we'd train with folds 1–8 and 10, then test with fold 9.
- Next, we'd train with folds 1–7 and 9–10, then test with fold 8.

This training and testing cycle continues until each fold has been used to test the model.

KFold Class

Scikit-learn provides the **KFold** class and the **cross_val_score** function (both in the module sklearn.model_selection) to help you perform the training and testing cycles described above. Let's perform k-fold cross-validation with the Digits dataset and the KNeighborsClassifier created earlier. First, create a KFold object:

```
In [46]: from sklearn.model_selection import KFold

In [47]: kfold = KFold(n_splits=10, random_state=11, shuffle=True)
```

The keyword arguments are:

- n_splits=10, which specifies the number of folds.
- random_state=11, which seeds the random number generator for *reproducibility*.
- shuffle=True, which causes the KFold object to randomize the data by shuffling it before splitting it into folds. This is particularly important if the samples might be ordered or grouped. For example, the Iris dataset we'll use later in this chapter has 150 samples of three *Iris* species—the first 50 are *Iris setosa*, the next 50 are *Iris versicolor* and the last 50 are *Iris virginica*. If we do not shuffle the samples, then the training data might contain none of a particular *Iris* species and the test data might be all of one species.

Using the KFold Object with Function cross_val_score

Next, use function cross_val_score to train and test your model:

```
In [48]: from sklearn.model_selection import cross_val_score

In [49]: scores = cross_val_score(estimator=knn, X=digits.data,
    ...:     y=digits.target, cv=kfold)
    ...:
```

The keyword arguments are:

- estimator=knn, which specifies the estimator you'd like to validate.
- X=digits.data, which specifies the samples to use for training and testing.
- y=digits.target, which specifies the target predictions for the samples.

- cv=kfold, which specifies the cross-validation generator that defines how to split the samples and targets for training and testing.

Function `cross_val_score` returns an array of accuracy scores—one for each fold. As you can see below, the model was quite accurate. Its *lowest* accuracy score was 0.97777778 (97.78%) and in one case it was 100% accurate in predicting an entire fold:

```
In [50]: scores
Out[50]:
array([0.97777778, 0.99444444, 0.98888889, 0.97777778, 0.98888889,
       0.99444444, 0.97777778, 0.98882682, 1.        , 0.98324022])
```

Once you have the accuracy scores, you can get an overall sense of the model's accuracy by calculating the mean accuracy score and the standard deviation among the 10 accuracy scores (or whatever number of folds you choose):

```
In [51]: print(f'Mean accuracy: {scores.mean():.2%}')
Mean accuracy: 98.72%

In [52]: print(f'Accuracy standard deviation: {scores.std():.2%}')
Accuracy standard deviation: 0.75%
```

On average, the model was 98.72% accurate—even better than the 97.78% we achieved when we trained the model with 75% of the data and tested the model with 25% earlier.

✓ Self Check

1. *(True/False)* Randomizing the data by shuffling it before splitting it into folds is particularly important if the samples might be ordered or grouped.
Answer: True.

2. *(True/False)* When you call `cross_val_score` to peform k-fold cross-validation, the function returns the best score produced while testing the model with each fold.
Answer: False. The function returns an array containing the scores for each fold. The mean of those scores is the estimator's overall score.

15.3.3 Running Multiple Models to Find the Best One

It's difficult to know in advance which machine learning model(s) will perform best for a given dataset, especially when they hide the details of how they operate from their users. Even though the `KNeighborsClassifier` predicts digit images with a high degree of accuracy, it's possible that other scikit-learn estimators are even more accurate. Scikit-learn provides many models with which you can quickly train and test your data. This encourages you to run *multiple models* to determine which is the best for a particular machine learning study.

Let's use the techniques from the preceding section to compare several classification estimators—`KNeighborsClassifier`, `SVC` and `GaussianNB` (there are more). Though we have not studied the `SVC` and `GaussianNB` estimators, scikit-learn nevertheless makes it easy for you to test-drive them by using their default settings.[6] First, let's import the other two estimators:

---

6. To avoid a warning in the current scikit-learn version at the time of this writing (version 0.20), we supplied one keyword argument when creating the SVC estimator. This argument's value will become the default in scikit-learn version 0.22.

```
In [53]: from sklearn.svm import SVC

In [54]: from sklearn.naive_bayes import GaussianNB
```

Next, let's create the estimators. The following dictionary contains key–value pairs for the existing `KNeighborsClassifier` we created earlier, plus new `SVC` and `GaussianNB` estimators:

```
In [55]: estimators = {
    ...:     'KNeighborsClassifier': knn,
    ...:     'SVC': SVC(gamma='scale'),
    ...:     'GaussianNB': GaussianNB()}
    ...:
```

Now, we can execute the models:

```
In [56]: for estimator_name, estimator_object in estimators.items():
    ...:     kfold = KFold(n_splits=10, random_state=11, shuffle=True)
    ...:     scores = cross_val_score(estimator=estimator_object,
    ...:         X=digits.data, y=digits.target, cv=kfold)
    ...:     print(f'{estimator_name:>20}: ' +
    ...:         f'mean accuracy={scores.mean():.2%}; ' +
    ...:         f'standard deviation={scores.std():.2%}')
    ...:
KNeighborsClassifier: mean accuracy=98.72%; standard deviation=0.75%
                 SVC: mean accuracy=99.00%; standard deviation=0.85%
          GaussianNB: mean accuracy=84.48%; standard deviation=3.47%
```

This loop iterates through items in the `estimators` dictionary and for each key-value pair performs the following tasks:

- Unpacks the key into `estimator_name` and value into `estimator_object`.
- Creates a `KFold` object that shuffles the data and produces 10 folds. The keyword argument `random_state` is particularly important here because it ensures that each estimator works with identical folds, so we're comparing "apples to apples."
- Evaluates the current `estimator_object` using `cross_val_score`.
- Prints the estimator's name, followed by the mean and standard deviation of the accuracy scores' computed for each of the 10 folds.

Based on the results, it appears that we can get slightly better accuracy from the SVC estimator—at least when using the estimator's default settings. It's possible that by tuning some of the estimators' settings, we could get even better results. The `KNeighborsClassifier` and `SVC` estimators' accuracies are nearly identical so we might want to perform hyperparameter tuning on each to determine the best.

Scikit-Learn Estimator Diagram
The scikit-learn documentation provides a helpful diagram for choosing the right estimator, based on the kind and size of your data and the machine learning task you wish to perform:

> https://scikit-learn.org/stable/tutorial/machine_learning_map/index.html

Self Check

1 *(True/False)* You should choose the best estimator before performing your machine learning study.
Answer: False. It's difficult to know in advance which machine learning model(s) will perform best for a given dataset, especially when they hide the details of how they operate from their users. For this reason, you should run multiple models to determine which is the best for your study.

2 *(Discussion)* How would you modify the code in this section so that it would also test a `LinearSVC` estimator?
Answer: You'd import the `LinearSVC` class, add a key–value pair to the `estimators` dictionary (`'LinearSVC': LinearSVC()`), then execute the `for` loop, which tests every estimator in the dictionary.

15.3.4 Hyperparameter Tuning

Earlier in this section, we mentioned that *k* in the k-nearest neighbors algorithm is a hyperparameter of the algorithm. Hyperparameters are set *before* using the algorithm to train your model. In real-world machine learning studies, you'll want to use hyperparameter tuning to choose hyperparameter values that produce the best possible predictions.

To determine the best value for *k* in the kNN algorithm, try different values of *k* then compare the estimator's performance with each. We can do this using techniques similar to comparing estimators. The following loop creates `KNeighborsClassifiers` with odd *k* values from 1 through 19 (again, we use odd *k* values in kNN to avoid ties) and performs k-fold cross-validation on each. As you can see from the accuracy scores and standard deviations, the *k* value 1 in kNN produces the most accurate predictions for the Digits dataset. You can also see that accuracy tends to decrease for higher *k* values:

```
In [57]: for k in range(1, 20, 2):
    ...:     kfold = KFold(n_splits=10, random_state=11, shuffle=True)
    ...:     knn = KNeighborsClassifier(n_neighbors=k)
    ...:     scores = cross_val_score(estimator=knn,
    ...:         X=digits.data, y=digits.target, cv=kfold)
    ...:     print(f'k={k:<2}; mean accuracy={scores.mean():.2%}; ' +
    ...:         f'standard deviation={scores.std():.2%}')
    ...:
k=1 ; mean accuracy=98.83%; standard deviation=0.58%
k=3 ; mean accuracy=98.78%; standard deviation=0.78%
k=5 ; mean accuracy=98.72%; standard deviation=0.75%
k=7 ; mean accuracy=98.44%; standard deviation=0.96%
k=9 ; mean accuracy=98.39%; standard deviation=0.80%
k=11; mean accuracy=98.39%; standard deviation=0.80%
k=13; mean accuracy=97.89%; standard deviation=0.89%
k=15; mean accuracy=97.89%; standard deviation=1.02%
k=17; mean accuracy=97.50%; standard deviation=1.00%
k=19; mean accuracy=97.66%; standard deviation=0.96%
```

Machine learning is not without its costs, especially as we head toward big data and deep learning. You must "know your data" and "know your tools." For example, compute time grows rapidly with *k*, because k-NN needs to perform more calculations to find the nearest neighbors. In an exercise, we'll ask you to try the function `cross_validate`, which does cross-validation *and* times the results.

Self Check

1 *(True/False)* When you create an estimator object, the default hyperparameter values that scikit-learn uses are generally the best ones for every machine learning study.
Answer: False. The default hyperparameter values make it easy for you to test estimators quickly. In real-world machine learning studies, you'll want to use hyperparameter tuning to choose hyperparameter values that produce the best possible predictions.

15.4 Case Study: Time Series and Simple Linear Regression

In the previous section, we demonstrated classification in which each sample was associated with a *distinct* class. Here, we continue our discussion of simple linear regression—the simplest of the regression algorithms—that began in Chapter 10's Intro to Data Science section. Recall that given a collection of numeric values representing an independent variable and a dependent variable, simple linear regression describes the relationship between these variables with a straight line, known as the regression line.

Previously, we performed simple linear regression on a time series of average New York City January high-temperature data for 1895 through 2018. In that example, we used Seaborn's `regplot` function to create a scatter plot of the data with a corresponding regression line. We also used the `scipy.stats` module's `linregress` function to calculate the regression line's slope and intercept. We then used those values to predict future temperatures and estimate past temperatures.

In this section, we'll

- use a *scikit-learn estimator* to reimplement the simple linear regression we showed in Chapter 10,

- use Seaborn's `scatterplot` function to plot the data and Matplotlib's `plot` function to display the regression line, then

- use the coefficient and intercept values calculated by the scikit-learn estimator to make predictions.

Later, we'll look at *multiple linear regression* (also simply called *linear regression*).

For your convenience, we provide the temperature data in the ch15 examples folder in a CSV file named ave_hi_nyc_jan_1895-2018.csv. Once again, launch IPython with the --matplotlib option:

```
ipython --matplotlib
```

Loading the Average High Temperatures into a DataFrame

As we did in Chapter 10, let's load the data from ave_hi_nyc_jan_1895-2018.csv, rename the 'Value' column to 'Temperature', remove 01 from the end of each date value and display a few data samples:

```
In [1]: import pandas as pd

In [2]: nyc = pd.read_csv('ave_hi_nyc_jan_1895-2018.csv')

In [3]: nyc.columns = ['Date', 'Temperature', 'Anomaly']

In [4]: nyc.Date = nyc.Date.floordiv(100)
```

```
In [5]: nyc.head(3)
Out[5]:
   Date  Temperature  Anomaly
0  1895         34.2     -3.2
1  1896         34.7     -2.7
2  1897         35.5     -1.9
```

Splitting the Data for Training and Testing

In this example, we'll use the **LinearRegression** estimator from **sklearn.linear_model**. By default, this estimator uses *all* the numeric features in a dataset, performing a **multiple linear regression** (which we'll discuss in the next section). Here, we perform *simple linear regression* using *one* feature as the independent variable. So, we'll need to select one feature (the Date) from the dataset.

When you select one column from a two-dimensional DataFrame, the result is a *one-dimensional* Series. However, scikit-learn estimators require their training and testing data to be *two-dimensional arrays* (or two-dimensional *array-like* data, such as lists of lists or pandas DataFrames). To use one-dimensional data with an estimator, you must transform it from one dimension containing *n* elements, into two dimensions containing *n rows* and one *column* as you'll see below.

As we did in the previous case study, let's split the data into training and testing sets. Once again, we used the keyword argument random_state for reproducibility:

```
In [6]: from sklearn.model_selection import train_test_split

In [7]: X_train, X_test, y_train, y_test = train_test_split(
   ...:     nyc.Date.values.reshape(-1, 1), nyc.Temperature.values,
   ...:     random_state=11)
   ...:
```

The expression nyc.Date returns the Date column's Series, and the Series' values attribute returns the NumPy array containing that Series' values. To transform this one-dimensional array into two dimensions, we call the array's **reshape** method. Normally, two arguments are the precise number of rows and columns. However, the first argument -1 tells reshape to *infer* the number of rows, based on the number of columns (1) and the number of elements (124) in the array. The transformed array will have only one column, so reshape infers the number of rows to be 124, because the only way to fit 124 elements into an array with one column is by distributing them over 124 rows.

We can confirm the 75%–25% train-test split by checking the shapes of X_train and X_test:

```
In [8]: X_train.shape
Out[8]: (93, 1)

In [9]: X_test.shape
Out[9]: (31, 1)
```

Training the Model

Scikit-learn does not have a separate class for simple linear regression because it's just a special case of multiple linear regression, so let's train a LinearRegression estimator:

```
In [10]: from sklearn.linear_model import LinearRegression
```

```
In [11]: linear_regression = LinearRegression()

In [12]: linear_regression.fit(X=X_train, y=y_train)
Out[12]:
LinearRegression(copy_X=True, fit_intercept=True, n_jobs=None,
         normalize=False)
```

After training the estimator, `fit` returns the estimator, and IPython displays its string representation. For descriptions of the default settings, see:

> http://scikit-learn.org/stable/modules/generated/
> sklearn.linear_model.LinearRegression.html

To find the best fitting regression line for the data, the `LinearRegression` estimator iteratively adjusts the slope and intercept values to minimize the sum of the squares of the data points' distances from the line. In Chapter 10's Intro to Data Science section, we gave some insight into how the slope and intercept values are discovered.

Now, we can get the slope and intercept used in the $y = mx + b$ calculation to make predictions. The slope is stored in the estimator's **coeff_** attribute (m in the equation) and the intercept is stored in the estimator's **intercept_** attribute (b in the equation):

```
In [13]: linear_regression.coef_
Out[13]: array([0.01939167])

In [14]: linear_regression.intercept_
Out[14]: -0.30779820252656265
```

We'll use these later to plot the regression line and make predictions for specific dates.

Testing the Model
Let's test the model using the data in X_test and check some of the predictions throughout the dataset by displaying the `predicted` and `expected` values for every fifth element—we discuss how to assess the regression model's accuracy in Section 15.5.8:

```
In [15]: predicted = linear_regression.predict(X_test)

In [16]: expected = y_test

In [17]: for p, e in zip(predicted[::5], expected[::5]):
    ...:     print(f'predicted: {p:.2f}, expected: {e:.2f}')
    ...:
predicted: 37.86, expected: 31.70
predicted: 38.69, expected: 34.80
predicted: 37.00, expected: 39.40
predicted: 37.25, expected: 45.70
predicted: 38.05, expected: 32.30
predicted: 37.64, expected: 33.80
predicted: 36.94, expected: 39.70
```

Predicting Future Temperatures and Estimating Past Temperatures
Let's use the coefficient and intercept values to predict the January 2019 average high temperature and to estimate what the average high temperature was in January of 1890. The `lambda` in the following snippet implements the equation for a line

$$y = mx + b$$

using the `coef_` as m and the `intercept_` as b.

15.4 Case Study: Time Series and Simple Linear Regression

```
In [18]: predict = (lambda x: linear_regression.coef_ * x +
   ...:                       linear_regression.intercept_)
   ...:

In [19]: predict(2019)
Out[19]: array([38.84399018])

In [20]: predict(1890)
Out[20]: array([36.34246432])
```

Visualizing the Dataset with the Regression Line
Next, let's create a scatter plot of the dataset using Seaborn's **scatterplot** function and Matplotlib's **plot** function. First, use scatterplot with the nyc DataFrame to display the data points:

```
In [21]: import seaborn as sns

In [22]: axes = sns.scatterplot(data=nyc, x='Date', y='Temperature',
   ...:      hue='Temperature', palette='winter', legend=False)
   ...:
```

The keyword arguments are:

- data, which specifies the DataFrame (nyc) containing the data to display.
- x and y, which specify the names of nyc's columns that are the source of the data along the *x*- and *y*-axes, respectively. In this case, x is the 'Date' and y is the 'Temperature'. The corresponding values from each column form *x-y* coordinate pairs used to plot the dots.
- hue, which specifies which column's data should be used to determine the dot colors. In this case, we use the 'Temperature' column. Color is not particularly important in this example, but we wanted to add some visual interest to the graph.
- palette, which specifies a Matplotlib color map from which to choose the dots' colors.
- legend=False, which specifies that scatterplot should not show a legend for the graph—the default is True, but we do not need a legend for this example.

As we did in Chapter 10, let's scale the y-axis range of values so you'll be able to see the linear relationship better once we display the regression line:

```
In [23]: axes.set_ylim(10, 70)
Out[23]: (10, 70)
```

Next, let's display the regression line. First, create an array containing the minimum and maximum date values in nyc.Date. These are the *x*-coordinates of the regression line's start and end points:

```
In [24]: import numpy as np

In [25]: x = np.array([min(nyc.Date.values), max(nyc.Date.values)])
```

Passing the array x to the predict lambda from snippet [16] produces an array containing the corresponding predicted values, which we'll use as the *y*-coordinates:

```
In [26]: y = predict(x)
```

Finally, we can use Matplotlib's `plot` function to plot a line based on the x and y arrays, which represent the *x*- and *y*-coordinates of the points, respectively:

```
In [27]: import matplotlib.pyplot as plt

In [28]: line = plt.plot(x, y)
```

The resulting scatterplot and regression line are shown below. This graph is nearly identical to the one you saw in Chapter 10's Intro to Data Science section.

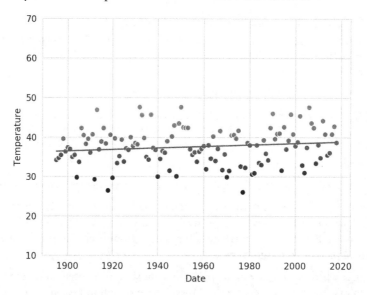

Overfitting/Underfitting

When creating a model, a key goal is to ensure that it is capable of making accurate predictions for data it has not yet seen. Two common problems that prevent accurate predictions are overfitting and underfitting:

- **Underfitting** occurs when a model is too simple to make predictions, based on its training data. For example, you may use a linear model, such as simple linear regression, when in fact, the problem really requires a non-linear model. For example, temperatures vary significantly throughout the four seasons. If you're trying to create a general model that can predict temperatures year-round, a simple linear regression model will underfit the data.

- **Overfitting** occurs when your model is too complex. The most extreme case, would be a model that memorizes its training data. That may be acceptable if your new data looks *exactly* like your training data, but ordinarily that's not the case. When you make predictions with an overfit model, new data that matches the training data will produce perfect predictions, but the model will not know what to do with data it has never seen.

For additional information on underfitting and overfitting, see

- https://en.wikipedia.org/wiki/Overfitting

- https://machinelearningmastery.com/overfitting-and-underfitting-with-machine-learning-algorithms/

✓ Self Check

1. *(Fill-In)* A LinearRegression object's _____ and _____ attributes can be used as *m* and *b*, respectively, in the equation $y = mx + b$ to make predictions.
Answer: coeff_, intercept_.

2. *(True/False)* By default, the LinearRegression estimator performs simple linear regression.
Answer: False. By default, the LinearRegression estimator uses all the numeric features in a dataset, performing a multiple linear regression.

3. *(IPython Session)* Use the predict lambda to estimate what the average January high temperature was in 1889 and to predict what it will be in 2020.
Answer:

```
In [29]: predict(1889)
Out[29]: array([36.34246432])

In [30]: predict(2100)
Out[30]: array([38.86338185])
```

15.5 Case Study: Multiple Linear Regression with the California Housing Dataset

In Chapter 10's Intro to Data Science section, we performed simple linear regression on a small weather data time series using pandas, Seaborn's regplot function and the SciPy's stats module's linregress function. In the previous section, we reimplemented that same example using scikit-learn's LinearRegression estimator, Seaborn's scatterplot function and Matplotlib's plot function. Now, we'll perform linear regression with a much larger real-world dataset.

The **California Housing dataset**[7] bundled with scikit-learn has 20,640 samples, each with eight numerical features. We'll perform a *multiple linear regression* that uses all eight numerical features to make more sophisticated housing price predictions than if we were to use only a single feature or a subset of the features. Once again, scikit-learn will do most of the work for you—LinearRegression performs multiple linear regression by default. In the exercises, we'll ask you to perform simple linear regressions with each individual feature and compare the results with this section's multiple linear regression. You should expect more meaningful results from the multiple linear regression.

We'll visualize some of the data using Matplotlib and Seaborn, so launch IPython with Matplotlib support:

```
ipython --matplotlib
```

---

7. http://lib.stat.cmu.edu/datasets. Pace, R. Kelley and Ronald Barry, Sparse Spatial Autoregressions, Statistics and Probability Letters, 33 (1997) 291-297. Submitted to the StatLib Datasets Archive by Kelley Pace (kpace@unix1.sncc.lsu.edu). [9/Nov/99].

15.5.1 Loading the Dataset

According to the California Housing Prices dataset's description in scikit-learn, "This dataset was derived from the 1990 U.S. census, using one row per census block group. A block group is the smallest geographical unit for which the U.S. Census Bureau publishes sample data (a block group typically has a population of 600 to 3,000 people)." The dataset has 20,640 samples—one per block group—with eight features each:

- median income—in tens of thousands, so 8.37 would represent $83,700
- median house age—in the dataset, the maximum value for this feature is 52
- average number of rooms
- average number of bedrooms
- block population
- average house occupancy
- house block latitude
- house block longitude

Each sample also has as its *target* a corresponding median house value in hundreds of thousands, so 3.55 would represent $355,000. In the dataset, the maximum value for this feature is 5, which represents $500,000.

It's reasonable to expect that more bedrooms or more rooms or higher income would mean higher house value. By combining these features to make predictions, we're more likely to get more accurate predictions.

Loading the Data

Let's load the dataset and familiarize ourselves with it. The `fetch_california_housing` function from the `sklearn.datasets` module returns a `Bunch` object containing the data and other information about the dataset:

```
In [1]: from sklearn.datasets import fetch_california_housing

In [2]: california = fetch_california_housing()
```

Displaying the Dataset's Description

Let's look at the dataset's description. The `DESCR` information includes:

- `Number of Instances`—this dataset contains 20,640 samples.
- `Number of Attributes`—there are 8 features (attributes) per sample.
- `Attribute Information`—feature descriptions.
- `Missing Attribute Values`—none are missing in this dataset.

According to the description, the target variable in this dataset is the *median house value*—this is the value we'll be trying to predict via multiple linear regression.

```
In [3]: print(california.DESCR)
.. _california_housing_dataset:
```

15.5 Multiple Linear Regression with the California Housing Dataset

```
California Housing dataset
--------------------------

**Data Set Characteristics:**

    :Number of Instances: 20640

    :Number of Attributes: 8 numeric, predictive attributes and
        the target

    :Attribute Information:
        - MedInc        median income in block
        - HouseAge      median house age in block
        - AveRooms      average number of rooms
        - AveBedrms     average number of bedrooms
        - Population    block population
        - AveOccup      average house occupancy
        - Latitude      house block latitude
        - Longitude     house block longitude

    :Missing Attribute Values: None

This dataset was obtained from the StatLib repository.
http://lib.stat.cmu.edu/datasets/

The target variable is the median house value for California districts.

This dataset was derived from the 1990 U.S. census, using one row per
census block group. A block group is the smallest geographical unit for
which the U.S. Census Bureau publishes sample data (a block group typi-
cally has a population of 600 to 3,000 people).

It can be downloaded/loaded using the
:func:`sklearn.datasets.fetch_california_housing` function.

.. topic:: References

    - Pace, R. Kelley and Ronald Barry, Sparse Spatial Autoregressions,
      Statistics and Probability Letters, 33 (1997) 291-297
```

Again, the Bunch object's data and target attributes are NumPy arrays containing the 20,640 samples and their target values respectively. We can confirm the number of samples (rows) and features (columns) by looking at the data array's shape attribute, which shows that there are 20,640 rows and 8 columns:

```
In [4]: california.data.shape
Out[4]: (20640, 8)
```

Similarly, you can see that the number of target values—that is, the median house values—matches the number of samples by looking at the target array's shape:

```
In [5]: california.target.shape
Out[5]: (20640,)
```

The Bunch's **feature_names** attribute contains the names that correspond to each column in the data array:

```
In [6]: california.feature_names
Out[6]:
['MedInc',
 'HouseAge',
 'AveRooms',
 'AveBedrms',
 'Population',
 'AveOccup',
 'Latitude',
 'Longitude']
```

15.5.2 Exploring the Data with Pandas

Let's use a pandas DataFrame to explore the data further. We'll also use the DataFrame with Seaborn in the next section to visualize some of the data. First, let's import pandas and set some options:

```
In [7]: import pandas as pd

In [8]: pd.set_option('precision', 4)

In [9]: pd.set_option('max_columns', 9)

In [10]: pd.set_option('display.width', None)
```

In the preceding set_option calls:

- 'precision' is the maximum number of digits to display to the right of each decimal point.

- 'max_columns' is the maximum number of columns to display when you output the DataFrame's string representation. By default, if pandas cannot fit all of the columns left-to-right, it cuts out columns in the middle and displays an ellipsis (...) instead. The 'max_columns' setting enables pandas to show all the columns using multiple rows of output. As you'll see momentarily, we'll have nine columns in the DataFrame—the eight dataset features in california.data and an additional column for the target median house values (california.target).

- 'display.width' specifies the width in characters of your Command Prompt (Windows), Terminal (macOS/Linux) or shell (Linux). The value None tells pandas to auto-detect the display width when formatting string representations of Series and DataFrames.

Next, let's create a DataFrame from the Bunch's data, target and feature_names arrays. The first snippet below creates the initial DataFrame using the data in california.data and with the column names specified by california.feature_names. The second statement adds a column for the median house values stored in california.target:

```
In [11]: california_df = pd.DataFrame(california.data,
    ...:                              columns=california.feature_names)
    ...:

In [12]: california_df['MedHouseValue'] = pd.Series(california.target)
```

15.5 Multiple Linear Regression with the California Housing Dataset

We can peek at some of the data using the head function. Notice that pandas displays the DataFrame's first six columns, then skips a line of output and displays the remaining columns. The \ to the right of the column head "AveOccup" indicates that there are more columns displayed below. You'll see the \ only if the window in which IPython is running is too narrow to display all the columns left-to-right:

```
In [13]: california_df.head()
Out[13]:
    MedInc  HouseAge  AveRooms  AveBedrms  Population  AveOccup  \
0   8.3252      41.0    6.9841     1.0238       322.0    2.5556
1   8.3014      21.0    6.2381     0.9719      2401.0    2.1098
2   7.2574      52.0    8.2881     1.0734       496.0    2.8023
3   5.6431      52.0    5.8174     1.0731       558.0    2.5479
4   3.8462      52.0    6.2819     1.0811       565.0    2.1815

   Latitude  Longitude  MedHouseValue
0     37.88    -122.23          4.526
1     37.86    -122.22          3.585
2     37.85    -122.24          3.521
3     37.85    -122.25          3.413
4     37.85    -122.25          3.422
```

Let's get a sense of the data in each column by calculating the DataFrame's summary statistics. Note that the median income and house values (again, measured in hundreds of thousands) are from 1990 and are significantly higher today:

```
In [14]: california_df.describe()
Out[14]:
             MedInc     HouseAge     AveRooms    AveBedrms    Population  \
count   20640.0000   20640.0000   20640.0000   20640.0000   20640.0000
mean        3.8707      28.6395       5.4290       1.0967    1425.4767
std         1.8998      12.5856       2.4742       0.4739    1132.4621
min         0.4999       1.0000       0.8462       0.3333       3.0000
25%         2.5634      18.0000       4.4407       1.0061     787.0000
50%         3.5348      29.0000       5.2291       1.0488    1166.0000
75%         4.7432      37.0000       6.0524       1.0995    1725.0000
max        15.0001      52.0000     141.9091      34.0667   35682.0000

            AveOccup     Latitude    Longitude  MedHouseValue
count     20640.0000   20640.0000   20640.0000     20640.0000
mean          3.0707      35.6319    -119.5697         2.0686
std          10.3860       2.1360       2.0035         1.1540
min           0.6923      32.5400    -124.3500         0.1500
25%           2.4297      33.9300    -121.8000         1.1960
50%           2.8181      34.2600    -118.4900         1.7970
75%           3.2823      37.7100    -118.0100         2.6472
max        1243.3333      41.9500    -114.3100         5.0000
```

Self Check

1   *(Discussion)* Based on the DataFrame's summary statistics, what was the average median household income across all block groups for California in 1990?
Answer: $38,707 (3.8707 * 10000—recall that the datasets median income is expressed in tens of thousands).

15.5.3 Visualizing the Features

It's helpful to visualize your data by plotting the target value against *each* feature—in this case, to see how the median home value relates to each feature. To make our visualizations clearer, let's use DataFrame method **sample** to randomly select 10% of the 20,640 samples for graphing purposes:

```
In [15]: sample_df = california_df.sample(frac=0.1, random_state=17)
```

The keyword argument frac specifies the fraction of the data to select (0.1 for 10%), and the keyword argument random_state enables you to seed the random number generator. The integer seed value (17), which we chose arbitrarily, is crucial for *reproducibility*. Each time you use the *same* seed value, method sample selects the *same* random subset of the DataFrame's rows. Then, when we graph the data, you should get the *same* results.

Next, we'll use Matplotlib and Seaborn to display scatter plots of each of the eight features. Both libraries can display scatter plots. Seaborn's are more attractive and require less code, so we'll use Seaborn to create the following scatter plots. First, we import both libraries and use Seaborn function set to scale each diagram's fonts to two times their default size:

```
In [16]: import matplotlib.pyplot as plt

In [17]: import seaborn as sns

In [18]: sns.set(font_scale=2)

In [19]: sns.set_style('whitegrid')
```

The following snippet displays the scatter plots.[8] Each shows one feature along the *x*-axis and the median home value (california.target) along the *y*-axis, so we can see how each feature and the median house values relate to one another. We display each scatter plot in a separate window. The windows are displayed in the order the features were listed in snippet [6] with the most recently displayed window in the foreground:

```
In [20]: for feature in california.feature_names:
    ...:     plt.figure(figsize=(16, 9))
    ...:     sns.scatterplot(data=sample_df, x=feature,
    ...:                     y='MedHouseValue', hue='MedHouseValue',
    ...:                     palette='cool', legend=False)
    ...:
```

For each feature name, the snippet first creates a 16-inch-by-9-inch Matplotlib Figure—we're plotting many data points, so we chose to use a larger window. If this window is larger than your screen, Matplotlib fits the Figure to the screen. Seaborn uses the current Figure to display the scatter plot. If you do not create a Figure first, Seaborn will create one. We created the Figure first here so we could display a large window for a scatter plot containing over 2000 points.

Next, the snippet creates a Seaborn scatterplot in which the *x*-axis shows the current feature, the *y*-axis shows the 'MedHouseValue' (*median house values*), and the 'MedHouseValue' determines the dot colors (hue). Some interesting things to notice in these graphs:

---

8. When you execute this code in IPython, each window will be displayed in front of the previous one. As you close each, you'll see the one behind it.

15.5 Multiple Linear Regression with the California Housing Dataset 631

- The graphs showing the latitude and longitude each have two areas of especially significant density. If you search online for the latitude and longitude values where those dense areas appear, you'll see that these represent the greater Los Angeles and greater San Francisco areas where house prices tend to be higher.

- In each graph, there is a horizontal line of dots at the *y*-axis value 5, which represents the median house value $500,000. The highest home value that could be chosen on the 1990 census form was "$500,000 or more."[9] So any block group with a median house value over $500,000 is listed in the dataset as 5. Being able to spot characteristics like this is a compelling reason to do data exploration and visualization.

- In the HouseAge graph, there is a vertical line of dots at the *x*-axis value 52. The highest home age that could be chosen on the 1990 census form was 52, so any block group with a median house age over 52 is listed in the dataset as 52.

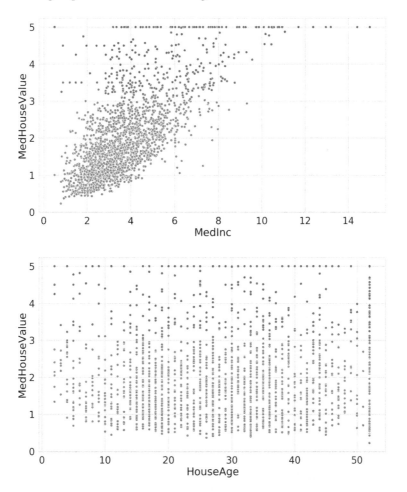

---

9. https://www.census.gov/prod/1/90dec/cph4/appdxe.pdf.

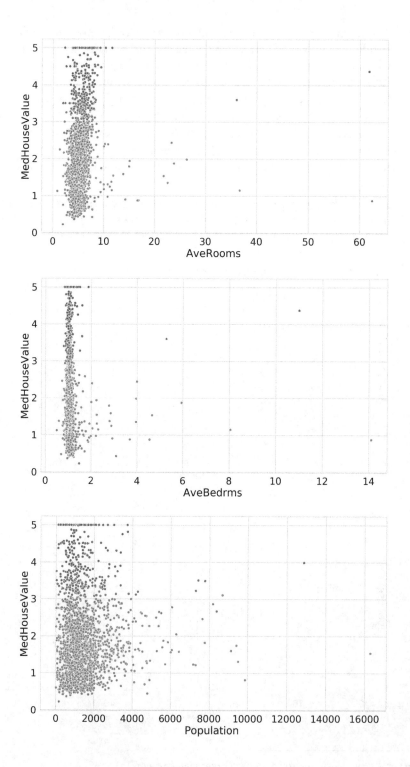

15.5 Multiple Linear Regression with the California Housing Dataset

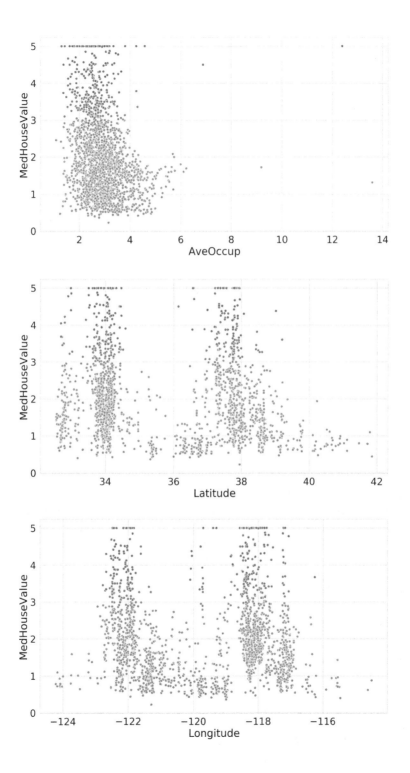

Self Check

1 *(Fill-In)* `DataFrame` method _____ returns a randomly selected subset of the `DataFrame`'s rows.
Answer: `sample`.

2 *(Discussion)* Why would it be useful in a scatter plot to plot a randomly selected subset of a dataset's samples?
Answer: When you are getting to know your data for a large dataset, there could be too many samples to get a sense of how they are truly distributed.

15.5.4 Splitting the Data for Training and Testing

Once again, to prepare for training and testing the model, let's break the data into training and testing sets using the `train_test_split` function then check their sizes:

```
In [21]: from sklearn.model_selection import train_test_split

In [22]: X_train, X_test, y_train, y_test = train_test_split(
    ...:     california.data, california.target, random_state=11)
    ...:

In [23]: X_train.shape
Out[23]: (15480, 8)

In [24]: X_test.shape
Out[24]: (5160, 8)
```

We used `train_test_split`'s keyword argument `random_state` to seed the random number generator for reproducibility.

15.5.5 Training the Model

Next, we'll train the model. By default, a `LinearRegression` estimator uses *all* the features in the dataset's `data` array to perform a multiple linear regression. An error occurs if any of the features are *categorical* rather than numeric. If a dataset contains categorical data, you either must preprocess the categorical features into numerical ones (which you'll do in the next chapter) or must exclude the categorical features from the training process. A benefit of working with scikit-learn's bundled datasets is that they're already in the correct format for machine learning using scikit-learn's models.

As you saw in the previous two snippets, `X_train` and `X_test` each contain 8 columns—one per feature. Let's create a `LinearRegression` estimator and invoke its `fit` method to train the estimator using `X_train` and `y_train`:

```
In [25]: from sklearn.linear_model import LinearRegression

In [26]: linear_regression = LinearRegression()

In [27]: linear_regression.fit(X=X_train, y=y_train)
Out[27]:
LinearRegression(copy_X=True, fit_intercept=True, n_jobs=None,
        normalize=False)
```

15.5 Multiple Linear Regression with the California Housing Dataset

Multiple linear regression produces separate coefficients for each feature (stored in coeff_) in the dataset and one intercept (stored in intercept_):

```
In [28]: for i, name in enumerate(california.feature_names):
    ...:     print(f'{name:>10}: {linear_regression.coef_[i]}')
    ...:
    MedInc: 0.4377030215382206
  HouseAge: 0.009216834565797713
  AveRooms: -0.10732526637360985
 AveBedrms: 0.611713307391811
Population: -5.756822009298454e-06
  AveOccup: -0.0033845664657163703
  Latitude: -0.419481860964907
 Longitude: -0.4337713349874016

In [29]: linear_regression.intercept_
Out[29]: -36.88295065605547
```

For positive coefficients, the median house value *increases* as the feature value *increases*. For negative coefficients, the median house value *decreases* as the feature value *increases*. Note that the population coefficient has a *negative exponent* (e-06), so the coefficient's value is actually -0.000005756822009298454. This is close to zero, so a block group's population apparently has little effect the median house value.

You can use these values with the following equation to make predictions:

$$y = m_1 x_1 + m_2 x_2 + \ldots m_n x_n + b$$

where

- $m_1, m_2, \ldots, m_n$ are the feature coefficients,
- b is the intercept,
- $x_1, x_2, \ldots, x_n$ are the feature values (that is, the values of the independent variables), and
- y is the predicted value (that is, the dependent variable).

✓ Self Check

1 *(True/False)* By default, a LinearRegression estimator uses *all* the features in the dataset to perform a multiple linear regression.
Answer: False. By default, a LinearRegression estimator uses all the *numeric* features in the dataset to perform a multiple linear regression. An error occurs if any of the features are *categorical* rather than numeric. Categorical features must be preprocessed into numerical ones or must be excluded from the training process.

15.5.6 Testing the Model

Now, let's test the model by calling the estimator's predict method with the test samples as an argument. As we've done in each of the previous examples, we store the array of predictions in predicted and the array of expected values in expected:

```
In [30]: predicted = linear_regression.predict(X_test)

In [31]: expected = y_test
```

Let's look at the first five predictions and their corresponding expected values:

```
In [32]: predicted[:5]
Out[32]: array([1.25396876, 2.34693107, 2.03794745, 1.8701254 ,
2.53608339])

In [33]: expected[:5]
Out[33]: array([0.762, 1.732, 1.125, 1.37 , 1.856])
```

With classification, we saw that the predictions were distinct classes that matched existing classes in the dataset. With regression, it's tough to get exact predictions, because you have continuous outputs. Every possible value of $x_1, x_2 \ldots x_n$ in the calculation

$$y = m_1x_1 + m_2x_2 + \ldots m_nx_n + b$$

predicts a value.

15.5.7 Visualizing the Expected vs. Predicted Prices

Let's look at the expected vs. predicted median house values for the test data. First, let's create a DataFrame containing columns for the expected and predicted values:

```
In [34]: df = pd.DataFrame()

In [35]: df['Expected'] = pd.Series(expected)

In [36]: df['Predicted'] = pd.Series(predicted)
```

Now let's plot the data as a scatter plot with the expected (target) prices along the *x*-axis and the predicted prices along the *y*-axis:

```
In [37]: figure = plt.figure(figsize=(9, 9))

In [38]: axes = sns.scatterplot(data=df, x='Expected', y='Predicted',
   ...:     hue='Predicted', palette='cool', legend=False)
   ...:
```

Next, let's set the *x*- and *y*-axes' limits to use the same scale along both axes:

```
In [39]: start = min(expected.min(), predicted.min())

In [40]: end = max(expected.max(), predicted.max())

In [41]: axes.set_xlim(start, end)
Out[41]: (-0.6830978604144491, 7.155719818496834)

In [42]: axes.set_ylim(start, end)
Out[42]: (-0.6830978604144491, 7.155719818496834)
```

Now, let's plot a line that represents *perfect predictions* (note that this is *not* a regression line). The following snippet displays a line between the points representing the lower-left corner of the graph (start, start) and the upper-right corner of the graph (end, end). The third argument ('k--') indicates the line's style. The letter k represents the color black, and the -- indicates that plot should draw a dashed line:

```
In [43]: line = plt.plot([start, end], [start, end], 'k--')
```

If every predicted value were to match the expected value, then all the dots would be plotted along the dashed line. In the following diagram, it appears that as the expected median house value increases, more of the predicted values fall below the line. So the model seems to *predict* lower median house values as the *expected* median house value increases.

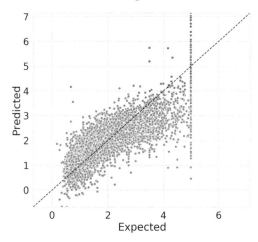

15.5.8 Regression Model Metrics

Scikit-learn provides many metrics functions for evaluating how well estimators predict results and for comparing estimators to choose the best one(s) for your particular study. These metrics vary by estimator type. For example, the sklearn.metrics functions confusion_matrix and classification_report used in the Digits dataset classification case study are two of many metrics functions specifically for evaluating *classification* estimators.

Among the many metrics for regression estimators is the model's **coefficient of determination**, which is also called the R^2 **score**. To calculate an estimator's R^2 score, call the sklearn.metrics module's **r2_score** function with the arrays representing the expected and predicted results:

```
In [44]: from sklearn import metrics

In [45]: metrics.r2_score(expected, predicted)
Out[45]: 0.6008983115964333
```

R^2 scores range from 0.0 to 1.0 with 1.0 being the best. An R^2 score of 1.0 indicates that the estimator perfectly predicts the dependent variable's value, given the independent variable(s) value(s). An R^2 score of 0.0 indicates the model cannot make predictions with any accuracy, based on the independent variables' values.

Another common metric for regression models is the **mean squared error**, which

- calculates the difference between each expected and predicted value—this is called the *error*,
- squares each difference and
- calculates the average of the squared values.

To calculate an estimator's mean squared error, call function **mean_squared_error** (from module sklearn.metrics) with the arrays representing the expected and predicted results:

```
In [46]: metrics.mean_squared_error(expected, predicted)
Out[46]: 0.5350149774449119
```

When comparing estimators with the mean squared error metric, the one with the value closest to 0 best fits your data. In the next section, we'll run several regression estimators using the California Housing dataset. For the list of scikit-learn's metrics functions by estimator category, see

> https://scikit-learn.org/stable/modules/model_evaluation.html

 Self Check

1 *(Fill-In)* An R^2 score of _____ indicates that an estimator perfectly predicts the dependent variable's value, given the independent variable(s) value(s).
Answer: 1.0.

2 *(True/False)* When comparing estimators, the one with the mean squared error value closest to 0 is the estimator that best fits your data.
Answer: True.

15.5.9 Choosing the Best Model

As we did in the classification case study, let's try several estimators to determine whether any produces better results than the `LinearRegression` estimator. In this example, we'll use the `linear_regression` estimator we already created as well as `ElasticNet`, `Lasso` and `Ridge` regression estimators (all from the `sklearn.linear_model` module). For information about these estimators, see

> https://scikit-learn.org/stable/modules/linear_model.html

```
In [47]: from sklearn.linear_model import ElasticNet, Lasso, Ridge

In [48]: estimators = {
    ...:     'LinearRegression': linear_regression,
    ...:     'ElasticNet': ElasticNet(),
    ...:     'Lasso': Lasso(),
    ...:     'Ridge': Ridge()
    ...: }
```

Once again, we'll run the estimators using k-fold cross-validation with a `KFold` object and the `cross_val_score` function. Here, we pass to `cross_val_score` the additional keyword argument `scoring='r2'`, which indicates that the function should report the R^2 scores for each fold—again, 1.0 is the best, so it appears that `LinearRegression` and `Ridge` are the best models for this dataset:

```
In [49]: from sklearn.model_selection import KFold, cross_val_score

In [50]: for estimator_name, estimator_object in estimators.items():
    ...:     kfold = KFold(n_splits=10, random_state=11, shuffle=True)
    ...:     scores = cross_val_score(estimator=estimator_object,
    ...:         X=california.data, y=california.target, cv=kfold,
    ...:         scoring='r2')
    ...:     print(f'{estimator_name:>16}: ' +
    ...:         f'mean of r2 scores={scores.mean():.3f}')
    ...:
LinearRegression: mean of r2 scores=0.599
      ElasticNet: mean of r2 scores=0.423
           Lasso: mean of r2 scores=0.285
           Ridge: mean of r2 scores=0.599
```

15.6 Case Study: Unsupervised Machine Learning, Part 1—Dimensionality Reduction

In our data science presentations, we've focused on getting to know your data. **Unsupervised machine learning** and visualization can help you do this by finding patterns and relationships among unlabeled samples.

For datasets like the univariate time series we used earlier in this chapter, visualizing the data is easy. In that case, we had two variables—date and temperature—so we plotted the data in two dimensions with one variable along each axis. Using Matplotlib, Seaborn and other visualization libraries, you also can plot datasets with three variables using 3D visualizations. But how do you visualize data with more than three dimensions? For example, in the Digits dataset, every sample has 64 features and a target value. In big data, samples can have hundreds, thousands or even millions of features.

To visualize a dataset with many features (that is, many dimensions), we'll first *reduce* the data to two or three dimensions. This requires an unsupervised machine learning technique called **dimensionality reduction**. When you graph the resulting information, you might see patterns in the data that will help you choose the most appropriate machine learning algorithms to use. For example, if the visualization contains *clusters* of points, it might indicate that there are distinct classes of information within the dataset. So a classification algorithm might be appropriate. Of course, you'd first need to determine the class of the samples in each cluster. This might require studying the samples in a cluster to see what they have in common.

Dimensionality reduction also serves other purposes. Training estimators on big data with significant numbers of dimensions can take hours, days, weeks or longer. It's also difficult for humans to think about data with large numbers of dimensions. This is called the **curse of dimensionality**. If the data has closely correlated features, some could be eliminated via dimensionality reduction to improve the training performance. This, however, might reduce the accuracy of the model.

Recall that the Digits dataset is already labeled with 10 classes representing the digits 0–9. Let's ignore those labels and use dimensionality reduction to reduce the dataset's features to two dimensions, so we can visualize the resulting data.

Loading the Digits Dataset
Launch IPython with:

    ipython --matplotlib

then load the dataset:

    In [1]: from sklearn.datasets import load_digits

    In [2]: digits = load_digits()

Creating a TSNE Estimator for Dimensionality Reduction
Next, we'll use the **TSNE estimator** (from the **sklearn.manifold** module) to perform dimensionality reduction. This estimator uses an algorithm called t-distributed Stochastic Neighbor Embedding (t-SNE)[10] to analyze a dataset's features and reduce them to the

---
10. The algorithm's details are beyond this book's scope. For more information, see https://scikit-learn.org/stable/modules/manifold.html#t-sne.

specified number of dimensions. We first tried the popular PCA (principal components analysis) estimator but did not like the results we were getting, so we switched to TSNE. We'll show PCA later in this case study.

Let's create a TSNE object for reducing a dataset's features to two dimensions, as specified by the keyword argument n_components. As with the other estimators we've presented, we used the random_state keyword argument to ensure the reproducibility of the "render sequence" when we display the digit clusters:

```
In [3]: from sklearn.manifold import TSNE

In [4]: tsne = TSNE(n_components=2, random_state=11)
```

Transforming the Digits Dataset's Features into Two Dimensions

Dimensionality reduction in scikit-learn typically involves two steps—training the estimator with the dataset, then using the estimator to transform the data into the specified number of dimensions. These steps can be performed separately with the TSNE methods **fit** and **transform**, or they can be performed in one statement using the **fit_transform** method:[11]

```
In [5]: reduced_data = tsne.fit_transform(digits.data)
```

TSNE's fit_transform method takes some time to train the estimator then perform the reduction. On our system, this took about 20 seconds. When the method completes its task, it returns an array with the same number of rows as digits.data, but only two columns. You can confirm this by checking reduced_data's shape:

```
In [6]: reduced_data.shape
Out[6]: (1797, 2)
```

Visualizing the Reduced Data

Now that we've reduced the original dataset to only two dimensions, let's use a scatter plot to display the data. In this case, rather than Seaborn's scatterplot function, we'll use Matplotlib's **scatter function**, because it returns a collection of the plotted items. We'll use that feature in a second scatter plot momentarily:

```
In [7]: import matplotlib.pyplot as plt

In [8]: dots = plt.scatter(reduced_data[:, 0], reduced_data[:, 1],
   ...:                    c='black')
   ...:
```

Function scatter's first two arguments are reduced_data's columns (0 and 1) containing the data for the *x*- and *y*-axes. The keyword argument c='black' specifies the color of the dots. We did not label the axes, because they do not correspond to specific features of the original dataset. The new features produced by the TSNE estimator could be quite different from the dataset's original features.

---

11. Every call to fit_transform trains the estimator. If you intend to reuse the estimator to reduce the dimensions of samples multiple times, use fit to once train the estimator, then use transform to perform the reductions. We'll use this technique with PCA later in this case study.

15.6 Unsupervised Machine Learning, Part 1—Dimensionality Reduction

The following diagram shows the resulting scatter plot:

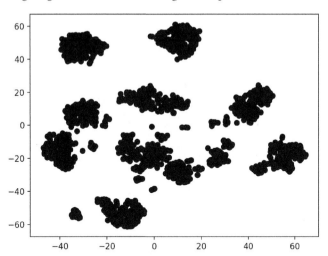

There are clearly *clusters* of related data points, though there appear to be 11 main clusters, rather than 10. There also are "loose" data points that do not appear to be part of specific clusters. Based on our earlier study of the Digits dataset this makes sense because some digits were difficult to classify.

Visualizing the Reduced Data with Different Colors for Each Digit

Though the preceding diagram shows clusters, we do not know whether all the items in each cluster represent the same digit. If they do not, then the clusters are not helpful. Let's use the known `targets` in the Digits dataset to color all the dots so we can see whether these clusters indeed represent specific digits:

```
In [9]: dots = plt.scatter(reduced_data[:, 0], reduced_data[:, 1],
   ...:        c=digits.target, cmap=plt.cm.get_cmap('nipy_spectral_r', 10))
   ...:
   ...:
```

In this case, `scatter`'s keyword argument `c=digits.target` specifies that the `target` values determine the dot colors. We also added the keyword argument

```
cmap=plt.cm.get_cmap('nipy_spectral_r', 10)
```

which specifies a color map to use when coloring the dots. In this case, we know we're coloring 10 digits, so we use `get_cmap` method of Matplotlib's `cm` object (from module `matplotlib.pyplot`) to load a color map (`'nipy_spectral_r'`) and select 10 distinct colors from the color map.

The following statement adds a color bar key to the right of the diagram so you can see which digit each color represents:

```
In [10]: colorbar = plt.colorbar(dots)
```

Voila! We see 10 clusters corresponding to the digits 0–9. Again, there are a few smaller groups of dots standing alone. Based on this, we might decide that a supervised-learning approach like k-nearest neighbors would work well with this data. In the exercises, you'll reimplement the colored clusters in a three-dimensional graph.

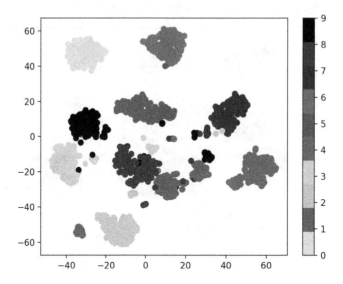

Self Check

1. *(Fill-In)* With dimensionality reduction training the estimator, then using the estimator to transform the data into the specified number of dimensions can be performed separately with the TSNE methods _____ and _____, or in one statement using the fit_transform method.
Answer: fit, transform.

2. *(True/False)* Unsupervised machine learning and visualization can help you get to know your data by finding patterns and relationships among unlabeled samples.
Answer: True.

15.7 Case Study: Unsupervised Machine Learning, Part 2—k-Means Clustering

In this section, we introduce perhaps the simplest unsupervised machine learning algorithms—**k-means clustering**. This algorithm analyzes *unlabeled* samples and attempts to place them in clusters that appear to be related. The k in "k-means" represents the number of clusters you'd like to see imposed on your data.

The algorithm organizes samples into the number of clusters you specify in advance, using distance calculations similar to the k-nearest neighbors clustering algorithm. Each cluster of samples is grouped around a **centroid**—the cluster's center point. Initially, the algorithm chooses k centroids at random from the dataset's samples. Then the remaining samples are placed in the cluster whose centroid is the closest. The centroids are iteratively recalculated and the samples re-assigned to clusters until, for all clusters, the distances from a given centroid to the samples in its cluster are minimized. The algorithm's results are:

- a one-dimensional array of labels indicating the cluster to which each sample belongs and
- a two-dimensional array of centroids representing the center of each cluster.

15.7 Unsupervised Machine Learning, Part 2—k-Means Clustering

Iris Dataset

We'll work with the popular **Iris dataset**[12] bundled with scikit-learn, which is commonly analyzed with both classification and clustering. Although this dataset is labeled, we'll ignore those labels here to demonstrate clustering. Then, we'll use the labels to determine how well the k-means algorithm clustered the samples.

The Iris dataset is referred to as a "toy dataset" because it has only 150 samples and four features. The dataset describes 50 samples for each of three *Iris* flower species—*Iris setosa*, *Iris versicolor* and *Iris virginica*. Photos of these are shown below. Each sample's features are the sepal length, sepal width, petal length and petal width, all measured in centimeters. The *sepals* are the larger outer parts of each flower that protect the smaller inside *petals* before the flower buds bloom.

Iris setosa: `https://commons.wikimedia.org/wiki/File:Wild_iris_KEFJ_(9025144383).jpg`.
Credit: Courtesy of Nation Park services.

Iris versicolor: `https://commons.wikimedia.org/wiki/Iris_versicolor#/media/File:IrisVersicolor-FoxRoost-Newfoundland.jpg`.
Credit: Courtesy of Jefficus,
`https://commons.wikimedia.org/w/index.php?title=User:Jefficus&action=edit&redlink=1`

---

12. Fisher, R.A., "The use of multiple measurements in taxonomic problems," Annual Eugenics, 7, Part II, 179-188 (1936); also in "Contributions to Mathematical Statistics" (John Wiley, NY, 1950).

Iris virginica: https://commons.wikimedia.org/wiki/File:IMG_7911-Iris_virginica.jpg.
Credit: Christer T Johansson.

 Self Check

1 *(Fill-In)* Each cluster of samples is grouped around a(n) _____—the cluster's center point.
Answer: centroid.

2 *(True/False)* The k-means clustering algorithm studies the dataset then automatically determines the appropriate number of clusters.
Answer: False. The algorithm organizes samples into the number of clusters you specify in advance.

15.7.1 Loading the Iris Dataset

Launch IPython with `ipython --matplotlib`, then use the `sklearn.datasets` module's `load_iris` function to get a Bunch containing the dataset:

```
In [1]: from sklearn.datasets import load_iris

In [2]: iris = load_iris()
```

The Bunch's `DESCR` attribute indicates that there are 150 samples (Number of Instances), each with four features (Number of Attributes). There are no missing values in this dataset. The dataset classifies the samples by labeling them with the integers 0, 1 and 2, representing *Iris setosa*, *Iris versicolor* and *Iris virginica*, respectively. We'll *ignore* the labels and let the k-means clustering algorithm try to determine the samples' classes. We show some key `DESCR` information in bold.:

```
In [3]: print(iris.DESCR)
.. _iris_dataset:

Iris plants dataset
--------------------

**Data Set Characteristics:**

    :Number of Instances: 150 (50 in each of three classes)
```

15.7 Unsupervised Machine Learning, Part 2—k-Means Clustering

```
:Number of Attributes: 4 numeric, predictive attributes and the class
:Attribute Information:
   - sepal length in cm
   - sepal width in cm
   - petal length in cm
   - petal width in cm
   - class:
         - Iris-Setosa
         - Iris-Versicolour
         - Iris-Virginica

:Summary Statistics:

===============  ====  ====  =======  =====  ====================
                 Min   Max   Mean     SD     Class Correlation
===============  ====  ====  =======  =====  ====================
sepal length:    4.3   7.9   5.84     0.83   0.7826
sepal width:     2.0   4.4   3.05     0.43   -0.4194
petal length:    1.0   6.9   3.76     1.76   0.9490   (high!)
petal width:     0.1   2.5   1.20     0.76   0.9565   (high!)
===============  ====  ====  =======  =====  ====================

:Missing Attribute Values: None
:Class Distribution: 33.3% for each of 3 classes.
:Creator: R.A. Fisher
:Donor: Michael Marshall (MARSHALL%PLU@io.arc.nasa.gov)
:Date: July, 1988
```

The famous Iris database, first used by Sir R.A. Fisher. The dataset is taken from Fisher's paper. Note that it's the same as in R, but not as in the UCI Machine Learning Repository, which has two wrong data points.

This is perhaps the **best known database to be found in the pattern recognition literature.** Fisher's paper is a classic in the field and is referenced frequently to this day. (See Duda & Hart, for example.) The data set contains 3 classes of 50 instances each, where each class refers to a type of iris plant. One class is linearly separable from the other 2; the latter are NOT linearly separable from each other.

.. topic:: References

  - Fisher, R.A. "The use of multiple measurements in taxonomic
    problems"
    Annual Eugenics, 7, Part II, 179-188 (1936); also in "Contributions
    to Mathematical Statistics" (John Wiley, NY, 1950).
  - Duda, R.O., & Hart, P.E. (1973) Pattern Classification and Scene
    Analysis.
    (Q327.D83) John Wiley & Sons. ISBN 0-471-22361-1. See page 218.
  - Dasarathy, B.V. (1980) "Nosing Around the Neighborhood: A New System
    Structure and Classification Rule for Recognition in Partially
    Exposed Environments". IEEE Transactions on Pattern Analysis and
    Machine Intelligence, Vol. PAMI-2, No. 1, 67-71.
  - Gates, G.W. (1972) "The Reduced Nearest Neighbor Rule". IEEE
    Transactions on Information Theory, May 1972, 431-433.
  - See also: 1988 MLC Proceedings, 54-64. Cheeseman et al"s AUTOCLASS
    II conceptual clustering system finds 3 classes in the data.
  - Many, many more ...

Checking the Numbers of Samples, Features and Targets

You can confirm the number of samples and features per sample via the `data` array's shape, and you can confirm the number of targets via the `target` array's shape:

```
In [4]: iris.data.shape
Out[4]: (150, 4)

In [5]: iris.target.shape
Out[5]: (150,)
```

The array `target_names` contains the names for the `target` array's numeric labels—`dtype='<U10'` indicates that the elements are strings with a maximum of 10 characters:

```
In [6]: iris.target_names
Out[6]: array(['setosa', 'versicolor', 'virginica'], dtype='<U10')
```

The array `feature_names` contains a list of string names for each column in the `data` array:

```
In [7]: iris.feature_names
Out[7]:
['sepal length (cm)',
 'sepal width (cm)',
 'petal length (cm)',
 'petal width (cm)']
```

15.7.2 Exploring the Iris Dataset: Descriptive Statistics with Pandas

Let's use a `DataFrame` to explore the Iris dataset. As we did in the California Housing case study, let's set the pandas options for formatting the column-based outputs:

```
In [8]: import pandas as pd

In [9]: pd.set_option('max_columns', 5)

In [10]: pd.set_option('display.width', None)
```

Create a `DataFrame` containing the `data` array's contents, using the contents of the `feature_names` array as the column names:

```
In [11]: iris_df = pd.DataFrame(iris.data, columns=iris.feature_names)
```

Next, add a column containing each sample's species name. The list comprehension in the following snippet uses each value in the `target` array to look up the corresponding species name in the `target_names` array:

```
In [12]: iris_df['species'] = [iris.target_names[i] for i in iris.target]
```

Let's use pandas' to look at a few samples. Once again notice that pandas displays a \ to the right of the column heads to indicate that there are more columns displayed below:

```
In [13]: iris_df.head()
Out[13]:
   sepal length (cm)  sepal width (cm)  petal length (cm)  \
0                5.1               3.5                1.4
1                4.9               3.0                1.4
2                4.7               3.2                1.3
3                4.6               3.1                1.5
4                5.0               3.6                1.4
```

```
        petal width (cm) species
0                    0.2  setosa
1                    0.2  setosa
2                    0.2  setosa
3                    0.2  setosa
4                    0.2  setosa
```

Let's calculate some descriptive statistics for the numerical columns:

```
In [14]: pd.set_option('precision', 2)

In [15]: iris_df.describe()
Out[15]:
       sepal length (cm)  sepal width (cm)  petal length (cm)  \
count             150.00            150.00             150.00
mean                5.84              3.06               3.76
std                 0.83              0.44               1.77
min                 4.30              2.00               1.00
25%                 5.10              2.80               1.60
50%                 5.80              3.00               4.35
75%                 6.40              3.30               5.10
max                 7.90              4.40               6.90

       petal width (cm)
count            150.00
mean               1.20
std                0.76
min                0.10
25%                0.30
50%                1.30
75%                1.80
max                2.50
```

Calling the describe method on the 'species' column confirms that it contains three unique values. Here, we know in advance of working with this data that there are three classes to which the samples belong, though this is not always the case in unsupervised machine learning.

```
In [16]: iris_df['species'].describe()
Out[16]:
count        150
unique         3
top       setosa
freq          50
Name: species, dtype: object
```

15.7.3 Visualizing the Dataset with a Seaborn pairplot

Let's visualize the features in this dataset. One way to learn more about your data is to see how the features relate to one another. The dataset has four features. We cannot graph one against the other three in a single graph. However, we can plot pairs of features against one another. Snippet [20] uses Seaborn function **pairplot** to create a grid of graphs plotting each feature against itself and the other specified features:

```
In [17]: import seaborn as sns
```

```
In [18]: sns.set(font_scale=1.1)

In [19]: sns.set_style('whitegrid')

In [20]: grid = sns.pairplot(data=iris_df, vars=iris_df.columns[0:4],
    ...:     hue='species')
    ...:
```

The keyword arguments are:

- data—The DataFrame[13] containing the data to plot.
- vars—A sequence containing the names of the variables to plot. For a DataFrame, these are the names of the columns to plot. Here, we use the first four DataFrame columns, representing the sepal length, sepal width, petal length and petal width, respectively.
- hue—The DataFrame column that's used to determine colors of the plotted data. In this case, we'll color the data by *Iris* species.

The preceding call to pairplot produces the following 4-by-4 grid of graphs:

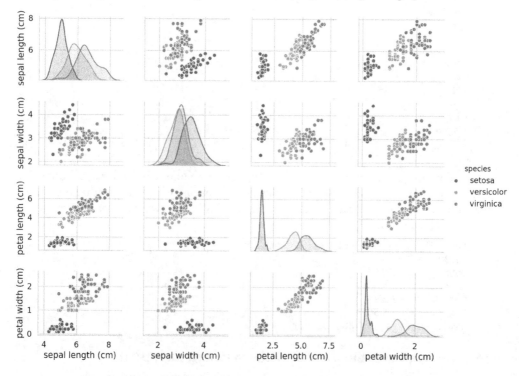

---

13. This also may be a two-dimensional array or list.

The graphs along the top-left-to-bottom-right diagonal, show the **distribution** of just the feature plotted in that column, with the range of values (left-to-right) and the number of samples with those values (top-to-bottom). For example, consider the sepal-length distributions:

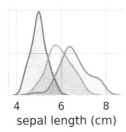

The blue shaded area indicates that the range of sepal length values (shown along the *x*-axis) for *Iris setosa* is approximately 4–6 centimeters and that most *Iris setosa* samples are in the middle of that range (approximately 5 centimeters). Similarly, the green shaded area indicates that the range of sepal length values for *Iris virginica* is approximately 4–8.5 centimeters and that the majority of *Iris virginica* samples have sepal length values between 6 and 7 centimeters.

The other graphs in a column show scatter plots of the other features against the feature on the *x*-axis. In the first column, the other three graphs plot the sepal width, petal length and petal width, respectively, along the *y*-axis and the sepal length along the *x*-axis.

Using separate colors for each *Iris* species, shows how the species relate to one another on a feature-by-feature basis. Interestingly, all the scatter plots clearly separate the *Iris setosa* blue dots from the other species' orange and green dots, indicating that *Iris setosa* is indeed in a "class by itself." We also can see that the other two species can sometimes be confused with one another, as indicated by the overlapping orange and green dots. For example, if you look at the scatter plot for sepal width vs. sepal length, you'll see the *Iris versicolor* and *Iris virginica* dots are intermixed. This indicates that it would be difficult to distinguish between these two species if we had only the sepal measurements available to us.

Displaying the `pairplot` in One Color

If you remove the hue keyword argument, `pairplot` function uses only one color to plot all the data because it does not know how to distinguish the species:

```
In [21]: grid = sns.pairplot(data=iris_df, vars=iris_df.columns[0:4])
```

As you can see in the resulting pair plot that follows, in this case, the graphs along the diagonal are histograms showing the distributions of all the values for that feature, regardless of the species. As you study each scatter plot, it appears that there may be only *two* distinct clusters, even though for this dataset we know there are *three* species. If you do not know the number of clusters in advance, you might ask a **domain expert** who is thoroughly familiar with the data. Such a person might know that there are three species in the dataset, which would be valuable information as we try to perform machine learning on the data.

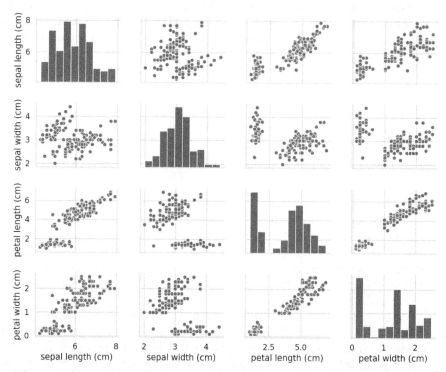

The preceding `pairplot` diagrams work well for a *small number of features* or a subset of features so that you have a small number of rows and columns, and for a relatively small number of samples so you can see the data points. As the number of features and samples increases, each scatter plot quickly becomes too small to read. For larger datasets, you may choose to plot a subset of the features and potentially a randomly selected subset of the samples to get a feel for your data.

✓ Self Check

1 *(Fill-In)* Seaborn's _____ function creates a grid of scatter plots showing features against one another.
Answer: `pairplot`.

2 *(True/False)* A plot of a feature's distribution shows the feature's range of values (left-to-right) and the number of samples with those values (top-to-bottom).
Answer: True.

15.7.4 Using a KMeans Estimator

In this section, we'll use k-means clustering via scikit-learn's **KMeans** estimator (from the `sklearn.cluster` module) to place each sample in the Iris dataset into a cluster. As with the other estimators you've used, the **KMeans** estimator hides from you the algorithm's complex mathematical details, making it straightforward to use.

15.7 Unsupervised Machine Learning, Part 2—k-Means Clustering

Creating the Estimator
Let's create the KMeans object:

```
In [22]: from sklearn.cluster import KMeans

In [23]: kmeans = KMeans(n_clusters=3, random_state=11)
```

The keyword argument n_clusters specifies the k-means clustering algorithm's hyperparameter k, which KMeans requires to calculate the clusters and label each sample. When you train a KMeans estimator, the algorithm calculates for each cluster a centroid representing the cluster's center data point.

The default value for the n_clusters parameter is 8. Often, you'll rely on domain experts knowledgeable about the data to help choose an appropriate k value. However, with hyperparameter tuning, you can estimate the appropriate k, as we'll do later. In this case, we know there are three species, so we'll use n_clusters=3 to see how well KMeans does in labeling the Iris samples. Once again, we used the random_state keyword argument for reproducibility.

Fitting the Model
Next, we'll train the estimator by calling the KMeans object's fit method. This step performs the k-means algorithm discussed earlier:

```
In [24]: kmeans.fit(iris.data)
Out[24]:
KMeans(algorithm='auto', copy_x=True, init='k-means++', max_iter=300,
    n_clusters=3, n_init=10, n_jobs=None, precompute_distances='auto',
    random_state=11, tol=0.0001, verbose=0)
```

As with the other estimator's, the fit method returns the estimator object and IPython displays its string representation. You can learn about the KMeans default arguments at:

> https://scikit-learn.org/stable/modules/generated/
>   sklearn.cluster.KMeans.html

When the training completes, the KMeans object contains:

- A labels_ array with values from 0 to n_clusters - 1 (in this example, 0–2), indicating the clusters to which the samples belong.
- A cluster_centers_ array in which each row represents a centroid.

Comparing the Computer Cluster Labels to the Iris Dataset's Target Values
Because the Iris dataset is labeled, we can look at its target array values to get a sense of how well the k-means algorithm clustered the samples for the three *Iris* species. With unlabeled data, we'd need to depend on a domain expert to help evaluate whether the predicted classes make sense.

In this dataset, the first 50 samples are *Iris setosa*, the next 50 are *Iris versicolor*, and the last 50 are *Iris virginica*. The Iris dataset's target array represents these with the values 0–2. If the KMeans estimator chose the clusters perfectly, then each group of 50 elements in the estimator's labels_ array should have a distinct label. As you study the results below, note that the KMeans estimator uses the values 0 through $k-1$ to label clusters, but these are not related to the Iris dataset's target array.

Let's use slicing to see how each group of 50 Iris samples was clustered. The following snippet shows that the first 50 samples were all placed in cluster 1:

```
In [25]: print(kmeans.labels_[0:50])
[1 1 1 1 1 1 1 1 1 1 1 1 1 1 1 1 1 1 1 1 1 1 1 1 1 1 1 1 1 1 1 1 1 1 1 1
 1 1 1 1 1 1 1 1 1 1 1 1 1 1]
```

The next 50 samples should be placed into a second cluster. The following snippet shows that most were placed in cluster 0, but two samples were placed in cluster 2:

```
In [26]: print(kmeans.labels_[50:100])
[0 0 2 0 0 0 0 0 0 0 0 0 0 0 0 0 0 0 0 0 0 0 0 0 0 0 0 0 0 2 0 0 0 0 0 0
 0 0 0 0 0 0 0 0 0 0 0 0 0 0]
```

Similarly, the last 50 samples should be placed into a third cluster. The following snippet shows that many of these samples were placed in cluster 2, but 14 of the samples were placed in cluster 0, indicating that the algorithm thought they belonged to a different cluster:

```
In [27]: print(kmeans.labels_[100:150])
[2 0 2 2 2 2 0 2 2 2 2 2 2 0 0 2 2 2 2 0 2 0 2 0 2 2 0 0 2 2 2 2 2 0 2 2
 2 2 0 2 2 2 0 2 2 2 0 2 2 0]
```

The results of these three snippets confirm what we saw in the `pairplot` diagrams earlier in this section—that *Iris setosa* is "in a class by itself" and that there is some confusion between *Iris versicolor* and *Iris virginica*.

✓ Self Check

1. *(IPython Session)* Try k-means clustering on the Iris dataset with two clusters, then display the first 50 and the last 100 elements of the estimator's `labels_` array.
Answer:

```
In [28]: kmeans2 = KMeans(n_clusters=2, random_state=11)

In [29]: kmeans2.fit(iris.data)
Out[29]:
KMeans(algorithm='auto', copy_x=True, init='k-means++', max_iter=300,
    n_clusters=2, n_init=10, n_jobs=None, precompute_distances='auto',
    random_state=None, tol=0.0001, verbose=0)

In [30]: print(kmeans2.labels_[0:50])
[1 1 1 1 1 1 1 1 1 1 1 1 1 1 1 1 1 1 1 1 1 1 1 1 1 1 1 1 1 1 1 1 1 1 1 1
 1 1 1 1 1 1 1 1 1 1 1 1 1 1]

In [31]: print(kmeans2.labels_[50:150])
[0 0 0 0 0 0 0 1 0 0 0 0 0 0 0 0 0 0 0 0 0 0 0 0 0 0 0 0 0 0 0 0 0 0 0 0
 0 0 0 0 0 0 1 0 0 0 0 1 0 0 0 0 0 0 0 0 0 0 0 0 0 0 0 0 0 0 0 0 0 0 0 0
 0 0 0 0 0 0 0 0 0 0 0 0 0 0 0 0 0 0 0 0 0 0 0 0 0 0 0 0]
```

In this case, you can see that all but three of the last 100 samples were placed in a single cluster.

15.7.5 Dimensionality Reduction with Principal Component Analysis

Next, we'll use the `PCA` estimator (from the `sklearn.decomposition` module) to perform dimensionality reduction. This estimator uses an algorithm called principal component

15.7 Unsupervised Machine Learning, Part 2—k-Means Clustering

analysis[14] to analyze a dataset's features and reduce them to the specified number of dimensions. For the Iris dataset, we first tried the TSNE estimator shown earlier but did not like the results we were getting. So we switched to PCA for the following demonstration.

Creating the PCA Object
Like the TSNE estimator, a PCA estimator uses the keyword argument n_components to specify the number of dimensions:

```
In [32]: from sklearn.decomposition import PCA

In [33]: pca = PCA(n_components=2, random_state=11)
```

Transforming the Iris Dataset's Features into Two Dimensions
Let's train the estimator and produce the reduced data by calling the PCA estimator's methods **fit** and **transform** methods:

```
In [34]: pca.fit(iris.data)
Out[34]:
PCA(copy=True, iterated_power='auto', n_components=2, random_state=11,
    svd_solver='auto', tol=0.0, whiten=False)

In [35]: iris_pca = pca.transform(iris.data)
```

When the method completes its task, it returns an array with the same number of rows as iris.data, but only two columns. Let's confirm this by checking iris_pca's shape:

```
In [36]: iris_pca.shape
Out[36]: (150, 2)
```

Note that we *separately* called the PCA estimator's fit and transform methods, rather than fit_transform, which we used with the TSNE estimator. In this example, we're going to *reuse* the trained estimator (produced with fit) to perform a second transform to reduce the cluster centroids from four dimensions to two. This will enable us to plot the centroid locations on each cluster.

Visualizing the Reduced Data
Now that we've reduced the original dataset to only two dimensions, let's use a scatter plot to display the data. In this case, we'll use Seaborn's scatterplot function. First, let's transform the reduced data into a DataFrame and add a species column that we'll use to determine the dot colors:

```
In [37]: iris_pca_df = pd.DataFrame(iris_pca,
    ...:                            columns=['Component1', 'Component2'])
    ...:

In [38]: iris_pca_df['species'] = iris_df.species
```

Next, let's scatterplot the data in Seaborn:

```
In [39]: axes = sns.scatterplot(data=iris_pca_df, x='Component1',
    ...:     y='Component2', hue='species', legend='brief',
    ...:     palette='cool')
    ...:
```

---

14. The algorithm's details are beyond this book's scope. For more information, see https://scikit-learn.org/stable/modules/decomposition.html#pca.

Each centroid in the KMeans object's cluster_centers_ array has the *same* number of features as the original dataset (four in this case). To plot the centroids, we must reduce their dimensions. You can think of a centroid as the "average" sample in its cluster. So each centroid should be transformed using the same PCA estimator we used to reduce the other samples in that cluster:

```
In [40]: iris_centers = pca.transform(kmeans.cluster_centers_)
```

Now, we'll plot the centroids of the three clusters as larger black dots. Rather than transform the iris_centers array into a DataFrame first, let's use Matplotlib's scatter function to plot the three centroids:

```
In [41]: import matplotlib.pyplot as plt

In [42]: dots = plt.scatter(iris_centers[:,0], iris_centers[:,1],
   ...:                     s=100, c='k')
   ...:
```

The keyword argument s=100 specifies the size of the plotted points, and the keyword argument c='k' specifies that the points should be displayed in black.

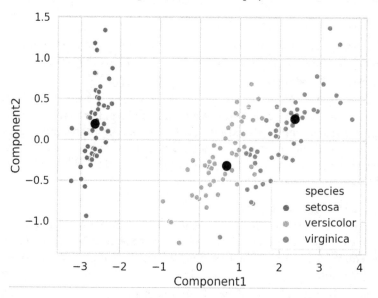

✓ Self Check

1. *(True/False)* Each centroid in a KMeans object's cluster_centers_ array has the *same* number of features as the original dataset.
Answer: True.

2. *(Discussion)* What is the purpose of the following statement?

    iris_centers = pca.transform(kmeans.cluster_centers_)

Answer: This statement reduces the centroids to the number of dimensions specified when the pca object was created. In the Iris case study, we were able to plot the reduced centroids in two dimensions at the centers of their corresponding clusters.

15.7.6 Choosing the Best Clustering Estimator

As we did in the classification and regression case studies, let's run multiple clustering algorithms and see how well they cluster the three species of Iris flowers. Here we'll attempt to cluster the Iris dataset's samples using the kmeans object we created earlier[15] and objects of scikit-learn's DBSCAN, MeanShift, SpectralClustering and AgglomerativeClustering estimators. Like KMeans, you specify the number of clusters in advance for the SpectralClustering and AgglomerativeClustering estimators:

```
In [43]: from sklearn.cluster import DBSCAN, MeanShift,\
    ...:     SpectralClustering, AgglomerativeClustering

In [44]: estimators = {
    ...:     'KMeans': kmeans,
    ...:     'DBSCAN': DBSCAN(),
    ...:     'MeanShift': MeanShift(),
    ...:     'SpectralClustering': SpectralClustering(n_clusters=3),
    ...:     'AgglomerativeClustering':
    ...:         AgglomerativeClustering(n_clusters=3)
    ...: }
```

Each iteration of the following loop calls one estimator's fit method with iris.data as an argument, then uses NumPy's unique function to get the cluster labels and counts for the three groups of 50 samples and displays the results. Recall that for the DBSCAN and MeanShift estimators, we did *not* specify the number of clusters in advance. Interestingly, DBSCAN correctly predicted three clusters (labeled -1, 0 and 1), though it placed 84 of the 100 *Iris virginica* and *Iris versicolor* samples in the same cluster. The MeanShift estimator, on the other hand, predicted only two clusters (labeled as 0 and 1), and placed 99 of the 100 *Iris virginica* and *Iris versicolor* samples in the same cluster:

```
In [45]: import numpy as np

In [46]: for name, estimator in estimators.items():
    ...:     estimator.fit(iris.data)
    ...:     print(f'\n{name}:')
    ...:     for i in range(0, 101, 50):
    ...:         labels, counts = np.unique(
    ...:             estimator.labels_[i:i+50], return_counts=True)
    ...:         print(f'{i}-{i+50}:')
    ...:         for label, count in zip(labels, counts):
    ...:             print(f'   label={label}, count={count}')
    ...:

KMeans:
0-50:
   label=1, count=50
50-100:
   label=0, count=48
   label=2, count=2
100-150:
   label=0, count=14
   label=2, count=36
```

---

15. We're running KMeans here on the *small* Iris dataset. If you experience performance problems with KMeans on larger datasets, consider using the MiniBatchKMeans estimator. The scikit-learn documentation indicates that MiniBatchKMeans is faster on large datasets and the results are almost as good.

```
DBSCAN:
0-50:
    label=-1, count=1
    label=0, count=49
50-100:
    label=-1, count=6
    label=1, count=44
100-150:
    label=-1, count=10
    label=1, count=40

MeanShift:
0-50:
    label=1, count=50
50-100:
    label=0, count=49
    label=1, count=1
100-150:
    label=0, count=50

SpectralClustering:
0-50:
    label=2, count=50
50-100:
    label=1, count=50
100-150:
    label=0, count=35
    label=1, count=15

AgglomerativeClustering:
0-50:
    label=1, count=50
50-100:
    label=0, count=49
    label=2, count=1
100-150:
    label=0, count=15
    label=2, count=35
```

Though these algorithms label every sample, the labels simply indicate the clusters. What do you do with the cluster information once you have it? If your goal is to use the data in supervised machine learning, typically you'd study the samples in each cluster to try to determine how they're related and label them accordingly. As we'll see in the next chapter, unsupervised learning is commonly used in deep-learning applications. Some examples of unlabeled data processed with unsupervised learning include tweets from Twitter, Facebook posts, videos, photos, news articles, customers' product reviews, viewers' movie reviews and more.

15.8 Wrap-Up

In this chapter we began our study of machine learning, using the popular scikit-learn library. We saw that machine learning is divided into two types. Supervised machine learning, which works with labeled data and unsupervised machine learning which works with unlabeled data. Throughout this chapter, we continued emphasizing visualizations using Matplotlib and Seaborn, particularly for getting to know your data.

We discussed how scikit-learn conveniently packages machine-learning algorithms as estimators. Each is encapsulated so you can create your models quickly with a small amount of code, even if you don't know the intricate details of how these algorithms work.

We looked at supervised machine learning with classification, then regression. We used one of the simplest classification algorithms, k-nearest neighbors, to analyze the Digits dataset bundled with scikit-learn. You saw that classification algorithms predicts the classes to which samples belong. Binary classification uses two classes (such as "spam" or "not spam") and multi-classification uses more than two classes (such as the 10 classes in the Digits dataset).

We performed the steps of a typical machine-learning case study, including loading the dataset, exploring the data with pandas and visualizations, splitting the data for training and testing, creating the model, training the model and making predictions. We discussed why you should partition your data into a training set and a testing set. You saw ways to evaluate a classification estimator's accuracy via a confusion matrix and a classification report.

We mentioned that it's difficult to know in advance which model(s) will perform best on your data, so you typically try many models and pick the one that performs best. We showed that it's easy to run multiple estimators. We also used hyperparameter tuning with k-fold cross-validation to choose the best value of k for the k-NN algorithm.

We revisited the time series and simple linear regression example from Chapter 10's Intro to Data Science section, this time implementing it using a scikit-learn `LinearRegression` estimator. Next, we used a `LinearRegression` estimator to perform multiple linear regression with the California Housing dataset that's bundled with scikit-learn. You saw that the `LinearRegression` estimator, by default, uses all the numerical features in a dataset to make more sophisticated predictions than you can with simple linear regression. Again, we ran multiple scikit-learn estimators to compare how they performed and choose the best one.

Next, we introduced an unsupervised machine learning and mentioned that it's typically accomplished with clustering algorithms. We used introduced dimensionality reduction (with scikit-learn's `TSNE` estimator) and used it to compress the Digits dataset's 64 features down to two for visualization purposes. This enabled us to see the clustering of the digits data.

We presented one of the simplest unsupervised machine learning algorithms, k-means clustering, and demonstrated clustering on the Iris dataset that's also bundled with scikit-learn. We used dimensionality reduction (with scikit-learn's `PCA` estimator) to compress the Iris dataset's four features to two for visualization purposes to show the clustering of the three *Iris* species in the dataset and their centroids. Finally, we ran multiple clustering estimators to compare their ability to label the Iris dataset's samples into three clusters.

In the next chapter, we'll continue our study of machine learning technologies with discussions of deep learning and reinforcement learning. We'll tackle some fascinating and challenging problems.

Exercises

15.1 *(Using PCA to Help Visualize the Digits Dataset)* In this chapter, we visualized the Digits dataset's clusters. To do so, we first used scikit-learn's TSNE estimator to reduce the dataset's 64 features down to two, then plotted the results using Seaborn. Reimplement that example to perform dimensionality reduction using scikit-learn's PCA estimator, then graph the results. How do the clusters compare to the diagram you created in the clustering case study?

15.2 *(Using TSNE to Help Visualize the Iris Dataset)* In this chapter, we visualized the Iris dataset's clusters. To do so, we first used scikit-learn's PCA estimator to reduce the dataset's four features down to two, then plotted the results using Seaborn. Reimplement that example to perform dimensionality reduction using scikit-learn's TSNE estimator, then graph the results. How do the clusters compare to the diagram you created in the clustering case study?

15.3 *(Seaborn pairplot Graph)* Create a Seaborn pairplot graph (like we showed for Iris) for the California Housing dataset. Try the Matplotlib features for panning and zooming the diagram. These are accessible via the icons in the Matplotlib window.

15.4 *(Human Recognition of Handwritten Digits)* In this chapter, we analyzed the Digits dataset and used scikit-learn's kNeighborsClassifier to recognize the digits with high accuracy. Can humans recognize digit images as well as the kNeighborsClassifier did? Create a script that randomly selects and displays individual images and asks the user to enter a digit from 0 through 9 specifying the digit the image represents. Keep track of the user's accuracy. How does the user compare to the k-nearest neighbors machine-learning algorithm?

15.5 *(Using TSNE to Visualize the Digits Dataset in 3D)* In Section 15.6, you visualized the Digits dataset's clusters in two dimensions. In this exercise, you'll create a 3D scatter plot using TSNE and Matplotlib's **Axes3D**, which provides x-, y- and z-axes for plotting in three dimensions. To do so, load the Digits dataset, create a TSNE estimator that reduces data to three dimensions and call the estimator's fit_transform method to reduce the dataset's dimensions. Store the result in reduced_data. Next, execute the following code:

```
from mpl_toolkits.mplot3d import Axes3D

figure = plt.figure(figsize=(9, 9))

axes = figure.add_subplot(111, projection='3d')

dots = axes.scatter(xs=reduced_data[:, 0], ys=reduced_data[:, 1],
    zs=reduced_data[:, 2], c=digits.target,
    cmap=plt.cm.get_cmap('nipy_spectral_r', 10))
```

The preceding code imports Axes3D, creates a Figure and calls its add_subplot method to get an Axes3D object for creating a three-dimensional graph. In the call to the Axes3D scatter method, the keyword arguments xs, ys and zs specify one-dimensional arrays of values to plot along the x-, y- and z-axes. Once the graph is displayed, be sure to drag the mouse on the image to rotate it left, right, up and down so you can see the clusters from various angles. The following images show the initial 3D graph and two rotated views:

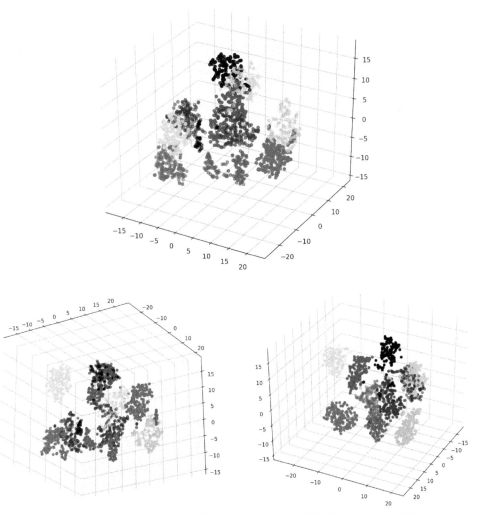

15.6 *(Simple Linear Regression with Average Yearly NYC Temperatures Time Series)* Go to NOAA's Climate at a Glance page (https://www.ncdc.noaa.gov/cag) and download the available time series data for the New York City average *annual* temperatures from 1895 through present (1895–2017 at the time of this writing). For your convenience, we provided the data in the file ave_yearly_temp_nyc_1895-2017.csv. Reimplement the simple linear regression case study of Section 15.4 using the average yearly temperature data. How does the temperature trend compare to the average January high temperatures?

15.7 *(Classification with the Iris Dataset)* We used unsupervised learning with the Iris dataset to cluster its samples. This dataset is in fact labeled so it can be used with scikit-learn's supervised machine learning estimators. Use the techniques you learned in the Digits dataset classification case study to load the Iris dataset and perform classification on it with the k-nearest neighbors algorithm. Use a KNeighborsClassifier with the default k value. What is the prediction accuracy?

15.8 *(Classification with the Iris Dataset: Hyperparameter Tuning)* Using scikit-learn's `KFold` class and `cross_val_score` function, determine the optimal *k* value for classifying Iris samples using a `KNeighborsClassifier`.

15.9 *(Classification with the Iris Dataset: Choosing the Best Estimator)* As we did in the digits case study, run multiple classification estimators for the Iris dataset and compare the results to see which one performs best.

15.10 *(Clustering the Digits Dataset with `DBSCAN` and `MeanShift`)* Recall that when using the `DBSCAN` and `MeanShift` clustering estimators you do not specify the number of clusters in advance. Use each of these estimators with the Digits dataset to determine whether each estimator recognizes 10 clusters of digits.

15.11 *(Using `%timeit` to Time Training and Prediction)* In the k-nearest neighbors algorithm, the computation time for classifying samples increases with the value of *k*. Use `%timeit` to calculate the run time of the `KNeighborsClassifier` cross-validation for the Digits dataset. Use values of 1, 10 and 20 for *k*. Compare the results.

15.12 *(Using `cross_validate`)* In this chapter, we used the `cross_val_score` function and the `KFold` class to perform k-fold cross-validation of the `KNeighborsClassifier` and the Digits dataset. In the k-nearest neighbors algorithm, the computation time for classifying samples increases with the value of *k*. Investigate the `sklearn.model_selection` module's `cross_validate` function, then use it in the loop of Section 15.3.4 both to perform the cross-validation and to calculate the computation times. Display the computation times as part of the loop's output.

15.13 *(Linear Regression with Sea Level Trends)* NOAA's Sea Level Trends website

> https://tidesandcurrents.noaa.gov/sltrends/

provides time series data for sea levels worldwide. Use their **Trend Tables** link to access tables listing sea-level time series for cities in the U.S. and worldwide. The date ranges available vary by city. Choose several cities for which 100% of the data is available (as shown in the **% Complete** column). Clicking the link in the **Station ID** column displays a table of time series data, which you can then export to your system as a CSV file. Use the techniques you learned in this chapter to load and plot each dataset on the same diagram using Seaborn's `regplot` function. In IPython interactive mode, each call to `regplot` uses the same diagram by default and adds data in a new color. Do the sea level rises match in each location?

15.14 *(Linear Regression with Sea Temperature Trends)* Ocean temperatures are changing fish migratory patterns. Download NOAA's global average surface temperature anomalies time series data for 1880–2018 from

> https://www.ncdc.noaa.gov/cag/global/time-series/globe/ocean/ytd/12/1880-2018

then load and plot the dataset using Seaborn's `regplot` function. What trend do you see?

15.15 *(Linear Regression with the Diabetes Dataset)* Investigate the Diabetes dataset bundled with scikit-learn

> https://scikit-learn.org/stable/datasets/index.html#diabetes-dataset

The dataset contains 442 samples, each with 10 features and a label indicating the "disease progression one year after baseline." Using this dataset, reimplement the steps of this chapter's multiple linear regression case study in Section 15.5.

15.16 *(Simple Linear Regression with the California Housing Dataset)* In the text, we performed multiple linear regression using the California Housing dataset. When you have meaningful features available and you have the choice between running simple and multiple linear regression, you'll generally choose multiple linear regression to get more sophisticated predictions. As you saw, scikit-learn's `LinearRegression` estimator uses all the numerical features by default to perform linear regressions.

In this exercise, you'll perform single linear regressions with each feature and compare the prediction results to the multiple linear regression in the chapter. To do so, first split the dataset into training and testing sets, then select one feature, as we did with the `DataFrame` in this chapter's simple linear regression case study. Train the model using that one feature and make predictions as you did in the multiple linear regression case study. Do this for each of the eight features. Compare each simple linear regression's R^2 score with that of the multiple linear regression. What produced the best results?

15.17 *(Binary Classification with the Breast Cancer Dataset)* Check out the Breast Cancer Wisconsin Diagnostic dataset that's bundled with scikit-learn

> https://scikit-learn.org/stable/datasets/index.html#breast-cancer-dataset

The dataset contains 569 samples, each with 30 features and a label indicating whether a tumor was malignant (0) or benign (1). There are only two labels, so this dataset is commonly used to perform **binary classification**. Using this dataset, reimplement the steps of this chapter's classification case study in Sections 15.2–15.3. Use the `GaussianNB` (short for Gaussian Naive Bayes) estimator. When you execute multiple classifiers (as in Section 15.3.3) to determine which one is best for the Breast Cancer Wisconsin Diagnostic dataset, include a `LogisticRegression` classifier in the `estimators` dictionary. Logistic regression is another popular algorithm for binary classification.

15.18 *(Project: Determine k in k-Means Clustering)* In the k-NN classification example, we demonstrated hyperparameter tuning to choose the best value of *k*. In k-means clustering, a challenge is determining the appropriate *k* value for clustering the data. One technique for determining *k* is called the elbow method. Investigate the elbow method, then use it with the Digits and Iris datasets to determine whether this technique yields the correct number of classes for each dataset.

15.19 *(Project: Automated Hyperparameter Tuning)* It's relatively easy to tune one hyperparameter using the looping technique we presented in Section 15.3.4 for determining *k* value in k-nearest neighbors algorithm. What if you need to tune more than one hyperparameter? Scikit-learn's `sklearn.model_selection` module provides tools for automated hyperparameter tuning to help you with this task. Class **GridSearchCV** uses a brute-force approach to hyperparameter tuning by trying every possible combination of the hyperparameters and value ranges for each that you specify. Class **RandomizedSearchCV** improves tuning performance by using random samples of the hyperparameter values you specify. Investigate these classes then reimplement the hyperparameter tuning in Section 15.3.4 using each class. Time the results of each approach.

15.20 *(Quandl Financial Time Series)* Quandl offers an enormous number of financial time series and a Python library for loading them as pandas `DataFrames`, making them easy to use in your machine learning studies. Many of the time series are free. Explore Quandl's financial data search engine at

    https://www.quandl.com/search

to see the range of time series data they offer. Investigate and install their Python module

    conda install -c conda-forge quandl

then use it to download their `'YALE/SPCOMP'` time series for the S&P Composite index (or another time series of your choice). Next, using time series data you downloaded, perform the steps in the linear regression case study of Section 15.5. Use only rows for which all the features have values.

15.21 *(Project: Multi-Classification of Digits with the MNIST Dataset)* In this chapter, we analyzed the Digits dataset that's bundled with scikit-learn. This is a subset and simplified version of the original MNIST dataset, which provides 70,000 digit-image samples and targets. Each sample represents a 28-by-28 image (784 features). Reimplement this chapter's digits classification case study using MNIST. You can download MNIST in scikit-learn using the following statements:

    from sklearn.datasets import fetch_openml
    mnist = fetch_openml('mnist_784', version=1, return_X_y=True)

Function `fetch_mldata` downloads datasets from `mldata.org`, which contains nearly 900 machine learning datasets and various ways to search them.

15.22 *(Project: Multi-Classification of Digits with the EMNIST Dataset)* The EMNIST dataset contains over 800,000 digit and character images. You can work with all 800,000 characters or subsets. One subset has 280,000 digits with approximately 28,000 of each digit (0–9). When the samples are divided evenly among the classes, the dataset is said to have balanced classes. You can download the dataset from

    https://www.nist.gov/itl/iad/image-group/emnist-dataset

in a format used with software called MATLAB, then use SciPy's `loadmat` function (module `scipy.io`) to load the data. The downloaded dataset contains several files—one for the entire dataset and several for various subsets. Load the digits subset, then transform the loaded data into a format usable for use with scikit-learn. Next, reimplement this chapter's digits classification case study using the 280,000 EMNIST digits.

15.23 *(Project: Multi-Classification of Letters with the EMNIST Dataset)* In the previous exercise, you downloaded the EMNIST dataset and worked with the digits subset. Another subset contains 145,600 letters with approximately 5600 of each letter (A–Z). Reimplement the preceding exercise using letter images rather than digits.

15.24 *(Try It: Clustering)* Acxiom is a marketing technology company. Their Personicx marketing software identifies clusters of people for marketing purposes. Try their "What's My Cluster?" tool

    https://isapps.acxiom.com/personicx/personicx.aspx

to see the marketing cluster to which they feel you belong.

15.25 *(Project: AutoML.org and Auto-Sklearn)* There are various ongoing efforts to simplify machine learning and make it available "to the masses." One such effort comes from AutoML.org, which provides tools for automating machine-learning tasks. Their **auto-sklearn** library at

> https://automl.github.io/auto-sklearn

inspects the dataset you wish to use, "automatically searches for the right learning algorithm" and "optimizes its hyperparameters." Investigate auto-sklearn's capabilities then:

- Reimplement the Digits classification case study (Sections 15.2–15.3) using the AutoSklearnClassifier in place of the KNeighborsClassifier estimator.
- Reimplement the California Housing dataset regression case study (Section 15.5) using the AutoSklearnRegressor in place of the LinearRegression estimator.

In each case, how do auto-sklearn's results compare to those in the original case studies? Does auto-sklearn choose the same models?

15.26 *(Research: Support Vector Machines)* Many books and articles indicate that support vector machines often yield the best supervised machine learning results. Research support vector machines vs. other machine learning algorithms. What are the primary reasons offered for why support vector machines perform best?

15.27 *(Research: Machine Learning Ethics and Bias)* Machine learning and artificial intelligence raise many ethics and bias issues. Should an AI algorithm be allowed to fire a company employee without human input? Should an AI-based military weapon, be allowed to make kill decisions without human input? AI algorithms often learn from data collected by humans. What if the data contains human biases regarding race, religion, gender and more? Some AI programs have already been proven to learn such human biases.[16] Research machine learning ethics and bias issues and make a top-10 list of the most common issues you encounter.

15.28 *(Project: Feature Selection)* Feature selection[17] involves choosing which dataset features to use when training a machine learning model. Research feature selection and scikit-learn's feature selection capabilities

> https://scikit-learn.org/stable/modules/feature_selection.html

Apply scikit-learn's feature selection capabilities to the Digits dataset, then reimplement the classification case study in Sections 15.2–15.3. Next, apply scikit-learn's feature selection capabilities to the California Housing dataset, then reimplement the linear regression case study in Section 15.5. In each case, do you get better results?

15.29 *(Research: Feature Engineering)* Feature engineering[18] involves creating new features based on existing features in a dataset. For example, you might transform a feature into a different format (such as transforming textual data to numeric data or transforming a date-time stamp into just a time of day), or you might combine multiple features into a single feature (such as combining latitude and longitude features into a location feature).

---

16. https://www.digitalocean.com/community/tutorials/an-introduction-to-machine-learning#human-biases.
17. https://en.wikipedia.org/wiki/Feature_selection.
18. https://en.wikipedia.org/wiki/Feature_engineering.

Research feature engineering and explain how it might be used to improve supervised machine learning prediction performance.

15.30 *(Project: Desktop Machine Learning Workbench—KNIME Analytics Platform)* There are many free and paid machine learning software packages (both web-based and desktop) for performing machine learning studies with little or no coding. Such tools are known as **workbenches**. KNIME is an open source desktop machine learning and analytics workbench available at

> https://www.knime.com/knime-software/knime-analytics-platform

Investigate KNIME, then install it and use it to implement this chapter's machine learning studies.

15.31 *(Project: Exploring Web-Based Machine Learning Tools—Microsoft Azure Learning Studio, IBM Watson Studio and Google Cloud AI Platform)* Microsoft's Azure Learning Studio, IBM's Watson Studio and Google's Cloud AI Platform are all web-based machine learning tools. Microsoft and IBM provide free tiers and Google provides an extended free trial. Research each of these web-based tools, then use one or more of interest to you to implement this chapter's machine learning studies.

15.32 *(Research Project: Binary Classification with the Titanic Dataset and the Scikit-Learn `DecisionTreeClassifier`)* Decision trees are a popular means of visualizing decision structures in business applications. Research "decision trees" online. Use the techniques you learned in the "Files and Exceptions" chapter to load the Titanic Disaster dataset from the RDatasets repository. One popular type of analysis on this dataset uses decision trees to predict whether a particular passenger survived or died in the tragedy. The `DecisionTreeClassifier` builds a decision tree internally which you can output in the DOT graphing language with the `export_graphviz` function (module `sklearn.tree`). You can use the open source Graphviz visualization software to create a decision-tree graphic from the DOT file.

Deep Learning

16

Objectives

In this chapter you'll:

- Understand what a neural network is and how it enables deep learning.
- Create Keras neural networks.
- Understand Keras layers, activation functions, loss functions and optimizers.
- Use a Keras convolutional neural network (CNN) trained on the MNIST dataset to recognize handwritten digits.
- Use a Keras recurrent neural network (RNN) trained on the IMDb dataset to perform binary classification of positive and negative movie reviews.
- Use TensorBoard to visualize the progress of training deep-learning networks.
- Learn what reinforcement learning, Q-learning and OpenAI Gym are and investigate them in exercises.
- Learn which pretrained neural networks come with Keras.
- Understand the value of using models pretrained on the massive ImageNet dataset for computer vision apps.

Deep Learning

16.1 Introduction	**16.8** ConvnetJS: Browser-Based Deep-Learning Training and Visualization
16.1.1 Deep Learning Applications	
16.1.2 Deep Learning Demos	**16.9** Recurrent Neural Networks for Sequences; Sentiment Analysis with the IMDb Dataset
16.1.3 Keras Resources	
16.2 Keras Built-In Datasets	
16.3 Custom Anaconda Environments	16.9.1 Loading the IMDb Movie Reviews Dataset
16.4 Neural Networks	16.9.2 Data Exploration
16.5 Tensors	16.9.3 Data Preparation
16.6 Convolutional Neural Networks for Vision; Multi-Classification with the MNIST Dataset	16.9.4 Creating the Neural Network
	16.9.5 Training and Evaluating the Model
	16.10 Tuning Deep Learning Models
16.6.1 Loading the MNIST Dataset	**16.11** Convnet Models Pretrained on ImageNet
16.6.2 Data Exploration	
16.6.3 Data Preparation	**16.12** Reinforcement Learning
16.6.4 Creating the Neural Network	16.12.1 Deep Q-Learning
16.6.5 Training and Evaluating the Model	16.12.2 OpenAI Gym
16.6.6 Saving and Loading a Model	
16.7 Visualizing Neural Network Training with TensorBoard	**16.13** Wrap-Up
	Exercises

16.1 Introduction

One of AI's most exciting areas is **deep learning**, a powerful subset of machine learning that has produced impressive results in computer vision and many other areas over the last few years. The availability of big data, significant processor power, faster Internet speeds and advancements in parallel computing hardware and software are making it possible for more organizations and individuals to pursue resource-intensive deep-learning solutions.

Keras and TensorFlow

In the previous chapter, scikit-learn enabled you to define machine-learning models conveniently with one statement. Deep learning models require more sophisticated setups, typically connecting multiple objects, called **layers**. We'll build our deep learning models with **Keras**, which offers a friendly interface to Google's **TensorFlow**—the most widely used deep-learning library.[1] François Chollet of the Google Mind team developed Keras to make deep-learning capabilities more accessible. His book *Deep Learning with Python* is a must read.[2] Google has thousands of TensorFlow and Keras projects underway internally and that number is growing quickly.[3,4]

---

1. Keras also serves as a friendlier interface to Microsoft's *CNTK* and the Université de Montréal's *Theano* (which ceased development in 2017). Other popular deep learning frameworks include Caffe (http://caffe.berkeleyvision.org/), Apache MXNet (https://mxnet.apache.org/) and PyTorch (https://pytorch.org/).
2. Chollet, François. *Deep Learning with Python*. Shelter Island, NY: Manning Publications, 2018.
3. http://theweek.com/speedreads/654463/google-more-than-1000-artificial-intelligence-projects-works.
4. https://www.zdnet.com/article/google-says-exponential-growth-of-ai-is-changing-nature-of-compute/.

Models

Deep learning models are complex and require an extensive mathematical background to understand their inner workings. As we've done throughout the book, we'll avoid heavy mathematics here, preferring English explanations.

Keras is to deep learning as Scikit-learn is to machine learning. Each encapsulates the sophisticated mathematics, so developers need only define, parameterize and manipulate objects. With Keras, you build your models from *pre-existing* components and quickly parameterize those components to your unique requirements. This is what we've been referring to as *object-based programming* throughout the book.

Experiment with Your Models

Machine learning and deep learning are empirical rather than theoretical fields. You'll experiment with many models, tweaking them in various ways until you find the models that perform best for your applications. Keras facilitates such experimentation.

Dataset Sizes

Deep learning works well when you have lots of data, but it also can be effective for smaller datasets when combined with techniques like transfer learning[5,6] and data augmentation[7,8]. Transfer learning uses existing knowledge from a previously trained model as the foundation for a new model. Data augmentation adds data to a dataset by deriving new data from existing data. For example, in an image dataset, you might rotate the images left and right so the model can learn about objects in different orientations. In general, though, the more data you have, the better you'll be able to train a deep learning model.

Processing Power

Deep learning can require significant processing power. Complex models trained on big-data datasets can take hours, days or even more to train. The models we present in this chapter can be trained in minutes to just less than an hour on computers with conventional CPUs. You'll need only a reasonably current personal computer. We'll discuss the special high-performance hardware called GPUs (Graphics Processing Units) and TPUs (Tensor Processing Units) developed by NVIDIA and Google to meet the extraordinary processing demands of edge-of-the-practice deep-learning applications.

Bundled Datasets

Keras comes packaged with some popular datasets. You'll work with two of these in the chapter's examples and several more in the exercises. You can find many Keras studies online for each of these datasets, including ones that take different approaches.

In the "Machine Learning" chapter, you worked with scikit-learn's Digits dataset, which contained 1797 handwritten-digit images that were selected from the much larger MNIST dataset (60,000 training images and 10,000 test images).[9] In this chapter you'll work with the full MNIST dataset. You'll build a Keras *convolutional neural network*

---

5. https://towardsdatascience.com/transfer-learning-from-pre-trained-models-f2393f124751.
6. https://medium.com/nanonets/nanonets-how-to-use-deep-learning-when-you-have-limited-data-f68c0b512cab.
7. https://towardsdatascience.com/data-augmentation-and-images-7aca9bd0dbe8.
8. https://medium.com/nanonets/how-to-use-deep-learning-when-you-have-limited-data-part-2-data-augmentation-c26971dc8ced.

(CNN or convnet) model that will achieve high performance recognizing digit images in the test set. Convnets are especially appropriate for computer vision tasks, such as recognizing handwritten digits and characters or recognizing objects (including faces) in images and videos. You'll also work with a Keras *recurrent neural network*. In that example, you'll perform sentiment analysis using the IMDb Movie reviews dataset, in which the reviews in the training and testing sets are labeled as positive or negative.

Future of Deep Learning

Newer automated deep learning capabilities are making it even easier to build deep-learning solutions. These include Auto-Keras[10] from Texas A&M University's DATA Lab, Baidu's EZDL[11] and Google's AutoML[12]. You'll explore Auto-Keras in the exercises.

Self Check

1. *(Fill-In)* _____ was developed by François Chollet of the Google Mind team as a friendly interface to Google's TensorFlow.
Answer: Keras.

2. *(Fill-In)* _____ are appropriate for computer vision tasks, such as recognizing handwritten digits and characters or recognizing objects (including faces) in images and video.
Answer: Convnets.

16.1.1 Deep Learning Applications

Deep learning is being used in a wide range of applications, such as:

- Game playing
- Computer vision: Object recognition, pattern recognition, facial recognition
- Self-driving cars
- Robotics
- Improving customer experiences
- Chatbots
- Diagnosing medical conditions
- Google Search
- Facial recognition
- Automated image captioning and video closed captioning
- Enhancing image resolution
- Speech recognition
- Language translation
- Predicting election results

---

9. "The MNIST Database." MNIST Handwritten Digit Database, Yann LeCun, Corinna Cortes and Chris Burges. http://yann.lecun.com/exdb/mnist/.
10. https://autokeras.com/.
11. https://ai.baidu.com/ezdl/.
12. https://cloud.google.com/automl/.

- Predicting earthquakes and weather
- Google Sunroof to determine whether you can put solar panels on your roof
- Generative applications—Generating original images, processing existing images to look like a specified artist's style, adding color to black-and-white images and video, creating music, creating text (books, poetry) and much more.

16.1.2 Deep Learning Demos

Check out these four deep-learning demos and search online for lots more, including practical applications like we mentioned in the preceding section:

- DeepArt.io—Turn a photo into artwork by applying an art style to the photo. `https://deepart.io/`.
- DeepWarp Demo—Analyzes a person's photo and makes the person's eyes move in different directions. `https://sites.skoltech.ru/sites/compvision_wiki/static_pages/projects/deepwarp/`.
- Image-to-Image Demo—Translates a line drawing into a picture. `https://affinelayer.com/pixsrv/`.
- Google Translate Mobile App (download from an app store to your smartphone)—Translate text in a photo to another language (e.g., take a photo of a sign or a restaurant menu in Spanish and translate the text to English).

16.1.3 Keras Resources

Here are some resources you might find valuable as you study deep learning:

- To get your questions answered, go to the Keras team's slack channel at `https://kerasteam.slack.com`.
- For articles and tutorials, visit `https://blog.keras.io`.
- The Keras documentation is at `http://keras.io`.
- If you're looking for term projects, directed study projects, capstone course projects or thesis topics, visit arXiv (pronounced "archive," where the X represents the Greek letter "chi") at `https://arXiv.org`. People post their research papers here in parallel with going through peer review for formal publication, hoping for fast feedback. So, this site gives you access to extremely current research.

16.2 Keras Built-In Datasets

Here are some of Keras's datasets (from the module `tensorflow.keras.datasets`[13]) for practicing deep learning. We'll use these in the chapter's examples, exercises and projects:

- MNIST[14] database of handwritten digits—Used for classifying handwritten digit images, this dataset contains 28-by-28 grayscale digit images labeled as 0

---

13. In the standalone Keras library, the module names begin with `keras` rather than `tensorflow.keras`.
14. "The MNIST Database." MNIST Handwritten Digit Database, Yann LeCun, Corinna Cortes and Chris Burges. `http://yann.lecun.com/exdb/mnist/`.

through 9 with 60,000 images for training and 10,000 for testing. We use this dataset in Section 16.6, where we study convolutional neural networks.

- **Fashion-MNIST[15] database of fashion articles**—Used for classifying clothing images, this dataset contains 28-by-28 grayscale images of clothing labeled in 10 categories[16] with 60,000 for training and 10,000 for testing. Once you build a model for use with MNIST, you'll be able to reuse that model with Fashion-MNIST by changing a few statements. You'll use this dataset in the exercises.

- **IMDb Movie reviews[17]**—Used for sentiment analysis, this dataset contains reviews labeled as positive (1) or negative (0) sentiment with 25,000 reviews for training and 25,000 for testing. We use this dataset in Section 16.9, where we study recurrent neural networks.

- **CIFAR10[18] small image classification**—Used for small-image classification, this dataset contains 32-by-32 color images labeled in 10 categories with 50,000 images for training and 10,000 for testing. You'll analyze this dataset with a convnet in the exercises.

- **CIFAR100[19] small image classification**—Also, used for small-image classification, this dataset contains 32-by-32 color images labeled in 100 categories with 50,000 images for training and 10,000 for testing. If you do the CIFAR10 exercise, you should be able to tweak your convnet model quickly for use with CIFAR100.

16.3 Custom Anaconda Environments

Before running this chapter's examples, you'll need to install the libraries we use. In this chapter's examples, we'll use the TensorFlow deep-learning library's version of Keras.[20] At the time of this writing, TensorFlow does not yet support Python 3.7. So, you'll need Python 3.6.x to execute this chapter's examples. We'll show you how to set up a *custom environment* for working with Keras and TensorFlow.

Environments in Anaconda

The Anaconda Python distribution makes it easy to create custom **environments**. These are separate configurations in which you can install different libraries and different library versions. This can help with *reproducibility* if your code depends on specific Python or library versions.[21]

The default environment in Anaconda is called the *base environment*. This is created for you when you install Anaconda. All the Python libraries that come with Anaconda are

---

15. Han Xiao and Kashif Rasul and Roland Vollgraf, Fashion-MNIST: a Novel Image Dataset for Benchmarking Machine Learning Algorithms, arXiv, cs.LG/1708.07747.
16. https://keras.io/datasets/#fashion-mnist-database-of-fashion-articles.
17. Andrew L. Maas, Raymond E. Daly, Peter T. Pham, Dan Huang, Andrew Y. Ng, and Christopher Potts. (2011). Learning Word Vectors for Sentiment Analysis. The 49th Annual Meeting of the Association for Computational Linguistics (ACL 2011).
18. https://www.cs.toronto.edu/~kriz/cifar.html.
19. https://www.cs.toronto.edu/~kriz/cifar.html.
20. There's also a standalone version that enables you to choose between TensorFlow, Microsoft's *CNTK* or the Université de Montréal's *Theano* (which ceased development in 2017).
21. In the next chapter, we'll introduce Docker as another reproducibility mechanism and as a convenient way to install complex environments for use on your local computer.

installed into the base environment and, unless you specify otherwise, any additional libraries you install also are placed there. Custom environments give you control over the specific libraries you wish to install for your specific tasks.

Creating an Anaconda Environment

The **conda create** command creates an environment. Let's create a TensorFlow environment and name it tf_env (you can name it whatever you like). Run the following command in your Terminal, shell or Anaconda Command Prompt:[22,23]

```
conda create -n tf_env tensorflow anaconda ipython jupyterlab
    scikit-learn matplotlib seaborn h5py pydot graphviz
```

This will determine the listed libraries' dependencies, then display all the libraries that will be installed in the new environment. There are many dependencies, so this may take a few minutes. When you see the prompt:

```
Proceed ([y]/n)?
```

press *Enter* to create the environment and install the libraries.[24]

Activating an Alternate Anaconda Environment

To use a custom environment, execute the **conda activate** command:

```
conda activate tf_env
```

This affects only the current Terminal, shell or Anaconda Command Prompt. When a custom environment is activated and you install more libraries, they become part of the activated environment, not the base environment. If you open separate Terminals, shells or Anaconda Command Prompts, they'll use Anaconda's base environment by default.

Deactivating an Alternate Anaconda Environment

When you're done with a custom environment, you can return to the base environment in the current Terminal, shell or Anaconda Command Prompt by executing:

conda deactivate

Jupyter Notebooks and JupyterLab

This chapter's examples are provided only as Jupyter Notebooks, which will make it easier for you to experiment with the examples. You can tweak the options we present and re-execute the notebooks. For this chapter, you should launch JupyterLab from the ch16 examples folder (as discussed in Section 1.10.3).

Self Check

1. *(Fill-In)* The default environment in Anaconda is called the _____ environment.
Answer: base.

---

22. Windows users should run the Anaconda Command Prompt as Administrator,
23. If you have a computer with an NVIDIA GPU that's compatible with TensorFlow, you can replace the tensorflow library with tensorflow-gpu to get better performance. For more information, see https://www.tensorflow.org/install/gpu. Some AMD GPUs also can be used with TensorFlow: http://timdettmers.com/2018/11/05/which-gpu-for-deep-learning/.
24. When we created our custom environment, conda installed Python 3.6.7, which was the most recent Python version compatible with the tensorflow library.

16.4 Neural Networks

Deep learning is a form of machine learning that uses artificial neural networks to learn. An **artificial neural network** (or just neural network) is a software construct that operates similarly to how scientists believe our brains work. Our biological nervous systems are controlled via *neurons*[25] that communicate with one another along pathways called *synapses*[26]. As we learn, the specific neurons that enable us to perform a given task, like walking, communicate with one another more efficiently. These neurons *activate* anytime we need to walk.[27]

Artificial Neurons
In a neural network, interconnected **artificial neurons** simulate the human brain's neurons to help the network learn. The connections between specific neurons are reinforced during the learning process with the goal of achieving a specific result. In **supervised deep learning**—which we'll use in this chapter—we aim to predict the target labels supplied with data samples. To do this, we'll train a general neural network model that we can then use to make predictions on unseen data.[28]

Artificial Neural Network Diagram
The following diagram shows a three-*layer* neural network. Each circle represents a neuron, and the lines between them simulate the synapses. The output of a neuron becomes the input of another neuron, hence the term neural network. This particular diagram shows a **fully connected network**—every neuron in a given layer is connected to *all* the neurons in the next layer:

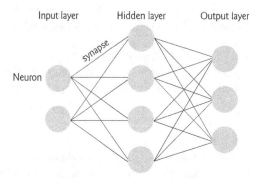

Learning Is an Iterative Process
When you were a baby, you did not learn to walk instantaneously. You learned that process *over time* with repetition. You built up the smaller components of the movements that enabled you to walk—learning to stand, learning to balance to remain standing, learning to lift your foot and move it forward, etc. And you got feedback from your environment. When you walked successfully your parents smiled and clapped. When you fell, you might have bumped your head and felt pain.

---

25. https://en.wikipedia.org/wiki/Neuron.
26. https://en.wikipedia.org/wiki/Synapse.
27. https://www.sciencenewsforstudents.org/article/learning-rewires-brain.
28. As in machine learning, you can create *unsupervised* deep learning networks—these are beyond this chapter's scope.

Similarly, we train neural networks iteratively over time. Each iteration is known as an **epoch** and processes every sample in the training dataset once. There's no "correct" number of epochs. This is a hyperparameter that may need tuning, based on your training data and your model. The inputs to the network are the features in the training samples. Some layers learn new features from previous layers' outputs and others interpret those features to make predictions.

How Artificial Neurons Decide Whether to Activate Synapses

During the training phase, the network calculates values called **weights** for every connection between the neurons in one layer and those in the next. On a neuron-by-neuron basis, each of its inputs is multiplied by that connection's weight, then the sum of those weighted inputs is passed to the neuron's **activation function**. This function's output determines which neurons to activate based on the inputs—just like the neurons in your brain passing information around in response to inputs coming from your eyes, nose, ears and more. The following diagram shows a neuron receiving three inputs (the black dots) and producing an output (the hollow circle) that would be passed to all or some of neurons in the next layer, depending on the types of the neural network's layers:

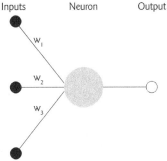

The values w_1, w_2 and w_3 are weights. In a new model that you train from scratch, these values are initialized randomly by the model. As the network trains, it tries to minimize the error rate between the network's predicted labels and the samples' actual labels. The error rate is known as the **loss**, and the calculation that determines the loss is called the **loss function**. Throughout training, the network determines the amount that each neuron contributes to the overall loss, then goes back through the layers and adjusts the weights in an effort to minimize that loss. This technique is called **backpropagation**. Optimizing these weights occurs gradually—typically via a process called **gradient descent**.

✓ Self Check

1. *(True/False)* Deep learning supports only supervised learning with labeled datasets.
Answer: False. As in machine learning, you can create unsupervised deep learning networks.

2. *(Fill-In)* In a(n) _____ neural network, every neuron in a given layer is connected to all the neurons in the next layer.
Answer: fully connected.

3 *(Fill-In)* We train neural networks iteratively over time. Each iteration is known as an _____ and processes every sample in the training dataset once.
Answer: epoch.

4 *(Fill-In)* As a neural network trains, it tries to minimize the error rate between the network's predicted labels and the samples' actual labels. The error rate is known as the _____.
Answer: loss.

16.5 Tensors

Deep learning frameworks generally manipulate data in the form of **tensors**. A "tensor" is basically a multidimensional array. Frameworks like TensorFlow pack all your data into one or more tensors, which they use to perform the mathematical calculations that enable neural networks to learn. These tensors can become quite large as the number of dimensions increases and as the richness of the data increases (for example, images, audios and videos are richer than text). Chollet discusses the types of tensors typically encountered in deep learning:[29]

- **0D (0-dimensional) tensor**—This is one value and is known as a *scalar*.
- **1D tensor**—This is similar to a one-dimensional array and is known as a *vector*. A 1D tensor might represent a sequence, such as hourly temperature readings from a sensor or the words of one movie review.
- **2D tensor**—This is similar to a two-dimensional array and is known as a *matrix*. A 2D tensor could represent a grayscale image in which the tensor's two dimensions are the image's width and height in pixels, and the value in each element is the intensity of that pixel.
- **3D tensor**—This is similar to a three-dimensional array and could be used to represent a color image. The first two dimensions would represent the width and height of the image in pixels and the *depth* at each location might represent the red, green and blue (RGB) components of a given pixel's color. A 3D tensor also could represent a *collection* of 2D tensors containing grayscale images.
- **4D tensor**—A 4D tensor could be used to represent a *collection* of color images in 3D tensors. It also could be used to represent one video. Each frame in a video is essentially a color image.
- **5D tensor**—This could be used to represent a collection of 4D tensors containing videos.

A tensor's *shape* typically is represented as a tuple of values in which the number of elements specifies the tensor's number of dimensions and each value in the tuple specifies the size of the tensor's corresponding dimension.

Let's assume we're creating a deep-learning network to identify and track objects in 4K (high-resolution) videos that have 30 frames-per-second. Each frame in a 4K video is 3840-by-2160 pixels. Let's also assume the pixels are presented as red, green and blue components

---

29. Chollet, François. *Deep Learning with Python*. Section 2.2. Shelter Island, NY: Manning Publications, 2018.

of a color. So *each frame* would be a 3D tensor containing a total of 24,883,200 elements (3840 * 2160 * 3) and each video would be a 4D tensor containing the sequence of frames. If the videos are one minute long, you'd have 44,789,760,000 elements *per tensor*!

Over 600 hours of video are uploaded to YouTube every minute[30] so, in just one minute of uploads, Google could have a tensor containing 1,612,431,360,000,000 elements to use in training deep-learning models—that's *big data*. As you can see, tensors can quickly become *enormous*, so manipulating them efficiently is crucial. This is one of the key reasons that most deep learning is performed on GPUs. More recently Google created TPUs (Tensor Processing Units) that are specifically designed to perform tensor manipulations, executing faster than GPUs.

High-Performance Processors

Powerful processors are needed for real-world deep learning because the size of tensors can be enormous and large-tensor operations can place crushing demands on processors. The processors most commonly used for deep learning are:

- NVIDIA GPUs (Graphics Processing Units)—Originally developed by companies like NVIDIA for computer gaming, GPUs are much faster than conventional CPUs for processing large amounts of data, thus enabling developers to train, validate and test deep-learning models more efficiently—and thus experiment with more of them. GPUs are optimized for the mathematical matrix operations typically performed on tensors, an essential aspect of how deep learning works "under the hood." NVIDIA's Volta Tensor Cores are specifically designed for deep learning.[31,32] Many NVIDIA GPUs are compatible with TensorFlow, and hence Keras, and can enhance the performance of your deep-learning models.[33]

- Google TPUs (Tensor Processing Units)—Recognizing that deep learning is crucial to its future, Google developed TPUs (Tensor Processing Units), which they now use in their Cloud TPU service, which "can provide up to 11.5 petaflops of performance in a single pod"[34] (that's 11.5 *quadrillion* floating-point operations per second). Also, TPUs are designed to be especially energy efficient. This is a key concern for companies like Google with already massive computing clusters that are growing exponentially and consuming vast amounts of energy.

✓ Self Check

1 *(Fill-In)* Deep learning frameworks generally manipulate data in the form of _____.
Answer: tensors.

2 *(True/False)* Tensors can always be processed on standard CPUs.
Answer: False. Tensors can quickly become *enormous* and place crushing demands on processors. For this reason, most deep learning is performed on GPUs or Google's TPUs (Tensor Processing Units).

---

30. https://www.inc.com/tom-popomaronis/youtube-analyzed-trillions-of-data-points-in-2018-revealing-5-eye-opening-behavioral-statistics.html.
31. https://www.nvidia.com/en-us/data-center/tensorcore/.
32. https://devblogs.nvidia.com/tensor-core-ai-performance-milestones/.
33. https://www.tensorflow.org/install/gpu.
34. https://cloud.google.com/tpu/.

3 *(Fill-In)* A(n) _____-dimensional tensor could represent a collection of grayscale images.
Answer: three.

16.6 Convolutional Neural Networks for Vision; Multi-Classification with the MNIST Dataset

In the "Machine Learning" chapter, we classified handwritten digits using the 8-by-8-pixel, low-resolution images from the Digits dataset bundled with Scikit-learn. That dataset is based on a subset of the higher-resolution MNIST handwritten digits dataset. Here, we'll use MNIST to explore deep learning with a **convolutional neural network**[35] (also called a **convnet** or **CNN**). Convnets are common in computer-vision applications, such as recognizing handwritten digits and characters, and recognizing objects in images and video. They're also used in non-vision applications, such as natural-language processing and recommender systems.

The Digits dataset has only 1797 samples, whereas MNIST has 70,000 labeled digit image samples—60,000 for training and 10,000 for testing. Each sample is a grayscale 28-by-28 pixel image (784 total features) represented as a NumPy array. Each pixel is a value from 0 to 255 representing the intensity (or shade) of that pixel—the Digits dataset uses less granular shading with values from 0 to 16. MNIST's labels are integer values in the range 0 through 9, indicating the digit each image represents.

The machine-learning model you used in the previous chapter produced as its output a digit image's predicted class—an integer in the range 0–9. The convnet model we'll build will perform **probabilistic classification**.[36] For each digit image, the model will output an *array* of 10 probabilities, each indicating the likelihood that the digit belongs to a particular one of the classes 0 through 9. The class with the *highest* probability is the predicted value.

Reproducibility in Keras and Deep Learning
We've discussed the importance of *reproducibility* throughout the book. In deep learning, reproducibility is more difficult because the libraries heavily parallelize operations that perform floating-point calculations. Each time operations execute, they may execute in a different order. This can produce differences in your results. Getting reproducible results in Keras requires a combination of environment settings and code settings that are described in the Keras FAQ:

> https://keras.io/getting-started/faq/#how-can-i-obtain-reproducible-results-using-keras-during-development

Basic Keras Neural Network
A Keras neural network consists of the following components:

- A **network** (also called a **model**)—A sequence of **layers** containing the neurons used to learn from the samples. Each layer's neurons receive inputs, process them (via an *activation function*) and produce outputs. The data is fed into the network via an **input layer** that specifies the dimensions of the sample data. This is fol-

---
35. https://en.wikipedia.org/wiki/Convolutional_neural_network.
36. https://en.wikipedia.org/wiki/Probabilistic_classification.

16.6 Multi-Classification with the MNIST Dataset

lowed by **hidden layers** of neurons that implement the learning and an **output layer** that produces the predictions. The more layers you *stack*, the deeper the network is, hence the term deep learning.

- A **loss function**—This produces a measure of how well the network predicts the target values. Lower loss values indicate better predictions.

- An **optimizer**—This attempts to minimize the values produced by the loss function to tune the network to make better predictions.

Launch JupyterLab

This section assumes that you've activated the tf_env Anaconda environment you created in Section 16.3 and launched JupyterLab from the ch16 examples folder. You can either open the MNIST_CNN.ipynb file in JupyterLab and execute the code in the cells we provided, or you can create a new notebook and enter the code on your own. If you prefer, you can work at the command line in IPython, however, placing your code in a Jupyter Notebook makes it significantly easier for you to *re-execute* this chapter's examples.

As a reminder, you can reset a Jupyter Notebook and remove its outputs by selecting **Restart Kernel and Clear All Outputs…** from JupyterLab's **Kernel** menu. This terminates the notebook's execution and removes its outputs. You might do this if your model is not performing well and you want to try different hyperparameters or possibly restructure your neural network.[37] You can then re-execute the notebook one cell at a time or execute the entire notebook by selecting **Run All** from JupyterLab's **Run** menu.

Self Check

1. *(Fill-In)* Convnets are common in _____ applications, such as recognizing handwritten digits and characters, and recognizing objects in images and video.
Answer: computer vision.

2. *(Fill-In)* _____ classification indicates the likelihoods that a sample belongs to each one of the classes the model predicts.
Answer: Probabilistic.

3. *(True/False)* An optimizer produces a measure of how well the network predicts the target values.
Answer: False. A *loss function* produces a measure of how well the network predicts the target values. An optimizer attempts to minimize the values produced by the loss function to tune the network to make better predictions.

16.6.1 Loading the MNIST Dataset

Let's import the tensorflow.keras.datasets.mnist module so we can load the dataset:

```
[1]: from tensorflow.keras.datasets import mnist
```

Note that because we're using the version of Keras built into TensorFlow, the Keras module names begin with "tensorflow.". In the standalone Keras version, the module names begin with "keras.", so keras.datasets would be used above. Keras uses *TensorFlow* to execute the deep-learning models.

---

37. We found that we sometimes had to execute this menu option twice to clear the outputs.

The mnist module's **load_data function** loads the MNIST training and testing sets:

```
[2]: (X_train, y_train), (X_test, y_test) = mnist.load_data()
```

When you call load_data it will download the MNIST data to your system. The function returns a tuple of two elements containing the training and testing sets. Each element is itself a tuple containing the samples and labels, respectively.

 Self Check

1 *(Fill-In)* By Default, Keras uses _____ as its backend to execute deep-learning models.
Answer: TensorFlow.

16.6.2 Data Exploration

Let's get to know the data before working with it. First, we check the dimensions of the training set images (X_train), training set labels (y_train), testing set images (X_test) and testing set labels (y_test):

```
[3]: X_train.shape
[3]: (60000, 28, 28)

[4]: y_train.shape
[4]: (60000,)

[5]: X_test.shape
[5]: (10000, 28, 28)

[6]: y_test.shape
[6]: (10000,)
```

You can see from X_train's and X_test's shapes that the images are higher resolution than those in Scikit-learn's Digits dataset (which are 8-by-8).

Visualizing Digits

Let's visualize some of the digit images. First, enable Matplotlib in the notebook, import Matplotlib and Seaborn and set the font scale:

```
[7]: %matplotlib inline

[8]: import matplotlib.pyplot as plt

[9]: import seaborn as sns

[10]: sns.set(font_scale=2)
```

The IPython magic

```
%matplotlib inline
```

indicates that Matplotlib-based graphics should be displayed *in the notebook* rather than in separate windows. For more IPython magics, you can use in Jupyter Notebooks, see:

https://ipython.readthedocs.io/en/stable/interactive/magics.html

Next, we'll display a randomly selected set of 24 MNIST training set images. Recall from the "Array-Oriented Programming with NumPy" chapter that you can pass a

16.6 Multi-Classification with the MNIST Dataset

sequence of indexes as a NumPy array's subscript to select only the array elements at those indexes. We'll use that capability here to select the elements at the same indexes in both the X_train and y_train arrays. This ensures that we display the correct label for each randomly selected image.

NumPy's **choice function** (from the numpy.random module) randomly selects the number of elements specified in its second argument (24) from the array of values in its first argument (in this case, an array containing X_train's range of indices). The function returns an array containing the selected values, which we store in index. The expressions X_train[index] and y_train[index] use index to get the corresponding elements from both arrays. The rest of this cell is the visualization code from the previous chapter's Digits case study:

```
[11]: import numpy as np
      index = np.random.choice(np.arange(len(X_train)), 24, replace=False)
      figure, axes = plt.subplots(nrows=4, ncols=6, figsize=(16, 9))

      for item in zip(axes.ravel(), X_train[index], y_train[index]):
          axes, image, target = item
          axes.imshow(image, cmap=plt.cm.gray_r)
          axes.set_xticks([])  # remove x-axis tick marks
          axes.set_yticks([])  # remove y-axis tick marks
          axes.set_title(target)
      plt.tight_layout()
```

You can see in the output below that MNIST's digit images have higher resolution than those in Scikit-learn's Digits dataset.

Looking at the digits, you can see why handwritten digit recognition is a challenge:

- Some people write "open" 4s (like the ones in the first and third rows), and some write "closed" 4s (like the one in the second row). Though each 4 has some similar features, they're all different from one another.

- The 3 in the second row looks strange—more like a merged 6 and 7. Compare this to the much clearer 3 in the fourth row.
- The 5 in the second row could easily be confused with a 6.
- Also, people write their digits at different angles, as you can see with the four 6s in the third and fourth rows—two are upright, one leans left and one leans right.

If you run the preceding snippet multiple times, you can see additional randomly selected digits.[38] You'll probably find that—if not for the labels displayed above each digit—it would be difficult for you to identify some of the digits. We'll soon see how accurately our first convnet will predict the digits in the MNIST test set.

16.6.3 Data Preparation

Recall from the "Machine Learning" chapter that Scikit-learn's bundled datasets were preprocessed into the shapes its models required. In real-world studies, you'll generally have to do some or all of the data preparation. The MNIST dataset requires some preparation for use in a Keras convnet.

Reshaping the Image Data

Keras convnets require NumPy array inputs in which each sample has the shape:

($width$, $height$, $channels$)

For MNIST, each image's *width* and *height* are 28 pixels, and each pixel has one *channel* (the grayscale shade of the pixel from 0 to 255), so each sample's shape will be:

(28, 28, 1)

Full-color images with RGB (red/green/blue) values for each pixel, would have three *channels*—one channel each for the red, green and blue components of a color.

As the neural network learns from the images, it creates many more channels. Rather than shade or color, the learned channels will represent more complex features, like edges, curves and lines, that will eventually enable the network to recognize digits based on these additional features and how they're combined.

Let's reshape the 60,000 training and 10,000 testing set images into the correct dimensions for use in our convnet and confirm their new shapes. Recall that NumPy array method `reshape` receives a tuple representing the array's new shape:

```
[12]: X_train = X_train.reshape((60000, 28, 28, 1))

[13]: X_train.shape
[13]: (60000, 28, 28, 1)

[14]: X_test = X_test.reshape((10000, 28, 28, 1))

[15]: X_test.shape
[15]: (10000, 28, 28, 1)
```

---

38. If you do run the cell multiple times, the snippet number next to the cell will increment each time, as it does in IPython at the command line.

Normalizing the Image Data

Numeric features in data samples may have value ranges that vary widely. Deep learning networks perform better on data that is scaled either into the range 0.0 to 1.0, or to a range for which the data's mean is 0.0 and its standard deviation is 1.0.[39] Getting your data into one of these forms is known as **normalization**.

In MNIST, each pixel is an integer in the range 0–255. The following statements convert the values to 32-bit (4-byte) floating-point numbers using the NumPy array method `astype`, then divide every element in the resulting array by 255, producing normalized values in the range 0.0–1.0:

```
[16]: X_train = X_train.astype('float32') / 255
```

```
[17]: X_test = X_test.astype('float32') / 255
```

One-Hot Encoding: Converting the Labels From Integers to Categorical Data

As we mentioned, the convnet's prediction for each digit will be an array of 10 probabilities, indicating the likelihood that the digit belongs to a particular one of the classes 0 through 9. When we evaluate the model's accuracy, Keras compares the model's predictions to the labels. To do that, Keras requires both to have the same shape. The MNIST label for each digit, however, is one integer value in the range 0–9. So, we must transform the labels into **categorical data**—that is, arrays of categories that match the format of the predictions. To do this, we'll use a process called **one-hot encoding**,[40] which converts data into arrays of 1.0s and 0.0s in which only one element is 1.0 and the rest are 0.0s. For MNIST, the one-hot-encoded values will be 10-element arrays representing the categories 0 through 9. One-hot encoding also can be applied to other types of data.

We know precisely which category each digit belongs to, so the categorical representation of a digit label will consist of a 1.0 at that digit's index and 0.0s for all the other elements (again, Keras uses floating-point numbers internally). So, a 7's categorical representation is:

```
[0.0, 0.0, 0.0, 0.0, 0.0, 0.0, 0.0, 1.0, 0.0, 0.0]
```

and a 3's representation is:

```
[0.0, 0.0, 0.0, 1.0, 0.0, 0.0, 0.0, 0.0, 0.0, 0.0]
```

The **tensorflow.keras.utils module** provides function **to_categorical** to perform one-hot encoding. The function counts the unique categories then, for each item being encoded, creates an array of that length with a 1.0 in the correct position. Let's transform y_train and y_test from one-dimensional arrays containing the values 0–9 into two-dimensional arrays of categorical data. After doing so, the rows of these arrays will look like those shown above. Snippet [21] outputs one sample's categorical data for the digit 5 (recall that NumPy shows the decimal point, but not trailing 0s on floating-point values):

---

39. S. Ioffe and Szegedy, C.. "Batch Normalization: Accelerating Deep Network Training by Reducing Internal Covariate Shift." https://arxiv.org/abs/1502.03167.
40. This term comes from certain digital circuits in which a group of bits is allowed to have only one bit turned on (that is, to have the value 1). https://en.wikipedia.org/wiki/One-hot.

```
[18]: from tensorflow.keras.utils import to_categorical

[19]: y_train = to_categorical(y_train)

[20]: y_train.shape
[20]: (60000, 10)

[21]: y_train[0]
[21]: array([ 0., 0., 0., 0., 0., 1., 0., 0., 0., 0.],
      dtype=float32)

[22]: y_test = to_categorical(y_test)

[23]: y_test.shape
[23]: (10000, 10)
```

✓ Self Check

1 *(Fill-In)* Deep learning networks perform better on data that is scaled either into the range 0.0 to 1.0, or to a range for which the data's mean is 0.0 and its standard deviation is 1.0. Getting your data into one of these forms is known as _____.
Answer: normalization.

2 *(What Does This Code Do?)* Assuming y_train contains integer labels 0–9 for the MNIST dataset's training data, what does the following statement do?

```
y_train = to_categorical(y_train)
```

Answer: This statement one-hot encodes the data in y_train, converting each element from an individual integer label in the range 0–9 to an array of 1.0s and 0.0s in which only the element representing the digit's label is 1.0 and the rest are 0.0s.

16.6.4 Creating the Neural Network

Now that we've prepared the data, we'll configure a convolutional neural network. We begin with the Keras **Sequential** model from the **tensorflow.keras.models** module:

```
[24]: from tensorflow.keras.models import Sequential

[25]: cnn = Sequential()
```

The resulting network will execute its layers sequentially—the output of one layer becomes the input to the next. This is known as a **feed-forward network**. As you'll see when we discuss *recurrent neural networks*, not all neural network operate this way.

Adding Layers to the Network

A typical convolutional neural network consists of several layers—an *input layer* that receives the training samples, *hidden layers* that learn from the samples and an *output layer* that produces the prediction probabilities. We'll create a basic convnet here. Let's import from the **tensorflow.keras.layers** module the layer classes we'll use in this example:

```
[26]: from tensorflow.keras.layers import Conv2D, Dense, Flatten,
      MaxPooling2D
```

We discuss each below.

Convolution

We'll begin our network with a **convolution layer**, which uses the relationships between pixels that are close to one another to learn useful features (or patterns) in small areas of each sample. These features become inputs to subsequent layers.

The small areas that convolution learns from are called **kernels** or **patches**. Let's examine convolution on a 6-by-6 image. Consider the following diagram in which the 3-by-3 shaded square represents the kernel—the numbers are simply position numbers showing the order in which the kernels are visited and processed:

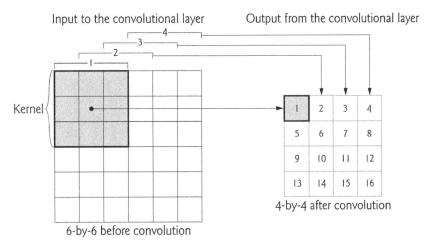

You can think of the kernel as a "sliding window" that the convolution layer moves one pixel at a time left-to-right across the image. When the kernel reaches the right edge, the convolution layer moves the kernel one pixel down and repeats this left-to-right process. Kernels typically are 3-by-3,[41] though we found convnets that used 5-by-5 and 7-by-7 for higher-resolution images. Kernel-size is a tunable hyperparameter.

Initially, the kernel is in the upper-left corner of the original image—kernel position 1 (the shaded square) in the input layer above. The convolution layer performs mathematical calculations using those *nine* features to "learn" about them, then outputs *one* new feature to position 1 in the layer's output. By looking at features near one another, the network begins to recognize features like edges, straight lines and curves.

Next, the convolution layer moves the kernel one pixel to the right (known as the *stride*) to position 2 in the input layer. This new position *overlaps* with two of the three columns in the previous position, so that the convolution layer can learn from all the features that touch one another. The layer learns from the nine features in kernel position 2 and outputs one new feature in position 2 of the output, as in:

---

41. https://www.quora.com/How-can-I-decide-the-kernel-size-output-maps-and-layers-of-CNN.

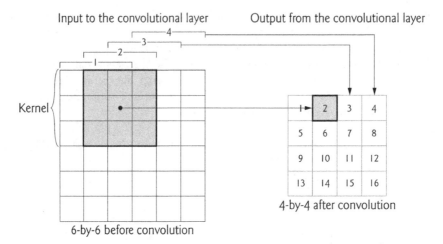

6-by-6 before convolution

4-by-4 after convolution

For a 6-by-6 image and a 3-by-3 kernel, the convolution layer does this two more times to produce features for positions 3 and 4 of the layer's output. Then, the convolution layer moves the kernel one pixel down and begins the left-to-right process again for the next four kernel positions, producing outputs in positions 5–8, then 9–12 and finally 13–16. The complete pass of the image left-to-right and top-to-bottom is called a **filter**. For a 3-by-3 kernel, the filter dimensions (4-by-4 in our sample above) will be *two less than the input dimensions* (6-by-6). For each 28-by-28 MNIST image, the filter will be 26-by-26.

The number of filters in the convolutional layer is commonly 32 or 64 when processing small images like those in MNIST, and each filter produces different results. The number of filters depends on the image dimensions—higher-resolution images have more features, so they require more filters. If you study the code the Keras team used to produce their pretrained convnets,[42] you'll find that they used 64, 128 or even 256 filters in their first convolutional layers. Based on their convnets and the fact that the MNIST images are small, we'll use 64 filters in our first convolutional layer. The set of filters produced by a convolution layer is called a **feature map**.

Subsequent convolution layers combine features from previous feature maps to recognize larger features and so on. If we were doing facial recognition, early layers might recognize lines, edges and curves, and subsequent layers might begin combining those into larger features like eyes, eyebrows, noses, ears and mouths. Once the network learns a feature, because of convolution, it can recognize that feature anywhere in the image. This is one of the reasons that convnets are used for object recognition in images.

Adding a Convolution Layer

Let's add a `Conv2D` convolution layer to our model:

```
[27]: cnn.add(Conv2D(filters=64, kernel_size=(3, 3), activation='relu',
                    input_shape=(28, 28, 1)))
```

The `Conv2D` layer is configured with the following arguments:

- `filters=64`—The number of filters in the resulting feature map.

---

42. https://github.com/keras-team/keras-applications/tree/master/keras_applications.

- `kernel_size=(3, 3)`—The size of the kernel used in each filter.
- `activation='relu'`—The **`relu`** (Rectified Linear Unit) activation function is used to produce this layer's output. `'relu'` is the most widely used activation function in today's deep learning networks[43] and is good for performance because it's easy to calculate.[44] It's commonly recommended for convolutional layers.[45]

Because this is the first layer in the model, we also pass the `input_shape=(28, 28,1)` argument to specify the shape of each sample. This automatically creates an input layer to load the samples and pass them into the `Conv2D` layer, which is actually the first *hidden* layer. In Keras, each subsequent layer infers its `input_shape` from the previous layer's output shape, making it easy to *stack* layers.

Dimensionality of the First Convolution Layer's Output
In the preceding convolutional layer, the input samples are 28-by-28-by-1—that is, 784 features each. We specified 64 filters and a 3-by-3 kernel size for the layer, so the output for each image is 26-by-26-by-64 for a total of 43,264 features in the feature map—a significant increase in dimensionality and an enormous number compared to the numbers of features we processed in the "Machine Learning" chapter's models. As each layer adds more features, the resulting feature maps' *dimensionality* becomes significantly larger. This is one of the reasons that deep learning studies often require tremendous processing power.

Overfitting
Recall from the previous chapter, that overfitting can occur when your model is too complex compared to what it is modeling. In the most extreme case, a model memorizes its training data. When you make predictions with an overfit model, they will be accurate if new data matches the training data, but the model could perform poorly with data it has never seen.

Overfitting tends to occur in deep learning as the dimensionality of the layers becomes too large.[46,47,48] This causes the network to learn *specific* features of the training-set digit images, rather than learning the *general* features of digit images. Some techniques to prevent overfitting include training for fewer epochs, data augmentation, dropout and L1 or L2 regularization.[49,50] We'll discuss dropout later in the chapter.

Higher dimensionality also increases (and sometimes explodes) computation time. If you're performing the deep learning on CPUs rather than GPUs or TPUs, the training could become intolerably slow.

---

43. Chollet, François. *Deep Learning with Python*. p. 72. Shelter Island, NY: Manning Publications, 2018.
44. https://towardsdatascience.com/exploring-activation-functions-for-neural-networks-73498da59b02.
45. https://www.quora.com/How-should-I-choose-a-proper-activation-function-for-the-neural-network.
46. https://cs231n.github.io/convolutional-networks/.
47. https://medium.com/@cxu24/why-dimensionality-reduction-is-important-dd60b5611543.
48. https://towardsdatascience.com/preventing-deep-neural-network-from-overfitting-953458db800a.
49. https://towardsdatascience.com/deep-learning-3-more-on-cnns-handling-overfitting-2bd5d99abe5d.
50. https://www.kdnuggets.com/2015/04/preventing-overfitting-neural-networks.html.

Adding a Pooling Layer

To reduce overfitting and computation time, a convolution layer is often followed by one or more layers that *reduce the dimensionality* of the convolution layer's output. A **pooling layer** *compresses* (or *down-samples*) the results by discarding features, which helps make the model more general. The most common pooling technique is called **max pooling**, which examines a 2-by-2 square of features and keeps only the maximum feature. To understand pooling, let's once again assume a 6-by-6 set of features. In the following diagram, the numeric values in the 6-by-6 square represent the features that we wish to compress and the 2-by-2 blue square in position 1 represents the initial pool of features to examine:

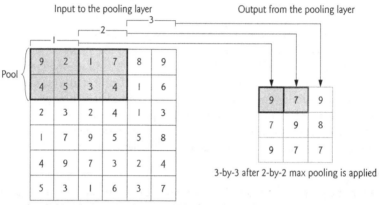

The max pooling layer first looks at the pool in position 1 above, then outputs the *maximum* feature from that pool—9 in our diagram. Unlike convolution, there's *no overlap* between pools. The pool moves by its width—for a 2-by-2 pool, the *stride* is 2. For the second pool, represented by the orange 2-by-2 square, the layer outputs 7. For the third pool, the layer outputs 9. Once the pool reaches the right edge, the pooling layer moves the pool down by its height—2 rows—then continues from left-to-right. Because every group of four features is reduced to one, 2-by-2 pooling *compresses* the number of features by 75%.

Let's add a `MaxPooling2D` layer to our model:

```
[28]: cnn.add(MaxPooling2D(pool_size=(2, 2)))
```

This reduces the previous layer's output from 26-by-26-by-64 to 13-by-13-by-64. In the exercises, we'll ask you to research and use `Dropout` layers, which provide another technique for reducing overfitting.

Though pooling is a common technique to reduce overfitting, some research suggests that additional convolutional layers which use larger strides for their kernels can reduce dimensionality and overfitting *without* discarding features.[51]

Adding Another Convolutional Layer and Pooling Layer

Convnets often have many convolution and pooling layers. The Keras team's convnets tend to double the number of filters in subsequent convolutional layers to enable the

---

51. Tobias, Jost, Dosovitskiy, Alexey, Brox, Thomas, Riedmiller, and Martin. "Striving for Simplicity: The All Convolutional Net." April 13, 2015. https://arxiv.org/abs/1412.6806.

model to learn more relationships between the features.[52] So, let's add a second convolution layer with 128 filters, followed by a second pooling layer to once again reduce the dimensionality by 75%:

```
[29]: cnn.add(Conv2D(filters=128, kernel_size=(3, 3), activation='relu'))
```

```
[30]: cnn.add(MaxPooling2D(pool_size=(2, 2)))
```

The input to the second convolution layer is the 13-by-13-by-64 output of the first pooling layer. So, the output of snippet [29] will be 11-by-11-by-128. For odd dimensions like 11-by-11, Keras pooling layers round *down* by default (in this case to 10-by-10), so this pooling layer's output will be 5-by-5-by-128.

Flattening the Results

At this point, the previous layer's output is three-dimensional (5-by-5-by-128), but the final output of our model will be a *one-dimensional* array of 10 probabilities that classify the digits. To prepare for the one-dimensional final predictions, we first need to *flatten* the previous layer's three-dimensional output. A Keras **Flatten** layer reshapes its input to one dimension. In this case, the Flatten layer's output will be 1-by-3200 (that is, 5 * 5 * 128):

```
[31]: cnn.add(Flatten())
```

Adding a Dense Layer to Reduce the Number of Features

The layers before the Flatten layer learned digit features. Now we need to take all those features and learn the relationships among them so our model can classify which digit each image represents. Learning the relationships among features and performing classification is accomplished with fully connected **Dense** layers, like those shown in the neural network diagram earlier in the chapter. The following Dense layer creates 128 neurons (units) that learn from the 3200 outputs of the previous layer:

```
[32]: cnn.add(Dense(units=128, activation='relu'))
```

Many convnets contain at least one Dense layer like the one above. Convnets geared to more complex image datasets with higher-resolution images like ImageNet—a dataset of over 14 million images[53]—often have several Dense layers, commonly with 4096 neurons. You can see such configurations in several of Keras's pretrained ImageNet convnets[54]—we list these in Section 16.11.

Adding Another Dense Layer to Produce the Final Output

Our final layer is a Dense layer that classifies the inputs into neurons representing the classes 0 through 9. The **softmax activation function** converts the values of these remaining 10 neurons into classification probabilities. The neuron that produces the highest probability represents the prediction for a given digit image:

```
[33]: cnn.add(Dense(units=10, activation='softmax'))
```

Printing the Model's Summary

A model's **summary method** shows you the model's layers. Some interesting things to note are the output shapes of the various layers and the number of parameters. The parameters

---

52. https://github.com/keras-team/keras-applications/tree/master/keras_applications.
53. http://www.image-net.org.
54. https://github.com/keras-team/keras-applications/tree/master/keras_applications.

are the *weights* that the network learns during training.[55,56] This is a relatively small network, yet it will need to learn nearly 500,000 parameters! And this is for tiny images that have less than one quarter of the resolution of the icons on most smartphone home screens. Imagine how many features a network would have to learn to process high-resolution 4K video frames or the super-high-resolution images produced by today's digital cameras. In the Output Shape, None simply means that the model does not know in advance how many training samples you're going to provide—this is known only when you start the training.

```
[34]: cnn.summary()
```

Layer (type)	Output Shape	Param #
conv2d_1 (Conv2D)	(None, 26, 26, 64)	640
max_pooling2d_1 (MaxPooling2	(None, 13, 13, 64)	0
conv2d_2 (Conv2D)	(None, 11, 11, 128)	73856
max_pooling2d_2 (MaxPooling2	(None, 5, 5, 128)	0
flatten_1 (Flatten)	(None, 3200)	0
dense_1 (Dense)	(None, 128)	409728
dense_2 (Dense)	(None, 10)	1290

```
Total params: 485,514
Trainable params: 485,514
Non-trainable params: 0
```

Also, note that there are no "non-trainable" parameters. By default, Keras trains *all* parameters, but it is possible to prevent training for specific layers, which is typically done when you're tuning your networks or using another model's learned parameters in a new model (a process called *transfer learning* that you'll explore in the exercises).[57]

Visualizing a Model's Structure

You can visualize the model summary using the **plot_model** function from the module tensorflow.keras.utils:

```
[35]: from tensorflow.keras.utils import plot_model
      from IPython.display import Image
      plot_model(cnn, to_file='convnet.png', show_shapes=True,
                 show_layer_names=True)
      Image(filename='convnet.png')
```

After storing the visualization in convnet.png, we use module IPython.display's **Image** *class* to show the image in the notebook. Keras assigns the layer names in the image:[58]

---

55. https://hackernoon.com/everything-you-need-to-know-about-neural-networks-8988c3ee4491.
56. https://www.kdnuggets.com/2018/06/deep-learning-best-practices-weight-initialization.html.
57. https://keras.io/getting-started/faq/#how-can-i-freeze-keras-layers.

16.6 Multi-Classification with the MNIST Dataset   **689**

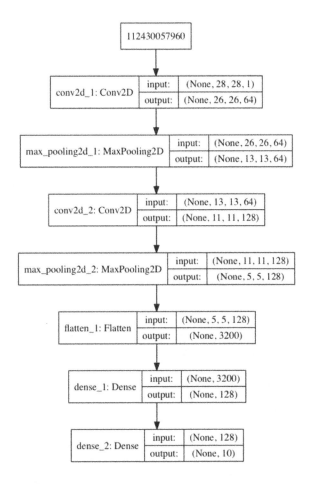

Compiling the Model

Once you've added all the layers you complete the model by calling its **compile** method:

```
[36]: cnn.compile(optimizer='adam',
                  loss='categorical_crossentropy',
                  metrics=['accuracy'])
```

The arguments are:

- optimizer='adam'—The *optimizer* this model will use to adjust the weights throughout the neural network as it learns. There are many optimizers[59]— 'adam' performs well across a wide variety of models.[60,61]

---

58. The node with the large integer value 112430057960 at the top of the diagram appears to be a bug in the current version of Keras. This node represents the input layer and should say "InputLayer".
59. For more Keras optimizers, see https://keras.io/optimizers/.
60. https://medium.com/octavian-ai/which-optimizer-and-learning-rate-should-i-use-for-deep-learning-5acb418f9b2.
61. https://towardsdatascience.com/types-of-optimization-algorithms-used-in-neural-networks-and-ways-to-optimize-gradient-95ae5d39529f.

- `loss='categorical_crossentropy'`—This is the *loss function* used by the optimizer in multi-classification networks like our convnet, which will predict 10 classes. As the neural network learns, the optimizer attempts to minimize the values returned by the loss function. The lower the loss, the better the neural network is at predicting what each image is. For binary classification (which we'll use later in this chapter), Keras provides `'binary_crossentropy'`, and for regression, `'mean_squared_error'`. For other loss functions, see https://keras.io/losses/.
- `metrics=['accuracy']`—This is a list of the *metrics* that the network will produce to help you evaluate the model. Accuracy is a commonly used metric in classification models. In this example, we'll use the accuracy metric to check the percentage of correct predictions. For a list of other metrics, see https://keras.io/metrics/.

✓ Self Check

1 *(Fill-In)* A(n) _____ network passes the output of one layer as the input to the next layer in sequence.
Answer: feed-forward.

2 *(Fill-In)* A(n) _____ layer uses the relationships between pixels that are close to one another to learn useful features (or patterns), such as edges, straight lines and curves.
Answer: convolution.

3 *(Fill-In)* A problem called _____ tends to occur in deep learning as the dimensionality of the layers becomes too large.
Answer: Overfitting.

4 *(True/False)* In Keras, you must specify the input shape for each new layer you add to your neural network model.
Answer: False. You specify the input shape only for the first layer. Keras *infers* the input shape for each subsequent layer, based on the output shape of the previous layer.

5 *(True/False)* In a convnet, learning the relationships among features and performing classification is accomplished with convolution layers.
Answer: False. Convolution layers learn features. Dense layers learn the relationships among features and perform classification.

6 *(Fill-In)* The _____ activation function converts neuron outputs into classification probabilities for multiple-classification.
Answer: `softmax`.

7 *(What Does This Code Do?)* Assuming `cnn` is a Keras convolutional neural network model, what does the following statement do?

```
cnn.add(MaxPooling2D(pool_size=(2, 2)))
```

Answer: This adds a new `MaxPooling2D` layer to an existing neural network model named `cnn`. Because the `pool_size` is specified as 2-by-2, each 2-by-2 square in the preceding layer's output will be reduced to a single value, compressing the previous layer's output by 75%.

8 *(What Does This Code Do?)* Assuming cnn is a Keras convolutional neural network model, what does the following statement do?

```
cnn.compile(optimizer='adam',
            loss='categorical_crossentropy',
            metrics=['accuracy'])
```

Answer: This statement configures the model to use the `'adam'` optimizer, the `categorical_crossentropy` loss function to perform multi-classification and the `'accuracy'` metric to indicate how well the network predicts the samples' classes.

16.6.5 Training and Evaluating the Model

Similar to Scikit-learn's models, we train a Keras model by calling its **fit** method:

- As in Scikit-learn, the first two arguments are the training data and the categorical target labels.

- **epochs** specifies the number of times the model should process the entire set of training data. As we mentioned earlier, neural networks are trained iteratively.

- **batch_size** specifies the number of samples to process at a time during each epoch. Most models specify a power of 2 from 32 to 512. Larger batch sizes can decrease model accuracy.[62] We chose 64. In the exercises, you'll try different values to see how they affect the model's performance.

- In general, some samples should be used to *validate* the model. If you specify validation data, after each epoch, the model will use it to make predictions and display the *validation loss and accuracy*. You can study these values to tune your layers and the `fit` method's hyperparameters, or possibly change the layer composition of your model. Here, we used the **validation_split argument** to indicate that the model should reserve the *last* 10% (0.1) of the training samples for validation[63]—in this case, 6000 samples will be used for validation. If you have separate validation data, you can use the `validation_data` argument (as you'll see in Section 16.9) to specify a tuple containing arrays of samples and target labels. In general, it's better to get *randomly selected validation data*. You can use scikit-learn's `train_test_split` function for this purpose (as we'll do later in this chapter), then pass the randomly selected data with the `validation_data` argument.

In the following output, we highlighted the training accuracy (acc) and validation accuracy (val_acc) in bold:

```
[37]: cnn.fit(X_train, y_train, epochs=5, batch_size=64,
              validation_split=0.1)
Train on 54000 samples, validate on 6000 samples
Epoch 1/5
54000/54000 [==============================] - 68s 1ms/step - loss:
0.1407 - acc: 0.9580 - val_loss: 0.0452 - val_acc: 0.9867
```

---

62. Keskar, Nitish Shirish, Dheevatsa Mudigere, Jorge Nocedal, Mikhail Smelyanskiy and Ping Tak Peter Tang. "On Large-Batch Training for Deep Learning: Generalization Gap and Sharp Minima." CoRR abs/1609.04836 (2016). https://arxiv.org/abs/1609.04836.
63. https://keras.io/getting-started/faq/#how-is-the-validation-split-computed.

```
Epoch 2/5
54000/54000 [==============================] - 64s 1ms/step - loss:
0.0426 - acc: 0.9867 - val_loss: 0.0409 - val_acc: 0.9878
Epoch 3/5
54000/54000 [==============================] - 69s 1ms/step - loss:
0.0299 - acc: 0.9902 - val_loss: 0.0325 - val_acc: 0.9912
Epoch 4/5
54000/54000 [==============================] - 70s 1ms/step - loss:
0.0197 - acc: 0.9935 - val_loss: 0.0335 - val_acc: 0.9903
Epoch 5/5
54000/54000 [==============================] - 63s 1ms/step - loss:
0.0155 - acc: 0.9948 - val_loss: 0.0297 - val_acc: 0.9927
[37]: <tensorflow.python.keras.callbacks.History at 0x7f105ba0ada0>
```

In Section 16.7, we'll introduce *TensorBoard*—a TensorFlow tool for visualizing data from your deep-learning models. In particular, we'll view charts showing how the training and validation accuracy and loss values change through the epochs. In Section 16.8, we'll demonstrate Andrej Karpathy's ConvnetJS tool, which trains convnets in your web browser and dynamically visualizes the layers' outputs, including what each convolutional layer "sees" as it learns. In the exercises, you'll run his MNIST and CIFAR10 models. These will help you better understand neural networks' complex operations.

As the training proceeds, the fit method outputs information showing you the progress of each epoch, how long the epoch took to execute (in this case, each took 63–70 seconds), and the evaluation metrics for that pass. During the last epoch of this model, the accuracy reached 99.48% for the training samples (acc) and 99.27% for the validation samples (val_acc). Those are impressive numbers, given that we have not yet tried to tune the hyperparameters or tweak the number and types of the layers, which could lead to even better (or worse) results. Like machine learning, deep learning is an empirical science that benefits from lots of experimentation.

Evaluating the Model

Now we can check the accuracy of the model on data the model has not yet seen. To do so, we call the model's **evaluate method**, which displays as its output, how long it took to process the test samples (four seconds and 366 microseconds in this case):

```
[38]: loss, accuracy = cnn.evaluate(X_test, y_test)
10000/10000 [==============================] - 4s 366us/step

[39]: loss
[39]: 0.026809450998473768

[40]: accuracy
[40]: 0.9917
```

According to the preceding output, our convnet model is 99.17% accurate when predicting the labels for unseen data—and, at this point, we have not tried to tune the model. With a little online research, you can find models that can predict MNIST with nearly 100% accuracy. The end-of-chapter exercises ask you to experiment with different numbers of layers, types of layers and layer parameters and observe how those changes affect your results.

16.6 Multi-Classification with the MNIST Dataset

Making Predictions

The model's **predict method** predicts the classes of the digit images in its argument array (X_test):

```
[41]: predictions = cnn.predict(X_test)
```

We can check what the first sample digit should be by looking at y_test[0]:

```
[42]: y_test[0]
[42]: array([0., 0., 0., 0., 0., 0., 0., 1., 0., 0.], dtype=float32)
```

According to this output, the first sample is the digit 7, because the categorical representation of the test sample's label specifies a 1.0 at index 7—recall that we created this representation via *one-hot encoding*.

Let's check the probabilities returned by the predict method for the first test sample:

```
[43]: for index, probability in enumerate(predictions[0]):
          print(f'{index}: {probability:.10%}')
0: 0.0000000201%
1: 0.0000001355%
2: 0.0000186951%
3: 0.0000015494%
4: 0.0000000003%
5: 0.0000000012%
6: 0.0000000000%
7: 99.9999761581%
8: 0.0000005577%
9: 0.0000011416%
```

According to the output, predictions[0] indicates that our model believes this digit is a 7 with *nearly* 100% certainty. Not all predictions have this level of certainty.

Locating the Incorrect Predictions

Next, we'd like to view some of the *incorrectly* predicted images to get a sense of the ones our model has trouble with. For example, if it's always mispredicting 8s, perhaps we need more 8s in our training data.

Before we can view incorrect predictions, we need to locate them. Consider predictions[0] above. To determine whether the prediction was correct, we must compare the index of the largest probability in predictions[0] to the index of the element containing 1.0 in y_test[0]. If these index values are the same, then the prediction was correct; otherwise, it was incorrect. NumPy's argmax function determines the index of the highest valued element in its array argument. Let's use that to locate the incorrect predictions. In the following snippet, p is the predicted value array, and e is the expected value array (the expected values are the labels for the dataset's test images):

```
[44]: images = X_test.reshape((10000, 28, 28))
      incorrect_predictions = []

      for i, (p, e) in enumerate(zip(predictions, y_test)):
          predicted, expected = np.argmax(p), np.argmax(e)

          if predicted != expected:
              incorrect_predictions.append(
                  (i, images[i], predicted, expected))
```

In this snippet, we first reshape the samples from the shape (28, 28, 1) that Keras required for learning back to (28, 28), which Matplotlib requires to display the images. Next, we populate the list incorrect_predictions using the for statement. We zip the rows that represent each sample in the arrays predictions and y_test, then enumerate those so we can capture their indexes. If the argmax results for p and e are different, then the prediction was incorrect, and we append a tuple to incorrect_predictions containing that sample's index, image, the predicted value and the expected value. We can confirm the total number of incorrect predictions (out of 10,000 images in the test set) with:

```
[45]: len(incorrect_predictions)
[45]: 83
```

Visualizing Incorrect Predictions
The following snippet displays 24 of the incorrect images labeled with each image's index, predicted value (p) and expected value (e):

```
[46]: figure, axes = plt.subplots(nrows=4, ncols=6, figsize=(16, 12))

      for axes, item in zip(axes.ravel(), incorrect_predictions):
          index, image, predicted, expected = item
          axes.imshow(image, cmap=plt.cm.gray_r)
          axes.set_xticks([])  # remove x-axis tick marks
          axes.set_yticks([])  # remove y-axis tick marks
          axes.set_title(
              f'index: {index}\np: {predicted}; e: {expected}')
      plt.tight_layout()
```

Before reading the expected values, look at each digit and write down what digit you think it is. This is an important part of getting to know your data:

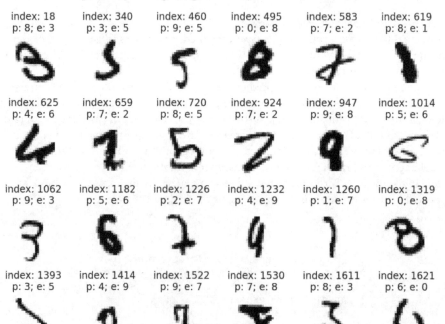

16.6 Multi-Classification with the MNIST Dataset

Displaying the Probabilities for Several Incorrect Predictions

Let's look at the probabilities of some incorrect predictions. The following function displays the probabilities for the specified prediction array:

```
[47]: def display_probabilities(prediction):
          for index, probability in enumerate(prediction):
              print(f'{index}: {probability:.10%}')
```

Though the 8 (at index 495) in the first line of the image output looks like an 8, our model had trouble with it. As you can see in the following output, the model predicted this image as a 0, but also thought there was 16% chance it was a 6 and a 23% chance it was an 8:

```
[48]: display_probabilities(predictions[495])
0: 59.7235262394%
1: 0.0000015465%
2: 0.8047289215%
3: 0.0001740813%
4: 0.0016636326%
5: 0.0030567855%
6: 16.1390662193%
7: 0.0000001781%
8: 23.3022540808%
9: 0.0255270657%
```

The 2 (at index 583) in the first row was predicted to be a 7 with 62.7% certainty, but the model also thought there was a 36.4% chance it was a 2:

```
[49]: display_probabilities(predictions[583])
0: 0.0000003016%
1: 0.0000005715%
2: 36.4056706429%
3: 0.0176281916%
4: 0.0000561930%
5: 0.0000000003%
6: 0.0000000019%
7: 62.7455413342%
8: 0.8310816251%
9: 0.0000114385%
```

The 6 (at index 625) at the beginning of the second row was predicted to be a 4, though that was far from certain. In this case, the probability of a 4 (51.6%) was only slightly higher than the probability of a 6 (48.38%):

```
[50]: display_probabilities(predictions[625])
0: 0.0008245181%
1: 0.0000041209%
2: 0.0012774357%
3: 0.0000000009%
4: 51.6223073006%
5: 0.0000001779%
6: 48.3754962683%
7: 0.0000000085%
8: 0.0000048182%
9: 0.0000785786%
```

Self Check

1. *(True/False)* The `validation_split` argument of a Keras model's `fit` method tells it to randomly select a percentage of the training samples to use as validation data.
Answer: False. The `fit` method takes the validation samples from the *end* of the training samples. For randomly selected validation samples, you can use `train_test_split` from Scikit-learn and pass the selected data to the `fit` method's `validation_data` argument.

2. *(Fill-In)* _____ tends to occur in deep learning as the dimensionality of the layers becomes too large.
Answer: Overfitting.

16.6.6 Saving and Loading a Model

Neural network models can require significant training time. Once you've designed and tested a model that suits your needs, you can save its state. This allows you to load it later to make more predictions. Sometimes models are loaded and further trained for new problems. For example, layers in our model already know how to recognize features such as lines and curves, which could be useful in handwritten character recognition (as in the EMNIST dataset) as well. So you could potentially load the existing model and use it as the basis for a more robust model. This process is called **transfer learning**[64,65]—you transfer an existing model's knowledge into a new model. A Keras model's **save** method stores the model's architecture and state information in a format called **Hierarchical Data Format (HDF5)**. Such files use the `.h5` file extension by default:

```
[51]: cnn.save('mnist_cnn.h5')
```

You can load a saved model with the **load_model function** from the `tensorflow.keras.models` module:

```
from tensorflow.keras.models import load_model
cnn = load_model('mnist_cnn.h5')
```

You can then invoke its methods. For example, if you've acquired more data, you could call `predict` to make additional predictions on new data, or you could call `fit` to start training with the additional data.

Keras provides several additional functions that enable you to save and load various aspects of your models. For more information, see

> https://keras.io/getting-started/faq/#how-can-i-save-a-keras-model

 Self Check

1. *(Fill-In)* You can load a previously saved model and use it as the basis for a more robust model. This process is called _____—you transfer an existing model's knowledge into a new model.
Answer: transfer learning.

---

64. https://towardsdatascience.com/transfer-learning-from-pre-trained-models-f2393f124751.
65. https://medium.com/nanonets/nanonets-how-to-use-deep-learning-when-you-have-limited-data-f68c0b512cab.

16.7 Visualizing Neural Network Training with TensorBoard

With deep learning networks, there's so much complexity and so much going on internally that's hidden from you that it's difficult to know and fully understand all the details. This creates challenges in testing, debugging and updating models and algorithms. Deep learning learns the features but there may be enormous numbers of them, and they may not be apparent to you.

Google provides the **TensorBoard**[66],[67] tool for visualizing neural networks implemented in TensorFlow and Keras. Just as a car's dashboard visualizes data from your car's sensors, such as your speed, engine temperature and the amount of gas remaining, a **TensorBoard dashboard** visualizes data from a deep learning model that can give you insights into how well your model is learning and potentially help you tune its hyperparameters. Here, we'll introduce TensorBoard. We encourage you to explore it more in the exercises.

Executing TensorBoard

TensorBoard monitors a folder on your system looking for files containing the data it will visualize in a web browser. Here, you'll create that folder, execute the TensorBoard server, then access it via a web browser. Perform the following steps:

1. Change to the `ch16` folder in your Terminal, shell or Anaconda Command Prompt.

2. Ensure that your custom Anaconda environment `tf_env` is activated:

    `conda activate tf_env`

3. Execute the following command to create a subfolder named `logs` in which your deep-learning models will write the information that TensorBoard will visualize:

    `mkdir logs`

4. Execute TensorBoard

    `tensorboard --logdir=logs`

5. You can now access TensorBoard in your web browser at

    `http://localhost:6006`

If you connect to TensorBoard before executing any models, it will initially display a page indicating "No dashboards are active for the current data set."[68]

The TensorBoard Dashboard

TensorBoard monitors the folder you specified looking for files output by the model during training. When TensorBoard sees updates, it loads the data into the dashboard:

---

66. https://github.com/tensorflow/tensorboard/blob/master/README.md.
67. https://www.tensorflow.org/guide/summaries_and_tensorboard.
68. TensorBoard does not currently work with Microsoft's Edge browser.

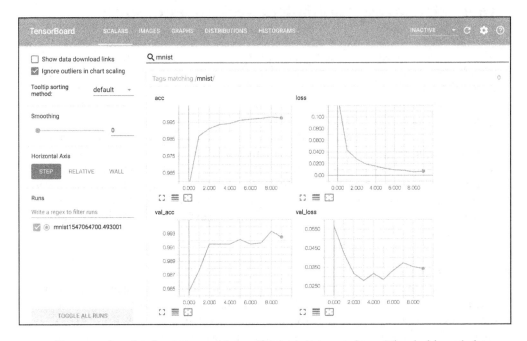

You can view the data as you train or after training completes. The dashboard above shows the TensorBoard **SCALARS** tab, which displays charts for individual values that change over time, such as the training accuracy (acc) and training loss (loss) shown in the first row, and the validation accuracy (val_acc) and validation_loss (val_loss) shown in the second row. The diagrams visualize a 10-epoch run of our MNIST convnet, which we provided in the notebook MNIST_CNN_TensorBoard.ipynb. The epochs are displayed along the x-axes starting from 0 for the first epoch. The accuracy and loss values are displayed on the y-axes. Looking at the training and validation accuracies, you can see in the first 5 epochs similar results to the five-epoch run in the previous section.

For the 10-epoch run, the training accuracy continued to improve through the 9th epoch, then decreased slightly. This might be the point at which we're starting to overfit, but we might need to train longer to find out. For the validation accuracy, you can see that it jumped up quickly, then was relatively flat for five epochs before jumping up then decreasing. For the training loss, you can see that it drops quickly, then continuously declines through the ninth epoch, before a slight increase. The validation loss dropped quickly then bounced around. We could run this model for more epochs to see whether results improve, but based on these diagrams, it appears that around the sixth epoch we get a nice combination of training and validation accuracy with minimal validation loss.

Normally these diagrams are stacked vertically in the dashboard. We used the search field (above the diagrams) to show any that had the name "mnist" in their folder name—we'll configure that in a moment. TensorBoard can load data from multiple models at once and you can choose which to visualize. This makes it easy to compare several different models or multiple runs of the same model.

16.7 Visualizing Neural Network Training with TensorBoard

Copy the MNIST Convnet's Notebook

To create the new notebook for this example:

1. Right-click the MNIST_CNN.ipynb notebook in JupyterLab's **File Browser** tab and select **Duplicate** to make a copy of the notebook.
2. Right-click the new notebook named MNIST_CNN-Copy1.ipynb, then select **Rename**, enter the name MNIST_CNN_TensorBoard.ipynb and press *Enter*.

Open the notebook by double-clicking its name.

Configuring Keras to Write the TensorBoard Log Files

To use TensorBoard, before you fit the model, you need to configure a **TensorBoard** object (module tensorflow.keras.callbacks), which the model will use to write data into a specified folder that TensorBoard monitors. This object is known as a **callback** in Keras. In the notebook, click to the left of snippet that calls the model's fit method, then type a, which is the shortcut for adding a new code cell *above* the current cell (use b for *below*). In the new cell, enter the following code to create the TensorBoard object:

```
from tensorflow.keras.callbacks import TensorBoard
import time

tensorboard_callback = TensorBoard(log_dir=f'./logs/mnist{time.time()}',
    histogram_freq=1, write_graph=True)
```

The arguments are:

- log_dir—The name of the folder in which this model's log files will be written. The notation './logs/' indicates that we're creating a new folder within the logs folder you created previously, and we follow that with 'mnist' and the current time. This ensures that each new execution of the notebook will have its own log folder. That will enable you to compare multiple executions in TensorBoard.
- histogram_freq—The frequency in *epochs* that Keras will output to the model's log files. In this case, we'll write data to the logs for every epoch.
- write_graph—When this is true, a graph of the model will be output. You can view the graph in the **GRAPHS** tab in TensorBoard.

Updating Our Call to fit

Finally, we need to modify the original fit method call in snippet 37. For this example, we set the number of epochs to 10, and we added the callbacks argument, which is a list of callback objects[69]:

```
cnn.fit(X_train, y_train, epochs=10, batch_size=64,
    validation_split=0.1, callbacks=[tensorboard_callback])
```

You can now re-execute the notebook by selecting **Kernel > Restart Kernel and Run All Cells** in JupyterLab. After the first epoch completes, you'll start to see data in TensorBoard.

---

69. You can view Keras's other callbacks at https://keras.io/callbacks/.

Self Check

1. *(Fill-In)* _____ visualizes Keras and TensorFlow neural networks, giving you insights into how well your model is learning and potentially helping you tune its hyperparameters.
Answer: TensorBoard.

16.8 ConvnetJS: Browser-Based Deep-Learning Training and Visualization

In this section, we'll overview Andrej Karpathy's JavaScript-based ConvnetJS tool for training and visualizing convolutional neural networks in your web browser:[70]

> https://cs.stanford.edu/people/karpathy/convnetjs/

You can run the ConvnetJS sample convolutional neural networks or create your own. We've used the tool on several desktop, tablet and phone browsers.

The ConvnetJS MNIST demo—which you'll run and explore in the exercises—trains a convolutional neural network using the MNIST dataset we presented in Section 16.6. The demo presents a scrollable dashboard that updates dynamically as the model trains and contains several sections:

Training Stats
This section contains a **Pause** button that enables you to stop the learning and "freeze" the current dashboard visualizations. Once you pause the demo, the button text changes to **resume**. Clicking the button again continues training. This section also presents training statistics, including the training and validation accuracy and a graph of the training loss.

Instantiate a Network and Trainer
In this section, you'll find the JavaScript code that creates the convolutional neural network. The default network has similar layers to the convnet in Section 16.6. The ConvnetJS documentation[71] shows the supported layer types and how to configure them. You can experiment with different layer configurations in the provided textbox and begin training an updated network by clicking the **change network** button.

Network Visualization
This key section shows one training image at a time and how the network processes that image through each layer. Click the **Pause** button to inspect all the layers' outputs for a given digit to get a sense of what the network "sees" as it learns. The network's last layer produces the probabilistic classifications. It shows 10 squares—9 black and one white, indicating the predicted class of the current digit image.

Example predictions on Test set
The final section shows a random selection of the test set images and the top three possible classes for each digit. The one with the highest probability is shown on a green bar and the other two are displayed on red bars. The length of each bar is a visual indication of that class's probability.

---

70. You also can download ConvnetJS from GitHub at https://github.com/karpathy/convnetjs.
71. https://cs.stanford.edu/people/karpathy/convnetjs/docs.html.

Exercises

In the exercises, experiment with the MNIST and CIFAR10 dataset demos, which classifies 32-by-32 color images in the 10 categories airplane, automobile, bird, cat, deer, dog, frog, horse, ship and truck.

16.9 Recurrent Neural Networks for Sequences; Sentiment Analysis with the IMDb Dataset

In the MNIST CNN network, we focused on stacked layers that were applied *sequentially*. Non-sequential models are possible, as you'll see here with *recurrent neural networks*. In this section, we use Keras's bundled IMDb (the Internet Movie Database) movie reviews dataset[72] to perform **binary classification**, predicting whether a given review's sentiment is positive or negative.

We'll use a **recurrent neural network (RNN)**, which processes sequences of data, such as time series or text in sentences. The term "recurrent" comes from the fact that the neural network contains *loops* in which the output of a given layer becomes the input to that same layer in the next **time step**. In a time series, a time step is the next point in time. In a text sequence, a "time step" would be the next word in a sequence of words.

The looping in RNNs enables them to learn and remember relationships among the data in the sequence. For example, consider the following sentences we used in the "Natural Language Processing" chapter. The sentence

        The food is not good.

clearly has negative sentiment. Similarly, the sentence

        The movie was good.

has positive sentiment, though not as positive as

        The movie was excellent!

In the first sentence, the word "good" on its own has positive sentiment. However, when *preceded by* "not," which appears earlier in the sequence, the sentiment becomes negative. RNNs take into account the relationships among the earlier and later parts of a sequence.

In the preceding example, the words that determined sentiment were adjacent. However, when determining the meaning of text there can be many words to consider and an arbitrary number of words in between them. In this section, we'll use a **Long Short-Term Memory (LSTM)** layer, which makes the neural network *recurrent* and is optimized to handle learning from sequences like the ones we described above.

RNNs have been used for many tasks including:[73,74,75]

- predictive text input—displaying possible next words as you type,

---

72. Maas, Andrew L. and Daly, Raymond E. and Pham, Peter T. and Huang, Dan and Ng, Andrew Y. and Potts, Christopher, "Learning Word Vectors for Sentiment Analysis," *Proceedings of the 49th Annual Meeting of the Association for Computational Linguistics: Human Language Technologies*, June 2011. Portland, Oregon, USA. Association for Computational Linguistics, pp. 142–150. http://www.aclweb.org/anthology/P11-1015.
73. https://www.analyticsindiamag.com/overview-of-recurrent-neural-networks-and-their-applications/.
74. https://en.wikipedia.org/wiki/Recurrent_neural_network#Applications.
75. http://karpathy.github.io/2015/05/21/rnn-effectiveness/.

- sentiment analysis,
- responding to questions with the predicted best answers from a corpus,
- inter-language translation, and
- automated closed captioning in video.

Self Check

1. *(Fill-In)* _____ is used to predict two possible classes—such as positive or negative sentiment in sentiment analysis.
Answer: binary classification.

2. *(True/False)* Neural networks always execute their layers sequentially, such that a given layer's output is passed immediately to the next layer.
Answer: False. The output of a recurrent layer in a recurrent neural network can be passed back as the input to that same layer, effectively creating a loop.

3. *(Fill-In)* A Long Short-Term Memory (LSTM) layer makes a neural network _____ and is optimized to handle learning from sequences.
Answer: recurrent.

16.9.1 Loading the IMDb Movie Reviews Dataset

The IMDb movie reviews dataset included with Keras contains 25,000 training samples and 25,000 testing samples, each labeled with its positive (1) or negative (0) sentiment. Let's import the **tensorflow.keras.datasets.imdb** module so we can load the dataset:

```
[1]: from tensorflow.keras.datasets import imdb
```

The imdb module's **load_data** function returns the IMDb training and testing sets. There are over 88,000 unique words in the dataset. The load_data function enables you to specify the number of unique words to import as part of the training and testing data. In this case, we loaded only the top 10,000 most frequently occurring words due to the memory limitations of our system and the fact that we're (intentionally) training on a CPU rather than a GPU (because most of our readers will not have access to systems with GPUs and TPUs). The more data you load, the longer training will take, but more data may help produce better models:

```
[2]: number_of_words = 10000

[3]: (X_train, y_train), (X_test, y_test) = imdb.load_data(
         num_words=number_of_words)
```

The load_data function returns a tuple of two elements containing the training and testing sets. Each element is itself a tuple containing the samples and labels, respectively. In a given review, load_data replaces any words outside the top 10,000 with a placeholder value, which we'll discuss shortly.

Self Check

1. *(Discussion)* What is the purpose of the num_words argument to the Keras IMDb dataset's load_data function?

Answer: This argument enables you to specify the number of most frequently occurring words you'd like to load and process. This can be helpful if you're training on a CPU and have memory limitations.

16.9.2 Data Exploration

Let's check the dimensions of the training set samples (X_train), training set labels (y_train), testing set samples (X_test) and testing set labels (y_test):

```
[4]: X_train.shape
[4]: (25000,)

[5]: y_train.shape
[5]: (25000,)

[6]: X_test.shape
[6]: (25000,)

[7]: y_test.shape
[7]: (25000,)
```

The arrays y_train and y_test are one-dimensional arrays containing 1s and 0s, indicating whether each review is positive or negative. Based on the preceding outputs, X_train and X_test also appear to be one-dimensional. However, their elements actually are *lists* of integers, each representing one review's contents, as shown in snippet [9]:[76]

```
[8]: %pprint
[8]: Pretty printing has been turned OFF

[9]: X_train[123]
[9]: [1, 307, 5, 1301, 20, 1026, 2511, 87, 2775, 52, 116, 5, 31, 7, 4,
91, 1220, 102, 13, 28, 110, 11, 6, 137, 13, 115, 219, 141, 35, 221, 956,
54, 13, 16, 11, 2714, 61, 322, 423, 12, 38, 76, 59, 1803, 72, 8, 2, 23,
5, 967, 12, 38, 85, 62, 358, 99]
```

Keras deep learning models require *numeric data*, so the Keras team preprocessed the IMDb dataset for you.

Movie Review Encodings

Because the movie reviews are numerically encoded, to view their original text, you need to know the word to which each number corresponds. Keras's IMDb dataset provides a dictionary that maps the words to their indexes. Each word's corresponding value is its frequency ranking among all the words in the entire set of reviews. So the word with the ranking 1 is the most frequently occurring word (calculated by the Keras team from the dataset), the word with ranking 2 is the second most frequently occurring word, and so on.

Though the dictionary values begin with 1 as the most frequently occurring word, in each encoded review (like X_train[123] shown previously), the ranking values are *offset by 3*. So any review containing the most frequently occurring word will have the value 4 wherever that word appears in the review. Keras reserves the values 0, 1 and 2 in each encoded review for the following purposes:

---

76. Here we used the %pprint magic to turn off pretty printing so the following snippet's output could be displayed horizontally rather than vertically to save space. You can turn pretty printing back on by re-executing the %pprint magic.

- The value 0 in a review represents *padding*. Keras deep learning algorithms expect all the training samples to have the same dimensions, so some reviews may need to be expanded to a given length and some shortened to that length. Reviews that need to be expanded are padded with 0s.
- The value 1 represents a token that Keras uses internally to indicate the start of a text sequence for learning purposes.
- The value 2 in a review represents an unknown word—typically a word that was not loaded because you called load_data with the num_words argument. In this case, any review that contained words with frequency rankings greater than num_words would have those words' numeric values replaced with 2. This is all handled by Keras when you load the data.

Because each review's numeric values are offset by 3, we'll have to account for this when we decode the review.

Decoding a Movie Review

Let's decode a review. First, get the word-to-index dictionary by calling the function **get_word_index** from the tensorflow.keras.datasets.imdb module:

```
[10]: word_to_index = imdb.get_word_index()
```

The word 'great' might appear in a positive movie review, so let's see whether it's in the dictionary:

```
[11]: word_to_index['great']
[11]: 84
```

According to the output, 'great' is the dataset's 84th most frequent word. If you look up a word that's not in the dictionary, you'll get an exception.

To transform the frequency ratings into words, let's first reverse the word_to_index dictionary's mapping, so we can look up every word by its frequency rating. The following dictionary comprehension reverses the mapping:

```
[12]: index_to_word = \
        {index: word for (word, index) in word_to_index.items()}
```

Recall that a dictionary's items method enables us to iterate through tuples of key–value pairs. We unpack each tuple into the variables word and index, then create an entry in the new dictionary with the expression index: word.

The following list comprehension gets the top 50 words from the new dictionary— recall that the most frequent word has the value 1:

```
[13]: [index_to_word[i] for i in range(1, 51)]
[13]: ['the', 'and', 'a', 'of', 'to', 'is', 'br', 'in', 'it', 'i',
       'this', 'that', 'was', 'as', 'for', 'with', 'movie', 'but', 'film', 'on',
       'not', 'you', 'are', 'his', 'have', 'he', 'be', 'one', 'all', 'at', 'by',
       'an', 'they', 'who', 'so', 'from', 'like', 'her', 'or', 'just', 'about',
       "it's", 'out', 'has', 'if', 'some', 'there', 'what', 'good', 'more']
```

Note that most of these are *stop words*. Depending on the application, you might want to remove or keep the stop words. For example, if you were creating a predictive-text application that suggests the next word in a sentence the user is typing, you'd want to keep the stop words so they can be displayed as predictions.

Now, we can decode a review. We use the index_to_word dictionary's two-argument method get rather than the [] operator to get value for each key. If a value is not in the dictionary, the get method returns its second argument, rather than raising an exception. The argument i - 3 accounts for the offset in the encoded reviews of each review's frequency ratings. When the Keras reserved values 0–2 appear in a review, get returns '?'; otherwise, get returns the word with the key i - 3 in the index_to_word dictionary:

```
[14]: ' '.join([index_to_word.get(i - 3, '?') for i in X_train[123]])
[14]: '? beautiful and touching movie rich colors great settings good
       acting and one of the most charming movies i have seen in a while i
       never saw such an interesting setting when i was in china my wife
       liked it so much she asked me to ? on and rate it so other would
       enjoy too'
```

We can see from the y_train array that this review is classified as positive:

```
[15]: y_train[123]
[15]: 1
```

16.9.3 Data Preparation

The number of words per review varies, but the Keras requires all samples to have the same dimensions. So, we need to perform some data preparation. In this case, we need to restrict every review to the *same* number of words. Some reviews will need to be *padded* with additional data and others will need to be *truncated*. The **pad_sequences** utility function (module **tensorflow.keras.preprocessing.sequence**) reshapes X_train's samples (that is, its rows) to the number of features specified by the maxlen argument (200) and returns a two-dimensional array:

```
[16]: words_per_review = 200

[17]: from tensorflow.keras.preprocessing.sequence import pad_sequences

[18]: X_train = pad_sequences(X_train, maxlen=words_per_review)
```

If a sample has more features, pad_sequences truncates it to the specified length. If a sample has fewer features, pad_sequences adds 0s to the beginning of the sequence to pad it to the specified length. Let's confirm X_train's new shape:

```
[19]: X_train.shape
[19]: (25000, 200)
```

We also must reshape X_test for later in this example when we evaluate the model:

```
[20]: X_test = pad_sequences(X_test, maxlen=words_per_review)

[21]: X_test.shape
[21]: (25000, 200)
```

Splitting the Test Data into Validation and Test Data

In our convnet, we used the fit method's validation_split argument to indicate that 10% of our training data should be set aside to validate the model as it trains. For this example, we'll manually split the 25,000 test samples into 20,000 test samples and 5,000 validation samples. We'll then pass the 5,000 validation samples to the model's fit method via the argument **validation_data**. Let's use Scikit-learn's train_test_split function from the previous chapter to split the test set:

```
[22]: from sklearn.model_selection import train_test_split
      X_test, X_val, y_test, y_val = train_test_split(
          X_test, y_test, random_state=11, test_size=0.20)
```

Let's also confirm the split by checking X_test's and X_val's shapes:

```
[23]: X_test.shape
[23]: (20000, 200)

[24]: X_val.shape
[24]: (5000, 200)
```

Self Check

1. *(What Does This Code Do?)* Assuming that X_train contains training samples that are variable-length sequences, what does the following statement do?

    ```
    X_test = pad_sequences(X_train, maxlen=500)
    ```

    Answer: This uses function pad_sequences from the tensorflow.keras.preprocessing.sequence module to ensure that every sample in X_train has the length 500. Any samples with sequences longer than 500 will be truncated to 500, and any samples with sequences shorter than 500 will be padded with leading 0s.

2. *(Write a Statement)* X_test and y_test represent 50,000 test samples from a dataset. Write a statement that uses train_test_split to randomly select 20,000 of these for use as validation data. Ensure that the same 20,000 elements are selected every time:
   Answer: X_test, X_val, y_test, y_val = train_test_split(
            X_test, y_test, random_state=11, test_size=0.40)

16.9.4 Creating the Neural Network

Next, we'll configure the RNN. Once again, we begin with a Sequential model to which we'll add the layers that compose our network:

```
[25]: from tensorflow.keras.models import Sequential

[26]: rnn = Sequential()
```

Next, let's import the layers we'll use in this model:

```
[27]: from tensorflow.keras.layers import Dense, LSTM

[28]: from tensorflow.keras.layers.embeddings import Embedding
```

Adding an Embedding Layer

Previously, we used *one-hot encoding* to convert the MNIST dataset's integer labels into *categorical* data. The result for each label was a vector in which all but one element was 0. We could do that for the index values that represent our words. However, this example processes 10,000 unique words. That means we'd need a 10,000-by-10,000 array to represent all the words. That's 100,000,000 elements, and almost all the array elements would be 0. This is not an efficient way to encode the data. If we were to process all 88,000+ unique words in the dataset, we'd need an array of nearly *eight billion* elements!

To *reduce dimensionality*, RNNs that process text sequences typically begin with an **embedding** layer that encodes each word in a more compact *dense-vector representation*.

The vectors produced by the embedding layer also capture the word's context—that is, how a given word relates to the words around it. So the embedding layer enables the RNN to learn word relationships among the training data.

There are also popular **predefined word embeddings**, such as **Word2Vec** and **GloVe**. These can be loaded into neural networks to save training time. They're also sometimes used to add basic word relationships to a model when smaller amounts of training data are available. This can improve the model's accuracy by allowing it to build upon previously learned word relationships, rather than trying to learn those relationships with insufficient amounts of data.

Let's create an `Embedding` layer (module `tensorflow.keras.layers`):

```
[29]: rnn.add(Embedding(input_dim=number_of_words, output_dim=128,
                        input_length=words_per_review))
```

The arguments are:

- `input_dim`—The number of unique words.
- `output_dim`—The size of each word embedding. If you load pre-existing embeddings[77] like *Word2Vec* and *GloVe*, you must set this to match the size of the word embeddings you load.
- `input_length=words_per_review`—The number of words in each input sample.

Adding an LSTM Layer

Next, we'll add an LSTM layer:

```
[30]: rnn.add(LSTM(units=128, dropout=0.2, recurrent_dropout=0.2))
```

The arguments are:

- `units`—The number of neurons in the layer. The more neurons the more the network can remember. As a guideline, you can start with a value between the length of the sequences you're processing (200 in this example) and the number of classes you're trying to predict (2 in this example).[78]
- `dropout`—The percentage of neurons to randomly disable when processing the layer's input and output. Like the pooling layers in our convnet, **dropout** is a proven technique[79,80] that reduces overfitting. Keras provides a **Dropout** layer that you can add to your models.
- `recurrent_dropout`—The percentage of neurons to randomly disable when the layer's output is fed back into the layer again to allow the network to learn from what it has seen previously.

---

77. https://blog.keras.io/using-pre-trained-word-embeddings-in-a-keras-model.html.
78. https://towardsdatascience.com/choosing-the-right-hyperparameters-for-a-simple-lstm-using-keras-f8e9ed76f046.
79. Yarin, Ghahramani, and Zoubin. "A Theoretically Grounded Application of Dropout in Recurrent Neural Networks." October 05, 2016. https://arxiv.org/abs/1512.05287.
80. Srivastava, Nitish, Geoffrey Hinton, Alex Krizhevsky, Ilya Sutskever, and Ruslan Salakhutdinov. "Dropout: A Simple Way to Prevent Neural Networks from Overfitting." *Journal of Machine Learning Research* 15 (June 14, 2014): 1929-1958. http://jmlr.org/papers/volume15/srivastava14a/srivastava14a.pdf.

The mechanics of how the LSTM layer performs its task are beyond the scope of this book. Chollet says: "you don't need to understand anything about the specific architecture of an LSTM cell; as a human, it shouldn't be your job to understand it. Just keep in mind what the LSTM cell is meant to do: allow past information to be reinjected at a later time."[81]

Adding a Dense Output Layer

Finally, we need to take the LSTM layer's output and reduce it to one result indicating whether a review is positive or negative, thus the value 1 for the **units** argument. Here we use the **'sigmoid' activation function**, which is preferred for binary classification.[82] It reduces arbitrary values into the range 0.0–1.0, producing a probability:

```
[31]: rnn.add(Dense(units=1, activation='sigmoid'))
```

Compiling the Model and Displaying the Summary

Next, we compile the model. In this case, there are only two possible outputs, so we use the **binary_crossentropy** loss function:

```
[32]: rnn.compile(optimizer='adam',
                  loss='binary_crossentropy',
                  metrics=['accuracy'])
```

The following is the summary of our model. Notice that even though we have fewer layers than our convnet, the RNN has nearly three times as many trainable parameters (the network's weights) as the convnet and more parameters means more training time. The large number of parameters primarily comes from the number of words in the vocabulary (we loaded 10,000) times the number of neurons in the Embedding layer's output (128):

```
[33]: rnn.summary()
```

Layer (type)	Output Shape	Param #
embedding_1 (Embedding)	(None, 200, 128)	1280000
lstm_1 (LSTM)	(None, 128)	131584
dense_1 (Dense)	(None, 1)	129

Total params: 1,411,713
Trainable params: 1,411,713
Non-trainable params: 0

Self Check

1 *(True/False)* When performing deep learning on text sequences, you must create your own word embeddings to learn relationships among the words in your dataset.
Answer: False. There are predefined word embeddings, such as Word2Vec and GloVe, that you can load into neural networks to save training time or to add basic word relationships to a model when smaller amounts of training data are available.

2 *(Write a Statement)* Assuming that your dataset has 100,000 unique words and that the sequences in the dataset are 500 words long, write a statement that creates a Keras

---

81. Chollet, François. *Deep Learning with Python.* p.204. Shelter Island, NY: Manning Publications, 2018.
82. Chollet, François. *Deep Learning with Python.* p.114. Shelter Island, NY: Manning Publications, 2018.

embedding layer with an output size of 256 and adds it to an existing `Sequential` object named `network`.
Answer: `network.add(Embedding(input_dim=100000, output_dim=256,`
  `input_length=500))`

3 *(Write a Statement)* Write a statement that creates a Keras `LSTM` with 256 units and 50% dropout.
Answer: `network.add(LSTM(units=256, dropout=0.5, recurrent_dropout=0.5))`

16.9.5 Training and Evaluating the Model

Let's train our model.[83] Notice for each epoch that the model takes significantly longer to train than our convnet did. This is due to the larger numbers of parameters (weights) our RNN model needs to learn. We bolded the accuracy (acc) and validation accuracy (val_acc) values for readability—these represent the percentage of training samples and the percentage of `validation_data` samples that the model predicts correctly.

```
[34]: rnn.fit(X_train, y_train, epochs=10, batch_size=32,
             validation_data=(X_test, y_test))
Train on 25000 samples, validate on 5000 samples
Epoch 1/5
25000/25000 [==============================] - 299s 12ms/step - loss:
0.6574 - acc: 0.5868 - val_loss: 0.5582 - val_acc: 0.6964
Epoch 2/5
25000/25000 [==============================] - 298s 12ms/step - loss:
0.4577 - acc: 0.7786 - val_loss: 0.3546 - val_acc: 0.8448
Epoch 3/5
25000/25000 [==============================] - 296s 12ms/step - loss:
0.3277 - acc: 0.8594 - val_loss: 0.3207 - val_acc: 0.8614
Epoch 4/5
25000/25000 [==============================] - 307s 12ms/step - loss:
0.2675 - acc: 0.8864 - val_loss: 0.3056 - val_acc: 0.8700
Epoch 5/5
25000/25000 [==============================] - 310s 12ms/step - loss:
0.2217 - acc: 0.9083 - val_loss: 0.3264 - val_acc: 0.8704
[34]: <tensorflow.python.keras.callbacks.History object at 0xb3ba882e8>
```

Finally, we can evaluate the results using the test data. Function `evaluate` returns the loss and accuracy values. In this case, the model was 85.99% accurate:

```
[35]: results = rnn.evaluate(X_test, y_test)
20000/20000 [==============================] - 42s 2ms/step

[36]: results
[36]: [0.3415240607559681, 0.8599]
```

Note that the accuracy of this model seems low compared to our MNIST convnet's results, but this is a much more difficult problem. If you search online for other IMDb sentiment-analysis binary-classification studies, you'll find lots of results in the high 80s. So we did reasonably well with our small recurrent neural network of only three layers. In the exercises, you'll be asked to study some online models and produce a better model.

---

83. At the time of this writing, TensorFlow displayed a warning when we executed this statement. This is a known TensorFlow issue and, according to the forums, you can safely ignore the warning.

16.10 Tuning Deep Learning Models

In Section 16.9.5, notice in the `fit` method's output that both the testing accuracy (85.99%) and validation accuracy (87.04%) were significantly less than the 90.83% training accuracy. Such disparities are usually the result of overfitting, so there is plenty of room for improvement in our model.[84,85] If you look at the output of each epoch, you'll notice both the training and validation accuracy continue to increase. Recall that training for too many epochs can lead to overfitting, but it's possible we have not yet trained enough. Perhaps one hyperparameter tuning option for this model would be to increase the number of epochs.

Some variables that affect your models' performance include:

- having more or less data to train with
- having more or less to test with
- having more or less to validate with
- having more or fewer layers
- the types of layers you use
- the order of the layers

In our IMDb RNN example, some things we could tune include:

- trying different amounts of the training data—we used only the top 10,000 words
- different numbers of words per review—we used only 200,
- different numbers of neurons in our layers,
- more layers or
- possibly loading pre-trained word vectors rather than having our Embedding layer learn them from scratch.

The compute time required to train models multiple times is significant so, in deep learning, you generally do not tune hyperparameters with techniques like k-fold cross-validation or grid search.[86] There are various tuning techniques,[87,88,89,90] but one particularly promising area is automated machine learning (AutoML). For example, the Auto-Keras[91] library is specifically geared to automatically choosing the best configurations for

---

84. https://towardsdatascience.com/deep-learning-overfitting-846bf5b35e24.
85. https://hackernoon.com/memorizing-is-not-learning-6-tricks-to-prevent-overfitting-in-machine-learning-820b091dc42.
86. https://www.quora.com/Is-cross-validation-heavily-used-in-deep-learning-or-is-it-too-expensive-to-be-used.
87. https://towardsdatascience.com/what-are-hyperparameters-and-how-to-tune-the-hyperparameters-in-a-deep-neural-network-d0604917584a.
88. https://medium.com/machine-learning-bites/deeplearning-series-deep-neural-networks-tuning-and-optimization-39250ff7786d.
89. https://flyyufelix.github.io/2016/10/03/fine-tuning-in-keras-part1.html and https://flyyufelix.github.io/2016/10/08/fine-tuning-in-keras-part2.html.
90. https://towardsdatascience.com/a-comprehensive-guide-on-how-to-fine-tune-deep-neural-networks-using-keras-on-google-colab-free-daaaa0aced8f.
91. https://autokeras.com/.

your Keras models. Google's Cloud AutoML and Baidu's EZDL are among various other automated machine learning efforts.

Self Check

1 *(True/False)* Overfitting occurs if you train your model for too few epochs.
Answer: False. Overfitting typically occurs when you train your model for too many epochs. This is one of the ways in which visualizing your model with TensorBoard can be helpful.

16.11 Convnet Models Pretrained on ImageNet

With deep learning, rather than starting fresh on every project with costly training, validating and testing, you can use **pretrained deep neural network models** to:

- make new predictions,
- continue training them further with new data or
- transfer the weights learned by a model for a similar problem into a new model—this is called *transfer learning*.

Keras Pretrained Convnet Models

Keras comes bundled with the following pretrained convnet models,[92] each pretrained on ImageNet[93]—a growing dataset of 14+ million images:

- Xception
- VGG16
- VGG19
- ResNet50
- Inception v3
- Inception-ResNet v2
- MobileNet v1
- DenseNet
- NASNet
- MobileNet v2

Reusing Pretrained Models

ImageNet is too big for efficient training on most computers, so most people interested in using it start with one of the smaller pretrained models.

You can reuse just the architecture of each model and train it with new data, or you can reuse the pretrained weights. For a few simple examples, see:

> https://keras.io/applications/

---

92. https://keras.io/applications/.
93. http://www.image-net.org.

ImageNet Challenge

In the end-of-chapter projects, you'll research and use some of these bundled models. You'll also investigate the *ImageNet Large Scale Visual Recognition Challenge* for evaluating object-detection and image-recognition models.[94] This competition ran from 2010 through 2017. ImageNet now has a continuously running challenge on the Kaggle competition site called the *ImageNet Object Localization Challenge*.[95] The goal is to identify "all objects within an image, so those images can then be classified and annotated." ImageNet releases the current participants leaderboard once per quarter.

A lot of what you've seen in the machine learning and deep learning chapters is what the Kaggle competition website is all about. There's no obvious optimal solution for many machine learning and deep learning tasks. People's creativity is really the only limit. On Kaggle, companies and organizations fund competitions where they encourage people worldwide to develop better-performing solutions than they've been able to do for something that's important to their business or organization. Sometimes companies offer prize money, which has been as high as $1,000,000 on the famous Netflix competition. Netflix wanted to get a 10% or better improvement in their model for determining whether people will like a movie, based on how they rated previous ones.[96] They used the results to help make better recommendations to members. Even if you do not win a Kaggle competition, it's a great way to get experience working on problems of current interest.

16.12 Reinforcement Learning

Reinforcement learning is a form of machine learning in which algorithms learn from their environment, similar to how humans learn—for example, a video game enthusiast learning a new game, or a baby learning to walk or recognize its parents.

The algorithm implements an **agent** that learns by trying to perform a task, receiving feedback about success or failure, making adjustments then trying again. The goal is to maximize the **reward**. The agent receives a **positive reward** for doing a right thing and a **negative reward** (that is, a **punishment**) for doing a wrong thing. The agent uses this information to determine the next action to perform and must try to maximize the reward.

Reinforcement learning was used in some key artificial-intelligence milestones that captured people's attention and imagination. In 2011, IBM's Watson beat the world's two best human Jeopardy! players in a $1 million match. Watson simultaneously executed hundreds of language-analysis algorithms to locate correct answers in 200 million pages of content (including all of Wikipedia) requiring four terabytes of storage.[97,98] Watson was trained with machine learning and used reinforcement learning techniques to learn the game-playing strategies (such as when to answer, which square to pick and how much money to risk on daily doubles).[99,100]

---

94. http://www.image-net.org/challenges/LSVRC/.
95. https://www.kaggle.com/c/imagenet-object-localization-challenge.
96. https://netflixprize.com/rules.html.
97. https://www.techrepublic.com/article/ibm-watson-the-inside-story-of-how-the-jeopardy-winning-supercomputer-was-born-and-what-it-wants-to-do-next/.
98. https://en.wikipedia.org/wiki/Watson_(computer).
99. https://www.aaai.org/Magazine/Watson/watson.php, *AI Magazine*, Fall 2010.
100. https://developer.ibm.com/articles/cc-reinforcement-learning-train-software-agent/.

Mastering the Chinese Board Game Go

Go—a board game created in China thousands of years ago[101]—is widely considered to be one of the most complex games ever invented with 10^{170} possible board configurations.[102] To give you a sense of how large a number that is, it's believed that there are (only) between 10^{78} and 10^{87} atoms in the known universe![103,104] In 2015, AlphaGo—created by Google's DeepMind group—used *deep learning* with two neural networks to beat the European Go champion Fan Hui. Go is considered to be a far more complex game than chess.

AlphaZero

More recently, Google generalized its AlphaGo AI to create AlphaZero—a game-playing AI that uses reinforcement learning to teach itself to play other games. In December 2017, AlphaZero learned the rules of and taught itself to play chess in *less than four hours*. It then beat the world champion chess program, Stockfish 8, in a 100-game match—winning or drawing every game. After training itself in Go for *just eight hours*, AlphaZero was able to play Go vs. its AlphaGo predecessor, winning 60 of 100 games.[105]

16.12.1 Deep Q-Learning

One of the most popular reinforcement learning techniques is Deep Q-Learning, which was originally described in the Google DeepMind team's paper "Playing Atari with Deep Reinforcement Learning."[106] Using Deep Q-Learning, they were able to develop an agent that learned to play Atari video games by observing changes in the pixels on the screen.

Deep Q-Learning combines Q-learning with deep learning. In Q-Learning a Q function determines the reward using a combination of the environment's current *state* and the action the agent performs. For example, if the agent is trying to learn how to avoid obstacles, every move the agent makes that does not hit an obstacle would get a positive reward and every move that collides with an obstacle would get a negative reward (that is, a punishment).

16.12.2 OpenAI Gym

Game playing is a key application of reinforcement learning. A tool called **OpenAI Gym** (https://gym.openai.com) has become popular for reinforcement learning research. It comes with several games environments that you can use to experiment with reinforcement learning and to develop your own algorithms. There are many additional environments (from Atari and others) that you can download and install into OpenAI Gym. In one of this chapter's project exercises, you'll research OpenAI Gym and experiment with its **CartPole environment** (shown below). This is a simple game with a *cart* (the black rect-

---

101. http://www.usgo.org/brief-history-go.
102. https://www.pbs.org/newshour/science/google-artificial-intelligence-beats-champion-at-worlds-most-complicated-board-game.
103. https://www.universetoday.com/36302/atoms-in-the-universe/.
104. https://en.wikipedia.org/wiki/Observable_universe#Matter_content.
105. https://www.theguardian.com/technology/2017/dec/07/alphazero-google-deepmind-ai-beats-champion-program-teaching-itself-to-play-four-hours.
106. Volodymyr, Koray, David, Alex, Ioannis, Daan, Riedmiller, and Martin. "Playing Atari with Deep Reinforcement Learning." December 19, 2013. https://arxiv.org/abs/1312.5602.

angle) that can move left or right on a track in one dimension and a *pole* (the vertical line) that's hinged to the cart. The goal of the game is to keep the pole vertical. As it falls, the algorithm moves the cart left or right to restore the pole to the vertical position.

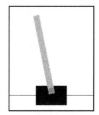 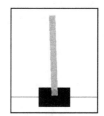 

16.13 Wrap-Up

In Chapter 16, you peered into the future of AI. Deep Learning and reinforcement learning have captured the imagination of the computer-science and data science-communities. This may be the most important AI chapter in the book.

We mentioned the key deep-learning platforms, indicating that Google's TensorFlow is the most widely used. We discussed why Keras, which presents a friendly interface to TensorFlow, has become so popular.

We set up a custom Anaconda environment for TensorFlow, Keras and JupyterLab, then used the environment to implement the Keras examples.

We explained what tensors are and why they're crucial to deep learning. We discussed the basics of neurons and multi-layered neural networks for building Keras deep-learning models. We considered some popular types of layers and how to order them.

We introduced convolutional neural networks (convnets) and indicated that they're especially appropriate for computer-vision applications. We then built, trained, validated and tested a convnet using the MNIST database of handwritten digits for which we achieved 99.17% prediction accuracy. This is remarkable, given that we achieved it by working with a only a basic model and without doing any hyperparameter tuning. In the exercises, you can try more sophisticated models and tune the hyperparameters to try to achieve better performance. We listed a variety of intriguing computer vision tasks, many of which you can investigate in the exercises.

We introduced TensorBoard for visualizing TensorFlow and Keras neural network training and validation. We also discussed ConvnetJS, a browser-based convnet training and visualization tool, which enables you to peek inside the training process.

Next, we presented recurrent neural networks (RNNs) for processing sequences of data, such as time series or text in sentences. We used an RNN with the IMDb movie reviews dataset to perform binary classification, predicting whether each review's sentiment was positive or negative. We also discussed tuning deep learning models and how high-performance hardware, like NVIDIA's GPUs and Google's TPUs, is making it possible for more people to tackle more substantial deep-learning studies.

Given how costly and time-consuming it is to train deep-learning models, we explained the strategy of using pretrained models. We listed various Keras convnet image-processing models that were trained on the massive ImageNet dataset, and discussed how transfer learning enables you to use these models to create new ones quickly and effectively.

We briefly introduced reinforcement learning, Deep Q-Learning and OpenAI Gym. In the exercises, you can investigate applications of each.

Deep learning and reinforcement learning are large, complex topics. We focused on the basics in the chapter. In the exercises, you can explore additional intriguing topics. Many of these would make nice term projects, directed study topics, capstone project topics and thesis topics at all levels.

In the next chapter, we present the big data infrastructure that supports the kinds of AI technologies we've discussed in Chapters 12 through 16. We'll consider the Hadoop and Spark platforms for big data batch processing and real-time streaming applications. We'll look at relational databases and the SQL language for querying them—these have dominated the database field for many decades. We'll discuss how big data presents challenges that relational databases don't handle well, and consider how NoSQL databases are designed to handle those challenges. We'll conclude the book with a discussion of the Internet of Things (IoT), which will surely be the world's largest big-data source and will present many opportunities for entrepreneurs to develop leading-edge businesses that will truly make a difference in people's lives.

Exercises

Convolutional Neural Networks

16.1 *(Image Recognition: The Fashion-MNIST Dataset)* Keras comes bundled with the Fashion-MNIST database of fashion articles which, like the MNIST digits dataset, provides 28-by-28 grayscale images. Fashion-MNIST contains clothing-article images labeled in 10 categories—0 (T-shirt/top), 1 (Trouser), 2 (Pullover), 3 (Dress), 4 (Coat), 5 (Sandal), 6 (Shirt), 7 (Sneaker), 8 (Bag), 9 (Ankle boot)—with 60,000 training samples and 10,000 testing samples. Modify this chapter's convnet example to load and process Fashion-MNIST rather than MNIST—this requires simply importing the correct module, loading the data then running the model with these images and labels, then re-run the entire example. How well does the model perform on Fashion-MNIST compared to MNIST? How do the training times compare?

16.2 *(MNIST Handwritten Digits Hyperparameter Tuning: Changing the Kernel Size)* In the MNIST convnet we presented, change the kernel size from 3-by-3 to 5-by-5. Re-execute the model. How does this change the prediction accuracy?

16.3 *(MNIST Handwritten Digits Hyperparameter Tuning: Changing the Batch Size)* In the MNIST convnet we presented, we used a training batch size of 64. Larger batch sizes can decrease model accuracy. Re-execute the model for batch sizes of 32 and 128. How do these values change the prediction accuracy?

16.4 *(Convnet Layers)* Remove the first `Dense` layer in this chapter's convnet model. How does this change the prediction accuracy? Several Keras pretrained convnets contain `Dense` layers with 4096 neurons. Add such a layer before the two `Dense` layers in this chapter's convnet model. How does this change the prediction accuracy?

16.5 *(Does the Size of the Training Data Set Matter?)* Rerun the MNIST convnet model with only 25% of the original training dataset, then 50%, then 75%. Use scikit-learn's `train_test_split` function to randomly select the training dataset items. Compare the results to when you trained the model with the complete training dataset.

16.6 *(Overfitting)* If you train and test on the *same* data, then the model might overfit the data. Train the MNIST model using all 70,000 training and testing samples, then evaluate the model on the test data and observe the results.

Recurrent Neural Networks

16.7 *(TensorBoard: Visualizing Deep Learning)* Use TensorBoard (Section 16.7) to visualize the training and validation accuracy and loss values for this chapter's recurrent neural network. Modify the example to try fewer and more epochs. How does this affect the prediction accuracy?

16.8 *(IMDb Sentiment Analysis: Removing Stop Words)* In Section 16.9's recurrent neural network example, use the techniques you learned in the "Natural Language Processing" chapter to remove the stop words from the reviews in the training and testing sets. Does this affect our RNN model's prediction accuracy?

16.9 *(IMDb Sentiment Analysis: Loading Pre-Trained Word Embeddings)* Modify the IMDb RNN in Section 16.9 to use pre-trained Word2Vec embeddings, rather than an Embedding layer. For details on how to do this, see https://blog.keras.io/using-pre-trained-word-embeddings-in-a-keras-model.html. Re-execute the RNN. Does loading pre-trained word embeddings improve the model's performance?

ConvnetJS Visualization

16.10 *(Run the Demo: Using the ConvnetJS Tool to Visualize an MNIST Convnet)* Section 16.8 introduced Andrej Karpathy's ConvnetJS browser-based deep-learning tool for training convolutional neural networks and observing their results. Visit the ConvnetJS website at https://cs.stanford.edu/people/karpathy/convnetjs/ and research its capabilities. Run the demo "Classify MNIST digits with a Convolutional Neural Network" and study the dashboard outputs described in Section 16.8 to get a better sense of what a convnet sees as it learns. Observe how changing the hyperparameters affects the model's statistics, by modifying the layer parameters in the **Instantiate a Network and Trainer** section, then click the **change network** button to begin training with the updated model.

16.11 *(Run the Demo: Using the ConvnetJS Tool to Visualize a CIFAR10 Convnet)* Keras's bundled CIFAR10 dataset contains 32-by-32 color images. Repeat the preceding exercise for the demo "Classify CIFAR-10 with Convolutional Neural Network" which classifies the color images into the categories airplanes, automobiles, birds, cats, deer, dogs, frogs, horses, ships and trucks.

Convolutional Neural Network Projects and Research

16.12 *(Project: Best MNIST Convnet Architectures)* Research the best MNIST convnet architectures and implement them using Keras. How do the results compare with this chapter's MNIST convnet?

16.13 *(Project: CIFAR10 Convnet)* Keras's bundled CIFAR10 dataset contains 32-by-32 color images labeled in 10 categories with 50,000 images for training and 10,000 for testing. Using the convnet techniques you learned in the MNIST case study, build, train and evaluate a convnet for CIFAR10. How accurate are the predictions compared to those you experienced with MNIST?

16.14 *(Research: Doppelganger—Find Someone Who Looks Just Like You)* It's often said that everyone has a doppelganger—that is, a look-alike. Research how deep learning convnets might be used to analyze images to find people who look alike. Find a Keras convnet that finds doppelgangers. Locate image datasets of people for training the model. Using a celebrity's photo see what image(s) the model predicts as that celebrity's doppelganger(s).

16.15 *(Project: Using Scikit-Learn to Evaluate the MNIST Model's Performance)* Use Scikit-learn's *classification report* and *confusion matrix* to check this chapter's MNIST model accuracy. Use Seaborn to visualize the confusion matrix.

16.16 *(Project: MNIST Handwritten Digits Model Tuning)* Try adding a third pair of `Conv2D` and `Pooling` layers to this chapter's convnet just before the `Flatten` layer. Use 256 neurons in the new Conv2D layer. How does this affect the model's performance?

16.17 *(Project: Convnets and Dropout)* Dropout layers have been shown to reduce overfitting and improve prediction performance. Generally they do this by randomly deactivating a percentage of the neurons in a given layer each time the weights are about to be updated. Dropout following convolutional layers is commonly set to 20–50%.[107,108] However, the optimal settings vary for each model and dataset.[109] Also, dropout can be applied to other layers. Research the optimal settings for dropout layers and where they're typically placed in a Keras model, then use at least one `Dropout` layer in this chapter's convnet. Does it improve the model's performance?

16.18 *(Project: Replacing Pooling Layers with Additional Convnet Layers)* Though pooling is a common technique to reduce overfitting, some research suggests that additional convolutional layers which use larger strides for their kernels can reduce dimensionality and overfitting *without* discarding features. Read the research paper "Striving for Simplicity: The All Convolutional Net" at https://arxiv.org/abs/1412.6806, then reimplement this chapter's convnet using only `Conv2D` and `Dense` layers. How does this affect the model's performance?

16.19 *(Project: EMNIST Handwritten Digits and Characters)* The EMNIST dataset (https://www.nist.gov/itl/iad/image-group/emnist-dataset) is a more recent version of MNIST. EMNIST has 814,255 digit and character images in 62 *unbalanced classes*, meaning the dataset's samples are not evenly split across the A–Z, a–z and 0–9 classes. The data is provided in a format used by software called Matlab. You can load it into Python via SciPy's `loadmat` function (module `scipy.io`). The downloaded dataset contains several files—one for the entire dataset and several for various subsets.

Research EMNIST and search for and study existing Python EMNIST deep-learning models. Load the EMNIST data and prepare it for use with Keras. Use scikit-learn's `train_test_split` function to split the data into training, validation and testing sets. Use 70% of the data for training, 10% for validation and 20% for testing. Reimplement this chapter's MNIST convnet for use with EMNIST and its 62 classes. What prediction accuracy do you get?

---

107. https://machinelearningmastery.com/dropout-regularization-deep-learning-models-keras/.
108. http://jmlr.org/papers/volume15/srivastava14a.old/srivastava14a.pdf.
109. http://micsymposium.org/mics2018/proceedings/MICS_2018_paper_27.pdf.

16.20 *(Project: Predicting EMNIST Digits with a Pretrained MNIST Model)* For this exercise, load the digits subset of EMNIST, which contains 280,000 digit images. Load the MNIST convnet model you trained in this chapter then use it to evaluate the prediction accuracy for the EMNIST digits. How accurate is your model with EMNIST?

16.21 *(Project: Predicting MNIST Digits with a Pretrained EMNIST Model)* For this exercise, load Keras's MNIST dataset and the EMNIST convnet model for both characters and digits you trained in Exercise 16.19. How accurate is your EMNIST model at predicting MNIST's digits?

16.22 *(Project: Transfer Learning with MNIST and EMNIST Digits)* Use scikit-learn's `train_test_split` function to split the *digits subset* of EMNIST into training (70%), validation (10%) and testing (20%) sets. Load the MNIST convnet model you trained in this chapter, then use its `fit` method to continue training the model with the EMNIST training set you created. Pass the validation set to `fit` via the `validation_data` argument. Evaluate the updated model with the testing data. How accurate is your model compared to the previous exercise?

16.23 *(Project: Binary Classification—Cats vs. Dogs)* Research and download Kaggle's Cats vs. Dogs dataset (`https://www.kaggle.com/c/dogs-vs-cats`) and study deep learning models that use it. Implement your own deep-learning convnet that performs binary-classification using the techniques presented in this chapter. How well does your convnet predict whether an image is a cat or a dog compared to the other Cats vs. Dogs convnets you studied?

16.24 *(Project: Predicting Image Classes with Pretrained Keras Convnet Models)* As we mentioned in Section 16.11, Keras comes with several pretrained convnet models. Investigate these online. Load one or more of the models as shown at `https://keras.io/` and use them to predict the classes of objects in your own images.

16.25 *(Research: Image Captioning with Keras)* Research how automated image captioning is accomplished. Investigate how to use Keras's pretrained models to create image captions. Locate, study and execute existing Keras image-captioning models. Try them with your own images.

16.26 *(Research: Video Closed Captioning with Keras)* Research how automated video closed captioning is accomplished. Investigate using Keras to implement a video closed-captioning system. Locate, study and execute existing Keras closed-captioning models. Try them with your own videos.

16.27 *(Research: OpenCV Object Detection, Face Detection and Facial Recognition)* Research how OpenCV is being used to implement computer-vision systems for object detection, face detection, facial recognition and more. Try several Python-based OpenCV examples. Try these models on your own image data.

16.28 *(Research: Lip Reading with a Convolutional Neural Network)* Research how deep learning and computer vision are being used to implement lip-reading systems. Locate, study and execute existing Keras lip-reading implementations. Try these on videos of your choice.

16.29 *(Research: Sign Language Recognition with Convnets)* Research how deep learning is being used to implement sign-language recognition systems. Locate, study and execute existing Keras sign-language recognition implementations.

Recurrent Neural Network Projects and Research

16.30 *(Project: Improving the IMDb RNN)* Our RNN example was 85.99% accurate. Try to improve our model's performance by increasing the number of words per review to 500 and increasing the number of neurons in the LSTM layer to 256. How do these changes affect the accuracy of the model? Research RNN models online for additional ways to potentially improve performance.

16.31 *(Project: Spam Detector with LSTM)* According to statista.com, over 50% of all emails are SPAM.[110] Research SPAM email detection with deep learning and Keras. Use the Spambase Dataset (https://archive.ics.uci.edu/ml/datasets/spambase), Keras and the recurrent-neural-network techniques you learned in this chapter to implement a deep-learning binary-classification model that predicts whether or not emails are SPAM. Investigate other SPAM email datasets and try them with your model.

16.32 *(Research: Recommendation Engines and Collaborative Filtering)* Companies like Amazon, Netflix and Spotify use recommendation engines and collaborative filtering to help consumers make decisions, such as which products to purchase, music to listen to or movies to watch. Research recommendation engines, collaborative filtering and how these techniques can be implemented using Keras. One popular dataset you'll encounter is the MovieLens 100K dataset, which has 100,000 ratings of 1700 movies from 1000 users. Locate Keras-based movie-recommendation models, study their code and try them.

16.33 *(Research: Anomaly Detection)* Credit-card companies, insurance companies, cyber security companies and others use machine-learning and deep-learning techniques to detect fraud and security breaches by looking for anomalies in data. Research anomaly-detection techniques and how they can be implemented using Keras. Look for sample anomaly detection datasets. Locate, study and try existing Keras anomaly detection model.

16.34 *(Research: Time Series Forecasting with Keras and LSTM)* Research time-series forecasting with Keras. Locate, study and run existing time-series forecasting examples.

16.35 *(Research: Text Summarization with RNNs)* Document summarization involves analyzing a document and extracting content to produce a summary. For example, with today's massive flow of information, this could be useful to busy doctors studying the latest medical advances in order to provide the best care. A summary could help them decide whether a paper is worth reading. Research how text summarization can be imlemented in Keras. Locate, study and try existing Keras text-summarization implementations.

16.36 *(Research: ChatBots and RNNs)* Research chatbots and recurrent neural networks. Locate, study and run chatbot examples implemented with Keras RNNs.

Automated Deep Learning Project

16.37 *(Project: Auto-Keras Automated Deep Learning)* Several of the preceding exercises ask you to tune your neural network architectures and hyperparameters. Research the deep-learning library Auto-Keras (https://autokeras.com/), which automates finding the appropriate deep-learning network configurations and hyperparameters. Then, use Auto-Keras to reimplement this chapter's MNIST and IMDb examples. Compare the accuracy of the Auto-Keras models to those we presented in the chapter.

---

110. https://www.statista.com/statistics/420391/spam-email-traffic-share/.

Reinforcement Learning Projects and Research

16.38 *(Research: Google's AlphaZero)* Research how Google's AlphaZero uses reinforcement learning to learn how to play games.

16.39 *(Project: Reinforcement Learning, Deep Q-Learning, OpenAI Gym and Game Playing with the CartPole Environment)* In Section 16.12, we briefly introduced reinforcement learning, Deep-Q Learning and OpenAI Gym. Research OpenAI Gym (https://gym.openai.com/), install it and execute its CartPole environment without any reinforcement learning implemented. Next, research solutions to the Cartpole problem using Deep-Q Learning and Keras, then study and run their code. Develop your own CartPole solution.

16.40 *(Research: Other OpenAI Gym Game-Playing Environments)* OpenAI Gym has many Atari video-game environments (https://gym.openai.com/envs/#atari). Research and execute several of these. Locate Keras Deep Q-Learning implementations that play these games, then study and run their code.

16.41 *(Research: Pong from Pixels)* Read Andrej Karpathy's blog post "Pong from Pixels" (http://karpathy.github.io/2016/05/31/rl/). Download and try his OpenAI Gym Pong reinforcement-learning implementation at:

> https://gist.github.com/karpathy/a4166c7fe253700972fcbc77e4ea32c5

16.42 *(Research: Solving Mazes with Reinforcement Learning, Open AI Gym and Keras)* Research how to solve mazes with reinforcement learning, OpenAI Gym and Keras. Download an OpenAI gym maze environment and try the maze reinforcement-learning solutions you find.

16.43 *(Research and Watch: Google DeepMind Agent Learning to Walk)* Read the Google DeepMind team's blog post https://deepmind.com/blog/producing-flexible-behaviours-simulated-environments/ in which they discuss teaching AI agents to walk. Watch the video of the process at: https://www.youtube.com/watch?v=hx_bgoTF7bs. For more details, see their research paper "Emergence of Locomotion Behaviours in Rich Environments" at https://arxiv.org/abs/1707.02286.

16.44 *(Research: Reinforcement Learning with Deep-Q Learning)* Read the Google DeepMind team's paper "Human-level control through deep reinforcement learning" at https://storage.googleapis.com/deepmind-media/dqn/DQNNaturePaper.pdf.

16.45 *(Research: Self-Driving Cars)* Research how deep-reinforcement learning is being used to help self-driving cars learn to drive.

16.46 *(Research: OpenAI Gym Retro)* An emulator is software that enables your computer to emulate how a different computer system works. There are many video game emulators, for example, that allow you to execute old video games on current computers. Research OpenAI's Gym Retro (https://github.com/openai/retro), which enables video game emulators to be used as OpenAI Gym environments. Gym Retro currently supports various Atari, NEC, Nintendo and Sega emulators. Try some of the environments. Look for and try Python reinforcement-learning solutions that use Gym Retro environments.

16.47 *(Research: Reinforcement Learning in Business and Industry)* There are many use-cases for reinforcement learning in business and industry including optimizing debt collection[111], self-training robots, optimizing warehouse space management, dynamic product pricing, stock trading, delivery route optimization, personalized shopping experiences, computing resource management, traffic light systems and more. Research how reinforcement learning is being used for each of these use-cases and investigate additional use-cases.

16.48 *(Research: Reinforcement Learning in Computational Neuroscience)* Research how reinforcement learning is being used in computational neuroscience.[112]

16.49 *(Research: 3D Tic-Tac-Toe)* Research and try Keras Deep Q-Learning implementations of three-dimensional tic-tac-toe. Can you beat the algorithm?

Generative Deep Learning

16.50 *(Watch: Sunspring Movie Generated By an AI Bot)* Research the movie Sunspring (https://en.wikipedia.org/wiki/Sunspring), which was written by an AI bot using a neural network. Watch the movie at http://www.thereforefilms.com/sunspring.html.

16.51 *(Demo: DeepDream—Psychedelic Art)* Research Google's DeepDream, which generates psychedelic images using information from their Inception convnet (which is one of the pretrained models bundled with Keras). Check out their online demo and gallery at https://deepdreamgenerator.com/. If you're interested in the source code, see https://github.com/google/deepdream. Try your hand at developing art with this approach.

16.52 *(Research: Creative Deep Learning—Generative Adversarial Neworks)* Generative Adversarial Networks (GANs)[113,114] are deep learning networks that can create realistic but fake images and video by using two competing deep-learning networks. Among their uses are creating elements and characters in video games, generating images of clothing models in different poses from an original image, applying different art styles to existing images, generating new art with the same style as existing art (neural style transfer), generating high-resolution images from low-resolution images and much more. Research applications of generative adversarial networks and try any demos you find. Investigate how such networks can be implemented in Keras.

16.53 *(Research: Creative Deep Learning—Converting Your Writing to Shakespearian Style)* Research how recurrent neural networks and LSTM can be used to generate text in different writing styles. Look for demos of converting your writing to Shakespearian style. Try any demos that you find.

Deep Fakes

16.54 *(Research: Detecting Deep Fakes)* Artificial intelligence technologies are making it possible to create images that look like original photos of people who do not even exist and *deep fakes*—realistic fake videos of people that capture their look, their voice, body motions and facial expressions. Research the deep learning techniques that are being used to detect deep fakes.

---

111. https://www.researchgate.net/publication/
220272023_Optimizing_debt_collections_using_constrained_reinforcement_learning.
112. http://www.princeton.edu/~yael/ICMLTutorial.pdf.
113. https://en.wikipedia.org/wiki/Generative_adversarial_network.
114. https://skymind.ai/wiki/generative-adversarial-network-gan.

16.55 *(Research: Ethics of Deep Fakes)* Research the many ethical issues surrounding deep fakes.

Additional Research

16.56 *(Research: Evolutionary Learning)* Research the recent developments in evolutionary learning—also called neuroevolution, evolutionary algorithms and evolutionary computation. Some people think these techniques might someday replace deep learning.

16.57 *(Research: Deep Learning in Poker)* Research how deep learning is being used to implement poker-playing agents that can beat the world's best poker players.

Big Data: Hadoop, Spark, NoSQL and IoT

17

Objectives
In this chapter you'll:
- Understand what big data is and how quickly it's getting bigger.
- Manipulate a SQLite relational database using Structured Query Language (SQL).
- Understand the four major types of NoSQL databases.
- Store tweets in a MongoDB NoSQL JSON document database and visualize them on a Folium map.
- Understand Apache Hadoop and how it's used in big-data batch-processing applications.
- Build a Hadoop MapReduce application on Microsoft's Azure HDInsight cloud service.
- Understand Apache Spark and how it's used in high-performance, real-time big-data applications.
- Use Spark streaming to process data in mini-batches.
- Understand the Internet of Things (IoT) and the publish/subscribe model.
- Publish messages from a simulated Internet-connected device and visualize its messages in a dashboard.
- Subscribe to PubNub's live Twitter and IoT streams and visualize the data.

Outline

17.1 Introduction
17.2 Relational Databases and Structured Query Language (SQL)
   17.3.1 A **books** Database
   17.3.2 **SELECT** Queries
   17.3.3 **WHERE** Clause
   17.3.4 **ORDER BY** Clause
   17.3.5 Merging Data from Multiple Tables: **INNER JOIN**
   17.3.6 **INSERT INTO** Statement
   17.3.7 **UPDATE** Statement
   17.3.8 **DELETE FROM** Statement
17.3 NoSQL and NewSQL Big-Data Databases: A Brief Tour
   17.3.1 NoSQL Key–Value Databases
   17.3.2 NoSQL Document Databases
   17.3.3 NoSQL Columnar Databases
   17.3.4 NoSQL Graph Databases
   17.3.5 NewSQL Databases
17.4 Case Study: A MongoDB JSON Document Database
   17.4.1 Creating the MongoDB Atlas Cluster
   17.4.2 Streaming Tweets into MongoDB
17.5 Hadoop
   17.5.1 Hadoop Overview
   17.6.2 Summarizing Word Lengths in *Romeo and Juliet* via MapReduce
   17.5.3 Creating an Apache Hadoop Cluster in Microsoft Azure HDInsight
   17.5.4 Hadoop Streaming
   17.5.5 Implementing the Mapper
   17.5.6 Implementing the Reducer
   17.5.7 Preparing to Run the MapReduce Example
   17.5.8 Running the MapReduce Job
17.6 Spark
   17.6.1 Spark Overview
   17.6.2 Docker and the Jupyter Docker Stacks
   17.6.3 Word Count with Spark
   17.6.4 Spark Word Count on Microsoft Azure
17.7 Spark Streaming: Counting Twitter Hashtags Using the **pyspark-notebook** Docker Stack
   17.7.1 Streaming Tweets to a Socket
   17.7.2 Summarizing Tweet Hashtags; Introducing Spark SQL
17.8 Internet of Things and Dashboards
   17.8.1 Publish and Subscribe
   17.8.2 Visualizing a PubNub Sample Live Stream with a Freeboard Dashboard
   17.8.3 Simulating an Internet-Connected Thermostat in Python
   17.8.4 Creating the Dashboard with Freeboard.io
   17.8.5 Creating a Python PubNub Subscriber
17.9 Wrap-Up
Exercises

17.1 Introduction

In Section 1.13, we introduced big data. In this capstone chapter, we discuss popular hardware and software infrastructure for working with big data, and we develop complete applications on several desktop and cloud-based big-data platforms.

Databases

Databases are critical big-data infrastructure for storing and manipulating the massive amounts of data we're creating. They're also critical for securely and confidentially maintaining that data, especially in the context of ever-stricter privacy laws such as **HIPAA** (**H**ealth **I**nsurance **P**ortability and **A**ccountability **A**ct) in the United States and **GDPR** (**G**eneral **D**ata **P**rotection **R**egulation) for the European Union.

First, we'll present **relational databases**, which store **structured data** in tables with a fixed-size number of columns per row. You'll manipulate relational databases via **Structured Query Language (SQL)**.

Most data produced today is **unstructured data**, like the content of Facebook posts and Twitter tweets, or **semi-structured data** like JSON and XML documents. Twitter processes each tweet's contents into a semi-structured JSON document with lots of *metadata*, as you saw in the "Data Mining Twitter" chapter. Relational databases are not geared to the unstructured and semi-structured data in big-data applications. So, as big data evolved, new kinds of databases were created to handle such data efficiently. We'll discuss the four major types of these **NoSQL databases**—key–value, document, columnar and graph databases. Also, we'll overview **NewSQL databases**, which blend the benefits of relational and NoSQL databases. Many NoSQL and NewSQL vendors make it easy to get started with their products through free tiers and free trials, and typically in cloud-based environments that require minimal installation and setup. This makes it practical for you to gain big-data experience before "diving in."

Apache Hadoop
Much of today's data is so large that it cannot fit on one system. As big data grew, we needed distributed data storage and parallel processing capabilities to process the data more efficiently. This led to complex technologies like Apache Hadoop for distributed data processing with massive parallelism among clusters of computers where the intricate details are handled for you automatically and correctly. We'll discuss Hadoop, its architecture and how it's used in big-data applications. We'll guide you through configuring a multi-node Hadoop cluster using the Microsoft Azure HDInsight cloud service, then use it to execute a Hadoop MapReduce job that you'll implement in Python. Though HDInsight is not free, Microsoft gives you a generous new-account credit that should enable you to run the chapter's code examples without incurring additional charges.

Apache Spark
As big-data processing needs grow, the information-technology community is continually looking for ways to increase performance. Hadoop executes tasks by breaking them into pieces that do lots of disk I/O across many computers. Spark was developed as a way to perform certain big-data tasks *in memory* for better performance.

We'll discuss Apache Spark, its architecture and how it's used in high-performance, real-time big-data applications. You'll implement a Spark application using functional-style filter/map/reduce programming capabilities. First, you'll build this example using a Jupyter Docker stack that runs locally on your desktop computer, then you'll implement it using a cloud-based Microsoft Azure HDInsight multi-node Spark cluster. In the exercises, you'll also do this example with the free Databricks Community Edition.

We'll introduce Spark streaming for processing streaming data in mini-batches. Spark streaming gathers data for a short time interval you specify, then gives you that batch of data to process. You'll implement a Spark streaming application that processes tweets. In that example, you'll use Spark SQL to query data stored in a Spark `DataFrame` which, unlike pandas `DataFrames`, may contain data distributed over many computers in a cluster.

Internet of Things
We'll conclude with an introduction to the Internet of Things (IoT)—billions of devices that are continuously producing data worldwide. We'll introduce the *publish/subscribe model* that IoT and other types of applications use to connect data users with data providers. First, without writing any code, you'll build a web-based dashboard using Freeboard.io and a sample live stream from the PubNub messaging service. Next, you'll simulate an Internet-connected thermostat which publishes messages to the free Dweet.io messaging service using the Python module Dweepy, then create a dashboard visualization of the data with Freeboard.io. Finally, you'll build a Python client that *subscribes* to a sam-

ple live stream from the PubNub service and dynamically visualizes the stream with Seaborn and a Matplotlib `FuncAnimation`.

End-of-Chapter Exercises
The rich exercise set encourages you to work with more big-data cloud and desktop platforms, additional SQL and NoSQL databases, NewSQL databases and IoT platforms. One exercise asks you to work with Wikipedia as another popular big-data source. Another asks you to implement an IoT application with the popular Raspberry Pi device simulator.

Experience Cloud and Desktop Big-Data Software
Cloud vendors focus on **service-oriented architecture** (SOA) technology in which they provide "as-a-Service" capabilities that applications connect to and use in the cloud. Common services provided by cloud vendors include:[1]

"As-a-Service" acronyms (note that several are the same)	
Big data as a Service (BDaaS)	Platform as a Service (PaaS)
Hadoop as a Service (HaaS)	Software as a Service (SaaS)
Hardware as a Service (HaaS)	Storage as a Service (SaaS)
Infrastructure as a Service (IaaS)	Spark as a Service (SaaS)

You'll get hands-on experience in this chapter with several cloud-based tools. In this chapter's examples, you'll use the following platforms:

- A *free* MongoDB Atlas cloud-based cluster.
- A multi-node Hadoop cluster running on Microsoft's Azure HDInsight cloud-based service—for this you'll use the *credit* that comes with a new Azure account.
- A *free* single-node Spark "cluster" running on your desktop computer, using a Jupyter Docker-stack container.
- A multi-node Spark cluster, also running on Microsoft's Azure HDInsight—for this you'll continue using your Azure new-account *credit*.

In the project exercises, you can explore various other options, including cloud-based services from Amazon Web Services, Google Cloud and IBM Watson, and the free *desktop* versions of the Hortonworks and Cloudera platforms (there also are cloud-based paid versions of these). You'll also explore and use a single-node Spark cluster running on the *free* cloud-based Databricks Community Edition. Spark's creators founded Databricks.

Always check the latest terms and conditions of each service you use. Some require you to enable credit-card billing to use their clusters. *Caution:* **Once you allocate Microsoft Azure HDInsight clusters (or other vendors' clusters), they incur costs. When you complete the case studies using services such as Microsoft Azure, be sure to delete your cluster(s) and their other resources (like storage). This will help extend the life of your Azure new-account credit.**

Installation and setups vary across platforms and over time. Always follow each vendor's latest steps. If you have questions, the best sources for help are the vendor's support

---

1. For more "as-a-Service" acronyms, see https://en.wikipedia.org/wiki/Cloud_computing and https://en.wikipedia.org/wiki/As_a_service.

capabilities and forums. Also, check sites such as `stackoverflow.com`—other people may have asked questions about similar problems and received answers from the developer community.

Algorithms and Data
Algorithms and data are the core of Python programming. The first few chapters of this book were mostly about algorithms. We introduced control statements and discussed algorithm development. Data was small—primarily individual integers, floats and strings. Chapters 5–9 emphasized *structuring* data into lists, tuples, dictionaries, sets, arrays and files. In Chapter 11, we refocused on algorithms, using Big-O notation to help us quantify how hard algorithms work to do their jobs.

Data's Meaning
But, what about the *meaning* of the data? Can we use the data to gain insights to better diagnose cancers? Save lives? Improve patients' quality of life? Reduce pollution? Conserve water? Increase crop yields? Reduce damage from devastating storms and fires? Develop better treatment regimens? Create jobs? Improve company profitability?

The data-science case studies of Chapters 12–16 all focused on AI. In this chapter, we focus on the big-data infrastructure that supports AI solutions. As the data used with these technologies continues growing exponentially, we want to learn from that data and do so at blazing speed. We'll accomplish these goals with a combination of sophisticated algorithms, hardware, software and networking designs. We've presented various machine-learning technologies, seeing that there are indeed great insights to be mined from data. With more data, and especially with big data, machine learning can be even more effective.

Big-Data Sources
The following articles and sites provide links to hundreds of free big data sources:

Big-data sources
"Awesome-Public-Datasets," GitHub.com, `https://github.com/caesar0301/awesome-public-datasets`.
"AWS Public Datasets," `https://aws.amazon.com/public-datasets/`.
"Big Data And AI: 30 Amazing (And Free) Public Data Sources For 2018," by B. Marr, `https://www.forbes.com/sites/bernardmarr/2018/02/26/big-data-and-ai-30-amazing-and-free-public-data-sources-for-2018/`.
"Datasets for Data Mining and Data Science," `http://www.kdnuggets.com/datasets/index.html`.
"Exploring Open Data Sets," `https://datascience.berkeley.edu/open-data-sets/`.
"Free Big Data Sources," Datamics, `http://datamics.com/free-big-data-sources/`.
Hadoop Illuminated, Chapter 16. Publicly Available Big Data Sets, `http://hadoopilluminated.com/hadoop_illuminated/Public_Bigdata_Sets.html`.
"List of Public Data Sources Fit for Machine Learning," `https://blog.bigml.com/list-of-public-data-sources-fit-for-machine-learning/`.
"Open Data," Wikipedia, `https://en.wikipedia.org/wiki/Open_data`.
"Open Data 500 Companies," `http://www.opendata500.com/us/list/`.

> **Big-data sources**
>
> "Other Interesting Resources/Big Data and Analytics Educational Resources and Research," B. Marr, http://computing.derby.ac.uk/bigdatares/?page_id=223.
>
> "6 Amazing Sources of Practice Data Sets,"
> https://www.jigsawacademy.com/6-amazing-sources-of-practice-data-sets/.
>
> "20 Big Data Repositories You Should Check Out," M. Krivanek,
> http://www.datasciencecentral.com/profiles/blogs/20-free-big-data-sources-everyone-should-check-out.
>
> "70+ Websites to Get Large Data Repositories for Free,"
> http://bigdata-madesimple.com/70-websites-to-get-large-data-repositories-for-free/.
>
> "Ten Sources of Free Big Data on Internet," A. Brown,
> https://www.linkedin.com/pulse/ten-sources-free-big-data-internet-alan-brown.
>
> "Top 20 Open Data Sources,"
> https://www.linkedin.com/pulse/top-20-open-data-sources-zygimantas-jacikevicius.
>
> "We're Setting Data, Code and APIs Free," NASA, https://open.nasa.gov/open-data/.
>
> "Where Can I Find Large Datasets Open to the Public?" Quora,
> https://www.quora.com/Where-can-I-find-large-datasets-open-to-the-public.

✓ Self Check for Section 17.1

1 *(Fill-In)* _____ databases store structured data in tables with a fixed-size number of columns per row and are manipulated via Structured Query Language (SQL).
Answer: Relational.

2 *(Fill-In)* Most data produced today is _____ data, like the content of Facebook posts and Twitter tweets, or _____ data like JSON and XML documents.
Answer: unstructured, semi-structured.

3 *(Fill-In)* Cloud vendors focus on _____ technology in which they provide "as-a-Service" capabilities that applications connect to and use in the cloud.
Answer: service-oriented architecture (SOA).

17.2 Relational Databases and Structured Query Language (SQL)

Databases are crucial, especially for big data. In Chapter 9, "Files and Exceptions," we demonstrated sequential text-file processing, working with data from CSV files and working with JSON. Both are useful when *most or all* of a file's data is to be processed. On the other hand, in transaction processing it is crucial to locate and, possibly, update an *individual* data item quickly.

A **database** is an integrated collection of data. A **database management system** (**DBMS**) provides mechanisms for storing and organizing data in a manner consistent with the database's format. Database management systems allow for convenient access and storage of data without concern for the internal representation of databases.

Relational database management systems (**RDBMSs**) store data in **tables** and define relationships among the tables. Structured Query Language (SQL) is used almost universally with relational database systems to manipulate data and perform **queries**, which request information that satisfies given criteria.[2]

Popular *open-source* RDBMSs include SQLite, PostgreSQL, MariaDB and MySQL. These can be downloaded and used *freely* by anyone. All have support for Python. We'll use SQLite, which is bundled with Python. Some popular proprietary RDBMSs include Microsoft SQL Server, Oracle, Sybase and IBM Db2.

Tables, Rows and Columns

A relational database is a logical table-based representation of data that allows the data to be accessed without consideration of its physical structure. The following diagram shows a sample Employee table that might be used in a personnel system:

```
         Number    Name      Department   Salary   Location
         23603    Jones        413        1100     New Jersey
         24568    Kerwin       413        2000     New Jersey
   Row { 34589    Larson       642        1800     Los Angeles
         35761    Myers        611        1400     Orlando
         47132    Neumann      413        9000     New Jersey
         78321    Stephens     611        8500     Orlando

         Primary key           Column
```

The table's primary purpose is to store employees' attributes. Tables are composed of **rows**, each describing a single entity. Here, each row represents one employee. Rows are composed of **columns** containing individual attribute values. The table above has six rows. The Number column represents the **primary key**—a column (or group of columns) with a value that's *unique* for each row. This guarantees that each row can be identified by its primary key. Examples of primary keys are social security numbers, employee ID numbers and part numbers in an inventory system—values in each of these are guaranteed to be unique. In this case, the rows are listed in ascending order by primary key, but they could be listed in descending order or no particular order at all.

Each column represents a different data attribute. Rows are unique (by primary key) within a table, but particular column values may be duplicated between rows. For example, three different rows in the Employee table's Department column contain number 413.

Selecting Data Subsets

Different database users are often interested in different data and different relationships among the data. Most users require only subsets of the rows and columns. Queries specify which subsets of the data to select from a table. You use Structured Query Language (SQL) to define queries. For example, you might select data from the Employee table to create a result that shows where each department is located, presenting the data sorted in increasing order by department number. This result is shown below. We'll discuss SQL shortly.

```
         Department    Location
            413        New Jersey
            611        Orlando
            642        Los Angeles
```

---

2. The writing in this chapter assumes that SQL is pronounced as "see-quel." Some prefer "ess que el."

SQLite

The code examples in the rest of Section 17.2 use the open-source SQLite database management system that's included with Python, but most popular database systems have Python support. Each typically provides a module that adheres to Python's **Database Application Programming Interface (DB-API)**, which specifies common object and method names for manipulating any database.

✓ Self Check

1. *(Fill-In)* A table in a relational database consists of _____ and _____.
   Answer: rows, columns.

2. *(Fill-In)* The _____ key uniquely identifies each record in a table.
   Answer: primary.

3. *(True/False)* Python's Database Application Programming Interface (DB-API) specifies common object and method names for manipulating any database.
   Answer: True.

17.2.1 A books Database

In this section, we'll present a books database containing information about several of our books. We'll set up the database in SQLite via the Python Standard Library's **sqlite3 module**, using a script provided in the ch17 example's folder's sql subfolder. Then, we'll introduce the database's tables. We'll use this database in an IPython session to introduce various database concepts, including operations that **create, read, update** and **delete** data—the so-called **CRUD** operations. As we introduce the tables, we'll use SQL and pandas DataFrames to show you each table's contents. Then, in the next several sections, we'll discuss additional SQL features.

Creating the books Database

In your Anaconda Command Prompt, Terminal or shell, change to the ch17 examples folder's sql subfolder. The following **sqlite3 command** creates a SQLite database named books.db and executes the books.sql SQL script, which defines how to create the database's tables and populates them with data:

```
sqlite3 books.db < books.sql
```

The notation < indicates that books.sql is input into the sqlite3 command. When the command completes, the database is ready for use. Begin a new IPython session.

Connecting to the Database in Python

To work with the database in Python, first call sqlite3's **connect function** to connect to the database and obtain a **Connection** object:

```
In [1]: import sqlite3

In [2]: connection = sqlite3.connect('books.db')
```

authors Table

The database has three tables—authors, author_ISBN and titles. The authors table stores all the authors and has three columns:

- **id**—The author's unique ID number. This integer column is defined as **autoincremented**—for each row inserted in the table, SQLite increases the id value by 1 to ensure that each row has a unique value. This column is the table's primary key.
- **first**—The author's first name (a string).
- **last**—The author's last name (a string).

Viewing the authors Table's Contents
Let's use a SQL query and pandas to view the authors table's contents:

```
In [3]: import pandas as pd

In [4]: pd.options.display.max_columns = 10

In [5]: pd.read_sql('SELECT * FROM authors', connection,
   ...:             index_col=['id'])
   ...:
Out[5]:
        first    last
id
1        Paul  Deitel
2      Harvey  Deitel
3       Abbey  Deitel
4         Dan   Quirk
5   Alexander    Wald
```

Pandas function **read_sql** executes a SQL query and returns a DataFrame containing the query's results. The function's arguments are:

- a string representing the SQL query to execute,
- the SQLite database's Connection object, and in this case
- an index_col keyword argument indicating which column should be used as the DataFrame's row indices (the author's id values in this case).

As you'll see momentarily, when index_col is not passed, index values starting from 0 appear to the left of the DataFrame's rows.

A SQL **SELECT** query gets rows and columns from one or more tables in a database. In the query:

```
SELECT * FROM authors
```

the **asterisk (*)** is a *wildcard* indicating that the query should get *all* the columns from the authors table. We'll discuss SELECT queries in more detail shortly.

titles Table
The titles table stores all the books and has four columns:

- **isbn**—The book's ISBN (a string) is this table's primary key. ISBN is an abbreviation for "International Standard Book Number," which is a numbering scheme that publishers use to give every book a unique identification number.
- **title**—The book's title (a string).
- **edition**—The book's edition number (an integer).
- **copyright**—The book's copyright year (a string).

Let's use SQL and pandas to view the `titles` table's contents:

```
In [6]: pd.read_sql('SELECT * FROM titles', connection)
Out[6]:
        isbn                            title  edition  copyright
0  0135404673      Intro to Python for CS and DS      1       2020
1  0132151006      Internet & WWW How to Program    5       2012
2  0134743350             Java How to Program     11       2018
3  0133976890                C How to Program      8       2016
4  0133406954  Visual Basic 2012 How to Program      6       2014
5  0134601548        Visual C# How to Program      6       2017
6  0136151574       Visual C++ How to Program      2       2008
7  0134448235              C++ How to Program     10       2017
8  0134444302          Android How to Program      3       2017
9  0134289366        Android 6 for Programmers     3       2016
```

author_ISBN Table
The `author_ISBN` table uses the following columns to associate authors from the `authors` table with their books in the `titles` table:

- `id`—An author's id (an integer).
- `isbn`—The book's ISBN (a string).

The `id` column is a **foreign key**, which is a column in this table that matches a *primary-key* column in another table—in particular, the `authors` table's `id` column. The `isbn` column also is a foreign key—it matches the `titles` table's `isbn` primary-key column. A database might have many tables. A goal when designing a database is to *minimize* data *duplication* among the tables. To do this, each table represents a specific entity, and foreign keys help link the data in *multiple* tables. The primary keys and foreign keys are designated when you create the database tables (in our case, in the `books.sql` script).

Together the `id` and `isbn` columns in this table form a *composite primary key*. Every row in this table *uniquely* matches *one* author to *one* book's ISBN. This table contains many entries, so let's use SQL and pandas to view just the first five rows:

```
In [7]: df = pd.read_sql('SELECT * FROM author_ISBN', connection)

In [8]: df.head()
Out[8]:
   id        isbn
0   1  0134289366
1   2  0134289366
2   5  0134289366
3   1  0135404673
4   2  0135404673
```

Every foreign-key value must appear as the primary-key value in a row of another table so the DBMS can ensure that the foreign-key value is valid. This is known as the **Rule of Referential Integrity**. For example, the DBMS ensures that the `id` value for a particular `author_ISBN` row is valid by checking that there is a row in the `authors` table with that `id` as the primary key.

Foreign keys also allow *related* data in *multiple* tables to be *selected* from those tables and combined—this is known as **joining** the data. There is a **one-to-many relationship** between a primary key and a corresponding foreign key—one author can write many books, and similarly one book can be written by many authors. So a foreign key can appear *many* times in its table but only *once* (as the primary key) in another table. For example, in the books database, the ISBN 0134289366 appears in several author_ISBN rows because this book has several authors, but it appears only once as a primary key in titles.

Entity-Relationship (ER) Diagram
The following **entity-relationship (ER) diagram** for the books database shows the database's *tables* and the *relationships* among them:

The first compartment in each box contains the table's name, and the remaining compartments contain the table's columns. The names in italic are primary keys. *A table's primary key uniquely identifies each row in the table.* Every row must have a primary-key value, and that value must be unique in the table. This is known as the **Rule of Entity Integrity**. Again, for the author_ISBN table, the primary key is the combination of both columns—this is known as a composite primary key.

The lines connecting the tables represent the *relationships* among the tables. Consider the line between authors and author_ISBN. On the authors end there's a 1, and on the author_ISBN end there's an infinity symbol (∞). This indicates a *one-to-many relationship*. For *each* author in the authors table, there can be an *arbitrary number* of ISBNs for books written by that author in the author_ISBN table—that is, an author can write *any* number of books, so an author's id can appear in multiple rows of the author_ISBN table. The relationship line links the id column in the authors table (where id is the primary key) to the id column in the author_ISBN table (where id is a foreign key). The line between the tables links the primary key to the matching foreign key.

The line between the titles and author_ISBN tables illustrates a *one-to-many relationship*—one book can be written by many authors. The line links the primary key isbn in table titles to the corresponding foreign key in table author_ISBN. The relationships in the entity-relationship diagram illustrate that the sole purpose of the author_ISBN table is to provide a **many-to-many relationship** between the authors and titles tables—an author can write *many* books, and a book can have *many* authors.

SQL Keywords
The following subsections continue our SQL presentation in the context of our books database, demonstrating SQL queries and statements using the SQL keywords in the following table. Other SQL keywords are beyond this text's scope:

SQL keyword	Description
SELECT	Retrieves data from one or more tables.
FROM	Tables involved in the query. Required in every SELECT.

SQL keyword	Description
WHERE	Criteria for selection that determine the rows to be retrieved, deleted or updated. Optional in a SQL statement.
GROUP BY	Criteria for grouping rows. Optional in a SELECT query.
ORDER BY	Criteria for ordering rows. Optional in a SELECT query.
INNER JOIN	Merge rows from multiple tables.
INSERT	Insert rows into a specified table.
UPDATE	Update rows in a specified table.
DELETE	Delete rows from a specified table.

Self Check

1 *(Fill-In)* A(n) _____ key is a field in a table for which every entry has a unique value in another table and where the field in the other table is the primary key for that table.
Answer: foreign.

2 *(True/False)* Every foreign-key value must appear as another table's primary-key value so the DBMS can ensure that the foreign-key value is valid—this is known as the Rule of Entity Integrity.
Answer: False. This is known as the Rule of Referential Integrity. The Rule of Entity Integrity states that every row must have a primary-key value, and that value must be unique in the table.

17.2.2 SELECT Queries

The previous section used SELECT statements and the * wildcard character to get all the columns from a table. Typically, you need only a subset of the columns, especially in big data where you could have dozens, hundreds, thousands or more columns. To retrieve only specific columns, specify a comma-separated list of column names. For example, let's retrieve only the columns first and last from the authors table:

```
In [9]: pd.read_sql('SELECT first, last FROM authors', connection)
Out[9]:
       first     last
0        Paul   Deitel
1      Harvey   Deitel
2       Abbey   Deitel
3         Dan    Quirk
4   Alexander     Wald
```

17.2.3 WHERE Clause

You'll often select rows in a database that satisfy certain **selection criteria**, especially in big data where a database might contain millions or billions of rows. Only rows that satisfy the selection criteria (formally called **predicates**) are selected. SQL's **WHERE clause** specifies a query's selection criteria. Let's select the title, edition and copyright for all books

17.2 Relational Databases and Structured Query Language (SQL)

with copyright years greater than 2016. String values in SQL queries are delimited by single (') quotes, as in '2016':

```
In [10]: pd.read_sql("""SELECT title, edition, copyright
   ...:                 FROM titles
   ...:                 WHERE copyright > '2016'""", connection)
Out[10]:
                           title  edition  copyright
0    Intro to Python for CS and DS        1       2020
1               Java How to Program      11       2018
2           Visual C# How to Program       6       2017
3                C++ How to Program      10       2017
4            Android How to Program       3       2017
```

Pattern Matching: Zero or More Characters

The WHERE clause may can contain the operators <, >, <=, >=, =, <> (not equal) and LIKE. Operator **LIKE** is used for **pattern matching**—searching for strings that match a given pattern. A pattern that contains the **percent (%)** wildcard character searches for strings that have *zero or more* characters at the percent character's position in the pattern. For example, let's locate all authors whose last name starts with the letter D:

```
In [11]: pd.read_sql("""SELECT id, first, last
   ...:                 FROM authors
   ...:                 WHERE last LIKE 'D%'""",
   ...:              connection, index_col=['id'])
   ...:
Out[11]:
     first    last
id
1     Paul  Deitel
2   Harvey  Deitel
3    Abbey  Deitel
```

Pattern Matching: Any Character

An **underscore (_)** in the pattern string indicates a single wildcard character at that position. Let's select the rows of all the authors whose last names start with any character, followed by the letter b, followed by any number of additional characters (specified by %):

```
In [12]: pd.read_sql("""SELECT id, first, last
   ...:                 FROM authors
   ...:                 WHERE first LIKE '_b%'""",
   ...:              connection, index_col=['id'])
   ...:
Out[12]:
     first    last
id
3    Abbey  Deitel
```

✓ Self Check

1 *(Fill-In)* SQL keyword _____ is followed by the selection criteria that specify the records to select in a query.
Answer: WHERE.

17.2.4 ORDER BY Clause

The **ORDER BY** clause sorts a query's results into ascending order (lowest to highest) or descending order (highest to lowest), specified with `ASC` and `DESC`, respectively. The default sorting order is ascending, so ASC is optional. Let's sort the titles in ascending order:

```
In [13]: pd.read_sql('SELECT title FROM titles ORDER BY title ASC',
    ...:              connection)
Out[13]:
                             title
0           Android 6 for Programmers
1              Android How to Program
2                    C How to Program
3                  C++ How to Program
4         Internet & WWW How to Program
5          Intro to Python for CS and DS
6                 Java How to Program
7     Visual Basic 2012 How to Program
8            Visual C# How to Program
9           Visual C++ How to Program
```

Sorting By Multiple Columns

To sort by multiple columns, specify a comma-separated list of column names after the ORDER BY keywords. Let's sort the authors' names by last name, then by first name for any authors who have the same last name:

```
In [14]: pd.read_sql("""SELECT id, first, last
    ...:                FROM authors
    ...:                ORDER BY last, first""",
    ...:             connection, index_col=['id'])
    ...:
Out[14]:
        first     last
id
3       Abbey    Deitel
2      Harvey    Deitel
1        Paul    Deitel
4         Dan     Quirk
5   Alexander     Wald
```

The sorting order can vary by column. Let's sort the authors in descending order by last name and ascending order by first name for any authors who have the same last name:

```
In [15]: pd.read_sql("""SELECT id, first, last
    ...:                FROM authors
    ...:                ORDER BY last DESC, first ASC""",
    ...:             connection, index_col=['id'])
    ...:
Out[15]:
        first     last
id
5   Alexander     Wald
4         Dan     Quirk
3       Abbey    Deitel
2      Harvey    Deitel
1        Paul    Deitel
```

17.2 Relational Databases and Structured Query Language (SQL)

Combining the WHERE and ORDER BY Clauses

The WHERE and ORDER BY clauses can be combined in one query. Let's get the isbn, title, edition and copyright of each book in the titles table that has a title ending with 'How to Program' and sort them in ascending order by title.

```
In [16]: pd.read_sql("""SELECT isbn, title, edition, copyright
    ...:                FROM titles
    ...:                WHERE title LIKE '%How to Program'
    ...:                ORDER BY title""", connection)
Out[16]:
         isbn                        title  edition  copyright
0  0134444302          Android How to Program        3       2017
1  0133976890                C How to Program        8       2016
2  0134448235              C++ How to Program       10       2017
3  0132151006    Internet & WWW How to Program        5       2012
4  0134743350             Java How to Program       11       2018
5  0133406954  Visual Basic 2012 How to Program        6       2014
6  0134601548        Visual C# How to Program        6       2017
7  0136151574       Visual C++ How to Program        2       2008
```

 Self Check

1 *(Fill-In)* SQL keyword _____ specifies the order in which records are sorted in a query.
Answer: ORDER BY.

17.2.5 Merging Data from Multiple Tables: INNER JOIN

Recall that the books database's author_ISBN table links authors to their corresponding titles. If we did not separate this information into individual tables, we'd need to include author information with each entry in the titles table. This would result in storing *duplicate* author information for authors who wrote multiple books.

You can merge data from multiple tables, referred to as joining the tables, with **INNER JOIN**. Let's produce a list of authors accompanied by the ISBNs for books written by each author—because there are many results for this query, we show just the head of the result:

```
In [17]: pd.read_sql("""SELECT first, last, isbn
    ...:                FROM authors
    ...:                INNER JOIN author_ISBN
    ...:                   ON authors.id = author_ISBN.id
    ...:                ORDER BY last, first""", connection).head()
Out[17]:
    first    last        isbn
0   Abbey   Deitel  0132151006
1   Abbey   Deitel  0133406954
2  Harvey   Deitel  0134289366
3  Harvey   Deitel  0135404673
4  Harvey   Deitel  0132151006
```

The INNER JOIN's **ON clause** uses a primary-key column in one table and a foreign-key column in the other to determine which rows to merge from each table. This query merges the authors table's first and last columns with the author_ISBN table's isbn column and sorts the results in ascending order by last then first.

738 Big Data: Hadoop, Spark, NoSQL and IoT

Note the syntax `authors.id` *(table_name.column_name)* in the `ON` clause. This **qualified name** syntax is required if the columns have the same name in both tables. This syntax can be used in any SQL statement to distinguish columns in different tables that have the same name. In some systems, table names qualified with the database name can be used to perform cross-database queries. As always, the query can contain an `ORDER BY` clause.

Self Check

1 *(Fill-In)* A(n) _____ specifies the fields from multiple tables that should be compared to join the tables.
Answer: qualified name.

17.2.6 INSERT INTO Statement

To this point, you've queried existing data. Sometimes you'll execute SQL statements that *modify* the database. To do so, you'll use a `sqlite3` **Cursor** object, which you obtain by calling the `Connection`'s **cursor** method:

```
In [18]: cursor = connection.cursor()
```

The pandas method `read_sql` actually uses a `Cursor` behind the scenes to execute queries and access the rows of the results.

The **INSERT INTO** statement inserts a row into a table. Let's insert a new author named Sue Red into the `authors` table by calling `Cursor` method **execute**, which executes its SQL argument and returns the `Cursor`:

```
In [19]: cursor = cursor.execute("""INSERT INTO authors (first, last)
    ...:                             VALUES ('Sue', 'Red')""")
    ...:
```

The SQL keywords `INSERT INTO` are followed by the table in which to insert the new row and a comma-separated list of column names in parentheses. The list of column names is followed by the SQL keyword **VALUES** and a comma-separated list of values in parentheses. The values provided must match the column names specified both in order and type.

We do not specify a value for the `id` column because it's an autoincremented column in the `authors` table—this was specified in the script `books.sql` that created the table. For every new row, SQLite assigns a unique `id` value that is the next value in the autoincremented sequence (i.e., 1, 2, 3 and so on). In this case, Sue Red is assigned `id` number 6. To confirm this, let's query the `authors` table's contents:

```
In [20]: pd.read_sql('SELECT id, first, last FROM authors',
    ...:             connection, index_col=['id'])
    ...:
Out[20]:
        first    last
id
1        Paul  Deitel
2      Harvey  Deitel
3       Abbey  Deitel
4         Dan   Quirk
5   Alexander    Wald
6         Sue     Red
```

Note Regarding Strings That Contain Single Quotes
SQL delimits strings with single quotes (`'`). A string containing a single quote, such as O'Malley, must have *two* single quotes in the position where the single quote appears (e.g., `'O''Malley'`). The first acts as an escape character for the second. Not escaping single-quote characters in a string that's part of a SQL statement is a SQL syntax error.

17.2.7 UPDATE Statement
An **UPDATE** statement modifies existing values. Let's assume that Sue Red's last name is incorrect in the database and update it to `'Black'`:

```
In [21]: cursor = cursor.execute("""UPDATE authors SET last='Black'
    ...:                             WHERE last='Red' AND first='Sue'""")
```

The **UPDATE** keyword is followed by the table to update, the keyword **SET** and a comma-separated list of *column_name = value* pairs indicating the columns to change and their new values. The change will be applied to *every* row if you do not specify a **WHERE** clause. The **WHERE** clause in this query indicates that we should update only rows in which the last name is `'Red'` and the first name is `'Sue'`.

Of course, there could be multiple people with the same first and last name. To make a change to only one row, it's best to use the row's unique primary key in the **WHERE** clause. In this case, we could have specified:

```
WHERE id = 6
```

For statements that modify the database, the Cursor object's rowcount attribute contains an integer value representing the number of rows that were modified. If this value is 0, no changes were made. The following confirms that the UPDATE modified one row:

```
In [22]: cursor.rowcount
Out[22]: 1
```

We also can confirm the update by listing the authors table's contents:

```
In [23]: pd.read_sql('SELECT id, first, last FROM authors',
    ...:             connection, index_col=['id'])
    ...:
Out[23]:
        first    last
id
1        Paul  Deitel
2      Harvey  Deitel
3       Abbey  Deitel
4         Dan   Quirk
5   Alexander    Wald
6         Sue   Black
```

17.2.8 DELETE FROM Statement
A SQL **DELETE FROM** statement removes rows from a table. Let's remove Sue Black from the authors table using her author ID:

```
In [24]: cursor = cursor.execute('DELETE FROM authors WHERE id=6')

In [25]: cursor.rowcount
Out[25]: 1
```

The optional WHERE clause determines which rows to delete. If WHERE is omitted, all the table's rows are deleted. Here's the authors table after the DELETE operation:

```
In [26]: pd.read_sql('SELECT id, first, last FROM authors',
    ...:             connection, index_col=['id'])
    ...:
Out[26]:
        first    last
id
1        Paul   Deitel
2      Harvey   Deitel
3       Abbey   Deitel
4         Dan    Quirk
5   Alexander     Wald
```

Closing the Database
When you no longer need access to the database, you should call the Connection's **close** method to disconnect from the database—not yet, though, as you'll use the database in the next Self Check exercises:

```
connection.close()
```

SQL in Big Data
SQL's importance is growing in big data. Later in this chapter, we'll use Spark SQL to query data in a Spark DataFrame for which the data may be distributed over many computers in a Spark cluster. As you'll see, Spark SQL looks much like the SQL presented in this section. You'll also use Spark SQL in the exercises.

Self Check for Section 17.2

1 *(IPython Session)* Select from the titles table all the titles and their edition numbers in descending order by edition number. Show only the first three results.
Answer:

```
In [27]: pd.read_sql("""SELECT title, edition FROM titles
    ...:              ORDER BY edition DESC""", connection).head(3)
Out[28]:
                  title  edition
0    Java How to Program      11
1     C++ How to Program      10
2       C How to Program       8
```

2 *(IPython Session)* Select from the authors table all authors whose first names start with 'A'.
Answer:

```
In [28]: pd.read_sql("""SELECT * FROM authors
    ...:              WHERE first LIKE 'A%'""", connection)
Out[28]:
   id      first    last
0   3      Abbey  Deitel
1   5  Alexander    Wald
```

3 *(IPython Session)* SQL's **NOT** keyword reverses the value of a WHERE clause's condition. Select from the titles table all titles that do *not* end with 'How to Program'.

Answer:

```
In [29]: pd.read_sql("""SELECT isbn, title, edition, copyright
    ...:                 FROM titles
    ...:                 WHERE title NOT LIKE '%How to Program'
    ...:                 ORDER BY title""", connection)
Out[29]:
         isbn                         title  edition  copyright
0  0134289366      Android 6 for Programmers       3       2016
1  0135404673  Intro to Python for CS and DS       1       2020
```

17.3 NoSQL and NewSQL Big-Data Databases: A Brief Tour

For decades, relational database management systems have been the standard in data processing. However, they require *structured data* that fits into neat rectangular tables. As the size of the data and the number of tables and relationships increases, relational databases become more difficult to manipulate efficiently. In today's big-data world, NoSQL and NewSQL databases have emerged to deal with the kinds of data storage and processing demands that traditional relational databases cannot meet. Big data requires massive databases, often spread across data centers worldwide in huge clusters of commodity computers. According to statista.com, there are currently over 8 million data centers worldwide.[3]

NoSQL originally meant what its name implies. With the growing importance of SQL in big data—such as SQL on Hadoop and Spark SQL—NoSQL now is said to stand for "Not Only SQL." NoSQL databases are meant for unstructured data, like photos, videos and the natural language found in e-mails, text messages and social-media posts, and semi-structured data like JSON and XML documents. Semi-structured data often wraps unstructured data with additional information called **metadata**. For example, YouTube videos are unstructured data, but YouTube also maintains metadata for each video, including who posted it, when it was posted, a title, a description, tags that help people discover the videos, privacy settings and more—all returned as JSON from the YouTube APIs. This metadata adds structure to the unstructured video data, making it semi-structured.

The next several subsections overview the four NoSQL database categories—key–value, document, columnar (also called column-based) and graph. In addition, we'll overview NewSQL databases, which blend features of relational and NoSQL databases. In Section 17.4, we'll present a case study in which we store and manipulate a large number of JSON tweet objects in a NoSQL document database, then summarize the data in an interactive visualization displayed on a Folium map of the United States. In the exercises, you can explore other types of NoSQL databases. You also can check out an implementation of the famous "six degrees of separation" problem in a NoSQL graph database.

17.3.1 NoSQL Key–Value Databases

Like Python dictionaries, **key–value databases**[4] store key–value pairs, but they're optimized for distributed systems and big-data processing. For reliability, they tend to repli-

---

3. https://www.statista.com/statistics/500458/worldwide-datacenter-and-it-sites/.
4. https://en.wikipedia.org/wiki/Key-value_database.

cate data in multiple cluster nodes. Some key–value databases, such as Redis, are implemented in memory for performance, and others store data on disk, such as HBase, which runs on top of Hadoop's HDFS distributed file system. Other popular key–value databases include Amazon DynamoDB, Google Cloud Datastore and Couchbase. DynamoDB and Couchbase are **multi-model databases** that also support documents. HBase is also a column-oriented database.

17.3.2 NoSQL Document Databases

A **document database**[5] stores semi-structured data, such as JSON or XML documents. In document databases, you typically add indexes for specific attributes, so you can more efficiently locate and manipulate documents. For example, let's assume you're storing JSON documents produced by IoT devices and each document contains a type attribute. You might add an index for this attribute so you can filter documents based on their types. Without indexes, you can still perform that task, it will just be slower because you have to search each document in its entirety to find the attribute.

The most popular document database (and most popular overall NoSQL database[6]) is **MongoDB**, whose name derives from a sequence of letters embedded in the word "hu**mongo**us." In an example, we'll store a large number of tweets in MongoDB for processing. Recall that Twitter's APIs return tweets in JSON format, so they can be stored directly in MongoDB. After obtaining the tweets we'll summarize them in a pandas `DataFrame` and on a Folium map. Other popular document databases include Amazon DynamoDB (also a key–value database), Microsoft Azure Cosmos DB and Apache CouchDB.

17.3.3 NoSQL Columnar Databases

In a relational database, a common query operation is to get a specific column's value for every row. Because data is organized into rows, a query that selects a specific column can perform poorly. The database system must get every matching row, locate the required column and discard the rest of the row's information. A **columnar database**[7,8], also called a **column-oriented database**, is similar to a relational database, but it stores structured data in columns rather than rows. Because all of a column's elements are stored together, selecting all the data for a given column is more efficient.

Consider our authors table in the books database:

```
        first    last
id
1         Paul   Deitel
2       Harvey   Deitel
3        Abbey   Deitel
4          Dan    Quirk
5    Alexander     Wald
```

In a relational database, all the data for a row is stored together. If we consider each row as a Python tuple, the rows would be represented as (1, 'Paul', 'Deitel'), (2, 'Harvey', 'Deitel'), etc. In a columnar database, all the values for a given column would be stored

---

5. https://en.wikipedia.org/wiki/Document-oriented_database.
6. https://db-engines.com/en/ranking.
7. https://en.wikipedia.org/wiki/Columnar_database.
8. https://www.predictiveanalyticstoday.com/top-wide-columnar-store-databases/.

together, as in (1, 2, 3, 4, 5), ('Paul', 'Harvey', 'Abbey', 'Dan', 'Alexander') and ('Deitel', 'Deitel', 'Deitel', 'Quirk', 'Wald'). The elements in each column are maintained in row order, so the value at a given index in each column belongs to the same row. Popular columnar databases include MariaDB ColumnStore and HBase.

17.3.4 NoSQL Graph Databases

A graph models relationships between objects.[9] The objects are called **nodes** (or **vertices**) and the relationships are called **edges**. Edges are *directional*. For example, an edge representing an airline flight points from the origin city to the destination city, but not the reverse. A **graph database**[10] stores nodes, edges and their attributes.

If you use social networks, like Instagram, Snapchat, Twitter and Facebook, consider your social graph, which consists of the people you know (nodes) and the relationships between them (edges). Every person has their own social graph, and these are interconnected. The famous "six degrees of separation" problem, which you'll explore in the exercises, says that any two people in the world are connected to one another by following a maximum of six edges in the worldwide social graph.[11] Facebook's algorithms use the social graphs of their billions of monthly active users[12] to determine which stories should appear in each user's news feed. By looking at your interests, your friends, their interests and more, Facebook predicts the stories they believe are most relevant to you.[13]

Many companies use similar techniques to create recommendation engines. When you browse a product on Amazon, they use a graph of users and products to show you comparable products people browsed before making a purchase. When you browse movies on Netflix, they use a graph of users and movies they liked to suggest movies that might be of interest to you.

One of the most popular graph databases is Neo4j. Many real-world use-cases for graph databases are provided at:

> https://neo4j.com/graphgists/

With most of the use-cases, sample graph diagrams produced by Neo4j are shown. These visualize the relationships between the graph nodes. Check out Neo4j's free PDF book, *Graph Databases*.[14]

17.3.5 NewSQL Databases

Key advantages of relational databases include their security and transaction support. In particular, relational databases typically use **ACID** (**Atomicity, Consistency, Isolation, Durability**)[15] transactions:

- *Atomicity* ensures that the database is modified only if *all* of a transaction's steps are successful. If you go to an ATM to withdraw $100, that money is not

---

9. https://en.wikipedia.org/wiki/Graph_theory.
10. https://en.wikipedia.org/wiki/Graph_database.
11. https://en.wikipedia.org/wiki/Six_degrees_of_separation.
12. https://zephoria.com/top-15-valuable-facebook-statistics/.
13. https://newsroom.fb.com/news/2018/05/inside-feed-news-feed-ranking/.
14. https://neo4j.com/graph-databases-book-sx2.
15. https://en.wikipedia.org/wiki/ACID_(computer_science).

removed from your account unless you have enough money to cover the withdrawal *and* there is enough money in the ATM to satisfy your request.

- *Consistency* ensures that the database state is always valid. In the withdrawal example above, your new account balance after the transaction will reflect precisely what you withdrew from your account (and possibly ATM fees).
- *Isolation* ensures that concurrent transactions occur as if they were performed sequentially. For example, if two people share a joint bank account and both attempt to withdraw money at the same time from two separate ATMs, one transaction must wait until the other completes.
- *Durability* ensures that changes to the database survive even hardware failures.

If you research benefits and disadvantages of NoSQL databases, you'll see that NoSQL databases generally do not provide ACID support. The types of applications that use NoSQL databases typically do not require the guarantees that ACID-compliant databases provide. Many NoSQL databases typically adhere to the **BASE (Basic Availability, Soft-state, Eventual consistency)** model, which focuses more on the database's availability. Whereas, ACID databases guarantee consistency when you write to the database, BASE databases provide consistency at some later point in time.

NewSQL databases blend the benefits of both relational and NoSQL databases for big-data processing tasks. Some popular NewSQL databases include VoltDB, MemSQL, Apache Ignite and Google Spanner.

✓ Self Check for Section 17.3

1 *(True/False)* Relational databases require unstructured or semi-structured data.
Answer: False. Relational databases require *structured data* that fits into rectangular tables.

2 *(Fill-In)* NoSQL is now said to stand for _____.
Answer: Not Only SQL.

3 *(True/False)* A NoSQL document database stores documents full of key–value pairs.
Answer: False. A NoSQL key–value database stores key–value pairs. A NoSQL document database stores semi-structured data, such as JSON or XML documents.

4 *(Fill-In)* Which NoSQL database type is similar to a relational database? _____.
Answer: A columnar (or column-oriented) database.

5 *(Fill-In)* Which NoSQL database type stores data in nodes and edges? _____.
Answer: A graph database.

17.4 Case Study: A MongoDB JSON Document Database

MongoDB is a document database capable of storing and retrieving JSON documents. Twitter's APIs return tweets to you as JSON objects, which you can write directly into a MongoDB database. In this section, you'll:

- use Tweepy to stream tweets about the 100 U.S. senators and store them into a MongoDB database,
- use pandas to summarize the top 10 senators by tweet activity and

- display an interactive Folium map of the United States with one popup marker per state that shows the state name and both senators' names, their political parties and tweet counts.

You'll use a free cloud-based MongoDB Atlas cluster, which requires no installation and currently allows you to store up to 512MB of data. To store more, you can download the MongoDB Community Server from:

> https://www.mongodb.com/download-center/community

and run it locally or you can sign up for MongoDB's *paid* Atlas service.

Installing the Python Libraries Required for Interacting with MongoDB

You'll use the **pymongo library** to interact with MongoDB databases from your Python code. You'll also need the dnspython library to connect to a MongoDB Atlas Cluster. To install these libraries, use the following commands:

```
conda install -c conda-forge pymongo
conda install -c conda-forge dnspython
```

keys.py

The ch17 examples folder's TwitterMongoDB subfolder contains this example's code and keys.py file. Edit this file to include your Twitter credentials and your OpenMapQuest key from the "Data Mining Twitter" chapter. After we discuss creating a MongoDB Atlas cluster, you'll also need to add your MongoDB connection string to this file.

17.4.1 Creating the MongoDB Atlas Cluster

To sign up for a free account go to

> https://mongodb.com

then enter your email address and click **Get started free**. On the next page, enter your name and create a password, then read their terms of service. If you agree, click **Get started free** on this page and you'll be taken to the screen for setting up your cluster. Click **Build my first cluster** to get started.

They walk you through the getting started steps with popup bubbles that describe and point you to each task you need to complete. They provide default settings for their free Atlas cluster (**M0** as they refer to it), so just give your cluster a name in the **Cluster Name** section, then click **Create Cluster**. At this point, they'll take you to the **Clusters** page and begin creating your new cluster, which takes several minutes.

Next, a **Connect to Atlas** popup tutorial will appear, showing a checklist of additional steps required to get you up and running:

- **Create your first database user**—This enables you to log into your cluster.
- **Whitelist your IP address**—This is a security measure which ensures that only IP addresses you verify are allowed to interact with your cluster. To connect to this cluster from multiple locations (school, home, work, etc.), you'll need to whitelist each IP address from which you intend to connect.
- **Connect to your cluster**—In this step, you'll locate your cluster's connection string, which will enable your Python code to connect to the server.

Creating Your First Database User
In the popup tutorial window, click **Create your first database user** to continue the tutorial, then follow the on-screen prompts to view the cluster's **Security** tab and click **+ ADD NEW USER**. In the **Add New User** dialog, create a username and password. Write these down—you'll need them momentarily. Click **Add User** to return to the **Connect to Atlas** popup tutorial.

Whitelist Your IP Address
In the popup tutorial window, click **Whitelist your IP address** to continue the tutorial, then follow the on-screen prompts to view the cluster's **IP Whitelist** and click **+ ADD IP ADDRESS**. In the **Add Whitelist Entry** dialog, you can either add your computer's current IP address or allow access from anywhere, which they do not recommend for production databases, but is OK for learning purposes. Click **ALLOW ACCESS FROM ANYWHERE** then click **Confirm** to return to the **Connect to Atlas** popup tutorial.

Connect to Your Cluster
In the popup tutorial window, click **Connect to your cluster** to continue the tutorial, then follow the on-screen prompts to view the cluster's **Connect to** *YourClusterName* dialog. Connecting to a MongoDB Atlas database from Python requires a connection string. To get your connection string, click **Connect Your Application**, then click **Short SRV connection string**. Your connection string will appear below **Copy the SRV address**. Click **COPY** to copy the string. Paste this string into the keys.py file as mongo_connection_string's value. Replace "<PASSWORD>" in the connection string with your password, and replace the database name "test" with "senators", which will be the database name in this example. At the bottom of the **Connect to** *YourClusterName*, click **Close**. You're now ready to interact with your Atlas cluster.

17.4.2 Streaming Tweets into MongoDB
First we'll present an interactive IPython session that connects to the MongoDB database, downloads current tweets via Twitter streaming and summarizes the top-10 senators by tweet count. Next, we'll present class TweetListener, which handles the incoming tweets and stores their JSON in MongoDB. Finally, we'll continue the IPython session by creating an interactive Folium map that displays information from the tweets we stored.

Use Tweepy to Authenticate with Twitter
First, let's use Tweepy to authenticate with Twitter:

```
In [1]: import tweepy, keys

In [2]: auth = tweepy.OAuthHandler(
   ...:     keys.consumer_key, keys.consumer_secret)
   ...: auth.set_access_token(keys.access_token,
   ...:     keys.access_token_secret)
   ...:
```

Next, configure the Tweepy API object to wait if our app reaches any Twitter rate limits.

```
In [3]: api = tweepy.API(auth, wait_on_rate_limit=True,
   ...:                  wait_on_rate_limit_notify=True)
   ...:
```

Loading the Senators' Data

We'll use the information in the file `senators.csv` (located in the ch17 examples folder's `TwitterMongoDB` subfolder) to track tweets to, from and about every U.S. senator. The file contains the senator's two-letter state code, name, party, Twitter handle and Twitter ID.

Twitter enables you to follow specific users via their numeric Twitter IDs, but these must be submitted as string representations of those numeric values. So, let's load `senators.csv` into pandas, convert the TwitterID values to strings (using Series method **astype**) and display several rows of data. In this case, we set 6 as the maximum number of columns to display. Later we'll add another column to the DataFrame and this setting will ensure that all the columns are displayed, rather than a few with ... in between:

```
In [4]: import pandas as pd

In [5]: senators_df = pd.read_csv('senators.csv')

In [6]: senators_df['TwitterID'] = senators_df['TwitterID'].astype(str)

In [7]: pd.options.display.max_columns = 6

In [8]: senators_df.head()
Out[8]:
  State           Name Party    TwitterHandle           TwitterID
0    AL  Richard Shelby     R        SenShelby            21111098
1    AL      Doug Jones     D     SenDougJones  941080085121175552
2    AK   Lisa Murkowski    R     lisamurkowski            18061669
3    AK    Dan Sullivan    R    SenDanSullivan          2891210047
4    AZ         Jon Kyl    R         SenJonKyl            24905240
```

Configuring the MongoClient

To store the tweet's JSON as documents in a MongoDB database, you must first connect to your MongoDB Atlas cluster via a pymongo **MongoClient**, which receives your cluster's connection string as its argument:

```
In [9]: from pymongo import MongoClient

In [10]: atlas_client = MongoClient(keys.mongo_connection_string)
```

Now, we can get a pymongo **Database** object representing the senators database. The following statement creates the database if it does not exist:

```
In [11]: db = atlas_client.senators
```

Setting up Tweet Stream

Let's specify the number of tweets to download and create the TweetListener. We pass the db object representing the MongoDB database to the TweetListener so it can write the tweets into the database. Depending on the rate at which people are tweeting about the senators, it may take minutes to hours to get 10,000 tweets. For testing purposes, you might want to use a smaller number:

```
In [12]: from tweetlistener import TweetListener

In [13]: tweet_limit = 10000

In [14]: twitter_stream = tweepy.Stream(api.auth,
    ...:     TweetListener(api, db, tweet_limit))
    ...:
```

Starting the Tweet Stream

Twitter live streaming allows you to track up to 400 keywords and follow up to 5,000 Twitter IDs at a time. In this case, let's track the senators' Twitter handles and follow the senator's Twitter IDs. This should give us tweets from, to and about each senator. To show you progress, we display the screen name and time stamp for each tweet received, and the total number of tweets so far. To save space, we show here only one of those tweet outputs and replace the user's screen name with XXXXXXX:

```
In [15]: twitter_stream.filter(track=senators_df.TwitterHandle.tolist(),
    ...:     follow=senators_df.TwitterID.tolist())
    ...:
Screen name: XXXXXXX
  Created at: Sun Dec 16 17:19:19 +0000 2018
Tweets received: 1
...
```

Class TweetListener

For this example, we slightly modified class `TweetListener` from the "Data Mining Twitter" chapter. Much of the Twitter and Tweepy code shown below is identical to the code you saw previously, so we'll focus on only the new concepts here:

```
 1  # tweetlistener.py
 2  """TweetListener downloads tweets and stores them in MongoDB."""
 3  import json
 4  import tweepy
 5
 6  class TweetListener(tweepy.StreamListener):
 7      """Handles incoming Tweet stream."""
 8
 9      def __init__(self, api, database, limit=10000):
10          """Create instance variables for tracking number of tweets."""
11          self.db = database
12          self.tweet_count = 0
13          self.TWEET_LIMIT = limit  # 10,000 by default
14          super().__init__(api)  # call superclass's init
15
16      def on_connect(self):
17          """Called when your connection attempt is successful, enabling
18          you to perform appropriate application tasks at that point."""
19          print('Successfully connected to Twitter\n')
20
21      def on_data(self, data):
22          """Called when Twitter pushes a new tweet to you."""
23          self.tweet_count += 1  # track number of tweets processed
24          json_data = json.loads(data)  # convert string to JSON
25          self.db.tweets.insert_one(json_data)  # store in tweets collection
26          print(f'    Screen name: {json_data["user"]["name"]}')
27          print(f'    Created at: {json_data["created_at"]}')
28          print(f'Tweets received: {self.tweet_count}')
29
30          # if TWEET_LIMIT is reached, return False to terminate streaming
31          return self.tweet_count != self.TWEET_LIMIT
32
```

17.4 Case Study: A MongoDB JSON Document Database

```
33      def on_error(self, status):
34          print(status)
35          return True
```

Previously, `TweetListener` overrode method `on_status` to receive Tweepy `Status` objects representing tweets. Here, we override the `on_data` method instead (lines 21–31). Rather than `Status` objects, `on_data` receives each tweet object's *raw JSON*. Line 24 converts the JSON string received by `on_data` into a Python JSON object. Each MongoDB database contains one or more **Collections** of documents. In line 25, the expression

```
self.db.tweets
```

accesses the `Database` object `db`'s `tweets` `Collection`, creating it if it does not already exist. Line 25 uses the `tweets` `Collection`'s **insert_one method** to store the JSON object in the `tweets` collection.

Counting Tweets for Each Senator

Next, we'll perform a full-text search on the collection of tweets and count the number of tweets containing each senator's Twitter handle. To text search in MongoDB, you must create a *text index* for the collection.[16] This specifies which document field(s) to search. Each text index is defined as a tuple containing the field name to search and the index type (`'text'`). MongoDB's wildcard specifier (`$**`) indicates that every text field in a document (a JSON tweet object in our case) should be indexed for a full-text search:

```
In [16]: db.tweets.create_index([('$**', 'text')])
Out[16]: '$**_text'
```

Once the index is defined, we can use the `Collection`'s **count_documents method** to count the total number of documents in the collection that contain the specified text. Let's search the database's `tweets` collection for every twitter handle in the `senators_df` DataFrame's `TwitterHandle` column:

```
In [17]: tweet_counts = []

In [18]: for senator in senators_df.TwitterHandle:
    ...:     tweet_counts.append(db.tweets.count_documents(
    ...:         {"$text": {"$search": senator}}))
    ...:
```

The JSON object passed to `count_documents` in this case indicates that we're using the index named `text` to `search` for the value of `senator`.

Show Tweet Counts for Each Senator

Let's create a copy of the `DataFrame` `senators_df` that contains the `tweet_counts` as a new column, then display the top-10 senators by tweet count:

```
In [19]: tweet_counts_df = senators_df.assign(Tweets=tweet_counts)

In [20]: tweet_counts_df.sort_values(by='Tweets',
    ...:     ascending=False).head(10)
    ...:
```

---

16. For additional details on MongoDB index types, text indexes and operators, see: https://docs.mongodb.com/manual/indexes, https://docs.mongodb.com/manual/core/index-text and https://docs.mongodb.com/manual/reference/operator.

```
Out[20]:
    State              Name   Party    TwitterHandle   TwitterID    Tweets
78     SC     Lindsey Graham      R   LindseyGrahamSC  432895323     1405
41     MA   Elizabeth Warren      D         SenWarren  970207298     1249
8      CA   Dianne Feinstein      D      SenFeinstein  476256944     1079
20     HI       Brian Schatz      D       brianschatz   47747074      934
62     NY      Chuck Schumer      D        SenSchumer   17494010      811
24     IL    Tammy Duckworth      D      SenDuckworth 1058520120      656
13     CT  Richard Blumenthal    D     SenBlumenthal  278124059      646
21     HI       Mazie Hirono      D       maziehirono   92186819      628
86     UT        Orrin Hatch      R     SenOrrinHatch  262756641      506
77     RI  Sheldon Whitehouse    D     SenWhitehouse  242555999      350
```

Get the State Locations for Plotting Markers

Next, we'll use the techniques you learned in the "Data Mining Twitter" chapter to get each state's latitude and longitude coordinates. We'll soon use these to place on a Folium map popup markers that contain the names and numbers of tweets mentioning each state's senators.

The file `state_codes.py` contains a `state_codes` dictionary that maps two-letter state codes to their full state names. We'll use the full state names with geopy's Open-MapQuest `geocode` function to look up the location of each state.[17] First, let's import the libraries we need and the `state_codes` dictionary:

```
In [21]: from geopy import OpenMapQuest
```

```
In [22]: import time
```

```
In [23]: from state_codes import state_codes
```

Next, let's get the geocoder object to translate location names into `Location` objects:

```
In [24]: geo = OpenMapQuest(api_key=keys.mapquest_key)
```

There are two senators from each state, so we can look up each state's location once and use the `Location` object for both senators from that state. Let's get the unique state names, then sort them into ascending order:

```
In [25]: states = tweet_counts_df.State.unique()
```

```
In [26]: states.sort()
```

The next two snippets use code from the "Data Mining Twitter" chapter to look up each state's location. In snippet [28], we call the `geocode` function with the state name followed by `', USA'` to ensure that we get United States locations,[18] since there are places outside the United States. with the same names as U.S. states. To show progress, we display each new `Location` object's string:

```
In [27]: locations = []
```

---

17. We use full state names because, during our testing, the two-letter state codes did not always return correct locations.
18. When we initially performed the geocoding for Washington state, OpenMapQuest returned Washington, D.C.'s location. So we modified `state_codes.py` to use "Washington State" instead.

17.4 Case Study: A MongoDB JSON Document Database

```
In [28]: for state in states:
    ...:     processed = False
    ...:     delay = .1
    ...:     while not processed:
    ...:         try:
    ...:             locations.append(
    ...:                 geo.geocode(state_codes[state] + ', USA'))
    ...:             print(locations[-1])
    ...:             processed = True
    ...:         except:  # timed out, so wait before trying again
    ...:             print('OpenMapQuest service timed out. Waiting.')
    ...:             time.sleep(delay)
    ...:             delay += .1
    ...:
Alaska, United States of America
Alabama, United States of America
Arkansas, United States of America
...
```

Grouping the Tweet Counts by State
We'll use the total number of tweets for the two senators in a state to color that state on the map. Darker colors will represent the states with higher tweet counts. To prepare the data for mapping, let's use the pandas DataFrame method **groupby** to group the senators by state and calculate the total tweets by state:

```
In [30]: tweets_counts_by_state = tweet_counts_df.groupby(
    ...:     'State', as_index=False).sum()
    ...:

In [31]: tweets_counts_by_state.head()
Out[31]:
   State  Tweets
0  AK         27
1  AL          2
2  AR         47
3  AZ         47
4  CA       1135
```

The `as_index=False` keyword argument in snippet [30] indicates that the state codes should be values in a column of the resulting **GroupBy** object, rather than the indices for the rows. The **GroupBy** object's **sum** method totals the numeric data (the tweets by state). Snippet [31] displays several rows of the GroupBy object so you can see some of the results.

Creating the Map
Next, let's create the map. You may want to adjust the zoom. On our system, the following snippet creates a map in which we initially can see only the continental United States. Remember that Folium maps are interactive, so once the map is displayed, you can scroll to zoom in and out or drag to see different areas, such as Alaska or Hawaii:

```
In [32]: import folium

In [33]: usmap = folium.Map(location=[39.8283, -98.5795],
    ...:     zoom_start=4, detect_retina=True,
    ...:     tiles='Stamen Toner')
    ...:
```

Creating a Choropleth to Color the Map

A **choropleth** shades areas in a map using the values you specify to determine color. Let's create a choropleth that colors the states by the number of tweets containing their senators' Twitter handles. First, save Folium's `us-states.json` file at

> https://raw.githubusercontent.com/python-visualization/folium/
> master/examples/data/us-states.json

to the folder containing this example. This file contains a JSON dialect called **GeoJSON** (**Geographic JSON**) that describes the boundaries of shapes—in this case, the boundaries of every U.S. state. The choropleth uses this information to shade each state. You can learn more about GeoJSON at http://geojson.org/.[19] The following snippets create the choropleth, then add it to the map:

```
In [34]: choropleth = folium.Choropleth(
    ...:     geo_data='us-states.json',
    ...:     name='choropleth',
    ...:     data=tweets_counts_by_state,
    ...:     columns=['State', 'Tweets'],
    ...:     key_on='feature.id',
    ...:     fill_color='YlOrRd',
    ...:     fill_opacity=0.7,
    ...:     line_opacity=0.2,
    ...:     legend_name='Tweets by State'
    ...: ).add_to(usmap)
    ...:

In [35]: layer = folium.LayerControl().add_to(usmap)
```

In this case, we used the following arguments:

- `geo_data='us-states.json'`—This is the file containing the GeoJSON that specifies the shapes to color.
- `name='choropleth'`—Folium displays the `Choropleth` as a layer over the map. This is the name for that layer that will appear in the map's layer controls, which enable you to hide and show the layers. These controls appear when you click the layers icon (![]) on the map
- `data=tweets_counts_by_state`—This is a pandas `DataFrame` (or `Series`) containing the values that determine the `Choropleth` colors.
- `columns=['State', 'Tweets']`—When the `data` is a `DataFrame`, this is a list of two columns representing the keys and the corresponding values used to color the `Choropleth`.
- `key_on='feature.id'`—This is a variable in the GeoJSON file to which the `Choropleth` binds the values in the `columns` argument.
- `fill_color='YlOrRd'`—This is a color map specifying the colors to use to fill in the states. Folium provides 12 colormaps: `'BuGn'`, `'BuPu'`, `'GnBu'`, `'OrRd'`, `'PuBu'`, `'PuBuGn'`, `'PuRd'`, `'RdPu'`, `'YlGn'`, `'YlGnBu'`, `'YlOrBr'` and `'YlOrRd'`.

---

19. Folium provides several other GeoJSON files in its examples folder at https://github.com/python-visualization/folium/tree/master/examples/data. You also can create your own at http://geojson.io.

17.4 Case Study: A MongoDB JSON Document Database

You should experiment with these to find the most effective and eye-pleasing ones for your application(s).

- `fill_opacity=0.7`—A value from 0.0 (transparent) to 1.0 (opaque) specifying the transparency of the fill colors displayed in the states.

- `line_opacity=0.2`—A value from 0.0 (transparent) to 1.0 (opaque) specifying the transparency of lines used to delineate the states.

- `legend_name='Tweets by State'`—At the top of the map, the `Choropleth` displays a color bar (the legend) indicating the value range represented by the colors. This `legend_name` text appears below the color bar to indicate what the colors represent.

The complete list of `Choropleth` keyword arguments is documented at:

> http://python-visualization.github.io/folium/
> modules.html#folium.features.Choropleth

Creating the Map Markers for Each State

Next, we'll create `Markers` for each state. To ensure that the senators are displayed in descending order by the number of tweets in each state's `Marker`, let's sort `tweet_counts_df` in descending order by the `'Tweets'` column:

```
In [36]: sorted_df = tweet_counts_df.sort_values(
    ...:     by='Tweets', ascending=False)
    ...:
```

The loop in the following snippet creates the `Markers`. First,

```
sorted_df.groupby('State')
```

groups `sorted_df` by `'State'`. A DataFrame's groupby method maintains the *original row order* in each group. Within a given group, the senator with the most tweets will be first, because we sorted the senators in descending order by tweet count in snippet [36]:

```
In [37]: for index, (name, group) in
enumerate(sorted_df.groupby('State')):
    ...:     strings = [state_codes[name]]   # used to assemble popup text
    ...:
    ...:     for s in group.itertuples():
    ...:         strings.append(
    ...:             f'{s.Name} ({s.Party}); Tweets: {s.Tweets}')
    ...:
    ...:     text = '<br>'.join(strings)
    ...:     marker = folium.Marker(
    ...:         (locations[index].latitude, locations[index].longitude),
    ...:         popup=text)
    ...:     marker.add_to(usmap)
    ...:
    ...:
```

We pass the grouped `DataFrame` to `enumerate`, so we can get an index for each group, which we'll use to look up each state's `Location` in the `locations` list. Each group has a name (the state code we grouped by) and a collection of items in that group (the two senators for that state). The loop operates as follows:

- We look up the full state name in the `state_codes` dictionary, then store it in the `strings` list—we'll use this list to assemble the `Marker`'s popup text.
- The nested loop walks through the items in the `group` collection, returning each as a named tuple that contains a given senator's data. We create a formatted string for the current senator containing the person's name, party and number of tweets, then append that to the `strings` list.
- The `Marker` text can use HTML for formatting. We join the `strings` list's elements, separating each from the next with an HTML `<br>` element which creates a new line in HTML.
- We create the `Marker`. The first argument is the `Marker`'s location as a tuple containing the latitude and longitude. The `popup` keyword argument specifies the text to display if the user clicks the `Marker`.
- We add the `Marker` to the map.

Displaying the Map
Finally, let's save the map into an HTML file

```
In [38]: usmap.save('SenatorsTweets.html')
```

Open the HTML file in your web browser to view and interact with the map. Recall that you can drag the map to see Alaska and Hawaii. Here we show the popup text for the South Carolina marker:

An exercise at the end of the chapter asks you to use the sentiment-analysis techniques you learned in previous chapters to rate as positive, neutral or negative the sentiment expressed by people who send tweets ("tweeters") mentioning each senator's handle.

Self Check for Section 17.4

1 *(Write a Statement)* Assuming that `atlas_client` is a pymongo `MongoClient` that's connected to a MongoDB Atlas cluster, write a statement that creates a new database with the name `football_players` and stores the resulting object in `db`.
Answer: `db = atlas_client.football_players`.

2 *(Fill-In)* A pymongo `Collection` object's _____ method inserts a new document into the `Collection`.
Answer: `insert_one`.

3 *(True/False)* Once you've inserted documents in a `Collection`, you can immediately text-search their contents.
Answer: False. To perform a text search, you must first create a text index for the collection specifying which document field(s) to search.

4 *(Fill-In)* A `folium` _____ uses GeoJSON to add color to a map.
Answer: `Choropleth`.

17.5 Hadoop

The next several sections show how Apache Hadoop and Apache Spark deal with big-data storage and processing challenges via huge clusters of computers, massively parallel processing, Hadoop MapReduce programming and Spark in-memory processing techniques. Here, we discuss Apache Hadoop, a key big-data infrastructure technology that also serves as the foundation for many recent advancements in big-data processing and an entire ecosystem of software tools that are continually evolving to support today's big-data needs.

17.5.1 Hadoop Overview

When Google was launched in 1998, the amount of online data was already enormous with approximately 2.4 million websites[20]—truly big data. Today there are now nearly two billion websites[21] (almost a thousandfold increase) and Google is handling over two trillion searches per year![22] Having used Google search since its inception, our sense is that today's responses are significantly faster.

When Google was developing their search engine, they knew that they needed to return search results quickly. The only practical way to do this was to store and index the entire Internet using a clever combination of secondary storage and main memory. Computers of that time couldn't hold that amount of data and could not analyze that amount of data fast enough to guarantee prompt search-query responses. So Google developed a **clustering** system, tying together vast numbers of computers—called **nodes**. Because having more computers and more connections between them meant greater chance of hardware failures, they also built in high levels of *redundancy* to ensure that the system would continue functioning even if nodes within clusters failed. The data was distributed across all these inexpensive "commodity computers." To satisfy a search request, all the computers in the cluster searched in parallel the portion of the web they had locally. Then the results of those searches were gathered up and reported back to the user.

---

20. http://www.internetlivestats.com/total-number-of-websites/.
21. http://www.internetlivestats.com/total-number-of-websites/.
22. http://www.internetlivestats.com/google-search-statistics/.

To accomplish this, Google needed to develop the clustering hardware and software, including distributed storage. Google published its designs, but did not open source its software. Programmers at Yahoo!, working from Google's designs in the "Google File System" paper,[23] then built their own system. They open-sourced their work and the Apache organization implemented the system as Hadoop. The name came from an elephant stuffed animal that belonged to a child of one of Hadoop's creators.

Two additional Google papers also contributed to the evolution of Hadoop—"MapReduce: Simplified Data Processing on Large Clusters"[24] and "Bigtable: A Distributed Storage System for Structured Data,"[25] which was the basis for Apache HBase (a NoSQL key–value and column-based database).[26]

HDFS, MapReduce and YARN

Hadoop's key components are:

- **HDFS (Hadoop Distributed File System)** for storing massive amounts of data throughout a cluster, and
- **MapReduce** for implementing the tasks that process the data.

Earlier in the book we introduced basic functional-style programming and filter/map/reduce. Hadoop MapReduce is similar in concept, just on a massively parallel scale. A MapReduce task performs two steps—**mapping** and **reduction**. The mapping step, which also may include *filtering*, processes the original data across the entire cluster and maps it into tuples of key–value pairs. The reduction step then combines those tuples to produce the results of the MapReduce task. The key is how the MapReduce step is performed. Hadoop divides the data into *batches* that it distributes across the nodes in the cluster—anywhere from a few nodes to a Yahoo! cluster with 40,000 nodes and over 100,000 cores.[27] Hadoop also distributes the MapReduce task's code to the nodes in the cluster and executes the code in parallel on every node. Each node processes only the batch of data stored on that node. The reduction step combines the results from all the nodes to produce the final result. To coordinate this, Hadoop uses **YARN** ("**yet another resource negotiator**") to manage all the resources in the cluster and schedule tasks for execution.

Hadoop Ecosystem

Though Hadoop began with HDFS and MapReduce, followed closely by YARN, it has grown into a large ecosystem that includes Spark (discussed in Sections 17.6–17.7) and many other Apache projects:[28,29,30]

---

23. http://static.googleusercontent.com/media/research.google.com/en//archive/gfs-sosp2003.pdf.
24. http://static.googleusercontent.com/media/research.google.com/en//archive/mapreduce-osdi04.pdf.
25. http://static.googleusercontent.com/media/research.google.com/en//archive/bigtable-osdi06.pdf.
26. Many other influential big-data-related papers (including the ones we mentioned) can be found at: https://bigdata-madesimple.com/research-papers-that-changed-the-world-of-big-data/.
27. https://wiki.apache.org/hadoop/PoweredBy.
28. https://hortonworks.com/ecosystems/.
29. https://readwrite.com/2018/06/26/complete-guide-of-hadoop-ecosystem-components/.
30. https://www.janbasktraining.com/blog/introduction-architecture-components-hadoop-ecosystem/.

- **Ambari** (https://ambari.apache.org)—Tools for managing Hadoop clusters.
- **Drill** (https://drill.apache.org)—SQL querying of non-relational data in Hadoop and NoSQL databases.
- **Flume** (https://flume.apache.org)—A service for collecting and storing (in HDFS and other storage) streaming event data, like high-volume server logs, IoT messages and more.
- **HBase** (https://hbase.apache.org)—A NoSQL database for big data with "billions of rows by[31] millions of columns—atop clusters of commodity hardware."
- **Hive** (https://hive.apache.org)—Uses SQL to interact with data in data warehouses. A **data warehouse** aggregates data of various types from various sources. Common operations include extracting data, transforming it and loading (known as ETL) into another database, typically so you can analyze it and create reports from it.
- **Impala** (https://impala.apache.org)—A database for real-time SQL-based queries across distributed data stored in Hadoop HDFS or HBase.
- **Kafka** (https://kafka.apache.org)—Real-time messaging, stream processing and storage, typically to transform and process high-volume streaming data, such as website activity and streaming IoT data.
- **Pig** (https://pig.apache.org)—A scripting platform that converts data analysis tasks from a scripting language called **Pig Latin** into MapReduce tasks.
- **Sqoop** (https://sqoop.apache.org)—Tool for moving structured, semi-structured and unstructured data between databases.
- **Storm** (https://storm.apache.org)—A real-time stream-processing system for tasks such as data analytics, machine learning, ETL and more.
- **ZooKeeper** (https://zookeeper.apache.org)—A service for managing cluster configurations and coordination between clusters.
- And more.

Hadoop Providers

Numerous cloud vendors provide Hadoop as a service, including Amazon EMR, Google Cloud DataProc, IBM Watson Analytics Engine, Microsoft Azure HDInsight and others. In addition, companies like Cloudera and Hortonworks (which at the time of this writing are merging) offer integrated Hadoop-ecosystem components and tools via the major cloud vendors. They also offer free *downloadable* environments that you can run on the desktop[32] for learning, development and testing before you commit to cloud-based hosting, which can incur significant costs. We introduce MapReduce programming in the example in the following sections by using a Microsoft cloud-based Azure HDInsight cluster, which provides Hadoop as a service.

---

31. We used the word "by" to replace "X" in the original text.
32. Check their significant system requirements first to ensure that you have the disk space and memory required to run them.

Hadoop 3
Apache continues to evolve Hadoop. Hadoop 3[33] was released in December of 2017 with many improvements, including better performance and significantly improved storage efficiency.[34]

17.5.2 Summarizing Word Lengths in *Romeo and Juliet* via MapReduce

In the next several subsections, you'll create a cloud-based, multi-node cluster of computers using Microsoft Azure HDInsight. Then, you'll use the service's capabilities to demonstrate Hadoop MapReduce running on that cluster. The MapReduce task you'll define will determine the length of each word in `RomeoAndJuliet.txt` (from the "Natural Language Processing" chapter), then summarize how many words of each length there are. After defining the task's mapping and reduction steps, you'll submit the task to your HDInsight cluster, and Hadoop will decide how to use the cluster of computers to perform the task.

17.5.3 Creating an Apache Hadoop Cluster in Microsoft Azure HDInsight

Most major cloud vendors have support for Hadoop and Spark computing clusters that you can configure to meet your application's requirements. Multi-node cloud-based clusters typically are *paid* services, though most vendors provide free trials or credits so you can try out their services.

We want you to experience the process of setting up clusters and using them to perform tasks. So, in this Hadoop example, you'll use Microsoft Azure's HDInsight service to create cloud-based clusters of computers in which to test our examples. Go to

> https://azure.microsoft.com/en-us/free

to sign up for an account. Microsoft requires a credit card for identity verification.

Various services are always free and some you can continue to use for 12 months. For information on these services see:

> https://azure.microsoft.com/en-us/free/free-account-faq/

Microsoft also gives you a credit to experiment with their *paid* services, such as their HDInsight Hadoop and Spark services. Once your credits run out or 30 days pass (whichever comes first), you cannot continue using paid services unless you authorize Microsoft to charge your card.

Because you'll use your new Azure account's credit for these examples,[35] we'll discuss how to configure a low-cost cluster that uses less computing resources than Microsoft allocates by default.[36] **Caution: Once you allocate a cluster, it incurs costs whether you're using it or not. So, when you complete this case study, be sure to delete your cluster(s) and other resources, so you don't incur additional charges.** For more information, see:

---

33. For a list of features in Hadoop 3, see https://hadoop.apache.org/docs/r3.0.0/.
34. https://www.datanami.com/2018/10/18/is-hadoop-officially-dead/.
35. For Microsoft's latest free account features, visit https://azure.microsoft.com/en-us/free/.
36. For Microsoft's recommended cluster configurations, see https://docs.microsoft.com/en-us/azure/hdinsight/hdinsight-component-versioning#default-node-configuration-and-virtual-machine-sizes-for-clusters. If you configure a cluster that's too small for a given scenario, when you try to deploy the cluster you'll receive an error.

```
https://docs.microsoft.com/en-us/azure/azure-resource-manager/
   resource-group-portal
```

For Azure-related documentation and videos, visit:

- `https://docs.microsoft.com/en-us/azure/`—the Azure documentation.
- `https://channel9.msdn.com/`—Microsoft's Channel 9 video network.
- `https://www.youtube.com/user/windowsazure`—Microsoft's Azure channel on YouTube.

Creating an HDInsight Hadoop Cluster

The following link explains how to set up a cluster for Hadoop using the Azure HDInsight service:

```
https://docs.microsoft.com/en-us/azure/hdinsight/hadoop/apache-
   hadoop-linux-create-cluster-get-started-portal
```

While following their **Create a Hadoop cluster** steps, please note the following:

- In *Step 1*, you access the Azure portal by logging into your account at

    `https://portal.azure.com`

- In *Step 2*, **Data + Analytics** is now called **Analytics**, and the HDInsight icon and icon color have changed from what is shown in the tutorial.

- In *Step 3*, you must choose a cluster name that does *not* already exist. When you enter your cluster name, Microsoft will check whether that name is available and display a message if it is not. You must create a password. For the **Resource group**, you'll also need to click **Create new** and provide a group name. Leave all other settings in this step as is.

- In *Step 5*: Under **Select a Storage account**, click **Create new** and provide a storage account name containing only lowercase letters and numbers. Like the cluster name, the storage account name must be unique.

When you get to the **Cluster summary** you'll see that Microsoft initially configures the cluster as **Head (2 x D12 v2), Worker (4 x D4 v2)**. At the time of this writing, the estimated cost-per-hour for this configuration was $3.11. This setup uses a total of 6 CPU nodes with 40 cores—far more than we need for demonstration purposes.

You can edit this setup to use fewer CPUs and cores, which also saves money. Let's change the configuration to a four-CPU cluster with 16 cores that uses less powerful computers. In the **Cluster summary**:

1. Click **Edit** to the right of **Cluster size**.
2. Change the **Number of Worker** nodes to 2.
3. Click **Worker node size**, then **View all**, select **D3 v2** (this is the minimum CPU size for Hadoop nodes) and click **Select**.
4. Click **Head node size**, then **View all**, select **D3 v2** and click **Select**.
5. Click **Next** and click **Next** again to return to the **Cluster summary**. Microsoft will validate the new configuration.
6. When the **Create** button is enabled, click it to deploy the cluster.

It takes 20–30 minutes for Microsoft to "spin up" your cluster. During this time, Microsoft is allocating all the resources and software the cluster requires.

After the changes above, our estimated cost for the cluster was $1.18 per hour, based on *average* use for similarly configured clusters. Our actual charges were less than that. If you encounter any problems configuring your cluster, Microsoft provides HDInsight chat-based support at:

```
https://azure.microsoft.com/en-us/resources/knowledge-center/
    technical-chat/
```

17.5.4 Hadoop Streaming

For languages like Python that are not natively supported in Hadoop, you must use **Hadoop streaming** to implement your tasks. In Hadoop streaming, the Python scripts that implement the mapping and reduction steps use the **standard input stream** and **standard output stream** to communicate with Hadoop. Usually, the standard input stream reads from the keyboard and the standard output stream writes to the command line. However, these can be *redirected* (as Hadoop does) to read from other sources and write to other destinations. Hadoop uses the streams as follows:

- Hadoop supplies the input to the mapping script—called the **mapper**. This script reads its input from the standard input stream.

- The mapper writes its results to the standard output stream.

- Hadoop supplies the mapper's output as the input to the reduction script—called the **reducer**—which reads from the standard input stream.

- The reducer writes its results to the standard output stream.

- Hadoop writes the reducer's output to the Hadoop file system (HDFS).

The mapper and reducer terminology used above should sound familiar to you from our discussions of functional-style programming and filter, map and reduce in the "Sequences: Lists and Tuples" chapter.

17.5.5 Implementing the Mapper

In this section, you'll create a mapper script that takes lines of text as input from Hadoop and maps them to *key–value pairs* in which each key is a word, and its corresponding value is 1. The mapper sees each word individually so, as far as it is concerned, there's only one of each word. In the next section, the reducer will summarize these key–value pairs by key, reducing the counts to a single count for each key. By default, Hadoop expects the mapper's output and the reducer's input and output to be in the form of key–value pairs separated by a *tab*.

In the mapper script (length_mapper.py), the notation #! in line 1 tells Hadoop to execute the Python code using python3, rather than the default Python 2 installation. This line must come before all other comments and code in the file. At the time of this writing, Python 2.7.12 and Python 3.5.2 were installed. Note that because the cluster does not have Python 3.6 or higher, you cannot use f-strings in your code.

```
 1  #!/usr/bin/env python3
 2  # length_mapper.py
 3  """Maps lines of text to key-value pairs of word lengths and 1."""
 4  import sys
 5
 6  def tokenize_input():
 7      """Split each line of standard input into a list of strings."""
 8      for line in sys.stdin:
 9          yield line.split()
10
11  # read each line in the the standard input and for every word
12  # produce a key-value pair containing the word, a tab and 1
13  for line in tokenize_input():
14      for word in line:
15          print(str(len(word)) + '\t1')
```

Generator function `tokenize_input` (lines 6–9) reads lines of text from the standard input stream and for each returns a list of strings. For this example, we are not removing punctuation or stop words as we did in the "Natural Language Processing" chapter.

When Hadoop executes the script, lines 13–15 iterate through the lists of strings from `tokenize_input`. For each list (`line`) and for every string (`word`) in that list, line 15 outputs a key–value pair with the word's length as the key, a tab (`\t`) and the value 1, indicating that there is one word (so far) of that length. Of course, there probably are many words of that length. The MapReduce algorithm's reduction step will summarize these key–value pairs, reducing all those with the same key to a single key–value pair with the total count.

17.5.6 Implementing the Reducer

In the reducer script (`length_reducer.py`), function `tokenize_input` (lines 8–11) is a generator function that reads and splits the key–value pairs produced by the mapper. Again, the MapReduce algorithm supplies the standard input. For each line, `tokenize_input` strips any leading or trailing whitespace (such as the terminating newline) and yields a list containing the key and a value.

```
 1  #!/usr/bin/env python3
 2  # length_reducer.py
 3  """Counts the number of words with each length."""
 4  import sys
 5  from itertools import groupby
 6  from operator import itemgetter
 7
 8  def tokenize_input():
 9      """Split each line of standard input into a key and a value."""
10      for line in sys.stdin:
11          yield line.strip().split('\t')
12
13  # produce key-value pairs of word lengths and counts separated by tabs
14  for word_length, group in groupby(tokenize_input(), itemgetter(0)):
15      try:
16          total = sum(int(count) for word_length, count in group)
17          print(word_length + '\t' + str(total))
18      except ValueError:
19          pass  # ignore word if its count was not an integer
```

When the MapReduce algorithm executes this reducer, lines 14–19 use the **groupby** function from the `itertools module` to group all word lengths of the same value:

- The first argument calls `tokenize_input` to get the lists representing the key–value pairs.
- The second argument indicates that the key–value pairs should be grouped based on the element at index 0 in each list—that is the key.

Line 16 totals all the counts for a given key. Line 17 outputs a new key–value pair consisting of the word and its total. The MapReduce algorithm takes all the final word-count outputs and writes them to a file in HDFS—the Hadoop file system.

17.5.7 Preparing to Run the MapReduce Example

Next, you'll upload files to the cluster so you can execute the example. In a Command Prompt, Terminal or shell, change to the folder containing your mapper and reducer scripts and the `RomeoAndJuliet.txt` file. We assume all three are in this chapter's ch17 examples folder, so be sure to copy your `RomeoAndJuliet.txt` file to this folder first.

Copying the Script Files to the HDInsight Hadoop Cluster

Enter the following command to upload the files. Be sure to replace *YourClusterName* with the cluster name you specified when setting up the Hadoop cluster and press *Enter* only after you've typed the entire command. The colon in the following command is required and indicates that you'll supply your cluster password when prompted. At that prompt, type the password you specified when setting up the cluster, then press *Enter*:

```
scp length_mapper.py length_reducer.py RomeoAndJuliet.txt
    sshuser@YourClusterName-ssh.azurehdinsight.net:
```

The first time you do this, you'll be asked for security reasons to confirm that you trust the target host (that is, Microsoft Azure).

Copying RomeoAndJuliet into the Hadoop File System

For Hadoop to read the contents of `RomeoAndJuliet.txt` and supply the lines of text to your mapper, you must first copy the file into Hadoop's file system. First, you must use ssh[37] to log into your cluster and access its command line. In a Command Prompt, Terminal or shell, execute the following command. Be sure to replace *YourClusterName* with your cluster name. Again, you'll be prompted for your cluster password:

```
ssh sshuser@YourClusterName-ssh.azurehdinsight.net
```

For this example, we'll use the following Hadoop command to copy the text file into the already existing folder `/examples/data` that the cluster provides for use with Microsoft's Azure Hadoop tutorials. Again, press *Enter* only when you've typed the entire command:

```
hadoop fs -copyFromLocal RomeoAndJuliet.txt
    /example/data/RomeoAndJuliet.txt
```

---

37. Windows users: If ssh does not work for you, install and enable it as described at https://blogs.msdn.microsoft.com/powershell/2017/12/15/using-the-openssh-beta-in-windows-10-fall-creators-update-and-windows-server-1709/. After completing the installation, log out and log back in or restart your system to enable ssh.

17.5.8 Running the MapReduce Job

Now you can run the MapReduce job for `RomeoAndJuliet.txt` on your cluster by executing the following command. For your convenience, we provided the text of this command in the file `yarn.txt` with this example, so you can copy and paste it. We reformatted the command here for readability:

```
yarn jar /usr/hdp/current/hadoop-mapreduce-client/hadoop-streaming.jar
    -D mapred.output.key.comparator.class=
       org.apache.hadoop.mapred.lib.KeyFieldBasedComparator
    -D mapred.text.key.comparator.options=-n
    -files length_mapper.py,length_reducer.py
    -mapper length_mapper.py
    -reducer length_reducer.py
    -input /example/data/RomeoAndJuliet.txt
    -output /example/wordlengthsoutput
```

The **yarn command** invokes the Hadoop's YARN ("yet another resource negotiator") tool to manage and coordinate access to the Hadoop resources the MapReduce task uses. The file `hadoop-streaming.jar` contains the Hadoop streaming utility that allows you to use Python to implement the mapper and reducer. The two `-D` options set Hadoop properties that enable it to sort the final key–value pairs by key (`KeyFieldBasedComparator`) in descending order numerically (`-n`; the minus indicates descending order) rather than alphabetically. The other command-line arguments are:

- `-files`—A comma-separated list of file names. Hadoop copies these files to every node in the cluster so they can be executed locally on each node.
- `-mapper`—The name of the mapper's script file.
- `-reducer`—The name of the reducer's script file
- `-input`—The file or directory of files to supply as input to the mapper.
- `-output`—The HDFS directory in which the output will be written. If this folder already exists, an error will occur.

The following output shows some of the feedback that Hadoop produces as the MapReduce job executes. We replaced chunks of the output with … to save space and bolded several lines of interest including:

- The total number of "input paths to process"—the 1 source of input in this example is the `RomeoAndJuliet.txt` file.
- The "number of splits"—2 in this example, based on the number of worker nodes in our cluster.
- The percentage completion information.
- `File System Counters`, which include the numbers of bytes read and written.
- `Job Counters`, which show the number of mapping and reduction tasks used and various timing information.
- `Map-Reduce Framework`, which shows various information about the steps performed.

```
packageJobJar: [] [/usr/hdp/2.6.5.3004-13/hadoop-mapreduce/hadoop-
streaming-2.7.3.2.6.5.3004-13.jar] /tmp/streamjob2764990629848702405.jar
tmpDir=null
...
18/12/05 16:46:25 INFO mapred.FileInputFormat: Total input paths to
process : 1
18/12/05 16:46:26 INFO mapreduce.JobSubmitter: number of splits:2
...
18/12/05 16:46:26 INFO mapreduce.Job: The url to track the job: http://
hn0-paulte.y3nghy5db2kehav5m0opqrjxcb.cx.internal.cloudapp.net:8088/
proxy/application_1543953844228_0025/
...
18/12/05 16:46:35 INFO mapreduce.Job:  map 0% reduce 0%
18/12/05 16:46:43 INFO mapreduce.Job:  map 50% reduce 0%
18/12/05 16:46:44 INFO mapreduce.Job:  map 100% reduce 0%
18/12/05 16:46:48 INFO mapreduce.Job:  map 100% reduce 100%
18/12/05 16:46:50 INFO mapreduce.Job: Job job_1543953844228_0025
completed successfully
18/12/05 16:46:50 INFO mapreduce.Job: Counters: 49
        File System Counters
            FILE: Number of bytes read=156411
            FILE: Number of bytes written=813764
...
        Job Counters
            Launched map tasks=2
            Launched reduce tasks=1
...
        Map-Reduce Framework
            Map input records=5260
            Map output records=25956
            Map output bytes=104493
            Map output materialized bytes=156417
            Input split bytes=346
            Combine input records=0
            Combine output records=0
            Reduce input groups=19
            Reduce shuffle bytes=156417
            Reduce input records=25956
            Reduce output records=19
            Spilled Records=51912
            Shuffled Maps =2
            Failed Shuffles=0
            Merged Map outputs=2
            GC time elapsed (ms)=193
            CPU time spent (ms)=4440
            Physical memory (bytes) snapshot=1942798336
            Virtual memory (bytes) snapshot=8463282176
            Total committed heap usage (bytes)=3177185280
...
18/12/05 16:46:50 INFO streaming.StreamJob: Output directory: /example/
wordlengthsoutput
```

Viewing the Word Counts

Hadoop MapReduce saves its output into HDFS, so to see the actual word counts you must look at the file in HDFS within the cluster by executing the following command:

```
hdfs dfs -text /example/wordlengthsoutput/part-00000
```

Here are the results of the preceding command:

```
18/12/05 16:47:19 INFO lzo.GPLNativeCodeLoader: Loaded native gpl library
18/12/05 16:47:19 INFO lzo.LzoCodec: Successfully loaded & initialized
native-lzo library [hadoop-lzo rev
b5efb3e531bc1558201462b8ab15bb412ffa6b89]
1       1140
2       3869
3       4699
4       5651
5       3668
6       2719
7       1624
8       1062
9       855
10      317
11      189
12      95
13      35
14      13
15      9
16      6
17      3
18      1
23      1
```

Deleting Your Cluster So You Do Not Incur Charges

Caution: Be sure to delete your cluster(s) and associated resources (like storage) so you don't incur additional charges. In the Azure portal, click **All resources** to see your list of resources, which will include the cluster you set up and the storage account you set up. Both can incur charges if you do not delete them. Select each resource and click the **Delete** button to remove it. You'll be asked to confirm by typing yes. For more information, see:

```
https://docs.microsoft.com/en-us/azure/azure-resource-manager/
    resource-group-portal
```

Self Check for Section 17.5

1 *(Fill-In)* Hadoop's key components are _____ for storing massive amounts of data throughout a cluster and _____ for implementing the tasks that process the data.
Answer: HDFS (Hadoop Distributed File System), MapReduce.

2 *(Fill-In)* For learning, development and testing before you commit to cloud-based services, the vendors _____ and _____ offer free downloadable environments with integrated Hadoop ecosystem components.
Answer: Cloudera, Hortonworks.

3 *(Fill-In)* The _____ command launches a MapReduce task.
Answer: yarn.

4 *(Fill-In)* To implement MapReduce tasks in languages like Python that are not natively supported, you must use Hadoop _____ in which the mapper and reducer communicated with Hadoop via the _____.
Answer: streaming, standard input and standard output streams.

5 *(True/False)* Hadoop MapReduce does not place requirements on the format of the mapper's output and the reducer's input and output.
Answer: False. MapReduce expects the mapper's output and the reducer's input and output to be in the form of key–value pairs in which each key and value are separated by a tab.

6 *(True/False)* Hadoop MapReduce keeps the task's final output in main memory for easy access.
Answer: False. In big-data processing, the results typically would not fit in main memory, so Hadoop MapReduce writes its final output to HDFS. To access the results, you must read them from the HDFS folder in which the output was written.

17.6 Spark

In this section, we'll overview Apache Spark. We'll use the Python **PySpark library** and Spark's functional-style filter/map/reduce capabilities to implement a simple word count example that summarizes the word counts in *Romeo and Juliet*.

17.6.1 Spark Overview

When you process truly big data, performance is crucial. Hadoop is geared to disk-based batch processing—reading the data from disk, processing the data and writing the results back to disk. Many big-data applications demand better performance than is possible with disk-intensive operations. In particular, fast streaming applications that require either real-time or near-real-time processing won't work in a disk-based architecture.

History

Spark was initially developed in 2009 at U. C. Berkeley and funded by DARPA (the Defense Advanced Research Projects Agency). Initially, it was created as a distributed execution engine for high-performance machine learning.[38] It uses an **in-memory architecture** that "has been used to sort 100 TB of data 3X faster than Hadoop MapReduce on 1/10th of the machines"[39] and runs some workloads up to 100 times faster than Hadoop.[40] Spark's significantly better performance on batch-processing tasks is leading many companies to replace Hadoop MapReduce with Spark.[41,42,43]

Architecture and Components

Though it was initially developed to run on Hadoop and use Hadoop components like HDFS and YARN, Spark can run standalone on a single computer (typically for learning

---

38. https://gigaom.com/2014/06/28/4-reasons-why-spark-could-jolt-hadoop-into-hyperdrive/.
39. https://spark.apache.org/faq.html.
40. https://spark.apache.org/.
41. https://bigdata-madesimple.com/is-spark-better-than-hadoop-map-reduce/.
42. https://www.datanami.com/2018/10/18/is-hadoop-officially-dead/.
43. https://blog.thecodeteam.com/2018/01/09/changing-face-data-analytics-fast-data-displaces-big-data/.

and testing purposes), standalone on a cluster or using various cluster managers and distributed storage systems. For resource management, Spark runs on Hadoop YARN, Apache Mesos, Amazon EC2 and Kubernetes, and it supports many distributed storage systems, including HDFS, Apache Cassandra, Apache HBase and Apache Hive.[44]

At the core of Spark are **resilient distributed datasets (RDDs)**, which you'll use to process distributed data using functional-style programming. In addition to reading data from disk and writing data to disk, Hadoop uses replication for fault tolerance, which adds even more disk-based overhead. RDDs eliminate this overhead by remaining in memory—using disk only if the data will not fit in memory—and by not replicating data. Spark handles fault tolerance by remembering the steps used to create each RDD, so it can rebuild a given RDD if a cluster node fails.[45]

Spark distributes the operations you specify in Python to the cluster's nodes for parallel execution. Spark streaming enables you to process data as it's received. Spark DataFrames, which are similar to pandas DataFrames, enable you to view RDDs as a collection of named columns. You can use Spark DataFrames with Spark SQL to perform queries on distributed data. Spark also includes **Spark MLlib** (the Spark Machine Learning Library), which enables you to perform machine-learning algorithms, like those you learned in the Chapters 15 and 16. We'll use RDDs, Spark streaming, DataFrames and Spark SQL in the next few examples. You'll explore Spark MLlib in the chapter exercises.

Providers

Hadoop providers typically also provide Spark support. In addition to the providers listed in Section 17.5, there are Spark-specific vendors like Databricks. They provide a "zero-management cloud platform built around Spark."[46] Their website also is an excellent resource for learning Spark. The paid Databricks platform runs on Amazon AWS or Microsoft Azure. Databricks also provides a free Databricks Community Edition, which is a great way to get started with both Spark and the Databricks environment. An exercise at the end of the chapter asks you to research Databricks Community Edition, then use it to reimplement the Spark examples in the upcoming sections.

17.6.2 Docker and the Jupyter Docker Stacks

In this section, we'll show how to download and execute a Docker stack containing Spark and the PySpark module for accessing Spark from Python. You'll write the Spark example's code in a Jupyter Notebook. First, let's overview Docker.

Docker

Docker is a tool for packaging software into **containers** (also called **images**) that bundle *everything* required to execute that software across platforms. Some software packages we use in this chapter require complicated setup and configuration. For many of these, there are preexisting Docker containers that you can download for free and execute locally on your desktop or notebook computers. This makes Docker a great way to help you get started with new technologies quickly and conveniently.

---

44. http://spark.apache.org/.
45. https://spark.apache.org/research.html.
46. https://databricks.com/product/faq.

Docker also helps with *reproducibility* in research and analytics studies. You can create custom Docker containers that are configured with the versions of every piece of software and every library you used in your study. This would enable others to recreate the environment you used, then reproduce your work, and will help you reproduce your results at a later time. We'll use Docker in this section to download and execute a Docker container that's preconfigured to run Spark applications.

Installing Docker
You can install Docker for Windows 10 Pro or macOS at:

> https://www.docker.com/products/docker-desktop

On Windows 10 Pro, you must allow the "Docker for Windows.exe" installer to make changes to your system to complete the installation process. To do so, click **Yes** when Windows asks if you want to allow the installer to make changes to your system.[47] Windows 10 Home users must use Virtual Box as described at:

> https://docs.docker.com/machine/drivers/virtualbox/

Linux users should install Docker Community Edition as described at:

> https://docs.docker.com/install/overview/

For a general overview of Docker, read the **Getting started** guide at:

> https://docs.docker.com/get-started/

Jupyter Docker Stacks
The Jupyter Notebooks team has preconfigured several Jupyter "Docker stacks" containers for common Python development scenarios. Each enables you to use Jupyter Notebooks to experiment with powerful capabilities without having to worry about complex software setup issues. In each case, you can open JupyterLab in your web browser, open a notebook in JupyterLab and start coding. JupyterLab also provides a **Terminal window** that you can use in your browser like your computer's Terminal, Anaconda Command Prompt or shell. Everything we've shown you in IPython to this point can be executed using IPython in JupyterLab's Terminal window.

We'll use the `jupyter/pyspark-notebook` Docker stack, which is preconfigured with everything you need to create and test Apache Spark apps on your computer. When combined with installing other Python libraries we've used throughout the book, you can implement most of this book's examples using this container. To learn more about the available Docker stacks, visit:

> https://jupyter-docker-stacks.readthedocs.io/en/latest/index.html

Run Jupyter Docker Stack
Before performing the next step, ensure that JupyterLab is not currently running on your computer. Let's download and run the `jupyter/pyspark-notebook` Docker stack. To ensure that you do not lose your work when you close the Docker container, we'll attach

---

47. Some Windows users might have to follow the instructions under **Allow specific apps to make changes to controlled folders** at https://docs.microsoft.com/en-us/windows/security/threat-protection/windows-defender-exploit-guard/customize-controlled-folders-exploit-guard.

a local file-system folder to the container and use it to save your notebook—Windows users should replace \ with ^. :

```
docker run -p 8888:8888 -p 4040:4040 -it --user root \
    -v fullPathToTheFolderYouWantToUse:/home/jovyan/work \
    jupyter/pyspark-notebook:14fdfbf9cfc1 start.sh jupyter lab
```

The first time you run the preceding command, Docker will download the Docker container named:

```
jupyter/pyspark-notebook:14fdfbf9cfc1
```

The notation ":14fdfbf9cfc1" indicates the specific jupyter/pyspark-notebook container to download. At the time of this writing, 14fdfbf9cfc1 was the newest version of the container. Specifying the version as we did here helps with *reproducibility*. Without the ":14fdfbf9cfc1" in the command, Docker will download the *latest* version of the container, which might contain different software versions and might not be compatible with the code you're trying to execute. The Docker container is nearly 6GB, so the initial download time will depend on your Internet connection's speed.

Opening JupyterLab in Your Browser
Once the container is downloaded and running, you'll see a statement in your Command Prompt, Terminal or shell window like:

```
Copy/paste this URL into your browser when you connect for the first
time, to login with a token:
    http://(bb00eb337630 or 127.0.0.1):8888/?token=
        9570295e90ee94ecef75568b95545b7910a8f5502e6f5680
```

Copy the long hexadecimal string (the string on your system will differ from this one):

```
9570295e90ee94ecef75568b95545b7910a8f5502e6f5680
```

then open http://localhost:8888/lab in your browser (localhost corresponds to 127.0.0.1 in the preceding output) and *paste* your token in the **Password or token** field. Click **Log in** to be taken to the JupyterLab interface. If you accidentally close your browser, go to http://localhost:8888/lab to continue your session.

In JupyterLab running in this Docker container, the work folder in the **Files** tab at the left side of the JupyterLab interface represents the folder you attached to the container in the docker run command's -v option. From this folder, you can open the notebook files we provide for you. Any new notebooks or other files you create will be saved to this folder by default. Because the Docker container's work folder is connected to a folder on your computer, any files you create in JupyterLab will remain on your computer, even if you decide to delete the Docker container.

Accessing the Docker Container's Command Line
Each Docker container has a command-line interface like the one you've used to run IPython throughout this book. Via this interface, you can install Python packages into the Docker container and even use IPython as you've done previously.

Open a separate Anaconda Command Prompt, Terminal or shell and list the currently running Docker containers with the command:

```
docker ps
```

The output of this command is wide, so the lines of text will likely wrap, as in:

```
CONTAINER ID        IMAGE                                    COMMAND
         CREATED           STATUS            PORTS
    NAMES
f54f62b7e6d5        jupyter/pyspark-notebook:14fdfbf9cfc1    "tini -g --
/bin/bash"  2 minutes ago     Up 2 minutes      0.0.0.0:8888->8888/tcp
    friendly_pascal
```

In the last line of our system's output under the column head NAMES in the third line is the name that Docker randomly assigned to the running container—friendly_pascal—the name on your system will differ. To access the container's command line, execute the following command, replacing *container_name* with the running container's name:

```
docker exec -it container_name /bin/bash
```

The Docker container uses Linux under the hood, so you'll see a Linux prompt where you can enter commands.

The app in this section will use features of the NLTK and TextBlob libraries you used in the "Natural Language Processing" chapter. Neither is preinstalled in the Jupyter Docker stacks. To install NLTK and TextBlob enter the command:

```
conda install -c conda-forge nltk textblob
```

Stopping and Restarting a Docker Container

Every time you start a container with docker run, Docker gives you a new instance that does *not* contain any libraries you installed previously. For this reason, you should keep track of your container name, so you can use it from another Anaconda Command Prompt, Terminal or shell window to stop the container and restart it. The command

```
docker stop container_name
```

will shut down the container. The command

```
docker restart container_name
```

will restart the container. Docker also provides a GUI app called Kitematic that you can use to manage your containers, including stopping and restarting them. You can get the app from https://kitematic.com/ and access it through the Docker menu. The following user guide overviews how to manage containers with the tool:

> https://docs.docker.com/kitematic/userguide/

17.6.3 Word Count with Spark

In this section, we'll use Spark's filtering, mapping and reducing capabilities to implement a simple word count example that summarizes the words in *Romeo and Juliet*. You can work with the existing notebook named RomeoAndJulietCounter.ipynb in the Spark-WordCount folder (into which you should copy your RomeoAndJuliet.txt file from the "Natural Language Processing" chapter), or you can create a new notebook, then enter and execute the snippets we show.

Loading the NLTK Stop Words
In this app, we'll use techniques you learned in the "Natural Language Processing" chapter to eliminate stop words from the text before counting the words' frequencies. First, download the NLTK stop words:

```
[1]: import nltk
     nltk.download('stopwords')
[nltk_data] Downloading package stopwords to /home/jovyan/nltk_data...
[nltk_data]   Package stopwords is already up-to-date!
[1]: True
```

Next, load the stop words:

```
[2]: from nltk.corpus import stopwords
     stop_words = stopwords.words('english')
```

Configuring a SparkContext
A **SparkContext** (from module pyspark) object gives you access to Spark's capabilities in Python. Many Spark environments create the SparkContext for you, but in the Jupyter pyspark-notebook Docker stack, you must create this object.

First, let's specify the configuration options by creating a **SparkConf** object (from module pyspark). The following snippet calls the object's **setAppName** method to specify the Spark application's name and calls the object's **setMaster** method to specify the Spark cluster's URL. The URL 'local[*]' indicates that Spark is executing on your local computer (as opposed to a cloud-based cluster), and the asterisk indicates that Spark should run our code using the same number of *threads* as there are cores on the computer:

```
[3]: from pyspark import SparkConf
     configuration = SparkConf().setAppName('RomeoAndJulietCounter')\
                                .setMaster('local[*]')
```

Threads enable a single node cluster to execute portions of the Spark tasks *concurrently* to simulate the parallelism that Spark clusters provide. When we say that two tasks are operating concurrently, we mean that they're both making progress at once—typically by executing a task for a short burst of time, then allowing another task to execute. When we say that two tasks are operating in *parallel*, we mean that they're executing simultaneously, which is one of the key benefits of Hadoop and Spark executing on cloud-based clusters of computers.

Next, create the SparkContext, passing the SparkConf as its argument:

```
[4]: from pyspark import SparkContext
     sc = SparkContext(conf=configuration)
```

Reading the Text File and Mapping It to Words
You work with a SparkContext using functional-style programming techniques, like filtering, mapping and reduction, applied to a **resilient distributed dataset (RDD)**. An RDD takes data stored throughout a cluster in the Hadoop file system and enables you to specify a series of processing steps to transform the data in the RDD. These processing steps are *lazy* (Chapter 5)—they do not perform any work until you indicate that Spark should process the task.

The following snippet specifies three steps:

- `SparkContext` method **textFile** loads the lines of text from `RomeoAndJuliet.txt` and returns it as an **RDD** (from module `pyspark`) of strings that represent each line.

- RDD method **map** uses its `lambda` argument to remove all punctuation with TextBlob's `strip_punc` function and to convert each line of text to lowercase. This method returns a new RDD on which you can specify additional tasks to perform.

- RDD method **flatMap** uses its `lambda` argument to map each line of text into its words and produces a single list of words, rather than the individual lines of text. The result of `flatMap` is a new RDD representing all the words in *Romeo and Juliet*.

```
[5]: from textblob.utils import strip_punc
     tokenized = sc.textFile('RomeoAndJuliet.txt')\
                   .map(lambda line: strip_punc(line, all=True).lower())\
                   .flatMap(lambda line: line.split())
```

Removing the Stop Words
Next, let's use RDD method **filter** to create a new RDD with no stop words remaining:

```
[6]: filtered = tokenized.filter(lambda word: word not in stop_words)
```

Counting Each Remaining Word
Now that we have only the non-stop-words, we can count the number of occurrences of each word. To do so, we first map each word to a tuple containing the word and a count of 1. This is similar to what we did in Hadoop MapReduce. Spark will distribute the reduction task across the cluster's nodes. On the resulting RDD, we then call the method **reduceByKey**, passing the `operator` module's `add` function as an argument. This tells method `reduceByKey` to *add* the counts for tuples that contain the same word (the key):

```
[7]: from operator import add
     word_counts = filtered.map(lambda word: (word, 1)).reduceByKey(add)
```

Locating Words with Counts Greater Than or Equal to 60
Since there are hundreds of words in *Romeo and Juliet*, let's filter the RDD to keep only those words with 60 or more occurrences:

```
[8]: filtered_counts = word_counts.filter(lambda item: item[1] >= 60)
```

Sorting and Displaying the Results
At this point, we've specified all the steps to count the words. When you call RDD method **collect**, Spark initiates all the processing steps we specified above and returns a list containing the final results—in this case, the tuples of words and their counts. From your perspective, everything appears to execute on one computer. However, if the `SparkContext` is configured to use a cluster, Spark will divide the tasks among the cluster's worker nodes for you. In the following snippet, sort in descending order (`reverse=True`) the list of tuples by their counts (`itemgetter(1)`).

The following snippet calls method `collect` to obtain the results and sorts those results in descending order by word count:

```
[9]: from operator import itemgetter
     sorted_items = sorted(filtered_counts.collect(),
                           key=itemgetter(1), reverse=True)
```

Finally, let's display the results. First, we determine the word with the most letters so we can right-align all the words in a field of that length, then we display each word and its count:

```
[10]: max_len = max([len(word) for word, count in sorted_items])
      for word, count in sorted_items:
          print(f'{word:>{max_len}}: {count}')
[10]:    romeo: 298
          thou: 277
        juliet: 178
           thy: 170S
         nurse: 146
       capulet: 141
          love: 136
          thee: 135
         shall: 110
          lady: 109
         friar: 104
          come: 94
      mercutio: 83
          good: 80
      benvolio: 79
         enter: 75
            go: 75
          i'll: 71
        tybalt: 69
         death: 69
         night: 68
      lawrence: 67
           man: 65
          hath: 64
           one: 60
```

17.6.4 Spark Word Count on Microsoft Azure

As we said previously, we want to expose you to both tools you can use for free and real-world development scenarios. In this section, you'll implement the Spark word-count example on a Microsoft Azure HDInsight Spark cluster.

Create an Apache Spark Cluster in HDInsight Using the Azure Portal
The following link explains how to set up a Spark cluster using the HDInsight service:

> https://docs.microsoft.com/en-us/azure/hdinsight/spark/apache-spark-jupyter-spark-sql-use-portal

While following the **Create an HDInsight Spark cluster** steps, note the same issues we listed in the Hadoop cluster setup earlier in this chapter and for the **Cluster type** select **Spark**.

Again, the default cluster configuration provides more resources than you need for our examples. So, in the **Cluster summary**, perform the steps shown in the Hadoop cluster setup to change the number of worker nodes to 2 and to configure the worker and head nodes to use **D3 v2** computers. When you click **Create**, it takes 20 to 30 minutes to configure and deploy your cluster.

Install Libraries into a Cluster
If your Spark code requires libraries that are not installed in the HDInsight cluster, you'll need to install them. To see what libraries are installed by default, you can use ssh to log into your cluster (as we showed earlier in the chapter) and execute the command:

    /usr/bin/anaconda/envs/py35/bin/conda list

Since your code will execute on multiple cluster nodes, libraries must be installed on *every* node. Azure requires you to create a Linux shell script that specifies the commands to install the libraries. When you submit that script to Azure, it validates the script, then executes it on every node. Linux shell scripts are beyond this book's scope, and the script must be hosted on a web server from which Azure can download the file. So, we created an install script for you that installs the libraries we use in the Spark examples. Perform the following steps to install these libraries:

1. In the Azure portal, select your cluster.
2. In the list of items under the cluster's search box, click **Script Actions**.
3. Click **Submit new** to configure the options for the library installation script. For the **Script type** select **Custom**, for the **Name** specify libraries and for the **Bash script URI** use:

    http://deitel.com/bookresources/IntroToPython/install_libraries.sh

4. Check both **Head** and **Worker** to ensure that the script installs the libraries on all the nodes.
5. Click **Create**.

When the cluster finishes executing the script, if it executed successfully, you'll see a green check next to the script name in the list of script actions. Otherwise, Azure will notify you that there were errors.

Copying RomeoAndJuliet.txt to the HDInsight Cluster
As you did in the Hadoop demo, let's use the scp command to upload to the cluster the RomeoAndJuliet.txt file you used in the "Natural Language Processing" chapter. In a Command Prompt, Terminal or shell, change to the folder containing the file (we assume this chapter's ch17 folder), then enter the following command. Replace *YourClusterName* with the name you specified when creating your cluster and press *Enter* only when you've typed the entire command. The colon is required and indicates that you'll supply your cluster password when prompted. At that prompt, type the password you specified when setting up the cluster, then press *Enter*:

    scp RomeoAndJuliet.txt sshuser@*YourClusterName*-ssh.azurehdinsight.net:

Next, use ssh to log into your cluster and access its command line. In a Command Prompt, Terminal or shell, execute the following command. Be sure to replace *YourClusterName* with your cluster name. Again, you'll be prompted for your cluster password:

    ssh sshuser@*YourClusterName*-ssh.azurehdinsight.net

To work with the RomeoAndJuliet.txt file in Spark, first use the ssh session to copy the file into the cluster's Hadoop's file system by executing the following command. Once again, we'll use the already existing folder /examples/data that Microsoft includes for use with HDInsight tutorials. Again, press *Enter* only when you've typed the entire command:

```
hadoop fs -copyFromLocal RomeoAndJuliet.txt
    /example/data/RomeoAndJuliet.txt
```

Accessing Jupyter Notebooks in HDInsight
At the time of this writing, HDInsight uses the *old* Jupyter Notebook interface, rather than the newer JupyterLab interface shown earlier. For a quick overview of the old interface see:

> https://jupyter-notebook.readthedocs.io/en/stable/examples/Notebook/Notebook%20Basics.html

To access Jupyter Notebooks in HDInsight, in the Azure portal select **All resources**, then your cluster. In the **Overview** tab, select **Jupyter notebook** under **Cluster dashboards**. This opens a web browser window and asks you to log in. Use the username and password you specified when setting up the cluster. If you did not specify a username, the default is `admin`. Once you log in, Jupyter displays a folder containing `PySpark` and `Scala` subfolders. These contain Python and Scala Spark tutorials.

Uploading the `RomeoAndJulietCounter.ipynb` Notebook
You can create new notebooks by clicking **New** and selecting PySpark3, or you can upload existing notebooks from your computer. For this example, let's upload the previous section's `RomeoAndJulietCounter.ipynb` notebook and modify it to work with Azure. To do so, click the **Upload** button, navigate to the ch17 example folder's `SparkWordCount` folder, select `RomeoAndJulietCounter.ipynb` and click **Open**. This displays the file in the folder with an **Upload** button to its right. Click that button to place the notebook in the current folder. Next, click the notebook's name to open it in a new browser tab. Jupyter will display a **Kernel not found** dialog. Select **PySpark3** and click **OK**. Do not run any cells yet.

Modifying the Notebook to Work with Azure
Perform the following steps, executing each cell as you complete the step:

1. The HDInsight cluster will not allow NLTK to store the downloaded stop words in NLTK's default folder because it's part of the system's protected folders. In the first cell, modify the call `nltk.download('stopwords')` as follows to store the stop words in the current folder (`'.'`):

    ```
    nltk.download('stopwords', download_dir='.')
    ```

    When you execute the first cell, `Starting Spark application` appears below the cell while HDInsight sets up a `SparkContext` object named `sc` for you. When this task is complete, the cell's code executes and downloads the stop words.

2. In the second cell, before loading the stop words, you must tell NLTK that they're located in the current folder. Add the following statement after the `import` statement to tell NLTK to search for its data in the current folder:

    ```
    nltk.data.path.append('.')
    ```

3. Because HDInsight sets up the `SparkContext` object for you, the third and fourth cells of the original notebook are not needed, so you can delete them. To do so, either click inside it and select **Delete Cells** from Jupyter's **Edit** menu, or click in the white margin to the cell's left and type dd.

4. In the next cell, you must specify the location of `RomeoAndJuliet.txt` in the underlying Hadoop file system. Replace the string `'RomeoAndJuliet.txt'` with the string

    `'wasb:///example/data/RomeoAndJuliet.txt'`

    The notation `wasb:///` indicates that `RomeoAndJuliet.txt` is stored in a Windows Azure Storage Blob (WASB)—Azure's interface to the HDFS file system.

5. Because Azure currently uses Python 3.5.x, it does not support f-strings. So, in the last cell, replace the f-string with the following older-style Python string formatting using the string method `format`:

    `print('{:>{width}}: {}'.format(word, count, width=max_len))`

You'll see the same final results as in the previous section.

Caution: Be sure to delete your cluster and other resources when you're done with them, so you do not incur charges. For more information, see:

    https://docs.microsoft.com/en-us/azure/azure-resource-manager/
    resource-group-portal

Note that when you delete your Azure resources, *your notebooks will be deleted as well.* You can download the notebook you just executed by selecting **File > Download as > Notebook (.ipynb)** in Jupyter.

Self Check for Section 17.6

1. *(Discussion)* How does Docker help with reproducibility?
Answer: Docker enables you to create custom Docker containers that are configured with the versions of every piece of software and every library used in your study. This enables others to recreate the environment you used, then prove your work.

2. *(Fill-In)* Spark uses a(n) _____ architecture for performance.
Answer: in-memory

3. *(True/False)* Hadoop and Spark both implement fault tolerance by replicating data.
Answer: False. Hadoop implements fault tolerance by replicating data across nodes. Spark implements fault tolerance by remembering the steps used to create each RDD so it can be rebuilt if a cluster node fails.

4. *(True/False)* Spark's significantly better performance on batch-processing tasks is leading many companies to replace Hadoop MapReduce with Spark.
Answer: True.

5. *(True/False)* You work with a `SparkContext` using functional-style filter, map and reduce operations, applied to a resilient distributed dataset (RDD).
Answer: True.

6. *(Discussion)* Assuming that `sc` is a `SparkContext`, what does the following code do? Are any results produced when the statement completes?

    ```
    from textblob.utils import strip_punc
    tokenized = sc.textFile('RomeoAndJuliet.txt')\
            .map(lambda line: strip_punc(line, all=True).lower())\
            .flatMap(lambda line: line.split())
    ```

Answer: This code first creates an RDD from a text file. Next it uses RDD method map to produce a new RDD containing the lines of text with punctuation removed and in all lowercase letters. Finally, it produces another new RDD representing the individual words in all the lines. This statement specifies only processing steps, which are lazy, so no results will be produced until you call an RDD method like collect that initiates the processing steps.

17.7 Spark Streaming: Counting Twitter Hashtags Using the pyspark-notebook Docker Stack

In this section, you'll create and run a Spark streaming application in which you'll receive a stream of tweets on the topic(s) you specify and summarize the top-20 hashtags in a bar chart that updates every 10 seconds. For this purpose of this example, you'll use the Jupyter Docker container from the first Spark example.

There are two parts to this example. First, using the techniques from the "Data Mining Twitter" chapter, you'll create a script that streams tweets from Twitter. Then, we'll use Spark streaming in a Jupyter Notebook to read the tweets and summarize the hashtags.

The two parts will communicate with one another via networking **sockets**—a low-level view of *client/server networking* in which a *client* app communicates with a *server* app over a network using techniques similar to file I/O. A program can read from a socket or write to a socket similarly to reading from a file or writing to a file. The socket represents one endpoint of a connection. In this case, the *client* will be a Spark application, and the *server* will be a script that receives streaming tweets and sends them to the Spark app.

Launching the Docker Container and Installing Tweepy

For this example, you'll install the Tweepy library into the Jupyter Docker container. Follow Section 17.6.2's instructions for launching the container and installing Python libraries into it. Use the following command to install Tweepy:

```
pip install tweepy
```

17.7.1 Streaming Tweets to a Socket

The script starttweetstream.py contains a modified version of the TweetListener class from the "Data Mining Twitter" chapter. It streams the specified number of tweets and sends them to a socket on the local computer. When the tweet limit is reached, the script closes the socket. You've already used Twitter streaming, so we'll focus only on what's new. Ensure that the file keys.py (in the ch17 folder's SparkHashtagSummarizer subfolder) contains your Twitter credentials.

Executing the Script in the Docker Container

In this example, you'll use JupyterLab's Terminal window to execute starttweetstream.py in one tab, then use a notebook to perform the Spark task in another tab. With the Jupyter pyspark-notebook Docker container running, open

```
http://localhost:8888/lab
```

in your web browser. In JupyterLab, select **File > New > Terminal** to open a new tab containing a **Terminal**. This is a Linux-based command line. Typing the ls command and pressing *Enter* lists the current folder's contents. By default, you'll see the container's work folder.

To execute starttweetstream.py, you must first navigate to the SparkHashtagSummarizer folder with the command[48]:

```
cd work/SparkHashtagSummarizer
```

You can now execute the script with the command of the form

```
ipython starttweetstream.py number_of_tweets search_terms
```

where *number_of_tweets* specifies the total number of tweets to process and *search_terms* one or more space-separated strings to use for filtering tweets. For example, the following command would stream 1000 tweets about football:

```
ipython starttweetstream.py 1000 football
```

At this point, the script will display "Waiting for connection" and will wait until Spark connects to begin streaming the tweets.

starttweetstream.py import Statements

For discussion purposes, we've divided starttweetstream.py into pieces. First, we import the modules used in the script. The Python Standard Library's **socket module** provides the capabilities that enable Python apps to communicate via sockets.

```
1  # starttweetstream.py
2  """Script to get tweets on topic(s) specified as script argument(s)
3     and send tweet text to a socket for processing by Spark."""
4  import keys
5  import socket
6  import sys
7  import tweepy
8
```

Class TweetListener

Once again, you've seen most of the code in class TweetListener, so we focus only on what's new here:

- Method __init__ (lines 12–17) now receives a connection parameter representing the socket and stores it in the self.connection attribute. We use this socket to send the hashtags to the Spark application.

- In method on_status (lines 24–44), lines 27–32 extract the hashtags from the Tweepy Status object, convert them to lowercase and create a space-separated string of the hashtags to send to Spark. The key statement is line 39:

  ```
  self.connection.send(hashtags_string.encode('utf-8'))
  ```

  which uses the connection object's **send** method to send the tweet text to whatever application is reading from that socket. Method send expects as its argument a sequence of bytes. The string method call encode('utf-8') converts the string to bytes. Spark will automatically read the bytes and reconstruct the strings.

---

48. Windows users should note that Linux uses / rather than \ to separate folders and that file and folder names are case sensitive.

17.7 Counting Twitter Hashtags Using the pyspark-notebook Docker Stack

```
 9  class TweetListener(tweepy.StreamListener):
10      """Handles incoming Tweet stream."""
11
12      def __init__(self, api, connection, limit=10000):
13          """Create instance variables for tracking number of tweets."""
14          self.connection = connection
15          self.tweet_count = 0
16          self.TWEET_LIMIT = limit  # 10,000 by default
17          super().__init__(api)  # call superclass's init
18
19      def on_connect(self):
20          """Called when your connection attempt is successful, enabling
21          you to perform appropriate application tasks at that point."""
22          print('Successfully connected to Twitter\n')
23
24      def on_status(self, status):
25          """Called when Twitter pushes a new tweet to you."""
26          # get the hashtags
27          hashtags = []
28
29          for hashtag_dict in status.entities['hashtags']:
30              hashtags.append(hashtag_dict['text'].lower())
31
32          hashtags_string = ' '.join(hashtags) + '\n'
33          print(f'Screen name: {status.user.screen_name}:')
34          print(f'   Hashtags: {hashtags_string}')
35          self.tweet_count += 1  # track number of tweets processed
36
37          try:
38              # send requires bytes, so encode the string in utf-8 format
39              self.connection.send(hashtags_string.encode('utf-8'))
40          except Exception as e:
41              print(f'Error: {e}')
42
43          # if TWEET_LIMIT is reached, return False to terminate streaming
44          return self.tweet_count != self.TWEET_LIMIT
45
46      def on_error(self, status):
47          print(status)
48          return True
49
```

Main Application

Lines 50–80 execute when you run the script. You've connected to Twitter to stream tweets previously, so here we discuss only what's new in this example.

Line 51 gets the number of tweets to process by converting the command-line argument sys.argv[1] to an integer. Recall that element 0 represents the script's name.

```
50  if __name__ == '__main__':
51      tweet_limit = int(sys.argv[1])  # get maximum number of tweets
```

Line 52 calls the socket module's **socket function**, which returns a socket object that we'll use to wait for a connection from the Spark application.

```
52      client_socket = socket.socket()  # create a socket
53
```

Line 55 calls the socket object's **bind method** with a tuple containing the hostname or IP address of the computer and the port number on that computer. Together these represent where this script will wait for an initial connection from another app:

```
54    # app will use localhost (this computer) port 9876
55    client_socket.bind(('localhost', 9876))
56
```

Line 58 calls the socket's **listen method**, which causes the script to *wait* until a connection is received. This is the statement that prevents the Twitter stream from starting until the Spark application connects.

```
57    print('Waiting for connection')
58    client_socket.listen()  # wait for client to connect
59
```

Once the Spark application connects, line 61 calls socket method **accept**, which accepts the connection. This method returns a tuple containing a new socket object that the script will use to communicate with the Spark application and the IP address of the Spark application's computer.

```
60    # when connection received, get connection/client address
61    connection, address = client_socket.accept()
62    print(f'Connection received from {address}')
63
```

Next, we authenticate with Twitter and start the stream. Lines 73–74 set up the stream, passing the socket object connection to the TweetListener so that it can use the socket to send hashtags to the Spark application.

```
64    # configure Twitter access
65    auth = tweepy.OAuthHandler(keys.consumer_key, keys.consumer_secret)
66    auth.set_access_token(keys.access_token, keys.access_token_secret)
67
68    # configure Tweepy to wait if Twitter rate limits are reached
69    api = tweepy.API(auth, wait_on_rate_limit=True,
70                    wait_on_rate_limit_notify=True)
71
72    # create the Stream
73    twitter_stream = tweepy.Stream(api.auth,
74        TweetListener(api, connection, tweet_limit))
75
76    # sys.argv[2] is the first search term
77    twitter_stream.filter(track=sys.argv[2:])
78
```

Finally, lines 79–80 call the **close** method on the socket objects to release their resources.

```
79    connection.close()
80    client_socket.close()
```

17.7.2 Summarizing Tweet Hashtags; Introducing Spark SQL

In this section, you'll use Spark streaming to read the hashtags sent via a socket by the script starttweetstream.py and summarize the results. You can either create a new notebook and enter the code you see here or load the `hashtagsummarizer.ipynb` notebook we provide in the `ch17` examples folder's `SparkHashtagSummarizer` subfolder.

Importing the Libraries

First, let's import the libraries used in this notebook. We'll explain the pyspark classes as we use them. From `IPython`, we imported the `display` module, which contains classes and utility functions that you can use in Jupyter. In particular, we'll use the `clear_output` function to remove an existing chart before displaying a new one:

```
[1]: from pyspark import SparkContext
     from pyspark.streaming import StreamingContext
     from pyspark.sql import Row, SparkSession
     from IPython import display
     import matplotlib.pyplot as plt
     import seaborn as sns
     %matplotlib inline
```

This Spark application summarizes hashtags in 10-second batches. After processing each batch, it displays a Seaborn barplot. The IPython magic

```
%matplotlib inline
```

indicates that Matplotlib-based graphics should be displayed in the notebook rather than in their own windows. Recall that Seaborn uses Matplotlib.

We've used several IPython magics throughout the book. There are many magics specifically for use in Jupyter Notebooks. For the complete list of magics see:

> https://ipython.readthedocs.io/en/stable/interactive/magics.html

Utility Function to Get the SparkSession

As you'll soon see, you can use **Spark SQL** to query data in resilient distributed datasets (RDDs). Spark SQL uses a Spark **DataFrame** to get a table view of the underlying RDDs. A **SparkSession** (module `pyspark.sql`) is used to create a `DataFrame` from an RDD.

There can be only one SparkSession object per Spark application. The following function, which we borrowed from the *Spark Streaming Programming Guide*,[49] defines the correct way to get a `SparkSession` instance if it already exists or to create one if it does not yet exist:[50]

```
[2]: def getSparkSessionInstance(sparkConf):
         """Spark Streaming Programming Guide's recommended method
            for getting an existing SparkSession or creating a new one."""
         if ("sparkSessionSingletonInstance" not in globals()):
             globals()["sparkSessionSingletonInstance"] = SparkSession \
                 .builder \
                 .config(conf=sparkConf) \
                 .getOrCreate()
         return globals()["sparkSessionSingletonInstance"]
```

---

49. https://spark.apache.org/docs/latest/streaming-programming-guide.html#dataframe-and-sql-operations.
50. Because this function was borrowed from the *Spark Streaming Programming Guide*'s **DataFrame and SQL Operations** section (https://spark.apache.org/docs/latest/streaming-programming-guide.html#dataframe-and-sql-operations), we did not rename it to use Python's standard function naming style, and we did not use single quotes to delimit strings.

Utility Function to Display a Barchart Based on a Spark DataFrame

We call function `display_barplot` after Spark processes each batch of hashtags. Each call clears the previous Seaborn barplot, then displays a new one based on the Spark DataFrame it receives. First, we call the Spark DataFrame's **toPandas** method to convert it to a pandas DataFrame for use with Seaborn. Next, we call the **clear_output** function from the IPython.display module. The keyword argument `wait=True` indicates that the function should remove the prior graph (if there is one), but only once the new graph is ready to display. The rest of the code in the function uses standard Seaborn techniques we've shown previously. The function call `sns.color_palette('cool', 20)` selects twenty colors from the Matplotlib `'cool'` color palette:

```
[3]: def display_barplot(spark_df, x, y, time, scale=2.0, size=(16, 9)):
         """Displays a Spark DataFrame's contents as a bar plot."""
         df = spark_df.toPandas()
         
         # remove prior graph when new one is ready to display
         display.clear_output(wait=True)
         print(f'TIME: {time}')
         
         # create and configure a Figure containing a Seaborn barplot
         plt.figure(figsize=size)
         sns.set(font_scale=scale)
         barplot = sns.barplot(data=df, x=x, y=y,
                              palette=sns.color_palette('cool', 20))
         
         # rotate the x-axis labels 90 degrees for readability
         for item in barplot.get_xticklabels():
             item.set_rotation(90)
         
         plt.tight_layout()
         plt.show()
```

Utility Function to Summarize the Top-20 Hashtags So Far

In Spark streaming, a **DStream** is a sequence of RDDs each representing a mini-batch of data to process. As you'll soon see, you can specify a function that is called to perform a task for every RDD in the stream. In this app, the function count_tags will summarize the hashtag counts in a given RDD, add them to the current totals (maintained by the SparkSession), then display an updated top-20 barplot so that we can see how the top-20 hashtags are changing over time.[51] For discussion purposes, we've broken this function into smaller pieces. First, we get the SparkSession by calling the utility function getSparkSessionInstance with the SparkContext's configuration information. Every RDD has access to the SparkContext via the context attribute:

```
[4]: def count_tags(time, rdd):
         """Count hashtags and display top-20 in descending order."""
         try:
             # get SparkSession
             spark = getSparkSessionInstance(rdd.context.getConf())
```

---

51. When this function gets called the first time, you might see an exception's error message display if no tweets with hashtags have been received yet. This is because we simply display the error message in the standard output. That message will disappear as soon as there are tweets with hashtags.

17.7 Counting Twitter Hashtags Using the pyspark-notebook Docker Stack

Next, we call the RDD's map method to map the data in the RDD to **Row** objects (from the pyspark.sql package). The RDDs in this example contain tuples of hashtags and counts. The Row constructor uses the names of its keyword arguments to specify the column names for each value in that row. In this case, tag[0] is the hashtag in the tuple, and tag[1] is the total count for that hashtag:

```python
# map hashtag string-count tuples to Rows
rows = rdd.map(
    lambda tag: Row(hashtag=tag[0], total=tag[1]))
```

The next statement creates a Spark DataFrame containing the Row objects. We'll use this with Spark SQL to query the data to get the top-20 hashtags with their total counts:

```python
# create a DataFrame from the Row objects
hashtags_df = spark.createDataFrame(rows)
```

To query a Spark DataFrame, first create a *table view*, which enables Spark SQL to query the DataFrame like a table in a relational database. Spark DataFrame method **createOrReplaceTempView** creates a temporary table view for the DataFrame and names the view for use in the from clause of a query:

```python
# create a temporary table view for use with Spark SQL
hashtags_df.createOrReplaceTempView('hashtags')
```

Once you have a table view, you can query the data using Spark SQL.[52] The following statement uses the SparkSession instance's sql method to perform a Spark SQL query that selects the hashtag and total columns from the hashtags table view, orders the selected rows by total in descending (desc) order, then returns the first 20 rows of the result (limit 20). Spark SQL returns a new Spark DataFrame containing the results:

```python
# use Spark SQL to get top 20 hashtags in descending order
top20_df = spark.sql(
    """select hashtag, total
       from hashtags
       order by total, hashtag desc
       limit 20""")
```

Finally, we pass the Spark DataFrame to our display_barplot utility function. The hashtags and totals will be displayed on the *x*- and *y*-axes, respectively. We also display the time at which count_tags was called:

```python
        display_barplot(top20_df, x='hashtag', y='total', time=time)
    except Exception as e:
        print(f'Exception: {e}')
```

Getting the SparkContext
The rest of the code in this notebook sets up Spark streaming to read text from the starttweetstream.py script and specifies how to process the tweets. First, we create the SparkContext for connecting to the Spark cluster:

```
[5]: sc = SparkContext()
```

---

52. For details of Spark SQL's syntax, see https://spark.apache.org/sql/.

Getting the `StreamingContext`

For Spark streaming, you must create a **`StreamingContext`** (module `pyspark.streaming`), providing as arguments the `SparkContext` and how often in seconds to process batches of streaming data. In this app, we'll process batches every 10 seconds—this is the *batch interval*:

```
[6]: ssc = StreamingContext(sc, 10)
```

Depending on how fast data is arriving, you may wish to shorten or lengthen your batch intervals. For a discussion of this and other performance-related issues, see the Performance Tuning section of the *Spark Streaming Programming Guide*:

> https://spark.apache.org/docs/latest/streaming-programming-
>   guide.html#performance-tuning

Setting Up a Checkpoint for Maintaining State

By default, Spark streaming does not maintain state information as you process the stream of RDDs. However, you can use Spark **checkpointing** to keep track of the streaming state. Checkpointing enables:

- fault-tolerance for restarting a stream in cases of cluster node or Spark application failures, and
- stateful transformations, such as summarizing the data received so far—as we're doing in this example.

`StreamingContext` method **`checkpoint`** sets up the checkpointing folder:

```
[7]: ssc.checkpoint('hashtagsummarizer_checkpoint')
```

For a Spark streaming application in a cloud-based cluster, you'd specify a location within HDFS to store the checkpoint folder. We're running this example in the local Jupyter Docker image, so we simply specified the name of a folder, which Spark will create in the current folder (in our case, the `ch17` folder's `SparkHashtagSummarizer`). For more details on checkpointing, see

> https://spark.apache.org/docs/latest/streaming-programming-
>   guide.html#checkpointing

Connecting to the Stream via a Socket

`StreamingContext` method **`socketTextStream`** connects to a socket from which a stream of data will be received and returns a `DStream` that receives the data. The method's arguments are the hostname and port number to which the `StreamingContext` should connect—these must match where the `starttweetstream.py` script is waiting for the connection:

```
[8]: stream = ssc.socketTextStream('localhost', 9876)
```

Tokenizing the Lines of Hashtags

We use functional-style programming calls on a `DStream` to specify the processing steps to perform on the streaming data. The following call to `DStream`'s **`flatMap`** method tokenizes a line of space-separated hashtags and returns a new `DStream` representing the individual tags:

```
[9]: tokenized = stream.flatMap(lambda line: line.split())
```

17.7 Counting Twitter Hashtags Using the pyspark-notebook Docker Stack

Mapping the Hashtags to Tuples of Hashtag-Count Pairs
Next, similar to the Hadoop mapper earlier in this chapter, we use `DStream` method `map` to get a new `DStream` in which each hashtag is mapped to a hashtag-count pair (in this case as a tuple) in which the count is initially 1:

```
[10]: mapped = tokenized.map(lambda hashtag: (hashtag, 1))
```

Totaling the Hashtag Counts So Far
`DStream` method `updateStateByKey` receives a two-argument `lambda` that totals the counts for a given key and adds them to the prior total for that key:

```
[11]: hashtag_counts = tokenized.updateStateByKey(
          lambda counts, prior_total: sum(counts) + (prior_total or 0))
```

Specifying the Method to Call for Every RDD
Finally, we use `DSteam` method `foreachRDD` to specify that every processed RDD should be passed to function `count_tags`, which then summarizes the top-20 hashtags so far and displays a barplot:

```
[12]: hashtag_counts.foreachRDD(count_tags)
```

Starting the Spark Stream
Now, that we've specified the processing steps, we call the `StreamingContext`'s `start` method to connect to the socket and begin the streaming process.

```
[13]: ssc.start()  # start the Spark streaming
```

The following shows a sample barplot produced while processing a stream of tweets about "football." Because football is a different sport in the United States and the rest of the world the hashtags relate to both American football and what we call soccer—we grayed out three hashtags that were not appropriate for publication:

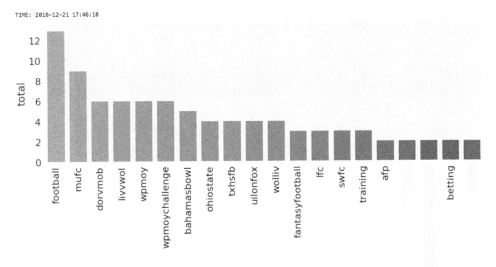

Self Check for Section 17.7

1. *(Fill-In)* Spark `DataFrame` method _____ returns a pandas `DataFrame`.
Answer: `toPandas`.

2. *(True/False)* You can use Spark SQL to query `RDD` objects using familiar Structured Query Language syntax.
Answer: False. Spark SQL requires a table view of a Spark `DataFrame`.

3. *(Discussion)* Assuming `hashtags_df` is a Spark `DataFrame`, what does the following code do?

    `hashtags_df.createOrReplaceTempView('hashtags')`

Answer: This statement creates (or replaces) a temporary table view for the `DataFrame` `hashtags_df` and names it `'hashtags'` for use in Spark SQL queries.

4. *(True/False)* By default, Spark streaming does not maintain state information as you process the stream of RDDs. However, you can use Spark checkpointing to keep track of the streaming state for fault-tolerance and stateful transformations, such as summarizing the data received so far.
Answer: True.

17.8 Internet of Things and Dashboards

In the late 1960s, the Internet began as the ARPANET, which initially connected four universities and grew to 10 nodes by the end of 1970.[53] In the last 50 years, that has grown to billions of computers, smartphones, tablets and an enormous range of other device types connected to the Internet worldwide. *Any* device connected to the Internet is a "thing" in the Internet of Things (IoT).

Each device has a unique Internet protocol address (IP address) that identifies it. The explosion of connected devices exhausted the approximately 4.3 billion available IPv4 (Internet Protocol version 4) addresses[54] and led to the development of IPv6, which supports approximately 3.4×10^{38} addresses (that's a lot of zeros).[55]

"Top research firms such as Gartner and McKinsey predict a jump from the 6 billion connected devices we have worldwide today, to 20–30 billion by 2020."[56] Various predictions say that number could be 50 billion. Computer-controlled, Internet-connected devices continue to proliferate. The following is a small subset IoT device types and applications.

---

53. https://en.wikipedia.org/wiki/ARPANET#History..
54. https://en.wikipedia.org/wiki/IPv4_address_exhaustion.
55. https://en.wikipedia.org/wiki/IPv6.
56. https://www.pubnub.com/developers/tech/how-pubnub-works/.

IoT devices		
activity trackers—Apple Watch, FitBit, ... Amazon Dash ordering buttons Amazon Echo (Alexa), Apple HomePod (Siri), Google Home (Google Assistant) appliances—ovens, coffee makers, refrigerators, ... driverless cars earthquake sensors	healthcare—blood glucose monitors for diabetics, blood pressure monitors, electrocardiograms (EKG/ECG), electroencephalograms (EEG), heart monitors, ingestible sensors, pacemakers, sleep trackers, ... sensors—chemical, gas, GPS, humidity, light, motion, pressure, temperature, ...	smart home—lights, garage openers, video cameras, doorbells, irrigation controllers, security devices, smart locks, smart plugs, smoke detectors, thermostats, air vents tsunami sensors tracking devices wine cellar refrigerators wireless network devices

IoT Issues

Though there's a lot of excitement and opportunity in IoT, not everything is positive. There are many security, privacy and ethical concerns. Unsecured IoT devices have been used to perform distributed-denial-of-service (DDOS) attacks on computer systems.[57] Home security cameras that you intend to protect your home could potentially be hacked to allow others access to the video stream. Voice-controlled devices are always "listening" to hear their trigger words. This leads to privacy and security concerns. Children have accidentally ordered products on Amazon by talking to Alexa devices, and companies have created TV ads that would activate Google Home devices by speaking their trigger words and causing Google Assistant to read Wikipedia pages about a product to you.[58] Some people worry that these devices could be used to eavesdrop. Just recently, a judge ordered Amazon to turn over Alexa recordings for use in a criminal case.[59]

This Section's Examples

In this section, we discuss the **publish/subscribe model** that IoT and other types of applications use to communicate. First, without writing any code, you'll build a web-based dashboard using Freeboard.io and subscribe to a sample live stream from the PubNub service. Next, you'll simulate an Internet-connected thermostat which publishes messages to the free Dweet.io service using the Python module Dweepy, then create a dashboard visualization of it with Freeboard.io. Finally, you'll build a Python client that subscribes to a sample live stream from the PubNub service and dynamically visualizes the stream with Seaborn and a Matplotlib `FuncAnimation`. In the exercises, you'll experiment with additional IoT platforms, simulators and live streams.

---

57. https://threatpost.com/iot-security-concerns-peaking-with-no-end-in-sight/131308/.
58. https://www.symantec.com/content/dam/symantec/docs/security-center/white-papers/istr-security-voice-activated-smart-speakers-en.pdf.
59. https://techcrunch.com/2018/11/14/amazon-echo-recordings-judge-murder-case/.

17.8.1 Publish and Subscribe

IoT devices (and many other types of devices and applications) commonly communicate with one another and with applications via **pub/sub (publisher/subscriber) systems**. A **publisher** is any device or application that sends a message to a cloud-based service, which in turn sends that message to all **subscribers**. Typically each publisher specifies a topic or channel, and each subscriber specifies one or more **topics** or **channels** for which they'd like to receive messages. There are many pub/sub systems in use today. In the remainder of this section, we'll use PubNub and Dweet.io. In the exercises, you can investigate Apache Kafka—a Hadoop ecosystem component that provides a high-performance publish/subscribe service, real-time stream processing and storage of streamed data.

17.8.2 Visualizing a PubNub Sample Live Stream with a Freeboard Dashboard

PubNub is a pub/sub service geared to real-time applications in which any software and device connected to the Internet can communicate via small messages. Some of their common use-cases include IoT, chat, online multiplayer games, social apps and collaborative apps. PubNub provides several live streams for learning purposes, including one that simulates IoT sensors (Section 17.8.5 lists the others).

One common use of live data streams is visualizing them for monitoring purposes. In this section, you'll connect PubNub's live simulated sensor stream to a Freeboard.io web-based dashboard. A car's dashboard visualizes data from your car's sensors, showing information such as the outside temperature, your speed, engine temperature, the time and the amount of gas remaining. A web-based dashboard does the same thing for data from various sources, including IoT devices.

Freeboard.io is a cloud-based dynamic dashboard visualization tool. You'll see that, without writing any code, you can easily connect Freeboard.io to various data streams and visualize the data as it arrives. The following dashboard visualizes data from three of the four simulated sensors in the PubNub simulated IoT sensors stream:

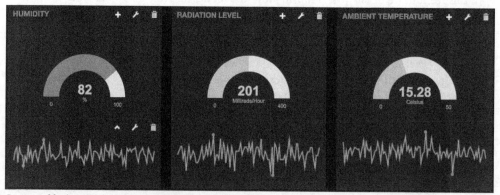

Courtesy of freeboard™, brought to you by Bug Labs, Inc. (https://freeboard.io, https://buglabs.net)

For each sensor, we used a **Gauge** (the semicircular visualizations) and a **Sparkline** (the jagged lines) to visualize the data. When you complete this section, you'll see the **Gauges** and **Sparklines** frequently moving as new data arrives multiple times per second.

In addition to their paid service, Freeboard.io provides an open-source version (with fewer options) on GitHub. They also provide tutorials that show how to add *custom plug-ins*, so you can develop your own visualizations to add to their dashboards.

Signing up for Freeboard.io
For this example, register for a Freeboard.io 30-day trial at

   https://freeboard.io/signup

Once you've registered, the **My Freeboards** page appears. If you'd like, you can click the **Try a Tutorial** button and visualize data from your smartphone.

Creating a New Dashboard
In the upper-right corner of the **My Freeboards** page, enter Sensor Dashboard in the **enter a name** field, then click the **Create New** button to create a dashboard. This displays the dashboard designer.

Adding a Data Source
If you add your data source(s) before designing your dashboard, you'll be able to configure each visualization as you add it:

1. Under **DATASOURCES**, click **ADD** to specify a new data source.
2. The **DATASOURCE** dialog's **TYPE** drop-down list shows the currently supported data sources, though you can develop plug-ins for new data sources as well.[60] Select **PubNub**. The web page for each PubNub sample live stream specifies the **Channel** and **Subscribe key**. Copy these values from PubNub's Sensor Network page at https://www.pubnub.com/developers/realtime-data-streams/sensor-network/, then insert their values in the corresponding **DATASOURCE** dialog fields. Provide a **NAME** for your data source, then click **SAVE**.

Adding a Pane for the Humidity Sensor
A Freeboard.io dashboard is divided into panes that group visualizations. Multiple panes can be dragged to rearrange them. Click the **+ Add Pane** button to add a new pane. Each pane can have a title. To set it, click the wrench icon on the pane, specify Humidity for the **TITLE**, then click **SAVE**.

Adding a Gauge to the Humidity Pane
To add visualizations to a pane, click its + button to display the **WIDGET** dialog. The **TYPE** drop-down list shows several built-in widgets. Choose **Gauge**. To the right of the **VALUE** field, click **+ DATASOURCE**, then select the name of your data source. This displays the available values from that data source. Click **humidity** to select the humidity sensor's value. For **UNITS**, specify %, then click **SAVE**. This displays the new visualization, which immediately begins showing values from the sensor stream.

Notice that the humidity value has four digits of precision to the right of the decimal point. PubNub supports JavaScript expressions, so you can use them to perform calculations or format data. For example, you can use JavaScript's function Math.round to round

---

60. Some of the listed data sources are available only via Freeboard.io, not the open source Freeboard on GitHub.

the humidity value to the closest integer. To do so, hover the mouse over the gauge and click its wrench icon. Then, insert `"Math.round("` before the text in the **VALUE** field and `")"` after the text, then click SAVE.

Adding a Sparkline to the Humidity Pane
A **sparkline** is a line graph without axes that's typically used to give you a sense of how a data value is changing over time. Add a sparkline for the humidity sensor by clicking the humidity pane's + button, then selecting Sparkline from the **TYPE** drop-down list. For the **VALUE**, once again select your data source and **humidity**, then click **SAVE**.

Completing the Dashboard
Using the techniques above, add two more panes and drag them to the right of the first. Name them **Radiation Level** and **Ambient Temperature**, respectively, and configure each pane with a **Gauge** and **Sparkline** as shown above. For the **Radiation Level** gauge, specify `Millirads/Hour` for the **UNITS** and 400 for the **MAXIMUM**. For the **Ambient Temperature** gauge, specify `Celsius` for the **UNITS** and 50 for the **MAXIMUM**.

17.8.3 Simulating an Internet-Connected Thermostat in Python
Simulation is one of the most important applications of computers. We used simulation with dice rolling in earlier chapters. With IoT, it's common to use simulators to test your applications, especially when you do not have access to actual devices and sensors while developing applications. Many cloud vendors have IoT simulation capabilities. In the exercises, you'll explore the IBM Watson IoT Platform and IOTIFY.io.

Here, you'll create a script that simulates an Internet-connected thermostat publishing periodic JSON messages—called dweets—to dweet.io. The name "dweet" is based on "tweet"—a dweet is like a tweet from a device. Many of today's Internet-connected security systems include temperature sensors that can issue low-temperature warnings before pipes freeze or high-temperature warnings to indicate there might be a fire. Our simulated sensor will send dweets containing a location and temperature, as well as low- and high-temperature notifications. These will be `True` only if the temperature reaches 3 degrees Celsius or 35 degrees Celsius, respectively. In the next section, we'll use freeboard.io to create a simple dashboard that shows the temperature changes as the messages arrive, as well as warning lights for low- and high-temperature warnings.

Installing Dweepy
To publish messages to dweet.io from Python, first install the Dweepy library:

```
pip install dweepy
```

The library is straightforward to use. You can view its documentation at:

```
https://github.com/paddycarey/dweepy
```

Invoking the `simulator.py` Script
The Python script `simulator.py` that simulates our thermostat is located in the `ch17` example folder's `iot` subfolder. You invoke the simulator with two command-line arguments representing the number of total messages to simulate and the delay in seconds between sending dweets:

```
ipython simulator.py 1000 1
```

Sending Dweets

The `simulator.py` is shown below. It uses random-number generation and Python techniques that you've studied throughout this book, so we'll focus just on a few lines of code that publish messages to `dweet.io` via Dweepy. We've broken apart the script below for discussion purposes.

By default, `dweet.io` is a public service, so any app can publish or subscribe to messages. When publishing messages, *you'll want to specify a unique name for your device.* We used `'temperature-simulator-deitel-python'` (line 17).[61] Lines 18–21 define a Python dictionary, which will store the current sensor information. Dweepy will convert this into JSON when it sends the dweet.

```
1   # simulator.py
2   """A connected thermostat simulator that publishes JSON
3   messages to dweet.io"""
4   import dweepy
5   import sys
6   import time
7   import random
8
9   MIN_CELSIUS_TEMP = -25
10  MAX_CELSIUS_TEMP = 45
11  MAX_TEMP_CHANGE = 2
12
13  # get the number of messages to simulate and delay between them
14  NUMBER_OF_MESSAGES = int(sys.argv[1])
15  MESSAGE_DELAY = int(sys.argv[2])
16
17  dweeter = 'temperature-simulator-deitel-python'  # provide a unique name
18  thermostat = {'Location': 'Boston, MA, USA',
19                'Temperature': 20,
20                'LowTempWarning': False,
21                'HighTempWarning': False}
22
```

Lines 25–53 produce the number of simulated message you specify. During each iteration of the loop, we

- generate a random temperature change in the range –2 to +2 degrees and modify the temperature,
- ensure that the temperature remains in the allowed range,
- check whether the low- or high-temperature sensor has been triggered and update the thermostat dictionary accordingly,
- display how many messages have been generated so far,
- use Dweepy to send the message to dweet.io (line 52), and
- use the `time` module's `sleep` function to wait the specified amount of time before generating another message.

---

61. To truly guarantee a unique name, `dweet.io` can create one for you. The Dweepy documentation explains how to do this.

```
23    print('Temperature simulator starting')
24
25    for message in range(NUMBER_OF_MESSAGES):
26        # generate a random number in the range -MAX_TEMP_CHANGE
27        # through MAX_TEMP_CHANGE and add it to the current temperature
28        thermostat['Temperature'] += random.randrange(
29            -MAX_TEMP_CHANGE, MAX_TEMP_CHANGE + 1)
30
31        # ensure that the temperature stays within range
32        if thermostat['Temperature'] < MIN_CELSIUS_TEMP:
33            thermostat['Temperature'] = MIN_CELSIUS_TEMP
34
35        if thermostat['Temperature'] > MAX_CELSIUS_TEMP:
36            thermostat['Temperature'] = MAX_CELSIUS_TEMP
37
38        # check for low temperature warning
39        if thermostat['Temperature'] < 3:
40            thermostat['LowTempWarning'] = True
41        else:
42            thermostat['LowTempWarning'] = False
43
44        # check for high temperature warning
45        if thermostat['Temperature'] > 35:
46            thermostat['HighTempWarning'] = True
47        else:
48            thermostat['HighTempWarning'] = False
49
50        # send the dweet to dweet.io via dweepy
51        print(f'Messages sent: {message + 1}\r', end='')
52        dweepy.dweet_for(dweeter, thermostat)
53        time.sleep(MESSAGE_DELAY)
54
55    print('Temperature simulator finished')
```

You do not need to register to use the service. On the first call to dweepy's **dweet_for** function to send a dweet (line 52), dweet.io creates the device name. The function receives as arguments the device name (dweeter) and a dictionary representing the message to send (thermostat). Once you execute the script, you can immediately begin tracking the messages on the dweet.io site by going to the following address in your web browser:

https://dweet.io/follow/temperature-simulator-deitel-python

If you use a different device name, replace "temperature-simulator-deitel-python" with the name you used. The web page contains two tabs. The **Visual** tab shows you the individual data items, displaying a sparkline for any numerical values. The **Raw** tab shows you the actual JSON messages that Dweepy sent to dweet.io.

17.8.4 Creating the Dashboard with Freeboard.io

The sites dweet.io and freeboard.io are run by the same company. In the dweet.io webpage discussed in the preceding section, you can click the **Create a Custom Dashboard** button to open a new browser tab, with a default dashboard already implemented for the temperature sensor. By default, freeboard.io will configure a data source named **Dweet**

17.8 Internet of Things and Dashboards

and auto-generate a dashboard containing one pane for each value in the dweet JSON. Within each pane, a text widget will display the corresponding value as the messages arrive.

If you prefer to create your own dashboard, you can use the steps in Section 17.8.2 to create a data source (this time selecting Dweepy) and create new panes and widgets, or you can you modify the auto-generated dashboard.

Below are three screen captures of a dashboard consisting of four widgets:

- A **Gauge** widget showing the current temperature. For this widget's **VALUE** setting, we selected the data source's `Temperature` field. We also set the **UNITS** to `Celsius` and the **MINIMUM** and **MAXIMUM** values to -25 and 45 degrees, respectively.

- A **Text** widget to show the current temperature in Fahrenheit. For this widget, we set the **INCLUDE SPARKLINE** and **ANIMATE VALUE CHANGES** to **YES**. For this widget's **VALUE** setting, we again selected the data source's `Temperature` field, then added to the end of the **VALUE** field

    `* 9 / 5 + 32`

    to perform a calculation that converts the Celsius temperature to Fahrenheit. We also specified `Fahrenheit` in the **UNITS** field.

- Finally, we added two **Indicator Light** widgets. For the first **Indicator Light**'s **VALUE** setting, we selected the data source's `LowTempWarning` field, set the **TITLE** to `Freeze Warning` and set the **ON TEXT** value to `LOW TEMPERATURE WARNING`—**ON TEXT** indicates the text to display when value is true. For the second **Indicator Light**'s **VALUE** setting, we selected the data source's `HighTempWarning` field, set the **TITLE** to `High Temperature Warning` and set the **ON TEXT** value to `HIGH TEMPERATURE WARNING`.

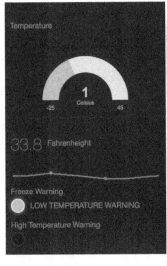

Courtesy of freeboard™, brought to you by Bug Labs, Inc. (`https://freeboard.io`, `https://buglabs.net`)

17.8.5 Creating a Python PubNub Subscriber

PubNub provides the `pubnub` Python module for conveniently performing pub/sub operations. They also provide seven sample streams for you to experiment with—four real-time streams and three simulated streams:[62]

- Twitter Stream—provides up to 50 tweets-per-second from the Twitter live stream and does not require your Twitter credentials.
- Hacker News Articles—this site's recent articles.
- State Capital Weather—provides weather data for the U.S. state capitals.
- Wikipedia Changes—a stream of Wikipedia edits.
- Game State Sync—simulated data from a multiplayer game.
- Sensor Network—simulated data from radiation, humidity, temperature and ambient light sensors.
- Market Orders—simulated stock orders for five companies.

In this section, you'll use the **pubnub** module to subscribe to their simulated Market Orders stream, then visualize the changing stock prices as a Seaborn barplot, like:

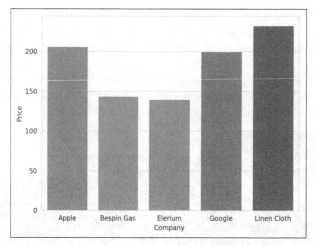

Of course, you also can publish messages to streams. For details, see the pubnub module's documentation at https://www.pubnub.com/docs/python/pubnub-python-sdk.

To prepare for using PubNub in Python, execute the following command to install the latest version of the pubnub module—the '>=4.1.2' ensures that at a minimum the 4.1.2 version of the pubnub module will be installed:

```
pip install "pubnub>=4.1.2"
```

The script `stocklistener.py` that subscribes to the stream and visualizes the stock prices is defined in the ch17 folder's pubnub subfolder. We break the script into pieces here for discussion purposes.

---

62. https://www.pubnub.com/developers/realtime-data-streams/.

Message Format
The simulated Market Orders stream returns JSON objects containing five key–value pairs with the keys `'bid_price'`, `'order_quantity'`, `'symbol'`, `'timestamp'` and `'trade_type'`. For this example, we'll use only the `'bid_price'` and `'symbol'`. The PubNub client returns the JSON data to you as a Python dictionary.

Importing the Libraries
Lines 3–13 import the libraries used in this example. We discuss the PubNub types imported in lines 10–13 as we encounter them below.

```
 1  # stocklistener.py
 2  """Visualizing a PubNub live stream."""
 3  from matplotlib import animation
 4  import matplotlib.pyplot as plt
 5  import pandas as pd
 6  import random
 7  import seaborn as sns
 8  import sys
 9
10  from pubnub.callbacks import SubscribeCallback
11  from pubnub.enums import PNStatusCategory
12  from pubnub.pnconfiguration import PNConfiguration
13  from pubnub.pubnub import PubNub
14
```

List and DataFrame Used for Storing Company Names and Prices
The list `companies` contains the names of the companies reported in the Market Orders stream, and the pandas `DataFrame` `companies_df` is where we'll store each company's last price. We'll use this `DataFrame` with Seaborn to display a bar chart.

```
15  companies = ['Apple', 'Bespin Gas', 'Elerium', 'Google', 'Linen Cloth']
16
17  # DataFrame to store last stock prices
18  companies_df = pd.DataFrame(
19      {'company': companies, 'price' : [0, 0, 0, 0, 0]})
20
```

Class SensorSubscriberCallback
When you subscribe to a PubNub stream, you must add a listener that receives status notifications and messages from the channel. This is similar to the Tweepy listeners you've defined previously. To create your listener, you must define a subclass of `SubscribeCallback` (module `pubnub.callbacks`), which we discuss after the code:

```
21  class SensorSubscriberCallback(SubscribeCallback):
22      """SensorSubscriberCallback receives messages from PubNub."""
23      def __init__(self, df, limit=1000):
24          """Create instance variables for tracking number of tweets."""
25          self.df = df  # DataFrame to store last stock prices
26          self.order_count = 0
27          self.MAX_ORDERS = limit  # 1000 by default
28          super().__init__()  # call superclass's init
29
```

```
30    def status(self, pubnub, status):
31        if status.category == PNStatusCategory.PNConnectedCategory:
32            print('Connected to PubNub')
33        elif status.category == PNStatusCategory.PNAcknowledgmentCategory:
34            print('Disconnected from PubNub')
35
36    def message(self, pubnub, message):
37        symbol = message.message['symbol']
38        bid_price = message.message['bid_price']
39        print(symbol, bid_price)
40        self.df.at[companies.index(symbol), 'price'] = bid_price
41        self.order_count += 1
42
43        # if MAX_ORDERS is reached, unsubscribe from PubNub channel
44        if self.order_count == self.MAX_ORDERS:
45            pubnub.unsubscribe_all()
46
```

Class SensorSubscriberCallback's __init__ method stores the DataFrame in which each new stock price will be placed. The PubNub client calls overridden method status each time a new status message arrives. In this case, we're checking for the notifications that indicate that we've subscribed to or unsubscribed from a channel.

The PubNub client calls overridden method message (lines 36–45) when a new message arrives from the channel. Lines 37 and 38 get the company name and price from the message, which we print so you can see that messages are arriving. Line 40 uses the DataFrame method at to locate the appropriate company's row and its 'price' column, then assign that element the new price. Once the order_count reaches MAX_ORDERS, line 45 calls the PubNub client's unsubscribe_all method to unsubscribe from the channel.

Function Update

This example visualizes the stock prices using the animation techniques you learned in Chapter 6's Intro to Data Science section. Function update specifies how to draw one animation frame and is called repeatedly by the FuncAnimation we'll define shortly. We use Seaborn function barplot to visualize data from the companies_df DataFrame, using its 'company' column values on the x-axis and 'price' column values on the y-axis.

```
47  def update(frame_number):
48      """Configures bar plot contents for each animation frame."""
49      plt.cla()  # clear old barplot
50      axes = sns.barplot(
51          data=companies_df, x='company', y='price', palette='cool')
52      axes.set(xlabel='Company', ylabel='Price')
53      plt.tight_layout()
54
```

Configuring the Figure

In the main part of the script, we begin by setting the Seaborn plot style and creating the Figure object in which the barplot will be displayed:

```
55  if __name__ == '__main__':
56      sns.set_style('whitegrid')  # white background with gray grid lines
57      figure = plt.figure('Stock Prices')  # Figure for animation
58
```

17.8 Internet of Things and Dashboards

Configuring the FuncAnimation and Displaying the Window

Next, we set up the `FuncAnimation` that calls function `update`, then call Matplotlib's `show` method to display the `Figure`. Normally, this method blocks the script from continuing until you close the `Figure`. Here, we pass the `block=False` keyword argument to allow the script to continue so we can configure the PubNub client and subscribe to a channel.

```
59      # configure and start animation that calls function update
60      stock_animation = animation.FuncAnimation(
61          figure, update, repeat=False, interval=33)
62      plt.show(block=False)  # display window
63
```

Configuring the PubNub Client

Next, we configure the PubNub subscription key, which the PubNub client uses in combination with the channel name to subscribe to the channel. The key is specified as an attribute of the `PNConfiguration` object (module `pubnub.pnconfiguration`), which line 69 passes to the new `PubNub` client object (module `pubnub.pubnub`). Lines 70–72 create the `SensorSubscriberCallback` object and pass it to the `PubNub` client's `add_listener` method to register it to receive messages from the channel. We use a command-line argument to specify the total number of messages to process.

```
64      # set up pubnub-market-orders sensor stream key
65      config = PNConfiguration()
66      config.subscribe_key = 'sub-c-4377ab04-f100-11e3-bffd-02ee2ddab7fe'
67
68      # create PubNub client and register a SubscribeCallback
69      pubnub = PubNub(config)
70      pubnub.add_listener(
71          SensorSubscriberCallback(df=companies_df,
72              limit=int(sys.argv[1] if len(sys.argv) > 1 else 1000))
73
```

Subscribing to the Channel

The following statement completes the subscription process, indicating that we wish to receive messages from the channel named `'pubnub-market-orders'`. The execute method starts the stream.

```
74      # subscribe to pubnub-sensor-network channel and begin streaming
75      pubnub.subscribe().channels('pubnub-market-orders').execute()
76
```

Ensuring the Figure Remains on the Screen

The second call to Matplotlib's show method ensures that the `Figure` remains on the screen until you close its window.

```
77      plt.show()  # keeps graph on screen until you dismiss its window
```

Self Check for Section 17.8

1 *(Fill-In)* IoT devices (and many other types of devices and applications) commonly communicate with one another and with applications via _____ systems.
Answer: pub/sub (publisher/subscriber)

2 *(Fill-In)* A(n) _____ is any device or application that sends a message to a cloud-based service, which in turn sends that message to all _____.
Answer: publisher, subscribers.

3 *(Fill-In)* A(n) _____ is a graph without axes that's typically used to give you a sense of how a data value is changing over time.
Answer: sparkline.

4 *(Fill-In)* In a PubNub Python client that subscribes to a channel, you must create a subclass of _____, then register an object of that class to receive status notifications and messages from the channel.
Answer: `SubscribeCallback`.

17.9 Wrap-Up

In this chapter, we introduced big data, discussed how large data is getting and discussed hardware and software infrastructure for working with big data. We introduced traditional relational databases and Structured Query Language (SQL) and used the `sqlite3` module to create and manipulate a books database in SQLite. We also demonstrated loading SQL query results into pandas `DataFrames`.

We discussed the four major types of NoSQL databases—key–value, document, columnar and graph—and introduced NewSQL databases. We stored JSON tweet objects as documents in a cloud-based MongoDB Atlas cluster, then summarized them in an interactive visualization displayed on a Folium map.

We introduced Hadoop and how it's used in big-data applications. You configured a multi-node Hadoop cluster using the Microsoft Azure HDInsight service, then created and executed a Hadoop MapReduce task using Hadoop streaming.

We discussed Spark and how it's used in high-performance, real-time big-data applications. You used Spark's functional-style filter/map/reduce capabilities, first on a Jupyter Docker stack that runs locally on your own computer, then again using a Microsoft Azure HDInsight multi-node Spark cluster. Next, we introduced Spark streaming for processing data in mini-batches. As part of that example, we used Spark SQL to query data stored in Spark `DataFrames`.

The chapter concluded with an introduction to the Internet of Things (IoT) and the publish/subscribe model. You used Freeboard.io to create a dashboard visualization of a live sample stream from PubNub. You simulated an Internet-connected thermostat which published messages to the free `dweet.io` service using the Python module Dweepy, then used Freeboard.io to visualize the simulated device's data. Finally, you subscribed to a PubNub sample live stream using their Python module.

The rich collection of exercises encourages you to work with more big-data cloud and desktop platforms, additional SQL and NoSQL databases, NewSQL databases and IoT platforms. You can work with Wikipedia as another big-data source, and you can implement IoT with the Raspberry Pi and Iotify simulators.

Thanks for reading *Intro to Python for Computer Science and Data Science: Learning to Program with AI, Big Data and the Cloud*. We hope that you enjoyed the book and that you found it entertaining and informative. Most of all we hope you feel empowered to apply the technologies you've learned to the challenges you'll face as you continue your education and pursue your career.

Exercises

SQL and RDBMS Exercises

17.1 *(Books Database)* In an IPython session, perform each of the following tasks on the books database from Section 17.2:
  a) Select all authors' last names from the authors table in descending order.
  b) Select all book titles from the titles table in ascending order.
  c) Use an INNER JOIN to select all the books for a specific author. Include the title, copyright year and ISBN. Order the information alphabetically by title.
  d) Insert a new author into the authors table.
  e) Insert a new title for an author. Remember that the book must have an entry in the author_ISBN table and an entry in the titles table.

17.2 *(Cursor Method fetchall and Attribute description)* When you use a sqlite3 Cursor's execute method to perform a query, the query's results are stored in the Cursor object. The Cursor attribute **description** contains **metadata** about the results stored as a tuple of tuples. Each nested tuple's first value is a column name in the query results. Cursor method **fetchall** returns the query result's data as a list of tuples. Investigate the description attribute and fetchall method. Open the books database and use Cursor method execute to select all the data in the titles table, then use description and fetchall to display the data in tabular format.

17.3 *(Contacts Database)* Study the books.sql script provided in the ch17 examples folder's sql subfolder. Save the script as addressbook.sql and modify it to create a single table named contacts. The table should contain an auto-incremented id column and text columns for a person's first name, last name and phone number. In an IPython session, insert contacts into the database, query the database to list all the contacts and contacts with a specific last name, update a contact and delete a contact.

17.4 *(Project: DB Browser for SQLite)* Investigate the open source DB Browser for SQLite (https://sqlitebrowser.org/). This tool provides a graphical user interface in which you can view and interact with a SQLite database. Use the tool to open the books.db database and view the contents of the authors table. In IPython, add a new author and remove it so you can see the table update live in DB Browser for SQLite.

17.5 *(Project: MariaDB)* Research the MariaDB relational database management system and its Python support, then use it to create a database and reimplement the IPython session in Section 17.2. You may need to update the SQL script that creates the database tables, as some features like auto-incremented integer primary keys vary by relational database management system.

NoSQL Database Exercises

17.6 *(MongoDB Twitter Example Modification: Sentiment Analysis Enhancement)* Using the sentiment analysis techniques you learned in the "Natural Language Processing" chapter, modify Section 17.4's case study as follows. Enable the user to select a senator, then use a pandas DataFrame to show a summary of the positive, negative and neutral tweets for that senator by state. Create a choropleth that colors each state by positive, negative and neutral sentiment. The popup map markers should show the number of tweets of each sentiment for that state.

17.7 *(Project: Six Degrees of Separation with Neo4j NoSQL Graph Database)* The famous "six degrees of separation" problem says that any two people in the world are connected to one another by six or fewer acquaintance connections.[63] A game based on this is called "Six degrees of Kevin Bacon"[64] in which any two movie stars in Hollywood can be connected to Kevin Bacon via the roles they've played in films (because he has appeared in so many films). Neo4j's Cypher language is used to query Neo4j databases. In their *Guide to Cypher Basics* (https://neo4j.com/developer/guide-cypher-basics/), they implement "Six degrees of Kevin Bacon" using a movie database. Install the Neo4j database on your system and implement their solution.

Hadoop Exercises

17.8 *(Project: Hadoop on the Desktop with Hortonworks HDP Sandbox)* Hortonworks Sandbox (https://hortonworks.com/products/sandbox/) is an open-source desktop platform for Hadoop, Spark and related technologies. Install a *desktop* version of the Hortonworks Data Platform (HDP) Sandbox, then use it to execute this chapter's Hadoop MapReduce example. *Caution:* Before installing HDP Sandbox, ensure that your system meets the substantial disk and memory requirements.

17.9 *(Project: Hadoop on the Desktop with Cloudera CDH Quickstart VM)* Cloudera CDH is an open-source desktop platform for Hadoop, Spark and related technologies. Install a Cloudera desktop Quick Start VM (search for "Cloudera CDH Quickstart VM" online), then use it to execute this chapter's Hadoop MapReduce example. *Caution:* Before installing a Cloudera CDH Quickstart VM, ensure that your system meets the substantial disk and memory requirements.

17.10 *(Research Project: Apache Tez)* Investigate Apache Tez—a high-performance replacement for MapReduce. How is it that Tez achieves its performance improvement over MapReduce?

Spark Exercises

17.11 *(Project: Spark on the Desktop with Hortonworks HDP Sandbox)* Hortonworks Sandbox (https://hortonworks.com/products/sandbox/) is an open-source desktop platform for Hadoop, Spark and related technologies. Install a desktop version of the Hortonworks Data Platform (HDP) Sandbox, then use it to execute this chapter's Spark examples. *Caution:* Before installing HDP Sandbox, ensure that your system meets the substantial disk and memory requirements.

17.12 *(Project: Spark on the Desktop with Cloudera Quickstart VM)* Cloudera CDH is an open-source desktop platform for Hadoop, Spark and related technologies. Install a Cloudera desktop Quick Start VM (search for "Cloudera CDH Quickstart VM" online), then use it to execute this chapter's Spark examples. *Caution:* Before installing a Quickstart VM, ensure that your system meets the substantial disk and memory requirements.

17.13 *(Project: Spark ML)* The "Machine Learning" chapter presented several popular machine-learning algorithms. These and many other algorithms are available in Spark via Spark ML and the PySpark library. Research Spark ML in PySpark, then reimplement one

---

63. https://en.wikipedia.org/wiki/Six_degrees_of_separation.
64. https://en.wikipedia.org/wiki/Six_Degrees_of_Kevin_Bacon.

of the "Machine Learning" chapter's examples using the Jupyter `pyspark-notebook` Docker container.

17.14 *(Project: IBM's Apache Spark Service)* Investigate IBM Watson's Apache Spark service (https://console.bluemix.net/catalog/services/apache-spark), which provides free Lite tier support for Spark streaming and Spark MLlib, then use it to implement one of the machine-learning studies from Chapter 15.

IoT and Pub/Sub Exercises

17.15 *(Watson IoT Platform)* Investigate the free Lite tier of the Watson IoT Platform (https://console.bluemix.net/catalog/services/internet-of-things-platform). They provide a live stream demonstration that receives sensor data directly from your smartphone, provides a 3D visualization of your phone and shows the sensor data. The visualization updates in real time as you move your phone. See https://developer.ibm.com/iotplatform/2017/12/07/use-device-simulator-watson-iot-platform for more information.

17.16 *(Raspberry Pi and Internet of Things)* IOTIFY is an IoT simulation service. Research IOTIFY, then follow their Hello IoT tutorial, which uses a simulated Raspberry Pi device.

17.17 *(Streaming Stock Prices Dashboard with IEX, PubNub and Freeboard.io)* Investigate the free stock-quote API provided by IEX (https://iextrading.com/) and Python modules on GitHub that enable you to use their APIs in Python applications. Create a Python IEX client that receives quotes for specific companies (you can look up their stock ticker symbols online). Research how to publish to a PubNub channel and publish the quotes to your channel. Use Freeboard.io to create a dashboard that subscribes to the PubNub channel you created and visualizes the stock prices as they arrive.

17.18 *(Project: Dweet.io and Dweepy)* Use Dweet.io and Dweepy to implement a text-based chat client script. Each person running the script would specify their own username. By default all clients will publish and subscribe to the same channel. As an enhancement, enable the user to choose the channel to use.

17.19 *(Project: Freeboard on GitHub)* Freeboard.io provides a free open-source version (with fewer options) on GitHub. Locate this version, install it on your system and use it to implement the dashboards we showed in Section 17.8.

17.20 *(Project: PubNub and Bokeh)* The Bokeh visualization library enables you to create dashboard visualizations from Python. In addition, it provides streaming support for dyanamically updating visualizations. Investigate Bokeh's streaming capabilities, then use them with PubNub's simulated sensor stream to create a Python client that visualizes the sensor data.

17.21 *(Research Project: IoT for the Entrepreneur)* If you're inclined to start a company, IoT presents many opportunities. Research IoT opportunities for entrepreneurs and create and describe an original idea for a business.

17.22 *(Research Project: Smart Watches and Activity Trackers)* Research the wearable IoT devices Apple Watch and Fitbit. List the sensors they provide and what they're able to monitor, and the dashboards they provide to help you monitor your health.

17.23 *(Research Project: Kafka Publish/Subscribe Messaging)* In this chapter you studied streaming and publish/subscribe messaging. Apache Kafka (`https://kafka.apache.org`) supports real-time messaging, stream processing and storage, and is typically used to transform and process high-volume streaming data, such as website activity and streaming IoT data. Research the applications of Apache Kafka and the platforms that use it.

Platform Exercises

17.24 *(Project: Spark with Databricks Community Edition)* Databricks[65] is an analytics platform created by the people who originally created Spark at U.C. Berkeley. In addition to being available through Amazon AWS and Microsoft Azure, they provide a free cloud-based Databricks Community Edition (`https://databricks.com/product/faq/community-edition`), which runs on AWS[66] and enables you to learn about and experiment with Spark without having to install any software locally. In fact, they implemented all the examples in their book *Spark: The Definitive Guide* using the free Databricks Community Edition.

Investigate the Databricks Community Edition's capabilities and follow their *Getting Started with Apache Spark* tutorial at `https://databricks.com/spark/getting-started-with-apache-spark`. Their notebook format and commands are similar but not identical to Jupyter's. Next, reimplement the Spark examples in Sections 17.6–17.7 using the Databricks Community Edition. To install Python modules into your Databricks cluster, follow the instructions at `https://docs.databricks.com/user-guide/libraries.html`. Like many of the data-science libraries we've used in the book, Databricks includes popular datasets you can use when learning Spark:

`https://docs.databricks.com/user-guide/faq/databricks-datasets.html`

17.25 *(Project: IBM Watson Analytics Engine)* You can access Hadoop, Spark and other tools in the Hadoop ecosystem via IBM's Watson Analytics Engine. To get started, the Watson Lite tier lets you create one cluster per 30-day period and use it for a maximum of 50 node hours[67] so that you can evaluate the platform or test Hadoop and Spark tasks. IBM also provides a separate Apache Spark service and various other big-data-related services. Research Watson Analytics Engine, then use it to implement and run this chapter's Hadoop and Spark examples. For a complete list of IBM's services, see their catalog at:

`https://console.bluemix.net/catalog/`

Other Exercises

17.26 *(Research Project: Big Data in Baseball)* Big data analytics techniques have been employed by some baseball teams and are credited with helping the 2004 Red Sox and the 2016 Cubs win World Series after long droughts. The books *Moneyball*[68] and *Big Data Baseball*[69] chronicle the data analytics successes of the 2002 Oakland Athletics and the

---

65. `http://databricks.com`.
66. `https://databricks.com/product/faq/community-edition`.
67. `https://console.bluemix.net/docs/services/AnalyticsEngine/faq.html#how-does-the-lite-plan-work-`.
68. Lewis, M., *Moneyball: The Art of Winning an Unfair Game*. W. W. Norton & Company. 2004.
69. Sawchik, T., *Big Data Baseball: Math, Miracles, and the End of a 20-Year Losing Streak*. Flatiron Books. 2016.

2013 Pittsburgh Pirates, respectively. On the downside, the *Wall Street Journal* reported that as a result of using data analytics, baseball games have become longer on average with less action.[70] Read either or both of those books to gain insights into how big-data analytics are used in sports.

17.27 *(Research Project: NewSQL Databases)* Research the NewSQL databases VoltDB, MemSQL, Apache Ignite and Google Spanner and discuss their key features.

17.28 *(Research Project: CRISPR Gene Editing)* Research how big data is being used with CRISPR gene editing. Research and discuss ethical and moral issues raised by CRISPR gene editing.

17.29 *(Research: Big-Data Ethics Conundrum)* Suppose big-data analytics predicts that a person with no criminal record has a significant chance of committing a serious crime. Should the police arrest that person? Investigate ethics issues with respect to big data.

17.30 *(Research Project: Privacy and Data Integrity Legislation)* In the chapter, we mentioned HIPAA (Health Insurance Portability and Accountability Act) in the United States and GDPR (General Data Protection Regulation) for the European Union. Laws like these are becoming more common and stricter. Investigate each of these laws and how they affect big-data analytics thinking.

17.31 *(Research Project: Cross-Referencing Databases)* Investigate and comment on the privacy issues caused by cross-referencing facts about individuals among various databases.

17.32 *(Research Project: Personally Identifiable Information)* Protecting users personally identifiable information (PII) is an important aspect of privacy. Research and comment on this issue in the context of big data.

17.33 *(Research Project: Wikipedia as a Big-Data Source)* Wikipedia is a popular big-data source. Investigate the capabilities they offer for accessing their information. Be sure to check out the `wikipedia` Python module and build an application that uses Wikipedia data.

---

70. "Baseball learns data's downside—analytics leads to longer games with less action," October 3, 2017. https://www.wsj.com/articles/the-downside-of-baseballs-data-revolutionlong-games-less-action-1507043924.

REVIEWER COMMENTS

"For a while, I have been looking for a book in Data Science using Python that would cover the most relevant technologies. Well, my search is over. A must-have book for any practitioner of this field. The machine learning chapter is a real winner!! The dynamic visualization is fantastic." —**Ramon Mata-Toledo, Professor, James Madison University**

"IBM Watson is an exciting chapter. I enjoyed running the code and using the Watson service. The code examples put together a lot of Watson services in a really nifty example. I enjoyed the OOP chapter—doctest unit testing is nice because you can have the test in the actual docstring so things are traveling together. The line-by-line explanations of the static and dynamic visualizations of the die rolling are just great." —**Daniel Chen, Data Scientist, Lander Analytics**

"A lucid exposition of the fundamentals of Python and Data Science. Excellent section on problem decomposition. Thanks for pointing out seeding the random number generator for reproducibility. I like the use of dictionary and set comprehensions for succinct programming. "List vs. Array Performance: Introducing %timeit" is convincing on why one should use ndarrays. Good defensive programming. Great section on Pandas Series and DataFrames—one of the clearest expositions that I have seen. The section on data wrangling is excellent. Natural Language Processing is an excellent chapter! I learned a tremendous amount going through it. Great exercises."
—**Shyamal Mitra, Senior Lecturer, University of Texas**

"My game programming students would appreciate these exercises." —**Pranshu Gupta, Assistant Professor, DeSales U.**

"I like the discussion of exceptions and tracebacks. I really liked the Data Mining Twitter chapter; it focused on a real data source, and brought in a lot of techniques for analysis (e.g., visualization, NLP). I like that the Python modules helped hide some of the complexity. Word clouds look cool." —**David Koop, Assistant Professor, U-Mass Dartmouth**

"I love the text! The right level for IT students. The examples are definitely a high point to this text. I love the quantity and quality of exercises. Avoiding heavy mathematics fits an IT program well." —**Dr. Irene Bruno, George Mason University**

"A great introduction to deep learning." —**Alison Sanchez, University of San Diego**

"I was very excited to see this textbook. I like its focus on data science and a general purpose language for writing useful data science programs. The data science portion distinguishes this book from most other introductory Python books."
—**Dr. Harvey Siy, University of Nebraska at Omaha**

"The collection of exercises is simply amazing. I've learned a lot in this review process, discovering the exciting field of AI. I liked the Deep Learning chapter, which left me amazed with the things that have already been achieved in this field. Many of the projects are really interesting." —**José Antonio González Seco, Consultant**

"An impressive hands-on approach to programming meant for exploration and experimentation."
—**Elizabeth Wickes, Lecturer, School of Information Sciences, University of Illinois at Urbana-Champaign**

"I was impressed at how easy it was to get started with NLP using Python. A meaningful overview of deep learning concepts, using Keras. I like the streaming example." —**David Koop, Assistant Professor, U-Mass Dartmouth**

"Really like the use of f-strings, instead of the older string-formatting methods. Seeing how easy TextBlob is compared to base NLTK was great. I never made word clouds with shapes before, but I can see this being a motivating example for people getting started with NLP. I'm enjoying the chapters in the latter parts of the book. They are really practical. I really enjoyed working through all the Big Data examples, especially the IoT ones." —**Daniel Chen, Data Scientist, Lander Analytics**

"A good overview of various neural networks with coding examples for classification problems for which neural networks are commonly used. The exercises in this chapter will give students insight into how changing the structure of neural networks and the amount of training/testing data affect performance. The Twitter examples covering trending topics, creating word clouds, and mapping the location of users are instructive and engaging. I like the real-world examples of data munging. Reviewing this book was enjoyable and even though I was fairly familiar with Python, I ended up learning a lot." —**Garrett Dancik, Associate Professor of Computer Science/Bioinformatics, Eastern Connecticut State University**

"I really liked the live input-output. The thing that I like most about this product is that it is a Deitel & Deitel book (I'm a big fan) that covers Python." —**Dr. Mark Pauley, University of Nebraska at Omaha**